DICTIONARY OF
ELECTRONICS
AND
NUCLEONICS

L E C HUGHES
R W B STEPHENS
L D BROWN

DICTIONARY OF

ELECTRONICS

AND

NUCLEONICS

BARNES & NOBLE, Inc.
NEW YORK
PUBLISHERS & BOOKSELLERS SINCE 1873

© L. E. C. HUGHES, R. W. B. STEPHENS

1969

First published in the United States, 1970
by Barnes & Noble, Inc.

SBN 389 01347 1

Printed in Great Britain by
T. & A. Constable, Ltd.,
Hopetoun Street, Edinburgh

PREFACE

Electronics and nucleonics are two of the most rapidly expanding fields of technology and the aim of this work is to give adequate coverage to both subjects within a single technical dictionary of reasonable size. The present volume was the inspiration of the late Dr L. E. C. Hughes, whose untimely death left his co-editors the formidable task of maintaining the high standard of his writing. Although we hope to have achieved this goal, there must inevitably be errors and omissions in any work covering such a wide and continuously developing field. In the words of Samuel Johnson, 'Every other author may aspire to praise, the lexicographer can only hope to escape reproach'. Suggestions from readers for improvements or additional material will always be appreciated.

The objective of the compilers has been to provide the most concise entry possible, consistent with conveying a complete and unambiguous meaning, and exact mathematical statements have been included only when the quantitative aspect of the reference is of fundamental importance. Entries are in word by word alphabetical order (single letters, as in **A-amplifier**, ranking as words), and there is an extensive system of cross-referencing, particularly under the basic electronic terms, e.g., **antenna, tube.** A comprehensive list of abbreviations, including those used in the main text and in electronics and associated fields generally, appears at the beginning of the book. During the later stages of production, some new terms, particularly in the computing field, together with a few older but less important definitions, were incorporated into a supplement to be found at the end of the book. The Appendices provide fuller statements of some of the fundamental theories of the increasingly complex fields of technology, particularly atomic and nuclear physics, as well as various useful and relevant tables of data.

Our thanks are due to the British Standards Institution, whose notation has been followed whenever possible, for use of their invaluable glossaries, and to other organizations noted in the Appendices, for permission to reproduce copyright material. Particular thanks are due to Mr J. H. Jupe and to the editor of *Industrial Electronics* for permission to reproduce material from the 'Trons Dictionary' which originally appeared in *British Communications and Electronics* (March 1962), and to Dr J. de Klerk who computed the Decibel Tables. The authors are deeply indebted to Mr Collocott, Mr Thorne and Miss Lynas for the care and attention they have given to the preparation of this volume for the press.

RWBS
LDB

PUBLISHER'S NOTE

Chambers Dictionary of Electronics and Nucleonics was begun under the editorship of Dr L. E. C. Hughes, sometime editor of *Chambers Technical Dictionary*. On the death of Dr Hughes, the task of completion was taken over by Dr Raymond W. B. Stephens, Reader, Physics Department, Imperial College, London, and Mr L. Denis Brown, M.Sc., Senior Lecturer, Physics Department, West Ham College of Science and Technology. To both of these highly experienced workers in the fields of electronics and nucleonics is due the very considerable extension of the original plan, including the provision of the invaluable Appendices.

CONTENTS

	PAGE
PREFACE	v-vi
ABBREVIATIONS, ACRONYMS AND SYMBOLS	1–18
THE DICTIONARY	19–301
APPENDICES	303–412

SYMBOLS, PREFIXES AND EQUIVALENTS
Greek Alphabet and Usual Meanings in Electronics,
  Telecommunications and Nuclear Science 304
Recommended Symbols for Electrical and Nuclear Quantities . . 305
Common U.K. and U.S. Abbreviations for Electrical Units
Verbal Interpretation of Letters
Expressions in Radio Communication
Electronic Term Equivalents
Contact Nomenclature Equivalents 306
International Morse Code
Electronic Gates performing Logical Operations
Electrical Supply Voltages and Frequencies in Various Countries . 307

SOME TERMS ENDING WITH 'TRON' 308–12

ELECTROMAGNETIC DATA
Electromagnetic Spectrum
Frequency Band Designation in Radio Spectrum 313
Allocation of HF Bands for Television and FM Radio in U.K.
TV Line Standards in Various Countries
Standard Frequency Transmissions in U.K. 314
Classification of Radio Transmissions
Navigation Transmission 315
British Waveguide Data 316–18

SMITH CHART and its Applications 319–20

COMPONENT CLASSIFICATION AND CODING
Colour Code for Resistors 321–22
Preferred Values of Resistances 322

PROPERTIES OF MATERIALS USED IN ELECTRONICS
Ferrite and Ferroelectric Materials 323–25
Piezomagnetic Materials 324–25
Piezoelectric Materials 326
Quartz Crystal Cuts 327
Insulating Materials 328–29

vii

	PAGE
Dielectric Heating	330
Electrical Properties of Gases and Semiconductors	331
Thermocouple Data	332–33
Wire Data (SWG)	334

ELECTROACOUSTIC DATA
Acoustic Spectrum	
Acoustic Absorption	335
Optimum Reverberation Times	
Acoustic Absorption Coefficients of some Common Materials .	336
Noise Measurement	337

SPACE VEHICLES AND SATELLITE COMMUNICATIONS
Space Research	338
Directory of Space Vehicles	338–40
Goonhilly Earth Stations	340

SUMMARY OF FUNDAMENTAL CONCEPTS OF ATOMIC AND NUCLEAR PHYSICS	341–56

IONIZING RADIATIONS
Radiation Hazards	357–58
Quality Factors of Commonly-encountered Radiations . . .	358
Chart of Radiation Effects	359
Biological and Chemical Effects of Radiation on Materials . .	360
Cross-sections of Nuclear Fuels	360–61
Calculations of Radiation Dose Rates and Shielding in Simple Cases	362–64
Data on Commonly-encountered Artificial Radioisotopes . .	365–73

DECIBEL TABLES	374–99

SYMBOLS, UNITS AND NOMENCLATURE IN PHYSICS
Physical Quantities	400
Units	400–01
Numbers	401
Symbols for Chemical Elements, Nuclides and Particles . . .	402
Quantum States	402–04
Nomenclature	404
International Symbols for Units	404–05
Physical Constants: Values in SI Units	406
Physical Concepts in Rationalized MKS Units	407–08
Standard Values and Equivalents	409

CONVERSION TABLES
Conversion Factors for British and Metric Mechanical Units . .	410
Illumination and Luminance Conversion Table	410–11
Conversion Factors required in Nuclear Reactor Physics . .	411
Unit Conversion Table	412

SUPPLEMENT	413–44

SMITH CHARTS	*Inside back cover*

ABBREVIATIONS, ACRONYMS AND SYMBOLS

N.B.—In many cases usage is not standardized, and both lower case and capital letter versions are current, also versions with and without points or hyphens. Certain groups of initial letters which form words have become by common usage generally accepted as terms in their own right and may be found in the main text. In accordance with BSI recommendations, symbols for physical quantities are in italic type; others in roman type. For Greek symbols, see Appendices.

A *Ampere; absolute* (temp.); *amplitude; ångström unit;* constant in Richardson-Dushmann equation; first Cauchy constant; Helmholtz free energy; refracting angle of prism.

A Symbol for *atomic weight; magnetic vector potential.*

Å *Ångström unit.*

a Linear acceleration; year.

a First van der Waal constant.

A0 Transmission with continuous carrier (unmodulated).

A1 Transmission with interrupted CW carrier.

A2 Transmission with carrier modulated by single frequency.

A3 Transmission with carrier modulated by speech or music.

AA *Artificial aerial.*

AAFC *Anti-aircraft fire control.*

AAGL *Anti-aircraft gun-laying* by radar.

AAM *Air-to-air missile.*

ABC *Automatic brightness* (or *bass*) *control.*

ABCSC *American-British-Canadian Stores Catalogue.*

ABK *Airborne identification kit* (transponder).

ABL *Atlas basic language.*

ABM *Automatic batch mixing.*

abs. *Absolute.*

AC *Automatic* (or *analogue*) *computer.*

a.c. *Alternating current.*

Ac *Actinium.*

ACA *Adjacent channel attenuation.*

ACC *Automatic chrominance* (or *contrast*) *control.*

ACE *Automatic computing engine.*

ACO *Admiralty Compass Observatory,* at Slough.

ACORN *Automatic checkout and recording* equipment.

ACR *Approach* (*airfield*) *control radar.*

ACRE *Automatic checkout and readiness* equipment.

ACSR *Aluminium conductor steel reinforced.*

AD *Average deviation.*

ADA *Action data automation.*

ADDAR *Automatic digital data acquisition and recording.*

ADF *Automatic direction finder.*

ADMA *Automatic drafting machine.*

ADP *Ammonium dihydrogen phosphate.*

a.d.p. *Automatic data processing.*

ADPC *Automatic data-processing centre.*

ADPS *Automatic display and plotting systems.*

A

AEA *Atomic Energy Authority* (U.K.).

AEC *Atomic Energy Commission* (U.S.).

AEE *Atomic Energy Establishment.*

AEI *Associated Electrical Industries.*

AERE *Atomic Energy Research Establishment,* at Harwell.

AEW *Airborne early warning* radar.

AF *Audio frequency; accumulation factor.*

AFC *Automatic frequency control.*

AFG *Analogue function generator.*

Ag *Silver.*

AGC *Automatic gain control.*

AGCA *Automatic ground-controlled approach.*

AGL *Automatic gun-laying* by radar.

AGR *Advanced gas-cooled reactor.*

AGS *Automatic gain stabilization; alternating gradient synchrotron.*

Ah *Ampere-hour.*

AI *Airborne interception.*

AID *Aeronautical Inspection Directorate.*

AIP *American Institute of Physics.*

AJ *Anti-jamming.*

Al *Aluminium.*

ALCOM *Algebraic computer.*

ALF *Automatic letter facing; absorption limiting frequency.*

ALPS *Advanced linear programming system.*

Alt. *Altitude* (angular in navigation).

ALU *Arithmetic and logic unit.*

AM *Amplitude modulation; ante meridiem.*

Am *Americium.*

AMC *Automatic modulation control.*

AMOS *Automatic* computer *Ministry of Supply; automatic meteorological observation station.*

amu *Atomic mass unit.*

ANACOM *Analogue computer.*

ANATRON *Analogue translator.*

ANL *Anti-* (or *automatic*) *noise limiter.*

AOS *Add or subtract.*

AP *Air Publication* of Air Ministry.

APC *Automatic phase control.*

APCHE *Automatic programmed checkout* equipment.

APEX *All-purpose electronic X-ray* computer (Birkbeck College).

API *Air position indicator.*

aprxly *Approximately* (computer language).

APT *Automatically-programmed* machine *tool; automatic picture transmission.*

APUHS *Automatic programme unit, high* speed.

APULS *Automatic programme unit, low* speed.

AQL *Acceptable quality level,* used in electronic component manufacture.

AR Radio DF Beacon (Admiralty publications).

Ar *Argon.*

ARC *Automatic relay calculator; automatic remote* (or *ratio*) *control.*

ARGUS *Automatic routine generating and updating system.*

ARL *Admiralty Research Laboratory,* at Poole; *acceptable reliability level.*

ARM *Automated route management.*

ARO *Automatic range only* in radar rangefinder.

ARRL *American Radio Relay League.*

ARSE *Air route surveillance* radar.

ART *Automatic reporting telephone.*

ARTCC *Air route traffic control centre.*

ARU *Audio response unit; acoustic resistance unit.*

As *Arsenic.*

ASA *American Standards Association.*

ASB *Automatic airborne search* radar.

ASC *Automatic selectivity* (or *synchronized*) *control.*

ASCC *Automatic sequence-controlled calculator.*

ASCII *American Standard Code for Information Interchange.*

ASD *Automatic synchronized discriminator.*

ASE *Association for Science Education; Admiralty Signals Establishment.*

ASEE *Association of Supervising Electrical Engineers.*

ASIST *Advanced Scientific Instruments Symbolic Translator.*

ASLT *Advanced solid logic technology.*

ASM *Air-surface missile.*

ASMI *Aerodrome-surface movement indicator.*

ASN *Average sample number.*

ASO *Advanced Solar Observatory.*

ASP *Automatic servo plotter.*

ASPPI *Azimuth-stabilized plan position indicator.*

ASR *Air-sea rescue; automatic send-receive.*

ASRE *Admiralty Signal and Radar Establishment,* at Portsdown.

AST *Atlantic standard time.*

ASTM *American Society for Testing Materials.*

ASV *Automatic self-verification.*

AT or **a.t.** *Ampere-turn.*

A/T *Action time.*

At *Astatine; attenuation.*

at. no. *Atomic number.*

at. wt. *Atomic weight.*

ATC *Antenna* (or *aerial*) *tuning control* (or *capacitor*); *air-traffic control.*

atm *Atmosphere.*

ATR *Anti-transmit-receive.*

ATS *Administrative terminal system.*

Au *Gold.*

AVC *Automatic volume control.*

AVE *Automatic volume expansion.*

AVO RTM for range of portable instruments measuring *amperes, voltage, ohms,* etc.

AWG *American wire gauge.*

AWRE *Atomic Weapons Research Establishment.*

awu *Atomic weight unit.*

Az *Azimuth.*

B Transmission with damped carrier (obsolete); *boron.*

B *Brightness;* symbol for *magnetic flux density; magnetic induction; susceptance;* second Cauchy constant; factor in Richardson-Dushmann equation.

b *Barn.*

b Second van der Waal constant; *bar.*

BA *British Association* for the Advancement of Science.

B/A *Beam approach.*

Ba *Barium.*

BABS *Blind approach beacon system.*

BAC *Binary asymmetric channel.*

BACE *Basic automatic checkout equipment.*

BAR *Buffer address register.*

BBC *British Broadcasting Corporation* (originally *Company*).

BCAC *British Conference on Automation and Computation.*

BCD/B *Binary-coded decimal/binary.*

BCD/Q *Binary-coded decimal/quaternary.*

BCFSK *Binary-code frequency-shift keying.*

BCI ITU symbol for *International Broadcasting Station.*

BCO *Binary-coded octal.*

BCRA *British Ceramic Research Association.*

BCRT *Bright cathode-ray tube.*

BCS *British Computer Society.*

BCW *Buffer control word.*

BDC *Binary decimal counter.*

BDI *Bearing deviation indicator.*

BDU *Basic display unit.*

Be *Beryllium.*

BEAMA *British Electrical and Allied Manufacturers Association.*

b.e.m.f. *Back electromotive force.*

BEPO *British Experimental Pile O* (large air-cooled, natural-uranium, graphite-moderated reactor at Harwell).

BFO *Beat-frequency oscillator.*

BH *Brinell hardness.*

BHP *Brake horsepower.*

Bi *Bismuth.*

BICEP *British Industrial Collaborative Exponential Programme*—a joint team to study physics of gas-moderated, gas-cooled reactor system.

BILE *Balanced inductor logical element.*

BIMAG *Bistable magnetic core.*

BIMCAM *British Industrial Measuring and Control Apparatus Manufacturers.*

BIOS *Biological investigation of space.*

BIPCO *Built-in place components.*

BIR *British Institute of Radiology.*

BIS *British Interplanetary Society.*

BIT *Built-in test.*

BIVAR *Bivariant function generator.*

BIX *Binary information exchange.*

BJCEB *British Joint Communications Electronics Board.*

Bk *Berkelium.*

BKS *British Kinematographic Society.*

BL *Blanking.*
BLEU *Blind-landing Experimental Unit.*
BLIP *Background-limited infrared photo-conduction.*
BMEWS *Ballistic missile early warning system.*
Bn *Beacon* (as in *W/T Bn*).
BNES *British Nuclear Energy Society.*
BOT *Beginning of tape.*
BP *Back projection; band pass.*
b.p. *Boiling-point.*
BPI *Bits per inch.*
BPL *Diode transistor logic.*
BPμL *Diode transistor micrologic.*
bps, b/s *Bits per second.*
Br *Bromine.*
BREMA *British Radio Equipment Manufacturers Association.*
Brit.I.R.E. *British Institute of Radio Engineers.* Now **IERE.**
BRPA *British Radiological Protection Association.*
BRS *Building research station; break request signal.*
BRVMA *British Radio Valve Manufacturers Association.*
BS *British Standard; binary subtract.*
BSDC *British Space Development Company.*
BSF *British Standard Fine* gauge for screws.
BSI *British Standards Institution.*
BSRA *British Sound Recording Association.*
BST *British standard time.*
BSW *British Standard Whitworth* thread of screws.
BTDL *Basic-transient diode logic.*
BTG *Beta thickness gauge.*
BTO *Bombing through overcast.*
B.t.u. *British thermal unit.*
BUIC *Back-up interceptor control.*
BVA *British Radio Valve Manufacturers Association.*
BW *Bandwidth; black-and-white* television.
BWG *Birmingham wire gauge.*
BWMS *British Wireless Marine Service.*
BWO *Backward-wave oscillator.*
BWR *Boiling-water reactor.*
BX Insulated wires in sleeving.

C *Coulomb; candle; candela; carbon.*
C *Capacitance; heat capacity;* Sutherland's constant.
c *Centi-* (prefix for ×10⁻²).
c Symbol for *velocity of light; concentration* of solution; Planck radiation law constants.
°C *Degree Centigrade;* originally and now internationally *Celsius.*
C₀ *Natural abundance.*
CA *Cascade amplifier.*
Ca *Calcium.*
CADF *Commutated aerial direction-finding.*
CAFD *Contact analogue flight display.*
CAI/O *Computer analogue input/output.*
cal *Calorie.*
CALDIC University of *California electronic* calculator.
CAM *Central address memory.*
C and C *Command and control.*

CANDU *Canadian deuterium uranium* reactor—a 200 MW nuclear power station near Lake Huron.
CAP *Civil Air Publication.*
Cap. *Capacitor; capacitance.*
CAR *Central apparatus room; channel address register.*
CART *Computerized automatic rating technique.*
CAS *Calibrated air speed.*
CAT *Cooled-anode transmitting* valve; *College of Advanced Technology.*
CAV *Cavity* resonance.
CAVALCADE *Calibrating, amplitude-variation and level-correcting analogue-digital equipment.*
CAW *Channel address word.*
CB *Central battery* system.
CBI *Compound Batch Identification.*
CBS *Columbia Broadcasting System; central battery signalling* system.
CC *Closed circuit* (transmission).
CCA *Carrier-controlled approach.*
CCD *Computer-controlled display.*
CCGCR *Closed-cycle gas-cooled reactor.*
CCIR *Comité Consultatif, International Radio.*
CCITT *Consultative Committee, International Telegraph and Telephone.*
CCR *Central control room.*
CCS *Collective call sign.*
cct *Circuit.*
ccw *Counter-clockwise.*
CD *Clock driver.*
Cd *Cadmium.*
c.d. *Current density.*
cd *Candela.*
CDA *Copper Development Association.*
CDC *Code-directed character.*
CDCE *Central data conversion equipment.*
CDF *Combined distribution frame.*
CDH *Command and data-handling.*
CDP *Communication data processor.*
CdS *Cadmium sulphide.*
CDT *Control data terminal.*
CDU *Coastal defence* radar against *U-boats; central display unit.*
CDX *Control differential transmitter* in U.S.
Ce *Cerium.*
CEGB *Central Electricity Generating Board* in U.K.
CEI *Communications electronics instructions.*
c.e.m.f. *Counter electromotive force.*
CEP *Circular error probability.*
CERL *Central Electricity Research Laboratory* in U.K.
CERN *Conseil Européen de Recherche Nucléaire,* near Geneva.
CET *Central European Time.*
CETEX *Committee for Extra-Terrestrial Exploration.*
Cf *Californium; compare with.*
cfm/s *Cubic feet per minute/second.*
c.g. *Centre of gravity.*
cg *Centigram.*
CGB *Convert gray to binary.*
CGS *Centimetre-gram(me)-second* system of units.

CH *Choke* coil.
CH(L) *Chain-home (low-flying).*
CHU *Centigrade heat unit.*
C/I *Carrier to interference* ratio.
Ci *Curie.*
CIC Radar *combat information centre.*
CIE *Commission Internationale de l'Éclairage.*
CIO *Central input/output* multiplex.
CISPR *International Special Committee on Radio Interference* of IEC.
CIT *Call-in time.*
CK *Check.*
CKO *Check operator.*
CL *Centre line* of plans, drawings; *central line; conversion loss.*
Cl *Chlorine.*
cl *Centilitre.*
CLD *Called* line.
CLG *Calling* line.
CLR *Computer language recorder.*
CLT *Computer language translator; communication line terminal.*
CM *Control mark; computer module.*
C/M *Communications.*
Cm *Curium.*
cm *Centimetre; circular mil.*
CMA *Cable Makers Association.*
CMC *Contact making/breaking clock.*
Cmd *Command Paper* from HMSO.
CMF *Cross-modulation factor.*
CML *Current mode logic.*
cmp *Computational.*
CMR *Continuous maximum rating* of equipment; *common mode rejection.*
C/N *Carrier/noise.*
CNL *Circuit net loss.*
CNR *Carrier/noise* power *ratio.*
CO *Coupled* (or *crystal*) *oscillator; change over.*
Co *Cobalt.*
CODIC *Computer-directed communications.*
CODIPHASE *Coherent digital phased-array* system.
CODIT *Computer direct to telegraph.*
COGB *Certified official government business.*
COHO *Coherent oscillator.*
COL *Computer-oriented language.*
COMET *Meteorological Office Computer.*
COMM *Communications.*
COMMZ *Communications zone.*
COMSAT *Communication satellite* (U.S.).
CONSORT A 10 kW light-water moderated and cooled research reactor designed by Imperial College (London) and G.E.C.
COP *Computer optimization package.*
COSMON *Component open/shut monitor.*
COSPAR *Committee for Space Research* of the International Council of Scientific Unions.
COZI *Communications zone indicator.*
CP *Constant potential; chemically pure; circular pitch; candle power; compare; cosmogenic; computer.*
CPA *Colour phase alternation.*
CPC *Card process* (or *programmed*) *electronic calculator; computer process control.*
CPDD *Command-post digital display.*

CPE *Central programme and evaluator; charged-particle equilibrium.*
CPIP *Computer pneumatic input panel.*
CPM *Cards per minute; critical path method.*
CPS *Cathode potential stabilization.*
cps *Counts per second.*
CPU *Central processing unit* for computers.
Cr *Chromium.*
CRAFT *Computerized relative allocation of facilities technique.*
CRAM *Card random-access memory.*
CRC *Carriage return contact.*
CRETE *Common radio and electronic test equipment* list of Naval test gear.
crit Mass of fissionable material which is *critical* under given conditions.
CRMR *Continuous-reading meter relay.*
CRO *Cathode-ray oscillograph.*
CRS *Command retrieval system.*
CRT *Cathode-ray tube.*
CRTU *Combined receiving and transmitting unit.*
CRYOSAR *Cryostatic switching-avalanche and recombination.*
Cs *Caesium.*
c/s *Cycles per second.*
CSA *Canadian Standards Association.*
CSIRO *Commonwealth Scientific and Industrial Research Organization.*
CSO *Chained sequential operation.*
CST *Central standard time* (U.S.).
CT *Centre tap* of a winding of an inductance, transformer, or battery; *control transformer* in a servo system.
CTB *Commonwealth Telecommunications Board.*
CTCA *Channel and traffic control agency.*
CT/N *Counter, n* stages.
CTP *Central transfer point.*
CTR *Controlled thermonuclear reaction.*
CTU *Centigrade thermal unit.*
CTV *Colour television.*
CU *Close-up; crosstalk unit; control unit.*
Cu *Copper; cubic.*
CV *Continuously variable.*
CVT *Constant-voltage transformer.*
CVU *Constant-voltage unit.*
CW *Continuous wave.*
cw *Clockwise; continuous-wound.*
CWO *Continuous-wave oscillator.*
CWV *Continuous-wave video.*
CX *Control transmitter.*
CYBORG *Cybernetic organism.*

D *Differential coefficient; deuterium; doublet lines* of ionized sodium vapour.
D Symbol for *electric flux* or *displacement;* coefficient of *fluid diffusion.*
d *Deci-* (prefix for $\times 10^{-1}$); *day.*
DAC *Data acquisition and control* system; *digital arithmetic centre.*
DAGC *Delayed automatic gain control.*
DAGMAR *Drift and ground-speed measuring airborne radar.*
DAME *Data acquisition and monitoring equipment* for computers.
DAP *Double-amplitude peak.*

DAPD *Directorate of Aircraft Production Development.*

DAS *Data acquisition system; digital attenuator system.*

DATACOM *Data communications.*

DAVC *Delayed automatic volume control.*

dB, db *Decibel.*

DBD *Double-base diode.*

dBk *Decibels* with reference to one *kilowatt.*

dBm *Decibels* with reference to one *milliwatt; decibel meter,* level referred to 1 mW in 600 ohm.

dBp *Decibels* with reference to one *picowatt.*

dBrap *Decibels* above *reference acoustic power.*

dBrn *Decibels* above *reference noise.*

dBv *Decibels* relative to 1 *volt.*

dBW *Decibels* relative to 1 *watt.*

DC *Data channel.*

d.c. *Direct current.*

DCC *Double cotton-covered* wire.

DCCU *Data communications control unit.*

DCR *Data conversion receiver.*

d.c.r. *Direct-current restorer.*

DCTL *Direct-coupled transistor logic.*

DCU *Decade counting unit.*

DCV *Direct-current voltage.*

DCX Oak Ridge thermonuclear device—*Direct-current Experiment* (U.S.).

DDA *Digital differential analyser.*

DDAS *Digital data acquisition system.*

DDC *Data distribution centre; digital data converter.*

DDCE *Digital data conversion equipment.*

DDG *Digital display generator.*

DDT *Digital data transmitter; dynamic debugging technique.*

Dec *Declination; decimal.*

decit *Decimal digit.*

DEF *Defence* specification.

Dem. *Demodulator.*

DERE *Dounreay Experimental Reactor Establishment,* in N. Scotland.

DES *Department of Education and Science; digital expansion system.*

DETAB *Decision tables.*

DEU *Data exchange unit.*

DEW *Distant early warning.*

DF *Direction-finding* (or *finder*); *degrees of freedom.*

DFG *Diode function generator.*

DH *Directly-heated.*

DHE *Data-handling equipment.*

DIAD *Drum information assembler and dispatcher.*

DIAMOD Form of rectifier modulator.

DIAN *Decca integrated air navigation system* (Decca-Dectra-Doppler).

DIDAS *Dynamic instrumentation data automobile system,* for telemetry transmission and control from mobile plants.

DIDO Heavy-water (DDO) moderated reactor with a high flux and using enriched uranium fuel.

DIGICOM *Digital communication* system.

DIMPLE *Deuterium moderated pile, low-energy* reactor, using heavy water.

DIN *Deutsche Industrien Normen* (German Standard).

DINA *Digital network analyser; direct-noise amplifier.*

dis. *Disconnect(ion).*

dist. *Distilled.*

DKT *Dipotassium tartrate* piezo crystal.

DL *Delay line; data link.*

dl *Decilitre.*

DMA *Direct memory address.*

DMC *Digital microcircuit.*

DME *Distance-measuring equipment.*

DMTR *Dounreay materials-testing reactor.*

dN *Decineper* attenuation unit.

DP *Data processing.*

dp *Deflection plate; double pole; dipole; dew point.*

DPC *Double paper-covered* cable.

DPDT *Double-pole double-throw* switch.

dpm/s *Disintegrations per minute/second.*

DR *Deduced* (*dead*) *reckoning* in navigation; *disk recorder.*

DRF *Dose reduction factor* (for protective agent).

DR3 Pluto-type reactor (Danish).

DS *Data synchronization.*

DSb *Double sideband.*

DSC *Double silk-covered.*

DSIF *Deep space instrumentation facilities.*

DSIR *Department of Scientific and Industrial Research.* (Obs., now **SRC.**)

DSR *Discriminating selector repeater.*

DT *Data transmission.*

dt *Double-throw.*

DTI *Dial test indicator; distortion transmission impairment.*

DVM *Digital voltmeter.*

DVST *Direct vision storage tube.*

Dwg *Drawing.*

DWI *Differential wave impedance.*

DX *Distant* radio communication.

Dy *Dysprosium.*

dyn. *Dynamo; dynamometer; dyne.*

DYSEAC Second National Bureau of *Standards Eastern Automatic Computer.*

E Symbol for *electric force; electric field strength; illumination; electromotive force; single electrode potential; Young's modulus of elasticity.*

e Symbol for base of natural logarithms (2·71828 . . .).

e Symbol for charge of electron; permittivity.

EAL *Electromagnetic amplifying lens.*

EASI *Electrical Accounting for the Security Industry.*

EAST *East Australian standard time.*

EAX *Electronic automatic exchange.*

EBR *Experimental breeder reactor.*

EBU *European Broadcasting Union* in West Europe.

EBWR *Experimental boiling-water reactor.*

ec *Electron-coupled.*

ECARS *Electronic Coordinatograph and Read-Out System.*

ECASS *Electronically-controlled automatic-switching system.*

ECC *Electrocardiocorder.*

ECCM *Electronic countermeasures.*
ECDS *Electrochemical diffused-collector transistor.*
ECG *Electrocardiogram (-graph).*
ECLO *Emitter-coupled logic operator.*
eco U.S. for *electron-coupled oscillator.*
ECX *Electronically-controlled* telephone *exchange.*
ED *Electrodynamic.*
EDA *Electrical Development Association.*
EDGE *Electronic data-gathering equipment.*
EDP *Electronic data processing.*
EDPM *Electronic data-processing machine.*
EDPS *Electronic data-processing system.*
EDSAC *Electronic* ⸢*delayed-storage automatic computer.*
EDT *Ethylenediamine tartrate* piezo crystal; *eastern daylight-saving time (EST +1 h).*
EDVAC *Electronic discrete variable automatic computer.*
EEA *Electronic Engineering Association.*
EEG *Electroencephalograph (-gram); electronephalograph.*
EF *Extra-fine* thread.
EFL *Effective focal length.*
EGO *Elliptically-orbiting Geophysical Observatory.*
EHF *Extremely high frequency.*
EHP *Effective horsepower.*
EHT *Extra high tension.*
EIA *Engineering Industries Association* (U.S.).
EKG *Electrocardiogram (-graph).*
ELCO *Electrolytic capacitor.*
ELDO *European Launcher Development Organization.*
ELEM *Element.*
ELF *Extremely low frequency.*
ELS *Electrostatic loudspeaker.*
ELT *Electrometer.*
EM *Electromagnetic.*
EMAR *Experimental memory address register.*
EMCCC *European Military Communications Coordinating Committee.*
e.m.f. *Electromotive force.*
EMIAC *Electrical and Musical Industries analogue computer.*
EMP *Electronic multiplying punches.*
EMTA *Electro Medical Trade Association.*
e.m.u. *Electromagnetic unit.*
ENEA *European Nuclear Energy Authority.*
ENI *Equivalent noise input.*
ENIAC *Electronic numeral integrator and calculator.*
ENSI *Equivalent noise sideband input.*
EOS *Electro-optical system.*
EOT *End of tape.*
EP *Electrically polarized; electropneumatic; extended-play; electroplate.*
EPI *Elevation-position indicator.*
EPIC *Electronic plotting equipment*, especially developed at the Royal Observatory, Herstmonceux, for recording star transits.
EPR *Equivalent parallel resistance* of a crystal; *electron paramagnetic resonance.*
EPTA *Electrophysiological Technologists Association.*
Er *Erbium.*

ERA *Electronic reading automaton,* which reads and codes printed figures.
ERMA *Electronic recording machine accounting.*
ERP *Effective* or *equivalent radiated power* from antenna.
ERTI *Electron-ray tuning indicator,* magic eye.
Es *Einsteinium.*
es *Electrostatic.*
ESC *Enamelled single-covered* (wire).
ESG *Electronic sweep generator.*
ESR *Equivalent series resistance* of resonating crystal; *effective signal radiated; electron spin resonance.*
ESRO *European Space Research Organization.*
ESS *Electronic switching system.*
EST *Eastern standard time* (U.S.).
e.s.u. *Electrostatic unit.*
e.s.v. *Electrostatic voltmeter.*
ET *Ephemeral time.*
ETA (or D) *Estimated time of arrival* or *departure.*
ETSPL *Equivalent threshold sound pressure level.*
Eu *Europium.*
eV *Electron-volt.*
EVATRON *Eccentric variable-angle thermionic rheostat.*
EWO *Electrical and Wireless Operators.*
EWR *Early warning radar.*
ExAM *Ex Air Ministry.*
Exp. *Exponential.*

F *Fahrenheit; farad; faraday; filament; fluorine; fuse.*
F Symbol for *Helmholtz function; magnetomotive force.*
f Symbol for *focal length/effective diameter* of a lens.
f Symbol for *frequency; acceleration.*
f$_\alpha$ *Alpha cutoff frequency.*
f$_c$ *Cutoff frequency.*
FA *French Army* valve.
FAB ITU abb. for *aeronautical broadcast station.*
FAC ITU abb. for *airport control station.*
FACT *Fully-automatic compiling technique.*
FAST *Flexible algebraic scientific translator.*
FAT ITU abb. for *flight test station.*
FAX ITU abb. for *fixed aeronautical station.*
fax *Facsimile* in U.S.
FB ITU abb. for *base station.*
FBC *Fully-buffered channel.*
FBI *Federation of British Industries.*
FC *Frequency changer* or *converter; fine control; front-connected;* ITU abb. for *coast station.*
f.c. *Foot-candles.*
FCB ITU abb. for *marine broadcasting station.*
FCC *Federal Communications Commission* (U.S.).
FCS *Fire control system* radar.
FCT *Filament centre-tap.*
FD *Frequency doubler.*
FDM *Frequency division multiplex.*
FDS *Fermi-Dirac-Sommerfeld.*

Fe *Iron.*
FEDAL *Failed element detector and locator* in nuclear reactors.
FEP *Financial evaluation programme.*
FET *Field-effect transistor.*
F-F *Flip-flop.*
FFF *Fission-fusion-fission* bomb.
ffrr *Full-frequency range recording.*
ffss *Full-frequency stereophonic sound.*
FIDO *Fog Investigation and Dispersal Organization.*
FIFO *Floating input-floating output.*
FL Electric wave *filter*; ITU abb. for *land station.*
FLE ITU abb. for *land telemetering station.*
FLF *Flip-flop.*
FLH ITU abb. for *land hydrological and metereological station.*
FM *Frequency modulation*; *feedback mechanism*; *floor manager* (of a studio).
Fm *Fermium.*
FMI *Flow measurement and indication.*
FNBR *Fast neutron breeder reactor.*
FO *Fast operation* of relay.
f.o.m. *Figure of merit.*
FOPT *Fibre optic photo transfer.*
FOT *Optimum working frequency* (abb. from French).
FP *Flame-proof* equipment;
f.p. *Freezing-point.*
fph *Full power hours*, of life for reactor fuel element.
f.p.m. *Feet per minute.*
FPS *Foot-pound-second.* System of legal units (Imperial) based on the *foot*, the *pound* (weight), and the *second*, as fundamental dimensions; used in ordinary commerce and mechanical engineering in U.K. and U.S.
f.p.s. *Feet* (or *frames*) *per second.*
FR *Fast release* of relay; *field resistance.*
Fr *Francium*; *French*; *frame.*
FRED *Figure reading electronic device.*
FS *Factor of safety*; *Federal Specification* in U.S.; *frame scan.*
FSD *Full-scale deflection.*
FSK *Frequency-shift keying.*
FSM *Field-strength meter.*
FSR *Feedback shift register.*
FTB *Frequency time base.*
FTC *Fast time constant.*
FTM *Frequency time modulation.*
FW *Full-wave* rectifier.
FWHM *Full width at half maximum amplitude.*
FWHP *Full width at half peak amplitude.*
FX ITU abb. for *fixed hydrological and meteorological station.*
Fx ITU abb. for *fixed station.*
FXBIN Decimal to *fixed binary* translation.
FXE ITU abb. for *fixed telemetering station.*

G *Gauss*; *giga-* (prefix for $\times 10^9$); *gilbert.*
G Symbol for *conductance*; *transconductance* in valves.
g *Gram(me).*
g Symbol for *acceleration due to gravity* at earth's surface.

gm Symbol for *transconductance* or *mutual conductance.*
GA *General arrangement*; *go-ahead* signal (or cue).
Ga *Gallium.*
Galv. *Galvanic*; *galvanometer.*
GAR *Guided aircraft rocket*, an air-to-air missile, self-homing by heat or radar.
GB *Gain bandwidth*; *grid bias.*
GC *Great Circle.*
GCA *Ground-controlled approach.*
GCI *Ground-controlled interception.*
GCL *Ground-controlled landing.*
Gc/s *Gigacycles per second*, unit of frequency in microwaves, equal to 10^9 Hz.
GCT *Greenwich civil time.*
GD *Ground detector.*
Gd *Gadolinium.*
Ge *Germanium.*
gen. Direct-current *generator.*
GG *Grounded grid* valve.
G/G *Ground to ground.*
GHA *Greenwich hour angle.*
GHz *Gigahertz.*
GL(T) *Gun-laying* (*turret*) by radar.
GM *Geiger-Müller* counter; *Gill-Morell* valve oscillator; *gramophone motor.*
Gm Same as **gm.**
GMT *Greenwich mean time.*
GMV *Guaranteed minimum value.*
gnd *Ground* (U.S.).
GOR *General operational requirement.*
GPAC *General purpose analogue computer.*
GPC *General purpose computer.*
GPDC *General purpose digital computer.*
GPI *Ground position indicator.*
GPS *Groups of pulses per second.*
GRF *Group repetition frequency.*
GSG *Glass-silicone-glass* laminate for electronic component packaging.
GST *Greenwich sidereal time.*
GT *Glass tube*; *game theory.*
GTS *Greenwich time signal* in broadcasting and television.
GW *Guided weapon.*
GZ *Ground-zero* of nuclear weapon.

H *Heater* (valves and tubes); *henry* (mutual inductance unit); *hydrogen.*
H Symbol for *magnetizing force*; for *magnetic field strength*; for *enthalpy*; for *Boltzmann's entropy constant.*
h *Hour.*
h Symbol for *Planck's constant.*
ħ Symbol for *Dirac's constant.*
HA *Hour angle* (navigation); *high angle* (radar); *half add.*
HAC *Heavy-aggregate concrete*, used for shielding.
H & D *Hurter and Driffield*, who derived procedure for relating density to exposure.
H & N *Hum and noise.*
HAWK *Homing all-the-way killer* missile.
HAZEL *Homogeneous aqueous zero energy level* reactor at Harwell.
HC *Handling capacity.*
HD *Hydrographic Dept.* Admiralty.
HDDS *High-density data system.*

HDM *Harmonic distortion meter.*
HDO *Hydrogen-deuterium-oxygen* (chemical symbol for heavy water).
He *Helium.*
HEC *Hollerith electronic computer.*
HECTOR *Heated experimental carbon thermal oscillator reactor* for test purposes at Winfrith Heath.
HELEN *Hydrogenous Experimental Liquid Experiment.* A light-water moderated sub-critical assembly at Winfrith used for extending uranium reactor calculations to plutonium-fuelled reactors.
HEM *Hybrid electromagnetic* wave.
HERALD *Highly-enriched reactor, Aldermaston.* A 5 MW light-water, moderated, cooled reactor using highly enriched fuel.
HERO *Hot experimental reactor of zero* power, at Windscale, anticipating **AGR.**
Het Supersonic *heterodyne* radio reception.
Hex Uranium *hexafluoride.*
HF *High frequency.*
Hf *Hafnium.*
HFRDF *High-frequency repeater distribution frame.*
Hg *Mercury.*
HHF *Hyper-high-frequency.*
HICAPOM *High-capacity communications.*
HIFAM *High-fidelity amplitude modulation.*
HIFAR *High-flux Australian reactor,* at Sydney, N.S.W.
HIG *Hermetic integrating gyroscope.*
HiK *High permittivity* material.
HILAC *Heavy ion linear accelerator.*
HMF *Hum modulation factor.*
HO *Hydrographic Office* (U.S.).
Ho *Holmium.*
HORACE *H_2O reactor Aldermaston critical experiment,* a zero-energy reactor at AWRE used for testing fuel elements arrangement in the HERALD reactor.
HP *High-pass* filter; *horsepower.*
HPI *Height-position indicator.*
HPRR *High-performance research reactor* (U.S.).
HR *High-range* radar.
HRE *Homogeneous reactor experiment* (U.S.)
HRI *Height range indicator.*
HRP *Horizontal radiation pattern* of a transmitting or receiving antenna.
HS *High-stability height* given by *sextant.*
HSAC *High-speed analogue computer.*
HSDA *High-speed data acquisition.*
HT(B) *High-tension (battery).*
HTGCR *High-temperature gas-cooled reactor,* with ceramic fuel elements.
HTO *Hydrogen-tritium-oxygen* (chemical symbol for tritiated water).
HV *High voltage; hard valve.*
HW *Half-wave* rectifier.
Hx *Hexode.*
HYCOTRAN *Hybrid computer translator.*
HYPO *High-power* water boiler reactor, sodium thiosulphate.
Hz *Hertz.*

I *Iodine; moment of inertia.*

I Symbol for *luminous intensity*; for *electric current*; for *acoustic intensity*; watts per sq. cm.; *ionic strength.*
i *Instantaneous current* $\sqrt{-1}$.
IAEA *International Atomic Energy Agency.*
IAGC *Instantaneous automatic gain control.*
I and C *Installation and checkout.*
IAS *Indicated air speed.*
IBM *International Business Machines* Ltd.
IBU One-time *International Broadcasting Union.*
IC *Internal connection* of valve; *intervalve coupling; information content; in charge; integrated circuit.*
ICBM *Intercontinental ballistic missile.*
ICCE *Imperial College computing element.*
ICI One time *International Commission on Illumination* (now **CIE**); *Imperial Chemical Industries.*
ICR *Institute of Cancer Research.*
ICRP *International Commission on Radiological Protection,* which formulates maximum permissible doses of radiation.
ICRU *International Commission on Radiological Units and Measurements.*
ICSE *Intermediate current stability experiment,* once proposed as a development from *zeta.*
ICT *International Computers and Tabulators* Ltd.
ICW *Interrupted continuous wave.*
ICWM *International Commission for Weights and Measures.*
IDF *Intermediate distribution frame* for circuits.
IDP *Integrated data processing,* that based on business management reorganization.
IDS *Integrated data store.*
IE *Institution of Electronics.*
IEC *International Electrotechnical Commission.*
IEE *Institution of Electrical Engineers.*
IEETE *Institution of Electrical and Electronics Technician Engineers.*
IEME *Inspectorate of Electrical and Mechanical Equipment.*
IERE *Institution of Electronic and Radio Engineers.*
IES *Illuminating Engineering Society.*
IF *Intermediate frequency,* especially of a supersonic heterodyne receiver; *information collector.*
IFF *Identification, friend or foe.*
IFPP *Irradiated fuel processing plant.*
IFR *Instrument Flight Rules; internal function register.*
IFRB *International Frequency Registration Board,* which advises ITU on the best use of radio frequencies.
IFRU *Interference frequency rejection unit.*
IG *Insulated gate.*
IGY *International Geophysical Year.*
IH *Indirectly-heated* cathode or valve.
ILLIAC University of *Illinois automatic* computer.
ILS *Instrument landing system.*
IMC *Image motion compensation.*
IMO *Interband magneto-optic effect.*
In *Indium.*
INE *Institute of Nuclear Engineers.*

Inst.P. *Institute of Physics.*
Int. *Internal; interior; international; integral.*
Intercom. *Intercommunication,* radio or telephone.
INTELSAT *International Satellite* Communications Organization.
inv *Inverter.*
I/O *Input/output.*
Io *Ionium.*
IOB *Input-output buffer.*
IOC *Input-output controller.*
IP *Imaginary part* (of complex number).
IPA *Intermediate power amplifier.*
IPC *Information processing centre; industrial process control.*
IPD *Insertion phase delay.*
IPE *Institution of Production Engineers.*
IPM/S *Interruptions per minute/second.*
IP(S) *Input primary* (*secondary*) of transformer.
IQSY *International quiet solar year.*
IR *India-rubber; information retrieval.*
Ir *Iridium.*
i.r. *Infrared.*
IRASER *Infrared laser.*
IRSU *International Radio Scientific Union.*
IS *Internal shield; insulating sleeving.*
ISB *Independent sideband* of carrier.
ISO *International Standardization Organization.*
ISR *Information storage and retrieval.*
IST *Indian standard time.*
IT *Input translator; item transfer.*
ITA *Independent Television Authority* (U.K.).
ITU *International Telecommunications Union,* which governs international telephone, telegraph and radio services.
I/T(V) *Inverted-T(V)* antenna.
IUPAC or P *International Union of Pure and Applied Chemistry or Physics.*
Iv *Inverter.*

J *Joule; Joule's equivalent;* inner quantum number.
J Symbol for *current density.*
j $\sqrt{-1}$ in electricity. Otherwise **i.**
JAN *Joint Army Navy* (equipment designation).
JASON A 10 kW light-water cooled, graphite-moderated research reactor.
JDC *Job description card.*
JERC *Joint Electronic Research Committee* of the Post Office and manufacturers.
JETEC *Joint Electron Tube Engineering Council* (U.S.).
JIE *Junior Institution of Engineers.*
J/S Ratio in dB of total interference power to signal-carrier power at the receiver.
JST *Japanese standard time.*

K *Kilohm; Kerr constant;* symbol for cathode (U.S.); *potassium.*
K25 Code name for Oak Ridge uranium separation plant.
K Degree *Kelvin* on absolute temperature scale.
A*

k *Kilo-* (prefix for $\times 10^3$); *kayser;* dielectric constant.
k Symbol for Boltzmann's constant; phase constant; *thermal conductivity; multiplication factor for chain reaction; bulk modulus of elasticity.*
kc *Kilocalorie; kilocycle/second* (U.S.).
kCi *Kilocurie.*
kc/s *Kilocycles per second.*
KERMA *Kinetic energy released* per unit *mass.*
kg *Kilogram(me).*
kHz *Kilohertz.*
k(ilo)var(h) *Kilovolt-ampere reactive* (*hour*).
kM *Kilomega-* (prefix for $\times 10^9$).
km *Kilometre.*
Kr *Krypton.*
kV *Kilovolt.*
kVA *Kilovolt-ampere.*
kVp *Kilovolts, peak.*
kW *Kilowatt.*
kWh *Kilowatt-hour,* unit of power (energy).
KY *Keying* device.

L Symbol for filter reaction consisting of series and shunt arm; *lambert; lumen.*
L Symbol for *inductance;* for *brightness* or *luminance.*
L$_{mn}$ *Mutual inductance.*
l *Litre;* azimuthal or orbital quantum number.
LA *Loop antenna; low angle* (radar).
La *Lanthanum.*
LAC *Load accumulator.*
LACE *Luton analogue computing engine.*
LADAR *Laser radar.*
Lam. *Lamination* for transformer.
LANAC *Laminar navigation anti-collision.*
Laq. *Lacquer* on wire.
LAT *Local apparent time.*
Lat. *Latitude.*
LB *Line buffer.*
LC *Inductance-capacitance* tuned circuit for oscillator; *loading coil* of circuit.
L/C *Inductance-capacitance* ratio for tuned circuit.
LCAO *Linear combination* of *atomic orbitals.*
LDDS *Low-density data system.*
LDRI *Low data rate input.*
LDX *Long-distance xeroxography.*
LEP *Lowest effective power.*
LET *Linear energy transfer.*
LF *Low frequency.*
LFC *Low-frequency choke.*
LHA *Local hour angle.*
Li *Lithium.*
LIDAR *Laser infrared radar.*
LIFT *Logically-integrated Fortran translator.*
LIM *Limit(er)* in frequency modulation.
LINAC *Linear accelerator.*
LITR *Low-intensity test reactor* (U.S.).
LL *Lower limit; lower level; loudness level.*
LLR *Load limiting resistor.*
lm *Lumen.*
LMFR *Liquid-metal fuel reactor* (U.S.).
LMS *Level measuring set,* which determines transmission levels relative to zero referenec level, 1 mW in 600 Ω.

LMTD *Logarithmic mean temperature differ-ence.*

ln Napierian or natural *logarithm* to base *e*.

LO *Local oscillator; locked open* relay.

Log Common *logarithm* to base 10.

Log dec *Logarithmic decrement* of decay of oscillation.

LOGANDS *Logical commands.*

LOGRAM *Logical programme.*

LOP *Line of position.*

LOPO *Low-power water boiler reactor* (U.S.).

LOS *Line of sight* for radio or radar transmission; *loss of signal.*

LP *Long-playing* record; *low-pass* filter; *linear programming.*

LPF *Low-pass* wave *filter.*

LpW *Lumens per watt* of a lamp.

LR *Level recorder.*

L+R, L−R Respectively the 'sum' and 'difference' elements in a stereo sum and difference signal complex.

LRD *Long-range data.*

LS *Loudspeaker; line scan.*

LSB *Least significant bit.*

LST *Local sidereal* (or *standard) time.*

Lt *Limit* (of summation or series).

LT(B) *Low-tension* (*battery*).

Lu *Lutecium.*

LUHF *Lowest useful high-frequency* radio wave.

LV *Low voltage.*

LVCD *Least voltage coincidence detection.*

LW *Long wave.*

Lw *Lawrencium.*

lx *Lux,* unit of illumination.

M *Mega-* (prefix for × 10^6).

M Symbol for *intensity of magnetization; magnetic moment; magnetic polarization; molecular weight; mutual inductance; m.m.f.; maxwell; Mach number.*

m *Milli-* (prefix for × 10^{-3}); *metre; minute.*

m Mass of electron.

MA *Mathematical Association.*

MAC *Magnetic automatic calculator; mechanical analogue computer.*

MADAM *Manchester automatic digital machine.*

MADRE *Magnetic drum receiving equipment* (U.S. Navy).

MADT *Microalloy diffused-base transistor.*

mag *Magnetron.*

magamp *Magnetic amplifier.*

MAGGI *Million ampere generator* at Aldermaston, used in thermonuclear studies.

MAGIC *Magnetic and germanium integer calculator.*

MANIAC *Mathematical analyser numerical indicator and computer* (U.S.).

MAR *Memory address register.*

MARS *Military affiliated radio system.*

MAT *Micro-alloy transistor,* suitable for VHF logical switching.

mA/V *Milliamperes per volt,* valve transconductance.

MAVAR *Mixed* (or *modulating) amplification by variable reactance;* see **parametric amplifier** in main text.

MB *Multi-band; mobile base; millibar; memory buffer.*

MBC *Mutual broadcasting system; miniature bayonet cap.*

MBR *Memory buffer register.*

MC *Mobile* (or *master) control; moving-coil.*

mc *Megacycle/second* (U.S.); *metre-candle.*

mCi *Millicurie.*

MCM *Monte Carlo method.*

MCR *Mobile control room.*

Mc/s, mcps *Megacycles per second.*

MCW *Modulated continuous wave.*

MD *Magnetic deflection* in CRT.

M-D *Modulation-demodulation.*

Md *Mendelevium.*

MDE ITU abb. for *mobile telemetering station.*

MDF *Main distribution frame* for circuits.

MDI *Magnetic direction indicator.*

MDR *Memory data register; multichannel data recorder.*

MDS *Minimum discernible signal.*

MDT *Mean down time.*

ME *Magic eye.*

MEA Beta *mercaptoethylamine* (cysteamine).

MEMISTOR *Memory resistor* storage device.

MERLIN *Medium-energy reactor light-water industrial neutron* source, a swimming-pool type of research reactor (AEI).

MES *Miniature Edison screw.*

MET *Meteorological; metal(lurgy).*

MEW *Microwave early warning.*

MF *Magnetic focus* of CRT; *medium frequency.*

mfd *Microfarad.* (Obs., now replaced by μF.)

mfp *Mean free path.*

mft *Mean free time.*

Mft L *Millifoot lamberts.*

MG *Motor generator.*

Mg *Magnesium.*

mg *Milligram(me).*

mH *Millihenry.*

MHCP *Mean horizontal candle-power.*

MHD *Magnetohydrodynamics.*

MHz *Megahertz.*

MICR *Magnetic ink character recognition.*

MIDAC *Michigan digital automatic computer.*

MIDAR *Microwave detection and ranging.*

MIG *Miniature integrating gyroscope* for inertial navigation.

mil *Military* (equipment designation).

min *Minute; minimum.*

MINNI A half-scale version of **MAGGI.**

Mintech *Ministry of Technology.*

MIR *Memory information register.*

MIT *Massachusetts Institute of Technology.*

mix Frequency *mixer.*

MK *Morse key.*

mk *Mark* of equipment.

MKG *Metre-kilogram(me).*

MKS *Metre-kilogram(me)-second* system of units.

MKSA *Metre-kilogram(me)-second-ampere,* system of units recently adopted by the International Electrotechnical Commission, in place of all other systems of units.

ML *Mutual inductance.*

ml *Millilitre.*
MLD *Minimum lethal dose* of radiation.
MLT *Maximum lethal time* of exposure to radiation, for 80% fatalities.
MM *Megamega-* (prefix for $\times 10^{12}$).
MMA *Multiple module access.*
m.m.f. *Magnetomotive force,* driving magnetic flux.
mmu *Milli* atomic *mass unit.*
Mn *Manganese.*
MO *Master oscillator; manually-operated.*
Mo *Molybdenum.*
mod. *Model; modulation; modification.*
MODEM Combined *modulator* and *demodulator.*
MOH ITU abb. for *mobile hydrological and meteorological station.*
mol Symbol for *mole.*
mon. *Monitor.*
MONECA *Motor network calculator.*
MOOSE *Man out of space easily,* U.S. space travel survival capsule.
MOPA *Master oscillator power amplifier.*
MOS *Metal oxide semiconductor; Ministry of Supply.*
MOSAIC *Ministry of Supply automatic integrator and computer.*
MOST *Metal oxide semiconductor transistor.*
MOUSE *Minimum orbital unmanned satellite earth.*
m.p. *Melting-point* in degrees.
MPC *Maximum permissible concentration.*
Mpc *Megaparsec.*
MPD *Maximum permissible dose* of gamma- or X-rays.
MPE *Maximum permissible exposure.*
MPL *Maximum permissible level* of gamma- or X-rays.
MPX *Multiplex.*
MR *Magnetic recorder.*
mr *Milliröntgen; millirad; millirem.*
MRC *Medical Research Council.*
MSAC *Moore School automatic computer.*
MSB *Most significant bit.*
MSC *Mile of standard cable.*
MSCP *Mean spherical candle-power.*
MSF *Medium standard frequency.*
MSL *Mean sea level.*
MST *Mountain standard time.*
MT *Mean time.*
Mt *Megaton.*
MTD *Mean temperature difference.*
MTI *Moving target indicator* radar.
MTR *Materials-testing reactor.*
Mu *Mutual conductance* in valve.
MUF *Maximum usable frequency.*
MUPO *Maximum undisturbed power output.*
MV *Medium voltage; megavolt.*
Mv *Mendelevium.* (obs.)
mV *Millivolt.*
MVC *Manual volume control.*
mV/m *Millivolts per metre.*
MW *Medium wave; megawatt.*
mW *Milliwatt.*
MWd/t *Megawatt days per tonne.*
MW(t) *Megawatt (thermal),* unit for heat power generated in a nuclear reactor, which has to be extracted by a coolant.

MWV *Maximum working voltage.*
Mx *Multiplex.*

N *Neper; Avogadro number; nitrogen; newton, shear modulus.*
N Symbol for *number of turns.*
n *Nano-* (prefix for $\times 10^{-9}$); *neutron; principal quantum number;* refractive index.
NA *Numerical aperture; not applicable; Nautical Almanac.*
Na *Sodium.*
NAB *National Association of Broadcasters* (U.S.).
NAFEC *National Aviation Facilities Experimental Centre* of the U.S. Federal Aviation Agency.
NAS *National Academy of Sciences* (U.S.).
NASA *National Aeronautics and Space Administration* (U.S.).
NATRON National Cash Register electronic data-processing system.
NAVAR Combined *navigation and radar* system.
NB *No bias.*
Nb *Niobium.*
NBC *Noise balancing circuit.*
NBFM *Narrow band frequency modulation.*
NBS *National Bureau of Standards.*
NC *Not connected* (on tube bases); *normally closed* contact of relay; *neutralizing capacitance* or *coil; no connection,* applicable to valve pins.
N/C *Numerical control.*
Nd *Neodymium.*
NDB Radio DF beacon (CAP).
NDT *Non-destructive testing.*
Ne *Neon.*
NEC *National Electronics Conference.*
NEMA *National Electrical Manufacturing Association.*
NEPTUNE An enriched-fuel, zero-energy experimental reactor (Harwell).
NERC *National Electronics Research Council.*
NERO *Sodium (Na) experimental reactor* of *zero* power (Winfrith)—a zero-energy graphite-moderated reactor system used to simulate a number of reactor systems.
NESTOR *Neutron source thermal reactor* (Winfrith), a modification of **JASON.**
NF *Noise* psophometric *factor.*
Nf *Nanofarad.*
NFB *Negative feedback.*
NI *Non-inductive.*
Ni *Nickel.*
NIC *Not in contact.*
NIMROD A code name for the 7 GeV proton synchrotron (Rutherford Laboratory).
NIO *National Institute* of *Oceanography.*
NIRNS *National Institute for Research in Nuclear Science,* which operates the high-energy Rutherford Laboratory at Harwell.
NLR *Noise load ratio.*
nm *Nautical mile.*
NMR *Nuclear magnetic resonance.*
No *Nobelium.*
n.o. *Normally open.*
NORC *Naval Ordnance research computer* (U.S.).

NOSMO *Norden optics setting mechanized operation.*
NP *Nickel-plated.*
Np *Neptunium.*
n.p.a. *Neutrons per absorption.*
NPD2 *Nuclear power demonstration,* 20 MW (E) generating plant (Canada).
n.p.f. *Neutrons per fission.*
NPL *National Physical Laboratory.*
NPM *Narrow-band phase modulation; counts per minute.*
NPR *Nuclear paramagnetic resonance.*
NQR *Nuclear quadruple resonance.*
NR *Noise ratio.*
nr *Non-reactive.*
NRC *National Research Council* (U.S.).
NRDC *National Research Development Corporation.*
NRM *Normalize.*
NRU A heavy-water moderated reactor 200 MW (Th) (Canada).
NRX Heavy-water reactor in Canada.
NRZ *Non-return-to-zero* in recording of analogue functions.
NSAA *National Space and Aeronautics Agency* (U.S.).
NSC *Noise suppression circuit.*
NSF *National Science Foundation* (U.S.).
nt *Nit.*
NTC *Negative temperature coefficient.*
NTI *Noise transmission impairment.*
n.t.p. or **s.t.p.** *Normal* or *standard temperature and pressure,* i.e., 0°C and 760 mm of mercury.
NTS *Not-to-scale.*
NTSC U.S. National Television System Committee.
NUDETS *Nuclear detonations.*
NUMARCOM *Nuclear* Power for *Marine* Purposes *Committee.*
nv *Neutron flux density.*
nVR *No voltage release.*
nvt *Total neutron flux* (time integral of flux density).

O *Oxygen.*
OABETA *Office Appliance and Business Equipment Trade Association.*
O and C *Operations and checkout.*
OAO *Orbiting astronomical observatory.*
OB *Outside broadcast.*
OBD *Omnibearing direction,* navigation.
OBI *Omnibearing indicator.*
OBS *Omnibearing direction selector; observed.*
O/C *Open circuit.*
OCC *Operations control centre.*
OCO *Open-close-open* contact.
OCP *Output control pulses.*
OCR *Optical character recognition.*
oct *Octal.*
OD *Outside diameter.*
ODO *Out of order.*
ODR *Omnidirectional range.*
OFHC *Oxygen-free high-conductivity copper.*
OGO *Orbiting geophysical observatory.*
OGRA Russian thermonuclear device, similar to **DCX.**
OI *Oil immersed* (or *insulated*).

OIRT *International Organization of Radio and Television* in East Europe.
O/L *Operations/logistics.*
OLMR *Organic liquid-moderated reactor.*
OLRT *On-line real time.*
OPCON *Optimizing control.*
OPTUL *Optical pulse transmitter using laser.*
OR *Outside radius; operations research.*
O/R *On request.*
ORACLE *Oak Ridge automatic computer and logical engine,* a high-speed U.S. digital computer.
ORB *Omnidirectional radio beacon* (**VOR** preferred).
ORD *Optical rotary dispersion.*
ORINS *Oak Ridge Institute* for nuclear studies.
ORNL *Oak Ridge National Laboratory.*
OS *Outsize; output secondary* of transformer.
Os *Osmium.*
Osc. *Oscillator.*
OSO *Orbiting solar observatory.*
OSWALD Early apparatus used on controlled thermonuclear reaction work at Aldermaston (now in Science Museum, Kensington).
OT *Overtime.*
OTU *Operational training unit.*
out *Output*
OWF *Optimum working frequency.*
OWR *Omega West reactor* in U.S.
OWU *Open window unit.*

P *Phosphorus; Poynting vector.*
P Symbol for *power;* for *polarization.*
p *Pico-* (prefix for $\times 10^{-12}$); *proton.*
p Symbol for *pressure;* momentum.
PA *Public-address* system; *power* (or *pulse*) *amplifier.*
Pa *Protactinium.*
PAC *Personal analogue computer.*
PAM *Pulse amplitude modulation.*
PAN Urgent signal preceding R/T message.
pan. *Panoramic; panchromatic.*
PANC *Power amplifier neutralizing capacitor.*
PAR *Precision-approach radar.*
par *Parameter.*
P-as-B or **P-as-R** *Programme as broadcast* or *recorded* in radio.
PAX *Private automatic exchange.*
PB *Pushbutton.*
Pb *Lead.*
PBF *Plates* or anodes for *beam forming.*
P(A)BX *Private (automatic) branch exchange.*
PC *Plate* or anode *circuit; programme* (or *pulse*) *counter; punched card; process controller; pulsating current.*
Pc *Printed circuits.*
PCC *Programme-controlled computer.*
PCM *Pulse code modulation.*
Pd *Palladium.*
p.d. *Potential difference* or voltage.
PDA *Post-deflection accelerator; probability distribution analyser.*
PDF *Probability distribution function.*
PDM *Pulse duration modulation.*

PDPS *Parts data-processing system.*
PDR *Power directional relay.*
PDU *Pilot's display unit.*
PE *Permanent echo.*
PEC *Photoelectric cell.*
PEM *Photoelectromagnetic effect.*
pen. *Pentode.*
PEP *Peak envelope power*; *Plessey electronic payroll.*
PERA *Production Engineering Research Association* of U.K.
PES *Photoelectric scanning.*
PF *Power factor*; *paper* and *foil* capacitor; *pulse frequency.*
pF *Picofarad.*
PFM *Pulse frequency modulation.*
pfr Tape *perforator.*
PG *Pulse generator.*
PGS *Pressure-gradient single-ended microphone.*
pH *Hydrogen* ion concentration *potential.*
ph *Phase*; *phot.*
PHA *Pulse height analyser.*
PhM *Phase modulation.*
phos. *Phosphor(escence).*
PI *Paper insulated*; *position indicator*; *programmed instruction.*
PIC *Pulse ionization chamber.*
PIM *Precision indication* of the *meridian.*
P in A *Parallax in Altitude* (navigation).
PIPPA *Pressurized pile for producing power and plutonium,* the protodesign for the Calder Hall type of reactor.
PIV *Peak inverse voltage.*
Pk *Peak*; *pack(age).*
PL *Proportional limit*; ITU abb. for *radio positioning land station.*
pl. *Plug*; *plate* or anode.
PLUTO *Plutonium* loop-testing reactor similar to **DIDO**, i.e., a heavy-water moderated reactor (at Harwell).
PM *Phase* or *pulse modulation*; *permanent magnet*; *potentiometer*; *post meridiem.*
Pm *Promethium.*
PMBX *Private manual branch exchange.*
PME *Photomagnetoelectric effect.*
PMH *Production per man hour.*
PO *Power output*; *polarity*; ITU abb. for *radio positioning mobile station*; *pulse type transmission* in radar.
Po *Polonium.*
POGO *Polar orbiting geophysical observatory.*
POLY *Polyethylene.*
POPI *Post Office position indicator,* a navigational radio aid, developed by the Post Office, in which sequential transmissions at a single frequency are made from two or more separated aerials at a fixed station.
POPOP 1,4 Di (2(5 phenyloxazolyl))–benzene.
Pot. *Potential*; *potentiometer.*
POV *Putting-on voltage* in Doppler navigation.
PP *Push-pull*; *pilot punch*; *pole piece.*
p-p *Peak-to-peak.*
p-ph-m *Pulse phase modulation.*
PPI *Plan position indicator.*

PPL *Plan position landing*; *transistor-transistor logic.*
PPM *Pulse position modulation*; *parts per million.*
PPμL *Transistor-transistor micrologic.*
PPO *Push-pull output*; 2, 5 Diphenyloxazole.
PPPI *Precision plan position indicator.*
PPPPI *Photographic projection plan position indicator.*
PPS *Pulses per second.*
P-pulse *Position pulse.*
Pr *Praseodymium.*
pref. *Prefix.*
PRF *Pulse repetition frequency.*
pri *Primary.*
PRN *Print numerically.*
PRO *Print octal.*
PROFIT *Programmed reviewing, ordering and forecasting inventory technique.*
PRONTO *Programme for numerical tool operation.*
PRP *Pseudo-random pulse.*
PRR *Pulse repetition (recurrence) rate.*
PRS *Pattern recognition system.*
PRV *Peak reverse voltage.*
PS *Pulses per second*; *periods per second* (obs.); *power supply* (U.S.).
PSEC *Picosecond* (10^{-12} sec.).
PSP *Peak sideband power.*
PST *Pacific standard time.*
PSW *Programme status word.*
Pt *Platinum.*
PTFE *Polytetrafluoroethylene.*
PTM *Pulse time modulation.*
PTP *Point to point.*
P to P *Plate to plate.*
PU *Pick-up.*
Pu *Plutonium.*
Puff *Picofarad.*
PV *Prime vertical.*
PVA *Polyvinyl acetate.*
PVC *Polyvinyl chloride.*
PVD *Paravisual director* system for instrument flying.
PW *Private* (third) *wire* in exchange circuit; *pulse width.*
PWM *Pulse width modulation.*
PWR *Pressurized-water reactor* (U.S.).

Q Symbol for *Q-factor*; *electric charge*; *reactive power*; *disintegration energy* or *heat.*
QA *Quick-acting relay*; *quiescent antenna.*
QAVC *Quiescent automatic volume control.*
QF *Quench frequency*; *quality factor.*
QMB *Quick make-and-break* contact.
QP *Quadruple play.*
QPP *Quiescent push-pull* valves or transistors.
QRA *Quality reliability assurance.*
QSG *Quasi-stellar galaxy.*
QSS *Quasi-stellar source.*
QT *Queueing theory.*

R Gas constant per mole; *resistance*; *reluctance*; conversion ratio; *Rydberg constant*; *Reynolds number.*
Ra *Differential anode resistance* of a valve.
Rp *Plate resistance*; U.S. for **Ra**.

r *Röntgen* unit; *radius.*
RA *Right angle; right ascension.*
Ra *Radium.*
RABAL *Radiosonde balloon.*
RACES *Radio Amateur Civil Emergency Service* (U.S.).
rad. *Radius; radian; radiator* (radio); dose of X or γ-rays; *radio.*
RADEX *Radiological exclusion* area.
RADIC *Redifon analogue-digital computing system,* a simulator for naval control equipment.
RAE *Royal Aircraft Establishment* at Farnborough, Hants.
RaG *Radium G.*
RAM *Random access memory.*
RAMAC *Random access method of accounting and control,* TN for a magnetic-disk store.
RAMPS *Resource allocation and multiproject scheduling.*
RAPCON *Radar approach control.*
RAPPI *Random access plan position indicator.*
RAS *Royal Astronomical Society.*
RAeS *Royal Aeronautical Society.*
RASCAL *Royal Aircraft Establishment sequence calculator.*
RATT *Radio teletypewriter.*
RAWIN *Radio wind* measurements.
RAX *Rural automatic exchange.*
RAYDAC *Raytheon digital computer* (U.S.).
RB *Radio bearing; rubber-based.*
Rb *Rubidium.*
rbe *Relative biological effectiveness* of radiation.
RBS *Radar bomb scoring,* in radar simulation.
RC *Record changer; radar* (or *ray*) *control; resistance-capacitance coupling.*
RCC *Resistor colour code; Radio Chemical Centre* at Amersham, Bucks, chief U.K. source of processed radioisotopes.
RCG *Reverberation controlled gain.*
RCM *Radar counter measure.*
RCO *Remote-control oscillator.*
RC(S) *Remote control (system).*
RCSC *Radio Component Standardization Committee.*
RD *Radar; recording demand (meter); research and development.*
rd *Rutherford.*
RdAc *Radioactinium.*
RDF *Radio direction-finding.*
RDFL *Reflection direction-finding, low-angle.*
RDT and E *Research, development, test and evaluation.*
RdTh *Radiothorium.*
RE *Real number.*
Re *Rhenium.*
recce *Reconnaissance.*
RECMF *Radio and Electronic Component Manufacturers' Federation.*
Reg. *Regulator; register.*
Rel. *Relay; release; reluctance; relief valve.*
REMC Joint Departmental *Radio and Electronics Measurements Committee* for Services.

REME *Royal Electrical and Mechanical Engineers,* a corps of the British Army.
Rep *Repeat.*
REPROD *Receiver protective device.*
Res. *Resistor; reserve.*
RETMA *Radio-Electronics-Television Manufacturers' Association* (U.S.).
RETSPL *Reference equivalent threshold sound pressure level.*
REX *Reactor experimental* in U.S.S.R.
RF *Radio frequency; range finder.*
RFC *Radio-frequency choke.*
RG *Radiogram(ophone); rectangular guide;* ITU abb. for *direction-finding radio station.*
RH *Relative humidity; Rockwell hardness.*
Rh *Rhodium.*
r/h *Röntgen, rad, rem* or *revolutions per hour.*
RHEL *Rutherford High Energy Laboratory.*
RHI *Range height indicator.*
rhm *Röntgen-hour-metre.*
RI *Royal Institution,* London; *read-in; reliability index.*
RIC *Radio Industry Council; Royal Institute of Chemistry.*
RID *Radio Intelligence Division* of the Federal Communications Commission (U.S.).
RISE *Research in supersonic environment* (U.S.).
RkVA, kVAR *Reactive-kilovolt-ampere,* capacitor or inductor loading.
RL *Radiolocation,* former term for *radar;* ITU abb. for *radio-navigation land station.*
RLA ITU abb. for *aeronautical marker beacon station.*
RLB ITU abb. for *aeronautical radio beacon station.*
RLC ITU abb. for **RACON** *station.*
RLG ITU abb. for *glide path station.*
RLL ITU abb. for *localizer station.*
RLM ITU abb. for *marine radio beacon station.*
RLN ITU abb. for *loran station.*
RLO ITU abb. for *omnidirectional range station.*
RLR ITU abb. for *radio-range station.*
RMI *Radio magnetic indicator.*
rms *Root-mean-square* of waveform.
rmse *Root-mean-square error.*
RMU *Radio maintenance unit.*
Rn *Radon.*
RNG *Radio range* (CAP).
RNSS *Royal Naval Scientific Service* of Admiralty.
RO ITU abb. for *radio-navigation mobile station; read out.*
ROA ITU abb. for *altimeter station.*
RP *Radar plot; real part* (of complex number).
RPC *Remote position control.*
RPMI *Revolutions per minute indicator.*
rpm/s *Revolutions per minute/second.*
RPS *Radiological Protection Service; Royal Photographic Society.*
RPT *Reactor for physical* and *technical* investigations (U.S.S.R.).
RR *Radio range; repetition rate.*
RRB *Radio Research Board.*

RRDE *Radio Research and Development Establishment*, now in **RRE**.

RRE *Radar Research Establishment* at Malvern.

RRS *Radio Research Station* of DSIR at Slough.

RS *Royal Society.*

RS and I *Rules, standards and institutions.*

RSGB *Radio Society of Great Britain.*

RSI *Royal Signals Institution.*

RSO *Radiological Safety Officer.*

RSS *Root-sum-squares.*

RT *Rise time* pulse.

R/T *Radio telephony.*

RTC *Radio tuned circuit.*

RTM *Registered trademark.*

RTP *Reference telephonic power.*

RTR *Repeater test rack.*

RTRA *Radio and Television Retailers Association.*

Ru *Ruthenium.*

R-unit *Röntgen* X- or γ-ray dosage *unit.*

RV *Radio vehicle; relief valve.*

RVA, VAR *Reactive-volt-ampere.*

RX *Receiver.*

ry *Relay.*

RZ *Return to zero*, in recording of analogue functions.

S *Sulphur; south; sabin; Stokes.*

S Symbol for *Poynting vector; apparent power; reluctance; sensitivity; entropy.*

s *Second; spin quantum number.*

SA *Slow acting* relay; *ship to aircraft.*

SAM *Surface-air missile.*

SAR *Submarine Advanced Reactor.*

SAT *Society of Acoustic Technology.*

SATCO *Signal automatic air traffic control.*

SB *Simultaneous broadcast.*

Sb *Antimony.*

sb *Stilb.*

SBA *Standard beam approach.*

SBC *Small bayonet cap.*

SBT *Surface barrier transistor.*

SC *Side contact* for valves; *single contact; sine-cosine; stop-continue* register.

S/C *Short-circuit.*

Sc *Scandium.*

SCA *Subcritical assembly.*

SCC *Single cotton-covered* wire.

SCE *Single cotton-covered enamelled* wire.

SCH *Schedule.*

SCORPIO *Subcritical carbon-moderated reactor assembly* for *plutonium* investigations —exponential and subcritical assemblies at Winfrith for experimental work.

SCR *Single-channel reception.*

SCRAM *Selective combat range artillery missile.*

SD *Semi-diameter.*

SDC *Single data converter.*

SDP *Slowing-down power.*

SDS *Special distress signal.*

SDV *Slowed-down video* radar.

Se *Selenium.*

SEAC National Bureau of *Standards Eastern automatic computer* (U.S.).

SEC *Simple electronic computer* (Birkbeck College, London).

sec *Second; secant; secondary.*

SECO *Sequential control* (Teletype system).

Sel. *Selector* in telephony.

sels. *Selsyn* motor.

selsyn *Self-synchronizing.*

SENN *Società Elettronucleare Nazionale* (Italy).

SEQUIN *Sequential quadrature inband* (CC colour TV system).

SERE *Services Electronic Research Establishment.*

SERT *Society of Electronic and Radio Technicians.*

SES *Small Edison screw* cap for lamps.

SF *Side frequency; signal frequency.*

SFERT *Système Fundamental Européen de Référence pour la Transmission Téléphonique* (European master telephone reference system).

SG, sg, s.g. *Screened-grid; spark gap; signal* (or *sweep*) *generator.*

SGHWR *Steam-generating heavy-water reactor.*

sgl *Signal.*

SH *Scleroscope hardness.*

SHA *Sidereal hour angle.*

SHF *Super-high-frequency* of electromagnetic waves, between 3000 and 30,000 MHz.

SHM *Ship heading marker* in radar.

s.h.m. *Simple harmonic motion.*

SI *Système International* d'Unités (system of units).

Si *Silicon.*

Si Co *Signal control.*

SIC *Specific inductive capacity* of dielectric.

SID *Sudden ionospheric disturbance.*

SIF *Sound intermediate frequency.*

sig *Signal.*

SIL *Speech interference level.*

SIMA *Scientific Instrument Manufacturers' Association* of Great Britain.

SIR *Submarine intermediate reactor.*

SIRA *Scientific Instrument Research Association* of Great Britain.

SISTER *Special Institute for Scientific* and *Technological Education* and *Research.*

SK *Socket; sketch.*

SLC *Searchlight control* radar; *straight-line* variation of *capacitance.*

SLEEP *Swedish low-energy experimental pile* reactor at Stockholm.

SLF *Straight-line frequency* (tuning capacitor)

SLM *Sound-level meter.*

SLT *Solid logic technology.*

SLW *Straight-line wavelength* (tuning capacitor).

Sm *Samarium.*

SMF *Special modifying factor.*

SML *Symbolic machine language.*

smp *Sampler.*

SMPTE *Society of Motion Picture and Television Engineers*, Hollywood, U.S.

SMT *Ship mean time.*

SN, S/N *Signal-to-noise* ratio in dB.

Sn *Tin.*

SNAP *Systems for nuclear auxiliary power.*

SO *Slow-operating* relay; *submarine oscillator.*
SOD *Sound-on-disk* for soundfilm.
SOF *Sound-on-film.*
Sol. *Solenoid; soluble.*
SOV *Shut-off valve.*
sp *Single-pole; special purpose.*
sp. *Specific; spark; spare.*
sp. gr. *Specific gravity.*
sp. vol. *Specific volume.*
SPA *Sudden phase anomaly.*
SPAT *Silicon precision alloy transistor.*
spdt *Single-pole double-throw* switch.
spec. *Specification; specimen.*
SPL *Sound pressure level.*
spst *Single-pole single-throw* switch.
SQC *Statistical quality control.*
SR *Ship-to-shore* radio or radar; *skywave synchronization* (loran); *slow release; starting relay; send-receive* switch.
Sr *Strontium.*
SRAEN *Système de Référence pour la détermination de l'Affaiblissement Equivalent pour la Netteté* (master telephone transmission reference system).
SRC *Science Research Council.*
SRDE *Signals Research and Development Establishment.*
SRE *Surveillance radar element,* which determines distance and direction of a landing aircraft; *sound reproduction equipment; sodium reactor experimental* (U.S.).
SRP *Society for Radiological Protection.*
SRRC *Scottish Reactor Research Centre.*
SRV *Surface recombination velocity.*
SS *Solid state; signal selector;* ITU abb. for *standard frequency station.*
SSB *Single sideband.*
SSC *Sea-state correction* in Doppler navigation.
SSEB *South of Scotland Electricity Board.*
SSEC *Selective sequence electronic calculator.*
SSI *Sector scan indicator.*
SS-LORAN *Sky-wave synchronized loran.*
SSM *Surface-surface missile.*
SSR *Secondary surveillance radar.*
SSS *Single-signal supersonic* heterodyne.
SSSC *Single sideband suppressed carrier.*
STALO *Stable local oscillator* for Doppler.
STANTEC *Standard Telephones electronic computer.*
STC *Sensitivity time control,* swept gain; *short-time constant; Standard Telephones and Cables.*
STD *Subscriber trunk dialling.*
std *Standard.*
STEP *Standard tape executive programme.*
STET *Specialized technique for efficient typesetting.*
STN *Station.*
s.t.p. See **n.t.p.**
STR *Scientific and Technological Research; submarine thermal reactor.*
STTO *Sawtooth timing oscillator.*
SU *Sensation unit* for hearing in dB; *strontium unit.*
SUN Commission on *Standard Units and Nomenclature* of IUPAP.

SUP *Suppressor grid* in a pentode valve; *suppress.*
SUPO *Super power* water boiler in U.S.
Supy *Supervisory.*
SW *Sandwich-wound; short wave.*
SWAC National Bureau of *Standards Western Automatic Computer* (U.S.).
SWAMI *Standing-wave area motion indicator.*
Swbd *Switchboard.*
SWG *Standard Wire Gauge.*
Swgr *Switchgear.*
S-wire *Sleeve*-connected *wire* of plug.
SWP *Safe working pressure.*
SWR *Standing-wave* (or *short-wave*) *ratio.* See **VSWR.**
SWTL *Surface wave transmission line.*
SX *Solvent extraction.*
Sync. *Synchronous, -ize, -ization.*

T Symbol for *filter section* consisting of shunt arm between two identical series arms; *tritium; Kelvin temperature; tesla; tera-* (prefix for $\times 10^{12}$),
T *Period.*
t *Celsius temperature; time; tonne; time constant; relaxation time; Fermi age; transmittance; radioactive mean life.*
t½ *Radioactive half-life.*
TA *Target* in CRT.
Ta *Tantalum.*
Tab. *Tabulate* in computer.
TAC *Television Advisory Committee* for development in U.K.
T and D *Transmission and distribution.*
TAS *True air speed.*
TASI *Time assigned speech interpolation,* a multichannel telephone system.
TAT 1 and 2 *Transatlantic telephone* cables.
TATG *Tuned anode tuned grid.*
TB *Transmitter-blocker; time base.*
Tb *Terbium.*
TC *Thermocouple; top contact* of valve; *trip coil; time closing.*
Tc *Technetium.*
TCG *Time-controlled gain.*
TD *Time delay; tunnel diode; turntable desk* for TV and SF.
TDF *Trunk distribution frame; two degrees of freedom.*
TDM *Time division multiplex; time duration modulation.*
TDR or **X** *Torque differential receiver* or *transmitter.*
TDTL *Tunnel diode transistor logic.*
TE *Transverse electric* waves in waveguides; *totally enclosed.*
Te *Tellurium.*
TE(S)FC *Totally enclosed (separately) fan-cooled.*
TELEX Automatic *tele*typewriter *ex*change service.
TEM *Transverse electric and magnetic.*
Tem. *Temperature; template; tempered.*
TEMA *Telecommunications Engineering and Manufacturing Association.*
term. *Terminal* of circuit.

TEWC *Totally-enclosed water-cooled.*
TFE Teflon-6, U.S. for **PTFE.**
TFT *Thin-film technology* (or *transistor*).
TG *Tuned grid.*
TG-LORAN *Traffic-guidance loran.*
TGTP *Tuned grid tuned plate.*
Th *Thorium.*
therm. *Thermal; thermostat; thermometer.*
TI *Technical Instructor; tuning indicator.*
Ti *Titanium.*
TIF *Telephone influence* (or *interference*)
  factor.
TIROS *Television and infrared observation
  satellite.*
TJ *Test* (or *telephone*) *jack.*
TL *Transmission level; time limit.*
Tl *Thallium.*
TLZ *Transfer on less than zero.*
TM *Technical manual* of instruction; *true
  motion* radar navigation; *transverse mag-
  netic waves* in waveguide; *temperature* (or
  tuning) *meter*; magnetic *tape module.*
Tm *Thulium.*
TMS *Transmission measuring set.*
TN *Trade name.*
TNF *Transfer on no overflow.*
TNZ *Transfer on non-zero.*
TO *Time opening; turn-over.*
tor. *Torque.*
TOS *Temporarily out of service.*
TP *Tuned plate.*
TPC *Triple paper-covered.*
tpdt *Triple-pole double-throw* wire.
TPI *Threads per inch* of screws.
t.p.i. *Tape-position indicator.*
tpr *Teleprinter.*
TPTG *Tuned plate tuned grid.*
TR *Transmitter-receiver; transmit-receive*
  (tube); *tape recorder; torque receiver.*
TRACE *Taxi-ing and routing of aircraft
  coordination equipment.*
tr(ans). *Transformer; transfer; transverse.*
TRE *Telecommunications Research Establish-
  ment,* now in **RRE.**
TRF *Tuned radio frequency.*
TRIDAC *Three-dimensional analogue com-
  puter* at RAE.
TRN *Transfer.*
TRS *Tough rubber-sheathed* cable.
TS *Terminal strip; temperature switch;
  Television Society.*
TSS *Time-sharing system.*
TST *Television signal tracer.*
tstr *Transistor.* See **X.**
TTC *Terminating toll centre.*
TTL *Transistor-transistor logic.*
TTM *Transit time modulation.*
TTR *Low-power thermal test reactor* in U.S.
tty *Teletypewriter.*
TU *Traffic unit* in telephony; *transmission
  unit,* replaced by **dB.**
TUF *Thermal utilization factor.*
TUN. *Tuning.*
TV(I) *Television (interference).*
TVC *Technical Valve Committee.*
TVM *Transistor voltmeter.*
TVR *Temperature variation* in *resistance.*
TW *Travelling wave.*

TW(I) *Tail warning (indicator)* radar.
T-wire *Tip-*connected *wire* of plug.
TWT *Travelling-wave tube.*
TWX *Teleprinter exchange* for Telex.
TX *Transmitter* of torque.

U *Uranium.*
U *Internal energy.*
u *Unified mass unit; group velocity.*
UAM *Underwater-air missile.*
UAX *Unit automatic exchange.*
UCC *University College computer,* London.
UD *Universal dipole.*
UDC *Universal decimal classification.*
UDOFT *Universal digital operational flight
  trainer.*
uF Modification of μF (*microfarad*).
UFO *Unidentified flying object.*
uH Modification of μH (*microhenry*).
UHF *Ultra-high frequency.*
UKAEA *United Kingdom Atomic Energy
  Authority.*
UNIVAC *Universal automatic computer* in
  U.S.
UPO *Undistorted power output.*
USAEC *United States Atomic Energy Com-
  mission.*
USM *Underwater-surface missile.*
USS *United States Sellers* standard threads.
USW *Ultra-short wave.*
UT *Universal time* (formerly **GMT**).
UUM *Underwater-underwater missile.*
u.v. *Ultraviolet.*

V *Valve; volt; vanadium.*
V Symbol for *electromotive force; potential
  difference.*
v *Velocity* (phase velocity).
VA *Volt-ampere.*
VAC, vac *Voltage of alternating current*
  (U.S.).
vac. *Vacuum.*
var. *Variable.*
Varicap. *Voltage-operated variable capacitor.*
VASCA *Valve and Semiconductor
  Manufacturers' Association.*
VB *Valence bond.*
VC *Variable capacitor; volt-coulomb; voice
  coil; volume control; video correlator;
  voltage compensator.*
VCD *Variable-capacitance diode.*
VCF *Variable crystal filter.*
VCM *Voltage-controlled multivibrator* in
  Doppler navigation.
v.d. *Vapour density.*
VDR *Voltage-dependent resistor.*
VDS *Variable depth sonar.*
VEB *Variable elevation beam* antenna.
vel. *Velocity.*
VERA *Versatile reactor assembly; vision
  electronic recording apparatus.*
VF *Voice* (or *video*) *frequency.*
VFO *Variable frequency oscillator.*
VGA *Variable gain amplifier.*
VGPI *Visual glide path indicator,* for landing
  aircraft.
VH *Vickers hardness.*
VHF *Very high frequency.*

VI   *Volume indicator.*
vib.   *Vibrator.*
Vid.   *Video.*
VIDAT   *Visual data acquisition.*
VIR   *Vulcanized india-rubber.*
vit.   *Vitreous.*
VLF   *Very low frequency*, i.e., below 30 kHz.
VLR   *Very long range* radar.
VM   *Vertical magnet; velocity modulation* of beam.
V/M, VPM   *Volts per metre* field strength; *volts per mil* dielectric strength.
VO   *Valve oscillator.*
VOC   *Variable output circuit.*
VODACOM   *Voice data communications.*
VOM   *Volt-ohm meter.*
v.p.   *Vapour pressure.*
VPC   *Voltage phasing control.*
VPM   *Volts per metre.*
VPR   *Virtual plan-position indicator reflectorscope.*
VR   *Voltage regulator* (or *reference*) tube; *voltage relay; vulcanized rubber.*
VRF   *Visual flight rules.*
VRR   *Visual radio range.*
VSB   *Vestigial sideband.*
VSG   *Variable speed gear*, continuously variable for mechanical integration in a computer.
VSR   *Very short range* radar.
VSW   *Very short waves.*
VSWR   *Voltage standing-wave ratio.*
VT(V)   *Vacuum tube (voltmeter).*
VTB   *Voltage time to breakdown.*
VTR   *Video tape recorder* or *recording.*
VTVM   *Vacuum tube voltmeter.*
VU   *Volume unit.*
VW   *Working voltage.*

W   *Watt; tungsten* or *wolfram.*
W   Symbol for *electrical energy.*
w   *Work function.*
WAC   *Working alternating current.*
WAF   *With all faults.*
waf.   *Wafer* switch.
WB   *Wet bulb* thermometer; *weather bureau.*
Wb   *Weber.*
WBNS   *Water boiler neutron source.*
WC   *Water-cooled.*
W/C, WPC   *Watts per candle-power.*
WDC   *Working direct current.*
WED   *War emergency dose.*
Wh   *Watt-hour.*
WL   *Wavelength.*
WO   *Wireless Operator; write out.*
WP   *Weather permitting; working pressure* (or *point*).
WPC   *Watts per candle-power.*
WPM   *Words per minute.*
WR   *Wire recorder; wave retardation* by electrode.

WRU   *Who are you?* on Telex.
W/T   *Wireless telegraphy; walkie-talkie.* See R/T.
WTR   *Westinghouse test reactor* in U.S.
WVAC   *Working voltage alternating current.*
WVDC   *Working voltage direct current.*
WW   *Wirewound* resistor.
WXD   ITU abb. for *meteorological radar station.*
WXR   ITU abb. for *radiosonde station.*

X   Symbol for *transistor* on Dwg; *crystal cut; horizontal deflection on CRT.*
X   Symbol for *reactance.*
Xe   *Xenon.*
X-former   *Transformer.*
XI/O   *Execute input/output.*
X-mitter   *Transmitter.*
XS   *Extra-strong.*
Xtal   *Crystal.*
XU   *X-unit* for X- or γ-rays.
X-wave   *Extraordinary wave* or *ray.*

Y   *Late operating contact* in relay; *yttrium; vertical deflection on CRT*; *3-phase star connection.*
Y   Symbol for *admittance.*
y   *Year.*
Yb   *Ytterbium.*
YIG   *Yttrium-iron-garnet.*
YP   *Yield point.*
YS   *Yield strength.*

Z   Symbol for *impedance;* for *figure of merit;* for *atomic number.*
z   Symbol for *valency of ion.*
ZA   *Zero and add.*
ZC   *Centre-zero* instrument.
ZD   *Zenith distance* (angular in navigation).
ZDA   *Zinc Development Association.*
ZEBRA   *Zero-energy breeder assembly* (Winfrith) for study of neutron physics of fast reactor fuel assemblies.
ZENITH   *Zero-energy nitrogen-heated thermal* reactor (Winfrith)—high-temperature gas-cooled reactor in Dragon project.
ZEPHYR   *Zero-energy, plutonium-fuelled, fast reactor* at Harwell.
ZERA   *Zero-energy critical assemblies* reactors at Harwell.
ZETR   *Zero-energy thermal reactor* (Harwell, now Dounreay) for studying homogeneous aqueous systems.
ZF   *Zero frequency.*
ZFB   *Signals fading badly.*
Zn   *Zinc.*
ZOE   *Zero energy.*
Zr   *Zirconium.*
ZST   *Zone standard time.*

DICTIONARY

A

α. See under **alpha**.

A-amplifier. One associated with, or immediately following, a high-quality microphone, as in broadcasting studios. *N.B.*—Not the same as **class-A amplifier.**

A and R display. Expansion of a part of a regular *A display* (q.v.).

A-battery. U.S. term for **low-tension battery.**

AB-battery. Combination of A (for heating cathodes) and B (anode supply) for portable equipment.

A-bomb. See **atomic bomb.**

A display. Coordinate display on a CRT in which a level time base represents distance and vertical deflections of beam indicate echoes.

AN direction finding. System in which two signals corresponding to morse A and N (. — and — .) are received simultaneously. When on course these are of equal intensity and blend into a continuous steady tone.

A-service area. Region around a broadcasting station where the electric field strength is greater than 10 mV/metre.

AT-cut. Special cut of a quartz crystal to obviate temperature effects; a cut such that the angle made with the Z-axis is 35·5°.

ATR tube. *Anti-transmit-receive* (q.v.) electronic switch, used in radar systems which have a common transmitting and receiving aerial.

AU diode. See **backward diode.**

ab-. Prefix to name of unit, indicating derivation in the CGS system.

abampere. The current which, when flowing round a single turn coil of 1 cm radius in vacuum, produces a field of 2π oersted at the centre. This is the CGS absolute electromagnetic unit of current and is equal to 10 absolute amperes.

aberration. In an image-forming system, e.g., optical or electronic lens, failure to produce a true image, i.e., a point object as a point image, etc. Five geometrical aberrations were recognized by von Seidel, viz., coma, spherical aberration, astigmatism, curvature of the field and distortion.

abnormal glow. Excess of current over that required to cover the cathode completely with a glow discharge.

abnormal polarization. Condition in an electromagnetic wave when the alternating magnetic field has a vertical component.

abnormal reflections. Those from the ionosphere for frequencies in excess of the *critical frequency*.

absolute address. Code designation of a specific storage register in a computer.

absolute ampere. The standard MKS unit of electric current; replaced the *international ampere* in 1948. See **ampere.**

absolute electrometer. High-grade attracted-disk electrometer in which an absolute measurement of potential can be made by 'weighing' the attraction between two charged disks against gravity.

absolute permeability. See **permeability.**

absolute temperature. Kelvin (Celsius or Centigrade) or Rankine (Fahrenheit) degree scale from absolute zero ($-273\cdot16°C$ or $-459\cdot69°F$), which is lowest possible, molecules then having no thermal energy.

absolute units. Those derived directly from the fundamental units of a system and not based on arbitrary numerical definitions. The internationally adopted fundamental units are the metre, kilogram, second, degree Kelvin, ampere and candela.

absolute wavemeter. One in which the frequency of an injected radio-frequency e.m.f. is determined by calculation of the elements of the circuit, or length of a resonant line (cymometer).

absorbed dose. See **dose.**

absorber. Any material which converts energy of radiation or particles into another form, generally heat. Energy transmitted is not absorbed. Scattered energy is often classed with absorbed energy. See **total absorption coefficient, true absorption coefficient.**

absorber valve. That in an absorption modulator which absorbs excess power during troughs of the modulation cycle. In a telegraph transmitter, it is used to stabilize voltages during keying.

absorbing material. Any medium used for absorbing energy from radiation of any type.

absorbing rod. See **control rod.**

absorption. (1) Reduction in intensity of any form of radiated energy resulting from energy conversion in medium. *Broad beam absorption*, relatively unaffected by scatter, is generally less than *narrow beam attenuation* where much energy is lost by scattering processes. See **absorber, attenuation.** (2) The process by which a fluid frequently penetrates the entire volume of a solid (cf. *adsorption*) or by which one chemical substance may dissolve into another. (3) The incorporation of a bombarding particle into a nucleus with which it interacts.

absorption capacitor. One connected across a spark gap to damp the discharge.

absorption coefficient. (1) At a discontinuity (*surface absorption coefficient*): (*a*) the fraction of energy which is absorbed or (*b*) the reduction of amplitude, for a beam of radiation or other wave system incident on a discontinuity in the medium through which it is propagated, or in the path along which it is transmitted. (2) In a medium (*linear absorption coefficient*), the natural logarithm of the ratio of incident and emergent energy or amplitude for a beam of radiation passing through unit thickness of a medium. (The *mass absorption coefficient* is defined in the same way but for a thickness of the medium corresponding to unit mass per unit area.) *N.B.—True absorption coefficients* exclude scattering losses. *Total absorption coefficients* include scattering losses.

absorption control modulation. Early system, in which a modulating valve absorbs power from the main radio-frequency oscillation in the radiating antenna.

absorption edge. The wavelength at which there is an abrupt discontinuity in the intensity of an absorption spectrum for electromagnetic waves, giving the appearance of a sharp edge in its photograph. It is due to one particular energy-dissipating process becoming possible or impossible at the limiting wavelength. In X-ray spectra of the chemical elements the K absorption edge for each element occurs at a wavelength slightly less than that for the K emission spectrum. Also **absorption discontinuity.**

absorption mesh. Metal grid used in waveguides to absorb unwanted component of energy propagated.

absorption meter. Instrument for measuring light transmitted through transparent sample (liquid or solid) using a detector such as a photocell. See also **absorption wavemeter.**

absorption spectrum. Graph relating absorption coefficient of medium for electromagnetic waves to frequency. Cf. *emission spectrum.*

absorption wavemeter. One which depends on resonance absorption in a tuned circuit, constructed with very stable inductance and capacitance.

abstat-. One-time U.S. prefix to names when used for absolute electrostatic units, e.g., *abstatampere, abstatfarad.*

abundance. For a specified element, the proportion or percentage of one isotope to the total, as occurring in nature. Also **abundance ratio, relative abundance.**

accelerating chain. That section of a vacuum system, e.g., cathode-ray tube or betatron, in which charged particles are accelerated by voltages on accelerating electrodes.

accelerating electrode. One in a thermionic valve or cathode-ray tube maintained at a high positive potential with reference to the electron source. It accelerates electrons in their flight to the anode but does not collect a high proportion of them.

accelerating machine. Machine used to accelerate charged particles to very high energies. See **betatron, Bevatron, cyclotron, linear accelerator, synchrocyclotron, synchrotron.**

accelerating potential. That applied to an electrode to accelerate electrons from a cathode.

accelerator See **accelerating machine.**

accelerator tube. A drift tube in which a beam of charged particles are accelerated by the action of a strong electric field.

accelerometer. Transducer used to give signal proportional to rate of acceleration of body. Used especially in vibration study.

accentuation. A technique for emphasizing particular bands in an audio-amplifier.

acceptance angle. The solid angle within which all incident light reaches the photocathode of a phototube.

acceptor. Impurity atoms introduced in small quantities into a crystalline semiconductor and having a lower valency than the semiconductor from which they attract electrons. In this way 'holes' or positive charge carriers are produced, the acceptor atom becoming negatively charged. The phenomenon is known as p-type conductivity, e.g., 100 parts of boron in 10^6 parts of silicon increase the conductivity of the latter one million fold. See also **donor, impurity.**

acceptor circuit. Tuned circuit responding to a signal of one specific frequency.

acceptor level. See **energy levels, donor and acceptor.**

access time. Maximum time to locate and extract *words* or digits from a *store*; from microseconds in a fast core or film store, to minutes in a long magnetic tape file store.

access to store. Location and extraction of data from a *store*. It can be *random* when data is anywhere in a large store, which has to be entirely examined (minutes in a large film store). It may be *fast* (microseconds) in cold stores or *slow* (milliseconds) in medium stores such as magnetic disks.

accumulator. (1) Voltaic cell which can be charged and discharged. On charge, when an electric current is passed through it, into the positive and out of the negative terminals (according to the conventional direction of flow of current), electrical energy is converted into chemical energy. The process is reversed on discharge, the chemical energy, less losses both in potential and current, being converted into useful electrical energy. Accumulators therefore form a useful portable supply of electric power, but have the disadvantage of being heavy. Also **reversible cell, secondary cell, storage battery.** See also **André-Venner-.** (2) Register in arithmetic unit of digital computer.

ace. See **quark.**

achromat. An optical lens system designed to the same focal length for two or more different wavelengths.

achromatic antenna. One with uniform radiation properties over a required range of frequencies.

achromatic prism. An optical prism with a minimum of dispersion but having a maximum of deviation.

Acmeola. TN for monitoring and editing equipment in which sound is reproduced and the picture projected only for observation and adjustment. **Editola** and **Moviola** are similar.

acorn. Colloquial name for a valve of very small dimensions, designed for operation at very high frequencies. The electrode capacitance and electron-transit time effects are reduced in proportion to the dimensions.

acoustic centre. The effective 'source' point of the spherically divergent wave system observed at distant points in the radiation field of an acoustic transducer.

acoustic delay. That introduced into sound reproduction of speech or music along a telephone line, by conversion to sound, which is caused to travel a suitable distance along a pipe before re-conversion into electric currents. Delay is also obtained through magnetic recording or through wave filters.

acoustic delay line. A device, magnetostrictive or piezoelectric, e.g., a quartz bar or plate of suitable geometry, which reflects an injected sound pulse many times within the body.

acoustic feedback. The excitation of the microphone in a public address system by sound generated by the reproducers of the same system. If this is excessive the system builds up a sustained oscillation. Also, in a radiogram or electric record-reproducer, reception through the pickup, while playing, of vibrations from the loudspeaker, producing a similar effect.

acoustic filter. One which uses tubes and resonating boxes in shunt and series as reactance elements, providing frequency cutoffs in acoustic wave transmission, as in an electric wave filter.

acoustic impedance, reactance and resistance. Impedance is given by complex ratio of sound pressure on surface to sound flux through surface, and has imaginary (reactance) and real (resistance) components respectively. Unit is the **acoustic ohm**.

acoustic interferometer. Instrument in which measurements are made by study of interference pattern set up by two sound or ultrasonic waves generated at the same source.

acoustic ohm. Unit of acoustic resistance, reactance and impedance.

acoustic reproduction. See **electrical reproduction.**

acoustic scattering. Irregular and multidirectional reflection and diffraction of sound waves produced by multiple reflecting surfaces, the dimensions of which are small, cf. a *wavelength*; or by certain discontinuities in the medium through which the wave is propagated.

acoustic spectrometer. An instrument designed to analyse a complex sound signal into its various wavelength components and measure their frequencies and relative intensities. See **optical spectrometer.**

acoustic spectrum. Graph showing frequency distribution of sound energy emitted by source.

acoustical compliance. The reciprocal of the *acoustical stiffness.*

acoustical mass (or **inertance**). Given by M,

where ωM is that part of the acoustical reactance which corresponds to the inductance of an electrical reactance. ω is the pulsatance $= 2\pi \times$ frequency.

acoustical stiffness (S_A). For an enclosure of volume V, given by $S_A = \rho c^2 / V$, where c is the velocity of propagation of sound and ρ is the density. It is assumed that the dimensions of the enclosure are small compared with the sound wavelength.

acoustics. The science of mechanical waves including production and propagation properties.

actinic radiation. Ultraviolet rays, which have enhanced biological effect by inducing chemical change; basis of the science of photochemistry.

actinides. A group name proposed for the series of elements of atomic numbers 90-101 (inclusive), which assumes that they have similar properties to actinium.

actinium (Ac). Radioactive element, at. no. 89, which gives its name to the actinium $(4n+3)$ series. Half-life 21·7 years. Produced from natural radioactive decay of the ^{235}U isotope or by neutron bombardment of ^{226}Ra.

activated cathode. Emitter in thermionic devices, comprising a filament of basic tungsten metal, alloyed with thorium, which is brought to the surface by process of activation, such as heating without electric field.

activated water. Water in which ions, radicals, atoms or molecules are made temporarily chemically reactive by the passage of ionizing radiation.

activation. Induction of radioactivity in otherwise non-radioactive atoms, e.g., in a cyclotron or reactor.

activation cross-section. Effective cross-sectional area of target nucleus undergoing bombardment by neutrons, etc. Measured in **barns.**

activation energy. Excess energy over that of the *ground state* which an atomic system must acquire to permit a particular process, such as emission, to occur.

activator. Impurity introduced to promote luminescence, i.e., a sensitizer such as copper in zinc sulphide.

active. Said of a transducer or filter section or a circuit which contains an effective source e.m.f., such as a transistor, valve, or modulator.

active area. The area of a metal rectifier which operates as the rectifying junction.

active deposit. Radioactive material deposited on a surface.

active lattice. Regular pattern of arrangement of fissionable and non-fissionable materials in the core of a lattice reactor.

active lines. Those which are effective in establishing a picture.

active materials. (1) General term for essential material required for the functioning of a device, e.g., iron or copper in a relay or machine, electrode materials in a primary or secondary cell, emitting surface material in a valve, electron, or photocell, phosphorescent and fluorescent material forming a phosphor in a cathode-ray tube, or that on the signal plate of a television camera. (2) Term applied to all types of radioactive isotopes.

active power. The time average over one cycle of the instantaneous input powers at the points of entry of a polyphase circuit. See **active volt-amperes.**

active transducer. Any transducer in which the applied power controls or modulates locally supplied power, which becomes the transmitted signal, as in a modulator, a radio transmitter, or a carbon microphone.

active volt-amperes. Product of the active voltage and the amperes in a circuit, or of the active current (amperes) and the voltage of the circuit; equal to the power in watts. Also termed **active power.**

active voltage. That component of a vector representing the voltage which is in phase with the current in a circuit.

activity. Measure of intensity of a radioactive source. Expressed as the number of disintegrating atoms per unit time, or a derived quantity such as number of scintillations per second on a screen, or rate of clicks in a Geiger-Müller counter. Often expressed in *curies*, or *röntgens/hour* at one metre from the isolated source. *Specific activity* is expressed in *curies/gram*, or a multiple or submultiple thereof.

actual decimal point. One stored in a computer in such a way that it appears as a printed character on read-out.

actuator. Transducer used to bring electronic equipment into operation.

acute exposure. Illness due to excessive single radiation dose.

adaptation kit. That which unites sections of a guided weapon, e.g., *safing*, *arming*, *fuzing* and *firing* divisions of system.

adaptive control. Self-adaptation of controlling system to approach overall optimum, according to a defined requirement.

Adcock antenna. Directional receiving antenna consisting of two spaced vertical dipoles or half-dipoles connected by a screened transmission line. It responds only to vertically-polarized waves, and is not subject to *night error.*

Adcock direction finder. One using Adcock antenna.

adder. Unit in a computer for summation of two quantities, applied to an amplifier in analogue computing or to an arithmetic unit of a digital computer. In mechanical computing, addition and subtraction can be by rotation of shafts, with differential gears.

Additron. TN of tube for high-speed addition and carrying of pulses, e.g., in computers.

address. Code indicating where data shall be stored in a register, store, or memory. In multi-address codes, there are several addresses which can move data between stores, e.g.,' to a *fast store* for quick access. See also **absolute-.**

adhesion. (1) Mutual forces between two magnetic bodies linked by magnetic flux, or between two charged non-conducting bodies, which keeps them in contact. (2) Intermolecular forces which hold matter together, particularly closely-contiguous surfaces of neighbouring media, e.g., liquid in contact with a solid. Equivalent U.S. term **bond strength.** See **bond energy.**

adiabatic process. One which occurs without interchange of heat with surroundings.

adiactinic. Said of a substance which does not transmit photochemically-active radiation.

adion. Ion which can move only on a surface, not away from it.

adjacent channel. One whose frequency is immediately above or below that of the required signal.

admittance. Property which permits the flow of current under the action of a potential difference. The reciprocal of *impedance.*

adsorption. The penetration of a fluid into the pores of a solid which bounds the volume it is occupying, or its adhesion to the surface of the solid. Adsorption is essentially a surface phenomenon and, if the solid was appreciably porous, *absorption* (see (2)) would take place.

adsorption potential. Change of potential in an ion in passing from a gas or solution phase on to the surface of an absorbent.

advance ball. Rounded point riding a surface to be cut by a stylus, to ensure uniformity of depth of cut.

advance wire. See **Eureka wire.**

advanced gas-cooled reactor. High-temperature reactor developed at Winfrith (U.K.), employing uranium-oxide fuel in beryllium or stainless steel cans. Now being used for commercial power generation.

advantage ratio. Ratio between the relative radiation dosage received at any point in a nuclear reactor and that of a reference position for the same time of exposure.

Advent. Project name for U.S. satellite communication system, using computer-directed antennae for transmission and reception.

Aeolian tones. Those produced when a stream of gas strikes a stretched wire at right angles to its length, e.g., wind on open telephone and telegraph wires.

Aeolight. Glowlamp in which luminous output is controlled by signal voltage applied to control electrode. Used in cinematography for sound recording.

aeolotropic. Same as **anisotropic.**

aerial. See **antenna.**

aerial efficiency. Same as **radiation efficiency.**

aerial resistance. See **antenna resistance.**

aerophare. See **radio beacon.**

aeroplane effect. Error in direction-finding by radio which arises from the tilt of the transmitting aerial on an aircraft, or from any horizontal component in the emitted wave.

aether. Same as **ether.**

affinity. Chemical ability to form molecules; measured by the valence properties of atoms. See **valence electrons.**

afterglow. Continued visibility or detection of luminous radiation from a gas-discharge tube or luminous screen when the exciting agency has been removed. Long-persistence cathode-ray tubes are used for retaining a display which has been momentarily excited. Also **persistence.** See **phosphorescence.**

afterheat. That which comes from fusion products in a reactor after it has been shut down.

afterimage. Formation of image on retina of eye after removal of a visual stimulus, in the colour complementary to this stimulus.

age theory. In nuclear reactor theory, the slowing-down of neutrons by elastic collisions. The *age equation* relates the spatial distribution of neutrons to their energy. The equation is given by

$$\nabla^2 q - \frac{\partial q}{\partial r} = 0,$$

where q is the slowing-down density, and r the spatial coordinate. It was first formulated by Fermi who assumed that the slowing-down process was continuous and so is least applicable to media containing light elements.

ageing. (1) Change in the properties of a substance with time. (2) A change in the magnetic properties of iron, e.g., increase of hysteresis loss of sheet-steel laminations. (3) Process whereby the subpermanent magnetism can be removed in the manufacture of permanent magnets. (4) The preliminary run of a valve, e.g., for 100 h, before taking it into service.

aggregate recoil. Nuclei recoiling after emission of α-particles may carry a small cluster of active atoms away from the surface with them. In strong-activity α-emitters (e.g., polonium, plutonium, etc.) this leads to a build-up of radioactive contamination on surrounding surfaces which may represent a serious health hazard.

agonic line. Line on a map joining points where the magnetic variation is zero.

air capacitor. One in which the dielectric is nearly all air, for tuning electrical circuits with minimum dielectric loss.

air-cooling. Blowing air through a chassis to reduce the temperature of components to safe values. The air can be drawn through a filter, e.g., of glass fibres, to avoid dust.

air dose. Radiation dose in röntgens delivered at a point in free air.

air equivalent. The thickness of an air column at 15°C and one atmosphere pressure which has the same absorption of a beam of radiation as a given thickness of a particular substance.

air gap. (1) Gap with points or knobs, adjusted to break down at specified voltage and hence limit voltages to this value. (2) Section of air, usually short, in a magnetic circuit, especially in a motor or generator, a relay, or a choke. The main flux passes through the gap, with leakage outside, depending on dimensions and permeability.

air insulation. At normal pressure, ordinary dust-free and dry air withstands a voltage-gradient of 35-38 kV/cm. This dielectric strength increases with pressure.

air-line. (1) Straight line drawn on the magnetization curve of a motor, or other electrical apparatus, expressing the magnetizing force necessary to maintain the magnetic flux across an air gap in the magnetic circuit. (2) Spectral line originated by a spark discharge in air.

air monitor. Radiation-measuring instrument used for monitoring contamination or dose rate in air.

air-spaced coil. Inductance coil in which the adjacent turns are spaced (instead of being wound close together) to reduce self-capacitance and dielectric loss.

air speed. Speed measured relative to the air in which aircraft or missile is moving, as distinct from *ground speed* (q.v.).

air terminal. Elevated structure acting as a lightning protector, collecting local charge and reducing electric field strength.

air wall. Wall of ionization chamber designed to give same ionization intensity inside chamber as in open space. This means the wall is made of elements with an atomic number similar to that for air.

airlift. General use of air or neutral gas to blow material (particularly radioactive) as solid or liquid in processing, to avoid pumps.

albedo. (1) Ratio of the light falling on a planet or satellite to that which is reflected. (2) Ratio of the neutron flow density out of a medium free from sources, to the neutron flow density into it, i.e., reflection factor of a surface for neutrons. (3) Secondary cosmic ray particles projected upwards.

Alcomax. TN for a material, of high coercivity used for permanent magnets. It is an alloy of aluminium, cobalt, iron, nickel, and copper.

Alford antenna. One comprising a vertical cylindrical sheet with a longitudinal slot, for FM broadcasting.

Algol. Universal symbolic machine language, useful for programming scientific calculations having no reference to the common code of a specific computer. (*Algebraically orientated language.*)

algorithm. Set of steps to be taken in operations to effect a desired calculation on a computer.

alien tones. Frequencies, harmonic and sum-and-difference products, introduced in sound reproduction because of non-linearity in some part of the transmission path.

aligned grid valve. See **beam-power valve.**

alignment. (1) Of superheterodyne receiver, adjustment of pre-set tuned circuits to give optimum performance. (2) Process of orientating, e.g., spin axes of atoms during magnetization and similar operations.

alignment chart. See **nomogram**.

aliquot. Small sample of radioactive material used for determination of level of activity.

all-pass network. One which introduces a specified phase-shift response without appreciable attenuation for any frequency.

Allegheny 4750. TN for a 50% nickel-iron isotropic magnetic alloy.

allobar. A form of element differing in isotopic composition from that occurring naturally.

allocation. Process of distributing stored data in a digital computer to keep the total access time required for a computation to a minimum.

allochromatic. Having photoelectric properties which arise from micro-impurities, or from previous specific irradiation.

allochromy. Fluorescent re-radiation of light of different wavelength from that incident on a surface. See **Stoke's law (1)**.

allowed band. Range of energy levels permitted to electrons in molecule or crystal. These may or may not be occupied.

allowed transition. Electron transition between energy levels which is not prohibited by any quantum selection rule.

alloy junction. One formed by alloying an impurity to an otherwise pure (better than 0·01 PPM impurity) semiconductor crystal.

Allström relay. Highly sensitive relay, using a photocell and light beam.

Alnico. TN for a high-energy permanent magnet material which is an alloy of aluminium, nickel, cobalt, iron and copper.

alpha. (1) Greek letter used to represent alpha-particles, etc. Symbol α. (2) Russian fusion device, similar to *zeta*.

alpha chamber. Ionization chamber for measurements of α-radiation intensity.

alpha counter. Tube for counting α-particles, with pulse selector to reject those arising from β- and γ-rays.

alpha counter tube. See **alpha chamber**.

alpha cutoff. Frequency at which the current amplification of a transistor has fallen by more than 3 dB (0·7) of its low-frequency value. Also **cutoff frequency**.

alpha decay. Radioactive disintegration resulting in emission of α-particle. Also **alpha disintegration**.

alpha decay energy (q_A). The sum of the kinetic energy of the alpha particle (E_A) and that of the recoil of the product atom (E_P), given by

$$E_A = \left(\frac{m_P}{m_P + m_A}\right) q_A \text{ and } E_P = \left(\frac{m_A}{m_P + m_A}\right) q_A,$$

where m_A and m_P are the respective masses of the alpha-particle and the recoil atom. Also **disintegration energy**.

alpha disintegration. See **alpha decay**.

alpha emitter. Natural or artificial radioactive isotope which disintegrates, through emission of α-rays.

alpha-particle. Particle identical in measured properties with high-speed atoms of helium,

stripped of their electrons and hence doubly positively charged, ejected from natural or artificial radioisotopes. It is composed, therefore, of two protons and two neutrons.

alpha radiation. Alpha-particles emerging from radioactive atoms.

alpha-rays. Streams of α-particles.

alpha-ray spectrometer. Instrument for investigating energy distribution of α-particles emitted by radioactive source.

alpha wave. The principal slow voltage wave produced in the human brain (frequency 10 Hz).

alphanumeric. Said of a system which uses character codes which include both alphabetical and numerical. Also **alphameric**.

Alphatron. TN for tube in which ionization is excited by α-rays from a radioactive source, obviating bombardment of thermionic cathode when there is relatively high gas-pressure.

alternate scanning. Method of scanning developed to compensate for the difference in frame frequencies when televising from cinematograph film.

alternating current. Electric current whose flow alternates in direction; the time of flow in one direction is a half-period, and the length of all half-periods is the same. The normal waveform of a.c. is sinusoidal, which allows simple vector or algebraic treatment. Provided by alternators and valve oscillators.

alternating-current bias. In magnetic tape or wire recording, the addition of a polarizing alternating current in signal recording to stabilize magnetic saturation.

alternating-current bridge. General term used for any null balancing network energized by an a.c. supply. The balance condition can always be derived by the application of Kirchhoff's laws.

alternating-current circuit. One which passes a.c., i.e., it can have a capacitor in series, which blocks direct current.

alternating-current magnet. Electromagnet excited by a.c., having normally a laminated magnetic circuit. See **shaded pole**.

alternating-current pickup. Interfering currents in one channel arising from e.m.fs. induced by currents in other channels, including power mains at power frequencies. Interference also occurs when d.c. is switched.

alternating-current resistance. See **differential anode resistance**.

alternating-current transformer. One designed for isolating electronic equipment from the mains, and for providing voltages suitable for the components. See **auto-transformer**.

alternating-gradient focusing. The principle follows the optical analogy whereby a series of alternate converging and diverging lenses may lead, under suitable conditions, to a net focusing effect since the rays will strike the diverging lenses nearer to the axis. Using magnetic or electrostatic lenses the idea has been used for the design of electron synchrotrons and ion linear accelerators.

alternator. A dynamo generating an alternating-current output.

aluminium (Al). Silver-white metal, forming a protective film of oxide, widely found in nature. At. no. 13, at. wt. 26·9815, m.p. 659·7°C, b.p. 1800°C, sp. gr. 2·70. Discovered by Wöhler in 1827. Much used in alloys for structural work; alloyed with iron and cobalt in many types of permanent magnet. Easily worked in sheets for chassis construction in electronic apparatus. Polished aluminium reflects well beyond the visible spectrum in both directions, and does not corrode in sea-water. Foil aluminium is much used for capacitors. The metal can be used as a window in X-ray tubes and as sheathing for reactor fuel rods. Aluminium is produced on a large scale where electric power is cheap, e.g., hydroelectric at Kitimat, when bauxite is electrolysed in fused cryolite. ^{28}Al and ^{29}Al are strong γ-ray emitters of very short life. As an electrode in gas-discharge tube, it does not sputter like other metals. U.S. **aluminum.**

aluminium anode cell. One with an aluminium anode immersed in an electrolyte which does not attack aluminium. The cathode may also be of aluminium or some other metal, e.g., lead. Such cells can be used as rectifiers or as high-capacitance capacitors. See **electrolytic capacitor.**

aluminium antimonide. A semiconducting material used for transistors up to temperature of 500°C.

aluminized screen. Cathode-ray tube fluorescent screen the conductivity of which has been improved by application of thin aluminium film. This gives greater contrast and avoids ion burn.

Alundum. TN for a form of aluminium oxide used for insulating, electrically and thermally, the cathode heater in a thermionic valve from the tubular cathode.

ambient illumination. Background uncontrollable light level at a location.

ambient noise. Acoustic noise existing in a room or any other environment, e.g., the ocean. Also called **room noise.**

ambient noise level. Random uncontrollable and irreducible noise level at a location or in a valve or circuit.

ambiguity. The condition of a servomechanism in which more than one null point is sought.

ambipolar. Said of any condition which applies equally to positive and negative ions (including electrons) in a plasma.

Amble. Technique for analogue computing certain differential equations, devised by O. Amble.

americium (Am). Transuranic element, at. no. 95, half-life 475 years. Of great value as a long life α-particle emitter free of criticality hazards and γ-radiation, e.g., in (αn) laboratory neutron sources.

amicrons. Particles of the order of 10^{-7} cm,

invisible in the ultramicroscope; they act as nuclei for larger submicron particles.

ammonium dihydrogen phosphate. Piezoelectric crystal used in microphones and other transducers; it can withstand a temperature higher than can Rochelle salt.

amperage. Current in amperes, more especially the rated current of an electrical apparatus, e.g., fuse or motor.

ampere. MKS unit of current flow, such that 1 amp in two parallel conductors 1 m apart repels itself with a force of 2×10^{-7} newton/m length, the conductors being infinitely thin, long, and in a vacuum. After Ampère of Paris. See also **absolute-.**

ampere-balance. See **Kelvin ampere-balance.**

ampere-hour. MKSA unit of charge, equal to 1 amp passing a point in a circuit for 1 h.

ampere-hour capacity. Capacity of an accumulator battery measured in ampere-hours, usually specified at a certain definite rate of discharge. Also applicable to primary cells.

ampere-hour efficiency. Ratio of the ampere-hours output during discharge to the ampere-hours input during charge, in an accumulator.

Ampère's law. That which states the magnetic field in the neighbourhood of a current-carrying conductor.

Ampère's rule. Simple rule for the direction of the magnetic field associated with a current. If a man is swimming with the current and is facing a magnet needle, the north-seeking pole of the magnet is deflected towards his left hand. Alternatively, the direction of the field is that of an advancing right-hand screw when turning with the current.

ampere-turn. MKSA unit of magnetomotive force, which drives flux through magnetic circuits, arising from 1 amp flowing round one turn of a conductor.

ampere-turn amplification (or gain). Ratio of the load ampere-turns to the control ampere-turns in a magnetic amplifier.

Amperite. TN for a type of **barretter.**

Ampex. System of magnetic recording of video signals, for storing and editing programmes. High-speed wide magnetic tape is scanned transversely by rotating magnetic heads.

amphoteric. Said of a chain molecule charged oppositely at the ends.

amplidyne. Rotary magnetic amplifier with a high power gain, useful in servomechanisms.

amplification factor. The number of volts by which the potential of the anode of a thermionic tube must be changed to counteract the effect upon the anode current of a change in grid potential of 1 volt. Symbol μ. See **mu-factor.**

amplifier. Device for controlling power from a source so that more is delivered at the output than is supplied at the input. Source of power may be mechanical, hydraulic, pneumatic, electric, etc. Electric amplifiers may be classified into (1) *valve*, which operates on voltage, (2) *repeater*, specially used for telephone circuits, (3) *transistor*, which operates on current, (4) *magnetic*, which operates on very low-frequency currents, (5) *solid state*, operated by transistor action in a single semiconductor block. See also:

A-	loaded push-pull-
audio-frequency-	lock-in-
B-	logarithmic-
band-pass-	low-frequency-
Black-	low-loading-
booster-	magnetic-
bridging-	maser
broadcasting-	microphone-
buffer-	parametric-
bullet-	paraphrase-
C-	photocell-
cascade-	power-
cascode-	pre-
cathode-follower-	pulse-
chopper-stabil-	push-pull-
ized-	quiescent push-pull-
class A-	radio-frequency-
class AB-	recording-
class B-	reflex-
class C-	rotary-
degenerative-	stabilized-feedback-
dielectric-	stagger-tuned-
direct-current-	step-down-
Doherty-	superconducting-
dynamo-electric-	television-
feedback-	thermionic-
gain-	time-shared-
grounded-cath-	torque-
ode-	transistor-
grounded-grid-	transmission-line-
head-	trap-
hydraulic-	travelling-wave-
intermediate-	trip-
frequency-	tuned-
K-	video-
line-	vogad
linear-	wideband-.

Amplistat. TN for a self-saturating magnetic amplifier circuit.

Amplitron. TN for amplifying tube operating at microwave frequencies.

amplitude. The magnitude of a vector, as contrasted with its *argument*. The latter defines the direction with reference to some standard direction. Also **modulus, tensor.** See **complex number, instantaneous-**.

amplitude discriminator. In a counting circuit, that which passes pulse amplitudes between specified limits, which may include zero and infinity.

amplitude distortion. Distortion of waveform arising from the non-linear static or dynamic response of a part of a communication system, the output amplitude of the signal at any instant not having a constant proportionality with the corresponding input signal. Can arise in networks, lines, or amplifiers.

amplitude filter. One which separates synchronizing signals in a television signal from the video (picture) signal.

amplitude modulation. System of modulation in which the amplitude of the transmitted carrier wave is varied in accordance with the impressed signal, the frequency and phase remaining unchanged.

amplitude peak. Maximum positive or negative excursion from zero of any periodic disturbance.

amplitude rms. Root-mean-square value of any periodic disturbance.

analogous measurements. Those made with models simulating difficult practical conditions, e.g., electrolytic troughs for plotting electric fields between odd-shape electrodes.

analog(ue). Referring to continuous variation of a magnitude in computation; cf. *digital*. Hence *analogue computer*, one which operates with continuously varying amplitudes, as contrasted with *digital computers*, which use *bits*.

analogue—digital converter. Circuit used to convert information in analogue form into digital form (or vice versa), e.g., in a digital voltmeter.

analogy. Correspondence of pattern or form, e.g., electric wave filter analogy for complex mechanical system, or sheet rubber models for investigation distribution of potentials between electrodes in valves, to facilitate conception and calculation from known techniques. U.S. analagy.

anchor ring. Old name for **toroid**.

and. An operator in symbolic logic.

and element. One producing an output signal only if all inputs are energized simultaneously, i.e., output signal is 1 when all input signals are 1.

Anderson bridge. An a.c. bridge used for measurement of inductance. It is not a four-arm bridge of the Wheatstone type and its balance conditions may be derived either by using Kirchhoff's laws or by the application of the star delta transformation.

André-Venner accumulator. One in which the anode is silver, the cathode zinc, with potassium hydroxide as electrolyte. It can be made in very small sizes, nearly dry, and for very high rates of charge and discharge.

anechoic room. One in which internal sound reflections are reduced to an ineffective value by lining with internally pointing pyramids of felt, foam plastic, or Fibreglass. Used for standard measurements on microphones, etc.

anelasticity. Any deviation from an ideal internal structure of a body which would dampen or attenuate an elastic wave therein.

anelectric. Term once used for a body which does not become electrified by friction.

anemometer. Transducer giving signal proportional to air velocity.

aneroid barometer. Transducer which can be used to give output signal proportional to barometric pressure for control systems.

angels. Stray reflections, e.g., from birds, etc., from the lower atmosphere.

angle modulation. Any system in which the transmitted signal varies the phase angle of an otherwise steady carrier frequency, i.e., phase and frequency modulation.

angle of arrival. Angle of elevation of a downcoming wave. See **musa**.

angle of deflection. That of the electron beam in a CRT.

angle of departure. Angle of elevation of maximum emission of electromagnetic energy from an antenna.

angle of dig or drag. Deviations from normal of angles for cutting stylus in disk sound recording.

angle of flow. Angle, or fraction of alternating cycle, during which current flows, e.g., in a thyratron.

angle of incidence. Angle between incidence ray (or beam) of radiation and normal to surface at the incidence point.

angle of tilt. Orientation of radar paraboloid to horizontal.

ångström. Unit of wavelength for electromagnetic radiation covering visible light and X-rays. Equal to 10^{-10} m; the international standard, once cadmium red = 6438·4696 Å, has been replaced by krypton orange = 6507·6373 Å. Also used for dimensions of crystal lattices.

angular correlations. The theory and analysis of the directional distributions of nuclear radiations together with experimental measurements, used to obtain information about the angular momenta and parities of the nuclear states and the transient fields which are involved.

angular distribution. The distribution relative to the incident beam of scattered particles or the products of nuclear reactions.

angular frequency. Frequency of a steady recurring phenomenon expressed in radians per second, i.e., frequency in Hz multiplied by 2π. Symbol p or ω; also called **radian frequency** or **pulsatance**.

angular height. Actual height in wavelengths of an aerial, multiplied by 2π radians or 360°.

angular momentum. In mechanics, the product of angular velocity and moment of inertia. The orbital angular momentum of an electron, measured as above, is quantized and can only have values which are exact multiples of the Dirac unit. The angular momentum of particles which (appear to) have spin energy is quantized to values that are multiples of half the Dirac unit. In some cases the physical reality of the spin has not been confirmed.

angular spacing. Spacing of aerials for direction-finding or broadcasting radiation, stated in terms of 360° per wavelength.

anharmonic. Said of any oscillation system in which the restoring force is non-linear with displacement, so that the motion is not simple harmonic.

anion. Negative ion, i.e., atom or molecule which has gained one or more electrons in an electrolyte, and is therefore attracted to an anode, the positive electrode. Anions include all non-metallic ions, acid radicals and the hydroxyl ion. In a primary cell, the deposition of anions on an electrode makes it the negative pole. Anions also exist in gaseous discharge.

aniseikon. Arrangement of two photocells, the same external object being focused on both but through separate chequerboard filters, so distorted that only movement is registered by unbalance in a bridge circuit. An electronic movement detector.

anisotropic. Said of crystalline material for which physical properties depend upon direction relative to crystal axes. These properties normally include elasticity, conductivity, permittivity, permeability, etc.

anisotropic coma. Distortion in a television image arising from inclination of the objective, which is the first electric lens adjacent to the electron-emitting cathode.

anisotropic conductivity. Body which has a different conductivity for different directions of current flow, electric or thermal.

anisotropic dielectric. One in which electric effects depend on the direction of the applied field, as in many crystals.

annealing. Process of maintaining a material at a known temperature to bleach-out dislocations, vacancies and other metastable conditions, e.g., in glass, magnetic material.

annihilation. Spontaneous conversion of a particle and corresponding anti-particle into radiation, e.g., positron and electron, which yield 2 γ-ray photons each of 0·511 MeV. See also **mass-energy equivalence**.

annihilation radiation. The radiation produced by the annihilation of a particle with its corresponding anti-particle.

annotation. Descriptive comments and explanations unconnected with machine instructions in a computer programme.

annunciator. Arrangement of indicators tripped by relays, for indicating which of a number of circuits has operated a bell, a machine, or one of many units in a plant.

anode. In a valve or tube, the electrode held at a positive potential with reference to cathode, and through which positive current generally enters the vacuum or plasma, through collection of electrons. Also positive electrode of battery or cell. In U.S. **plate**. See

carbon-	rotating-
carbonized-	tuned-
disk-	ultor.
holding-	

anode battery. See **high-tension battery**.

anode bend. The more or less abrupt curve in the anode-current versus grid-voltage characteristic of a triode, which occurs at small values of anode current. Also called **bottom bend**.

anode-bend detector. See **plate-bend detector**.

anode-bend rectification. That dependent on the curvature of the anode-current versus grid-voltage characteristic.

anode characteristic. Graph relating anode current and anode voltage for an electron tube.

anode circuit. Closed circuit formed by the anode-cathode path in the valve, the B-battery, and the coupling impedance or transformer and its load.

anode conductance. Anode current divided by the anode potential. Frequently, though incorrectly, used for the slope of the anode-current versus anode-voltage characteristic.

anode current. Current flowing to the anode of a thermionic tube.

anode-current characteristic. Curve relating the anode current of a multi-electrode tube to the potential of one of the electrodes, e.g., anode current versus grid voltage.

anode-current surface. Surface geometrically relating the anode current to the potential of two electrodes, usually the anode and grid, of a multi-electrode tube. The z (vertical) coordinate represents the anode current, whilst the x and y coordinates represent the grid and anode voltages.

anode dark space. Dark zone near the anode in a glow-discharge tube.

anode dissipation. Generally the energy produced at the anode of a thermionic tube and wasted as heat owing to the bombardment by electrons; specifically, the maximum permissible power which may be dissipated at the anode.

anode drop or **fall.** Component of the anode-to-cathode potential difference in a gas-filled discharge tube which is independent of the anode current.

anode effect. In electrolysis the sudden drop in current due to the formation of a film of gas on the surface of the anode.

anode efficiency. Ratio of a.c. power in the load to d.c. power supplied to the anode. This may vary between 20% and 80%, depending on conditions of drive; also called **conversion efficiency, plate efficiency.**

anode feed. Supply of direct current to anode of valve, generally decoupled so that the supply circuit does not affect the condition of operation of the valve.

anode glow. Luminous zone on anode side of positive column in a gas discharge tube.

anode impedance. Complex a.c. impedance between anode and cathode of thermionic valve, including interelectrode capacitance. Misapplied to **differential anode resistance.**

anode load impedance. That between anode and cathode outside a valve, not affected by the feed.

anode load resistance. External load between anode and steady polarizing voltage.

anode modulation. Insertion of the modulating signal into the anode circuit of a valve, which is oscillating or rectifying the carrier. Also called **plate modulation.**

anode polishing. Same as **electrolytic polishing.**

anode-ray current. Current in a partial vacuum, represented by the movements of positively-charged particles.

anode rays. See **positive rays.**

anode resistance. The anode potential divided by the anode current; frequently, though incorrectly, applied to **differential anode resistance.**

anode saturation. Limitation of current through the anode of a valve, arising from current, voltage, temperature, or space charge.

anode stopper. Small resistance joined directly to anode contact on thermionic valve and intended to inhibit parasitic oscillation.

anode strap. Conducting strip connecting segments of magnetron anode.

anode tap. Tapping point on the inductance coil of a tuned-anode circuit, to which the anode is connected. The position of the tap is adjusted so that the tube operates into the optimum impedance.

anodized. Said of a metal surface protected by chemical or electrolytic action.

anolyte. Ionized liquid in neighbourhood of an anode during electrolysis.

anomalous dispersion. Departure from the normal observation that the refractive index of a medium decreases with increase in wavelength, whereby there is a discontinuity or even reversal for particular wavelengths.

anomalous magnetization. Irregular distribution of magnetization, e.g., when consequent poles exist as well as main poles on a magnetic circuit.

anomalous scattering. See **scattering.**

Anotron. TN for cold-cathode glow valve, with sodium cathode and copper anode.

answer print. First print from the edited negative, shown to the producers of the soundfilm for final approval before release.

antenna. Exposed wire from which electromagnetic wave energy is radiated or received. Also **aerial.**
See also:

achromatic-	director
Adcock-	discone-
Alford-	dish
anti-fading-	doublet-
anti-static-	dumb aerial
aperiodic-	earthed aerial
artificial-	end-on aerial
balancing-	array
beam-	equivalent height
beam array	fan-
beavertail-	ferrite-rod-
Bellini-Tosi-	fishbone-
broadside-	flat-top-
buried-	frame-
cage-	Franklin-
Cassegrain-	H-
cheese-	harmonic-
Chireix-Mesny-	Hertz-
coaxial-	horizontal-
cosecant-	inverted-L-
counterpoise	inverted-V-
cubical-	J-
current feed	Kooman's array
curtain-	lens-
dielectric-	Marconi-
dipole-	marker-
directional-	musa

omnidirectional-	stub aerial
omnidirectional	T-
radiator	tiered array
parasitic-	travelling-wave-
periodic-	tuned-
pinetree-	turnstile-
polyrod-	unidirectional-
quarter-wave-	unipole-
rhombic-	unloaded-
sausage-	untuned-
Schwartzschild-	variable elevation
shunt-excited-	beam-
skirt dipole	vee-
slot-	voltage-fed-
spaced-	wave-
steerable-	Yagi-
sterba-	zepp.

antenna array. System consisting of a number of aerials, arranged to obtain directional properties in transmitting or receiving. See **beam array.**

antenna changeover switch. Switch used for transferring an antenna from the transmitting to the receiving equipment, and vice versa, protecting the receiver.

antenna download. Wire running from the elevated part or conductor of an antenna down to the transmitting or receiving equipment.

antenna earthing switch. One used for disconnecting the antenna from the transmitting or receiving apparatus and connecting it directly to earth, as a protection against lightning.

antenna effect. (1) Error in direction-finding arising from non-symmetry of antenna or interference by downcoming waves when antenna acts as an open aerial. (2) Action of a loop antenna in picking up signals from directions in which it is not normally responsive, due to asymmetrical distribution of capacitance to earth.

antenna feeder. The transmission line or cable by which energy is fed from the transmitter to the antenna.

antenna field. Map showing electromagnetic field strength produced by antenna in the form of contour lines joining points of equal field intensity.

antenna gain. Ratio of maximum flux of energy from an antenna to that which would have been received from a single dipole radiating the same power.

antenna impedance. Complex ratio of voltage to current at the point where the feeder is connected.

antenna load. (1) Same as **dummy load.** (2) The load impedance offered by an antenna to a transmitter.

antenna resistance. Total power supplied to an antenna system divided by the square of a specified current, e.g., in the feeder, or at the earth connection of an open-wire antenna.

antenna-shortening capacitor. That connected in series with an antenna operated at a frequency higher than its first natural frequency, in order to lower the impedance

between the base of the antenna and earth.

anthracene. Organic crystal ($C_{14}H_{10}$), useful as a scintillator in photoelectric detection of beta-particles.

anti-baryon. Anti-particle of a baryon. The term *baryon* is often used generically to include both.

anti-bonding orbital. Orbital electron of two atoms, which increases in energy when the atoms are brought together, so acting against the closer bonding of a molecule.

anti-capacitance switch. One designed to have very little capacitance between the terminals when in the open condition.

anti-cathode. Target electrode of X-ray tube (normally also the anode).

anti-clutter. Automatic gain control to diminish display of locally-returned echoes.

anti-coherer. See **decoherer.**

anti-coincidence circuit. (1) One which delivers a pulse if one of two pulses is independently applied, but not when both are applied together, or non-simultaneously within an assigned time interval. (2) Unit of a computer with '0' output only when all inputs agree on '1'.

anti-coincidence counter. System of counters and circuits which record only if an ionizing particle passes through particular counters but not through the others.

anti-cyclotron. A type of travelling-wave tube.

anti-fading antenna. One designed for the optimum ratio of low elevation to high elevation radiation; usually a vertical mast or wire about 0·6 wavelength in height for medium broadcast wavelengths.

anti-ferromagnetism. Reversed spin of neighbouring atoms in certain lattices, exhibiting paramagnetism with certain properties of ferromagnetism, usually at low temperatures.

anti-hunting. In a d.c. closed loop feedback system, the use of a transformer in series with the load current, so that a rate-of-change signal can be injected into the system to ensure stability and avoid self-oscillation.

anti-lepton. Positron, positive muon or antineutrino. *Lepton* (q.v.) is often used as a generic term for leptons and anti-leptons together.

anti-microphonic holder. Valve-holder in which the valve is supported on springs or resilient material, to reduce the effects of mechanical shock.

antimony (Sb). Metallic element, at. no. 51, at. wt. 121·75, m.p. 630°C, sp. gr. 6·6. Used in alloys for cable covers, etc., and as a donor impurity in germanium. Has several radioactive isotopes which emit very penetrating gamma-radiation. These are employed in (γn) laboratory neutron sources.

anti-neutrino. Particle, the emission of which is assumed to accompany radioactive decay brought about by electron capture or positron (i.e., $\beta+$) emission. An apparently different anti-neutrino is associated in an exactly analogous way with positive muon decay.

anti-neutron. Anti-particle with spin and magnetic moment oppositely orientated to those of a neutron.

antinode. At certain positions in a standing-wave system of acoustic or electric waves or vibrations, the location of maxima of some wave characteristic, e.g., amplitude, displacement, velocity, current, pressure, voltage. At the *nodes* these properties would have minimum values.

anti-noise microphone. One in which the diaphragm is exposed to noise on both sides, and on one side to a wanted speech sound, thus improving the speech/noise ratio in the output.

anti-parallel. Said of vectors which are parallel but operate in opposite directions.

anti-particle. One which has all the properties of another particle, but reversed in one respect, e.g., the *negatron* (*negative* elec*tron*) and the *positron* (*posi*tive elec*tron*). Interaction between particle and anti-particle means simultaneous *annihilation*, with production of great energy, which is radiated.

anti-polarizing winding. One on a transformer or choke which carries a d.c. to neutralize the magnetizing effect of another d.c., e.g., anode current to a valve.

anti-proton. Short-lived particle, half-life 0·05 microseconds, identical with proton, but with negative charge; annihilating with normal proton, it yields mesons.

anti-resonance frequency. Frequency at which the parallel impedance of a tuned circuit rises to a maximum.

anti-static antenna. One in which the receptive portion is placed outside the interfering field as much as possible, and is connected to the receiver by screened leads.

anti-static fluid. That applied to a direct-recording disk before cutting a record, to obviate swarf adhering to the surface, because of generation of electric charge. Also applied to finished gramophone disks to avoid the attraction of dust, and to sound-film to avoid static discharges and consequent scratches on the sound-track.

anti-Stokes lines. Those in scattered or fluorescent light with frequencies greater than that in the incident radiation, because of departure of atoms or molecules from their normal states.

anti-symmetric. Pattern or waveform in which symmetry is complete except for one particular, e.g., sign of electric charge, direction of current, or of components in waveform.

anti-transmit-receive. Gas discharge which blocks a pulse transmitter when pulses are being received. See ATR tube, TR tube.

aperiodic. Said of circuit or system when it responds equally to all frequencies of interest.

aperiodic (or non-resonant) antenna. One with useful efficiency over a range of radio frequencies, terminated to minimize resonance by reflection, e.g., *rhombic antenna*.

aperiodic regeneration. That employing direct- or battery-coupled amplifiers in which the degree of regeneration is independent of frequency.

aperture. (1) Space or solid angle through which most of the radiation energy from an antenna passes, including flare of horn. (2) Ratio of focal length to diameter of a lens.

aperture distortion. That arising from the scanning spot having finite, instead of infinitely small, dimensions.

aperture lens. Electron lens formed from holes in diaphragms, which are maintained at differing potentials.

apertured disk. See Nipkow disk.

apochromat. A compound lens corrected so that it has the same focal length for at least three different wavelengths. Cf. *achromat*.

apostilb. Unit of luminance equal to 1/10,000 lambert.

apparent power. The volt-amperes, i.e., the product of volts and amperes in an a.c. circuit or system.

apparent resistance. Obsolete term for impedance.

apple. Colloquialism for a type of valve used in an audio-frequency amplifier.

apple-and-biscuit microphone. Omni-directional microphone in which a moving-coil element is located on a sphere, with an adjacent plate to regulate its response in all directions normal to the axis; also ball-and-biscuit microphone, thistle microphone.

Apple tube. TN for cathode-ray tube, with single electron gun, the phosphor on the screen having red, green and blue stripes of fluorescence.

Appleton layer. Same as F-layer.

applicator. Electrode used in industrial high-frequency heating or medical diathermy; often specially shaped to fit the sample or body. See also heating inductor.

applied power. That *applied* to an electrical transducer is not equal to the actual power received, because of the reflection arising from non-equality of impedance matching. The *applied power* is the power which would be received if the load matched the source in impedance.

Aquadag. RTM for commercial colloidal graphite in water dispersion for internal coating of cathode-ray tubes; also Dixonag.

aquarium. Colloquialism for the booth or sound-proof enclosure in which, in sound-film production, mixing is executed.

aquarium reactor. See swimming-pool reactor.

aquo (or aqueous) ion. In water solution, an ionized molecule which is combined with one or more molecules of water.

arc. Ionic gaseous discharge maintained between electrodes, characterized by low voltage and high current. See arc furnace, arc lamp, mercury-arc rectifier.

arc absorber. Same as *spark absorber* (q.v.), but referring to a discharge likely to be destructive if not extinguished.

arc-back. In a mercury-arc rectifier, flow of

electrons opposite to that intended. Caused by a heated spot on the anode acting as a cathode, leading to possible damage.

arc baffle. Means of preventing liquid mercury contacting an anode in a mercury-arc rectifier. Also called **splash baffle.**

arc chute. Deflector assisting the blowout of an air arc by convection. Also **arc shield.**

arc converter. One which generates a.c. in a circuit when fed with d.c., especially in high-voltage d.c. transmission of power.

arc crater. Depression formed in electrodes between which an electric arc has been maintained. In arc welding, the depression which occurs in the weld metal.

arc deflector. Magnetic arrangement for controlling the position of the arc in an arc lamp. Also **arc shield.**

arc duration. Time during which an arc exists between the contacts of an opening switch or circuit-breaker. In a.c. circuits, usually measured in cycles, varying between half a cycle and perhaps 20 cycles.

arc furnace. One using electric arc for heating purposes. Very high temperatures are obtained with mirror focusing.

arc lamp. One using electric arc for lighting.

arc modulation. In a mechanical scanning system, modulation of the intensity of the light source by variation of the current in an arc discharge.

arc shield. See **arc chute, arc deflector.**

arc spectrum. Emission spectrum from substance excited by an electric arc. Cf. *spark spectra.*

arc-stream voltage. Voltage drop along the arc stream of an electric arc, excluding the voltage drops at the anode and cathode.

arc suppression. Any scheme, such as Petersen coil or magnetic field, for obviating or blowing out an arc to avoid damage.

arc-through. Overflow of electron stream into an intended non-conducting period.

arc transmitter. One in which the source of the high-frequency current is an arc discharge, which maintains oscillations in an oscillatory circuit because of effective negative differential resistance. Now obsolete.

arcing contact (or tips). An auxiliary contact fitted to a switch or circuit-breaker, arranged so that it opens after and closes before the main contact, thereby bearing the brunt of any burning due to the arc which occurs when a circuit is interrupted. Designed for easy replacement.

arcing ring. Circular or oval ring conductor, placed concentrically with a pin insulator or a string of insulators, for deflecting an arc from the insulator surface which could be damaged.

arcing voltage. That below which a voltage cannot be maintained between two electrodes.

area. Space around a transmitter and antenna defined functionally, i.e., grade of service, fading, mush, interference.

areal density. That of a layer or film of material, expressed as mass/unit-area, for absorption data of rays.

B

Argand diagram. The vector diagram for showing the magnitude and phase angle of a vector quantity with reference to some other vector quantity. See **complex number.**

argon (Ar). Inert gas, at. no. 18, at. wt. 39·948, m.p. −189·3°C, b.p. −185·8°C. Filling gas in radiation counters, fluorescent tubes, etc.

argument. See **amplitude.**

arithmetic unit. That unit which performs the *logical operations* in an electronic computer, under the control of the *programme.*

arm. See **branch.**

armature. (1) Piece of low-reluctance ferromagnetic material (*keeper*) for temporarily bridging the poles of a permanent magnet, to reduce the leakage field and preserve magnetization. (2) Moving part which closes a magnetic circuit and which indicates the presence of electric current as the agent of actuation, as in all relays, electric bells, sounders, and telephone receivers. (3) The rotator of a d.c. motor or generator.

armature ratio. Ratio of distance moved by the spring buffer of an electromagnetic relay to that moved by the armature.

armature reaction. Distortion of magnetic field in armature core resulting from rotational motion.

armature relay. A relay operated electromagnetically, thus causing the armature to be magnetically attracted.

armed lodestone. One fitted with iron pole-pieces for concentrating flux.

Armstrong circuit. The original super-regenerative receiving circuit; also the supersonic heterodyne circuit.

Armstrong oscillator. The original oscillator, in which tuned circuits in the anode and grid circuits of a valve are coupled.

array. See **antenna array.**

arsenic (As). Element, at. no. 33, at. wt. 74·9216, m.p. 820°C, sp. gr. 5·7 (crystallized). Important as donor impurity in germanium semiconductor devices.

articulation. Percentage intelligibility of a telephone communication system, usually tested with meaningless syllables of vowel-consonant type.

artificial antenna. Combination of resistances, capacitors and inductances with the same characteristics as an antenna except that it does not radiate energy. Also **dummy antenna, phantom antenna.** See **dumb aerial.**

artificial ear. Device for testing earphones which presents an acoustic impedance similar to the human ear and includes facilities for measuring the sound pressure produced at the ear.

artificial earth. See **counterpoise.**

artificial horizon. A free gyroscope connected to an indicator to simulate the horizon.

artificial line. Repeated network units which have collectively some or all of the transmission properties of a line; also **simulated line.**

artificial load. See **dummy load.**

artificial radioactivity. Radiation of α-, β-, γ-rays and positrons from isotopes after high-energy bombardment. Discovered by Irène Curie in 1933. Effected generally by neutron bombardment in reactors.

artificial voice. Loudspeaker and baffle for simulating speech in testing of microphones.

asdic. Underwater ultrasonic detecting system which transmits a pulse and receives a reflection from underwater objects, particularly submarines, at a distance. Also used by trawlers to detect shoals of fish. Equivalent U.S. term **sonar.** (*A*llied *S*ubmarine *D*etection *I*nvestigation *C*ommittee.)

asmodular. Modular method of assembly. See **module.**

aspect ratio. See **picture ratio.**

asperity. Actual region of contact between two surfaces, elastically and plastically flattened to take the load (normal force).

assembler. Unit which produces a set of computer machine instructions from a set of symbolic input data. An assembler normally converts specific single-step instructions into machine language and is therefore more restricted than a *compiler* (2, q.v.) which can convert single general instructions into complete subroutines for carrying them out.

assigned frequency. That assigned as centre frequency of a class of transmission, with tolerance, by authority.

associated emission. That which brings about equilibrium between incident photons and secondary electrons in ionization.

astable circuit. Valve or transistor circuit (with two quasi-stable states) which is free-running and self-sustaining in oscillation, e.g., multivibrator. Also **free-running circuit.**

astatic galvanometer. Moving-magnet galvanometer in which adjustable magnets form an astatic system.

astatic microphone. Same as **omnidirectional microphone.**

astatic system. Ideally an arrangement of two or more magnetic needles on a single suspension so that in a uniform magnetic field, such as the earth's field, there is no resultant torque on the suspension.

astatine (At). Unstable element, at. no. 85. Half-life of most stable isotope about 8 h.

Aston dark space. That in the immediate neighbourhood of a cathode, in which the emitted electrons have velocities insufficient to ionize the gas.

Aston whole-number rule. Atomic weights of isotopes are approximately whole numbers when expressed in awu. See **mass spectrograph.**

astron. Type of chamber used in thermo-nuclear research.

astronomical unit. Unit of distance based on that between sun and earth, equal to $149 \cdot 5 \times 10^6$ km.

asymmetric conductor. Conductor which has a different conductivity for currents flowing in different directions through it, e.g., valve or transistor diode.

asymmetric top. A model of a molecule having no threefold or higher-fold axis of symmetry.

asymmetrical, dissymmetrical, non-symmetrical. Said of circuits, networks, or transducers when the impedance (image impedance or iterative impedance) differs in the two directions.

asymmetrical conductivity. Phenomenon whereby a substance, or a combination of substances such as in a rectifier, conducts electric current differently in opposite directions.

asymmetry potential. The potential difference between the inside and outside surface of a hollow electrode.

asynchronous (or non-synchronous) computer. One in which operations are not all timed by a master clock. The signal to start an operation is provided by the completion of the previous operation.

Atlas. Very powerful I.C.T. digital computer.

Atmite. TN of a non-linear resistance material comprising silicon carbide.

atmosphere microphone. One used for background noise to a transmission, the adjustable output being added to the main transmission.

atmospheric acoustics. Branch of acoustics concerned with the propagation of sound in the atmosphere, of importance in sound-ranging and aircraft noise.

atmospheric carbon. An equilibrium amount of ^{14}C having been built up in the universe over the centuries, and its atom becoming oxidized relatively quickly to carbon dioxide, the atmospheric carbon dioxide is therefore slightly radioactive.

atmospheric electricity. That causing increasing potential with height, about 100 V/m, in calm conditions, altered considerably by thunderclouds. See **lightning.**

atmospheric radio wave. One which is propagated by reflections in the atmosphere using either, or both, the ionosphere and the troposphere.

atmospheric radioactivity. Natural background radioactivity of air due to active gases and suspended particles.

atmospheric waveguide duct. Under certain atmospheric conditions, a layer of the atmosphere which acts as a true waveguide conductor for radio-frequency waves, even down to 20 MHz.

atmospherics. Interfering or disturbing signals of natural origin. Also X's, **sferics, spherics, strays, static.**

atom. Smallest unit of an element which can enter into chemical combination.

atom bomb. See **atomic bomb.**

atomic absorption coefficient (μ_a). For an element, fractional decrease in intensity per number of atoms per unit area, i.e.,

$$\mu_a = \frac{\text{linear absorption coefficient}}{\text{number of atoms per unit volume}}$$
$$= \frac{\text{mass absorption coefficient}}{\text{number of atoms per unit mass}}.$$

When the medium contains only one nuclide then μ_a is the equivalent of the *total cross-section* (q.v.) for the radiation concerned.

atomic arc welding. Same as **atomic hydrogen welding.**

atomic bomb. Bomb in which the explosive power (20,000 tons TNT equivalent or more) is provided by nuclear fissionable material such as uranium-235 and plutonium-239. The bombs dropped on Hiroshima and Nagasaki (1945) were of this type. Also called **A-bomb, atom bomb, fission bomb.** See also **hydrogen bomb.**

atomic bond. Link between adjacent atoms formed by interaction of valence electron from each.

atomic clock. One which depends for its frequency of operation on the change of spin of the valency electron of caesium. The latest value for this frequency is 9, 192, 631, 770 ± 20 Hz.

atomic core (or kernel). That part of the atom other than the valence electrons.

atomic disintegration. Natural decay of radioactive atoms, with radiation, into chemically different atomic products.

atomic energy. Strictly the energy (chemical) obtained from changing the combination of atoms originally in fuels. Misapplied to energy obtained from breakdown of fissile atoms in nuclear reactors. See **nuclear energy.**

atomic frequency. A natural vibration frequency in an atom—used in the atomic clock.

atomic heat. Product of specific heat and atomic weight in grams; approximately the same for most solid elements at high temperatures.

atomic hydrogen welding. Process in which an electric arc is drawn between tungsten electrodes placed in a jet of hydrogen, which becomes ionized and releases its energy on impact. Also called **atomic arc welding.**

atomic mass unit (amu). Exactly 1/12th the mass of a neutral atom of the most abundant isotope of carbon ^{12}C.

$$1 \text{ amu} = 1 \cdot 6600 \times 10^{-24} \text{ g.}$$

Before 1960 the amu was defined in terms of the mass of the ^{16}O isotope and 1 amu was $1 \cdot 6599 \times 10^{-24}$ g.

atomic number. The order of an element in the periodic (Mendeleev) chemical classification, and identified with the number of unit positive charges in the nucleus (independent of the associated neutrons). Equal to the number of external electrons in the neutral state of the atom, and determines its chemistry. Symbol Z.

atomic polarizability. Electrical susceptibility per atom, defined as

$$a = \chi/N,$$

where χ is the susceptibility, and N is number of atoms per unit volume.

atomic scattering. That of radiation (usually cathode rays or X-rays) by the individual atoms in the medium through which it passes.

atomic spectrum. Emission spectrum arising from electron transitions inside an atom and characteristic of the element concerned.

atomic structure. Model of nucleus and surrounding orbiting electrons which conforms with observations, particularly spectral. See **Bohr, Rutherford, Sommerfeld atoms.**

atomic transmutation. Change of atomic number of atom due to bombardment by high-energy radiation or particles—most easily produced by neutron irradiation. This produces change of chemical nature of element, e.g., gold can be transmuted into mercury—the converse of what ancient alchemists attempted.

atomic volume. Ratio for an element of the atomic weight to the density; this shows a remarkable periodicity with respect to Z.

atomic (or equivalent) weight. The weight of atoms of an element, formerly expressed in atomic weight units (awu), but now more correctly given on the *unified scale* where 1u is $1 \cdot 6600 \times 10^{-24}$ g. For natural elements with more than one isotope, it is the average for the mixture of isotopes.

atomic weight unit (awu). Exactly 1/16th of the weighted mean of the masses of the neutral atoms of oxygen of isotopic composition found in rain water or fresh-water lakes.

Atomichron. TN for frequency-establishing system based on resonance in caesium vapour. See **atomic clock.**

attenuation. General term for reduction in magnitude, amplitude, or intensity of a physical quantity arising from absorption, scattering, or geometrical dispersion. The latter, arising from diminution by the inverse-square law, is not generally considered as attenuation proper. See also **absorption.**

attenuation coefficient. See **total absorption coefficient.**

attenuation compensation. The use of networks to correct for varying attenuation, e.g., in transmission lines. See **pre-emphasis.**

attenuation constant. Defined by

$$\rho = \rho_0 \epsilon^{-\alpha x},$$

where α is the constant, which may be complex, x is a distance, and ρ is a physical quantity, e.g., acoustic pressure in a sound wave, vector amplitude in an electromagnetic wave, etc. A more fundamental quantity is

$$\mu = \alpha\lambda,$$

the loss per wavelength distance of propagation. See **decibel, neper, propagation constant, transfer constant.**

attenuation distortion. Distortion of a complex waveform resulting from the differing attenuation of each separate frequency component in the signal. This form of distortion is difficult to avoid, e.g., in transmission lines.

attenuator. (1) Arrangement of resistors, capacitors, etc., which introduces known attenuation into a measuring circuit or line. A *variable attenuator* uses switching to vary the attenuation introduced into the circuit. This is often standardized at 600 or 75 ohms impedance level in ladder attenuators consisting of a series of T-sections. See pad (2). (2) Section of waveguide for diminishing intensity of wave transmitted. Geometry may include vane, piston, flap, disk, or a rotary member, in association with absorbing material to avoid reflections. (3) Gas tube used to absorb and hence control a radio-frequency oscillation. See also:

chimney-   potential-
piston-   waveguide-.

attenuator card. Thin-film attenuator produced by vacuum deposition of a thin rectangular sheet of resistive material on a dielectric substrate.

attracted-disk electrometer. Fundamental instrument in which potential is measured by the attraction between two oppositely-charged disks.

audibility of sound. Absolute intensity or ratio of the intensity of a sound stimulus to the threshold intensity at the same frequency; see phon, threshold of sound.

audio frequency. One which, in an acoustic wave, makes it audible. In general, any wave motion including frequencies in the range of, e.g., 30 Hz to 20 kHz.

audio-frequency amplifier. Amplifier for frequencies within the audible range, or some section of this.

audio-frequency choke. Inductor with appreciable reactance at audio frequencies.

audio-frequency shift modulation. Method of facsimile transmission in which tone values from black to white are represented by a graded system of audio frequencies.

audio-frequency transformer. Transformer for insertion into a communication channel, so designed that it has a specified, normally uniform, response for signals at all frequencies required for sound reproduction.

audiogram. Standard graph or chart which indicates the hearing loss (in *bels*) of an individual ear in terms of frequency. See noise-, sound-level meter.

audiometer. Instrument for measurement of acuity of hearing.

audion. Obsolete name for triode valve.

audition limits. Extreme frequencies of sound waves perceivable by the normal ear, and the extent of perception between the maximum tolerable loudness and the minimum perceptible. See phon, threshold of sound.

auditory canal. Duct connecting the eardrum with the external ear (*pinna*).

auditory perspective. Same as stereophony.

auditory sensation area. Area between the limits of audition expressed as curves on a chart.

Auger effect. Non-radiative transition of an atom from an electronic excited state to a lower level, with the emission of an electron. Arises when two atomic energy levels belonging to different series lie close enough together for mutual interaction, so that repulsion gives a shift of the energy levels.

Auger yield. For a given excited state of an atom of a given element, the probability of

de-excitation by Auger process instead of by X-ray emission.

Aurama. System of colour and sound control by cues added to stereophonically-recorded music and sound-effects on magnetic tape.

aureole. Luminous glow from outer portion of electric arc. This has different spectral distribution to that from the highly-ionized core.

aurora. Luminous ionization in thin atmosphere above magnetic poles, excited by particles from sun spiralling in earth's magnetic field.

auroral zone. Zone where radio transmission is limited by aurora.

Austen-Cohen formula. Semi-empirical formula for the field strength at a distance of r kilometres from the transmitting antenna of effective height H metres, and carrying a current of I amperes, the wavelength being λ kilometres. The field strength in microvolts per metre is:

$$\frac{377HI}{\lambda r}\exp.\left(\frac{-0 \cdot 0015r}{\sqrt{\lambda}}\right) \text{ for } \lambda > 1 \text{ km.}$$

autoalarm. See automatic call device.

auto-capacitance coupling. Coupling of two circuits by a capacitor included in series with a common branch.

autocode. Procedure for operating digital computer so that it helps to prepare its own programme.

autocorrelation. Technique for detecting weak signals against strong background level. Signal is subjected to controlled delay, the original delay signals then being fed to the autocorrelation unit which responds strongly only if delay is exact multiple of signal period.

autodyne. Of an electrical circuit in which same elements and valves are used both as oscillator and detector. Also endodyne, self-heterodyne.

autodyne oscillator. The valve used in an autodyne receiver which functions as a generator and for reception (or amplification).

autodyne receiver. One utilizing the principle of beat reception and including an autodyne oscillator.

auto-electric effect. See auto-emission.

auto-emission. That arising at normal temperatures by a high voltage gradient, stripping electrons from surface atoms; also called cold emission, field emission.

autoheterodyne. See under autodyne.

auto-inductive coupling. Coupling of two circuits by an inductance included in series with a common branch.

autoland. Electronic system for blind aircraft landings.

automatic alarm. Same as automatic call device.

automatic bias. Grid bias from a resistor in the cathode circuit of a valve, the grid circuit being returned to the end of the resistor remote from the cathode.

automatic brightness control. Circuit used

in some television receivers to keep average brightness level of screen constant.

automatic call device. System of relays, responsive to a prearranged set of signals, connected to an unattended receiver, so that an alarm is sounded on operation. Frequently used on ships for detection of distress signals. International alarm signal is 12 4-sec. dashes at 1-sec. intervals followed by SOS call repeated 3 times. Also **auto-alarm, automatic alarm.**

automatic check. Circuits built into a computer which check that it is functioning correctly.

automatic contrast control. Form of automatic gain control used in video signal channel of television receiver.

automatic control. (1) Switching system which operates control switches in correct sequence and at correct intervals automatically. (2) Control system incorporating servomechanism or similar device so that feedback signal from output of system is used to adjust the controls and maintain optimum operating conditions.

automatic data processing. Complete processing of data by computer, *read out* for immediate use.

automatic direction-finding. System in which servomotors, controlled by the incoming signal, cause the rotatable loop antenna or goniometer to hunt for the direction of maximum or minimum signal response.

automatic frequency control. Electronic or mechanical means for automatically compensating (in a receiver) frequency drifts in transmission carrier or local oscillator.

automatic gain control. Auto-adjustment of the gain of the radio amplifier of a radio receiver, through feedback of the rectified carrier to the grid bias of the valves, so that the output of the demodulator is very nearly constant, although the received carrier varies over a wide range.

automatic grid bias. Use of potential drop across a cathode resistor for biasing grid or other circuit.

automatic letter facing. System using one or two phosphorescent ink lines on stamps which are detected photoelectrically.

automatic meteorological observation station. Transistorized and packaged apparatus which transmits weather data for electronic computation.

automatic monitor. Apparatus which compares the quality of transmission at different parts of a system, and raises an alarm if there is appreciable variation.

automatic phase control. In reproducing colour television images, the circuit which interprets the phase of the chrominance signal as a signal to be sent to a matrix.

automatic phase control loop. Feedback circuit in which the phase of a local oscillator is controlled by a comparison with that of a reference signal, to obtain a correction voltage for application to the controlled source.

automatic picture control. See **automatic contrast control.**

automatic pilot. Servo system for control of aircraft in flight.

automatic programming. System in which a computer is used for preparing programme instructions in machine form.

automatic quiet gain control. Joint use of automatic gain control and muting.

automatic shutter. In a film projector, the shutter which cuts off the light from the arc if the film should stop, instead of maintaining intermittent motion; without this safeguard, the intense heat from the arc would ignite nitrate film.

automatic tracking. Servo control of radar system operated by received signal, to keep antenna aligned on target.

automatic tuning. (1) System of tuning in which any of a number of predetermined transmissions may be selected by means of pushbuttons or similar devices. (2) Fine tuning of receiver circuits by electronic means, following rough tuning by hand.

automatic volume compression. Reduction of signal voltage range from sounds which vary widely in volume, e.g., orchestral music. This is necessary before they can be recorded or broadcast but requires corresponding expansion in the reproducing system in order to compensate.

automatic volume control. Alteration of the contrast (dynamics) of sound during reproduction by any means. By compression (*compounder*) a higher level of average signal is obtained for modulation of a carrier, the expansion (*expander*) performing the reverse function at the receiver. In high-fidelity reproduction, arbitrary expansion can be disturbing because of variation in background noise, if present.

automatic volume expansion. Expansion of contrast (dynamics), e.g., keeping the maximum level constant and automatically diminishing the lower levels.

automation. Industrial closed-loop control system in which manual operation of controls is replaced by servo operation.

autonomics. Study of self-regulating systems for process control, optimizing performances.

Autophotic. Same as **Photronic.**

autoradiograph. Photograph made of radioactive material with self-emitted radiation, e.g., by placing photographic emulsion near to surface of object.

Autosyn. TN for class of synchros (U.S.).

auto-transductor. One in which the same winding is used for power transfer and control.

auto-transformer. Single winding on a laminated core, the coil being tapped to give desired voltages.

auxiliary grid. In a pentode, the second grid, maintained at high positive potential.

auxiliary store. See **secondary memory** (2).

auxochrome. A chromophore (or group of atoms) having a selective absorption frequency for radiation and functioning as a colour centre.

auxometer. An apparatus for measuring the magnifying power of an optical system.

availability ratio. See **operating ratio**.

available line. Percentage of total length of scanning line on a CRT screen on which information can be displayed.

available power. Maximum obtained from source by adjusting load. Limited by source impedance or linearity.

available power efficiency. The ratio of the electrical power available at the terminals of an electro-acoustic transducer to the acoustical power output of the transducer. The latter should conform with the reciprocity principle so that the efficiency in sound reception is equal to that in transmission.

available power gain. The ratio of the available output power of a transducer to the available signal power at the input.

available power response. For an electro-acoustic transducer, the ratio of the mean square sound pressure at a distance of one metre, in a defined direction from the 'acoustic centre' of the transducer, to the available electrical power input. The response will be expressed in dB above the reference response of (1 microbar)^2 per watt of available electrical power.

avalanche. Self-augmentation of ionization. See **Townsend avalanche, Zener effect**.

avalanche diode. See **Zener diode**.

average deviation (AD). For a number of like quantities X_J, the AD is given by

$$\frac{1}{n} \sum_{j=1}^{n} \left| X_J - \bar{x} \right|$$

where the mean

$$\bar{x} = \frac{1}{n} \sum_{j=1}^{n} X_J.$$

average life. See **mean life**.

average power output. For an amplitude-modulation transmitter, the instantaneous output averaged over one cycle of the *modulation* signal.

Avogadro number. The number of molecules in a *mol* (q.v.) of any pure chemical substance. It is also the number of atoms in a gram-atom for any element, and has the value $6 \cdot 02257 \times 10^{23}$ mole^{-1}. Also **Loschmidt number**, especially on the Continent.

axial ratio. Ratio of major to minor axis of polarization ellipse for wave propagated in waveguide, polarized light, etc.

axiotron. Valve in which the electron stream to the anode is controlled by the magnetic field of the heating current.

axonometry. Measurement of the axes of crystals.

Ayrton-Mather shunt. Design of shunt which can be used with almost any current-sensitive device, to reduce its response by a series of accurately known factors.

azimuth. The angle between the plane of the meridian and the vertical plane containing an object. See **bearing**.

azimuth error. That found in a transit instrument if the horizontal axis is not aligned exactly east-west.

azimuth marker. Line on radar display made to pass through target in order that the bearing may be determined.

azimuth-stabilized PPI. Form of plan-position indicator display which is stabilized by a gyro-compass so that the top of the screen always corresponds to north.

azimuthal quantum number. Quantum number associated with eccentricity of elliptic electron orbits in atom. Now normally replaced by orbital quantum number.

azusa. Radio tracking system for missile guidance.

B

β. See under **beta**.

B-amplifier. Amplifier following mixers or faders associated with microphone circuits in broadcasting studios, the faders and mixers following the A-amplifiers. *N.B.*— Not the same as **class-B amplifier.**

B-battery. U.S. term for **high-tension battery.**

B display. Rectangular radar display with target bearing indicated by horizontal co-ordinate and target distance by the vertical coordinate, the targets appearing as bright spots.

B-layer. Weakly reflecting and scattering layer or region 10-30 km above the earth's surface, possibly associated with water-vapour or ice in the stratosphere; postulated to explain short-period return signals when these are projected vertically.

B-service area. Region surrounding a broadcasting transmitter where the field strength is between 5 and 10 millivolts per metre.

BT-cut. Special cut of a quartz crystal to obviate temperature effects, such that the angle made with the Z-axis is −40°.

B-Y signal. Component of colour television chrominance signal. Combined with luminance (Y) signal, it gives primary blue component.

Babinet's principle. The radiation field beyond a screen which has apertures, added to that produced by a complementary screen (in which metal replaces the holes and spaces the metal), is identical with the field which would be produced by the unobstructed beam of radiation, i.e., the two diffraction patterns will be complementary.

back bias. See **back lighting.**

back contact. Contact in a relay assembly which is isolated when a moving contact separates from it on operation of the relay.

back coupling. Any form of coupling which permits the transfer of energy from the output circuit of an amplifier to its input circuit. See **feedback, regeneration** (1).

back electromotive force. That which arises in an inductance (because of rate of change of current), in an electric motor (because of conversion of energy), in a primary cell (because of polarization), or in a secondary cell (when being charged). Also **counter-electromotive force.**

back emission. That of electrons from the anode.

back heating. Excess heat in cathode due to electron bombardment, as in a magnetron.

back lighting. Of television camera tubes, illumination of rear surface of mosaic, giving greater sensitivity. Also **back bias.**

back lobe. Lobe of polar diagram for antenna, microphone, etc., which points in the reverse direction to that required.

back porch. See **porch.**

back-porch effect. The prolonging of the collector current in a transistor for a brief time after the input signal (particularly if large) has decreased to zero.

back projection. The use of a translucent background on which synchronized cinematographic pictures or a static scene are projected, so that both this and the action can be photographed together.

back scatter. The deflection of radiation or particles by scattering through angles greater than 90° with reference to the original direction of travel.

back-shunt keying. Keying a radio transmitter between the radiating aerial and a dummy aerial.

back-to-back. (1) Parallel connection of valves, such that the anode of one is connected to the cathode of the other, for rectification, frequency-doubling, etc. (2) A connection used in thyratron (or ignitron) rectifiers to control the alternating current to a load.

back wave. See **spacing wave.**

backfire. See **arc-back.**

background. A general problem in physical measurements which limits the sensitivity of detecting any given phenomenon. Background consists of extraneous signals arising from any cause which might be confused with the required measurements, e.g., in electrical measurements of nuclear phenomena and of radioactivity, it would include counts emanating from amplifier noise, cosmic rays, insulator leakage, etc. Cf. *noise ratio.*

background controls. Beam current (brightness) controls for the three electron guns of a colour TV tube.

background count. General extraneous count, in addition to that desired from a source.

background noise. Unwanted frequencies entering a wanted frequency band, which cannot be separated from wanted signals. Residual output from microphones, pickups, lines, etc., giving a signal/noise ratio. Also **ground noise.**

backing coil. See **bucking coil.**

backing pump. Roughing pump to establish a low pressure, at which a vapour diffusion pump can operate efficiently.

backlash. Property of most regenerative and oscillator circuits by which oscillation is maintained with a smaller positive feedback than is required for inception.

backplate. In a television camera tube, the plate behind, and insulated from, the mosaic, which transmits the video signal by capacitance.

backstop. The structure of a relay which limits the travel of the armature away from the pole-piece or core.

backwall cell. Semiconductor photovoltaic cell in which light passes through a polarized grid to active layer carried on a metal plate.

backward diode. One with characteristic of reverse shape to normal. Also sometimes known as AU diode.

backward scatter. Scattering backwards of radio signals tangent to lower level of ionosphere. Used to monitor effectiveness of transmission.

backward wave. In a travelling-wave tube, a wave with group velocity in the opposite direction to the electron beam. See also **forward wave.**

backwash diode. A diode valve which is connected across the line of a pulse modulator to absorb energy during the voltage pulse reversals. Also **overswing diode.**

Badger rule. An empirical relationship between the force constants and the vibrational frequencies of the electronic states of diatomic molecules.

baffle. (1) Extended surface surrounding a diaphragm of a loudspeaker, so that an acoustic short-circuit is prevented and the diaphragm loaded to transmit acoustic energy, especially at low frequencies. (2) Internal structure in a tube for controlling discharge or its decay.

baffle blankets. Blankets temporarily placed about a set, to regulate the distribution of reflected sound during sound-film production.

baffle plate. Plate inserted into waveguide to produce change in mode of transmission.

bake-out. Preliminary heating of the electrodes and container of a mercury-arc rectifier or any other type of tube or valve, to ensure freedom from the later release of gases.

Bakelite. TN for a phenolic resin plastic of good insulating properties. The transparent form is very valuable for polariscopes.

balance. (1) Adjustment of sources of sound in studios so that the final transmission adheres to an artistic standard. (2) A balance in bridge measurements is said to be obtained when the various impedances forming the arms of the bridge have been adjusted, so that no current flows through the detector.

balanced amplifier. Same as **push-pull amplifier.**

balanced-beam relay. One having two coils arranged to exert their forces on plungers at each end of a beam pivoted about its central point.

balanced circuit. For a.c. and d.c., one which is balanced to earth potential, i.e., the two conductors are at equal and opposite potentials with reference to earth at every instant. See **unbalanced circuit.**

balanced line. One in which the impedances to earth of the two conductors are, or are made to be, equal; also **balanced system.**

balanced mixer. A waveguide technique to minimize the effect of local oscillator noise.

balanced modulator. Matched pair of valves, operated as modulators, with their anodes connected in push-pull. The carrier voltage is applied to the two grids in phase and to the modulating voltage in antiphase, so that the carrier components in the anode currents cancel. Used in suppressed-carrier systems.

balanced network. One arranged for insertion into a *balanced circuit* and therefore symmetrical electrically about the midpoints of its input and output pairs of terminals.

balanced-pair cable. One with two conductors forming a loop circuit, the wires being electrically balanced to each other and earth (shield), e.g., an open-wire antenna feeder. Cf. *coaxial cable.*

balanced system. See **balanced line.**

balanced termination. Terminating impedance, the midpoint of which is at earth potential, or connected to earth.

balancing. See **neutralization.**

balancing antenna. Auxiliary reception antenna which responds to interfering but not to the wanted signals. The interfering signals thus picked up are balanced against those picked up by the main antenna, leaving signals more free from interference.

balancing capacitance. That connected between appropriate points in the grid and anode circuits of a valve amplifier to neutralize effects of internal grid-anode capacitance. Also called **neutralizing** or **neutrodyning capacitance.**

ball-and-biscuit microphone. See **apple-and-biscuit microphone.**

ball resolver. Computer device in which a ball is in contact with three wheels at points, radii of which are mutually perpendicular. Rotation of one wheel rotates the ball and hence the other two wheels at speeds $R \cos \theta$ and $R \sin \theta$, depending on its own speed R and inclination of its axis θ.

ballast lamp. Normal incandescent lamp used as a ballast resistor, current limiter, alarm, or to stabilize a discharge lamp.

ballast resistor. One inserted into a circuit to swamp or compensate changes, e.g., those arising through temperature fluctuations. One similarly used to swamp the negative resistance of an arc or gas discharge. See also **barretter.**

ballast tube. Same as **barretter** or **ballast resistor.**

ballistic galvanometer. One of long period which can be used to measure the total charge passing through the coil.

Balmer series. A group of lines in the hydrogen spectrum named after the discoverer and given by the formula:

$$\nu = R_H\left(\frac{1}{2^2} - \frac{1}{n^2}\right),$$

where n has various integral values, ν is the wave number and R_H is the hydrogen Rydberg number ($= 109,677 \cdot 576$ cm^{-1}).

balun. Abb. for *balance to unbalance transformer*, usually a resonating section of transmission line, for coupling balanced to unbalanced lines. Also **bazooka.**

banana plug. A single conductor plug which has a spring metal tip in the shape of a banana. The corresponding socket or jack is termed a **banana jack.**

Banana tube. Trade name (cf. U.S. Apple tube) for a single-gun cathode-ray tube developed especially for colour television, in which the tube has the shape of a long cylinder with phosphor stripes parallel to the axis in the three primary colours.

band. See under:

allowed-	rejection-
conduction-	service-
energy-	sidebands
filter attenuation-	transmission-
filter transmission-	V-
frequency-	valence-
guard-	X-.

band-edge energy. The band of energy between two defined limits in a semi-conductor. The lower limit corresponds to the lowest energy required by an electron to remain free, while the upper one is the maximum permissible energy of a freed electron.

band elimination filter. Filter which highly attenuates currents having frequencies within a specified nominal range and freely passes currents having frequencies outside this range. Also called **band rejection filter, band stop filter.**

band ignitor tube. A valve of mercury pool type in which the control electrode is a metal band outside the glass envelope. Also **capacitron.**

band merit. Parameter of a valve, the product of the bandwidth and maximum gain (as a ratio) possible; alternatively the mutual conductance divided by 2π times the sum of the grid and anode capacitances.

band-pass amplifier. Amplifier with band-pass frequency characteristic.

band-pass filter. One which freely passes currents having frequencies within specified nominal limits, and highly attenuates currents with frequencies outside these limits.

band-pass tuning. Arrangement of two coupled circuits, tuned to same frequency and having substantially uniform response to range of frequencies, instead of marked variation of response with frequency, which characterizes a single-tuned circuit.

band rejection filter. See **band elimination filter.**

band spectrum. Molecular optical spectrum of numerous very closely spaced lines spread through a limited band of frequencies.

band-spreading. Use of a relatively large fixed capacitor in parallel with a smaller variable capacitor, to reduce the band of frequencies covered by variation of the latter.

band stop filter. See **band elimination filter.**

band-switching. System of inductance or capacitance switching, allowing a number of frequency bands to be covered by the same tuning dial.

bandwidth. (1) Section of frequency spectrum within which component frequencies of a

B*

signal can pass, ideally with zero attenuation. (2) Arbitrary measure of a radio transmission, the width in Hz within which 99% of the energy is contained.

bang-bang. Control mechanism of servo system with two set points, operation being between these, with no gradation.

bar-and-yoke. Method of magnetic testing in which the sample is in the form of a bar, clamped into a yoke of relatively large cross-section, which forms a low-reluctance return path for the flux.

bar generator. Source of pulse signals, which gives a bar pattern for testing TV CRTs.

bar relay. A relay in which a bar mechanism operates simultaneously several contacts.

barium (Ba). Metallic element, at. no. 56, at. wt. 137·34, m.p. 850°C. Barium salts are used in coating thermionic cathodes.

barium getter. Barium metal is supported in a nickel capsule and heated with eddy currents.

barium titanate. $BaTiO_3$, a ceramic-type material. It is an electret which has piezo-electric properties and a higher Curie point than Rochelle salt.

Barkhausen criterion. The feedback condition which must be satisfied in the design of an oscillator. The product of the complex gain and the feedback factor must be -1.

Barkhausen effect. The tendency for magnetization to take place in discrete steps rather than by continuous change.

Barkhausen-Kurz oscillator. Triode valve with anode and grid fed negatively and positively through a lecher system, the oscillation depending on the retarding field on both sides of the grid.

barn. Unit of effective cross-sectional area of nucleus, equal to 10^{-24} cm².

Barnett effect. Magnetization of ferro-magnetic material caused by rapid rotation. Also **Einstein-de Haas effect, gyromagnetic effect.**

barrel. In a system of *transposition*, the section in which all possible transpositions are effected in order, before repetition in the next section, to effect balance of circuits, i.e., side impedances to earth.

barrel distortion. Distortion of image on CRT screen such that the sides of a rectangle bulge outwards.

barretter. Iron-wire resistor mounted in a glass bulb containing hydrogen, and having a temperature variation, so arranged that the change of resistance ensures that the current in the circuit in which it is connected remains substantially constant over a wide range of voltage.

barrier. In cinematograph film, thin black line which separates adjacent frames in the projection print, and from the sound-track.

barrier layer. Double electrical layer formed at the surface of substances which have differing work functions, there being a diffusion of electrons up the work-function gradient.

barrier-layer cell. One in which illumination

of the barrier layer results in a small electro-motive force because of differing work functions of dissimilar materials.

bars. Artificial appearance of bars or a cross on a television screen, generated electronically and used for testing television systems and circuits. See **bar generator.**

Bartlett force. Force acting between two nucleons due to spin exchange. See **short-range forces.**

barye. Less commonly-used term for **microbar.**

baryon. Nucleon or hyperon; see **antibaryon.**

base. (1) In a navigation chain, the line which joins two of the stations. (2) Middle region of a transistor into which minority carriers are injected from the external circuit. (3) Part of valve where leads from electrodes are connected to rigid pins which fit into sockets in a holder, so that a valve can be changed quickly. Applicable also to lamps, capacitors, crystals, etc. (4) See **radix.**

base resistance. Total resistance to base current, including spread.

base-spreading resistance. That between active contact and external terminal in transistor.

base transmission factor. Complex ratio of minority-carrier current arriving at collector to that leaving emitter; also called **base transport factor.**

baseline. (1) Line on CRT screen along which echo displacements are measured. (2) Great circle line joining two navigational stations, e.g., on loran systems.

basic cycle. In a computer, the time necessary to complete a set of operations for the execution of each instruction.

basket coil. One with criss-cross layers, so designed to minimize self-capacitance. Also **duolateral coil.**

bass boost. Amplifier circuit adjustment which regulates the attenuation of the lowest frequencies in the audio scale, to offset the progressive loss towards low frequencies.

bass compensation. Differential attenuation introduced into a sound-reproducing system when the loudness of the reproduction is reduced below normal, to compensate for the diminishing sensitivity of the ear towards the lowest frequencies reproduced.

bass frequency. One towards the lower limit of frequency in an audio-frequency signal or a channel for such, e.g., below 250 Hz.

bass reflex cabinet. Loudspeaker cabinet in which sound wave from back of cone is delayed by half a period and then radiated in phase with that from the front.

bathochromes. Particular groups of atoms in organic compounds which have the effect of lowering frequency of the radiation absorbed by these compounds.

battery. General term for a number of objects cooperating together, e.g., a number of accumulator cells, dry cells, capacitors, radars. See:

AB- grid-bias-
high-tension- plate-.
low-tension-

battery coupling. Interstage coupling in amplifiers required to transmit very low frequencies or direct currents; the grid of one stage is connected to the anode of the preceding stage by a battery, of the requisite voltage, to maintain the grid at its correct operating potential.

baud. Unit of speed of telegraphic code transmission; equal to twice the number of dots evenly sent per second. See **bit.**

Baudot code. One comprising impulses in the time frame of five units, devised by J. M. Baudot for mechanical transmission of signals. Used in teleprinters, computers, process controllers.

bay. (1) Unit of racks designed to accommodate numbers of standard-sized panels, e.g., repeaters or logical units. (2) Unit of horizontally extended antenna, e.g., between masts.

Bayard and Alpert gauge. One for measuring very low gas pressure by collecting ions on a fine wire inside a helical grid.

Bayhurst curve. Graph showing relationship between efficiency of beta-particle counter assembly and the particle energy.

bazooka. See **balun.**

beacon. See **radio beacon.**

beam. A collimated or approximately unidirectional flow of electromagnetic radiation (radio, light, X-rays) or of particles (atoms, electrons, molecules). The angular beam width is defined by the half-intensity points.

beam angle. (1) Angle of response curve of an antenna in which the bulk of energy is transmitted or received. (2) The solid angle subtended by the cone of electrons emerging from the first focus of the beam in an electron gun.

beam antenna. One with very marked directional properties. Original high curtain of vertical wires with a similar reflector.

beam array. Beam antenna composed of a number of spaced radiators, as distinct from one dependent upon reflectors for its directional properties.

beam bender. U.S. colloquialism for **ion trap.**

beam convergence. Focusing of the three electron beams of a colortron-type triple gun colour television tube, so that they converge at the shadow-mask aperture.

beam coupling. That provided between circuits by an electron beam passing between electrodes, as in a beam tetrode.

beam current. That portion of the gun current which passes through the aperture in the anode and impinges on the fluorescent screen.

beam-deflection valve. One in which an electron beam is deflected by side plates and so switched between push-pull anodes.

beam-forming electrode. Electrode to which a potential is applied to concentrate the electron stream into one or more beams. Used in beam tetrodes and cathode-ray tubes.

beam hole. Hole in shield of reactor, or that

around a cyclotron, for extracting a beam of neutrons or gamma-rays.

beam-indexing colour TV tube. Tube in which a signal, generated by beam after deflection, is used to control image colour.

beam jitter. (1) Random movements of radar beam due to imperfections or wear in mechanical drive of antenna. (2) Random movements of electron beam in cathode-ray tube due to electronic noise.

beam loading. Coupling of two electrodes by an electron beam leading to an admittance between them.

beam pentode. See **beam-power valve.**

beam-power valve. One in which the control grid and screen are aligned so that a bunching of electrons is equivalent to a suppressor grid, without the power loss of the latter.

beam rider (or riding). System in which a guided missile maintains and returns to a course of maximum signal on a radio beam.

beam suppression. Application of a large negative potential to the control electrode of a cathode-ray tube, in order to suppress the beam during flyback period between successive scanning lines.

beam switching. See **lobe switching.**

beam tetrode. Tetrode having an additional pair of plates normally connected internally to the cathode, so designed as to concentrate the electron beam between the screened grid and anode and thus reduce secondary emission effects. See also **beam-power valve.**

beam transadmittance. See **forward transadmittance.**

beam trap. Bucket-formed electrode mounted in a cathode-ray tube, to catch the electron beam when it is not required to excite fluorescence in the screen.

bearing. Angle of direction (horizontal plane) in degrees from true north, e.g., of an arriving radio wave as determined by a direction-finding system. See **azimuth.**

bearing classification. Classes considered by operator to be within:

Class A, $\pm 2°$; B, $\pm 5°$; C, $\pm 10°$.

beat. Periodic variation in the amplitude of a wave containing two sinusoidal summation components of nearly equal frequencies.

beat frequency. Generally, the difference frequency produced by the intermodulation of two frequencies. Specifically, the supersonic (intermediate) frequency in a supersonic heterodyne receiver.

beat frequency oscillator. Same as heterodyne oscillator.

beat frequency wavemeter. Same as heterodyne wavemeter.

beat (or beatnote) receiver. See **supersonic heterodyne receiver.**

beat reception. See heterodyne reception.

beating oscillator. See local oscillator.

beavertail antenna. One producing a broad flat radar beam.

beavertail beam. A broad flat radar beam due to imperfections or wear in mechanical drive of antenna.

Becquerel cell. See **photochemical cell.**

Becquerel effect. Flow of current between two similar metallic electrodes immersed in an electrolyte which is produced when one of the electrodes is illuminated.

Becquerel rays. The penetrating alpha (α), beta (β) and gamma (γ) radiations from uranium, discovered by Becquerel.

bel. A non-dimensional unit used for expressing ratio of power units (P_1 and P_2).

$$N = \log_{10}(P_1/P_2) \text{ bels.}$$

Ten times the size of the more frequently used *decibel* (q.v.). On the Continent **neper** is used instead of bel.

Bellini-Tosi antenna. Directional antenna comprising two crossed loops. The direction of maximum reception is controlled by a goniometer, which varies the relative couplings of the two loops to the receiver.

bench. Fixed rails with adjustable and slidable supports for a waveguide system.

bend. Alteration of direction of a rigid or flexible waveguide. It is E or *minor* when electric vector is in plane of arc of bending and H or *major* when at right angles to this. Also **corner.** See **anode-, H-plane-.**

benito. A CW navigation system giving bearing and range relative to ground station.

berkelium (Bk). Element, at. no. 97, discovered by S. G. Thompson in 1950 from bombarding ^{241}Am with helium ions.

Bernoulli's theorem. That the sum of the pressure, potential energy and kinetic energy is constant at any point in a tube through which liquid is flowing, pressure being smallest at points of the greatest velocity.

beryl. A silicate of beryllium and aluminium, which is the chief ore of beryllium.

beryllium (Be). Steely uncorrodible white metal. Gk. and Lat. *beryl* = old mineral name. Discovered by Wöhler in 1828. At. no. 4, at. wt. 9·0122, m.p. 1280°C, b.p. 2450°C, sp. gr. 2·7. Main use is for windows in X-ray tubes and as an alloy for hardening copper. Used as a powder for fluorescent tubes until found poisonous. The metal can be evaporated on to glass, forming a mirror for ultraviolet light. As a slight alloy with nickel, it has the highest coefficient for secondary electron emission, 12·3. Alpha-particles projected into beryllium make it a useful source of neutrons, from which they were discovered by Chadwick in 1932. The oxide is also a good reflector of neutrons. Formerly called **glucinum.**

beta. Greek letter used principally to represent beta-particles, and complex feedback factor of electronic circuits. Symbol β.

beta circuit. Colloquialism for *feedback path* in an amplifier.

beta decay. See beta disintegration.

beta detector. A radiation detector specially designed to record or monitor beta-radiation.

beta disintegration. Radioactive transformation of a nuclide in which the mass number remains unchanged but the atomic number changes by (a) $+1$, with negative β-particle emission or (b) -1, with positron emission or electron capture. Also known as **beta decay.**

beta-disintegration energy. For negatron emission it equals the sum of the kinetic energies of the beta-particle, the neutrino and the recoil atom. In the case of positron emission there is in addition the energy equivalence of two electron rest masses.

beta-ell or βl. Continental expression for the total attenuation, in *nepers*, of a line of length l and attenuation constant β nepers per unit length (km or mile) for any frequency.

beta-particle. A negatron or a positron emitted from a nucleus during beta decay.

beta-ray spectrometer. One which determines the spectral distribution of energies of beta-particles from radioactive substances or secondary electrons.

beta-rays. Streams of β-particles.

beta thickness gauge. Thickness-measuring instrument, based on absorption of beta-particles from radioactive source.

beta wave. High-frequency wave (15-60 Hz) produced in human brain.

betatopic. Said of atoms differing in atomic number by one unit. One atom can be considered as ejecting an electron (β-particle) to produce the other one.

betatron. Accelerator for high-energy beams of electrons and consequently very highly penetrating X-rays. Electrons are accelerated by a rapidly changing magnetic field, orbit being of constant radius. Also **induction accelerator, rheotron.** See **cyclotron.**

Bethe hole. Arrangement for tapping off power from a waveguide by attaching a tube at a reverse angle.

Bethenod-Latour alternator. High-frequency alternator in which alternating currents in the stator generate currents of higher frequency in the rotor. By repeating the process, frequencies of the order of 100 kHz have been attained in a single machine.

BeV. U.S. abb. for *billion-electron-volt*; see GeV.

Bevatron. Accelerator at Berkeley University in California, used to accelerate protons and other atomic particles up to 6 GeV.

Beverage antenna. See **wave antenna.**

bias. (1) Adjustment of a relay so that it operates for currents greater than a given current (against which it is *biased*), or for a current of one polarity. (2) In a computer, the average of random errors when these are not balanced about zero error. See:

alternating-current-	direct-current-
automatic-	grid-
automatic grid-	line-.
cathode-	

bias current. Non-signal current supplied to electrode of semiconductor device, magnetic amplifier, tape recorder, etc., to control operation at optimal working point.

bias resistor. That used in cathode-bias circuit.

bias voltage. Generally non-signal or mean potential of any electrode in a thermionic tube, measured with reference to the cathode. Specially applied to that of control grid.

bias winding. Transformer or choke winding, current in which controls the operating point on the magnetic circuit.

biasing. Polarization of a recording head in magnetic-tape recording, to improve linearity of amplitude response, using d.c. or a.c. much higher than the maximum audio frequency to be reproduced.

Biax. Minute computer element of ferrite, utilizing flux interaction between normal magnetic fields, established by wires through orthogonal holes.

biaxial crystal. One which has two optical axes and relevant physical properties.

biconical horn. Two flat cones apex to apex for radiating uniformly in horizontal directions when driven from a coaxial line.

bidirectional microphone. One, such as the normal open ribbon pressure-gradient microphone, which is most sensitive in both directions along one axis.

bifilar resistor. One wound with two wires in parallel, to reduce inductance and to balance capacitive effects at high frequencies.

bifilar winding. One used for non-inductive coils, in which the current passes through two wires side-by-side, in opposite directions, so that their outer magnetic field is largely balanced. Used for resistors in radio circuits.

bigrid valve. Four-electrode thermionic tube with two control grids, each having approximately the same control on the anode current. Used as a modulating or mixing valve, or as an amplifier operating with low anode voltages.

bilateral impedance. Any electrical or electro-mechanical device in which power can be transmitted in both directions.

billicapacitor. Variable capacitor having a maximum capacitance of a few micro-microfarads, used for fine tuning adjustments.

billion-electron-volt. See **BeV.**

bimetallic strip. Strip of two metals having different temperature coefficients, so arranged that the strip deflects when subjected to a change in temperature; used in thermal switches.

bimorph. Unit in microphones and vibration detectors in which two piezoelectric plates are cemented together in such a way that application of p.d. causes one to contract and the other to expand, so that the combination bends, as in a bimetallic strip.

binary. Involving the integer 2; see **binary scale.**

binary arithmetic. Arithmetical operations carried out on *binary scale* (q.v.).

binary cell. An information storage element used in computer work which can have one or other of two stable states.

binary-coded decimal system. Scheme of computation whereby decimal numbers are individually coded and used as binary numbers.

binary counter. Flip-flop or toggle circuit which gives an output pulse for two input pulses, thus dividing by two.

binary digit. See bit.

binary scale. In computers, scale of counting with radix 2; used because the figures 1 and 0 can be represented by *on* and *off*, *pulse* or *no pulse*, in electronic circuits.

binaural. Listening with two ears, the result of which is a sense of directivity of the arrival of a sound wave. Said of a stereophonic system with two channels (matched) applying sound to a pair of ears separately, e.g., by earphones. The effect arises from relative phase delay between wavefronts arriving at the two ears.

binding energy. (1) That required to remove a particle from a system, e.g., electron, when it is the *ionization potential* (q.v.). (2) That required to overcome forces of cohesion and disperse a solid into constituent atoms.

Binistor. *N-p-n* silicon bistable device for registering or controlled switching.

binode. Three-electrode thermionic tube having one cathode and two anodes. Used for full-wave rectification; also **double diode**, **duo-diode**.

binomial distribution. Statistical distribution applicable to processes involving a number of individual observations with each of which a specific event may or may not be associated. A special case of *multinomial distribution* where several events may be associated with each observation. See also **Poisson distribution**.

biological (rbe) dose. See dose, rem.

biological half-life. Time interval required for half of a quantity of radioactive material absorbed by a living organism to be eliminated naturally.

biological hole. A cavity within a nuclear reactor in which biological specimens are placed for irradiation experiments.

biological shield. Screen required in a reactor or other high-energy machine to nullify the effect of intensity of radiation on humans.

bioluminescence. The production of light by living organisms, e.g., glow-worms, some deep-sea fish, some bacteria, some fungi.

biophysics. The study of the physical phenomena associated with living organisms, or of live organic phenomena, by physical techniques.

biotron. Two-valve amplifying circuit, in which high amplification is obtained by aperiodic regeneration.

bipolar transistor. One making use of both negative and positive charge carriers.

birdie. Transistorized electronic device for detecting enemy aircraft and directing the fire of local missile batteries to them. *B*attery *I*ntegration and *R*adar *D*isplay *E*quipment.

birefringence. See **double refraction**.

bismuth (Bi). Element, at. no. 83, at. wt. 208·980, m.p. 271°C. ^{209}Bi is the heaviest stable nuclide. It is strongly diamagnetic

and has a low capture cross-section for neutrons, hence its possible use as a liquid metal coolant for nuclear reactors. A high absorption for γ-rays makes it a useful filter or window for these while transmitting neutrons.

bismuth spiral. Flat coil of bismuth wire used in magnetic flux measurements; the change of flux is measured by observing the change in resistance of the bismuth wire, which increases with increasing fields.

bistable circuit. Valve or transistor circuit which has two stable states which can be decided by input signals; much used in counters and scalers.

bit. The smallest unit of *information* in a computer, a unit of storage capacity. Abb. for *binary digit*, i.e., 1 or 0, the normal units for expressing information, corresponding with *on* and *off* or *pulse* and *no pulse*.

Bitter pattern. Pattern showing boundaries of magnetic domains in ferromagnetic material, formed by powder concentrated along boundaries. Also **powder pattern**.

Black amplifier. Original amplifier with reversed retroaction (negative feedback) for degeneration.

black body. A theoretically ideal thermal radiator for which the spectral distribution of the radiation depends only upon the temperature.

black box. Colloquialism for computer control unit.

Black feedback. Same as **bridge feedback**.

black level. That percentage of maximum amplitude possible in a positive video signal which corresponds to black in a transmitted picture, a lesser amplitude being concerned with synchronizing. Usually between 30% and 40% in a positive video signal.

black negative or positive. Television picture signals in which the voltage corresponding to black is respectively negative or positive in relation to the voltage corresponding to white.

black-out effect. Temporary loss of sensitivity in vacuum tube after subjection to a strong short pulse.

black peak. Maximum excursion of a video signal in the black direction, which may be positive or negative.

blank groove. Unmodulated groove on disk recording.

blanket. Surround of fertile material to absorb stray neutrons from the core of a reactor and so breed in a small way further fissile material, e.g., by absorption of excess neutrons, thorium becomes uranium.

blanking. (1) Blocking or disabling a circuit for a required interval of time. (2) In a CRT, shutting off the beam during flyback.

blanking interval. Time during which there is no video signal.

blanking level. Reference level in a video signal, the demarcation between the synchronizing signals and the picture information.

blast wave. See **shock wave**.

Blattnerphone. Early type of magnetic-wire sound recorder.

bleeder resistor. Resistor placed across secondary of transformer to regulate its response curve, especially when the transformer is not loaded with a proper terminating resistance. One placed in a power supply or rectifier circuit to control its regulation.

blemish. A mosaic imperfection which affects the transmission of an image in a camera tube.

blind approach beacon system. Aircraft navigation system in which switched beams from a transmitter indicate to the pilot his azimuth in relation to the landing runway.

blind monitoring. Control of microphone outputs in broadcasting, particularly in outside broadcasts, when the operator is out of sight of the persons originating the transmission.

blind spot. Point, within normal range of a transmitter, at which field strength is abnormally small. Usually results from interference pattern produced by surrounding objects, or geographical features, e.g., valleys.

blinking. Modification of a loran transmission, so that a fluctuation in display indicates incorrect operation.

blip. Spot on CRT screen indicating radar reflection.

blister. See radome.

Bloch wall. Transition layer (few hundred lattice constants thick) in which the spin directions change slowly from the orientation in one ferromagnetic domain to that of a neighbouring domain of differing magnetic orientation.

block. Group of *words* arranged sequentially on magnetic tape, which form a unit in operations in computers.

blocked impedance. That of the input of a transducer when the output load is infinite, e.g., when the moving coil of a loudspeaker is prevented from moving.

blocking. Cutoff of anode current in a valve because of the application of a high negative voltage to the grid; used in *gating* or *blanking*.

blocking capacitor. One in signal path to prevent d.c. continuity. Also called **buffer capacitor**.

blocking oscillator. Valve or transistor tuned oscillator, which has more than sufficient positive feedback for oscillation, but in which a condition periodically supervenes to suspend normal oscillation, e.g., by integration of grid current by a capacitor. Discharge of capacitor removes condition and restores initial state. Cycle is continuously repeated, producing sawtooth and pulse waveforms up to very high frequencies.

blood count. Count of concentration of red and/or white blood corpuscles. The ratio of red to white is upset by ionizing radiation.

blooming. (1) Spread of spot on CRT phosphor due to excessive beam current. (2)

Coating of dielectric surfaces to reduce reflection of EM waves.

bloop. Dull thud in sound-film reproduction caused by joints in negative sound-track before printing the positive projection prints.

blooping patch. Black patch painted on the negative sound-track to give a gradual change in the exposure area and so prevent a bloop on projecting the positive print.

blowout magnet. A permanent magnet or electromagnet used to extinguish more rapidly the arc (in a switch, etc.) due to breaking an electric circuit.

blue glow. Visible evidence of ionization in thermionic valve, due to gas.

blue light. That produced by discharge lamps, which operate with low heat dissipation, for stage or studio; also called **cold light**.

Bode equalizer. Variable-response adjuster, in which one control regulates the equalization for all frequencies.

Bode plot (or diagram). One in which gain in dB and phase are plotted against frequency, to study margins of control in a servo system.

body burden. Radioactive material retained in the body at any time after absorption.

body capacitance. Change in capacitance because of proximity of the hand or body. Used for changing frequency of oscillator in electronic musical instruments, for triggering control devices, burglar alarms, etc.

body-centred cubic. A crystal whose structure unit cell is a cube with an atom located at each corner and one at the centre of each cube. See **face-centred cubic**.

body-section radiography. See **tomography**.

Bohr atom. Concept of the atom, with electrons moving in a limited number of circular orbits about the nucleus. These are *stationary states*. Emission or absorption of electromagnetic radiation results only in a *transition* from one orbit (state) to another.

Bohr magneton. Unit of magnetic moment, for electron defined in CGS by

$$\mu_e = eh/4\pi m_e c,$$

where e = charge (e.s.u.)

h = Planck's constant

m_e = rest mass

c = velocity of light,

so that $\mu_e = 9 \cdot 27 \times 10^{-21}$ erg gauss^{-1}.

The **nuclear Bohr magneton** is defined by

$$\mu_p = eh/4\pi Mc = \frac{\mu_e}{1836} = 5 \cdot 05 \times 10^{-24} \text{ erg gauss}^{-1},$$

M being the rest mass of the proton.

Bohr radius. The radius of any permissible orbit in the Bohr model of the hydrogen atom, i.e.,

$$R_n = \frac{n^2 h^2}{4\pi^2 m_e e^2} = 0 \cdot 529 \times 10^{-8} n^2 \text{ cm.,}$$

where n may have any positive integral value (n = 1 for the orbit of lowest energy, which is usually what is meant by Bohr radius), h is Planck's constant, me is the rest mass of the electron, and e is the electronic charge.

Bohr-Sommerfeld atom. Atom obeying modifications of Bohr's laws suggested by Sommerfeld and allowing for possibility of elliptic electron orbits.

boiling-water reactor. Reactor cooled by allowing water to boil.

bolometer. Early instrument for measuring radiant energy and for detecting radiofrequency currents, generally by unbalancing a bridge. Modern forms make use of very large changes in resistance of metals at very low temperatures, and are very sensitive. Erroneously applied to *barretter* or *thermistor* when used in microwave work.

Boltzmann's constant. Given by

$$k = R/N = 1 \cdot 3803 \cdot 10^{-23} \text{ joule deg}^{-1} \text{ C,}$$

where R = ideal gas constant per g mol.

N = Avogadro number.

Boltzmann's equation. The fundamental particle conservation diffusion equation based on the description of individual collisions, and expressing the fact that the time rate of change of the density of particles in the medium is equal to the rate of production less the rate of leakage and the rate of absorption.

Boltzmann's principle. Statistical distribution of large numbers of small particles when subjected to thermal agitation and acted upon by electric, magnetic or gravitational fields. In statistical equilibrium number of particles n per unit volume in any region is given by

$$n = n_0 \cdot \exp \cdot (-E/kT),$$

where k = Boltzmann's constant

T = absolute temperature

E = potential energy of a particle in given region

n_0 = number per unit volume when $E = 0$.

bombardment. Process of directing a beam of neutrons or high-energy charged particles onto a target material in order to produce nuclear reactions.

bond. Link between atoms, considered to be electrical attraction arising from electrons as distributed around the nucleus of atoms thus bonded, controlling properties of such compounds. Represented by a dot (·) or a line (—) between atoms, e.g.,

H—O—H or H·O·H.

See:

atomic-	ionic-
covalent-	metallic-
hydrogen-	molecular-.

bond angle. That between axes of atoms in valence combination, e.g., that between hydrogen bonds in water is $109 \cdot 5°$.

bond energy. That associated with a chemical bond between atoms, measured by energy required to break bond. See **adhesion** (2).

bond length. Internuclear distance of two atoms joined by chemical bond.

Bond notation. That describing the way in which a piezoelectric crystal is cut.

bond strength. See **adhesion** (2).

Bondacust. TN for a type of sound-absorbing soft felt.

bonded wire. Enamelled insulated wire also coated with a thin plastic; after forming a coil, it is heated by a current or in an oven or both, for the plastic to set and the coil to attain a solid permanent form.

bone conduction. Conduction of sound energy to inner ear through cranial bones.

bone seeker. Radioelement similar to calcium, e.g., Sr, Ra, Pu, which can pass into bone where it continues to radiate.

bone tolerance dose. The maximum radioactive dose which can safely be given in treatment without bone damage.

Boolean algebra. That which is found useful in analysing and synthesizing binary circuits, using analogous logical elements.

Boolean calculus. Boolean algebra into which time variations have been introduced.

boom. Mechanical arrangement for swinging the microphone clear of artists and cameras in sound-film and television studios.

boost transformer. Same as **buck transformer**.

booster amplifier. One used specially to compensate loss in mixers and volume controls, in order to obviate reduction in signal/noise ratio.

booster response. An automatic controller method of operation in which there exists a continuous linear response between rate of change of the controlled variable and the position of the final control element. Also called **rate action, time response.**

booster station. One which rebroadcasts a transmission received directly on the same wavelength.

Boothroyd-Creamer system. A form of time division multiplex transmission.

bootstrap. A self-sustaining system in liquid rocket engines by which the main propellants are transferred by a turbo-pump which is driven by hot gases. In turn the gas generator is fed by propellants from the pump.

bootstrap circuit. Thermionic valve in which drive between cathode and grid is through a high resistance, output being taken from a resistor and shunted capacitor in the cathode circuit. A step signal applied to grid results in the potential of the source with reference to earth rising regularly with time. This linear sweep has wide application for accurate timing measurements in radar, etc.

bornite detector. Early demodulator consisting of a steel point in contact with a bornite crystal, with marked rectifying properties.

boro-carbon resistor. Resistor constructed of boron and carbon to be substantially independent of temperature.

boron (B). Amorphous brown-yellow element. Discovered by Davy 1808, also Gay-Lussac and Thenard. At. no. 5, at. wt. $10 \cdot 811$, m.p. 2300°C, b.p. 2550°C, sp. gr. $2 \cdot 5$. Can be formed into a conducting metal. Boron carbide is often used as a cutting tool, having extreme hardness. Most important in reactors, because of great cross-section (absorption) for neutrons; thus, boron steel is used for control rods. The isotope ^{10}B, on absorbing neutrons, breaks into two charged particles, ^7Li and He, which are easily detected, and is therefore most useful for detecting and measuring neutrons.

boron chamber. Counter tube containing boron trifluoride, or boron-covered electrodes, for the detection and counting of slow-speed neutrons, which eject α-particles from the isotope ^{10}B.

Bose-Einstein statistics. Statistical mechanics laws obeyed by a system of particles whose wave function is unchanged when two particles are interchanged.

boson. A fundamental particle described by Bose-Einstein statistics and having angular momentum nh, where n is an integer and h is Planck's constant.

bottle. Colloquialism for **valve (1)**.

bottom bend. Same as **anode bend**.

bottoming. The operation of a device with a non-linear voltage current characteristic at or below the 'knee' of the curve.

Boucherot circuit. An arrangement of inductances and capacitances, whereby a constant-current supply is obtained from a constant-voltage circuit.

bounce. Colloquialism in sound recording, implying that reverberation is relatively high for high-frequency components.

bound charge. That induced static charge which is 'bound' by the presence of the charge of opposite polarity, which induces it. In a dielectric, the charge arising from polarization. Also **surface charge**.

boundary. The surface in the transition region between p-type and n-type semiconductor material at which the donor and accepted concentration are equal.

boundary layer. The region within a fluid near to an interface in which the velocity differs significantly from that of the main fluid stream. The presence of this layer is of considerable importance in heat transfer problems, e.g., in nuclear reactors.

Bourdon gauge. Form of pressure-sensitive transducer for fluids.

box antenna. See **travelling-wave antenna**.

box baffle. Box, with or without apertures and damping, fitted in a side with an open diaphragm loudspeaker unit, generally coil-driven.

boxcar. Converter of digital or sample data to analogue form, output forming a step waveform, used in computers.

Brackett series. A group of spectral lines of atomic hydrogen in the infrared given by the formula

$$\nu = R_H \left(\frac{1}{n_1^2} - \frac{1}{n_2^2} \right),$$

in which ν is the wave number, R_H is 109,677·591 cm^{-1}, $n_1 = 4$ and n_2 has various integral values.

Bragg angle. Angle of incidence of a beam of X-rays which results in wave reflection from the surface of a crystal lattice.

Bragg curve. Graph giving average number of ions per unit distance along beam of initially monoenergetic α-particles (or other ionizing particles) passing through a gas.

Bragg law. Conditions for reflecting beam of X-rays with maximum intensity. Equation is:

$$\sin \theta = n\lambda/2d,$$

where θ = angle which the incident and reflected waves make with the crystal planes (Bragg angle)
n = integer
λ = wavelength of X-rays
d = spacing of planes or layers of atoms.

The law also applies for the reflection of de Broglie waves associated with protons, electrons, neutrons, etc.

Bragg rule. An empirical relationship according to which the mass stopping power of an element for α-particles (also applicable to other charged particles) is proportional to (at. wt.)$^{-\frac{1}{2}}$.

brain voltage. Electrical signal waves generated in human brain. Usually classed as alpha, beta and delta waves according to frequency.

brake field. Same as **retarding field**.

branch or arm. (1) Electric components comprising a conducting path between junction points of common connection in a network. (2) Alternative modes of radioactive decay. (3) Same as **conditional jump**, for which see **jump**.

branch jack. See **jack**.

branching. The existence of two or more modes by which a radionuclide can undergo radioactive decay, e.g., ^{64}Cu can undergo β^-, β^+ and electron capture decay.

Branly coherer. Original form consisting of two electrodes immersed in iron filings contained in a glass tube; these cohered with high-frequency current, and decohered with a tapper.

Braun tube. Original name for **cathode-ray tube**, after Karl Ferdinand Braun (1850–1918) the inventor.

Bravais lattice. The space lattice of a crystal structure.

break-before-make. Classification of switch and relay wipers where existing contacts are opened before new ones close.

break delay. The time required after de-energizing a coil for a relay to release the contacts.

break-in. Attachment of an operator's circuit to a telegraph, telephone line, or radio channel for transmitting on to or taking over control of a circuit already established.

breakdown diode. See **Zener diode**.

breakdown transfer characteristic. Relationship between breakdown voltage of discharge tube and current flowing to third electrode.

breakdown voltage. Voltage at which a marked increase in the current through an insulator or semiconductor occurs. See also **disruptive voltage**.

breathing. Noise arising in noise-reduction systems when timed to operate too fast. See **hush-hush**.

breeder reactor. One which produces more fissile material than is consumed in establishing the requisite neutron flux. See **converter reactor**.

breeding ratio (b_r). The number of fissionable atoms produced per fissionable atom destroyed in a nuclear reactor. ($b_r - 1$) is known as the **breeding gain**.

Breit-Wigner formula. An equation which relates the cross-section σ of a particular nuclear reaction to the energy E if the incident particle (A) and the energy E_r of a resonance level is involved, then

$$\sigma(A,B) = (2l+1)\frac{\lambda^2}{4\pi}\ \frac{W_A W_B}{(E-E_r)^2+(\frac{1}{2}W)^2},$$

where $\sigma(A, B)$ is the cross-section for the reaction which involves the capture of particle A and the emission of particle B. W is the total width of the energy level and W_A, W_B are the partial widths respectively of the energy levels for the modes of disintegration in which particle A is re-emitted and B is emitted. λ is the *de Broglie wavelength* (q.v.) of particle A, and l is the orbital angular momentum quantum number of incident particle A.

bremsstrahlung. Electromagnetic radiation arising from collision or deviation between fast-moving electrons and atoms (Ger. *bremsen*=to brake and *strahl*=a stream).

brevium. Name sometimes used for uranium $X_2(UX_2)$.

Brewster (or polarizing) angle. Particular angle of incidence for which a wave on a perfect dielectric, polarized parallel to the plane of incidence, is totally transmitted.

Brewster law. That relating the Brewster angle (θ) to the refractive index (n) of the medium for a particular wavelength, viz.,

$$\tan\theta = n.$$

For sodium light incident on particular glass, $n = 1\cdot66$, θ is $51°$.

bridge. Circuit in which two potentials, each divided from a common source, can be made equal, e.g., Wheatstone bridge for d.c., and analogous developments for a.c. Widely used in measurements. See:

alternating-current-	Owen-
Anderson-	permeability-
Campbell-	reverberation-
Carey-Foster-	Schering-
comparison-	thermistor-
discharge-	transformer-ratio-
Hay-	Wheatstone-
Kelvin-	Wien-.

bridge feedback. Original negative feedback amplifier in which a balanced bridge provides a feedback voltage, independent of

load impedance; also called **Black feedback**.

bridge network. Same as **lattice network**.

bridge neutralizing. Method for overcoming the adverse effects of interelectrode capacitances in thermionic valve amplifiers. Two valves are connected in push-pull with the anodes and grids cross-connected through balancing capacitors, the whole forming a balanced bridge.

bridge oscillator. One in which positive feedback and limitation of amplitude is determined by a bridge, which contains a quartz crystal for determining the frequency of oscillation. Devised by Meachan for high stability of operation in crystal clocks, etc.

bridge receiver. One used when 2-way working is carried out on one wavelength. By the use of the bridge principle it is sensitive to the distant, but not to the local, transmitter.

bridge rectifier. Type of full-wave rectifier using 4 rectifiers in the form of a bridge. The alternating supply is connected across one diagonal and the direct-current output is taken from the other.

bridged-T filter. One consisting of a T-network, with a further arm bridging the two series arms; used for phase compensation.

bridging, non-bridging. (1) Said of wipers in step relays when they do or do not bridge adjacent contacts when passing from one to the next. (2) Bridging is also undesirable wear in relay contacts due to a protuberance building up on one of the contacts leading to a false closure of the circuit.

bridging amplifier. One for monitoring (supervising) or tapping a channel without abstracting appreciable power; also **monitoring amplifier**. See **trap amplifier**.

bright emitter. A thermionic valve with pure tungsten cathode, for emitting electrons at ca. 2600°K. Originally used of any thermionic valve, but now restricted to those of high power.

brightness. See **luminance**. As a quantitative term *brightness* is deprecated.

brightness control. Electrical control which alters the brightness and/or contrast (gamma factor) on a cathode-ray tube screen.

brilliance control. Control of the average illumination over the whole of a picture.

Brillouin formula. A quantum mechanical analogue in paramagnetism of the Langevin equation in classical theory of magnetism.

Brillouin zone. Polyhedron in k-space, k being position wave vector of the groups or bands of electron energy states in the band theory of solids.

British Standard. A publication giving recommendations respecting materials, their use and measurements, issued by British Standards Institution.

broadcast. Said of signals radiated from antennae for general reception.

broadcast channel. Frequency band used for interference-free and widespread reception from a single transmitting source.

broadcast receiver. One whose tuning ranges

cover those normally used for broadcasting.

broadcast transmitter. Radio transmitter specially designed for broadcasting, the requirements for faithful transmission being higher than those of an ordinary commercial transmitter.

broadcasting amplifier or **repeater.** One of superior performance for programmes in a broadcast or *programme channel.*

broadside antenna. Array in which the main direction of the reception or radiation of electromagnetic energy is normal to the line of radiating elements.

Broca tube. Early form of acoustic detector.

Broglie wavelength. See de Broglie wavelength.

bromine (Br). Element no. 35, at. wt. 79·909, b.p. 58·8°C. One of the halogens used in halogen-quenched Geiger tubes.

Bronson resistance. The resistance between two electrodes in a gas when exposed to a constant source of ionization.

Brown loudspeaker. Early type of loud-speaker in which the sound power, generated by a reed-driven unit, is more effectively and efficiently radiated by a metal horn.

Brownian movement. Small movements of light suspended bodies such as galvanometer coils, due to statistical fluctuations in the bombardment by surrounding air molecules.

brush. (1) A rubbing contact on a commutator switch or relay. Also **wiper.** (2) The wiper which 'reads' punched cards, producing an electric contact through the aperture.

brush(ing) discharge. Discharge from a con-ductor when the p.d. between it and its surroundings exceeds a certain value but is not high enough to cause a spark or an arc.

bubble chamber. Development of *cloud chamber* (q.v.), in which a vessel is filled with a transparent liquid (for photographing tracks), which is so highly superheated that an ionizing particle passing through starts violent boiling by initiating the development of a string of bubbles along its path.

Buchmann-Meyer pattern. See optical pattern.

buck. A voltage is said to buck another voltage in series with it, if of opposite polarity.

buck transformer. One with secondary in the main circuit to regulate voltage according to a controlling circuit feeding the primary; also called **boost transformer.**

bucking coil. A winding on an electromagnet to oppose the magnetic field of the main winding. Such a device, known as the *hum-bucking coil,* is sometimes used in electromagnetic loudspeakers to smooth out voltage pulsations in the power supply.

Buckley gauge. Sensitive pressure gauge depending on gas ionization.

buckling. In reactor diffusion theory, buckling gives a measure of the curvature of the neutron density distribution. In a homo-geneous reactor the buckling factor for the reactor to go critical will depend upon its geometrical configuration.

buffer. (1) An electronic circuit to decouple the output of the buffer from its input,

thus avoiding reaction between a driving and a driven circuit. (2) A memory type of device in a computing system which compensates for any difference in the rates of flow of information in various parts of the system.

buffer amplifier. Amplifier between two units, used principally to isolate them.

buffer capacitor. See blocking capacitor.

buffer memory. Same as **temporary memory.**

buffer storage. A temporary means of coupling between two forms of data storage.

bug. (1) Colloquialism for a code-transmitting key which is semi-automatic in its operation. (2) Elusive fault in equipment or computer programme.

build-up. Increase in radiation dose rate below surface of an absorber, due to release of secondary electrons.

build-up time. See rise time.

building-out network. One which is connected to a basic network, to match more exactly line impedance over the frequency range of interest.

bulb. The gas-tight envelope, usually glass, which encloses the electrodes of a thermionic valve.

bulk test. The large sample usually required for a shield radiation test of a material having a high attenuation.

bullet amplifier. A colloquialism for the amplifier mounted in a cylinder and associ-ated with a condenser microphone hanging therefrom.

buncher. Arrangement which velocity-modu-lates and thereby introduces bunches in electron space current passed through it. Bunching would be *ideal* if the bunches contained electrons all having the same velocity. Also called **buncher gap, input gap.** See catcher, debunching, rhumbatron.

bunching. Collective grouping of electrons along the beam of electrons passing through a rhumbatron.

bunching angle. That transit or phase angle between modulation and extraction of energy in a bunched beam of electrons.

Burger's vector. Translation vector of crystal lattice representing a displacement which creates a lattice dislocation.

buried antenna. One in which the wires are buried under the ground, the e.m.f. depend-ing on tilt of the ground wavefront.

burn. See ion burn.

burnable poison. A substance to influence the long-term reactivity variations of a fission reactor. It should have a high neutron capture cross-section and give rise to a capture reaction product of low capture cross-section.

burning voltage. The minimum voltage between the anode and cathode to maintain the discharge of a cold-cathode electron tube or lamp. This burning or 'running' voltage is less than the starting or ignition voltage needed to indicate the discharge.

burnout. Sudden and protracted change in

crystal rectifier characteristics as a result of excess voltage.

burnup. In a reactor, consumption (by fission) of fuel rods or other materials.

burst. (1) Sudden increase in strength in received radio signals caused by sudden changes in ionosphere. (2) Unusually large pulse arising in an ionization chamber, caused by a cosmic-ray shower. See also **colour burst.**

burst-can detector. An instrument for the early detection of ruptures of the sheaths of fuel elements inside a reactor. Also **leak detector.**

burst mode. Mode of communications between the processor and in/out devices.

burst pedestal. Part of *colour burst* (q.v.) in a colour television signal.

burst signal. Component of transmitted signal, which acts as reference for chrominance components. These operate circuits to establish correctly colour elements in the reproduced image.

burst slug. Fuel element with a small leak emitting fission products. Also **cartridge.**

bus. See **highway.**

bus(-bar). (1) Sectionalizing switch in power circuits. (2) Supply *rail* maintained at a constant potential (including zero or earth) in electronic equipment.

bust. The bad performance of a programmer or a machine operator.

Butex. TN for dioxydiethylether, used for separating U and Pu from fission products.

butt-jointed. See **joint.**

butterfly circuit. One in which both the inductance and capacitance between opposite re-entrant sections of a punched disk are varied by rotating an insulated coaxial vane of the same shape as the cavity. See **differential capacitor.**

Butterworth filter. Constant-k filter designed to give response of maximum flatness through pass band. Cf. *Chebyshev filter.*

button microphone. Small microphone which can be fitted in the buttonhole.

buzzer. An electric device of similar principle to the electric bell but minus a gong or hammer. It gives a buzzing note on excitation.

by-pass. Alternative signal path, usually of lower impedance, e.g., by-pass capacitor providing a low shunting impedance for a.c. currents in a circuit.

byte. A group of bits shorter than a word and operated on as a unit.

C

C-amplifier. Line or distribution amplifier following a B-amplifier in broadcasting studios. Not the same as **class-C amplifier.**

C-battery. U.S. term for **grid-bias battery.**

C-core. Strip-wound magnetic core, coiled to shape, cut into two C-shaped pieces and clamped over a coil for transformers, chokes, or magnetic amplifiers.

C display. Radar display in which bright spot represents the target, with horizontal and vertical displacements representing bearing and elevation respectively.

CIE coordinates. Set of colour coordinates specifying proportions of three theoretical additive primary colours required to produce any hue. These theoretical primaries were established by the CIE and form basis of all comparative colour measurement. See **chromaticity diagram.**

C-layer. Reflecting or scattering region between about 35 and 70 km above the Earth's surface, postulated to explain return signals sometimes obtained with vertically-radiated waves.

CR-law. Relates to exponential rise or decay of charge on capacitor in series with a resistor, and, by extension, to signal distortion on long submarine cables.

C-ring. Method of tuning of magnetrons.

cable. See under:

| coaxial- | mile of standard- |
| composite- | television-. |

cable film. Substandard TV film, scanned at low speed for signal transmission up to 4·5 kHz through cables, for reconstruction and video transmission. Cable time is 1000 times viewing time.

cable form. The normal scheme of cabling between units of apparatus. The bulk of the cable is made up on a board, using nails at the appropriate corners, each wire of the specified colour identification being stretched over its individual route with adequate *skinner* (q.v.). When the cable is bound with twine and waxed, it is fitted to the apparatus on the racks and the skinners connected, by soldering, to the *tag blocks* (q.v.).

cable splice. Junction formed in multicore cable.

cadmium (Cd). White metallic element, at. wt. 112·40, m.p. 320·9°C, b.p. 767°C, sp. gr. 8·648. Cadmium plating is used as corrosion protective to aluminium and its alloys. It is a powerful absorber of neutrons and is used in steel, for the detail control of nuclear reactors, and always for an emergency shut-down. Films of cadmium are photosensitive in the ultraviolet between 2500 and 2950, with a peak at 2600Å.

cadmium cell. (1) Vacuum photocell having a cadmium or cadmium-coated cathode, with maximum spectral sensitivity in the ultraviolet range. (2) Old name for **Weston standard cadmium cell.**

cadmium red line. That originally proposed as a reproducible standard for length $=6438\cdot4696$Å. See **krypton.**

caesium (Cs). Metallic element, at. no. 55, at. wt. 132·905, m.p. 28·6°C, b.p. 713°C, sp. gr. 1·88. As a photosensor it has a peak response at 8000Å in the infrared, thermal and photo-emission being high. Caesium, when alloyed with antimony, gallium, indium, and thorium, is generally photosensitive. Fission products from a nuclear reactor are ^{37}Cs with half-life of 27 y, and ^{34}Cs with half-life 2·5 y, in such abundance that this source is likely to be used for irradiation of foodstuffs on a large scale.

caesium cell. One having a cathode consisting of a thin layer of caesium deposited on minute globules of silver; particularly sensitive to infrared radiation, but generally approximating to that of the eye.

caesium clock. Frequency-determining apparatus based on caesium ion resonance of 9. 192 631 770 0 GHz.

caesium-oxygen cell. One in which the vacuum is replaced by an atmosphere of oxygen at very low pressure. It is more sensitive to red light than the caesium cell.

cage antenna. One comprising a number of wires connected in parallel, and arranged in the form of a cage, to reduce the copper losses and increase the effective capacitance.

calcium (Ca). Metallic element, at. no. 20, at. wt. 40·08, m.p. 850°C, b.p. 1440°C, sp. gr. 1·58. Used as a getter in low-noise valves.

calibration error. That in the bearings given by a ship's direction-finder, due to currents in the hull, masts, and rigging. The error is corrected in the *initial calibration.*

californium (Cf). Manmade element, at. no. 98, produced in a cyclotron. Its longest lived isotope is ^{251}Cf with 800 years half-life.

call-in. The transfer control, from the main to a subroutine, during a subsidiary operation in computer programming.

call number or **word.** Computer coding for a subroutine.

call sign. Letter and/or numeral for a transmitting and/or receiving station, or one of its authorized channels. Used for calling or identification.

calomel cell. A half-cell comprising a mercury electrode in a solution of potassium chloride saturated with mercurous chloride (calomel). Also **calomel electrode.**

calorie. The quantity of energy which as heat would raise the temperature of 1 gm of water by 1°C. See **kilocalorie** (also called **Calorie**).

calutron. American-designed electromagnetic isotope separator. See also **race track**.

cam. Sliding mechanical device used to convert rotational to linear motion or vice versa. Widely used in servo systems.

camera. Television picture-transmitting instrument consisting of lens, tube and controls, usually mounted on dolly for easy control.

camera channel. In a television studio, the camera, with all its supplies, monitor, control position, and communication to the operator, which forms a unit, with others, for supplying video signal to the control room.

camera signal. The video signal output of a television camera.

camera signal characteristic. The sensitivity with reference to wavelength, in colour television, of each of the colour separation channels. Also **camera spectral characteristic.**

camera tube. One which converts an image of an external scene into a video signal. Essential component in a television camera channel. In U.S. **pickup tube.**

camouflage. Treatment of objects so that there is ineffective reflection of radar waves.

Campbell bridge. An a.c. measuring bridge for the comparison of mutual inductances.

Campbell gauge. Electrical bridge, one arm being the filament of a lamp located in the low gas pressure to be measured.

Campbell's formula. That which gives the effective attenuation of a coil-loaded transmission line in terms of the constants of the line and the magnitude of the loading.

can. Cover for reactor fuel rods, often of aluminium or magnox. Also called **jacket.**

canal. Water-filled trench into which highly-active elements from reactor core can be discharged. The water acts as a shield against radiation but allows objects to be easily inspected.

canal rays. Same as **positive rays.**

canaries. Extraneous high-frequency noises reproduced from a recording channel.

candela. Luminous intensity unit, superseding the *international candle*; it is 1/60 of the luminous intensity of 1 cm² of melting platinum.

candle power. Luminous intensity provided by a candle of specific composition burning at a known rate. Unit, formerly *international candle*, now **candela.**

cannibalize. To dismantle apparatus so as to recover parts for new models or prototypes.

capacitance (self-). Assuming that all other conductors are sufficiently remote from a conducting body, capacitance of latter is defined as the total electric charge on it divided by its potential. See:

body-	inter-electrode-
direct-	internal-
distributed-	partial-
geometric-	stray-
input-	total-.

capacitance bridge. A.c. bridge network for the measurement of capacitance, e.g., *Schering bridge, Wien bridge.*

capacitance coefficients (C_{ij}). Charges $(q_1 \ldots q_n)$ of system of conductors can be expressed in terms of coefficients of electric induction (C_{ij}) by the following equations:

$$q_1 = C_{1\infty}V_1 + C_{12}(V_1 - V_2) + \ldots + C_{1n}(V_1 - V_n)$$

$$q_2 = C_{21}(V_2 - V_1) + C_{2\infty}V_2 + \ldots + C_{2n}(V_2 - V_n)$$

$$\cdots \quad \cdots \quad \cdots \quad \cdots$$

$$q_n = C_{n1}(V_n - V_1) + C_{n2}(V_n - V_2) + \ldots + C_{n\infty}V_n$$

where $C_{km} = -C_{mk}(m \neq k)$,

and $C_{m\infty} = C_{m1} + C_{m2} + \ldots + C_{m(n-1)} + C_{mn}$.

Fundamental relation for partial capacitances of a number of conductors, e.g., electrodes in valves, conductors in cables, variable air capacitors.

capacitance coupling. That between circuits or valves, which is effected either by a capacitor in a common arm or branch, or by a capacitor between appropriate points in the circuits.

capacitance reaction. Reaction from the output to the input circuit of an amplifier, through a path which includes a capacitor.

capacitive load. Terminating impedance which is markedly capacitive, taking an a.c. *leading* in phase on the source e.m.f., e.g., capacitor loudspeaker.

capacitive reactance. Negative of the reciprocal of the product of the angular frequency ($\omega = 2\pi$ frequency) and capacitance. Measured in ohms when the capacitance is in farads.

capacitor. Electric component having *capacitance*; formed by conductors (usually thin and extended) separated by a dielectric, which may be vacuum, paper (waxed or oiled), mica, glass, plastic foil, fused ceramic, air, etc. Maximum p.d. which can be applied depends on electrical breakdown of dielectric used. Modern construction uses sheets of metal foil and insulating material wound into a compact assembly. Air capacitors, of adjustable parallel vanes, are used for tuning high-frequency oscillators. Formerly **condenser.** See:

absorption-	electrolytic-
air-	gang-
antenna-shortening-	HiK-
billicapacitor	load-
blocking-	paper-
ceramic-	parallel-plate-
coupling-	reaction-
differential-	self-sealing-
disk-	shortening-
dry electrolytic-	solid-state-

square-law-
straight-line-
straight-line
frequency-
straight-line wave-
length-

tantalum-
telephone-
trimming-
tuning-
vibrating-
wet electrolytic-.

capacitor loudspeaker. See **electrostatic loudspeaker.**

capacitor microphone. One with a stretched or slack conducting foil diaphragm, which is polarized at a steady potential, or at a high-frequency voltage, both of which are modulated by variations in capacitance due to varying sound pressure.

capacitor modulator. Capacitor microphone or similar transducer, which, by variation in capacitance, modulates an oscillation either in amplitude or frequency. See **Vibron.**

capacitor pickup. One in which the tracking stylus moves an electrode of a polarized capacitor, thereby generating an e.m.f. for subsequent amplification and acoustic reproduction.

capacitor start. Starting unit for electric motor using series capacitance to advance phase of current.

capacitron. See **band-ignitor tube.**

capacity. (1) The output of any electrical apparatus, e.g., that of a motor in HP, or a generator in kW. In an accumulator (secondary battery), *capacity* is measured by the ampere-hours of charge it can deliver. Capacity of a switch is the current it can break under specified circuit conditions. (2) The volume of a tank, etc., or delivery of a pump, etc. (3) Range of numbers which can be handled by a computer, limited by *storage, register,* or *accumulator,* without *overflow.*

capacity earth. See **counterpoise.**

Capenhurst plant. UKAEA plant for production of enriched uranium by gaseous diffusion of uranium hexafluoride.

capstan. Rotating synchronous drive on a tape deck, which determines the linear speed of the tape, as contrasted with the feed and take-up rotations, both of which vary during recording.

capture. (1) Any process in which an atomic or nuclear system acquires an additional particle. In a nuclear radiative capture process there is an emission of electromagnetic radiation only, e.g., in reactor physics there may be emission of a γ-ray subsequent to the capture of a neutron by the nucleus. (2) In frequency and phase modulation the diminution to zero of a weak signal (noise) by a stronger signal.

carbon (C). Amorphous or crystalline (graphite and diamond) element, at. no. 6, at. wt. 12·011, m.p. 3550°C, sp. gr. (diamond) 3·51, sp. gr. (graphite) 2·25, sp. gr. (amorphous) 1·8-2·1. Lat. *carbo* =charcoal. Near black body, hence good radiator of heat, especially from valve anodes. Widely known as electrode in Leclanché cells, in rods for arcs in cinema projectors and searchlights, and in brushes for electric generators and motors. Alloyed with iron for steels. Colloidal carbon or graphite (Aquadag) is used to coat glass in cathode-ray tubes and electrodes in valves, to inhibit photoelectrons and secondary electrons. High-purity carbon, crystallized to graphite in a coke furnace for many days, is used in many types of nuclear reactors, particularly for moderation of neutrons.

carbon anode. One constructed of carbon, usually in graphite form, to resist the high temperatures encountered under some conditions of operation of valves.

carbon button. See **carbon microphone.**

carbon cycle. (1) The series of thermonuclear reactions by which four protons are fused into a helium nucleus in the presence of a carbon-12 nucleus. The source of solar and stellar energy. Reaction temp. $\sim 15 \times 10^{6}$°C. (2) The biological circulation of carbon from the atmosphere into living organisms and, after their death, back again.

carbon dating. Atmospheric carbon dioxide contains a constant proportion of radioactive ^{14}C, formed by cosmic radiation. Living organisms absorb this isotope in the same proportion. After death it decays with a half-life $5·57 \times 10^{3}$ years. The proportion of ^{12}C to the residual ^{14}C indicates the period elapsed since death.

carbon dioxide. Gas, formula CO_2, solid at -79°C, a convenient temperature for testing electronic components. High-pressure carbon dioxide has found a considerable use as the coolant in carbon-moderated nuclear reactors for power generation.

carbon microphone. A microphone in which a normally d.c. energizing current is modulated by changes in the resistance of a cavity filled by granulated carbon which is compressed by the movement of the diaphragm. The diameter of the cavity is frequently very much less than that of the diaphragm and it is then known as a **carbon button.**

carbon resistor. Negative temperature coefficient, non-inductive resistor, formed of powdered carbon with ceramic binding material. Used for low temperature measurements because of the large increase of resistance as temperature decreases.

carbonized anode. Metallic anode coated with carbon, in the form of lampblack, to assist in the radiation of heat and reduce the secondary emission of electrons.

carborundum detector. One consisting of a point contact between steel and a carborundum crystal, most sensitive when a small steady voltage is maintained across the contact.

carcel unit. Unit of luminous flux, formerly used in France.

Carcinotron. TN for a wide-tuning and high-efficiency oscillator tube, depending on backward waves and generating frequencies of the order of 10^{11} Hz.

card. (1) Standard size card, printed with

numbers up to 80 in columns, with holes correspondingly punched, for operation of process controls and computers. (2) A board containing a circuit with a plug or connector for attachment to a card box or tray.

card box. Portion of a complete unit for accommodation of jacked-in replaceable electronic units. Also **tray.** See **card** (2).

card printer. Component of a computing system which prints in tabular form the information carried by the presence of holes in punched cards.

card punch. A machine for punching holes in a card to follow a given code, thus providing an information register.

card reader. Component part of a computing system which scans punched cards and delivers signals or words corresponding to the information recorded in the holes.

cardan mount. Type of *gimbal mount* (q.v.) used for compasses and gyroscopes.

cardiogram. Trace produced by electro-cardiograph showing voltage waveform generated during heartbeats.

cardioid diagram. Heart-shaped polar diagram showing response of the combination of loop and vertical antennae used in direction-finding systems.

cardioid microphone. One comprising an electrodynamic unit and a ribbon unit backed with an acoustic termination; combination of the outputs results in a wide control of the polar response curve, which can be made reasonably independent of frequency.

cardioid reception. Reception with antenna, horizontal response of which is determined by a vertical loop and open vertical wire. Also **heart-shaped reception.**

cardiotachometer. An electronic amplifying instrument for recording and timing the heart rate.

cardiotron. A portable form of electro-cardiograph.

Carey-Foster bridge. A form of *Wheatstone bridge* (q.v.) of particular use in measuring difference between two nearly equal low resistances.

carpet. Airborne electronic apparatus for radar jamming.

carriage return. That of an automatic typewriter or page teleprinter, usually coinciding with line feed, initiated by special code.

carrier. (1) Vehicle for communicating a signal, when the medium cannot convey the signal but can convey the carrier, as in radio transmission, the signal being speech, video, or coded words or figures. (2) Circuit transmission on cable or open-wire circuits in which, after balanced modulation, the carrier and one sideband are filtered out; at the receiving end *demodulation* is effected after the insertion of a similar carrier. (3) Constant frequency in an amplitude-modulation radio transmission. It is *quiescent* if present only when modulated; it is *controlled* or *floating* when it varies to a

lesser extent with the modulation. (4) Non-active material mixed with, and chemically identical to, radioactive compound. (5) See **carriers,** and also:

controlled-	modulated-
exalted-	pilot-
floating-	reconditioned-
majority-	sound-.

carrier condition. That of voltages and currents in an amplifier for a modulated signal when there is, at the time considered, no modulation.

carrier-controlled approach. Radar system used for landings on aircraft carriers.

carrier density. The concentration of electrons and/or holes in semiconductors.

carrier filter. Electric wave filter suitable for discriminating between currents used in carrier telephony according to their frequency, particularly when they are combined with or separated from currents of normal telephonic frequency.

carrier-free. Said of radioisotopes when substantially free from a chemical carrier.

carrier frequency. The steady frequency of current or voltage which is modulated by speech or telegraphic signals, resulting in no change in the carrier but the addition of side frequencies in perfect amplitude modulation.

carrier metal. Metal, usually a thin film of silver, on which, after surface oxidation, molecular layers of caesium are deposited, when making photosensitive surfaces.

carrier mobility. The mean drift velocity in unit electric field.

carrier noise. Radiated noise which has been introduced into the carrier of a transmitter before modulation.

carrier shift. (1) Change in carrier amplitude because of non-linearity in modulation. (2) Telegraphy transmission where closing of key changes frequency of radiated signal.

carrier suppression. (1) Transmission of a modulated carrier wave with the carrier suppressed, with reinsertion of the carrier at the receiving end. (2) Suppression of the carrier from the radiating system when not required for modulation, especially on board ship to protect the receivers from noise arising from variability of eddy-current paths in the rigging, etc.

carrier (or voice-frequency) telegraphy. Use of modulated frequencies, usually in the five-unit code originated by teleprinters, and transmitted, with others, as a voice-frequency signal in telephone circuits. Also called **wired wireless.**

carrier-to-noise ratio. Ratio of received carrier signal to noise voltage immediately before demodulation or limiting stage.

carrier wave. That component of a modulated wave which is independent of the modulation. This appears as one or both sidebands of frequencies, which represent the information being transmitted and received.

carriers. Electrons in semiconductors and their effective *holes*, which give *n* (negative)-

conduction or *p* (positive)-conduction in rectifier and transistor units respectively. These comprising more than half the total number are the *majority carriers*; the remaining are the *minority carriers*.

carry. In arithmetic and computing, the process of replacing digits whose sum equals the base by a single digit in the column immediately to the left.

cartridge. Sealed electromechanical transducer unit in gramophone pickup. See also **burst slug, slug.**

cascade. Number of devices connected in such a way that each operates the next one in turn, e.g., valves in an amplifier.

cascade amplifier. Series of thermionic valves so connected that the output of one stage is amplified by the succeeding stage.

cascade control. System in which one or more control systems depend on another central system, which determines the index (or desired) point in operation.

cascade particle. Hyperon found in cosmic-ray tracks, and now normally known as **Xi particle.**

cascade shower. Manifestation of cosmic rays in which high-energy mesons, protons and electrons create high-energy photons, which produce further electrons and positrons, thus increasing the number of particles until the energy is dissipated.

cascade amplifier. Thermionic valve circuit in which a grounded-cathode triode followed by a grounded-grid triode provides a low-noise amplifier for very high radio frequencies. Also **Wallman amplifier.**

Cassegrain antenna. One in which radiation from a focus is collimated from one surface, e.g., a parabola, and reflected by another surface, e.g., a plane, or in reverse.

cassette. (1) Holder for reels of magnetic tape, which can be easily clamped and detached from the tape deck of a recorder. (2) Holder of X-ray plate before, during, and after exposure. (3) Lightproof holder for photographic film.

cassieopeium. See **lutecium.**

castle. Temporary or permanent structure surrounding radioactive materials and their detector, to obviate interference from external sources or cosmic rays, e.g., a leadbox.

cataphoresis. See preferred term **electrophoresis.**

catcher. Element in a thermionic valve which abstracts or catches the energy in a bunched electron stream as it passes through it. See **buncher.**

catcher foil. Aluminium sheet used for measuring power levels in nuclear reactor by absorption of fission fragments.

catching diode. One used to *clamp* a voltage or current at a preset value. When its anode becomes positive, it prevents the potential rising above any potential applied to its cathode.

cathelectrotonus. Physiological excitability produced in muscle tissue by passage of electric current.

cathode. (1) Negative terminal of an electrolytic cell, i.e., one at which the positively-charged ions (cations) are discharged. Electrode at which electrons enter the cell. It is the positive terminal of a cell or battery, i.e., one at higher potential, so that the positive current in the external circuit flows from the terminal to the other terminal. See **anode.** (2) Source of electrons, in valves and tubes, produced by heat or by applying a high electric field (*cold emission*). In a mass spectrometer, secondary emission from a cold cathode is sometimes used as an ion source.

See also:

activated-	equipotential-
anti-	indirectly-heated-
cold-	ionic-heated-
directly-heated-	S-type-
dispenser-	virtual-.
dull-emitter-	

cathode bias. Grid bias obtained from a resistor in the cathode circuit, which takes the anode current. Negative feedback also results, obviated by a large shunting capacitor.

cathode coating. Low work-function surface layer applied to cathode. The cathode coating impedance is between the base metal and this layer.

cathode coupling. That effected to or from a thermionic valve by an impedance connected between the cathode and negative terminal of the high-tension supply.

cathode current. Total current from the cathode to all other electrodes in a thermionic valve.

cathode dark space. Dark sheath, noticed by Crookes and by Hittorf, surrounding a cathode in a gas-discharge tube.

cathode disintegration. That arising from positive-ion bombardment.

cathode drop. Potential difference over the cathode dark space, entailing great loss of efficiency in cold discharge tubes.

cathode efficiency. Ratio of emission current to energy supplied to cathode. Also **emission efficiency.**

cathode follower. A valve network in which the output load is in the cathode circuit, the input being applied between the grid and the end of the load remote from the cathode; particularly used for very high frequencies, when the valve becomes an impedance-changer.

cathode-follower amplifier. Amplifier with load impedance connected between cathode and ground. Resulting intrinsic negative feedback limits maximum voltage gain to unity.

cathode glow. That only near the surface of a cathode, of colour depending on the gas or vapour in the tube. If an arc occurs in a partial vacuum, it may fill the greater part of the discharge tube.

cathode heating time. That for heating of cathode to give full emission before accelerating voltages are applied. Especially important in thyratron tubes and large valves.

cathode interface impedance. That between the cathode surface and its support, which may increase with age or through bad coating.

cathode keying. Operation of telegraphy transmitter by key inserted in cathode lead.

cathode luminous sensitivity. Ratio of cathode current of photoelectric cell to luminous intensity.

cathode modulation. Modulation produced by signal applied to cathode of valve through which carrier wave passes.

cathode potential stabilized camera tube. A type of image orthicon television camera tube in which the cathode potential is used to stabilize the mosaic potential and prevent secondary emission. Also called **cathode potential stabilized emitron.**

cathode ray. Stream of negatively-charged particles (electrons) emitted normally from the surface of the cathode in a rarefied gas. The velocity of the electrons is proportional to the square root of the potential difference through which they pass, and is equal to 595 km/sec for 1 volt. The beam can be deflected in the direction of an applied electric field, or at right angles to an applied magnetic field. Widely used applications are for oscilloscopes and in television systems.

cathode-ray camera. Combination of cathode-ray tube and moving-film camera. The photographic film may be internal or external to the vacuum chamber.

cathode-ray direction-finding. Arrangement by which the direction of arrival of an incoming signal is shown by the inclination of a line on the screen of a cathode-ray tube; this has a circular scale and is actuated by the outputs of receiving antennae, usually Bellini-Tosi loops or a double Adcock.

cathode-ray furnace. One in which a small specimen is raised to a very high temperature by focusing on it an intense beam of cathode rays.

cathode-ray indicator. Engine indicator using a cathode-ray tube for recording the diagram. The electron beam is deflected by voltages proportional to cylinder pressure and to time respectively, giving an indicator diagram on a time base; suitable for the highest speeds, being free from all inertia.

cathode-ray lamp. One in which light is obtained from a refractory source by high-energy electron bombardment.

cathode-ray oscillograph. Complete equipment, including high-voltage supply, for *registering* transient waveforms on a photographic plate within the vacuum of a cathode-ray tube, continuously evacuated.

cathode-ray oscilloscope. Complete equipment, including a cathode-ray tube, amplifiers, time-base generators, power supply, for *observing* repeated and transient waveforms of current or voltage, which present a *display* on a *phosphor*.

cathode-ray tube. An electronic tube in which a well-defined and controllable beam of electrons is produced and directed on to a surface to give a visible or otherwise detectable display or effect.

cathode-ray tuning indicator. See electric eye.

cathode-ray voltmeter. One of known deflectional sensitivity, employed as a voltmeter. It can indicate crest values of the voltage applied to the deflector plates.

cathode resistor. Resistor connected between cathode and earth, often to provide bias.

cathode spot. Area on a cathode where electrons are emitted into an arc, the current density being much higher than with simple thermionic emission.

cathodic etching. Erosion of a cathode by a glow discharge through positive-ion bombardment, for showing microstructure.

cathodoluminescence. Excitation of dislodged metal from an anode by a cathode-ray beam, with consequent characteristic radiation and with possible afterglow.

cathodophone. Microphone utilizing the silent discharge between a heated oxide-coated filament in air and another electrode. This discharge is modulated directly by the motion of the air particles in a passing sound wave. Also **ionophone.**

cation. Ion in an electrolyte which carries a positive charge and which migrates towards the cathode under the influence of a potential gradient in electrolysis. It is the deposition of the cation in a primary cell which determines the *positive terminal.*

catkin. Small receiving valve in which anode also serves as envelope.

cat's whisker. A colloquial term for the fine wire making contact with the crystal in some early forms of crystal detector.

Cauer filter. One which has sections designed from impedance considerations according to Cauer's filter theory.

Cavendish experiment. Experiment to demonstrate that all charges reside on the surface of a conductor; first performed by Henry Cavendish (1731-1810).

cavitation. Formation of local cavities in liquids, e.g., by propeller in hydrodynamic flow, or by ultrasonic waves.

cavity frequency meter. One attached to a radar waveguide system, with a calibrated adjustment for resonance.

cavity magnetron. One in which the circular anode has a number of radial slots with openings facing the cathode, for stabilizing the frequency of oscillation and increasing efficiency, the slots forming a travelling-wave system.

cavity modes. Stable EM or acoustic fields in a cavity which can exhibit resonance. They are *degenerate* if, having similar frequencies, they have fields which differ in pattern to that of main resonance.

cavity resonator. Any nearly closed section of waveguide or coaxial line in which a pattern of electric and magnetic fields can be established. Also applies to sound fields.

cell. (1) Electric battery or accumulator. (2) Unit of storage in computer. (3) Unit of homogeneous reactivity in reactor core. (4) Photoelectric tube. (5) Small storage or work space for 'hot' radioactive preparations. (6) Small volume unit on mathematical coordinate system. (7) Small item forming part of experimental assembly, e.g., Kerr cell, dielectric test cell, etc. See also:

aluminium anode-	Kerr-
backwall-	Leclanché-
barrier-layer-	photochemical-
binary-	primary-
cadmium-	Reuben-Mallory-
caesium-	sea-
caesium-oxygen-	selenium-
calomel-	solar-
Clark-	TR-
conductivity-	tuned-
Daniell-	voltaic-
electrolytic-	Weston standard
frontwall-	cadmium-.
half-	

cell call. (1) A call number in computer programming. (2) A set of symbols used in computer programming to identify a subroutine.

Celotex. TN for wall and ceiling board with good sound insulating properties.

censor key (or **switch**). One used in sound or television broadcasting to cut off a channel or change to another channel. Also **cut key.**

cent. (1) Unit of reactivity equal to one hundredth of *dollar* (q.v.). (2) Frequency interval equal to one twelve-hundredth root of two (i.e., 1200 cents =octave).

centi-. Prefix for one-hundredth (10^{-2}). Symbol c.

centre frequency. (1) Geometric or arithmetic midpoint of the cutoff frequencies of a wave filter. (2) Carrier frequency, when modulated symmetrically.

centre holes. Line of small holes in centre of punched tape for feeding by star wheel, punched holes for code being in line across tape; also called **feed holes.**

centre of mass. The point in an assembly of mass particles which moves as if the total mass of the assembly were concentrated there and resultant of forces acted there.

centring controls. Controls used to centre display on face of CRT.

Ceramac. TN for a ceramic ferrite magnetic core material.

ceramic capacitor. One using barium titanate, which has a high dielectric constant but lower leakage resistance.

ceramic pickup. One employing a piezoelectric element of ceramic material, e.g., barium titanate.

ceramic reactor. Reactor with ceramic (usually uranium oxide) fuel elements, suitable for use at high temperatures.

ceraunograph. Electronic instrument for recording thunderstorms.

Cerenkov. See **Cherenkov.**

cerium (Ce). A rare earth metallic element, at. no. 58, at. wt. 140·12, sp. gr. at 20°C 6·9, m.p. 635°C, specific electrical resistivity 78 microhms cm. Alloyed with iron and several rare elements, it is used as the sparking component in automatic lighters, etc. It is also a constituent (0·15%) of the aluminium base alloy ceralumin, and is photosensitive in the ultraviolet region. It is also used on tracer bullets, for flashlight powders, and in gas mantles. It is a getter for noble gases in vacuum apparatus. There are a number of isotopes which are fission products, ^{144}Ce of 290 days half-life being a pure electron emitter of importance.

cermet. See **metal ceramic.**

Cerrobend, Cerroseal. TNs for low m.p. solders.

chad. Disk of paper removed by punching. Hence *chad tape,* one punched for computer or process control; *chadless tape,* one partially punched, so that there is room for corresponding character printing.

Chadwick-Goldhaber effect. Early name for nuclear reactions due to bombardment with γ-radiation (photodisintegration). Obs.

chaff. Same as **window.**

chain. Any series whose members are related by a specific phenomenon. See **accelerating-, radioactive-.**

chain code. One in which a set of equal words have a unit set of binary digits formed by side displacement and added bits.

chain reaction. Self-maintaining fission reaction, i.e., one for which each nucleus undergoing fission releases at least one neutron that initiates a further fission reaction.

Chambers-Imrie-Sharpe curve. Graph of radiation dose against depth of penetration for human body in fall-out field.

channel. (1) General term for a unique transmission path, as for programme, vision, recording, picture, camera, shared, two-way. (2) One or more similar highways for transmission of computer data. (3) Range of radio frequencies occupied by a modulated transmission. With double sideband amplitude modulation, its width in Hz is twice the highest modulating frequency. A *clear channel* is one occupied by a single transmission, free from interference from other transmissions. (4) Passage through reactor core for coolant, fuel rod, or control rod. See **adjacent-, broadcast-, camera-, chrominance-, luminance-, sound-.**

channel capacity. The maximum possible information rate for a channel.

channel circuit. Same as **forked circuit.**

channel effect. In a transistor, by-passing base component by leakage due to surface conduction.

channel width. (1) Frequency band allocated to service or transmission. (2) See **pulse height selector.**

channelling effect. In radiation, the greater transparency of an absorbing material with voids, relative to similar homogeneous material. Expressed numerically by the ratio of the attenuation coefficients.

Chaperon resistor. Wirewound resistor of low residual reactance.

character. One of a unique set of letters and figures representing data for computer processing. *Ignore, marker* (*sentinel* in U.S.), *key, tag, sign,* are special characters.

character reading or **recognition.** Peripheral component in a computing system which recognizes printed or written figures or letters, and displays or provides codings correspondingly. Special alphabets may be required, or magnetic ink.

characteristic. Graph illustrating performance of a particular device, taken under specified conditions, e.g., for a valve, anode current against variable grid voltage, the anode voltage remaining constant. See **dynamic-**.

characteristic impedance. (1) Of a line or filter, that at a point in an infinite line, or an infinite number of cascaded filter sections. The ratio of voltage to current is the same for each point on the line. Also **surge impedance**. See **image impedance**. (2) Of an antenna, that with which certain forms of antennae (e.g., wave and rhombic antennae) must be terminated to prevent standing waves through end reflection.

characteristic spectrum. Ordered arrangement of radiated (optical or X-ray) wavelengths related to the atomic structure of the material giving rise to them.

characteristic X-rays. Those arising from the transitions of electrons in the inner shells.

Charactron. TN for cathode-ray tube which displays letters and numbers.

charge. (1) Quantity of unbalanced electricity in a body, i.e., excess or deficiency of electrons, giving the body negative or positive electrification respectively. See **negative** (1) and **positive**. That of an ion, one or more times that of an electron, of either sign. (2) Electrical energy stored in chemical form in secondary cell. (3) Load connected to RF heater. (4) Fuel material in nuclear reactor.

charge density. Value of quotient of distributed charge over the surface (or volume) of a body and the relevant area (or volume) of the body.

charge exchange. Exchange of charge between neutral atom and ion in a plasma. After its charge is neutralized, high-energy ions can normally escape from plasma—hence this process reduces plasma temperature.

charge-independent. Said of nuclear forces between particles the magnitude and sign of which does not depend on whether the particles are charged. See **nuclear force, short-range forces**.

charge-mass ratio. Ratio of electric charge to mass of particle, of great importance in physics of all particles and ions.

charge-storage. Principle whereby image information is stored on a mosaic of photocells or piezoelectric crystals, especially in camera tubes for television.

charged. Said of a capacitor when a working potential difference is applied to its electrodes. Said of a secondary cell or battery (accumulator) when it stores the maximum (rated) energy in chemical form, after passing the necessary ampere-hours of charge through it. Said of a conductor when it is held at an operating potential, e.g., traction conductors or rails, or mains generally.

charged-particle equilibrium. See **electronic equilibrium**.

charging current. (1) Impulse of current passing into a capacitor when a steady voltage is applied suddenly, the actual current being limited by the total resistance of the circuit. (2) A.c. which flows through a capacitor, when an alternating voltage is applied. (3) That for charging an accumulator.

charging voltage. Electromotive force required to pass the correct charging current through an accumulator, about 2.5 V for each lead-acid cell.

chassis. Mounting for circuit components of an electronic or electric unit. A printed-circuit board, or a perforated and bent metal box. Detachable electric couplings may be plug and socket, springs, or rubbing contacts. Any major part of an assembly to which smaller parts are attached.

chatter. Rapid closing and opening of contacts on a relay, which reduces their life.

Chebyshev filter. Third order constant-k filter of which insertion loss is characterized by a Chebyshev polynomial of the sixth order. These show some variation of the residual attenuation in the nominal pass band; but have a more rapid increase of attenuation than equivalent *Butterworth filters* (q.v.) outside the pass band.

Chebyshev response. Minimum fluctuation of response of a network from specification, obtained by a design procedure.

check programme. One arranged to reveal faults, or incipient faults, in operation of a computer. Also **test programme.**

check receiver. Radio or TV receiver for verifying quality or content of a programme.

cheese antenna. Covered parabolic antenna, suitable for ship navigation, with narrow vertical aperture for spreading the emitted wave sideways.

chemical binding effect. Variation of effective cross-section of nucleus for neutron absorption with type of chemical bond.

chemical bond. The electric forces linking atoms in molecules or single crystals, classified according to the electron configuration of the atoms. Thus *covalent* bonds are exchanged between the atoms, *heteropolar* bonds produce molecules with resulting dipole moment whilst *homopolar* bonds do not; *ionic* (or *electrostatic*) bonds are due to electrostatic attraction following ionization, *metallic* bonds are formed by electrons in lower energy levels leaving the valence electrons free to produce metallic conduction, and *molecular* (or *dative covalent*) bonds are due to two electrons contributed from the same atom.

chemiluminescence. Luminescence arising from chemical processes, e.g., glow-worm.

Cherenkov counter. Radiation counter which operates through the detection of Cherenkov radiation.

Cherenkov radiation. Electromagnetic radiation which arises when high-energy particles are ejected into a medium in which the radiation velocity is less than that of the particles. Occurs when the medium has a high refractive index much greater than unity, as in water.

chickens. Said of the signal corresponding to a line of scanning when the amplitude of the signal spuriously wanders towards black (with positive modulation), arising from accumulation of charges on the mosaic of the camera tube.

Child-Langmuir-Schottky equation. This gives the anode current in a space-charge-limited valve as

$$I = G \cdot V^{\frac{3}{2}},$$

where I=current, G=perveance (a constant for a given material and design), and V= maintained anode voltage.

chimney attenuator. Type of coaxial line attenuator.

chip. See swarf.

chirality. Relationship between spin vector and momentum vector of certain nuclear particles, especially neutrinos.

Chireix-Mesny antenna. Broadside antenna, with similar reflector, consisting of a square and diagonals.

Chireix transmitter. One with parallel stages which amplify sidebands shifted $\pm 90°$ in phase as class-C, with high efficiency, before combining to form an amplitude-modulated transmission.

chirp. Variable frequency note produced by instability of radio receiver.

chlorine (Cl). One of the halogens, a poisonous gaseous element, at. no. 17, at. wt. 35·453.

choke. (1) In a waveguide, an inductance coil designed to prevent the passage of high-frequency currents, e.g., lightning surges, in electrical power circuits. Often exhibiting high reactance because of self-resonance. See audio frequency. (2) Component in radar waveguide, with two quarter-wavelength sections, one short-circuited. See audio-frequency-, high-frequency-, smoothing-, swinging-.

choke coupling. (1) Use of the impedance of a choke for coupling the successive stages of a multi-stage amplifier, at a frequency below the resonant frequency of the choke. (2) In a waveguide, coupling flange with quarter-wavelength groove which breaks surface continuity at current node.

choke feed. Use of a high inductance path for the d.c. component of the anode current of a valve, the a.c. signal being fed through a capacitor.

choke-jointed. See joint.

choke modulation. System of anode modulation in which the modulating and modulated valves are coupled by means of a high-inductance choke included in the common anode-voltage supply lead. Also called constant-current modulation.

chopper. (1) Light interrupter, wheel, or vibrator, such that photocell output is alternating and so can be easily amplified. See Vibron. (2) Tone wheel or ticker interrupter, generally a rotating commutator, used to break up the continuous oscillations generated by a valve oscillator into trains for transmission in interrupted continuous-wave telegraphy. This renders audible the rectified signals from a continuous-wave telegraph transmitter without heterodyning.

chopper amplifier. See vibrating-reed amplifier.

chopper-stabilized amplifier. D.c. amplifier in which a mechanical chopper is used to stabilize gain and minimize drift.

Christmas-tree pattern. See optical pattern.

chroma control. Control which adjusts colours in television pictures. Abb. for chromaticity control.

chromatic aberration. An enlargement of the focused spot caused (1) in a cathode-ray tube, by the differences in the electron velocity distribution through the beam, and (2) in an optical lens system using white light, by the refractive index of the glass varying with the wavelength of the light.

chromaticity. Colour quality of light, as defined by its chromaticity coordinates, or alternatively by its purity (saturation) and dominant wavelength.

chromaticity control. See chroma control.

chromaticity coordinate. Ratio of any one of the tri-stimulus values of a colour sample to the sum of the three tri-stimulus values.

chromaticity diagram. Plane diagram in which one of three chromaticity coordinates is plotted against another. Generally applied to the CIE (x,y)-diagram in rectangular coordinates for colour television.

chromaticity flicker. Flicker resulting from periodic variations in chromaticity only.

Chromaton. TN for colour TV receiver tube similar to chromatron but with smaller deflection angle.

chromatron. Cathode-ray tube for viewing colour television images.

chrome. Term used in Munsell colour system to indicate saturation.

chrominance. Colorimetric difference between any colour and a reference colour of equal luminance and specified chromaticity. The reference colour is generally a specified white, e.g., CIE illuminant C for artificial daylight.

chrominance channel. Circuit path carrying the chrominance signal in a colour television system.

chrominance signal. In the NTSC colour television system, the carrier whose modulation sidebands are added to the monochrome signal in the upper part of the video-frequency band to convey colour information.

chrominance subcarrier regenerator. In a NTSC colour television receiver, the circuit which performs the function of generating a local subcarrier locked in phase with the transmitter burst. See **automatic phase control loop, quadricorrelator.**

chromium (Cr). Metallic element, at. no. 24, at. wt. 51·996, used in stainless steel and as a decorative plating over conventional steel.

chromophoric electrons. Valence electrons of elements with excitation levels which lead them to absorb one or more wavelengths of visible light.

chronic exposure. Excessive exposure to radiation acquired gradually over a long period; cf. *acute exposure.*

chronometer. Precision time-indicating device.

chronoscope. Electronic instrument for precision measurement of very short time intervals.

chronotron. Device which superimposes pulses on a power line to determine times of phenomena.

CinemaScope. TN for cinematography in which a wide field is compressed in distorted form on to normal-size film by an anamorphic lens, and rectified by a similar lens in the projector, giving a picture up to 60 ft wide.

Cinerama. TN for cinematography in which three cameras cover adjacent fields of view, giving a total horizontal field of 146°. Using three projectors, the combined image is formed on a curved screen 51 ft wide.

circle diagram. Graphical representation of complex impedances at different points in a transmission system on an orthogonal network. An example is the **Smith chart.**

circuit. Series of conductors, forming a partial branched path, which substantially confine the flow of electrons forming a current because of great contrast in conductivity. See also:

alternating-	feedback-
current-	Flewelling-
anode-	forked-
Armstrong-	inductive-
astable-	integrating-
balanced-	intermediate-
bistable-	magnetic-
bootstrap-	Meissner-
Boucherot-	muting-
butterfly-	non-inductive-
clipping-	open-
closed-	phantom-
control-	phase-shifting-
copper-	printed-
coupled-	radiating-
delay-	radio-
differentiating-	Reinartz-
discharge-	rejector-
divider	reset-
double-	resonant-
tuned-	safety-
doubler-	scaler
earth-	scaling-
Eccles-Jordan-	selector-
equivalent-	separator-

short-	tone-control-
shunt-	transistor equivalent-
single-	trigger-
single-wire-	tuned-
sneak-	tuned-grid-
solid-	two-way-
sweep-	unbalanced-
talk-back-	untuned-
tank-	Wagner earth.

circuit breaker. Device for opening electric circuit under abnormal operating conditions, e.g., excessive current, high ambient radiation level, etc. Also called **constant breaker.**

circuit cheater. Circuit which for test purposes simulates a component or load. Cf. *dummy load.*

circuit diagram. Conventional representation of wiring system of electrical or electronic equipment.

circuit noise. See **thermal noise.**

circuit parameters. Relevant values of physical constants associated with circuit elements.

circular magnetization. Magnetization of cylindrical magnetic material in such a way that the lines of force are circumferential.

circular mil. U.S. unit for wire sizes, equal to area of wire one mil (0·001 in) diameter.

circular polarization. State of polarization of an electromagnetic wave when its electric and magnetic fields each contain two equal components, at right angles in space and in phase quadrature.

circular scanning (or **sweep**). System in which the scanning element follows a circular (or a spiral) path.

circular time base. Circuit for causing the spot on the screen of a cathode-ray tube to traverse a circular path at constant angular velocity.

circulating current. That which flows round the loop of a complete circuit, as contrasted with *longitudinal current*, which flows along the two sides or *legs* of the same circuit, in parallel.

circulating-fuel reactor. Reactor using liquid fissile material continuously circulating through reactor core.

circulating memory. One in which the impulses in a sonic delay-line store are taken from the output, re-shaped, and re-inserted in the input. The delay line may be a column of mercury, or a fine nickel wire (magneto-striction).

circulation. Line integral of a vector, e.g., representing an electric or magnetic field intensity or fluid flow, round a closed contour.

cladding. Thin covering, e.g., of reactor fuel units to contain fission products and of dielectric rods to form light guides.

clamp. Circuit in which a waveform is adjusted and maintained at a definite level when recurring after intervals.

clamping diode. A diode valve used to clamp a voltage at some point in a circuit.

Clapp oscillator. Low drift *Colpitts oscillator* (q.v.).

Clark cell. A standard cell for calibrating sources of e.m.f. as in a potentiometer. This mercury-zinc cell has an e.m.f. of 1·434 volts at 15°C but is more temperature-sensitive than the Weston cadmium cell.

class-A, -B, -C, or -O insulating materials. Classification of insulating materials according to the temperature which they will withstand.

class-A amplifier. One in which the polarizing voltages are adjusted for operation on the linear portion of the characteristic curves of the valve, without grid current. The suffix 1 is added if grid current does not flow during any part of the cycle, and 2 if it does (this also applies to class -AB, -B and -C amplifiers).

class-AB amplifier. One in which the valve has its grid bias so adjusted that the operation is intermediate between classes A and B, i.e., the anode current is shut off during part of the excitation cycle, but the quiescent anode current is not reduced to a small value.

class-B amplifier. One in which the grid bias is adjusted to give the lower cutoff in anode current. Applied, colloquially, also to the combination of two such valves in one envelope, the two valves being designed to operate with grid current and substantially zero grid bias, and in antiphase.

class-C amplifier. One in which the grid bias is greatly in excess of that in class-B, and in which the anode output power becomes proportional to the anode voltage for a given grid excitation. In a given valve the anode current will flow for less than one-half of each cycle of an alternating voltage applied to the grid.

classical. Said of theories based on concepts established before relativity and quantum mechanics, i.e., largely in conformity with Newton's mechanics and Maxwell's electromagnetic theory. U.S. term **non-quantized.**

classical scattering. See **Thomson scattering.**

Clausius-Mosotti equation. One relating electrical polarizability to permittivity, principally for fluid dielectrics.

clean start. Commencement of transmission of a unit of programme when the *gain* has been previously adjusted to transmit normal volume or level, and without subsequent adjustment.

clean-up. (1) Improvement in vacuum which occurs in an electric discharge tube or vacuum lamp consequent upon absorption of the residual gases by the glass. (2) The removal of residual gas by a *getter*.

clear. (1) To remove data from computer stores so that fresh data can be recorded. (2) Radio message 'in plain', i.e., not *coded* or *scrambled*, so that it is immediately understood.

clear channel. See **channel.**

clearance rate. Rate of elimination of a particular isotope by a given organ.

clearing field. (1) One applied across cloud chamber to prevent or eradicate ionization tracks. (2) One applied to eliminate trace from storage oscilloscope tube screen.

click. (1) Short impulse of sound, with wide frequency spectrum, with no perceptible concentration of energy-giving characterization. (2) Atmospheric disturbance of very short duration, arising from lightning.

click filter. Simple filter for eliminating surges arising from key clicks.

click method. That for determining the resonant frequency of oscillatory circuit, depending upon click produced in telephones of a heterodyne wavemeter when oscillatory circuit coupled thereto is brought into resonance.

C-line. Fraunhofer line in spectrum of sun at 6562·8Å, arising from ionized hydrogen in its atmosphere.

clipping. (1) Loss of initial or final speech sounds in telephone transmission, due to the operation of voice-switching apparatus, which is required to prevent the circuit from singing. Obviated by the use of *delay networks*, so that the switching has time to operate the transmission circuits before the speech currents pass their contacts. (2) A type of distortion arising from severe overloading so that the tips of a sound waveform become clipped, due to the fact that the output cannot be increased despite a greater input to the system.

clipping circuit. That for removing the peaks, tails, or frequency components of pulses in electrical circuits.

clock frequency. That provided by a stable oscillator to regulate the digits (*bits*) and to keep them synchronous throughout a computer system.

close coupling. See **overcoupling.**

closed circuit. (1) One in which there is zero impedance to the flow of any current, the voltage dropping to zero. (2) One composed of inductors, capacitors and resistors, all of relatively small dimensions, to load a transmitter instead of an extended antenna and consequent radiation. (3) System in which transmission (pickup) and reproduction of television sound and scene are fixed, the question of broadcast or free reception not arising. Used in industry, medicine, theatre, etc., the scanning system being most common. Also **radiovision.**

closed-circuit recording. That for subsequent dissemination in broadcasting, and not of a performance being broadcast.

closed cycle. (1) In thermodynamics, heat engine in which a working substance circulates continuously without replenishment. (2) Cooling system for nuclear reactor in which coolant is circulated continuously through core and heat exchanger.

closed-cycle control system. One in which the controller is worked by a change in the quantity being controlled, e.g., an automatic voltage regulator in which a field current is actuated by a deviation of the voltage from a desired value, the *reference voltage*.

closed diaphragm. Diaphragm or cone which is not directly open to the air, but communicates with the latter through a horn, which serves to match the high mechanical impedance of the diaphragm with the low radiation impedance of the outer air.

closed-loop system. A control system, involving one or more feedback control loops, which combines functions of controlled signals and of commands, in order to keep relationships between the two stable. Also **feedback control system.**

closed subroutine. One which is entered through a jump instruction in the programme; cf. *open subroutine*.

cloud chamber. Chamber in which saturated gas or air can be suddenly cooled by expansion, revealing the paths of rays of particles which cause ionization and hence nuclei for condensation.

Clusius column. Device used for isotope separation by method of thermal diffusion.

cluster. See **ion cluster.**

clutter. Any irregular interference or noise arising from echoes, sea, rain, clouds, buildings. Gives rise to radar echoes.

co-altitude. See **zenith distance.**

coarse scanning. Rapid scan carried out to determine approximate location of any radar target.

coastal refraction. Refraction, towards the normal, of waves arriving from sea to land at their incidence with the shore-line, resulting in an appreciable error in radio direction-finding for bearings making a small angle with the shore-line.

coastline effect. Same as **shore effect.**

coated cathode. One sprayed or dipped with electron-emitting compounds.

coaxial antenna. Exposed λ/4 length of coaxial line, with reversed metal cover, acting as a dipole. λ = wavelength.

coaxial cable. One in which a central conductor is surrounded by an outer tubular conductor. If the intervening dielectric is air, the low attenuation makes it suitable for TV transmission of video-frequency currents.

coaxial feeder. Same as **concentric feeder.**

coaxial filter. One in which a section of coaxial line is fitted with re-entrant elements to provide the inductance and capacitance of a filter section.

coaxial line. See **coaxial cable.**

coaxial-line resonator. One in which standing waves are established in a coaxial line, short- or open-circuited at the end remote from the drive. Of very high Q, these are used for stabilizing oscillators, or for selective coupling between valves. See **coaxial stub.**

coaxial-line tube. See **Heil oscillator.**

coaxial relay. Switching device in which the coaxial circuit on both sides of the contact is maintained at its correct impedance level, thus avoiding wave reflection in the current path.

coaxial stub. Section of coaxial line short-circuited at one end and functioning as a high impedance at quarter-wave resonance.

cobalt (Co). Hard, grey, metallic element, at. no. 27, at. wt. 58·9332, m.p. 1480°C, sp. gr. 8·9. It is used in permanent magnets and hardened steel alloys. The radioactive isotope ^{60}Co emits 1·17 and 1·33 MeV γ-rays, and is widely used in radiography and industrial irradiation.

cobalt bomb. Theoretical nuclear weapon loaded with ^{59}Co. The long life radioactive ^{60}Co formed on explosion would make the surrounding area uninhabitable. Also a radioactive source comprising ^{60}Co capsule in lead shield with shutters.

Cobol. Business machine language, using automatic conversion from normal language to machine code. (*C*ommon *b*usiness *o*riented *l*anguage.)

cochlea. Inner ear forming sensitive part of hearing organ.

cochlear potentials. Electric potentials within the cochlear structures resulting from acoustic stimulation.

Cockcroft and Walton accelerator. High-voltage machine in which rectifiers charge capacitors in series. The discharge of these drives charged particles through an accelerating tube.

cocked hat. Navigational term for intersection of three bearing lines on chart.

codan. Circuit for silencing a receiver in the absence of a signal. (*C*arrier-*o*perated *d*evice, *a*nti-*n*oise.)

code. Variations in a unified pattern which represent relevant pieces of information on a one-to-one basis. Thus numbers in a *binary* or *denary* scale of numbers can represent letters or numerals, money, stocks, etc., as data for processing. Such numbers can then be coded into, e.g., a 5-unit code for manipulation. See **computer-.**

code delay. Arbitrary time interval introduced between pulses from master and slave navigational transmitters.

coder. In a PM system, the sampler that tests the signal at specified intervals.

coding. (1) Alteration of width or gap, or chopping in an IFF system. (2) Programming of a computer, whether by written instructions on a coding form, by pseudo-code, or in machine language.

coefficient. Numerical constant prefixed as a multiplier to a variable quantity in calculating the magnitude of a physical property. Thus, if the *coefficient of expansion* of brass is 0·000018 per deg. C, the expansion of a brass rod of length *l* heated through *t* deg. C would be 0·000018 × *lt*.
See also:

absorption-	reflection-
Hall-	total absorption-
inductance-	Townsend-
Peltier-	transmission-
potential-	true absorption-.
recombination-	

coercimeter. Instrument for measurement of coercive force.

coercive force. Reverse magnetizing force required to bring magnetization to zero, after ferromagnetic material has been saturated and left with appreciable *residual magnetization* (q.v.).

coercivity. Coercive force when the cyclic magnetization reaches saturation.

coffin. Transportable chamber of lead for radioactive materials. Also **flask**.

coherence. Opposite of randomness, especially with reference to radio, light and acoustic waves.

coherent. Said of waves controlled in a very narrow frequency band.

coherent oscillator. One which is stabilized by being locked to the transmitter of a radar set for beating with a reflected incoming pulse signal, and used with radar-following circuits.

coherent pulse. Said when individual trains of high-frequency waves are all in the same phase.

coherent sources. Those between which definite phase relationships are maintained, so making it possible for interference effects to occur.

coherer. Early form of detector of electromagnetic waves, in which the resistance of an imperfect contact is abruptly reduced by the passage of high-frequency currents. See **Branly-.**

coherer effect. On an attempted contact between metals, the breakdown of a film by sufficient voltage and the build-up of a conducting bridge of metal by melting.

coil. (1) Roll of tape, said to be *blank* when *feed holes* alone have been punched. (2) A winding of conducting wire, with a core of air or magnetic material for providing inductance. (3) A coil of wire to carry a current to provide a magnetic field, usually with an iron core (then called **winding**).

See also:

air-spaced-	load-
basket-	loading-
bucking-	pancake-
compensating-	reaction-
convergence-	repeating-
coupling-	retardation-
deflector-	Ruhmkorff-
focusing-	scanning-
Helmholtz-	search-
honeycomb-	spark-
inductance-	Tesla-
induction-	voice-.

coil antenna. See frame antenna.

coin analysis. Breakdown, according to a *programme*, in which the output of a wages-processing is also expressed in specified coins or notes.

coincidence. (1) Unit with '1' output only when all inputs agree on '1'. (2) When counting nuclear particles, operation of counters apparently simultaneously, or within an assigned time interval.

coincidence counter. Combination of counters and circuits which record only when all the counters are operated simultaneously or within a specified interval of time.

coincidence gate. Electronic circuit producing an output pulse only when each of two (or more) input circuits receive pulses at the same instant, or within a prescribed time interval.

cold cathode. Electrode from which electron emission results from high-potential gradient at the surface at normal temperatures.

cold-cathode rectifier. Rectifier for low-frequency a.c., comprising a convex cathode and a concave anode placed close together in low-pressure gas.

cold emission. See auto-emission.

cold light. See blue light.

cold store. (1) One constructed from an array of units, each of which depends on persistence of current in a bridge in a hole in a sheet of lead or tin, maintained at a low temperature by liquid helium. (2) Use of cryotron as register. At liquid helium temperature superconductivity is removed in a straight current-carrying wire by an axial magnetic field, thus greatly changing its impedance and operating it as a switch. Sn and Pb are the most suitable conductors.

cold tests. Measurements on electronic circuits, when there is no emission from heated cathodes, to obtain parameters.

Cole-Cole plot. Graph of real against imaginary part of complex permittivity, theoretically a semicircle.

collapse. Term in plasma physics for shock heating produced by sudden application of magnetic field.

collateral. Term used for nuclides which decay into one of the main radioactive series although initially not part of it.

collator. Apparatus for matching or checking cards in separate packs.

collective electron theory. Assumption that ferromagnetism arises from free electrons. Fermi-Dirac statistics identify the Curie point with the transition from ferro- to para-magnetic states.

collector. (1) Any electrode which collects electrons, which have already completed and fulfilled their function, e.g., screen grid. (2) Outer section of a transistor, which delivers a primary flow of carriers.

collector capacitance. The capacitance of the depletion layer forming the collector of a transistor.

collector current. The current which flows at the collector of a transistor on applying a suitable bias.

collector-current runaway. The continued increase of the collector current arising from an increase of temperature in the collector junction when the current grows. See thermal runaway.

collector efficiency. The ratio of the useful power output to the d.c. power input of a transistor.

collector junction. One biased in the high-

collision resistance direction, current being controlled by *minority carriers*. The semiconductor junction between the collector and base electrodes of a transistor.

collision. With particles, contact is not apparent unless there is *capture*. Collision means such nearness of approach that there is more or less mutual interaction because of the forces associated with the particles, although no impact.

collision frequency. See **frequency of collision.**

colloidal electrolyte. One formed from long-chain hydrocarbon compounds, end radicals of which can ionize, thus providing some properties of electrolytes.

colorimeter. Instrument for quantitative measurement of hue.

colour balance. Suitable selection of phosphor efficiencies and electrical and optical properties of a colour TV display for the achromatic reproduction of a grey scale of varying luminance.

colour bar. Bar of colour produced on a colour TV display. Also the corresponding video waveform.

colour break-up. Transitory separation of colours in an intermittent TV picture arising from motion of the eye of the observer.

colour burst. In NTSC colour TV system, that part of the composite colour signal consisting of a sine wave at subcarrier frequency present during part of the back porch of the line-synchronizing pulse; used to establish a phase reference for demodulating the chrominance signal.

colour cast. Predominance of one-colour primary in colour image resulting from lack of colour balance.

colour cell. The smallest area in a colour picture tube which includes a complete set of all the primary colours in the repeating pattern.

colour centre. (1) Unit region of CRT phosphor, determined by the electron beam and perforated screen in establishing a colour television image. Also called **U-centre.** (2) Lattice defects in crystal which interact with electrons in order to produce intense local light absorption. A few such defects give coloured crystals their characteristic hue.

colour coder. Apparatus in colour TV to generate chrominance subcarrier and composite colour signal from the camera signals. Also **colourflexer.**

colour contamination. Error in colour reproduction caused by incomplete separation of primaries.

colour coordinates. Set of numbers representing the location of a hue on a *chromaticity diagram.*

colour decoder. Circuit in a TV receiver which extracts, decodes, and separates the 3 constituent colours.

colour difference signal. NTSC colour TV transmissions comprise a monochrome signal, and 3 colour difference signals. The latter are combined with the monochrome signal in the receiver to produce one of the required primary colour signals.

colour disk. Filter disk used in field sequential colour TV systems.

colour edging. See **fringing.**

colour field. Field in one of the colours selected for transmission.

colour filter. Film of material selectively absorbing certain wavelengths, and hence changing spectral distribution of transmitted radiation.

colour flicker. Flicker observed as a result of fluctuations in chromaticity.

colour gate. Circuit in colour TV receiver which allows only primary colour signal corresponding to excited phosphor to reach modulation electrode of tube.

colour killer. Circuit rendering the chrominance channel of a colour TV receiver inoperative during reception of monochrome signals.

colour mixture curve. Representation of the specified three colours which match a given colour.

colour negative. Image in which the correct hues have been replaced by their corresponding complementary colour.

colour phase alternation. Sequence of the colour signals in a video signal.

colour picture signal. Monochrome video signal, plus a subcarrier conveying the colour information, which is transmitted with synchronizing signals.

colour plane. In a multibeam colour picture tube, the (near) plane containing all the colour centres.

colour primaries. Set of (usually three) colours from which multicolour images are built up in printing, photography and TV. See **primary additive colours, primary subtractive colours.**

colour purity magnet. Magnet placed near to neck of colour picture tube to modify path of electron beam and thus improve purity of the displayed colour.

colour reference signal. Continuous signal which determines the phase of the *burst signal* (q.v.).

colour response. Output of a television camera tube with reference to the colour of the light incident on it.

colour subcarrier. Signal, conveying the colour information as a modulation, added to the monochrome video and synchronizing signals.

colour television. The transmission of pictures in colour. See **television.**

colour television mask. A perforated material disk which directs a particular colour electron beam in a colour picture tube to the appropriate phosphor dot in each cluster on the screen.

colour television tube. Any CRT suitable for production of a colour TV picture.

colour temperature. That temperature of a black body which radiates with the same dominant wavelengths as those apparent from a source being described. See **Planck's law.**

C

colour threshold. The luminance level below which colour differences are indiscernible.

colour triangle. That drawn on a chromaticity diagram, representing entire range of chromaticities obtainable as additive mixtures of three prescribed primaries, represented by the corners of the triangle.

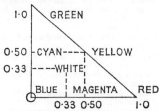

(The sum of the red, green and blue colour coordinates is always unity. The red and green coordinates are plotted along the x and y axes respectively.)

colourflexer. See **colour coder.**

Colpitts oscillator. Triode valve in which the shunt-tuned circuit is connected between anode and grid, the cathode connection being taken from the midpoint of two series capacitors which tune the inductance.

columbium. See **niobium.**

column. (1) Position in a number corresponding to a power of the *base* or *radix* (usually 10 or 2). Also *place.* (2) Loudspeaker enclosure in which the back of the drive unit gets a resonant bass loading from a column of air. The speaker would be mounted at the end of the column, or, if this is tapered, it is located at a distance from the narrow end.

coma. Plumelike distortion of spot arising from misalignment of focusing elements of gun. See **anisotropic-.**

comb filter. One containing a number of *pass* or *rejection* bands of frequency.

command. (1) Set of signals which bridge computer programme instruction and intended operation of circuits. (2) U.S. term for **instruction.**

common. (1) Said of points in a circuit which are common to the input and output circuits of a transistor. (2) Joined by a *rail.*

common-base connection. The operation of a transistor in which the signal is fed between base and emitter, the output being between collector and the base with the latter earthed. Also **grounded base.**

common-collector connection. The operation of a transistor in which the collector is earthed and the signal is fed between collector and base. The output is between emitter and collector. Also **grounded collector.**

common-emitter connection. The operation of a transistor in which the signal is fed between base and emitter with the latter earthed. The output is between emitter and collector. Also **grounded emitter.**

common-frequency broadcasting. The use of the same carrier frequency by two or more broadcast transmitters, sufficiently separated for their useful service areas not to overlap. Also **shared-channel broadcasting.**

communications receiver. Exceptionally sensitive and selective receiver designed for signal communications.

commutating reactor or reactance. One in the cathode lead of a mercury-arc rectifier, to keep current passing while the arc jumps from one anode to the next anode.

commutation factor. Product of rate of current decay and rate of voltage rise after a gas discharge, both expressed per microsecond.

commutative rule. Process in which the result is correctly obtained irrespective of the order of the steps taken, e.g., $pq=qp$.

commutator. The part of a motor or generator armature through which electrical connections are made by rubbing brush contacts.

compander. Device for compressing the volume range of the transmitted signal and re-expanding it at the receiver, thus reducing the signal/noise ratio. (*Com*presser . . . ex*pander.*)

comparator. Circuit which compares two sets of impulses, and acts on their matching or otherwise.

comparison bridge. One designed to measure small differences from a standard magnitude.

compass. See **gyrocompass, magnetic-, radio-.**

compatibility. (1) That a transmitted signal carrying chrominance elements can be received correctly, but without coloration, on a television receiving equipment which normally receives black-and-white images. (2) That different computers can accept and process data prepared in the same form or by the same peripheral equipment.

compensated semiconductor. Material in which there is a balanced relation between *donors* and *acceptors,* by which their opposing electrical effects are partially cancelled.

compensating coils. Current-carrying coils to adjust distribution of magnetic flux as desired.

compensation. In a sound-reproducing system, adjustment of an actual frequency response to one specified.

compensation theorem. That change in current produced in a network by a small change in any impedance Z carrying a current I is the result of an apparent e.m.f. of $-I . \delta Z$.

compiler. (1) See **programmer.** (2) Automatic computer programming unit which transcribes the operator's instructions from a programme-orientated language into the machine language of the specific computer. See also **assembler.**

complement. Because most computers cannot subtract, negative numbers are made positive by deducting each numeric from radix minus one; after operations of adding, one is added to the least significant digit.

complementarity. The correspondence in quantum mechanics between particles of momentum p and energy E, and wave trains of frequency f and wavelengths λ. This is given by equations: $p=\dfrac{h}{\lambda}$ and $E=hf$, where h is Planck's constant.

complementary colours. Pairs of colours which combine to give spectral white.

complementary symmetry. That shown by otherwise identical p-n-p and n-p-n transistors.

complementary transistors. A n-p-n and a p-n-p transistor pair used to produce a push-pull output using a common signal input.

complex number. One with a directional property which can be expressed in the following forms:
$$z=x+jy$$
$$=r(\cos\theta+j\sin\theta),$$
where $j=\sqrt{-1}$, θ is the *amplitude, argument,* or *phase* and r is the *modulus.*

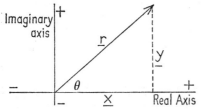

The operator $\sqrt{-1}$ means an anti-clockwise rotation of a vector by 90° on an Argand diagram, because a double operation is denoted by $j^2=-1$, i.e., a reversal on the diagram. The component x of the complex number is known as the *real part* and y as the *imaginary part.* A *real* or *imaginary number* is one for which the other component has a value of zero.

complex Poynting vector. See **Poynting vector.**

compliance. 2π (frequency) times the reciprocal of the negative imaginary part of acoustic or mechanical impedance. See **acoustical-.**

component. Any part of a signal (including a space) which is uniform in frequency and amplitude during its allotted duration.

Composertron. Complex magnetic tape recording assembly, in which elements of recorded sound can be synthesized and reproduced as a composition.

composite cable. One containing different purpose conductors inside a common sheath.

composite conductor. One in which strands of different metals are used in parallel.

composite filter. Combination of a number of filter sections, or half-sections, all having the same cutoff frequencies and specified impedance levels.

composite resistor. One formed of solid rod of carbon compound.

composite signal. Total TV signal, comprising video, blanking, synchronizing, burst, and chrominance component signals.

compound horn. Rectangular horn for electromagnetic waves in which flaring occurs in both planes.

compound modulation. Use of an already modulated wave as a further modulation envelope. Also called **double modulation.**

compound motor or generator. One which has both series and shunt field windings.

compound nucleus. Highly excited unstable nucleus formed by absorption of bombarding particle.

compression. (1) Reduction of gain of amplifiers for high signal levels. Normally expressed by: $10\log_{10}G_H/G_L$, where G_H and G_L are respectively power gains for high and low power levels. (2) Increase of density along high-pressure wavefront of sound wave.

Compton effect. Elastic scattering of photons by electrons, i.e., collisions in which both energy and momentum are conserved. If λ_s and λ_i are respectively the wavelengths associated with scattered and incident photons, Compton shift is given by
$$\lambda_s-\lambda_i=\lambda_0(1-\cos\theta),$$
where $\theta=$ angle between directions of the incident and scattered photons, and λ_0 is the Compton wavelength of the electron.

Compton electrometer. A sensitive form of quadrant electrometer.

Compton recoil electron. An electron which has been set in motion following an interaction with a photon (Compton effect).

Compton wavelength. Wavelength associated with the mass of any particle, such that
$$\lambda=h/mc,$$
where
$\lambda=$ wavelength $\qquad m=$ rest mass
$h=$ Planck's constant $\quad c=$ velocity of light.

computer. Any mechanical, electric, or electronic apparatus or assembly which performs calculations, simple or including integration and differentiation, or simultaneous equations. There are three main types: (1) *analogue* (q.v.), which deals with continuous correspondences; (2) *digital* (q.v.), which operates on units; (3) *hybrid* (q.v.) or *quantized,* in which operations are carried out on the quantities processed using both analogue and digital techniques. An *analogue—digital converter* (q.v.) is used to change continuous amplitudes into digital signals and vice versa.

See also:

asynchronous-	poison-
dynamic storage-	real-time-
parallel-	serial-.

computer code. A system of rules and characters to convey instructions.

concentration. General term for quantity of substance or energy per unit volume.

concentration cell. One with similar electrodes in common electrolyte, the e.m.f. arising from differences in concentration at the electrodes.

concentric. Term replaced by *coaxial* for all telecommunications and radar purposes.

concentric feeder. Transmission line in which two conductors are concentric tubes. The absence of external fields and freedom from pickup make it specially suitable for use with directional antenna systems; also called **coaxial feeder**.

condenser. Original name of **capacitor**. Obsolete in U.K. and U.S., but retained on the Continent.

conditional jump. See **jump** (1).

conditionally stable. Amplification system which satisfies Nyquist stability criterion until its gain is reduced.

condor. Radio-navigation system related to *benito* (q.v.).

conductance. Property of material by which it allows current to flow through when a p.d. is applied; reciprocal of resistance, measured in mhos. See also **anode-, conversion-, differential anode-, electrode-, equivalent-, molar-**.

conductance ratio. That between equivalent conductance of a given solution and its value at infinite dilution.

conduction. General term for transmission of energy (electrical, thermal, acoustical) without bodily movement of transmitting medium. *Convection* implies motion of medium, and *radiation* means transmission regardless of medium.

conduction angle. The fraction of a cycle (expressed as an angle) for which a diode rectifier passes current. With an efficient smoothing filter this will be a very short period coincident with the peak input voltage.

conduction band. In band theory of solids, band which is only partially filled, so that electrons can move freely in it, hence permitting conduction of current.

conduction by defect. In a doped semiconductor, conduction by *holes* in the valency electron band.

conduction current. That resulting from flow of electrons or ions in a conducting medium.

conduction electrons. The electrons situated in the conduction band of a solid which are free to move under the influence of an electric field.

conduction hole. In a crystal lattice of semiconductor, conduction obtained by electrons filling holes in sequence, equivalent to a positive current.

conductivity. (1) Conductance at a specified temperature between opposite faces of a cube of material, 1 cm edge. Reciprocal of *resistivity*. (2) Ratio of air volume current to acoustic potential difference for an acoustic orifice. (3) Rate of heat flow across unit area normal to the direction of flow, per unit temperature gradient. See also **anisotropic-, asymmetrical-, electrical-, *p*-type-**.

conductivity cell. Any cell with electrodes for measuring conductivity of liquids or molten metals or salts.

conductivity modulation. That effected in a semiconductor by varying a charge-carrier density.

conductor. Material substance (solid, liquid,

gas) which can conduct electric charges (current) continuously. See **asymmetric-, composite-, ionic-, lightning-**.

cone diaphragm. One of felted, doped or Bakelized paper, driven by a circular coil carrying speech currents near its apex; widely used for radiating sound in loudspeaking receivers. Also **cone loudspeaker**.

cone of silence. Cone-shaped space in which signals from a transmitter are virtually undetectable.

cones. Perceptive elements on periphery of human retina, which are primarily used for daylight and colour vision as opposed to *rods* (q.v.).

configuration control. Control of nuclear reactivity by alterations to the configuration of the fuel, reflector and moderator assembly.

confinement. See **containment**.

conic section. General name for class of curves obtained by taking a plane section of a cone. These cover all possible orbits for a particle moving under the influence of any form of inverse square law. See **eccentricity**.

conical scanning. Scanning similar to *lobe switching*, but circular, the direction of maximum response generating a conical surface. Used in missile guidance.

conjugate branches. Any two branches of an electrical network such that an e.m.f. in one branch produces no current in the conjugate branch.

conjugate impedance. Two impedances are conjugate when their resistance components are equal and their reactances are equal but opposite in sign.

Conpernik. RTM for an equal nickel-iron isotropic alloy.

consequent pole. An effective pole not at the end of ferromagnetic material, e.g., at the ends of a diameter of a ring magnet; must occur in pairs. Used in Broca galvanometer.

conservation of energy law. That which states that energy cannot be lost or gained in a closed system, but interchanged in form at specific rates of conversion. Relativistically, addition of energy to a mass, when approaching the speed of light, is manifest in an increase of mass, according to Einstein's law, which is completely reversible.

conservative field. One for which the potential function is single-valved.

consistency. Maximum deviation of a number of readings from a mean on a measuring instrument, when a constant magnitude is applied.

consol. Long-range navigational system which can determine a line of position within close limits of accuracy. U.S. **consolan**; German **sonne**.

console. (1) Location of playing controls in an organ, electronic or pipe, includings swells, pedals, several manuals, speaking stops and couplers. (2) Central controlling desk in power station, process plant, computer, or reactor, where an operator can supervise operations and give instructions.

consonance. The acoustical condition where two pure tones blend pleasingly.

constant. In a formula, a multiplying factor, independent of all variables. If *absolute*, it has always the same value, if *arbitrary* (parameter), it has one particular value in a given case.

constant breaker. See **circuit breaker.**

constant-current characteristic. Graph which exhibits performance of electrical equipment when variables are adjusted to keep its operating current a specified value.

constant-current modulation. See **choke modulation.**

constant-current transformer. One designed to supply steady secondary current under varying conditions of load.

constant-frequency oscillator. One in which special precautions are taken to ensure that the frequency remains constant under varying conditions of load, supply voltage, etc.

constant-*k* filter. Simple filter formed from a constant-*k* network only.

constant-*k* network. Iterative network for which product of series and shunt impedances is frequency independent.

constant-luminance transmission. Type of colour television transmission in which transmission primaries consist of one luminance and two chrominance primaries.

constant-resistance network. One in which iterative impedance in at least one direction is resistive and independent of the applied signal voltage.

constant-speed motor. Electric motor designed to operate at same speed under widely varying conditions of torque.

constant-voltage transformer. One with or without extra components which gives a constant voltage with varying load current, or with varying input voltage over a specified range.

constant-volume amplifier. See **vogad.**

constantan wire. See **Eureka wire.**

constriction resistance. That across the actual area of contact through which current passes from one metal to another, equal to

$$(\rho_1 + \rho_2)/4a,$$

where a = radius of circular contact area ρ_1 and ρ_2 = resistivities.

contact bounce (or chatter). Intermittent opening and closing of relay contacts.

contact breaker. An automatic device for interrupting a contact.

contact electromotive force. That which arises at the contact of dissimilar metals at the same temperature, or the same metals at different temperatures.

contact metal. That used for contacts on springs of relays, generally silver, platinum, tungsten, Elkonite, etc.

contact-modulated amplifier. See **vibrating-reed amplifier.**

contact noise. Noise voltage arising across a contact, with or without adsorbed gases, arising from differences in work function of

contacting conductors, one of which may be a semiconductor.

contact nomenclature. Of switches and relays. See table in Appendices.

contact potential. Potential difference between any substances, metal contacts, electrolytes and electrodes, arising from pressure of free electrons in these substances.

contact rectifier. One depending on contact potential between different solids, the relation between current and voltage not being linear.

contact resistance. Resistance at surface of contact between two conductors; mainly determined by constriction of current in the materials near the point of contact, increasing resistance. Contact area is extended by pressure.

contactor. Switch operated other than manually.

containment. Holding together of sufficient density of plasma in strong magnetic fields at a very high temperature, so that fusion nuclear reactions from deuterons and tritons are promoted. Also **confinement.**

continuity. (1) Supervision of succession of items from various sources in making up a broadcast programme. (2) The existence of an uninterrupted path for current in a circuit.

continuous control. System in which controller is supplied with continuously varying actuating signal; cf. *on-off* control.

continuous current. Earlier name for **direct current**, now obsolete in U.K. and U.S. but retained on Continent.

continuous oscillations. Those generated by alternator, arc, or oscillating valve, which are not broken into individually decaying groups, as opposed to the *damped oscillations* of a spark transmitter, the latter now being obsolete. Also **undamped oscillations.**

continuous rating. Maximum power dissipation which could be allowed to continue indefinitely.

continuous spectrum. One which shows continuous non-discrete changes of intensity with wavelengths or particle energy. See also **heterogeneous radiation.**

continuous wave. Steady radiated oscillation, modulated on-and-off by a coded signal; it must be heterodyned to give an audio note for reception by ear, and further rectified to operate a recorder or tape machine.

continuous-wave telegraphy. That in which each signal (dot or dash) consists of a train of continuous oscillations.

contour recording. Same as **hill-and-dale recording.**

contra wire. See **Eureka wire.**

contrast. Ratio of brightness of two acceptable levels in an image. Range of contrast is maximum contrast acceptable or possible. See **point gamma.**

contrast amplification. That in which dynamic contrast in sound reproduction is increased by electronic means, compensating precisely for contrast reduction (*control*) which is necessary in most communication systems to avoid intrusion of noise.

contrast control. Detailed control of highlights and shadows in a reproduction image, e.g., as in TV.

contrast range. See **dynamic range** (2).

contrast sensitivity. Liminal ratio of $\delta B/B$ of a small spot which differs from brightness B of the larger surroundings.

control. (1) General term for manual or automatic adjustment (usually by potentiometer, fader, or attenuator) of power level in a transmission within its dynamic range. (2) Said to be *sequential* when the instructions are fed to an electronic digital computer in order during an operation. It is said to be *dynamic* when the instructions are varied or altered in sequence during, or as a result of, an operation. (3) Regulation of relation between upper and lower levels of transmission in circuits, to keep it within the overload and noise level of the system. (4) Necessity in sound reproduction to reduce range between useful maximum and minimum levels in a transmission. Usually determined by hand, but in a transmission system where there can be exact compensation at the receiving end, reduction can be performed automatically to a set rule. (5) Maintenance of power level at desired setting by adjustments to reactivity of reactor core through absorbing rods or other means.

control action. Type of control in a regulating system, distinguished as *proportional, integral,* or *derivative,* when depending on displacement, summation, or velocity, or a combination of these, as these vary with time.

control ampere-turns. Magnetomotive force applied to a magnetic amplifier.

control characteristic. Curve connecting output quantity against control quantity under determined conditions in a magnetic amplifier.

control circuit. (1) Separate circuit or line in parallel with a transmission path, for use of operators. (2) All windings and components which determine control current in a magnetic amplifier.

control current or **voltage.** One which, by its magnitude, direction, or relative phase, determines operation of functional apparatus, e.g., that which operates lamps through local relays, traction motors distributed along a train, that in a mercury-arc rectifier, or process control.

control electrode. One, e.g., a grid, the primary function of which is to control flow of electrons between two other electrodes, without necessarily taking current itself, control being by voltage which regulates electrostatic fields.

control hysteresis. Ambiguous control, depending on previous conditions. Jump or snap action arising in valve or magnetic amplifiers because of excessive positive feedback under certain conditions of load.

control panel. Panel containing full set of indicating devices and remote control units required for operation of industrial plant,

reactor, chemical works, etc. Cf. *console* (2).

control point. Value of controlled variable, departure from which causes a controller to operate in such a sense as to reduce the *error* and restore an intended steady state. Also **set point.**

control ratio. In discharge tubes, ratio of change in anode voltage to a compensating change in grid voltage for striking.

control register. That which stores the current instructions which govern the computer for a given cycle.

control relay. See **relay types.**

control rod. Rod moved in and out of reactor core to vary rate of reactivity. May be neutron-absorbing rod, e.g., cadmium, or, less often, fuel rod. See **regulating rod, shim rod.**

control system. Servo controller unit operated by error-sensing head through closed loop.

control turns (or **windings**). Those wires on the core of a magnetic amplifier or transductor which carry the control current. U.S. term **signal windings.**

control valve. A valve (i.e., tube) in an automatic gain control circuit.

control voltage. See **control current.**

controlled area. Interior or exterior area where radiation exposure may reach level at which continuous records must be kept, and which is under supervision of Radiological Safety Officer.

controlled carrier. Transmission in which the magnitude of the carrier is controlled by the signal, so that the depth of modulation is nearly independent of the magnitude of the signal.

controlled variable. The quantity or condition which is measured and controlled.

convection current. Current in which the charges are carried by moving masses appreciably heavier than electrons.

convergence. (1) In a colour TV display tube (tricolour kinescope), meeting or crossing of the three electron beams at a common point. For proper reproduction of colour pictures convergence must be maintained over the whole scanning area. (2) If an algorithm involves using methods of successive approximation, these converge when each result is nearer to the true value than the last.

convergence coils (or **magnets**). Coils used to ensure convergence of the electron beams in a triple gun colour TV tube.

converse magnetostriction. The change of magnetic properties, e.g., induction, in a ferromagnetic material when subjected to mechanical stress or pressure.

conversion coefficient. See **internal conversion.**

conversion conductance. Ratio of component of current of intermediate frequency flowing in the output circuit of a conversion detector when working into zero impedance, to the signal frequency input voltage; expressed in mA/V.

conversion detector. See **mixer** (2).

conversion efficiency. See **anode efficiency.**

conversion electron. One ejected from inner shell of atom when excited nucleus returns to ground state, and the energy released is given to orbital electron instead of appearing as quantum.

conversion factor. Generally, ratio of two measures of the same physical quantity, expressed in different units.

conversion gain. Effective amplification of a conversion detector, measured as the ratio of output voltage of intermediate frequency to input voltage of signal frequency. Expressed in dB, the gain is twenty times the logarithm of this ratio. See also **conversion ratio.**

conversion mixer. Same as frequency changer (2).

conversion ratio (R). Number of fissionable atoms produced per fissionable atom destroyed in a reactor. Corresponding conversion gain is defined as $R-1$.

conversion transconductance. Ratio of output current of one frequency to corresponding voltage of different frequency in a superheterodyne converter.

conversion transducer. Same as **frequency changer** (2).

converter. (1) U.S. for **frequency changer.** (2) A machine or device for changing a.c. to d.c. or the converse, particularly transistor oscillator for supplying high tension from low tension. (3) A device for changing information coded in one form into the same information coded in another, e.g., analogue-digital converter. See **arc-, direct-current-, thermal-, transistor power-pack.**

converter reactor. One in which fertile material in reactor core is converted into fissile material different from the fuel material; cf. *breeder reactor.*

'cookie cutter' tuning. Tuning of cavity magnetrons involving changing capacitance of cavities by the introduction of conducting rings between straps; cf. *'crown of thorns' tuning.*

cooled-anode valve. Large thermionic valve in which special provisions are made for dissipating the heat generated at the anode, effected by circulating water, oil, or air around the anode, or by radiation from its surface.

Coolidge tube. High-voltage self-rectifying X-ray tube, the first to have a heated cathode as source of electrons.

cooling. The decay of activity of highly radioactive waste before it is processed or disposed of.

Cooper-Hewitt lamp. A form of mercury vapour ultraviolet lamp.

coordinate potentiometer. One in which two linear potentiometers carry a.c. currents 90° apart in phase, so that the resultant voltage between tappings can be adjusted both in phase and magnitude.

coplanar-grid valve. Four-electrode valve containing two control grids intermeshed with, but insulated from, each other, and exercising equal control effects on the anode

current. Used as a detector and as a high output amplifier where the anode supply voltage is limited.

copper (Cu). Bright, reddish, metallic element, at. no. 29, at. wt. 63·54, sp. gr. 8·95, m.p. 1083°C, resistivity at 0°C $1·65 \times 10^{-8}$ ohm/m, much used as electrical conductor. Nickel-iron wires with a copper coating are frequently used for *lead-ins* through glass seals, and so form vacuum-tight joints. Copper-oxide photosensitors are red-sensitive in the backwall type, but the copper-oxide rectifier, much used in d.c. electric power generation, is being superseded by selenium, germanium, and silicon types, except for small rectifying instruments, where the characteristic curves are most suitable. ^{64}Cu is a mixed radiator, of half-life 13 h.

copper circuit. One suitable for signals with d.c. component, e.g., video signals. See **direct-current restoration.**

copper-oxide rectifier. One depending on the potential barrier between copper and copper oxide or copper sulphide. Small copper-oxide rectifiers have suitable characteristics for a.c. indicating meters.

copy. The transfer of information from one register to another in a computer without changing the information in the original register.

core. (1) Magnetic material increasing inductance of a coil; it may be wound from tape, moulded from ferro-particles and binder, or of punched laminations; it may be a complete magnetic circuit (divided to contain coil) or simply a rod. (2) In an atom the nucleus and all complete shells of electrons. (3) That part of a nuclear reactor containing the fissile material, dispersed or in cans.

Coriolis effects. Those arising on the surface of the earth because of rotation, resulting in tangential and centrifugal acceleration, particularly in liquids and gyroscopes. Similarly with particles. Allowed for in calculating paths of satellites.

corkscrew rule. That which relates the direction of the magnetic field to the current direction in a conductor. The corkscrew is driven in direction of current and sense of rotation of the handle gives direction of magnetic field.

corner. See **bend.**

corner admittance. In coaxial-line resonators, the admittance which a reflex klystron presents to the line.

corner reflector. Metal structure, of three mutually perpendicular sheets, for returning radar signals.

cornucopia. Cross-section used for horns (named after traditional horn of plenty).

corona. Phenomenon of air breakdown when electric stress at the surface of a conductor exceeds a certain value. At higher values, stress results in luminous discharge. See **critical voltage.**

corona tube. One utilizing a corona effect, as in voltage reference or stabilizer tubes.

corona voltmeter. Instrument for measuring high voltages by observing the conditions under which a corona discharge takes place on a specially designed wire.

correction time. Time required for a controller to correct an error, and bring the controlled function within a specified range of operation.

correlation. The extent to which two variables are related. Correlation techniques are applied in many scientific and engineering fields.

correspondence principle. The principle that, in the limit of high quantum numbers, the predictions of quantum and classical physics always correspond.

corrosion voltmeter. Instrument which locates and estimates corrosion of materials by measuring e.m.f. arising from electrochemical action between material and corrosive agent.

Corti organ. Part of *cochlea* in which auditory nerve fibres terminate.

cosecant antenna. Radiator comprising a surface so shaped that the radiation amplitude pattern is described by a cosecant curve over a wide angle. This gives approximately the same intensity for near and far objects. The method also applies to electron beams.

cosine law (Lambert's). The energy emitted from a perfectly diffusing surface in any direction is proportional to the cosine of the angle which that direction makes with the normal.

cosine potentiometer. Voltage divider in which the output of an applied direct voltage is proportional to the cosine of the angular displacement of a shaft.

cosmic abundance. Quantity of an isotope (estimated) in the universe, relative to silicon or as a percentage of the whole.

cosmic noise. Radio interference due to extraterrestial phenomena, e.g., sun spots.

cosmic-ray shower. The simultaneous appearance of a number of downward directed light ionizing particles as detected by a cloud chamber.

cosmic rays. Highly penetrating rays from outer space; according to Millikan and others, *primary cosmic rays* consist approximately as follows:

two-thirds: protons
one-third: alpha-particles 90 %
heavier nuclei 10 %.

Collision with atmospheric particles results in *secondary cosmic rays* and particles of many kinds, including neutrons, mesons, and hyperons.

cosmic-rays altitude effect. This gives a maximum for cosmic-ray intensity where atmospheric pressure is about 4·5cm Hg.

cosmic rays — longitude and magnetic latitude effects. These give rise to cosmic-ray intensity variation due to earth's magnetic field.

cosmotron. Large proton-synchrotron, using frequency modulation of an electric field, which accelerates the protons up to energies greater than 1 BeV.

Cotton balance. An instrument for measuring the intensity of a magnetic field by finding the vertical force on a current-carrying wire placed at right angles to the field.

Cotton-Mouton effect. Effect occurring when a dielectric becomes double-refracting on being placed in a magnetic field H. The retardation δ of the ordinary over the extraordinary ray in traversing a distance l in the dielectric is given by $\delta = C_m \lambda l H^2$, where λ is the light wavelength and C_m is the Cotton-Mouton constant.

coulomb. MKSA unit of electric charge, realized by one ampere passing a point in a circuit or across a surface in 1 sec. Previous international coulomb is that which deposits 0·00111800 g. of Ag when passed through defined solution $AgNO_3$ in water in 1 sec. Equals charge on $1/1·601864 \cdot 10^{-19}$ electrons.

coulomb energy. Fraction of binding energy arising from simple electrostatic forces between electrons and ions.

coulomb force. Electrostatic attraction or repulsion between two charged particles.

coulomb potential. One calculated from Coulomb's inverse-square law and from known values of electric charge. The term is used particularly in nuclear physics to indicate that component of the potential energy of a particle which varies with position in accordance with an inverse-square law and should be contrasted with, e.g., *Yukawa potential.*

coulomb scattering. Scattering of particles by action of coulomb force.

Coulomb's law. Fundamental law which states that the force of attraction or repulsion between two point charges is proportional to the product of their charges and inversely proportional to the square of the distance between times the permittivity of the medium between them. If $Q_1 \, Q_2$ are point charges at distance d apart, force is

$$F \propto \frac{Q_1 \, Q_2}{\varepsilon d^2},$$

where ε = relative permittivity of medium.

Coulomb's law for magnetism. The force between two isolated point magnetic poles (theoretical abstractions) would be proportional to the product of their strengths and inversely proportional to the square of their distance apart times the permeability of the medium between them $F \propto \dfrac{M_1 \, M_2}{\mu \, d^2}$,

where M_1 and M_2 are the strengths of the two poles, d is their distance apart, and μ is the relative permeability of the medium.

coulometer. Voltameter or electrolytic cell—especially when designed for use in measurement of the quantity of electricity passed.

count. Summation of photons or ionized particles with a counting tube, which passes pulses to counting circuits, including dividers. See **background**.

count ratemeter. One which gives a continuous indication of the rate of count of ionizing radiation, e.g., for radiac survey. Also **ratemeter.**

counter. (1) Equipment for counting individual events which can be detected, e.g., in a *G-M counting tube* or a *boron chamber* for neutrons. (2) Loosely a complete counting assembly, e.g., counter, scaler and register.

See also:

alpha-	gas-flow-
anti-coincidence-	Geiger-Müller-
Cherenkov-	proportional-
coincidence-	radiation-
crystal-	ring-
electro-	scintillation-
mechanical-	wall effect
end-window-	well-.

counter efficiency. Ratio of counts recorded by counter to number of incident particles or photons reaching detector. Counts may be lost due to (i) absorption in window, (ii) passage through detector without initiating ionization, (iii) passage through detector during *dead time* following previous count.

counter electromotive force. See **back electromotive force.**

counter life. The total number of counts a nuclear counter can be expected to make without serious deterioration of efficiency.

counter range. See **start-up procedure.**

counter recovery time. The minimum time between the start of a recorded pulse and a subsequent one which attains a specified percentage of the amplitude of the first.

counter tube. (1) One for detecting ionizing radiation by electric discharge resulting from Townsend avalanche and operating in proportional or Geiger region. (2) Loosely applied to individual scaling tube (e.g., dekatron) in complete counting assembly.

counterpoise. A network of conductors placed a short distance above the surface of the ground but insulated from it, and used for the earth connection of an antenna. It will have a large capacity to earth and serves to greatly reduce the earth current losses that would otherwise take place. Also **artificial earth, capacity earth, counterpoise antenna.**

coupled-circuit effect. Property, exhibited by any two resonant circuits coupled by mutual reactance greater than a critical value, of showing maxima of response at two frequencies, neither of which coincides with resonant frequencies of the separate circuits.

coupled circuits. Two or more circuits electrically connected so that they react on each other, leading to energy transfer.

coupled oscillator. Valve in which positive feedback from output circuit to input circuit by mutual inductance is sufficient to initiate or maintain oscillation.

coupling. Interaction between different systems or different components of a system.

C*

See also:

back-	j.j.-
battery-	line-
beam-	loop-
capacitance-	parafeed-
cathode-	paraphase-
choke-	reactance-
critical-	resistance-
cross-	resistance-
direct-	capacitance-
directional-	Russell-Saunders-
electron-	Schwinger-
electrostatic-	tight-
halo-	transformer-
inductance-	tuned-anode-
inductive-	undercoupling-
intermediate-	valve-.
intervalve-	

coupling capacitor. Any capacitor for coupling two circuits, particularly that for coupling antenna to a transmitter or receiver.

coupling coefficient or **factor.** Ratio of total effective positive (or negative) impedance common to two resonant circuits to geometric mean of total positive (or negative) reactances of two separate circuits.

coupling coil. One whose inductance is a small fraction of the total for circuit of which it forms a part; used for inductive transfer of energy to or from the circuit.

coupling element. The component through which energy is transferred in a coupled system.

coupling resistance. Common resistance between two circuits for transference of energy from one circuit to the other.

coupling spin-orbit. Interaction between the spin and the orbital angular momentum of an electron.

coupling transformer. A transformer used as a coupling element.

course error. See **drift angle.**

course line. Horizontal projection of intended path.

course-line deviation. The angular difference between track and course.

course-line indicator. Cross-pointer navigational instrument registering *course error.*

course push (or **pull**). Error on course indicator of aircraft produced by polarization error in received signal.

covalent bond. An ordinary chemical bond between atoms in which each provides one electron to a shared pair. When one atom provides both electrons it is said to be a **dative bond.**

crab angle. Angle between track and direction in which vessel or aircraft is heading. See **leeway.**

crack detector. An electromagnetic device for detecting flaws by the gathering of fine magnetic powder along the flaw lines in an iron specimen when magnetized, or by the reflection of ultrasonic waves.

crater lamp. Discharge tube so designed that a concentrated light source arises in a crater in a solid cathode.

crest factor. See **peak factor.**

crest voltmeter. See **peak voltmeter.**

Crestatron. TN for travelling-wave tube, using the beating-wave principle.

crispening current. One which uses non-linear means to improve definition in a reproduced picture.

crit. See **critical mass.**

crith. Unit of mass, that of 1 litre of hydrogen at STP.

critical angle. (1) The angle of radiation of a radio wave which will just not be reflected from the ionosphere. (2) See **total internal reflection.**

critical coupling. That between two circuits or systems tuned to the same frequency, which gives maximum energy transfer without *overcoupling* (q.v.).

critical damping. Damping in an oscillatory electric circuit or in an oscillating mechanical system (such as the movement of an indicating instrument) just sufficient to prevent free oscillations from arising (ringing).

critical field. In the case of a *magnetron*, the smallest steady magnetic flux density (at a constant anode voltage) which would prevent an electron (assumed to be emitted at zero velocity from the cathode) from reaching the anode. It can also mean the magnetic field applied to a conductor below which the *superconducting* transition occurs at a given temperature. Also called **cutoff field.**

critical frequency. (1) Frequency of a radio wave which is just sufficient to penetrate an ionized layer in the upper atmosphere. (2) Lowest frequency which can be propagated as a periodic wave motion in a waveguide. It will have infinite guide wavelength and phase velocity. The value depends upon the guide dimensions and mode. Also **cutoff frequency.**

critical mass. The minimum mass of fissionable material which can sustain a chain reaction. Also **crit.**

critical potential. A measure of the energy (in electron-volts) required to ionize a given atom or raise it to an excited state.

critical size. The minimum size for a nuclear reactor core of given configuration.

critical voltage. That which, when applied to a gas-discharge tube, just initiates discharge.

critical wavelength. Free space wavelength corresponding to *critical frequency* for waveguide.

criticality. State in nuclear reactor when multiplication factor for neutron flux reaches unity and external neutron supply is no longer required to maintain power level.

Crookes tube. Original gas-discharge tube, illustrating striated positive column, Faraday dark space, negative glow, cathode glow, etc.

cross bombardment. A method of identification of radioactive nuclides through their production by differing reactions.

cross close-up. One which gives alternately the heads of two persons facing each other.

cross coupling. Undesired transfer of interfering power from one circuit to another by induction, leakage, etc.

cross cut, intercut. Cutting so that sections of a scene are reproduced alternately.

cross-fade. To fade in one channel while fading out another in order to substitute gradually the output of one for that of the other, e.g., to create the impression of a change of scene. Also **dissolve, mix.**

cross-field multiplier. One where a feedback voltage representing the product restores to zero deflection a beam simultaneously deflected by electric and magnetic fields, both representing a magnitude to be multiplied.

cross-hatch pattern. A test pattern of vertical and horizontal lines used on a television picture tube.

cross magnetizing. The effect of the armature reaction on the magnetic field of a current generator.

cross modulation. Impression of the envelope of one modulated carrier upon another carrier, due to non-linearity in the medium transmitting both carriers. See **Luxemburg effect.**

cross neutralization. A method of neutralization in push-pull amplifiers.

cross-over. In an electron lens system, location where streams of electrons from the object pass through a very small area, substantially a point, before forming an image.

cross-over area. The point at which the electron beam comes to a focus inside the accelerating anode of a cathode-ray tube.

cross-over frequency. The frequency in a two-channel loudspeaker system at which the high and low frequency units deliver equal acoustic power; alternatively it may be more generally applied to electric dividing networks when equal electric powers are delivered to each of adjacent frequency channels.

cross-over network. Same as **dividing network.**

cross product. Same as **vector product.**

cross-section. The measure of the probability of a particular process. The value of the cross-section for any particular process will depend upon the element under bombardment and upon the nature and energy of the bombarding particles. For example, if A is the *capture cross-sectional area* of an element for neutrons and a volume containing n atoms of the element is exposed to a uniform beam (of sectional area a) of neutrons, then the number of neutrons captured is $\dfrac{nA}{a} \times N$, where N is total number of neutrons in the beam.

cross-talk. Interference between two telephone channels arising from coupling between them.

crossbar transformer. Coupling device between a coaxial cable and a waveguide, the latter having a short transverse rod, to the centre of which the central conductor is connected.

crossfire. Interference of signals in one circuit with those in another circuit.

'crown of thorns' tuning. Tuning of cavity magnetrons involving changing inductance of cavities by the introduction of conducting rods along their axes; cf. *'cookie cutter' tuning.*

crushing. Distortion of contrast in a reproduced picture by diminishing contrast at upper or lower ends of contrast scale (black-white). See **gamma** (4).

cryogenic gyro. One depending on electron spin in atoms at very low temperature.

cryogenics. The study of material at very low temperature.

cryometer. A thermometer for measuring very low temperatures.

cryosar. Low-temperature germanium switch with on-off time of a few nanoseconds.

cryostat. Low-temperature thermostat.

cryotron. Miniature electronic switch, operating in liquid helium, consisting of a short wire wound with a very fine control wire. When a magnetic field is induced *via* the control wire, the main wire changes from superconductive to resistive. Used as a memory, characterized by exceedingly short access time. See **superconductivity.**

crystal. (1) Solid substance showing some marked form of geometrical pattern, to which certain physical properties, angle and distance between planes, refractive index, etc., can be attributed. (2) Piezoelectric element, shaped from a crystal in relation to crystallographic axes, e.g., quartz, tourmaline, Rochelle salt, ammonium dihydrogen phosphate, to give *facets* to which electrodes are fixed or deposited for use as transducers or frequency standards. See **biaxial-, ideal-, ionic-, piezoelectric-, pulling by-, twin-, uniaxial-, zero-cut-.**

crystal analysis. Usually *X-ray crystallography* (q.v.).

crystal anisotropy. In general, the directional variations of any physical property, e.g., elasticity, thermal conductivity, etc., in crystalline materials. Leads to existence of favoured directions of magnetization, related to lattice structure in some ferromagnetic crystals.

crystal axes. The axes of the natural co-ordinate system formed by the crystal lattice. These axes are perpendicular to the natural faces for many common crystals.

crystal burn-out. Permanent structural damage due to exposure to excessive RF power.

crystal cell. Kerr cell using a quartz or other suitable crystal in place of the more usual nitrobenzine.

crystal control. Control of frequency of a radio transmitter by an oscillating piezoelectric crystal, usually of quartz.

crystal-controlled transmitter. A radio transmitter whose carrier frequency is controlled directly by a crystal oscillator.

crystal counter. One in which an operating pulse is obtained from a crystal when made conducting by an ionizing particle or wave.

crystal cut. A plane section of a crystal, the cut being classified by its orientation relative to a *crystal axis.*

crystal cutter. Device used for cutting grooves in gramophone recording, a piezoelectric crystal being the means of initiating the mechanical displacements of the stylus.

crystal detector. Demodulator incorporating a stable contact, e.g., 'cat's whisker' between semiconductors, e.g., germanium, silicon.

crystal diamagnetism. The property of negative susceptibility shown by silver, bismuth, etc.

crystal diode. Crystal of high purity, to which is attached a fine wire, yielding a high-frequency rectifier of low capacitance. Larger units are used for power rectification in traction.

crystal drive. System in which oscillations of low power are generated in a crystal oscillator, being subsequently amplified to a level requisite for transmission.

crystal electrostriction. The dimensional changes of a dielectric crystal under an applied electric field; cf. *magnetostriction.*

crystal filter. Band-pass filter in which piezoelectric crystals provide very sharp frequency-discriminating elements, especially for group modulation of multi-channels over coaxial lines.

crystal gate receiver. Supersonic heterodyne receiver in which one (or more) piezoelectric crystals is included in the intermediate-frequency circuit, in order to obtain a high degree of selectivity.

crystal goniometer. See **goniometer.**

crystal growing. Technique of forming semiconductors by extracting crystal slowly from molten state. Also called **crystal pulling.**

crystal indices. See **Miller indices.**

crystal lattice. Three-dimensional repeating array of points used to represent structure of crystal, and classified into 14 groups by Bravais.

crystal loudspeaker. One depending on vibrations excited in a *bimorph* piezocrystal which drives, e.g., a small cone diaphragm, acting as a *tweeter.*

crystal microphone. One depending on the generation of an e.m.f. in a bimorph crystal when flexed by the varying pressure in an applied sound wave. Also **piezoelectric microphone.**

crystal mixer. Frequency changer in a waveguide system, fed from a local oscillator.

crystal oscillator. Valve or transistor oscillator in which frequency is held within very close limits by rapid change of mechanical impedance (coupled piezoelectrically) when passing through resonance.

crystal oven. Chamber containing a piezoelectric crystal; maintained at very constant temperature to ensure constancy of frequency of resonance.

crystal pickup. Piezoelectric transducer, producing an e.m.f. due to mechanical drive arising from vibration. Widely used for record reproduction and vibration measurements.

crystal pulling. See **crystal growing.**

crystal receiver. One using a simple crystal demodulator and hence no external supplies.

crystal rectifier. One which depends on differential conduction in semiconductor crystals, suitably 'doped', such as Ge or Si.

crystal spectrometer. An instrument using a crystal to measure the wavelengths of X-rays, γ-rays, and particle waves.

crystal triode. Early name for **transistor.**

crystallization. Slow formation of a metal crystal by withdrawing it from the surface of a melt, particularly Cu, Se, Ge.

crystallogram. Diffraction pattern from a crystal, whereby its structure can be calculated.

crystallography. The scientific study of crystal structure.

crystalloid. Crystalline substance which dissolves in water and is not a colloid. (Obs.)

Cubex Steel. TN for silicon-iron alloy with special crystal orientation developed in manufacture.

cubical antenna. One with radiating elements arranged as sides of adjacent cubes.

cumulative dose. Integrated radiation dose resulting from repeated exposure.

cumulative excitation. Successive absorption of energy by electrons in collision, leading to ionization.

cumulative-grid detector. See **grid-leak detector.**

cuprous oxide. Substance formed by partial oxidation of copper, used in copper oxide rectifiers.

curie. Unit of radioactivity, defined as $3 \cdot 700 \cdot 10^{10}$ disintegrations/s, roughly equal to the activity of 1 g of radium.

Curie balance. A torsion balance for measuring the magnetic properties of non-ferromagnetic materials by the force exerted on the specimen in a non-uniform magnetic field.

Curie point (or **temperature**). (1) Temperature at which a ferromagnetic material loses its magnetization. Also **magnetic transition temperature.** (2) Temperature (*upper Curie point*) above which a ferroelectric material loses its polarization. (3) Temperature (*lower Curie point*) below which some ferroelectric materials lose their polarization.

Curie-Weiss effect. Change of ferromagnetism to paramagnetism at the Curie point.

Curie's law. The susceptibility of a paramagnetic material is inversely proportional to its absolute temperature.

curium (Cm). Man-made radioactive element, at. no. 96, produced from americium. There are several long-lived isotopes (up to $1 \cdot 7 \times 10^7$ years half-life), all alpha-emitters.

current. Rate of flow of charge in a substance, solid, liquid, or gas. Conventionally it is opposite to the flow of (negative) electrons, this having been fixed before the nature of the electric current had been determined. Practical unit of current is the *ampere*, equal to $6 \cdot 25 \cdot 10^{18}$ electrons/s.

See also:

alternating-	magnetizing-
anode-	Maxwell's
anode-ray-	circulating-
beam-	polarization-
circulating-	pre-conduction-
collector-	primary-
control-	pulsating-
dielectric-	quiescent-
direct-	reactive component of-
feed-	reverse (or inverse)-
filament-	reverse grid-
gas-	reversible absorption-
gate-	saturation-
gun-	sinusoidal-
hole-	telluric-
induced-	thermionic-
ionic-	transfer-
ionization-	unidirectional-
Leduc-	Zener-.
light-	

current amplification. Ratio of output current to input current of valve amplifier, magnetic amplifier, transistor amplifier, or photomultiplier, often expressed in dB.

current antinode. A point of maximum current in a standing-wave system along a transmission line or aerial.

current attenuation. Ratio of output to input currents of a transducer, expressed in dB.

current density. Current flowing per unit cross-sectional area of conductor or plasma; expressed in amperes (or milliamperes)/cm².

current feed. Delivery of radio power to a current maximum (loop or antinode) in a resonating part of an antenna.

current feedback. (1) In thermionic circuits, a negative feedback voltage proportional to the output current. Also **voltage feedback.** (2) A feedback current in magnetic amplifiers and transistors.

current gain. In a transistor, ratio of ultimate change in collector current at constant voltage, resulting from a change in emitter current.

current generator. Two terminals from which a constant current flows independently of the potential difference, internal admittance being zero. Approximated in screened-grid and pentode valves.

current limiter. A component which sets an upper limit to the current which can be passed.

current margin. In a relay, difference between steady-state currents corresponding to values used for signalling and for just operating the relay.

current node. Opposite of *current antinode*, a point of zero electric current.

current regulator. Circuit employed to control the current supplied to a unit.

current (or voltage) resonance. Condition of a circuit when the magnitude of a current (or voltage) passes through a maximum as the frequency is changed through resonance. Obs. syntony or **tuning** (Lodge).

current saturation. Condition when the anode current in a triode valve has reached its maximum value.

current transformer. (1) One designed to be connected in series with circuit, drawing predetermined card into *upper* and *lower* areas. instrument when the line current is transformed down to changed range.

curtain. Shield, usually made of cadmium, which absorbs neutrons.

curtain antenna Large number of vertical radiators or reflectors in a plane, as in original beam system.

curtate. Horizontal division of a standard punched card into *upper* and *lower* areas.

Curtis winding. The winding of low-capacitance and low-inductance resistors in which the wire is periodically reversed.

curvature of field. Said of an optical system in which the surface of maximum definition of the image is not plane but curved.

cut. (1) In filming, sudden ending of a shot, either in the original *take* or *recording*, or in the subsequent *editing*. (2) Commencement and the groove in making a disk recording. (3) Removal of a section of a programme or a recording. (4) Proportion of input material to any stage of an isotope separation plant, which forms useful product. Also called **splitting ratio.**

cut key. See censor key.

cutoff field. Same as **critical field.**

cutoff frequency. The theoretical frequency is that at which the attenuation constant changes from zero to a positive value (or vice versa). See also **alpha cutoff, critical frequency.**

cutoff voltage. Negative value which must be applied to the grid of a thermionic valve to reduce the anode current substantially to zero. Approximately equal to anode voltage divided by amplification factor.

cut-out. Off-switch operated automatically if safe operating conditions are not maintained, e.g., water flow cut-out.

Cutler feed. Resonant cavity at end of waveguide used to feed mirror antennae.

cutting head. Electrodynamical or electromechanical transducer which deflects the cutting stylus in *direct* (lacquer) disk recording. See also **crystal cutter.**

Cx-band. See **X-band.**

cybernetics. Application of information theory to comparison of mechanical or electrical controls with biological equivalents.

cycle. (1) Complete series of changes in a periodically varying quantity, e.g., an alternating current, during one period. (2) Sequence of computer operations which continues until a criterion is reached for stoppage. (3) Interval which must be allowed for minimum (*minor*) or maximum (*major*) length of computer *words*, or a recurrence in a rotating (cyclic) store.

cycle criterion. The number of times in computer programming that a particular cycle must be performed.

cycle reset. In computer programming, setting a cycle index back to its original value.

cycles per second (c/s). Number of complete repetitions, in 1 s of periodic disturbance. Replaced as standard U.K. term by **hertz.**

cycling. Periodic change imposed on controlled variable or a function by a controller. See also **hunting.**

cyclogram. Graph showing characteristics of negative resistance oscillator and constructed on isocline diagram. It forms a closed loop for any stable oscillator, the loop being a circle when the output is a sine wave.

cyclograph. An electron optical system in which the electron beam moves in two perpendicular directions.

cyclophon. Tube which uses the fundamental principle of electron beam switching.

cyclotron. Machine in which positively-charged particles are accelerated in a spiral path within dees in a vacuum between poles of a magnet, energy being provided by a high-frequency voltage across the dees.

cyclotron frequency. In a magnetron, frequency with which an electron orbits in magnetic field in the absence of electric field.

cylindrical wave. One where equiphase surfaces form coaxial cylinders.

cymograph. See **kymograph.**

cymometer. Original (Lodge) wavemeter, which measured wavelength when sliding tubes were adjusted for resonance.

cymoscope. (Obs.) Any detector of electric oscillations.

D

D display. *C display* (q.v.) in which targets appear as bright spots, elongated vertically approximately in proportion to the range.

D-layer. The lowest layer or region of absorbing ionization up to 90 km above the earth's surface considered to exist as a consequence of particle radiation of hydrogen bursts from the sun. It brings about complete inhibition of short-wave communication but some improvement in long-wave communication. See also **Dellinger fade-out.**

d-levels. See **diffuse series.**

d-state. That of an orbital electron when the orbit has angular momentum 2 Dirac units.

damped oscillation. One which dies away rapidly, having been excited by an impulse, as in breaking a current by opening a contact. See **continuous oscillations, decay factor.**

damped waves. Electromagnetic waves radiated from a damped oscillatory circuit, excited by a timed spark. (Obs.)

damper. Vibration-absorbing pad or isolator for reducing the transmission of vibrational energy from a disturbing source.

damping. Extent of reduction of amplitude of oscillation in an oscillatory system, due to energy dissipation, e.g., friction and viscosity in mechanical system, and resistance in electrical system. With no supply of energy, the oscillation dies away at a rate depending on the *degree of damping.* The effect of damping is to increase slightly period of vibrations. It also diminishes sharpness of resonance for frequencies in the neighbourhood of natural frequency of vibrator. See **logarithmic decrement.**

damping factor. See **decay factor.**

danger coefficient. Expression of change in reactivity when sample is inserted into reactor.

Daniell cell. Primary cell with zinc and copper electrodes, the zinc rod being inserted in sulphuric acid contained within a porous pot which is itself immersed in a copper pot containing copper sulphate solution.

daraf. Unit of elastance, the reciprocal of capacitance in farads. (*Farad* backwards).

dark current. Residual current, depending on temperature, in a photocell when there is no incident illumination. Cf. *light current.*

dark resistance. Resistance of a selenium or other photocell in the dark.

dark space. See **anode-, Aston-, cathode-, Faraday-.**

dark spot. Reproduced spot arising from cloud of electrons released from a mosaic in a television camera. See **shading** (1).

dark-trace screen. One which yields a dark trace under electron beam bombardment.

d'Arsonval galvanometer. A moving-coil galvanometer in which a light rectangular coil is suspended in the strong field between the poles of a horseshoe magnet. A similar movement is used in commercial ammeters and voltmeters.

dash pot. A damping device using the viscous property of a fluid to minimize the oscillatory motion of a system.

data. All the *operands* and results of computer operations directed by the detailed *instructions* comprising the *programme.*

data-handling capacity. (1) For a telemetering system, the maximum amount of information which can be transmitted and received over a given link. (2) The number of bits of information which can be stored at any one time in a computing system and the rate at which these may be fed to the input.

data-handling system. Automatic or semi-automatic equipment for collecting, receiving, transmitting and storing numerical data. It may be handled continuously (as analogue or position signals) or in discrete steps (as digital or binary signals). If necessary the system should be able to carry out mathematical operations on the stored data and to record the results.

data hold. Computer equipment for reconstructing a continuous function (analogue) from sampled data by extrapolation.

data processing. Various operations of adding, subtracting, etc., performed by an electronic computer on data supplied, either as punched cards or tape, or on magnetic tape, according to a programme. See **automatic-.**

data reduction. Rearranging or reducing the quantity of the experimentally obtained data to permit the more rapid extraction of conclusions.

data storage. See **memory capacity.**

Datamatic 1000. U.S. commercial electronic digital computer and data-handling system with a very large storage capacity and fast operation.

dative bond. See **covalent bond.**

daughter product. Of a given nuclide, any nuclide that originates from it by radioactive decay. Also **decay product.**

Davisson chart. Skew coordinate system for showing logarithm of the emission current density against logarithm of the cathode-power dissipation as a straight line.

Davisson-Germer experiment. Demonstration of wavelike diffraction patterns from electrons passing through a nickel crystal. Confirmed independently by G. P. Thomson, who shared a Nobel Prize with Davisson for this work.

Davo. Speech-synthesizing computer developed at MIT.

Day modulation. Transmission of two quadrature carriers, separately modulated with different signals, thus doubling the use of the radio channel.

de-accentuator. Circuit arrangement for de-emphasis in frequency-modulation reception.

dead. Said of a circuit from which there is no output when a signal is applied to the input.

dead band. (1) Permitted tolerance in controlled function without operation of automatic control. (2) Range of signal input to a magnetic amplifier which does not effect a change in the output.

dead-beat. Said of an instrument or other oscillating system when it is critically damped, i.e., when, after receiving some disturbing impulse, it takes up its final position without oscillations and in minimum time.

dead-end effect. Increase in effective resistance of a coil because of currents in unused section arising from self-capacitance. Obviated by short-circuiting unused turns.

dead-end switch. Multi-point switch arranged to short-circuit the unused end turns of an inductance coil.

dead load. See dummy load.

dead reckoning (derived from *deduced reckoning*). Estimate of position in navigation, made from the last known location.

dead room. See anechoic room.

dead short. A short-circuit of extremely small resistance.

dead spot. A region where the reception of radio transmissions over a particular frequency range is extremely weak or practically zero.

dead time. (1) Time after ionization during which further operation is not possible; reduced by a *quench*, as in G-M counters. When dead time of detector is variable subsequent circuits may incorporate a fixed electronic dead time. Also called **insensitive time.** (2) In computers, time delay in a controlling system between error signal and response.

dead-time correction. That applied to the observed rate in a radiation counter to allow for probable events during the dead time.

dead zone. Range of error signal value, within which controlling action does not take place.

deathnium centre. Crystal lattice imperfection in semiconductor at which (it is believed) electron hole pairs are produced or recombine.

de Broglie wavelength. That assignable to any particle because of its momentum in motion, i.e.,

$$\lambda = h/mv,$$

where λ = wavelength
h = Planck's constant
v = velocity
m = relativistic mass.

debug. To remove all defects from a programme or from operation of a system.

debunching. Tendency for a beam of electrons

or a velocity-modulated beam of electrons to spread because of their mutual repulsion. In gas cathode-ray tube, the maintenance of focus arises from relatively static positive ions in the path of beam counteracting debunching. See **buncher, bunching.**

Debye length. Maximum distance at which coulomb fields of charged particles in a plasma may be expected to interact.

Debye unit. The unit of dipole moment, i.e., 10^{-18} statcoulomb cm.

decade. Any ratio of 10 : 1. Specifically the interval between frequencies of this ratio.

decade box. A resistance (capacitance or inductance) box divided into sections so that each section has ten switched positions and ten times the value of the preceding section. The switches can therefore be set to any integral value within the range of the box.

decade scaler. One having a scaling factor of 10.

decametric waves. The radio waveband from 10 to 100 m.

decay. See **alpha-, beta-, double beta-, dual beta-, Fermi-.**

decay cable. A length of concentric cable between a test object and surge generator to delay the arrival of the surge.

decay constant. See **disintegration constant.**

decay factor. That which expresses rate of decay of oscillations in damped oscillatory system, given by natural logarithm of ratio of two successive amplitude maxima divided by the time interval between them. Calculated from ratio of resistance coefficient to twice mass in a mechanical system, and ratio of resistance to twice inductance in an electrical system. Also called **damping factor.** See **logarithmic decrement.**

decay law. When the phenomenon is random, and a change depends on a momentary condition, the law of decay is *exponential*, e.g., sound intensity in an enclosure, activity in a radioactive substance, charge into and out of a capacitor, and rise and fall of a current in an inductor (air core).

decay product. See **daughter product.**

Decca. High-precision radio-navigation system depending on detected interference between signals locked in phase from two or more stations. Resulting pattern on map is hyperbolic.

decelerating electrode. One which is intended to reduce the velocity of electrons.

deci-. Prefix for one-tenth (10^{-1}). Symbol *d*.

decibel. Unit of power-level difference, measured by $10 \log_{10} W_2/W_1$, where W_2 is a power controlled by W_1. Used as a measure of response in all types of electrical communication circuits. Abb. **dB.** See **bel.**

decibel meter. Meter which has a scale calibrated approximately uniformly in logarithmic steps and labelled with decibel units; used for determining the power levels in communication circuits, relative to a datum power level, now 1 mW in 600 ohms.

decilog. Unit of ratio, one-tenth of logarithm

to base-10. Same as **decibel,** but universal in application.

decimetric waves. The radio waveband from 10 cm to 1 m.

decineper. Unit of voltage and current attenuation in lines and amplifiers, of magnitude one-tenth of the *neper,* defined by $d = 10 \log_e(x_1/x_2)$, where d is number of decinepers, x_1 and x_2 are currents (voltages or acoustic pressures). In properly terminated networks, one decineper equals 0·8686 dB.

deck. Series of units mounted on common horizontal platform, or one tier of multibank assembly.

declination. (1) Angle between the magnetic meridian and the geographic meridian at a point on the earth's surface. (2) The angular distance between a celestial body and the celestial equator, measured in degrees north or south.

declinimeter. An apparatus for determining the direction of a magnetic field with respect to astronomical or survey coordinates.

decode. Extraction of a signal from a pulse-code modulation. Any mechanism which *reads out* coded signals into their normal expression.

decoherer. Device for restoring a *coherer* to its sensitive, non-conducting condition after the arrival of a signal; also **anti-coherer.** See **tapper.**

Decometer. Interpolating phase indicator to locate the path of an aircraft within a *Decca* (q.v.) lane.

decomposition voltage. The minimum voltage which will cause continuous electrolysis in an electrolytic cell.

decontamination. Removal or neutralization of radioactive bacteriological or chemical contamination.

decontamination factor. Ratio of initial to final contamination for a given process.

decoupling filter. Simple resistor-capacitor section(s), which decouple feedback circuits and so prevent oscillation or *motor-boating* (relaxation oscillation).

decrement. (1) Ratio of successive amplitude in a damped harmonic motion. See **logarithmic-.** (2) The amount by which a variable is reduced during a mathematical operation; cf. *increment.*

Dectra. Long-range long-wave navigational aid, using a phase comparison of radio waves from fixed shore transmitting stations, giving a pattern of hyperbolic position lines and drawing a pictorial plot of a course on a chart.

de-emphasis. Correction of frequency response distortion originated by *pre-emphasis* by introducing a complementary frequency response; necessary in a form of frequency modulation. Also **post-emphasis.**

de-energize. To disconnect a circuit from its source of power.

dees. Pair of hollow half-cylinders, i.e., D-shaped, in the vacuum of a cyclotron for accelerating charged particles in a spiral, a high-frequency voltage being applied to them in antiphase.

defect. Lattice imperfection which may be due to the introduction of a minute proportion of a different element into a perfect lattice, e.g., indium into germanium crystal to form an extrinsic semiconductor for a transistor. A 'point' defect is a 'vacancy' or an 'interstitial atom' while a 'line' defect relates to a dislocation in the lattice.

definition. Extent to which a television or phototelegraphic system is capable of reproducing the detail of the transmitted image; defined as the number of picture elements, or the number of scanning lines, into which the picture is divided. For TV in U.K., 405; in U.S., 525; Continent and UHF stations in U.K., 625 horizontal lines.

deflagrator. Primary cell (of low internal resistance) capable of high peak output currents.

deflection sensitivity. (1) Ratio of displacement of spot or angle of an electron beam to the voltage producing it. (2) Ratio of displacement of spot or angle of beam to the magnetic field producing it, or the current in the deflecting coils. Also applies to galvanometer or other instrument for measuring current, voltage, or quantity of electricity.

deflector coil(s). Coil(s) so arranged that a current passing through produces a magnetic field which deflects the beam in a cathode-ray tube using magnetic deflection. Usually applied around the neck of the tube.

deflector plates (or deflecting electrodes). Electrodes so arranged in a cathode-ray tube that the electrostatic field produced by a p.d. deflects the beam.

deformation potential. Potential barrier formed in semiconducting materials by lattice deformation.

degassing. Removal of last traces of gas from valve envelopes, by pumping or *gettering,* with or without heat, the former with eddy currents, electron bombardment, or baking.

degaussing. Neutralization of the magnetization of a mass of magnetic material, e.g., a ship, by an encircling current-carrying conductor. Used to avoid the detonation of magnetic mines.

degeneracy. The condition in a resonant system when two or more modes have the same frequency.

degenerate. Said of particles or electrons when their energy states are low and equal; also when an atom is stripped of all electrons, as in a star.

degenerate electron gas. An electron gas which is far below its Fermi temperature, so that a large fraction of the electrons completely fill the lower energy levels and have to be excited out of these levels in order to take part in any physical processes.

degenerate gas. (1) Gas which is so concentrated, e.g., electrons in the crystal lattice of a conductor, that the Maxwell-Boltzmann law is inapplicable. (2) Gas at very high temperature in which most of the electrons are stripped from the atoms.

degenerate semiconductor. One in which the conduction approaches that of a simple metal.

degeneration. Same as **negative feedback**.

degenerative amplifier. One for wide frequency bands and minimized amplitude distortion, these being obtained by reversed or negative retroaction (feedback).

degradation. Loss of energy of motion solely by collision. Deliberate slowing of neutrons in a reactor is *moderation*. In an isolated system the *entropy* increases.

degrees of freedom. The number of independent variables defining the state of a system.

de-ionization. Process whereby an ionized gas returns to its normal neutral condition when sources of ionization have been removed.

de-ionization grid. That in a gas-discharge tube, to assist de-ionization of plasma.

de-ionization potential. Voltage below which ionization and consequent current flow cannot take place.

de-ionization time. (1) Period necessary for the substantially complete recombination of electrons and positive ions in a gas after the removal of the ionizing agent. (2) In a gas tube, e.g., thyratron, the time for the grid to regain control after interruption of the anode current.

Dekatron. Multi-electrode neon tube in which a set of 10 electrodes function in turn on the application of voltage pulses, thus indicating the number of pulses. Such tubes in cascade constitute a visible counter of impulses in the decimal system.

del. The differential operator used in vector analysis, also known as **nabla** and written as ∇.

delay. (1) That which can be introduced into the transmission of a signal, by recording it magnetically on tape, disk, or wire, and reproducing it at a point traversed later by the record. Used in public-address systems, to give illusion of distance and to coalesce contributions from original source and reproducers. (2) In computers, unit which delays signals, e.g., by exactly one *bit*. See **acoustic-, ionization-, phase-.**

delay cable. A length of concentric cable between a test object and surge generator to delay the arrival of the surge.

delay circuit. One which delays the wavefront of a signal (telephonic or telegraphic) by acoustic tubes, recording and reproduction, or wave filters. Also transmission lines for phasing in feeders between antennae, radio receivers and transmitters.

delay distortion. Change in waveform during transmission because of non-linearity of delay with frequency, which is $d\beta/d\omega$, where β = phase delay in radians, and $\omega = 2\pi \times$ frequency in Hz.

delay equalizer. Network or wave filter which makes linear delay in radians against frequency, and so maintains the wavefront of a transmission.

delay line. (1) Column of mercury, quartz plate, or length of nickel wire, in which impressed sonic signals travel at a finite speed and which, by the delay in travelling, can act as a *store*, the signals being constantly recirculated and abstracted (*gated*) as required. (2) Real or artificial transmission line used to delay a propagated electrical signal.

delay network. Artificial line of electrical networks, designed to give a specified phase delay in the transmission of currents over a frequency band represented by speech, so that time may be allowed for switches or relays to be operated.

delay tank. Reservoir for collecting radioactive liquid waste and retaining until activity has decayed to acceptable level for disposal.

delay-time store. One which uses the circulation of impulses along a mercury tube, quartz plate, or nickel-wire transmission line, with form correction and amplifiers.

delayed automatic gain control. That operative above a threshold voltage signal in a radio receiver. Provides full amplification for very weak signals and constant output from detector for signals above the threshold. See also **automatic gain control, quiet automatic volume control.**

delayed critical. Assembly of fissile material critical only after release of delayed neutrons; cf. *prompt critical*.

delayed neutrons. Those arising from fission but not released instantaneously. Fission neutrons are all *prompt*; those apparently delayed (up to seconds) arise from breakdown of fission products, not primary fission. Such delay eases control of reactors.

delayed-speech feedback test. A test used to detect malingering by persons apparently showing symptoms of deafness. The speech of the subject is delayed by approximately 0·2 s using a tape recorder, and is fed to one ear by an earphone. When the speech is heard, this delay normally causes hesitancy, stuttering and raising of the voice.

delayed sweep. Time-base circuit in which start of deflecting waveform is delayed by a known period relative to the trigger voltage.

Dellinger fade-out. Complete fade-out (which may last for minutes or hours) and inhibition of short-wave radio communication because of the formation of a highly absorbing D-layer, lower than the regular E- and F-layers of the ionosphere, on the occasion of a burst of hydrogen particles from an eruption associated with a sun spot.

delta connection. The connection of a 3-phase electrical system such that the corresponding windings of the transformers form a triangle.

delta impulse function. Infinitely narrow pulse of great amplitude, such that the product of its height and duration is unity.

delta-matching transformer. A matching network between 2-wire transmission lines and half-wave antennae.

delta network. One with three branches all in series; a **loop or mesh.**

delta-particle. Very short-lived hyperon which decays almost instantaneously through the strong interaction.

delta-ray. Any particle ejected by recoil action from passage of ionizing particles, e.g., in a Wilson cloud chamber.

delta-ray spectrometer See **spectrometer**.

delta waves. The lower frequency brain waves (1–8 Hz).

Deltamax. TN for an orientated alloy of equal percentage iron and nickel.

demagnetization. (1) *Removal* of magnetization of ferromagnetic materials by the use of diminishing saturating alternating magnetizing forces. (2) *Reduction* of magnetic induction by the internal field of a magnet, arising from the distribution of the primary magnetization of the parts of the magnet. (3) *Removal* by heating above the Curie point. (4) *Reduction* by vibration.

demagnetization factor (*N*). Diminution factor applied to the intensity of magnetization (*I*) of a ferromagnetic material, to obtain the demagnetizing field (ΔH), i.e., $\Delta H = NI$. *N* depends primarily on the geometry of the body concerned.

demodulation. Inverse of *modulation*. Generally effected by passing the modulated carrier, or the high-frequency signal with an added carrier, through a non-linear system, so that the output currents or voltages contain difference frequencies between the carrier and side frequencies which can be extracted and reformed into the original modulating signal. Previously denoted by **detection.** Also previously applied to the reduction of the depth of modulation in a carrier when the latter is partially rectified in a high-frequency amplifier, with or without the presence of a relatively strong alien carrier.

demodulation (or detection or rectification) by leaky grid. The modulated signal is applied through a capacitor to the grid of a triode valve, making use of grid curvature, the mean current (*envelope*) passing to earth *via* a resistor, at the same time controlling the anode current.

demodulation of an exalted carrier. Same as homodyne reception.

demodulator. See **detector, thermionic rectifier**.

demountable. Said of large thermionic valves when they can be taken apart for cleaning and filament replacement, and are continuously pumped during operation.

Dempster positive-ray analysis. A technique of separating particles having different charge-to-mass ratios using a magnetic field.

denaturant. Isotope added to fissile material to render it unsuitable for military use.

densitometer. Any instrument for measuring the optical transmission or reflecting properties of a material.

density effect. The variation of mass absorption with density which occurs for particles of relativistic energies.

density modulation. Time variation in density imposed on an electron beam.

depleted uranium. Sample of uranium having less than its natural content of ^{235}U.

depletion. Reduction in the proportion of a specific isotope in a given mixture.

depletion layer. In semiconductor materials, location where mobile electrons do not neutralize the charge of the donors and acceptors taken together. Has been applied for use as an ultrasonic transducer.

depletion-layer capacitance. That relating to the depletion layer of a semiconductor diode.

depletion-layer transistor. One depending on the movement of carriers through a depletion layer. See **spacistor**.

Deplistor. Three-terminal component, which can exhibit a negative differential resistance under certain circuit conditions.

depolarization. Reduction of polarization, usually in electrolytes, but sometimes in dielectrics. In the former it may refer to the removal of the gas collected at the plates of a cell during charge or discharge.

depth dose. Ratio or percentage of the surface X-ray dose which is received within the body. See **Chambers-Imrie-Sharpe curve**.

depth finder. Radar or ultrasonic depth-sounding equipment. See **asdic**.

depth of penetration. That thickness of hollow conductor which would give the same loss for the effective current as a solid conductor of the same dimensions. See **skin depth**.

de-rating. The reduction of the maximum performance ratings of equipment when used under unusual conditions in order to maintain an adequate margin of safety.

derivative equalizer. One designed to improve the definition (resolution) in a reproduced picture by adjusting both response and phase relations.

derivative feedback. Feedback signal in control system proportional to time derivative error.

derived units. Units derived from the three fundamental units of a system by consideration of the dimensions of the quantity to be measured. See **dimensions**.

desampling. Process of extrapolation in reconstructing a continuous analogue function from samples of a variable.

desaturation. See **saturation**.

desensitizing. See **muting**.

desired value. See **index value**.

desorption. The reverse process to *adsorption*.

detail file. Data-storage file whose contents are relatively transient.

detection. See **demodulation**.

detector. (1) Original name for circuit or apparatus for making received radio signals manifest, e.g., crystal detectors, etc. Replaced by *demodulator*, which extracts the signal from a carrier with minimum distortion. (2) Circuit component concerned with *demodulation*, i.e., extracting signal which has *modulated* a wave. It is *linear* if there is exact proportionality between output voltage or current and depth of modulation, and *square-law* if output voltage or current is proportional to square of depth of modulation. It is substantially linear for small depths of modulation. (3) Device in which presence of radiation induces observable physical change.

See also:

beta-	mixer (2)
bornite-	phase-sensitive-
burst-can-	plate-bend-
carborundum-	pyron-
crack-	rectifying-
crystal-	regenerative-
grid-leak-	retarding-field-
infinite-	second-
impedance-	silicon-
ionized-gas-	thermionic-.
leak-	

detuning. Adjustment of a resonant circuit so that its resonant frequency does not coincide with that of the applied e.m.f.

Deuce. TN for first-generation electronic digital computer. (*Digital electronic universal computing engine.*)

deuterium (D). Isotope of element hydrogen, having one neutron and one proton in nucleus, symbol D, when required to be distinguished from natural hydrogen, which is both ^{1}H and ^{2}H. This heavy hydrogen is thus twice as heavy as ^{1}H, but similarly ionized in water. As a chemical, lithium deuteride is the essential element in H-bombs. See **deuteron.**

deuterium reactor. See **heavy-water reactor.**

deuteron. Charged particle, D+, the nucleus of *deuterium*, a stable but lightly-bound combination of 1 proton and 1 neutron. Obtained from *heavy water* by electrolysis, mainly used as a bombarding particle accelerated in cyclotrons. Used in *zeta* (q.v.).

deviation. (1) Difference between the instantaneous quantity and the *index* (or *desired*) *value*, e.g., difference between the magnetic north and the setting of a compass needle due to a local magnetic disturbance or between the actual value of the controlled variable and that to which the controlling mechanism is set in an automatic control system. Also **error.** (2) The error in a vessel's compass, i.e., the angle between the direction it points and the actual direction of magnetic north. A deviation table is normally prepared showing the deviation for various headings. (3) In FM radio transmission the amount the carrier increases or decreases when modulated.

deviation absorption. That of radio waves at frequencies near the critical frequency of the ionosphere.

deviation distortion. Consequence of any restriction of bandwidth or linearity of discrimination in the transmission and reception of a frequency-modulated signal.

deviation ratio. Frequency departure of a frequency-modulated carrier from its mean value (that when there is no modulation) divided by this value.

deviation sensitivity. Ratio of change in navigational-course indication to deviation from course line.

dew line. Line of radar missile warning stations along 70th parallel of latitude. (*Distant early-warning line.*)

diacritical current. The current in a coil to produce a flux equal to half that required for saturation.

diagonalizing. Practice of radiating the same programme at different times, and on different wavelengths.

diamagnetism. Phenomenon in some materials, in which the susceptibility is negative, i.e., the magnetization opposes the magnetizing force, the permeability being less than unity. Arises from precession of spinning charges in a magnetic field and is common to all materials but is masked by paramagnetism in paramagnetic materials. See **crystal-.**

diamond antenna. Same as **rhombic antenna.**

diaphragm. A vibrating membrane as in a loudspeaker, telephone and similar sound sources; also in receivers, e.g., the human eardrum.

diathermy. Heat therapy applied to human body by high-frequency currents, generated by specially designed electronic oscillators.

dice. Small rectangular or circular pieces of semiconductor material for fabrication of devices.

dichroic. Said of materials, such as solution of chlorophyll, which exhibit one colour by reflected light and another colour by transmitted light.

dichroic mirror. In television, colour-selective mirror which reflects a particular band of spectral energy and transmits all others.

Dicke's radiometer. An instrument for precise measuring of microwave noise power by reference to a standard source in a waveguide.

Dictaphone. TN for dictating machine, originally using as a recording medium a wax cylinder, the incised grooves being subsequently shaved off. Present-day Dictaphone machines use plastic belts, also magnetic tape.

dielectric. Substance, solid, liquid or gas, which can sustain a steady electric field, and hence an insulator. It can be used for cables, terminals and capacitors. See **anisotropic-.**

dielectric absorption. Phenomenon that the charging or discharging current of a dielectric does not die away according to the normal exponential law, due to absorbed energy in the medium.

dielectric amplifier. One which operates through a capacitor, the capacitance of which varies with applied voltage.

dielectric antenna. Antenna in which required radiation field is principally produced from a non-conducting dielectric.

dielectric breakdown. Passage of large current through normally non-conducting medium at sufficiently intense field strengths, accompanied by a reduction of resistance to a relatively low value.

dielectric constant. See **permittivity.**

dielectric current. A changing electric field applied to a dielectric may give rise to a displacement current (always), an absorption current, a conduction current.

dielectric diode. A capacitor whose negative plate can emit electrons into, e.g., CdS crystals, so that current flows in one direction.

dielectric dispersion. Variation of dielectric constant with frequency.

dielectric fatigue. Breakdown of a dielectric subjected to a repeatedly applied electric stress, insufficient to break down dielectric if applied once or a few times.

dielectric guide. Possible transmission path of very-high-frequency electromagnetic energy functionally realized in a dielectric channel, the dielectric constant of which differs from its surroundings.

dielectric heating. Radio-frequency heating in which energy is released in a non-conducting medium through dielectric hysteresis.

dielectric hysteresis. Phenomenon in which the polarization of a dielectric depends not only on the applied electric field, but also on its previous variation. This leads to power loss with alternating electric fields.

dielectric lens. By analogy with optics, a lens of dielectric material used for beaming electromagnetic waves from a radar horn.

dielectric loss. Dissipation of power in a dielectric under alternating electric stress; equal to

$$W = \omega C V^2 \delta,$$

where W = power loss
         V = rms voltage
         C = capacitance
         δ = power factor
         ω = $2\pi \times$ frequency.

dielectric polarization. Phenomenon explained by formation of doublets (dipoles) of elements of dielectric under electric stress.

dielectric radiation. That measured by dipole moment/unit volume for an antenna, depending on dielectric material.

dielectric relaxation. Time delay, arising from dipole moments in a dielectric when an applied electric field varies.

dielectric strain. See **displacement.**

dielectric strength. Electric stress (volts per cm or mm) required to puncture thin dielectric; stress, steady or alternating, is normally maintained for one minute.

dielectric viscosity. Phenomenon in which the polarization lags behind the changes in the applied field, depending on its rate of change.

dielectric wedges. Such as are used for matching in waveguides, e.g., air to dielectric.

difference limen. Increase in physiological stimulus which is just detectable in a specified percentage of trials. The ratio of this to the original magnitude is the *relative difference limen.*

differential. An electronic device or circuit whose operation depends on the difference of two opposing effects.

differential absorption ratio. Ratio of concentration of radioisotope in different tissues or organs, at a given time after its active material has been ingested or injected.

differential analyser. Analogue computer used for solving differential equations.

differential anode conductance. Reciprocal of *differential anode resistance.*

differential anode resistance. The slope of the anode voltage versus anode current curve of a multi-electrode valve, when taken with all other electrodes maintained at constant potentials with reference to the cathode. At high frequencies resistance values are larger than for d.c. (see **skin effect**) and a separate value may be quoted. Also called **a.c. resistance, incremental resistance, slope resistance.**

differential capacitor. One with one set of moving plates and two sets of fixed plates so arranged that, as capacitance of the moving plates to one set of fixed plates is increased, that to the other set is decreased. Used for balancing and for control of regeneration.

differential ionization chamber. Two-compartment system in which the resultant ionization current recorded is the difference between the currents in the two chambers.

differential permeability. Ratio of a small change in magnetic flux density of magnetic material, to change in the magnetizing force producing it, i.e., slope of the magnetization loop at the point in question.

differential relay. See **relay types.**

differential resistance. Ratio of a small change in the voltage drop across a resistance to the change in current producing the drop, i.e., the slope of the voltage current characteristic for the material.

differential selsyn. One whose stator and rotor are wound to produce rotating magnetic fields so that a differential angle is introduced into the control system by a change in the rotor position.

differential susceptibility. Ratio of a small change in the intensity of magnetization of a magnetic material to the change of magnetizing force producing it, i.e., the slope of the intensity-magnetization (hysteresis) loop.

differential winding. A winding in a compound motor which is in opposition to the action of another winding.

differentiating circuit (or **differentiator).** Electric circuit comprising resistors, capacitors and/or inductors, used, e.g., with servomechanisms, output voltage being proportional to rate of change (differential) of input voltage. Cf. *integrating circuit.*

diffraction. Property, exhibited by electromagnetic waves and all other waves, of bending around edges of obstacles in their path; one factor which accounts for the propagation of radio waves over the curved surface of the earth, also for the audibility of sound around corners. See **electronic-.**

diffraction angle. That between the direction of an incident beam of light, sound or electrons, and any resulting diffracted beam.

diffuse reflection. Same as **non-specular reflection.**

diffuse series. Series of optical spectrum lines observed in the spectra of alkali metals. Energy levels for which the orbital quantum number is two are designated **d-levels.**

diffuse sound. That in a (nearly) closed space, the totality of all intensities, phases and directions arising from multiple reflections from the walls, so that the sound at any point does not appear to come from any particular direction; *reverberant component* in sound reproduction.

diffuse transmission. Transmission of energy through a medium, such that a large proportion of that emitted has been scattered.

diffuse transmittance. See **transmittance.**

diffused junction. One formed by the diffusion of an impurity into a semiconductor crystal.

diffuser. (1) Translucent material in front of studio lamp to diffuse light and soften shadows. (2) Irregular structure, pyramids and cylinder, to break up reflections of sound waves in a studio.

diffusion. General transport of matter whereby molecules or ions mix through normal thermal agitation. Migration of ions may be directed and accelerated by electric fields, as in *dialysis*.

diffusion area. See **diffusion length.**

diffusion capacitance. The rate of change of injected charge with the applied voltage in a semiconductor diode.

diffusion constant. The ratio of diffusion current density to the gradient of charge carrier concentration in a semiconductor.

diffusion length. (1) Average distance travelled by carriers in semiconductor between generation and recombination. (2) A length significant to neutron diffusion theory. This length squared (termed the *diffusion area*) is one sixth of the mean square distance travelled by a thermalized neutron.

diffusion of solids. In semiconductors, the migration of atoms into pure elements to form surface alloy for providing minority carriers.

diffusion plant. Plant used for isotope separation by diffusion or thermal diffusion.

diffusion potential. The p.d. across the boundary of an electrical double layer in a liquid.

diffusion pump. See **Gaede diffusion pump.**

diffusion theory. Simplified neutron migration theory based on *Fick's law* (q.v.); less accurate than the more detailed *transport theory*.

Digilock. Telemetry system, using pulse-coded digits.

digit. Epoch occupied by any character, which may be any letter, figure or sign; one of a limited number in a set for counting, or in which *data* can be *coded*.

digit check. Redundant digit added to *words* in a process to reveal faults in that process.

digit sign. The + or − associated with a *word* read into a computer.

digital computer. One which performs simple calculations at high speed by impulses, related to digits in a code.

digital frequency meter. Class of *wavemeter* in which cycles of the input signal are counted over a known time interval, so that digital read-out can be provided.

digital read-out. Feature incorporated in some modern instruments, providing direct numerical presentation of the measured quantity.

digital voltmeter. One which indicates a voltage by its nearest numerical magnitude, e.g., by moving-coil optical deflection, switching of electroluminescent elements, dekatrons, digitrons.

digitization. Conversion of an analogue signal to a digital signal, with steps between specified levels. Also **quantization.**

digitron. Glow tube used for numerical read-out. The cathodes are usually shaped to form the digits 0 to 9.

diheptal. Referring to 14 in number, e.g., pins on base of a tube or valve.

dimensions. Measurements of physical quantities need only involve measurements of the fundamental units of the system. The dimensions of a quantity are the powers to which these units are involved in measurements of the quantity, e.g., the dimensions of force are mass × length × time^{-2} (MLT^{-2}).

diode. (1) Simplest electron tube, with a heated cathode and anode; used because of unidirectional and hence rectification properties. (2) Semiconductor device with similar properties, evolved from primitive crystal rectifiers for radio reception. See also:

backward-	equivalent-
catching-	parametric-
crystal-	tunnel-
dielectric-	Zener-.
efficiency-	

diode characteristic. A graph showing the anode current as a function of the anode voltage.

diode clipper. A clipping circuit which employs a diode.

diode mixer. Device performing same function in a radio-frequency line as a crystal mixer.

diode-pentode, diode-triode. Thermionic diode in same envelope as a pentode and triode respectively.

diode voltmeter. Valve voltmeter in which measured voltage is rectified by a valve or semiconductor diode, the rectified voltage then usually being amplified and displayed on a moving-coil meter.

dip. The angle measured in a vertical plane between the direction of the resultant earth's magnetic field and the horizontal. Also **inclination.**

diplex. See **simplex.**

diplexer. A means of coupling which permits a radar and a communication transmitter to operate with the same aerial.

dipole. Close combination of equal positive and negative charges of electricity or magnetism, not translated by a uniform field, but restored to alignment by a torque proportional to the sine of the angle of deflection. Certain large dielectric molecules also exhibit dipole effects, reflected in the polarization and temperature trends in the dielectric constant of substance. See **doublet.**

dipole antenna. One comprising a straight conductor, up to half-wavelength long, attached to a feeder at its centre. Maximum response is for waves polarized along the axis of the conductor.

dipole molecule. One which has a permanent moment due to the permanent separation of the effective centres of the positive and negative charges.

dipole moment. Product of the magnitude of an electric or magnetic charge and the distance between it and its opposite charge; this determines the distant potential, which falls off as the inverse square of the distance. See **induced-.**

Dirac's constant. Planck's constant divided by 2π (usually termed **h-bar** and written $\hbar$). The unit in which *electron spin* is measured.

direct capacitance. That between two conductors, as if there are no other conductors.

direct coupling. Intervalve and transistor coupling so that zero-frequency currents are amplified to the same extent as higher frequency currents. The correct grid bias for succeeding valves is obtained either by a potentiometer or by a reversed polarity battery; also **direct-current coupling.**

direct current. Current which flows in one direction only, and which does not have any appreciable pulsations in its magnitude.

direct-current amplifier. Amplifier which uses direct coupling between stages in order to amplify very low frequency changes in voltage or current (zero-frequency signals). Chief problem involved is drift stabilization.

direct-current balancer. The coupling and connecting of two or more similar direct-current machines so that the conductors connected to the junction points of the machines are maintained at constant potentials.

direct-current bias. In magnetic tape or wire recording, the addition of a polarizing direct current in signal recording to stabilize magnetic saturation.

direct-current bridge. A four-arm null bridge energized by a d.c. supply. The prototype is the Wheatstone bridge, other examples the metre bridge and the Post Office box.

direct-current component. That part of the picture signal which determines the average or datum brightness of the reproduced picture.

direct-current converter. A converter which changes direct current from one voltage to another.

direct-current coupling. See **direct coupling.**

direct-current/direct-current converter. High-voltage d.c. obtained by rectifying the output of an oscillating transistor fed by low-voltage d.c. Also called **direct-current transformer.**

direct-current dump. The intentional or accidental removal of all d.c. power from a computer system.

direct-current erasing. The removal of signals in a tape by a unidirectional magnetic field using a d.c. current.

direct-current generator. A rotary machine to convert mechanical into direct-current power.

direct-current meter. One which responds only to d.c. component of a signal, e.g., *moving-coil* instruments.

direct-current motor. One which operates on direct current.

direct-current picture transmission. Television transmission in which the d.c. component of the signal represents the average illumination.

direct-current quadricorrelator. Noise-resistant two-mode type of automatic frequency and phase control circuit.

direct-current restoration. Use of a *catch* or *clamp* diode to restore or hold black *or* white level at the designated pedestal level. Also **reinsertion.**

direct-current restorer. The means of restoring a d.c. (or low-frequency) component of a signal after transmission in a line which only easily transmits high frequencies.

direct-current transformer. (1) Device to measure large d.c. currents by means of associated magnetic field. (2) Colloquial name for **direct-current/direct-current converter.**

direct-current transmission. Inclusion of d.c. or very low frequency in the transmitted video signal in TV. If omitted, it has to be *restored* in relation to the *pedestal* in the receiver.

direct drive. Transmitting system in which the antenna circuit is directly coupled to the oscillator circuit.

direct ray. That portion of the wave from a transmitter which proceeds directly to the receiver, without reflection from the Heaviside layer; also **direct wave, ground ray, ground wave.**

direct-recorded disk. One which, after a groove has been cut into its plastic surface, is immediately ready for *playback* (reproduction). Widely used for broadcast transcription; *records* available to the public can be *processed* from direct-recorded disks.

direct viewing. Television reception in which the received image is viewed directly on the screen of a cathode-ray tube without projection or reflecting devices.

direct wave. See **direct ray.**

direction-finding. Principle and practice of determining a bearing by radio means, using a discriminating antenna system and a radio receiver, so that direction of an arriving wave (ostensibly direction or bearing of a distant transmitter) can be determined. See AN-, automatic-, cathode-ray-, frame-.

directional antenna. One in which the transmitting and receiving properties are concentrated along certain directions.

directional coupling. That afforded by a device in which power can be transmitted in specified paths and not in others, e.g., line-balancing toroid, valve amplifier, and certain waveguide configurations.

directional filter. Combination of high- and low-pass filters, which determines route or direction of transmission of signals. Separate filters are not directional, being passive.

directional gain. Ratio (expressed in decibels) of the response (generally along the axis, where it is a maximum) to the mean-spherical (or hemispherical with reflector or baffle) response, of an antenna, loudspeaker or microphone; also **directivity index.**

directional gyroscope. A free gyroscope which holds its position in azimuth and therefore indicates any angular deviation from the reference position.

directional homing. Following a course so that the objective is always at the same relative bearing.

directional microphone. One which is directional in response, either inherently (as in the ribbon microphone) or with a parabolic reflector of adequate dimensions.

directional radio. System using directional antennae at the transmitting and/or receiving ends, usually to minimize interference.

directional receiver. Receiving system using a directional antenna for discrimination against noise and other transmissions.

directional transmitter. Transmitting system using a directional antenna, to minimize power requirements and to diminish effect of interference.

directive efficiency. Ratio of maximum to average radiation or response of an antenna; the *gain*, in dB, of antenna over a dipole being fed with the same power.

directive gain. For a given direction, 4π times the ratio of the radiation intensity in that direction to the total power which is radiated by the aerial.

directivity. Measurement, in dB, of the extent to which a directional antenna, loudspeaker, or microphone, concentrates its radiation or response in specified directions.

directivity angle. Angle of elevation of direction of maximum radiation or reception of electromagnetic wave by an antenna.

directivity index. Same as **directional gain.**

directly-heated cathode. Metallic (-coated) wire heated to a temperature such that electrons are freely emitted; also called **filament cathode.**

director. Free resonant dipole element in front of antenna array which assists the directivity of array in the same direction. See **Yagi array.**

disadvantage factor. Ratio of average neutron flux in reactor lattice to that within actual fuel element.

discharge. (1) Abstraction of energy from a cell by allowing current to flow through a load. (2) Reduction of the p.d. at the terminals

(plates) of a capacitor to zero. (3) Flow of electric charge through gas or air due to ionization, e.g., lightning, or at reduced pressure, as in fluorescent tubes. (4) Unloading of fuel from reactor.

discharge bridge. Measurement of the ionization or discharge, in dielectrics or cables, depending on the amplification of the high-frequency components of the discharge.

discharge circuit. Discharge of a capacitor by making a parallel valve or tube conducting by altering its grid bias or by using a neon lamp.

discharge lamp. Lamp in which luminous output arises from ionization in gaseous discharge; see **discharge** (3).

discharge tube. Any device in which conduction arises from ionization, initiated by electrons of sufficient energy.

discomposition effect. See **Wigner effect.**

discone antenna. Radiator for VHF transmission, comprising a vertical cone, which acts as a vertical dipole with very constant driving impedance over a wide frequency range.

discrimination. The selection of a signal having a particular characteristic, e.g., frequency, amplitude, etc., by the elimination of all the other input signals at the discriminator.

discriminator. (1) Circuit which converts a frequency or phase modulation into amplitude modulation for subsequent demodulation. (2) Circuit which rejects pulses below a certain amplitude level and shapes the remainder to standard amplitude and profile. See **amplitude-, frequency-.**

dish. Colloquialism for *parabolic reflector*, solid or mesh, used for directive radiation and reception, especially for radar or radio-telescopes.

disintegration. A process in which a nucleus ejects one or more particles, applied especially, but not only, to spontaneous radioactive decay. See **alpha decay, atomic-, beta-, cathode-.**

disintegration constant. The probability of radioactive decay of a given unstable nucleus per unit time. Statistically it is the constant λ expressing the exponential decay $\exp(-\lambda t)$ of activity of a quantity of this isotope with time. It is also the reciprocal of the mean life of an unstable nucleus. Also **decay constant, transformation constant.**

disintegration energy. See **alpha decay energy.**

disintegration voltage. In a hot-cathode gas tube, the lowest anode potential at which occurs the destructive positive ion bombardment of the cathode.

disk anode. Final anode in an electrostatically focused cathode-ray tube, usually a circular plate containing a central aperture through which the beam passes. See **ultor.**

disk capacitor. One in which the variation in capacitance is effected by the relative axial motion of disks.

disk scanner. Rotating disk carrying apertures, lenses, prisms, or other picture-scanning elements used in mechanical scanning systems.

disk-seal tube. Same as **megatron**.

dislocation. A lattice imperfection in a crystal structure, classified according to type, e.g., edge dislocation, screw dislocation.

dispatcher. The section of a digital computer performing the switching operations which determine the sources and destinations for the transfer of words.

dispenser cathode. One which is not coated, but is continuously supplied with suitable emissive material from a separate electrode element.

dispersive medium. One in which the phase velocity is a function of frequency.

displacement. Vector representing the electric flux in a medium and given by the product of the absolute permittivity ϵ and the field strength E. In unrationalized units $D = 1/4\pi \times \epsilon E$. Also called **dielectric strain, electric flux density.**

displacement current. Integral of the *displacement current density* through a surface. The time rate of change of the electric flux, D. Current postulated in a dielectric when the electric stress or potential gradient is varied. Distinguished from a normal or conduction current in that it is not accompanied by motion of current carriers in the dielectric. Concept introduced by Maxwell for the completion of his electromagnetic equations.

displacement flux. The integral of the normal component of displacement over any surface in a dielectric. See **displacement.**

displacement law. Soddy and Fajans formulation that radiation of an alpha-particle *reduces* the atomic number by two and the mass number by four, and that radiation of a beta-particle *increases* the atomic number by one, but does not change the mass number. It was later found that emission of a positron *decreases* the atomic number by one, but does not change the mass number. Gamma emission and isomeric transition change neither mass nor atomic number. Displacement laws are summarized as follows:

Type of disintegration	Change in atomic number	Change in mass number
alpha emission	-2	-4
beta electron emission	$+1$	0
,, positron emission	-1	0
,, electron capture	-1	0
isomeric transition	0	0
gamma emission	0	0

A change in atomic number means displacement in the periodic classification of the chemical elements; a change in mass number determines the radioactive series.

display. Mode of showing information on a cathode-ray tube screen, especially in radar

and navigation. See A-, A & R-, B-, C-, chart comparison unit-, D, E, F, G, H, I-, J, K, L, M, N, P-, plan-position indicator.

disruptive discharge. That of a capacitor when the discharge arises from breakdown (puncture) of the dielectric by an electric field strength which it cannot withstand.

disruptive voltage. That which is just sufficient to puncture the dielectric of a capacitor. A test voltage is normally applied for one minute. See also **breakdown voltage.**

dissector tube. One in which the electron pattern of emission from a continuous photoemissive surface is scanned over an aperture, which creates the video signal.

dissipation. (1) Loss or diminution of power, leading to flat tuning for steady-state oscillations and damping in free oscillations. Removes sharpness of cutoff in filters. Dependent on resistance, measured at high-frequency, including radiation. It is the dissipation of eddy-current energy in conductors and of displacement polarization energy in dielectrics that makes industrial high-frequency heating possible. (2) Power released from any electrode, depending on electron collection, radiation or conduction of heat received, or ion bombardment, controlled by graphite, fins, or water-cooling. See **anode-.**

dissipation factor. The co-tangent of the phase angle (δ) for an inductor or capacitor. For low loss components the dissipation factor is approximately equal to the power factor

$$\tan \delta \simeq \frac{\sigma}{\omega \epsilon}$$ for a low loss dielectric, where σ

is electrical conductivity, ϵ is permittivity of the medium, and ω is $2\pi \times$ frequency.

dissipationless line. A hypothetical transmission line in which there is no energy loss. Also **lossless line.**

dissipative network. One designed to absorb power, as contrasted with networks which attenuate power by impedance reflection. All networks dissipate to some slight extent, because neither capacitors nor inductors can be made entirely loss-free.

dissociation, Arrhenius theory of. In electrolytic dissociation, the original idea that an acid, base or salt on solution ionizes to some extent, an applied field causing a drift up or down the grade of potential, with associated changes of boiling and freezing points.

dissolve. See **cross-fade.**

dissonance. A displeasing combination of two pure tones; cf. *consonance.*

dissymmetrical. See **asymmetrical.**

distance mark. One on the screen of a cathode-ray tube to denote distance in radar target.

distant-reading compass. Gyro flux-gate compass in which the indicator is remote from the sensing device.

distorting network. A network altering the response of a part of a system, and anticipating the correction of response required to restore a signal waveform before actual distortion has occurred, e.g., owing to the inevitable frequency distortion in a line, or to minimize noise interference.

distortion. (1) Change of waveform or spectral content of any wave or signal due to any cause. (2) Errors in geometry of image due to imperfections in the focusing system. In facsimile reproduction, electrical and optical distortion are normally both present.

See also:

amplitude-	linear-
aperture-	modulation-
attenuation-	non-linear-
barrel-	origin-
coma	phase-delay-
delay-	phase-
deviation-	phase-intercept-
frequency-	pincushion-
harmonic-	transient-
intermodulation-	trapezium-.
keystone-	

distortion factor. Ratio of the rms harmonic content to the total rms value of the distorted sine wave.

distortion set. Instrument which measures the extent of a specified type of distortion in a communication system, e.g., distortion and noise monitor which measures the noise, hum and audio-frequency distortion in an amplitude- (or frequency-) modulated signal.

distortionless line. One with constants such that the attenuation (a minimum value) and the delay time are constant in magnitude with variation in frequency. The characteristic impedance is purely resistive. For such a line

$$L \cdot G = R \cdot C,$$

where R = resistance
G = leakance
L = inductance
C = capacitance,
all being distributed values per unit length. Also **distortionless condition.**

distributed amplifier. Same as **transmission-line amplifier.**

distributed capacitance. (1) That distributed along a transmission line, which, with distributed resistance and/or inductance, reduces the velocity of transmission of signals. (2) That between the separate parts of a coil, lowering its inductance; represented by an equivalent lumped capacitor across the terminals, giving the same frequency of resonance.

distributed constants. Those applicable to real or artificial transmission lines and waveguides because dimensions are comparable with the wavelength of transmitted energy.

distributed inductance. Said of a circuit which has an inductance distributed uniformly along it, e.g., a power transmission line, a loaded telephone circuit, or a travelling-wave valve or tube.

distribution cable. A communication cable extending from a feeder cable into a defined service area.

distribution coefficients. Chromaticity coordinates for spectral (monochromatic) radiations of equal power, i.e., for the component radiations forming an equal energy spectrum.

distribution control. The means by which the distribution of scanning speeds is varied during the trace interval.

ditch. Groove designed to assist choking action for high attenuation.

dither. Small continuous signal supplied to servo motor operating hydraulic valve or similar device, and producing a continuous mechanical vibration which prevents sticking.

divergence angle. Angle of spread of electron beam, arising from mutual repulsion or debunching.

divergent. Description of reactor or critical experiment, when multiplication constant exceeds unity.

diversity reception. System designed to reduce fading; several antennae, each connected to its own receiver, are spaced several wavelengths apart from one another, the demodulated outputs of the receivers being combined. Alternative systems use antennae orientated for oppositely polarized waves (*polarized diversity*) or independent transmission channels on neighbouring frequencies (*frequency diversity*).

divertor. Trap used in thermonuclear device to divert impurity atoms from entering plasma, and fusion products from striking walls of chamber.

divider. Thermionic circuit which reduces by an integral factor the number of pulses or alternations/s passing through it.

dividing network. A frequency-selective network which arranges for the input to be fed into the appropriate loudspeakers, usually two, covering high and low frequencies respectively. Also **cross-over network, loudspeaker dividing network.**

Dixonag. Same as **Aquadag.**

Doba's network. Shaping circuit used in pulse amplifiers when rise times of a few nanoseconds are required.

dog house. Enclosure protecting feed equipment at base of large transmitting antenna.

Doherty amplifier. One in which one section takes low signal amplitudes up to half-maximum, the other section coming in up to full power, the signal being divided and united with 90° phase-shift networks. The result is high efficiency and linearity.

Doherty transmitter. One in which high efficiency of amplification of amplitude-modulated wave is obtained by two valves connected to the load, one directly, the other through a 90° retarding network.

Dolezalek electrometer. The original quadrant electrometer used for measurement of very small currents. Consists of a suspended paddle (usually connected to a high voltage) which rotates between four quadrants, two earthed and two being charged by the current under measurement. (This is *hetero-*

static operation. Idiostatic connection has the paddle joined to one pair of quadrants.)

dollar. U.S. unit of reactivity (contributed by the delayed neutrons) corresponding to the prompt critical condition in a reactor. See also **cent**.

domain. In ferroelectric or ferromagnetic material, region where there is saturated polarization, depending only on temperature. The transition layer between adjacent domains is the Bloch wall, and the average size of the domain depends on the constituents of the material and its heat treatment. Domains can be seen under the microscope when orientated by strong electric or magnetic fields.

dominant mode. (1) The mode with the lowest cutoff frequency in waveguide propagation. (2) In a cavity resonator, the dominant mode is that corresponding to the lowest excitation frequency. This is also true of microwave oscillators (magnetrons, klystrons, etc.) using resonant cavities.

dominant wavelength. The wavelength of monochromatic light which matches a specified colour when combined in suitable proportions with a reference standard light.

donor. Impurity atoms which add to the conductivity of a semiconductor by contributing electrons to a nearly empty conduction band, so making *n*-type conduction possible. See also **acceptor, impurity**.

donut. See **doughnut**.

doorknob transformer. Device for coupling a coaxial line to a waveguide through a circular hole at the side, the inner conductor terminating on the far side.

doorknob valve (or tube). A thermionic valve in which the electrodes are very small and closely spaced. It was designed for VHF transmission.

dope additive. Specific impurity added to ultra-pure semiconductor to give required electrical properties, e.g., indium to germanium.

doped junction. One with a semiconductor crystal which has had impurity added during a melt.

doping. Addition of known impurities to a semiconductor, to achieve the desired properties in diodes and transistors, i.e., *donor* in *p*-type and *acceptor* in *n*-type.

Doppler broadening. Frequency spread of radiation single spectral lines because of Maxwell distribution of velocities in the molecular radiators.

Doppler effect. Change in apparent frequency (or wavelength) because of relative motion of a source of radiation and an observer. Thus, for sound:

$$\text{frequency} \atop \text{observed} = \left(\frac{V+W-V_o}{V+W-V_s}\right)\left(\text{frequency} \atop \text{of source}\right),$$

where V_s, V_o = velocities of source and observer

W = velocity of wind in same direction

V = phase velocity of wave.

Effect is observed in the in-line motion of stars, leading to the idea of an expanding universe; in acoustics, e.g., drop in pitch of a whistle on a passing train.

Doppler navigation. System using ground reflection and dependent on *Doppler effect*.

Doppler radar. Any means of detection by reflection of electromagnetic waves which depends on measurement of change of frequency of a continuous wave after reflection by a target having relative motion.

doran. Doppler ranging system for tracking missiles. (*Doppler range*.)

dose. A general expression for the quantity of ionizing radiation to which a sample has been exposed. The following terms are preferred:

    absorbed dose. Measured in *rad*.

    biological (rbe) dose. Measured in *rem*.

    exposure dose. Measured in *röntgens*.

    integral dose. (*a*) For tissue, etc., the total absorbed dose measured in *gm. rad*. (*b*) For neutron irradiation in reactors, the time integral of the neutron flux through the sample.

See also:

	maximum
air-	permissible-
bone tolerance-	skin-
cumulative-	standing-off-
depth-	tissue-
	tolerance-.

dose rate. The *absorbed* or *exposure dose* received per unit time.

dose ratemeter. Instrument for measuring radiation dose rate.

dosemeter or **dosimeter**. Instrument for measuring *dose* (q.v.). Commonly takes the form of a pocket electroscope used in radiation surveys, hospitals and Civil Defence. A quartz fibre, after being charged, is viewed against a scale, and by its movement across the scale on discharge it gives a measure of the radiation field and dosage experienced.

dot. Unit digital impulse, the basis of telegraphic codes, and electronic computing. See **band, bit**.

dot cycle. Minimum period of an *on-off* or *mark-space*. See **band**.

dot frequency. Number of dots/s in a continuous train of dots. A measure of speed of telegraphic transmission.

dot generator. One which provides impulses for a test pattern for adjusting television images on a cathode-ray tube screen.

dot-sequential. Said of colour television based on the association of individual primary colours in sequence with successive picture elements.

double amplitude. The sum of the maximum values of the positive and negative half-waves of an alternating quantity; also **peak-to-peak amplitude**.

double-beam cathode-ray tube. One containing two complete sets of beam-forming and beam-deflecting electrodes operated from the same cathode, thus allowing two separate waveforms to appear on the screen simultaneously (double-gun tube). Also can refer to use of a single gun with a beam-switching circuit, i.e., two inputs continuously interchanged, and to single beam tube in which the beam is split and the two parts are separately deflected, usually in one plane only.

double beta decay. An energetically possible process involving the emission of two beta-particles simultaneously. Not to be confused with **dual beta decay.**

double bridge. Network for measuring low resistances. See **Kelvin bridge.**

double camera. One in which the film registering the outside scene is accompanied by a parallel film for optical or magnetic synchronous recording of the accompanying sound.

double-coil loudspeaker. Electrodynamic loudspeaker with two driving coils separated by a compliance, the coils driving one or two open cone diaphragms, thus operating more effectively over a wide range of frequency.

double-cone loudspeaker. Large open coil-driven cone loudspeaker, with a smaller free-edge cone fixed to the coil former, thus assisting radiation for high audio frequencies.

double detection receiver. Same as **supersonic heterodyne receiver.**

double diode. See **binode.**

double-frequency oscillator. One in which two sets of oscillations, having different frequencies, are simultaneously generated.

double-gun cathode-ray tube. See **double-beam cathode-ray tube.**

double-hump effect. Property of two coupled resonant circuits, each separately resonant to the same frequency, of showing maximum response to two frequencies disposed about the common resonance frequency.

double image. See **ghost.**

double-layer. Screen with separate layers of phosphors with differing properties, i.e., colour or persistence.

double limiting. The process of limiting in radar the positive and the negative amplitudes of a wave.

double moding. Irregular switches of frequency by magnetron oscillator, due to changing mode.

double modulation. See **compound modulation.**

double precision. Computer operation using *words* of double normal length in *bits.*

double reception. Simultaneous reception of two signals on different wavelengths by two receivers connected to the same antenna.

double refraction. Division of electromagnetic wave in anisotropic media into oppositely polarized components propagated with different velocities. These components are termed the *ordinary ray,* where the wavefronts are spherical so that the normal laws of refraction are obeyed, and the *extraordinary ray,* where the wavefronts are not spherical so that the velocity

is a function of the direction of propagation. Also **birefringence.**

double sideband transmitter. One which transmits both the sidebands produced in the modulation of a radio-frequency carrier wave. Cf. *single sideband system.*

double throw switch. One which enables connections to be made with either of two sets of contacts.

double triode. Thermionic valve with two triode assemblies in the same envelope.

double-tuned circuit. Circuit containing two elements which may be tuned separately.

double vibration. Obsolete for **cycle** (1).

doubler circuit. (1) One with over-driven valve, so that a frequency can be filtered from the output which is double the frequency of the input. (2) A form of self-saturating magnetic amplifier having an a.c. output.

doublet. Non-polar valence bond between two atoms, formed by one electron from each.

doublet antenna. One comprising two short straight conductors, essentially less than a quarter-wavelength long, in line and fed at the centre by a transmission line.

doubling time. (1) The time required for the neutron flux in a reactor to double. (2) In a *breeder reactor,* the time required for the amount of fissile material to double.

doughnut. (1) Anchor-ring shape used in circular-path accelerator tubes of glass or metal, e.g., zeta. (2) Traditional shape of a pile of annular laminations for magnetic testing, since there is no external field. (3) Traditional shape of loading coils on transmission lines, permitting exact balance in addition to inductance; also **donut.** See **toroid.**

dovap. Doppler system for tracking missiles, in which the missile carrier or transponder returns a double-frequency signal. (*D*oppler *v*elocity *a*nd *p*osition.)

Dow oscillator. See **electron-coupled oscillator.**

down-time. Period during which a computer is not available for normal operation.

downlead. See **antenna downlead.**

downward modulation. That in which the instantaneous amplitude of the carrier is always smaller than the amplitude of the unmodulated carrier.

drag-cup generator. A servo unit used to generate a feedback signal proportional to the time derivative of the error.

Dragon project. The international cooperative research programme on the design of high-temperature gas-cooled reactors, being carried out at Winfrith Heath in Dorset.

Dreadnought. RN first nuclear-powered submarine fitted with American S5W pressurized water reactor.

drift. (1) In a toroid for plasma studies, tendency for the plasma to drift from several causes, mainly because of non-uniformity of steady magnetic field across area of tube. (2) Slow variation in performance, e.g., of thermionic amplifier during warming up or after a long period, or of a magnetic amplifier. (3) In computers, change in output of directly coupled analogue amplifier, arising from external causes and which cannot be separated from signal after amplification.

drift angle. The angle between the planned course and the track. Also sometimes used for angle between heading and track.

drift mobility. The average drift velocity of the carriers per unit electric field in a homogeneous semiconductor. In general, the mobilities of electrons and holes will be different.

drift rate. Time variation of output of voltage regulator or reference tube.

drift space. (1) Space between the buncher and reflector in a klystron, in which the electrons congregate in longitudinal waves. (2) Space in an electron tube which is free of electric or magnetic fields.

drift transistor. One in which resistivity increases continuously between emitter and collector junctions, improving high-frequency performance.

drift tube. Section (space) of any tube, e.g., accelerator or klystron, in which electrons pass without significant change in velocity, while the accelerating RF voltage changes.

drift velocity. The average velocity (in direction of applied electric field) of electrons in a conductor (or of ions in a gas).

drive. (1) Voltage applied to a vacuum device, especially the grid in a transmitting valve, or base of a transistor. (2) In radio, that which controls a master resonator in an oscillator, e.g., *quartz crystal, tuning-fork, resonating line* or *cavity*. (3) In circuits, the alternating voltage applied to the grid of an amplifying valve. Specially the master oscillator circuit and its subsequent amplifying stages in a transmitter using independent drive. Cf. *exciter, excitation.*

drive pattern. Pattern in facsimile transmission, which results from a variation of density due to periodic errors in the position of the recording spot.

driven elements. Those in an antenna which are fed by the transmitter, as compared with reflector, director, or parasitic elements.

driver stage. Bank of valves which drives the final stage in a radio transmitter.

driver unit. Same as **exciter.**

driving-point impedance. The ratio of the e.m.f. at a particular point in a system to the current at that point.

driving potential. The positive potential applied to the anode of a photocell to drive the electrons to the anode after they have been released from the cathode by the incident light.

driving signals. The original line and frame pulses which synchronize the whole television system.

drone. Guided missile used as a target or for reconnaissance.

drop. Computer digits are said to *drop in* if they are recorded without a signal, and *drop out* if not recorded from a signal.

drop-out. Said of a relay when it de-operates,

i.e., contacts revert to de-energized condition. See also **drop.**

dropping resistor. One whose purpose is to reduce a given voltage by the voltage drop across the resistance itself.

dry box. Sealed box for handling material in low humidity atmosphere. Not synonymous with **glove box.**

dry cell. A primary cell in which the contents are in the form of a paste. See **Leclanché cell.**

dry electrolytic capacitor. One in which the negative pole takes the form of a sticky paste, which is sufficiently conducting to maintain a gas and oxide film on the positive aluminium electrode.

dry flashover voltage. The breakdown voltage between electrodes in air of a clean dry insulator.

dry joint. Faulty solder joint giving high resistance contact due to residual oxide film.

dry rectifier. Solid-state device, in contrast to a thermionic or electrolytic rectifier.

dual amplifier. See **reflex amplifier.**

dual beta decay. *Branching* (q.v.), where a radioactive nuclide may decay by either electron or positron emission. Not to be confused with **double beta decay.**

dual channel sound. A technique used in television receivers in which a separate intermediate frequency stage is employed for sound and video signals after the common first detector stage.

dual modulation. Simultaneous amplitude and frequency modulation of a carrier.

dual track. Use of two tracks on a magnetic tape, so that recording and subsequent reproduction can proceed along one track and return along the other, thus obviating re-winding.

Duane and Hunt's law. The maximum photon energy in an X-ray spectrum is equal to the kinetic energy of the electrons producing the X-rays, so that the maximum frequency, as deduced from quantum mechanics, is

$$eV/h,$$

where V = applied voltage
e = electronic charge
h = Planck's constant.

dubbing. (1) The combination of two sources of sound into one recording. (2) The replacement of the original sound-track of a ciné film, e.g., with one in a different language.

duct. Layer in the atmosphere which, because of its refractive properties, keeps electromagnetic radiated (or acoustic) energy within its confines. It is *surface-* or *ground-based* when the surface of the earth is one confining plane.

Duddell arc. A d.c. arc which gives rise to an audio-frequency current. If a suitable tuned circuit is connected across the arc then the arc will sing acoustically. Also **singing arc.**

dull-emitter cathode. One from which electrons are emitted in large quantities at temperatures at which incandescence is barely visible. The emitting surface is the oxide of one or more of the alkaline-earth metals.

dumb aerial. Non-radiating resistive network, used for absorbing the output power from a transmitter during the spacing periods, in absorber keying. Similar to **artificial antenna.**

dummy antenna. Same as **artificial antenna.**

dummy load. One which matches a feeder or transmitter; so designed to absorb the full load without radiation, particularly for testing; also called **antenna load.**

dump. Holding store for data required for future operations.

dump condenser. Condenser which allows steam from heat exchangers to be by-passed from the turbines in a nuclear power station, if the electrical load is suddenly taken off.

duode. Electrodynamic open diaphragm loudspeaker driven by eddy currents in a metal former, the *voice coil* being wound over a rubber compliance; the arrangement gives enhanced width of response, with damping of diaphragm resonances.

duodecal. Twelve-contact tube base.

duodecimal. Arithmetic system based on radix twelve and with twelve distinct digits.

duo-diode. See **binode.**

duodynatron. System comprising two resonant circuits, connected to the inner grid and anode of a tetrode, the outer grid being maintained at a higher potential than any of the other electrodes; oscillations of different frequencies are maintained in the two resonant circuits owing to secondary emission from the inner grid and anode.

duolateral coil. See **basket coil.**

duophase. The use of a choke in the cathode or anode circuit of a valve in an amplifier, to obtain a reversed-phase voltage for driving a push-pull output stage.

duo-triode. Two triode valves in a single glass envelope.

duplet. Single valence bond between atoms arising from a shared pair of electrons.

duplex. (1) U.S. for separate electron systems in a common vacuum, with or without a common cathode, e.g., double diode or duo-diode pentode. (2) See **simplex.**

duplexer. Circuit using TR switches, which permit the same antenna to be used for transmission and reception, the switch protecting the receiver from the high-powered transmission.

duplication check. The checking in a computer that the results of two independent performances of the same operation are identical.

duplicator. Machine tool with a mechanical or electronic controlled cutting head for contour reproduction.

Duralumin. TN for corrosion-resistant aluminium alloy of high tensile strength, widely used for antennae.

durchgriff. *Penetration factor* (q.v.) or inverse of *amplification factor* (q.v.) of a triode valve.

Dushman equation. See **Richardson-Dushman equation.**

dust core. Magnetic circuit embracing or threading a high-frequency coil, made of ferromagnetic particles compressed into an insulating matrix binder, thus obviating losses at high frequency because of eddy currents.

dust monitor. Instrument which separates airborne dust, and tests for radioactive contamination.

duty. The cycle of operations which an apparatus is called upon to perform whenever it is used; e.g., with a motor, it is the starting, running for a given period, and stopping; or with a circuit-breaker, it may be closing and opening for a given number of times with given time intervals between; or the prescription of a process timer.

duty factor. Pulse signals giving ratio of pulse duration to space interval.

dwarf waves. Those of length less than one centimetre, produced by electronic oscillations in the inter-electrode space of a valve.

dyadic operation. One on two operands.

Dykanol. TN for dielectric material used in paper capacitors.

Dynamax. TN for high-permeability iron alloy suitable for use in toroidal cores.

dynamic. General term applied to non-steady operating conditions.

dynamic-capacity electrometer. See **vibrating-reed electrometer.**

dynamic characteristic. Generally, any characteristic curve taken under normal working conditions. Specifically applied to the anode current/grid voltage relationship, when the effect of the anode load impedance is included.

dynamic focusing. The automatic process of varying the voltage on the focusing electrode of a colour television picture tube so that the beam spots remain always in focus as they are swept across the flat screen.

dynamic loudspeaker. Open diaphragm, driven by a *voice coil* on a former; intended to be used in a plane (Rice-Kellog) baffle, from the side of a box (box baffle), or at the neck of a large horn (flare).

dynamic microphone. See **moving-coil microphone.**

dynamic noise suppressor. One which automatically reduces the effective audio bandwidth, depending on the level of the required signal to that of the noise.

dynamic range. (1) Range of intensities in a sample of radio programme, as measured by a programme meter, a peak meter or volumeunit (VU) meter, and expressed in dB. (2) Ratio of maximum to minimum brightness in the original or reproduced TV image. Also **contrast range.**

dynamic regulator. A regulator in a transmission system which requires contro power to adjust at all but a few settings.

dynamic resistance. Effective differential resistaicc over a section of a static characteristic, e.g., arc, valve, tunnel diode. Can be positive or negative, the latter being essential for sustaining self-oscillation.

dynamic sensitivity. The alternating component of phototube anode current divided by the alternating component of incident radiant flux.

dynamic storage computer. One in which signals are circulating and from which they can be extracted, e.g., delay line.

dynamic subroutine. One which must be modified by the computer in accordance with some feature of the parameters to be used.

dynamo. Electromagnetic machine which converts mechanical energy into a.c. or d.c. electrical supply. See **alternator.**

dynamo-electric amplifier. Low and zero frequency mechanically rotating armature in a controlled magnetic field; used in servo systems.

dynamometer. Instrument for measurement of torque. Electric dynamometers can be used for measurement of a.c. current, voltage or power.

dynamotor. A rotary converter for transforming a.c. to d.c.

Dynatrol. TN for type of servo motor. (*Dyna*mic con*trol*.)

dynatron. Circuit in which steady-state oscillations are set up in a tuned circuit between screen and anode of a tetrode, the latter exhibiting negative resistance when the anode potential is below the potential of the screen.

dynatron oscillator. One which uses the normal negative slope resistance of a screened grid valve to sustain oscillations in a shunt-tuned circuit.

dyne. The absolute CGS unit of force which will accelerate 1 gm. at 1 cm. per sec. per sec. ($=10^{-5}$ newton).

Dynistor. TN for *p-n-p-n* transistor type component with negative resistance characteristic. (*Dyna*tron trans*istor*.)

dynode. Intermediate electrodes (between cathode and final anode) in photomultiplier or electron multiplier tube. Dynode electrodes are those which emit secondary electrons and provide the amplification.

dynode chain. Resistance potential divider used to supply increasing potential to successive dynodes of an electron multiplier.

dynode spots. Spurious signals which may be produced in image orthicons.

Dyotron. TN for single cavity microwave oscillator.

dysprosium (Dy). Element in rare earth group, at. no. 66, at. wt. 162·50.

E

E-bend. A smooth bend in the axis direction of a waveguide, the axis being maintained in a plane parallel to the polarization direction.

E display. Radar display in which target range and elevation are plotted as horizontal and vertical coordinates of the blip.

E-layer. Most regular of the ionized regions in the ionosphere, which reflects waves from a transmitter back to earth. Its effective maximum density increases from zero before dawn to its greatest at noon, and decreases to zero after sunset, at heights varying between 110 and 120 km. There are at least two such layers; also called **Heaviside layer** or **Kennelly-Heaviside layer.**

E-transformer. An electric sensing device in an automatic control system which gives an error voltage in response to linear motion. Consists of coils for detecting small displacement of a magnetic armature, which affects balance of currents when off-centre.

E-wave. See TM-wave.

ear microphone. A contact microphone which is shaped to fit into the ear of the wearer, whose voice is picked up for consequent transmission.

early effect. Variation of junction capacitance and effective base thickness of a transistor with supply potentials.

early-warning radar. A radar system for the detection of approaching aircraft or missiles at greatest possible distances.

earphone. Transducer for close application to the ear, in hearing-aids, bridge measurements, for monitoring and for the reception of radio signals.

earphone coupler. A suitably shaped cavity with an incorporated microphone used for acoustical testing of earphones.

earth. System of plates or wires buried in the ground, to which connection is made to provide a path to ground for currents flowing in the antenna circuit. U.S. **ground.** See also **earth potential.**

earth (or ground) capacitance. The capacitance between an electrical circuit and earth (or conducting body connected to earth).

earth circuit. That part of a radio transmitter or receiver circuit which includes an earth lead or counterpoise.

earth currents. (1) Currents in the earth which, by electromagnetic induction, cause irregular currents to flow in submarine cables and so interfere with the reception of the transmitted signals. (2) Direct currents in the earth, which are liable to cause corrosion of the lead sheaths of cables; they can be the earth-return currents of power systems.

earth electrode system. The totality of conductors, conduits, shields, and screens which

are connected to the main earth by low-impedance conductors.

earth lead. Connection between a radio transmitting or receiving apparatus and its earth.

earth potential. (1) The electric potential of the earth; usually regarded as zero, so that all other potentials are referred to it. (2) The potential of any point in a circuit when it does not vary at a radio frequency, no matter what the steady or low-frequency potential may be. See **earthy.**

earth-return circuit. One which comprises an insulated conductor between two points, the circuit being completed through the earth.

earth system. System of wires, usually buried underneath an antenna system, to increase conductivity of the ground around antenna, making a nearly perfect reflector. Also **ground system.**

earth wire. One leading to the earth connection.

earth's field. That which arises from natural magnetization of the earth, exhibiting considerable variation in magnitude and direction. Directed from a region (magnetic N pole) off New Zealand towards an area (magnetic S pole) in the north of Canada. These points are not exactly diametrical and the effective poles are only about a third of the way from the earth's centre to the surface.

earth's radius. Effective radius for calculations on radio transmission, allowing for a refractive index increasing linearly with height, is 4/3 actual radius for radio waves.

earthed aerial. Marconi aerial, in which an elevated wire is earthed at its lower end.

earthing switch. One for connecting an antenna to earth when not in use, as a protection against lightning and/or the accumulation of static charge. See **antenna-.**

earthy. Said of (1) circuits when they are connected to earth, either directly (as for direct currents) or through a capacitor (with alternating currents); (2) any point in a communicating system (e.g., the midpoint of a shunting resistance across a balanced line) which is at earth potential, although not connected to earth through zero impedance; (3) points of a bridge when reduced to earth potential by a Wagner earth.

ebonite. A hard insulating material of rubber which has been vulcanized, i.e., the latex molecules have been cross-linked through sulphur atoms.

ebony resin. A synthetic resin having little strength and used for encapsulating and embedding electronic components.

eccentric groove. See **locked groove.**

Eccles-Jordan circuit. Original bistable multivibrator, using two triodes or transistors. See also **flip-flop.**

echo. (1) Received acoustic wave, distinct from a directly received wave, because it has travelled a greater distance due to reflection or refraction. (2) Return signal in radar, whether from wanted object or from side or back lobe radiation.

echo box. Adjustable test resonator (of high Q) for returning a signal to the receiver from the transmitter.

echo chamber. Same as **reverberation chamber.**

echo checking. Verification of computer data transmission by reference back to and matching with original data.

echo flutter. A rapid sequence of reflected radar pulses arising from one initial pulse.

echo (or sound) ranging sonar. Determination of distance and direction of objects, such as submarines, by the reception of the reflection of an ultrasonic pulse under water. See **asdic.**

echo sounding. Use of echoes of pressure waves sent down to the bottom of the sea and reflected, the delay between sending and receiving times giving a measure of the depth; used also to detect wrecks and shoals of fish.

echo suppression. (1) In navigational apparatus, the desensitizing of equipment for a short period after the reception of a pulse. By this means, pulses travelling in indirect paths are rejected. (2) In telephone 2-way circuits, the attenuation of echo currents in one direction which is due to telephone currents in the other direction.

echo suppressor. Circuit used to eliminate reflected signals arising from mismatch in line transmission, or duplicate signals received by indirect transmission path in radio communication.

eddy-current heating. See **induction heating.**

eddy-current loss. The loss of energy due to eddy-current heating which becomes more important at high radio frequencies. Hence care in design of cores of inductances and use of powdered iron and of ferrite cores.

eddy currents. Those arising through varying electromotive forces consequent on varying magnetic fields, resulting in diminution of the latter and dissipation of power. Used for mechanical damping and braking (as in electricity meters) and for induction heating as applied in case-hardening. Also called **Foucault currents.**

edge effect. (1) Deviation from parallelism in fields at the edge of parallel plate capacitors, or between poles of permanent or electromagnets, thus leading to non-uniformity of the field at the edges. (2) In acoustic absorption measurements, the variations which arise from the size, shape or division of the areas of material being tested—a diffraction effect.

edge tones. Tones produced by the impact of an air jet on a sharply-edged dividing surface, as in an organ pipe mouth.

edges. The rising and falling portions of a pulse signal, termed the **leading** and **trailing edges.**

Edison cell. See **nickel-iron secondary cell.**

Edison effect. The phenomenon of electrical conduction between an incandescent filament and an independent cold electrode contained in the same envelope, when the second electrode is made positive with reference to a part of the filament. Precursor of *Fleming diode.* Also **Richardson effect.**

edit. To prepare computer data for initial or further processing.

Editola. See **Acmeola.**

effective antenna height. Height (in metres) which, when multiplied by the field strength (in volts per metre) incident upon the antenna, gives the e.m.f. (in volts) induced therein. It is less than the physical height and differs from the equivalent height in that it is also a function of the direction of arrival of the incident wave.

effective atomic number. That of a compound or mixture, given by Spier as

$$Z_e = [a_1 Z_1^{2 \cdot 94} + a_2 Z_2^{2 \cdot 94} \ldots \ldots]^{1/2 \cdot 94},$$

where $a_1, a_2 \ldots$ are the fractional electron contents of elements of at. no. $Z_1, Z_2 \ldots$.

effective bandwidth. The bandwidth of an ideal (rectangular) band-pass filter, which would pass the same proportion of the signal energy as the actual filter.

effective energy (or wavelength). The quantum energy (or wavelength) of a monochromatic beam of X-rays or γ-rays with the same penetrating power as a given heterogeneous beam. Its value depends upon the nature of the absorbing medium.

effective half-life. The time required for the activity of a radioactive nuclide in the body to fall to half its original value as a result of both biological elimination and radioactive decay. Its value is given by

$$\frac{\tau(b\tfrac{1}{2}) \times \tau(r\tfrac{1}{2})}{\tau(b\tfrac{1}{2}) + \tau(r\tfrac{1}{2})},$$

where $\tau(b\tfrac{1}{2})$ and $\tau(r\tfrac{1}{2})$ are the biological and radioactive half-lives respectively.

effective radiated power. Actual maximum or unmodulated power delivered to a transmitting antenna, multiplied by the factor of *gain* in a specified direction in the horizontal plane.

effective reactance. Quotient of the component of the voltage in quadrature with the current (sinusoidal), and that current.

effective resistance. The total a.c. resistance covering eddy-current losses, iron losses, dielectric and corona losses, and transformed power, as well as conductor loss. For a sinusoidal current it is the component of the voltage in phase with current divided by that current. Measured in ohms.

effective value. Same as **rms value**, $1/\sqrt{2}$ times the amplitude of a sine wave of current or voltage or of an acoustic wave.

effective wavelength. See **effective energy.**

efficiency (or booster) diode. One used in television receivers to increase the beam deflection voltage of a CRT during the forward part of cycle of operation.

effluent monitor. Instrument for measuring level of radioactivity in fluid effluent.

effluve. The corona discharge from an electrostatic machine or high-frequency generator, to stimulate the human skin.

egg-box lens. Same as slatted lens.

eidophor system. A projection television system in which diffraction effects are used to produce the modulation of light for the picture.

eigentones. The natural frequencies of vibration of a system.

eigenvalues. Possible values for a parameter of an equation for which the solutions will be compatible with the boundary conditions. In quantum mechanics the energy eigenvalues for the Schrödinger equation are possible *energy levels* for the system.

eightfold way. See supermultiplet.

Einstein-Bose statistics. See Bose-Einstein statistics.

Einstein-de Haas effect. See Barnett effect.

Einstein energy. See mass-energy equivalence.

Einstein photoelectric equation. That which gives the energy of an electron, just ejected photoelectrically from a surface by a photon,

i.e., $E = h\nu - w$,

where E = kinetic energy

h = Planck's constant

ν = frequency of photon

w = work function.

einsteinium (Es). Manmade element, at. no. 99, produced by bombardment in a cyclotron, but also recognized in H-bomb débris, it having been produced by beta decay of uranium which had captured a large number of neutrons. The longest-lived isotope is ^{254}Es, with a half-life of greater than 2 y.

elastance. The reciprocal of the capacitance of a capacitor, so termed because of electromechanical analogy with a spring. Unit is the *daraf*.

elastic collision. Such collision that no change in the total energy or momentum takes place.

elastic scattering. See scattering.

elasticity. The reciprocal of the *permittivity* of a dielectric.

electret. Permanently-polarized dielectric material, formed by cooling barium titanate from above a Curie point, or waxes, e.g., carnauba, in a strong electric field.

electric. Said of any phenomena which depend essentially on a peculiarity of electric charges. Cf. *electrical*.

electric axis. Direction in a crystal which gives the maximum conductivity to the passage of an electric current, e.g., the X-axis of a piezoelectric crystal. See circuit.

electric (or electrostatic) component. That component of an electromagnetic wave which produces a force on an electric charge, and along the direction of which currents in a conductor exposed to the field are urged to flow.

electric conduction. Transmission of energy by flow of charge along a conductor.

electric dipole. See electric doublet.

electric discharge. Same as field discharge.

electric double layer. A positive and a negative layer distribution of electric charge very close together so that effectively the total charge is zero but the two layers form an assembly of dipoles thus giving rise to an electric field.

electric doublet. System with a definite electric moment, mathematically equivalent to two equal charges of opposite sign at a very small distance apart.

electric eye. Miniature cathode-ray tube in a radio receiver which exhibits a pattern determined by the rectified output voltage obtained from the received carrier, thus assisting in tuning the receiver. Also used for balancing a.c. bridges. Also cathode-ray tuning indicator, magic eye.

electric field. Region in which attracting or repelling forces are exerted on any electric charge present.

electric field strength. The strength of an electric field is measured by the force exerted on a unit charge at a given point. Expressed in volt/metre.

electric flux. Surface integral of the electric field intensity normal to the surface. The electric flux is conceived as emanating from a positive charge and ending on a negative charge without loss.

electric flux density. See displacement.

electric intensity. See electric field strength.

electric length. Effective length of an electric path, i.e., transmission line expressed in wavelengths, degrees, or radians.

electric moment. Product of the magnitude of either of two equal electric charges and the distance between their centres, with axis direction from the negative to the positive charge. See also magnetic moment.

electric motor. Any device for converting electrical energy into mechanical torque.

electric oscillations. Electric currents which periodically reverse their direction of flow, at a frequency determined by the constants of a resonant circuit. See also continuous oscillations, electronic oscillations.

electric polarization. The dipole moment per unit volume of a dielectric.

electric potential. Measured by the energy of a unit positive charge at a point, expressed relative to that at infinite distance or at the surface of the earth (*zero potential*).

electric shielding. See Faraday cage.

electric spectrum. The colour spectrum of an electric arc.

electric storm. A meteorological disturbance in which the air becomes highly charged with static electricity. In the presence of clouds this leads to thunderstorms.

electric strength. Maximum voltage which can be applied to an insulator or insulating material without sparkover or breakdown taking place. The latter arises when the applied voltage gradient coincides with a

D

breakdown strength at a temperature which is attained through normal heat dissipation.

electric susceptibility. The amount by which the relative permittivity of a dielectric exceeds unity—or the ratio of the polarization produced by unit field, to the permittivity of free space.

electric wave filter. One in which there is a phase retardation for those currents which are passed and an effective time delay for a signal, comprising these frequencies, in getting through the filter. Also **frequency-discriminating filter.**

electric wind. Stream of air caused by the repulsion of charged particles from a sharply pointed portion of a charged conductor.

electrical. Descriptive of means related to, pertaining to, or associated with electricity, but not inherently functional. Cf. *electric.*

electrical chain. Number of circuits (e.g., tuned circuits) coupled together so that energy is transferred from one to the next.

electrical communication. See **telecommunication.**

electrical conductivity. Ratio of current density to applied electric field. Expressed in ohm^{-1} cm^{-1} in CGS units and ohm^{-1} metre^{-1} in MKSA units. Conductivity of metals at high temperatures varies as T^{-1}, where T is absolute temperature. At very low temperatures, variation is complicated but increases (at one stage proportional to T^{-5}), until finally limited by material defects of structure. See **conductance.**

electrical degrees. Angle, expressed in degrees, of phase difference of vectors, representing currents or voltages arising in different parts of a circuit.

electrical heating. Heating by electrical means, such as current flow through a resistance, induction currents in a conductor, displacement currents in a dielectric, etc. Cf. *arc.*

electrical hygrometer. An instrument which measures humidity using electrical techniques.

electrical interference. Interference due to the operation of electrical apparatus other than that arising from radio transmissions.

electrical recording. Use of amplified currents from microphones for operating electromagnetic or electrodynamic drives for the cutting stylus in wax recording, in contrast to the previous acoustic (mechanical) method of recording, now obsolete. See **tape recording.**

electrical reproduction. Reproduction from gramophone records by piezo- or electromagnetic devices operated by the tracking needle, as contrasted with *acoustic reproduction*, in which the tracking needle drives the centre of the diaphragm of a sound box connected to a horn, and now obsolescent.

electrical reset. Restoration of a magnetic device, e.g., relay or circuit-breaker, by auxiliary coils or relays.

electrical resonance. Condition arising when a maximum of current or voltage occurs when the frequency of the electrical source is varied, e.g., when the length of a transmission line approximates to multiples of a quarter-wavelength and the current or voltage becomes abnormally large.

electrical technology. That covering the practical applications of electricity as accepted and established. Electrical engineering includes new applications of electric science for definite purposes, e.g., trade or military.

electricity. The manifestation of a form of energy associated with static or dynamic electric charges. See **atmospheric-, negative-, positive-, unit quantity of-.**

electrification. (1) Charging a network to a high potential. (2) Conversion of any motive power to electric drive, e.g., trains. (3) Charging a conductor by electric induction from another charged conductor.

electro-acoustics. Branch of technology dealing with the interchange of electric and acoustic energy, e.g., as in a transducer.

electro-arteriograph. Instrument for recording blood flow rates.

electro-ballistics. The study of velocities of projectiles using electronic means.

electrobiology. The study of electrical phenomena associated with the function of living organisms.

electrocapillarity. Change of surface tension at the interface between two conductors, e.g., mercury and acid, when a current is passed across. The effect forms the basis of the capillary electrometer, now obsolescent.

electro-cardiograph. Instrument for study of voltage waves produced by action of heart muscles of living animals.

electrocataphoresis. See **electrophoresis.**

electrochemical equivalent. The weight of a substance deposited at the cathode of a voltameter per coulomb of electricity passing through it.

electrochemical series. The classification of elements in the order of the electrode potential which is developed when an element is immersed in a solution of normal ionic concentration.

electrochemistry. Branch of technology dealing with the electrical and electronic aspects of chemical laws or processes.

electrochronograph. The combination of an electrically driven clock and an electromagnetic recorder for recording short time intervals.

electroculture. Stimulation of plant growth through electricity.

electrocution. Death resulting from electric shock.

electrode. (1) Conductor whereby an electric current is led into or out of a liquid (as in an electrolytic cell), or a gas (as in an electric discharge lamp or gas tube), or a vacuum (as in a valve). (2) In a semiconductor, emitter or collector of electrons or holes.

See also:

accelerating-	cathode
anode	collector
applicator	control-
beam-forming-	decelerating-

> dynode
> gas-
> grid
> guard ring
> hydrogen-
> intensifier-
> modulating-
> negative-
> pilot-
> plate (1)
> positive-
> signal-
> target.

electrode admittance. Admittance measured between an electrode and earth when all other potentials on electrodes are maintained constant.

electrode characteristic. Graph relating current in electronic device to potential of one electrode, that of all others being maintained constant.

electrode conductance. In-phase or real component of an electrode admittance.

electrode current. The net current flowing in a valve (or tube) from an electrode into the surrounding space.

electrode dark current. The current which flows in a camera tube or phototube when there is no radiation incident on the photocathode, given certain specified conditions of temperature and shielding from radiation. It limits the sensitivity of the device.

electrode dissipation. Power released at an electrode, usually an anode, because of electron or ion impact. In large valves the temperature is held down by radiating fins, graphiting to increase radiation, or by water- or oil-cooling.

electrode drop. The potential drop arising from the finite resistance of an electrode.

electrode impedance. Ratio of a small sinusoidal voltage on an electrode to the corresponding sinusoidal current, all other electrodes being maintained at constant potential.

electrode resistance. Reciprocal of *electrode conductance.*

electrodeless discharge. That arising in a discharge tube because of a sufficiently intense externally applied high-frequency electromagnetic field.

electrodeposition. The deposition electrolytically of a substance on an electrode, as in electroplating or electroforming.

electrodermal effect. Change in skin resistance consequent upon emotional reactions.

electrodialysis. Removal of electrolytes from a colloidal solution by an electric field between electrodes in pure water outside the two dialysing membranes between which is contained the colloidal solution.

electro-disintegration. Disintegration of nucleus under electron bombardment.

electrodynamic loudspeaker. Open-diaphragm loudspeaker, in which the radiating cone is driven by current in a coil fixed to the apex of the cone and moving axially in the radial magnetic field of a pot magnet.

electrodynamic microphone. See moving-coil microphone.

electrodynamic wattmeter. One for low frequency measurements. It depends on the torque exerted between currents carried by fixed and movable coils.

electrodynamics. Science dealing with the interaction, or forces between, currents, or the forces on currents in independent magnetic fields.

electroencephalograph. Instrument for study of voltage waves associated with the brain. It effectively comprises a sensitive detector (voltage or current), a d.c. amplifier of very good stability and an electronic recording system.

electro-endosmosis. Movement of liquid, under an applied electric field, through a fine tube or membrane. Also **electro-osmosis.**

electrofluor. A transparent material which has the property of storing electrical energy and releasing it as visible (fluorescent) light.

electrofluorescence. See **electroluminescence.**

electroforming. Electrodeposition of copper on stainless steel formers, to obtain components of waveguides with closely dimensioned sections.

electrogen. A molecule that emits electrons when it is illuminated.

electrography. Direct electric recording on to teledeltos paper.

electrokinetics. Science of electric charges in motion, without reference to the accompanying magnetic field.

electrokymograph. An electronic recording instrument used in electrobiology.

electroluminescence. That produced by the application of an electric field to a dielectric phosphor. Also termed **electrofluorescence.**

electrolysis. Chemical change, generally decomposition, effected by a flow of current through a solution of the chemical, or in its molten state, based on ionization.

electrolyte. Chemical, or its solution in water, which conducts current through ionization. See **colloidal-.**

electrolyte strength. Extent towards complete ionization in a dilute solution. When concentrated, the ions join in groups, as indicated by lowered mobility.

electrolytic capacitor. Effectively an electrolytic cell in which a very thin layer of non-conducting material has been deposited on one of the electrodes by an electric current. This is known as 'forming' the capacitor, the deposited layer providing the dielectric. Because of its thinness a larger capacitance is achieved in a smaller volume than in the normal construction of a condenser. In the so-called *dry* electrolytic capacitor the dielectric layer is a gas which however is actually 'formed' from a moist paste within the capacitor.

electrolytic cell. An assembly of electrodes in an electrolyte, such as a voltameter.

electrolytic conduction. Conduction by electrolytic ions.

electrolytic detector. Demodulator consisting of a metal electrode of very small area immersed in an electrolyte. When this is the cathode, polarization increases the resistance leading to partial rectification of the signal. (Obsolete.)

electrolytic dissociation. The splitting-up

(which is reversible) of substances into oppositely-charged ions.

electrolytic ion. Charged current carriers formed by dissociation of ionic compound in polar liquid such as water. See electrostatic bonding.

electrolytic polarization. Change in the potential of an electrode when a current is passed through it. As the current rises, polarization reduces the p.d. between the two electrodes of the system.

electrolytic polishing. By making a metal surface an anode and passing a current under certain conditions, there is a preferential solution so that the microscopic irregularities vanish, leaving a smoother surface; also termed **anode polishing, electropolishing.**

electrolytic rectifier. One with electrolyte and electrodes, such that one electrode becomes polarized and opposes the flow of current in one direction.

electrolytic tank. A device used to simulate electrical focusing systems, e.g., cathode-ray tubes or thermionic valves. The tank is filled with a poorly conducting fluid in which is immersed a scale model in metal of the desired system. Appropriate voltages are applied between the various parts and the equipotentials are traced out using a suitable probe.

electromagnet. Ferromagnetic core, embraced by a current-carrying coil, and which exhibits appreciable magnetic effects only when current passes.

electromagnetic braking. The use of an electromagnet to apply or remove a brake.

electromagnetic deflection. Deflection of the beam in a cathode-ray tube by a magnetic field produced by a system of coils carrying currents, e.g., for scanning a television image or for providing a time-base deflection.

electromagnetic focusing. The focusing of a beam of charged particles by magnetic fields associated with current-carrying coils. Used in CRTs and electron microscopes.

electromagnetic horn. Metal horn designed to radiate a beam of ultra-short-wave energy originated by a dipole within the horn.

electromagnetic induction. Transfer of electrical power from one circuit to another by varying the magnetic linkage. See Faraday's law of induction.

electromagnetic inertia. The energy required to stop or start a current in an inductive circuit. An electrical inductance behaves like a mass in a mechanical system.

electromagnetic lens. One using current-carrying coils to focus electron beams.

electromagnetic loudspeaker. One involving the motion of a magnetic armature.

electromagnetic mirror. A reflecting surface for electromagnetic waves; cf. radar *dish*.

electromagnetic pickup. One in which the motion of the needle, in following the recorded track, causes a fluctuation in the magnetic flux carried in any part of a magnetic circuit, with consequent electromotive forces in any coil embracing such magnetic circuit.

electromagnetic pole piece. In a U-shaped core, the pole pieces are attached at the free end and are often conical in shape to concentrate the magnetic field in the air gap.

electromagnetic pump. Pump designed for conducting fluids (e.g., liquid metals) and maintaining circulation without use of moving parts.

electromagnetic radiation. The emission and propagation of electromagnetic energy from a source, including long (radio) waves, heat rays, light, X-rays, and (hard) gamma-rays.

electromagnetic reaction. Reaction between the anode and grid circuits of a valve obtained by electromagnetic coupling; also called **inductive reaction, magnetic reaction.**

electromagnetic separator. Isotope separation by *electromagnetic focusing*, as in a mass spectrometer.

electromagnetic spectrum. See electromagnetic wave.

electromagnetic units. Any system of units based on assigning arbitrary value to permeability of free space. Unity on CGS EM system, $4\pi \times 10^{-7}$ on MKSA system. See unit conversion table in Appendices.

electromagnetic wave. One formed of time-dependent mutually perpendicular electric and magnetic fields. The *spectrum* of electromagnetic waves includes radio waves, heat and light rays, ultraviolet rays, X-rays and gamma-rays.

electromagnetism. Science of the properties of, and relations between, magnetism and electric currents.

electromechanical counter. One which records mechanically the number of electric pulses fed to a solenoid.

electromechanical recorder. One which changes electrical signals into a mechanical motion of a similar form and cuts the shape of the motion in appropriate medium.

electrometer. Fundamental instrument for measuring potential difference, depending on the attraction or repulsion of charges on plates or wires. See also absolute-, attracted-disk-, Compton-, Dolezalek-, Hoffman binant-, Lindemann-, string-, vibrating-reed-, Wulf string-.

electrometer tube or valve. A valve designed to have a very high insulation resistance to the grid, and to be suitable for use with very high grid input resistance circuits. Currents measured are in the range of $10^{-10} - 10^{-14}$ amp.

electromotive force. Difference of potential produced by sources of electrical energy which can be used to drive currents through external circuits. Abb. e.m.f., symbol *E*. See back-.

electromotive intensity. Same as potential gradient.

electromyograph. Instrument for study of voltages produced by muscular contractions.

electron. A fundamental particle with negative electric charge of $1 \cdot 601 \times 10^{-19}$ coulomb and mass $9 \cdot 107 \times 10^{-28}$ gm. Electrons are grouped round the nuclei of atoms in several possible shells, but also exist independently and produce the various electric effects observable in different materials. Due to their small size the wave properties of electrons and relativistic effects are particularly significant. An equivalent particle but with a positive electric charge is known as the *positron* and is the anti-particle to the electron. The term electron is sometimes used generically to cover both electrons and positrons. See **de Broglie wavelength, relativistic mass; also Auger effect, beta-particle, cathode ray, Compton effect, electron spin** (see **spin** (1)), **electron-volt, K, L, M, N, O, P,** and **Q shells, photoelectron, secondary-, valence-, *n*-type semiconductor.**

electron affinity. (1) Tendency of certain substances, notably oxidizing agents, to capture an electron. (2) See **work function.**

electron attachment. Formation of negative ion by attachment of free electron to neutral atom or molecule.

electron beam. A stream of electrons moving with the same velocity and direction in neighbouring paths and usually emitted from a single source such as a cathode.

electron-beam valve (or tube). One in which several electrodes control one or more electron beams. See **cathode-ray tube, klystron,** etc.

electron binding energy. Same as **ionization potential.**

electron camera. Generic term for a device which converts an optical image into a corresponding electric current directly by electronic means, without the intervention of mechanical scanning. See **emitron, iconoscope, image dissector.**

electron capture. That of a shell electron (K or L) by its own nucleus, decreasing the atomic number of the atom without change of mass.

electron charge/mass ratio. A fundamental physical constant, the mass being the rest mass of the electron.

electron cloud. Region in the inter-electrode space of an electron discharge tube containing large numbers of relatively stationary electrons. See also **space charge, virtual cathode.**

electron concentration. Ratio of number of valence electrons to number of atoms in a molecule.

electron conduction. That which arises from the drift of free electrons in metallic conductors when an electric field is applied. See *p*-type and *n*-type **semiconductors.**

electron-coupled oscillator. Valve with four (or more) electrodes, the first three electrodes being used for the production of the oscillations, the output being taken from a subsequent electrode which is electron-coupled to the oscillator circuit proper. Characterized by the very small effect of the load on the frequency of oscillation.

electron coupling. That between two circuits, due to an electron stream controlled by the one circuit influencing the other circuit. Such coupling tends to be unidirectional, the second circuit having little influence on the first.

electron density. The number of electrons per gram of a material. It is approx. 3×10^{23} for most light elements. In an ionized gas the equivalent electron density is the product of the ionic density and the ratio of the mass of an electron to that of a gas ion.

electron device. One which depends on the conduction of electrons through a vacuum, gas, or semiconductor.

electron diffraction. The investigation of crystal structure by the patterns obtained on a screen from electrons diffracted from the surface of crystals or as a result of transmission through thin metal films.

electron discharge. Current produced by the passage of electrons through otherwise empty space.

electron-discharge tube. Highly-evacuated tube containing two or more electrodes between which electrons pass.

electron drift. The actual transfer of electrons in a conductor as distinct from energy transfer arising from encounters between neighbouring electrons.

electron emission. The liberation of electrons from a surface.

electron gas. The 'atmosphere' of free electrons in vacuo, in a gas, or in a conducting solid. The laws obeyed by an electron gas are governed by *Fermi-Dirac statistics*, unlike ordinary gases to which *Maxwell-Boltzmann statistics* apply.

electron gun. Assembly of electrodes in a cathode-ray tube which produces the electron beam, comprising a cathode from which electrons are emitted, an apertured anode, and one or more focusing diaphragms and cylinders.

electron jet. Narrow stream of electrons, similar to a beam, but not necessarily focused.

electron lens. A composite arrangement of magnetic coils and charged electrodes to focus or divert electron beams in the manner of an optical lens. See also **electromagnetic focusing, electrostatic focusing.**

electron mass. A result of relativity theory, that mass can be ascribed to kinetic energy, is that the effective mass (m) of the electron should vary with its velocity according to the experimentally confirmed expression:

$$m = \frac{m_0}{\sqrt{1 - \left(\dfrac{v}{c}\right)^2}},$$

where m_0 is the mass for small velocities, c is the velocity of light, and v that of the electron.

electron microscope. Tube in which electrons emitted from the cathode are focused, by suitable magnetic and electrostatic fields, to form an enlarged image of the cathode on a fluorescent screen. By passing the electrons through an object, such as a virus, a vastly enlarged image can be obtained on a photographic plate. The instrument has very high resolving power compared with the optical microscope, due to the shorter wavelength associated with electron waves.

electron mirror. A 'reflecting' electrode in an electron tube, e.g., reflex klystron.

electron multiplier. Electron tube in which anode is replaced by a series of auxiliary electrodes, maintained at successively increasing positive potentials up to the final anode. Electrons emitted from the cathode impinge on the first of the auxiliary electrodes, from which secondary electrons are ejected, and travel to the next electrode, where the process is repeated. With suitable materials for the auxiliary electrodes, the number of secondary electrons emitted at each stage is greater than the number of incident electrons, so that very high overall amplification of the original tube current results. See **dynode.**

electron octet. The (up to) eight valency electrons in an outer shell of an atom or molecule. Characterized by great stability, in so far as the complete shell around an atom makes it chemically inert, and around a molecule (by sharing) makes a stable chemical compound.

electron optics. Control of free electrons by curved electric and magnetic fields, leading to focusing and formation of images. See **electron microscope.**

electron pair. Two valence electrons shared by adjacent nuclei, so forming non-polar bond.

electron paramagnetic resonance. Resonance arising from conduction electrons in metals and semiconductors.

electron radius. The classical theoretical value is $2 \cdot 82 \times 10^{-13}$ cm., but experimentally the effective value varies greatly with the interaction concerned.

electron relay. Three- (or more) electrode valve operating relays or circuits without taking any input energy.

electron runaway. Condition in a plasma when the electric fields are sufficiently large for an electron to gain more energy from the field than it loses, on average, in a collision.

electron scanning. Scanning or establishing a television image by an electron beam in a television camera tube or a cathode-ray tube (kinescope), normally using a rectangular raster with horizontal lines.

electron sheath. Electron space charge around an anode, when the supply of electrons is greater than demanded by the anode circuit; see also **electron cloud.**

electron shell. Group of electrons surrounding a nucleus of an atom, and having adjacent energy levels. Radiation is emitted or absorbed by electrons jumping between shells, according to certain rules, the radiation wavelengths being determined by changes of level (energy).

electron sink. In a toroid for plasma studies, the cold electrons which absorb energy from the injected high-speed ions.

electron spin. See **spin** (1).

electron-spin resonance. A branch of microwave spectroscopy in which orbital electrons of a molecular sample resonate with RF fields in a waveguide cavity—the resonance being detected by the resulting energy absorption from the signal.

electron telescope. Astronomical optical instrument, including electronic image converter, associated with a normal telescope.

electron trap. An acceptor impurity in a semiconductor.

electron tube or **valve.** In U.S., any electron device in which the electron conduction is in a vacuum or gas inside a gas-tight enclosure.

electron-volt. General unit of energy of moving particles, equal to the kinetic energy acquired by an electron losing one volt of potential, equal to $1 \cdot 601 . 10^{-12}$ erg; abb. **eV.** Larger units are the million-electron-volt (MeV), and the billion-electron-volt (BeV) or giga-electron-volt (GeV).

electron-wave tube. See **travelling-wave tube.**

Electrone. TN for electronic music generation using electrostatic generators for pure frequencies, which are synthesized into desired timbres for replacing church and theatre organs.

electronegative. (1) Said of acid-forming elements, generally non-metallic, whose atoms have a relatively large affinity for electrons in joining other atoms in forming chemical compounds. (2) Carrying a negative charge of electricity. (3) Tending to form negative ions, i.e., having a relatively positive electrode potential.

electronic. Pertaining essentially to devices depending on electrons, e.g., *electronic computer, electronic control.*

electronic charge. The unit in which all nuclear charges are expressed. It is equal to $1 \cdot 601 \times 10^{-19}$ coulomb.

electronic configuration. Arrangement of atoms or electrons in various states or orbits in a molecule or crystal.

electronic control. General description of a control of machinery, traction, lifts, etc., which includes essential electronic switching and timing of operations, with or without tape triggering.

electronic efficiency. For a valve, the ratio of the power at the desired frequency delivered by the electron stream to the circuit, to the average power supplied to the stream.

electronic equilibrium. That between ionization inside and outside the volume of gas in an air-wall ionization chamber when accurate radiation dose measurements are possible. Also **charged-particle equilibrium.**

electronic flash. Battery or mains device which charges a capacitor, the latter discharging through a tube containing neon (stroboscope) or xenon (photography) when triggered.

electronic formula. Formula which gives the

electronic charges available in radicals and atoms due to the valency electrons in the outer shell. Also called **ionic formula.**

electronic keying. Production of telegraphic signals by all-electronic system.

electronic microphone. One in which the acoustic pressure is applied to an electrode of a valve.

electronic music. See **electrosonic music.**

electronic oscillations. Those of very-high-frequency, generated by electrons moving between electrodes in a valve, the frequency being determined by the transit time. See **Barkhausen-Kurz oscillator.**

electronic pickup. One in which external vibration affects the grid of a valve and thereby modulates the anode current.

electronic sky screen equipment. See **elsse.**

electronic switch. Device for opening or closing electric circuit by electronic means.

electronic tuning. Changing the operating frequency of a system by changing the characteristics of a coupled electron beam.

electronic voltmeter. One which depends on the amplifying and rectifying properties of valves or transistors.

electronic wattmeter. One which uses valves or transistors.

electronics. Branch of science that deals with the study and application of electron devices, e.g., electron tubes, transistors, magnetic amplifiers, etc.

electronogen. Photosensitive molecule which may emit an electron when illuminated.

electro-optical effect. Interaction between a a strong electrical field and the refracting properties of a dielectric. See **Kerr effect, Pockel's effect.**

electro-osmosis. Same as **electro-endosmosis.**

electropathology. The study of human diseases using electrical apparatus.

electrophonic effect. Perception of sound when relevant currents are passed through the human body.

electrophonic music. Same as **electrosonic music.**

electrophoresis. Motion of colloidal particles under an electric field in a fluid, positive groups to the cathode and negative groups to the anode.

electrophorus. Simple electrostatic machine for repeatedly generating charges. A resinous plate, after rubbing, exhibits a positive charge, which displaces a charge through an insulated metal plate placed in partial contact. Earthing the upper surface of this plate leaves a net negative charge on the metal plate when it is removed, a process which can be repeated indefinitely.

electroplane camera. An electronic device applied to the lens elements of a motion picture camera to improve depth of field.

electroplating. Deposition of one metal on another by electrolytic action on passing a current through a cell (for decoration or for protection from corrosion). Metal is taken from the anode and deposited on the cathode through a solution containing the metal as an ion.

electroplating bath. Tank in which objects to be electroplated are hung. It is filled with electrolyte at the correct temperature, with anodes of the metal to be deposited on articles which are made cathodes.

electropneumatic. Said of control system using both electronic and pneumatic elements.

electropolar. Possessing magnetic poles or positive and negative charges.

electropolishing. See **electrolytic polishing.**

electropositive. (1) Carrying a positive charge of electricity. (2) Tending to form positive ions, i.e., having a relatively negative electrode potential.

electroscope. Indicator and measurer of small electric charges, usually two gold leaves which diverge because of repulsion of like charges; with one gold leaf and a rigid brass plate indication is more precise. See **gold-leaf-, Lauritsen-, pith-ball-.**

electrosonic music. Music or other sounds produced by electronic means (e.g., by oscillators, photocells, or generators), then combined electrically and reproduced through loudspeakers; also called **electronic music, electrophonic music.**

electrostatic accelerator. One which depends on the electrostatic field due to large d.c. potentials.

electrostatic actuator. Apparatus used for absolute calibration of a microphone through application of a known electrostatic force.

electrostatic adhesion. That between two substances, or surfaces, due to electrostatic attraction between opposite charges.

electrostatic bonding. Valence linkage between atoms arising from transfer of one or more electrons from outer shell of one atom to outer shell of another, transfer leading to more near completion of outer shells of both atoms, producing *ions* on dissociation.

electrostatic component. See **electric component.**

electrostatic coupling. That between valves, one anode applying a voltage to the next grid through a capacitor.

electrostatic deflection. Deflection of the beam of a cathode-ray tube by an electrostatic field produced by two plates between which the beam passes on its way to the fluorescent screen.

electrostatic field. Electric field associated with stationary charges.

electrostatic focusing. Focusing in high-vacuum cathode-ray tubes by the electrostatic field produced by two or more electrodes maintained at suitable potentials.

electrostatic generator. Generator operated by electrostatic induction, e.g., Van de Graaff generator.

electrostatic induction. Movement and manifestation of charges in a conducting body by the proximity of charges in another body. Also the separation of charges in a dielectric by an electric field.

electrostatic instrument. See **electroscope, electrometer.**

electrostatic Kerr effect. Dispersion of the plane of polarization experienced by a beam of plane-polarized light on its passage through a transparent medium subjected to an electrostatic strain. The basis of action of some light-modulation systems.

electrostatic lens. An arrangement of tubes and diaphragms at different electric potentials. See **electron lens.**

electrostatic loudspeaker. One in which the mechanical forces are due to action of electrostatic fields. Also **capacitor loudspeaker.**

electrostatic machine. See **electrostatic generator.**

electrostatic memory. A device in which the information is stored as electrostatic energy, e.g., storage tube, Williams tube. Also **electrostatic storage.**

electrostatic microphone. See **capacitor microphone.**

electrostatic precipitation. Use of an electrostatic field to precipitate solid (or liquid) particles in a gas, e.g., in dust removal.

electrostatic reaction. Capacitance reaction through a separate capacitor. See also **Miller effect.**

electrostatic screen. (1) Conducting shield surrounding instruments or other apparatus, to prevent their being influenced by external electric fields. (2) An earthed conducting plate interposed between two circuits to prevent unwanted capacitance coupling between them; also **electrostatic shield.**

electrostatic separator. Apparatus in which materials having different permittivities are deflected by different amounts when falling between charged electrodes, and therefore fall into different receptacles.

electrostatic shield. Same as **electrostatic screen** (2).

electrostatic storage. See **electrostatic memory.**

electrostatic units. Units for electric and magnetic measurements in which the permittivity of a vacuum is taken as unity, with no dimensions on CGS system.

electrostatic voltmeter. One depending for its action upon the attraction or repulsion between charged bodies; usual unit of calibration *kilovolt.* See also **electrometer.**

electrostatic wattmeter. One which utilizes electrostatic forces to measure a.c. power at high voltages.

electrostatics. Section of science of electricity which deals with the phenomena of electric charges substantially at rest.

electrostriction. Change in the dimensions of a dielectric accompanying the application of an electric field. See **crystal-.**

electrotherapy. The treatment of diseases by electricity. Also **electrotherapeutics.**

electrothermic (or electrothermal) instrument. One depending on the Joulean heating of a current for its operation. See **bolometer, hot-wire instrument.**

electrotyping. The production (or reproduction) of printing plates by means of electroforming.

electrovalence. Chemical bond in which an electron is transferred from one atom to another, the resulting ions being held together by electrostatic attraction.

electroviscosity. Minor change of viscosity when an electric field is applied to certain polar liquids.

element. (1) Chemically simple substance which cannot be resolved into simpler substances by normal chemical means. Because of the existence of *isotopes* (q.v.) of elements, an element cannot be regarded as a substance which has identical atoms, but of atoms which have the same atomic number. Before 1940, 92 elements had been discovered, in 1968 the number was 105, the more recently discovered elements being the artificial products, often very short-lived, of nuclear fission in a cyclotron, or, as in the case of einsteinium (99) and fermium (100), of large-scale military nuclear explosions. The total probable number of elements may be 110. (2) In computing, unit of a code pattern, e.g., an impulse or no impulse, or a dot or dash, etc. (3) Unit of an assembly, especially the detailed parts of electron tubes which affect its operation. (4) Component part of a system when considered theoretically. (5) In mathematics, a unit in a matrix or determinant.

elementary particle. Particle believed incapable of subdivision. At the 1968 International Conference on High Energy Physics it was reported that the existence of 68 elementary particles grouped in isotopic spin multiplets had been confirmed. Detection of some 30 others, including the intermediate vector boson, was awaiting confirmation.

elliptic polarization. That of an electromagnetic wave in which the electric and magnetic fields each contain two unequal components, at right-angles in space and in phase quadrature.

Elsie. Letter-sorting system in which the address is presented to an operator who operates keys which print phosphorescent dots for photoelectric routing on the envelope. Coding is by first three and last two letters of the office. (*E*lectronic *l*etter *s*orting and *i*ndicating *e*quipment.)

elsse. Telemetric system for tracking the flight-path of ground-launched missiles. (*E*lectronic *s*ky *s*creen *e*quipment.)

emanations. Heavy isotopic inert gases resulting from decay of natural radioactive elements, i.e., *radon* from radium, *thoron* from thorium, *actinon* from actinium; these gases are short-lived and decay to other radioactive elements. They are radioisotopes 222, 220, 219 of element 86 or radon (q.v.), sometimes (deprecated) called *emanation.*

emission. Release of electrons from parent atoms on absorption of energy in excess of normal average. This can arise from (1) *thermal* (thermionic) agitation, as in valves, Coolidge X-ray tubes, cathode-ray tubes; (2) *secondary* emission of electrons, which are ejected by impact of higher energy primary electrons; (3) *photoelectric* release on absorption of quanta above a certain energy level; (4) *field* emission by actual stripping from parent atoms by high electric field. See **associated-, auto-, nuclear-, secondary-, secondary grid-.**

emission current. The total electron flow from an emitting source.

emission efficiency. See **cathode efficiency.**

emission spectrum. Wavelength distribution of electromagnetic radiation emitted by self-luminous source.

emissive power. The total energy emitted per unit area of a surface at a specified temperature and under defined surroundings. Also applies to emission of electrons. See **emission.**

emissivity. The ratio of emissive power of a surface at a given temperature to that of a black body at the same temperature and with the same surroundings. Values range from 1·0 for lampblack down to ·02 for polished silver.

emitron. Early British TV camera tube similar to *iconoscope* (q.v.). See **cathode potential stabilized camera tube.**

emitter. Outer section of a transistor, by which minority carriers enter the *base* region.

emitter junction. One biased in the low-resistance direction, to inject minority carriers into the *base* region.

emphasizer. An audio-frequency circuit which selects and amplifies specific frequencies or frequency bands.

empire cloth. An insulating fabric which is impregnated with linseed oil.

empty band. See **energy band.**

emulsion technique. Study of nuclear particles by means of tracks formed in photographic emulsion.

enabling pulse. One which opens a *gate* (q.v.) which is normally closed.

enantiotropy. The property of a substance of existing in two crystal forms, one being stable below, and the other above, a transition temperature.

encephalogram. Trace produced by electro-encephalograph showing voltage waves generated in the brain.

encoder. A network in a computer in which only one input is excited at any one time and each input gives rise to a combination of outputs.

end-hats. Metal fitments at the ends of the cathode in certain magnetrons to reduce space charge.

end-of-selection signal. One which, normally given by a register, may be transmitted in either direction and can be used to prepare the circuit for speech.

end-on (or end-fire) aerial array. A linear aerial array such that maximum radiation is along the axis of the array.

end plate. Disk-shaped electrode, partially or completely closing the end of a concentric cylindrical arrangement of anode and cathode in a thermionic tube, to which a

D*

negative potential is applied to prevent the escape of electrons from the end of the tubular anode.

end-plate magnetron. Magnetron fitted with end plates which, in conjunction with a magnetic field directed exactly parallel with the cathode, performs in the same manner as one without end plates, in which the magnetic field is slightly inclined to the axis of the cathode.

end product. The stable nuclide forming the final member of a radioactive decay series.

end-window counter. GM counter designed so that radiation of low penetrating power can enter one end. This is usually covered with a thin mica sheet.

endodyne. Same as **autodyne.**

endoergic process. Nuclear process in which energy is consumed.

energy. The capacity for doing work. An electrical capacitance of magnitude C, holding a charge Q, has an electrical energy $\frac{1}{2}\frac{Q^2}{C}$, which is resident in the field between the plates. On discharge an amount of work equal to $\frac{1}{2}\frac{Q^2}{C}$ will be obtained. *Units of energy*: in the metric system, the *erg* is the work done by a force of one dyne moving through 1 cm in the direction of force; and the *joule*, equal to 10^7 erg, is the work done by a force of 1 newton moving through 1 m in the direction of the force.

energy band. In a solid the energy levels of individual atoms interact to form bands of permitted levels with gaps between. Normally there is a valence band with a full complement of electrons and a conduction band which is *empty*. When these overlap, metallic conduction is possible. In semiconductors there is a small gap and intrinsic conduction occurs only when some electrons acquire the energy necessary to surmount this gap and enter the conduction band. In insulators the gap is huge and cannot normally be surmounted.

energy gap. Range of forbidden energy levels between two permitted bands.

energy levels. Electron energies in atoms are limited to a fixed range of values termed *permitted energy levels* and represented by horizontal lines drawn against a vertical energy scale. See **eigenvalues, Schrödinger equation.**

energy levels—donor and acceptor. In semiconductors a *donor level* is an intermediate level close to the conduction band; being filled at absolute zero, electrons in this level can acquire energies corresponding to the conduction level at other temperatures. An *acceptor level* is an intermediate level close to the normal band, but empty at absolute zero; electrons corresponding to the normal band can acquire energies corresponding to the intermediate level at other temperatures. See **Fermi characteristic energy level.**

energy/mass equivalence. Einstein formula,
$$E = mc^2$$
where E = energy
m = mass
c = velocity of light,
gives the energy liberated when mass is annihilated, as in A- and H-bombs and the stars. Reversal also occurs.

enriched uranium. Uranium in which the proportion of the fissile isotope ^{235}U has been increased above its natural abundance.

enriched-uranium reactor. Reactor using enriched uranium as fuel.

enrichment. (1) Of reactor fuel, raising the proportion of fissile nuclei above that for natural uranium. (2) Of isotope separation, raising the proportion of the desired isotope above that present initially.

enrichment factor. (1) In the U.K., the abundance ratio of a product divided by that of the raw material. The *enrichment* is the enrichment factor less unity. (2) In the U.S., same as **separation factor.**

enthalpy. Thermodynamic property of a working substance, defined as $h = u + pv$, where u = internal energy, p = pressure, and v = volume of system. Useful in studying flow process, replacing total heat.

entropy. (1) Deprecated term for *information* rate, the number of *bits* required to represent a message. (2) In thermodynamic processes, it is concerned with the probability of a given distribution of momentum among molecules. Any 'free' physical system will distribute its energy so that the entropy increases but the available energy diminishes. Any process in which no change of entropy occurs is said to be *isentropic*. An example of this is the propagation of a normal sound wave.

envelope. The modulation waveform within which the carrier of an amplitude-modulated signal is contained, i.e., the curve connecting the peaks of successive cycles of the carrier wave.

envelope delay. The propagation time between two fixed points of the envelope of a modulated wave.

envelope-delay distortion. This arises when the rate of change of phase shift with frequency for a transmission system is variable over the required frequency range.

epicadmium neutrons. Neutrons with epithermal energies just above the cadmium cutoff.

episcotizer. A rotating disk with alternate transparent and opaque sections used with a photocell to reduce the intensity of the light falling on the sensitive surface. See **sector disk.**

epitaxial transistor. One made from a semiconductor material by the deposition of the material on a suitable monocrystalline support.

epitaxy. Unified crystal growth or deposition of one crystal layer on another.

epithermal. Having energy just above thermal agitation level; comparable with *chemical bond energy*.

epithermal neutrons. See neutron.

epithermal reactor. See intermediate reactor.

Eputmeter. TN for a counting device for measuring the number of *events per unit time*.

equal energy source. An electromagnetic or acoustic source whose radiated energy is distributed equally over its whole frequency spectrum.

equal signal system. One in which two signals are emitted for radio range, an aircraft receiving equal signals only when on the indicated course.

equalization. Electronically, the reduction of distortion by compensating networks which allow for the particular type of distortion over the requisite band.

equalizer. See **equalizing network.**

equalizing network. (1) Network incorporating any inductance, capacitance or resistance, which is deliberately introduced into a transmission circuit to alter the response of the circuit in a desired way; particularly to equalize a response over a frequency range. (2) A similar arrangement incorporated in the coupling between valves in an amplifier. Also **equalizer.**

equalizing pulse. A pulse used in TV at twice the line frequency, which is applied immediately before and after the vertical synchronizing pulse. This is done to reduce any effect of line-frequency pulses on the interlace.

equalizing signals. Those added to ensure triggering at the exact time in a frame cycle.

equation of time. See mean solar day.

equation solver. A computer of specific design to solve mathematical equations.

equilibrium time. The time required for steady-state working of an isotope separation plant to be established.

equipartition of energy. The Maxwell-Boltzmann law, which states that the available energy in a closed system eventually distributes itself equally among the *degrees of freedom* (q.v.) present.

equipotential cathode. One in which the electron-emitting surface is carried on a metal or ceramic body, which is heated by an internal wire.

equipotential surface or region. One where there is no difference of potential, and hence no electric field. See **Faraday cage.**

equivalent absorption. Of an object in a room (or the room itself), the equivalent surface area of unity absorption factor which absorbs acoustic energy at the same rate. The *sabin* is the acoustical unit.

equivalent binary digits. The number required to express a given number of decimal digits or other characters.

equivalent circuit. One consisting of resistances, inductances, and capacitances, which behaves, as far as the current and voltage at its terminals are concerned, exactly as some other circuit or piece of apparatus, e.g., a transformer may be represented by an arrangement of resistances and inductances, and an effective self-capacitance.

equivalent conductance. Electrical conductance of a solution which contains one gram-equivalent weight of solute at a specified concentration, measured when placed between two plane-parallel electrodes, 1 cm apart.

equivalent diode. One which would draw the same cathode current as a given valve which has more electrodes.

equivalent electrons. These which occupy the same orbit in an atom, and therefore have the same principal and orbital quantum numbers.

equivalent height. That of a perfect antenna, erected over a perfectly conducting ground, which, when carrying a uniformly distributed current equal to the maximum current in the actual antenna, radiates the same amount of power.

equivalent network. One identical to another network either in general or at some specified frequency. The same input applied to each would produce outputs identical in both magnitude and phase, generated across the same internal impedance.

equivalent reactance. The value which the reactance of an equivalent circuit must have in order that it shall represent the system of magnetic or dielectric linkages present in the actual circuit.

equivalent resistance. The value which the resistance of an equivalent circuit must have in order that the loss in it shall represent the total loss occurring in the actual circuit.

equivalent sine wave. One which has the same frequency and the same rms value as a given wave.

equivalent T-networks. T- (or π-) networks equivalent in electrical properties to sections of transmission line, provided these are short in comparison with the wavelength.

equivalent weight. See **atomic weight**.

erase. To remove data in the store or register of a computer, e.g., electrostatic, magnetic, or any other form.

erase head. In magnetic tape recording, the head which saturates the tape with high-frequency magnetization, in order to remove any previous recording.

erasing. The removal of a recording from a magnetic tape by saturating the magnetic material with d.c. or by high-frequency a.c. in an erasing head or by the field of a permanent magnet.

erbium (Er). Element in rare earth group, at. no. 68, at. wt. 167·26.

erg. The unit of work or energy. See **energy**.

ergometer. An instrument for measuring the work done by a device.

error. See **deviation**.

error-correcting code. One constructed with redundant elements so that certain types of error can be detected *after* reception and corrected.

error in bearing. The difference between the corrected and the true (or relative) bearing.

error signal. Feedback signal in control system representing deviation of controlled variable from set value.

Esaki diode. See **tunnel diode**.

escape factor. Fraction of the tracks in a nuclear emulsion which are not complete, due to the escape of the particle from the emulsion.

Estiatron. Travelling-wave tube with electrostatic focusing.

etchant. Chemical for removing copper from laminate during production of printed circuits.

ether. A hypothetical, non-material entity supposed to fill all space whether 'empty' or occupied by matter. The theory that electromagnetic waves need such a medium for propagation is no longer tenable.

ethylenediamine tartarate. A chemical in crystal form, exhibiting marked piezoelectric phenomena; used in narrow-band carrier filters.

Ettinghausen effect. Effect analogous to the *Hall effect*, concerned with a metal strip carrying an electric current longitudinally and placed in a magnetic field perpendicular to the plane of the strip. Differential temperatures are created between the opposite edges of the strip.

etude. Projected improved *stellarator*, characterized by a variable twist angle on the end loops. Magnetohydrodynamic theory suggests this should be the most stable form of controlled thermonuclear device so far developed.

eudiometer. (1) Voltameter-like instrument in which quantity of electricity passing can be found from volume of gas produced by electrolysis. (2) Similar system to determine volume changes in a gas mixture due to combustion.

eureka. See **rebecca-eureka**.

Eureka wire. Wire of an alloy of copper and nickel, approx. 55% Cu, 45% Ni. It has a very small temperature coefficient and does not deteriorate at high temperatures. Used for winding resistance coils, and can be soft-soldered.

europium (Eu). Element in rare earth group, at. no. 63, at. wt. 151·96.

Eurovision. System for relays of television programmes through the EBU between various countries, with or without change of resolution (lines per frame).

eutectic. Said of alloys with composition giving lowest possible melting-point.

evaporation. Escape of electrons or other particles having by chance a sufficiently enhanced outward velocity.

even-even nuclei. Nuclei for which the numbers of protons and neutrons are both even. Normally stable.

even-odd nuclei. Nuclei with an even number of protons and an odd number of neutrons.

exalted carrier. Addition of a synchronized carrier before demodulation, to improve linearity and to mitigate the effects of fading during transmission.

except. Operator in symbolic logic equivalent to *and not*. In Boolean algebra 'A except B' is written 'A.B.'

excess code. In computers, one which increases decimal digits before conversion to binary digits.

excess conduction. That arising from excess electrons provided by donor impurities.

excess electron. One added to a semiconductor by, e.g., a donor impurity.

excess pressure. Same as **sound pressure.**

excess-3 code. Binary coding of a decimal number, to which has been added 3. Complements are formed by 1 changed to 0, and 0 changed to 1, in all numerics.

exchange. (1) The interchange of one particle between two others, e.g., a pion between two nucleons, leading to establishment of *exchange forces.* (2) Possible interchange of state between two indistinguishable particles, which involves no change in the wave function of the system.

exchange force. Force arising from exchange of one particle between two others—see **exchange** (1). Both covalent molecular bond (*electron exchange*) and nuclear force between nucleons (*pion exchange*) are examples.

excitation. (1) Signal which drives any valve stage in a transmitter or receiver. (2) Addition of energy to a system, such as an atom or nucleus, raising it above the *ground state.* (3) Current in a coil which gives rise to a m.m.f. in a magnetic circuit, especially in a generator or motor. (4) The m.m.f. itself. (5) The magnetizing current of a transformer.

excitation anode. An auxiliary anode used to maintain the cathode spot of a mercury-pool cathode valve.

excitation potential. Potential required to raise orbital electron in atom from one energy level to another. Also **resonant potential**; see **ionization potential.**

excited atom. One with more energy than in the normal or ground state. The excess may be associated with the nucleus or an orbital electron.

excited nucleus. One with an excess of energy over its normal or ground state.

exciter. An original source of high-frequency oscillations in an independent drive transmitter, comprising the master oscillator and its immediately subsequent amplifying stages; also called **driver unit.**

exciter lamp. That which provides the light to be modulated for recording sound photographically on a sound-track, or the light-source for modulation by the sound-track in the sound-head of a projector.

exciton. Bound hole-electron pair in semiconductor. These have a definite half-life during which they migrate through crystal. Their eventual recombination releases energy as a photon, or less often, several phonons.

excitron. Single-anode steel-tank mercury-arc rectifier, the arc being initiated by a jet of mercury from a pool.

exclusion principle. See **Pauli-.**

exoergic process. Nuclear process in which energy is liberated.

expanded sweep. A technique for speeding-up the motion of the electron beam in a CRO during a part of the sweep.

expander. Amplifying apparatus for automatically increasing the contrast in speech modulation, particularly after reception of speech which has had its contrast compressed by a *compressor.*

expansion. A technique by which the effective amplification applied to a signal depends on the magnitude of the signal being larger for the bigger signals; see **automatic volume-.**

exploring coil. One used to measure magnetic field strengths by finding the change of magnetic linkages on removing the coil from a position in the field to a remote point out of the influence of the field. Also **search coil.**

exponential horn. One for which the taper or flare follows an exponential law. Ideal acoustically, but not fully realisable in practice. See **logarithmic horn.**

exponential reactor. One with insufficient fuel to make it diverge, but excited by an external source of neutrons for the determination of its properties.

exposure dose. See **dose.**

exposure meter. A combination of photocell and current meter for measuring light intensity as guide for correct exposure time in photography.

extended tube. See **remote cutoff.**

extensometer. In general, an instrument for measuring dimensional changes of a material. Specifically, this can be done electrically by measuring the changes of capacitance in a capacitor which are directly influenced by the changes in dimensions of this specimen.

external feedback. Feedback in magnetic amplifiers in which the feedback voltage is derived from the rectified output current.

external radiation. Radiation received by a body from sources outside itself.

external store or **memory.** Supplementary computer store, outside the main unit.

extinction coefficient. Constant corresponding to *absorption coefficient* for electromagnetic waves, divided by *refractive index.*

extinction voltage (or potential). (1) Lowest anode potential which sustains a discharge in a gas at low pressure. (2) In a gas-filled tube, the p.d. across the tube which will extinguish the arc.

extraction. Isolation of required information *bit* from *word* in computer register, according to machine instruction.

extraction instruction. In a digital computer, an instruction to form a new word by a suitable arrangement of selected segments of given words.

extradop. Doppler system for tracking missiles.

extraordinary ray. See **double refraction.**

extrapolation. Extension of data of variables by a function over a range for which its validity has not been verified, but for which there is some justification. See **interpolation.**

extremely high frequencies. Band between 3×10^4 and 3×10^5 MHz. Abb. **EHF.** See also **ultra high frequencies.**

extrinsic. Said of conduction properties arising from impurities in the crystal.

extrinsic semiconductor. See **semiconductor.**

F

F display. Type of radar display, used with directional antenna, in which the target appears as a bright spot which is off-centre when the aim is incorrect.

F-layer. Upper ionized layer in the ionosphere resulting from the ultraviolet radiation from the sun and capable of reflecting radio waves back to earth at frequencies up to 50 MHz. At a regular height of 300 km during the night, it falls to about 200 km during the day. During some seasons, this remains as the F_1 layer while an extra F_2 layer rises to a maximum of 400 km at noon. Considerable variations are possible during particle bombardment from the sun, the layer rising to great heights or vanishing. Also known as the **Appleton layer.**

f-levels. See **fundamental series.**

f-state. That of an orbital electron when the orbit has angular momentum of three Dirac units.

face. Readable side of a punched computer card, which can be passed from a *pack* in a *hopper* or *magazine*, and read *face-up* or *face-down.*

face-centred cubic. A crystal structure whose unit cell structure has equivalent points occurring at corners as well as at centres of a six-faced shape, such as a cube. See **body-centred cubic.**

face plate. That part of a cathode-ray tube which carries the phosphor screen, independently of the glass envelope, but mounted therein.

facsimile. The scanning of any still graphic material to convert the image into electrical signals, for subsequent reconversion into a likeness of the original.

facsimile bandwidth. The frequency difference between the highest and lowest components necessary for the adequate transmission of the facsimile signals.

facsimile baseband. The frequency band occupied by the signal before modulation of the carrier frequency to produce the radio or transmission line frequency.

facsimile density. Measure of the light transmission (or reflection) properties of an area, given by

$$\log_{10}\left(\frac{\text{incident intensity}}{\text{transmitted (or reflected) intensity}}\right).$$

facsimile modulation. The process by which the amplitude, phase or frequency of the wave in facsimile transmission is varied with time, in accord with a signal. The device for changing the type of modulation from receiving to transmitting, or vice versa, is known as a **converter.**

facsimile receiver. One for translating signals from communication channel into facsimile record of original copy.

facsimile telegraphy. Transmission of still pictures over telegraph circuits by photoelectric scanning, modulating a carrier, and consequent reconstruction of the picture by synchronous scanning. A radio link may be included in the transmission circuit. Also **picture telegraphy.**

facsimile transmitter. The means for translating the subject copy into signals suitable for the communication channel. The copy is rotated on a drum and the differences in the brightness of the light reflected from the copy modulate the output of a receiving photocell.

fade, fading. Phenomenon represented by more or less periodic reductions in the received field strength of a distant station, usually as a result of interference between reflected and direct waves from source. Mitigated by using a specific downcoming wave (*musa*), or by *diversity reception.*

fader. Potentiometer device or variable attenuator; used, in a communication channel, for varying a signal level continuously from zero to maximum, or vice versa. It is thus a means of maintaining a constant signal level while one signal is faded out and another is faded in.

fading area. Area in which fading is experienced in night reception of radio waves, between the primary and secondary areas surrounding a station transmitting on medium and long waves.

fail safe. Design in which power supply, control or structural failure leads to automatic operation of protective devices.

fall-out. Airborne radioactive contamination resulting from distant nuclear explosion, inadequately filtered reactor coolant, etc.

fall time. The decaying portion of a wave pocket or pulse; usually the time taken for the amplitude to decrease from 90% to 10% of the peak amplitude.

family. The group of radioactive nuclides which form a decay series.

fan antenna. Antenna in which a number of vertically inclined wires are arranged in a fanwise formation, the apex being at the bottom.

fan marker beacon. A form of marker beacon radiating a vertical fan-shaped pattern.

farad. The practical and absolute MKSA unit of electrostatic capacitance, defined as that which, when charged by a p.d. of one volt, carries a charge of one coulomb. Equal to 10^{-9} electromagnetic units and 9×10^{11} electrostatic units. Symbol F. This unit is in practice too large, and the subdivisions *microfarad* (μF), *nanofarad* (nF), and *picofarad* (pF) are in general usage.

faraday. Quantity of electricity associated

109

with one gram-equivalent of chemical charge, i.e., 96,500 coulomb.

Faraday cage. Earthed wire screen (e.g., a number of parallel wires joined at one end which is earthed) completely surrounding a piece of equipment in order to shield it from external electric fields and so that there can be no electric field within. Also called **Faraday shield.**

Faraday dark space. Dark region in a gas-discharge column between the negative glow and the positive column.

Faraday disk. Rotating disk in the gap of an electromagnet, so that a low e.m.f. is generated across a radius. Used in the Lorentz machine to generate a calculated electromotive force to balance against the drop across a resistance due to a steady current, thus establishing the latter in absolute terms from the calculation of the mutual inductance between the disk and the exciting air-cored coil.

Faraday effect. Rotation of the plane of polarization of (1) a plane-polarized light beam when propagated through a transparent isotropic medium, (2) a plane-polarized microwave passing through a ferrite along the direction of a magnetic field. If l is the length of path traversed, H is the strength of the magnetic field, and θ is the angle of rotation then $\theta = ClH$, where C is *Verdet's constant.*

Faraday shield. See **Faraday cage.**

Faraday tube. Tube of force in an electric field, of such magnitude that unit charge gives rise to one tube.

Faraday's ice-pail experiment. Classical experiment which consists in lowering a charged body into a metal pail connected to an electroscope, in order to show that charges reside only on the outside surface of conductors.

Faraday's laws of electrolysis. (1) The amount of chemical change produced by a current is proportional to the quantity of electricity passed. (2) The amounts of different substances liberated or deposited by a given quantity of electricity are proportional to the chemical equivalent weights of those substances.

Faraday's law of induction. The e.m.f. induced in any circuit is proportional to the rate of change of the number of magnetic lines of force linked with the circuit. Principle used in every motor. *Maxwell's field equations* (q.v.) involve a more general mathematical statement of this law.

faradic currents. Induced currents obtained from secondary winding of an induction coil and used for curative purposes.

Fast. System for testing transistors, incorporating a memory for subsequent read-out of parameters. (Facility for *a*utomatically *s*orting and *t*esting.)

fast fission. ^{238}U has a fission threshold for neutrons of energy about 1 MeV, and the fission cross-section increases rapidly with energy. Fission of this isotope by fast neutrons may cause a substantial increase in the reactivity of a thermal reactor (the **fast effect**).

fast neutron. See **neutron.**

fast reaction. Nuclear reaction involving strong interaction and occurring in a time of the order of 10^{-23} sec. Due to the strong interaction forces.

fast reactor. Reactor without moderator in which chain reaction is maintained almost entirely by fast fission.

fast store. One of small capacity, and hence of quick access, for programme or data often required in a calculation, thus increasing overall speed of operation, e.g., short mercury line or short nickel wire, in which coded information is continuously circulated.

fathometer. An ultrasonic depth-finding device.

fault current. That caused by defects in electrical circuit or device, such as short-circuit in system. The peak value of current is the accepted measure.

fault-finding. General description of locating and diagnosing faults, according to a pre-arranged schedule, generally arranged in a chart or table, with or without special instruments.

fault rate. See **reliability.**

fax. Abb. for **facsimile.**

Feather analysis. An approximate method of determining the range of β-rays forming part of a combined β-γ spectrum, by comparison of the absorption curve with that for a pure β-emitter.

Fechner colours. The visual sensations of colour which are induced by intermittent achromatic stimuli.

Fechner law. See **Weber-Fechner law.**

feed. To offer a programme or signal at some point in a communication network. See **anode-.**

feed current. Direct-current component of anode current of a thermionic valve when separated from alternating components.

feed holes. Same as **centre holes.**

feedback. Transfer of some output energy of an amplifier to its input in order to modify its characteristics. *Current* (or *voltage*) *feedback* (q.v.) when feedback signal depends on current (or voltage) in output respectively. *Bridge feedback* (q.v.) depends on their combination. See also:

acoustic-   positive-
back coupling   stabilized-feedback
derivative-   amplifier.
negative-

feedback admittance. The short-circuit transadmittance from output to input electrode of a thermionic valve.

feedback circuit. (1) Circuit conveying current or voltage feedback from the output to the input of an amplifier. (2) Circuit for using personnel from a programme source elsewhere.

feedback control loop. A closed transmission path including an active transducer, forward and feedback paths and one or more mixing points. The system is such that a given relation is maintained between the input and output signals of the loop.

feedback control system. U.S. for **closed loop system**.

feedback factor. In an amplifier using negative feedback it is the product of the fraction of the output voltage fed back and the gain of the amplifier before applying feedback.

feedback oscillator. Any oscillator in which the frequency is controlled by the parameters of a positive or negative feedback loop.

feedback path. That from the loop output signal to the feedback signal in a feedback control loop.

feedback ratio. That of the number of turns in the feedback winding to the number of the output winding in a simple series magnetic amplifier using current feedback.

feedback signal. That which is responsive in an automatic controller to the value of the controlled variable.

feedback transducer. One which generates a signal, generally electrical, depending on quantity to be controlled, e.g., for rotation potentiometer, synchro or tacho, giving proportional, derivative or integral signals respectively.

feedback windings. Those control windings in a saturable reactor to which are made the feedback connections.

feeder. (1) Conductor, or system of conductors, connecting the radiating portion of an antenna to the transmitter or receiver. It may be balanced pair, a quad, or coaxial. (2) In electrical circuits, the lines running from the main switchboard to the branch panels in an installation. See also **antenna-**, **concentric-**, **open-wire-**, **single-wire-**, **twin-**.

feedthrough. A conductor used to connect patterns on opposite sides of the board of a printed circuit, or an insulated conductor for connection between two sides of a metal earthing screen.

feeling threshold. Sound level applied to the ear, so loud that it results in a painful perception; about 120-130 on the *phon scale*.

fenestration. A surgical operation to improve hearing which involves a new 'window' being opened to the inner ear.

Fermat's principle. See **least time, principle of**.

fermi. A very small length unit, i.e., 10^{-3} cm. Used in nuclear physics, being of the order of the radius of the proton, viz., $1·2 \times 10^{-13}$ cm.

Fermi age. Slowing-down area for neutron calculated from *Fermi age theory*, which assumes that neutrons on being slowed down lose energy continuously and not in finite discrete amounts. (This name is deprecated as Fermi age has dimensions of cm^2.)

Fermi characteristic energy level. The highest occupied energy level in a partly filled band at a temperature of absolute zero. All energy levels below this will be occupied. In a semiconductor it will be the energy level

for which the Fermi-Dirac distribution function has a value of 0·5.

Fermi constant. A universal constant which indicates the coupling between a nucleon and a lepton field. Its value is $1·4 \times 10^{-49}$ erg cm^3, and is important in β-decay theory.

Fermi decay. Theory of ejection of electrons as β-particles.

Fermi-Dirac gas. An assembly of particles which obey Fermi-Dirac statistics.

Fermi-Dirac statistics. Statistical mechanics laws obeyed by a system of particles whose wave function changes sign when two particles are interchanged.

Fermi distribution. The energy distribution of electrons in a metal in which all levels are empty above the Fermi level except for a thin layer ($\sim kT$ in thickness) about the level.

Fermi level. See **Fermi characteristic energy level**.

Fermi plot. See **Kurie plot**.

Fermi potential. The equivalence of the energy of the Fermi level to an electric potential.

Fermi resonance. Degeneracy in polyatomic molecules when two vibrational levels belonging to different vibrations (or combinations) may have nearly the same energy.

Fermi selection rules. A particular set of *nuclear selection rules*, which follow from a definite assumption about the coupling term.

Fermi temperature. The degeneracy temperature of a Fermi-Dirac gas, which is defined by EF/k, where EF is the energy of the Fermi level and k is Boltzmann's constant. This temperature is of the order of tens of thousands of degrees for the free electrons in a metal.

fermion. Generalized particle treated by Fermi-Dirac statistical theory. Fermions have total angular moments of $(n+\frac{1}{2})\hbar$, where n is 0 or an integer and $\hbar$ (the Dirac constant) is the quantum or unit of angular momentum. Both baryons and leptons are fermions and are subject to the Pauli exclusion principle. Their number always appears to be conserved in any type of interaction. See **baryon number**, **lepton number**.

fermium (Fm). Manmade element, at. no. 100, produced in a cyclotron. The longest lived isotope appears to be ^{257}Fm with half-life 79 days, undergoing self-fission. It is also produced by β-decay of einsteinium and found in H-bomb débris.

ferractor. Magnetic amplifier with ferrite core.

ferrimagnetism. Anti-parallel alignment of magnetic moments of neighbouring ions, observed in ferrites and similar compounds.

Ferristor. TN for a wire-wound coil on a ferromagnetic core, encapsulated in epoxy resin.

ferrite. A ceramic iron oxide compound having ferromagnetic properties. It has formula MFe_2O_4, where M is generally a metal such as cobalt, nickel or zinc.

ferrite-bead memory. A memory device in which a mixture of ferrite powders is fused into small beads directly to the current-carrying wires of a memory matrix.

ferrite core. A magnetic core, usually in the form of a small toroid, made of ferrite material such as nickel ferrite, nickel-cobalt ferrite, manganese-magnesium ferrite, yttrium-iron-garnet, etc. These materials have high resistance and make eddy-current losses very low at high frequencies.

ferrite-rod antenna. Small reception antenna, using ferrite rod to accept electromagnetic energy, output being from an embracing coil. Also called **loopstick antenna.**

Ferrocart. TN for early ferromagnetic material suitable for use at high frequencies on account of its small hysteresis and eddy-current losses, achieved by subdivision of material into fine particles.

ferroelectric materials. Dielectric materials (usually ceramics) with domain structure, which exhibit spontaneous electric polarization. Analogous to *ferromagnetic materials* (see *ferromagnetism*). Have relative permittivities of up to 10^5, and show dielectric hysteresis. Rochelle salt was first to be discovered. Others include barium titanate and potassium dihydrogen phosphate.

ferromagnetic amplifier. A paramagnetic amplifier which depends on the non-linearity in ferroresonance phenomena at high RF power levels.

ferromagnetic resonance. A special case of paramagnetic resonance, exhibited by ferromagnetic materials — sometimes termed **ferroresonance**. It involves unpaired spins of electrons which are very close together and so subject to large exchange forces.

ferromagnetism. Phenomenon in which there is marked increase in magnetization in an independently established magnetic field. Some magnetism is retained in some ferromagnetic substances but can be removed by *demagnetization*. Certain elements (iron, nickel, cobalt) and alloys with other elements (titanium, aluminium) exhibit relative permeabilities much in excess of unity, up to 10^6 (*ferromagnetic materials*); some have marked hysteresis, leading to permanent magnets, storage devices for computers, magnetic amplifiers, etc.

ferromanganese. A ferromagnetic alloy of iron and manganese.

ferrometer. An a.c. instrument for measuring the magnetization of a ferromagnet.

ferronickel. A ferromagnetic alloy of iron and nickel.

ferroresonance. See **ferromagnetic resonance.**

ferrospinel. A crystalline material which has the equivalent function to that of the M in a ferrite. See **ferrite.**

Ferroxcube. TN for a ferrite, particularly useful for rod aerials and memory units.

fertile. Capable of conversion into fissile material in a reactor. ^{232}Th and ^{238}U are the most important examples.

Fessenden oscillator. A low-frequency underwater sound source of the moving-coil type.

Fick's law. The rate of molecular diffusion is proportional to the negative of the concentration gradient. This law holds for mass and energy transfer and also for neutron diffusion.

fidelity. The exactness of reproducibility of the input signal at the output end of a system transmitting information. See **high-.**

fiducial points. Points on the scale of an indicating instrument located by direct calibration, as contrasted with the intervening points, which are inserted by interpolation or subdivision.

field. (1) Space in which there are electromagnetic oscillations associated with a radiator. That component which represents the interchange of energy between radiator and space (practically negligible beyond a few wavelengths) is termed *induction field*; *radiation field* represents energy which is lost from the radiator to space. Region where components radiated by antenna elements are parallel is called **Fraunhofer region**, and where not the **Fresnel region.** The latter will exist between the antenna and the Fraunhofer region, and is usually taken to extend a distance $2D^2/\lambda$, where λ is the wavelength of the radiation and D is the aerial aperture in a given aspect. (2) In computers, area obtained by a vertical division of a punched card. (3) A set of scanning lines which, when interlaced with other such sets, constructs the complete TV picture. (4) The interaction of bodies or particles is explained in terms of fields, viz., electric, magnetic, gravitational, acoustic fields (see **field theory**). See also:

antenna-	meson-
clearing-	nuclear-
colour-	retarded-
conservative-	retarding-
critical-	rotating-
electrostatic-	solenoidal-
free-	sound-
guide-	uniform-.
magnetic-	

field control. The adjustment of the field current of a generator or motor to control the voltage and speed respectively.

field density. The number of lines of force passing normally through unit area of an electric or magnetic field.

field discharge. The passage of electricity through a gas as a result of ionization of the gas; it takes the form of a brush discharge, an arc, or a spark. Also **electric discharge.**

field emission. Same as **auto-emission.**

field-emission microscope. A device for investigating the effect of adsorption at a metal point on the value of the work function. A high voltage ($\sim$10-20kV) is applied between the single crystal metal point and a curved fluorescent screen. The electrons emitted from the point form a fluorescent image on the screen.

field-enhanced. Said of electron emission when a very strong field is effective at the emission surface.

field-free. Said of electron emission when there is no electric field at the emitting surface.

field frequency. Product of the *frame frequency* and number of *fields* per frame.

field intensity. Same as **field strength.**

field of force. Principle of *action at a distance*, i.e., mechanical forces experienced by an electric charge, a magnet, or a mass, at a distance from an independent electric charge magnet or mass, because of fields established by these, and described by uniform laws.

field oscillator. Same as **framing oscillator.**

field-sequential. In colour television, pertaining to the association of individual primary colours with successive fields.

field strength. (1) Vector representing the quotient of a force and the *charge* (or *pole*) in an electric (or magnetic) field, with the direction of the force; also called **field intensity.** (2) In radio, electromagnetic wave in volt/metre, i.e., that which induces an e.m.f. of 1 volt in an antenna of 1 metre effective height. Usually measured by frame or loop antenna.

field-synchronizing impulse. Impulse transmitted at the end of each field-scanning cycle, to synchronize time-base generator of the receiver with that at the transmitter.

field theory. As yet unverified attempt to link the properties of all fields (electric, magnetic, gravitational, nuclear) into a unified system.

Fieldistor. TN for transistor which uses an external field for the control of the electron flow.

figure of merit. General parameter which can be derived specifically for the performance of instruments, e.g., current or power required to deflect the light beam of a reflecting galvanometer 1 millimetre on a scale at 1 metre, or power amplification of a magnetic amplifier.

filament. The dissipative metallic element forming the heater of a thermionic valve.

filament cathode. Same as **directly-heated cathode.**

filament current. That required (a.c. or d.c.) to heat a filament in a thermionic valve.

filament efficiency. Ratio of the current emitted from a filament to that required to heat it.

filament getter. One which absorbs gas readily when hot, and is used for this purpose in a high-vacuum assembly.

filament limitation. The limitation of anode current in a thermionic tube by the finite emission from the filament, as distinct from *space-charge limitation*; also called **filament saturation.**

filament reactivation. Process of restoring the emission from used thoriated tungsten directly-heated valves, by operating for a short time at a temperature well above normal.

filamentary transistor. One whose length is much greater than its transverse dimensions.

film. (1) Any thin layer of substance, e.g., that which carries a light-sensitive emulsion for photography, that which carries iron-oxide particles in a matrix for sound recording. (2) Thin layer of material deposited, formed or adsorbed on another, down to monomolecular dimensions, e.g., electroplated films, oxide on aluminium, sputtered depositions on glass or microcomponents.

film badge. Small photographic film used as radiation monitor and dosimeter. Normally worn on lapel, wrist or finger and sometimes partly covered by cadmium and tin screens so that exposure to neutrons, and to β- and γ-rays, can be estimated separately.

film recording. Process of recording sound on sound-track on the edge of cinematograph film, for synchronous reproduction with the picture. The track is exposed so as to vary the density in the direction of motion of the film according to the recorded sounds, in order that subsequent scanning by a finely focused slit and photocell may reproduce the original modulation of the density.

film resistor. One formed by deposition of a metallic film on an insulator.

film store. In computers, store in which magnetic effect is produced in magnetic alloy deposited on a film, when currents in grid wires on both sides coincide.

filter. (1) Any device, or chain of similar devices, which discriminates between single frequencies or bands of frequencies of vibrations of specific phenomena, e.g., *crystal* filter in supersonic heterodyne receivers, *piezoelectric* or *magnetostriction* couplers, plain mechanical *resonators*, acoustic and electric *wave filters*, and *light filters*. (2) Device which discriminates between particles of different energies in the same beam.

See also:

acoustic-	high-pass-
amplitude-	interference-
band elimination-	lattice-
band-pass-	light-
bridged-T-	m-derived-
Butterworth-	magnetostrictive-
carrier-	neutral-
Cauer-	octave-
Chebyshev-	pi-section-
coaxial-	prototype-
colour-	scratch-
composite-	T-section-
crystal-	tared-
decoupling-	voice-
directional-	wave-
electric wave-	waveguide-.
harmonic-	

filter attenuation. The loss of signal power in its passage through a filter due to absorption, reflection or radiation. It is usually given in decibels.

filter attenuation band. A frequency band in which there is appreciable attenuation.

filter network. Usually a combination of inductors and capacitors for frequency separation of the input waves.

filter transmission band. A frequency band in

which there is negligible attenuation. Also called **pass band**.

fine structure. Splitting of optical spectrum lines into multiplets due to interaction between spin and orbital angular momenta of electrons in the emitting atoms.

finger. Probe used to sense presence or absence of a perforation in punched card or tape system.

fire control. System whereby gunfire is controlled by radar, which directs the aim after detection of target and computation of course with regard to all conditions and corrections.

fired. Said of gas tubes when discharging, particularly TR pulse tubes during the transmission condition.

Firestreak. De Havilland supersonic air-to-air rocket missile, based on infrared homing.

firing. (1) Sudden change from unsaturation to saturation of the core in a magnetic amplifier. (2) Establishment of discharge through gas tube.

firing angle. Angle of phase of an applied voltage which makes a vacuum or gas valve conducting, as in a *gating* valve or thyratron.

firing power. The minimum RF power required to start a discharge in a switching tube for a specified ignitor current.

firing time. The interval between applying a d.c. voltage to the ignitor electrode of a switching tube and the beginning of the discharge.

first detector. See **mixer** (2).

fish-paper. A flexible insulating material, usually of varnished cambric, for separating coil windings, etc.

fishbone antenna. End-fire array of vertical resonators spaced along a transmission line.

fissile. Capable of fission, i.e., breakdown into lighter elements of certain heavy isotopes (^{232}U, ^{235}U, ^{239}Pu), when these capture neutrons of suitable energy. Associated with Einstein release of energy; also **fissionable**. See **reactor**.

fission. The spontaneous or induced disintegration of a heavy atom into two or more lighter ones. The process involves a loss of mass which is converted into nuclear energy.

fission bomb. Same as **atomic bomb**.

fission chain. Atoms formed by uranium or plutonium fission have too high a neutron proton ratio for stability. This is corrected either by neutron emission (the delayed neutrons) or more usually by the emission of a series of beta-particles, so forming a short radioactive decay chain.

fission chamber. Ionization chamber lined with a thin layer of uranium. This can experience fission by slow neutrons, which are thereby detected by the consequent ionization.

fission neutrons. Those released by fission, having a continuous spectrum of energy with a maximum of ca. 10^6 eV.

fission poisons. Fission products with abnor-

mally high thermal neutron absorption cross-sections, which reduce the reactivity of nuclear reactors. Principally xenon and samarium, ^{135}Xe having an absorption cross-section of 3·5 million barn for slow neutrons.

fission products. Lower mass, highly radioactive atoms and particles resulting from fission, e.g., strontium-90, which is the major contributor to radiation in *fall-out* from nuclear explosions.

fission spectrum. The energy distribution of neutrons released by fission.

fission yield. The percentage of fissions for which one of the products has a specific mass number. Fission yield curves show two peaks of approximately 6% for mass numbers of about 97 and 138. The probability of fission dividing into equal mass products falls to about 0·01%.

fissionable. See **fissile**.

fissioning distribution. The modification of the neutron energy spectrum which results from allotting to each neutron a weight which is equal to the probability that its next collision will result in fission.

fix. Geographical location determined by intersection of radio bearings from 2 or 3 radio transmitters in known positions.

fixed lead. Navigation in which the missile flight path leads the line of sight by a constant angle.

fixed-point notation. Representation of a number by a fixed set of digits so that decimal or radix point has predetermined location.

flag. (1) Shield to protect lenses of television cameras from unwanted light. (2) Supporter of the *getter*, which is fired by eddy currents after the valve is evacuated. (3) An information bit used to indicate boundary of a field or end of a word in computing.

flap attenuator. One consisting of a strip of absorbing material which is introduced through a non-radiating slot in a waveguide.

flare. (1) Excess brightness in an image, especially at edges. (2) The law of increase of sectional area of horn on proceeding to the mouth.

flare out. (1) Controlled approach path of aircraft immediately prior to landing. (2) Increase in cross-section termination to open-ended waveguide or speaking tube. When the increase is linear in one plane on y, a *sectoral horn* is formed.

flash arc. See **Rocky Point effect**.

flash radiography. High-intensity, short-duration, X-ray exposure from an X-ray tube fed by a Marx high-voltage generator or betatron.

flasher. Switching circuit controlled by electrical relay with resistance and capacitance or by mechanical cams, for regularly applying power to lamps, e.g., for light-buoys or advertisements.

flashing. Operation in manufacture of thermionic cathode, in which it is raised to a very high temperature for a short period. Also used with carbon filaments in carbon atmosphere to obtain uniform cross-section.

flashover. An electric discharge over the surface of an insulator.

flashover voltage. The highest value of a voltage impulse which just produces flashover.

flask. See **coffin.**

flat lighting. That resulting from diffuse sources in front of the object, obviating depth or moulding.

flat random noise. See **white noise.**

flat region. Portion of reactor core over which neutron flux (and hence power level) is approximately uniform.

flat-top antenna. One in which the uppermost wires are horizontal; also called **roof antenna.**

flat tuning. Inability of a tuning system to discriminate sharply between signals having different frequencies.

flattening material. Neutron absorber or depleted fuel rod used in centre of reactor core to give larger *flat region.*

Fleming diode. That of 1904, used as a detector of received radio signals, with an incandescent filament and a separate anode. Also **Fleming valve.** See **Edison effect.**

Fletcher-Munson contours. Equal-loudness curves for aural perception, measured just outside the ear, extending from 20 to 20,000 Hz, and from the threshold of hearing to the threshold of pain. The basis of the *phon scale* of equivalent loudness level, and of the weighting networks used in sound-level indicating meters.

Flewelling circuit. Original super-regenerative receiving circuit in which quenching oscillations are generated by the same valve as is used for super-regeneration.

flexible resistor. One resembling a flexible cable.

Flexiguide. TN for *flexi*ble wave*guide.*

flexion point. U.S. for upper bend in the characteristic of a diode, depending on temperature limitation of emission, if stable.

Flexowriter. Keyboard perforator which provides simultaneous page-printing and punched tape. Page-printing is also possible from paper tape, or a new tape from an existing tape.

flexure crystal. Quartz-bar crystal, the bending of which is used for stabilizing valve oscillators or exhibiting resonance through ionization of surrounding neon gas.

flicker. Visual perception of fluctuation of brightness at frequencies lower than that covered by persistence of vision. Threshold of flicker depends on brightness and angle from optic axis. See also **chromaticity flicker.**

flicker effect. Irregular emission of electrons from a thermionic cathode due to spontaneous changes in condition of emitting surface; it results in an electronic noise.

flight control. Control of vehicle, e.g., aircraft or missile, so that it attains its target, taking all conditions and corrections into account.

Generally done by computer, controlled by signals representing actual and intended path.

flight path. The path in space of an aircraft or projectile. (Its *course* is the horizontal projection of this path.)

flip-flop. Originally (and still in U.S.), bistable pair of valves or transistors, two stable states being switched by pulses. In U.K., similar circuit with one stable state and one unstable state temporarily achieved by pulse. Circuit for binary counting, similar to *multivibrator* without coupling capacitors. See **Eccles-Jordan circuit, toggle.**

floating action. Description of control when rate of control is related to prescribed performance.

floating address. A label used to identify a particular word in a computer programme regardless of its specific location at any given time. Also **symbolic address.**

floating battery. An electrical supply system in which a storage battery and electrical generator are connected in parallel to share load, so that the former carries the whole load if the generator fails.

floating carrier. System in which a carrier is varied in amplitude to accommodate modulating signals.

floating-carrier wave. Transmission in which the carrier is reduced or even suppressed to zero during intervals of no modulation by signals, to economize in power or to avoid interference with reception.

floating grid. One which is unconnected in a valve and so is free to accumulate electrons.

floating gyro. One spinning in enclosure containing fluid to reduce friction, mainly by buoyancy.

floating point. Multiplying or dividing by radix, analogous to moving the decimal point when radix=10.

floating-point notation. That which uses additional digits to indicate the position of the decimal or other radix point in computers.

floating potential. That appearing on an isolated electrode when all other potentials on electrodes are held constant.

flocculation. Separation of radioactive waste products from water by coagulation.

flop-over. U.S. for a *flip-flop* (U.K. meaning) or trigger circuit.

flow counter. See **gas-, liquid-.**

flowchart. Chart giving a pictorial representation of the nature and sequence of operations to be carried out according to programme, in a computer or automatic control system.

fluctuation noise. Noise produced in the output circuit of an amplifier by shot and flicker effects.

Fluon. TN for **polytetrafluoroethylene.**

fluorescence. Emission of light, generally visible, from materials *irradiated* (generally from a higher frequency source), or from impact of electrons, as in a *phosphor.* See **Stokes' law.**

fluorescent screen. One coated with a layer of

luminescent material so that it fluoresces when excited, e.g., by X-rays or cathode rays. See **fluoroscopy, phosphor** (2).

fluorescent yield. Probability of a specific excited atom emitting a photon, in preference to an Auger electron.

fluorine (F). Pale greenish-yellow gas, the most electronegative (non-metallic) of the elements and the first of the halogens. Highly corrosive and never found free. Lat. *fluo* =to flow. Discovered by Scheele in 1771 and isolated by Moissan in 1886. At. no. 9, at. wt. 18·9984, m.p. −223°C, b.p. −187°C. Essential in separating the isotopes of uranium, in the compound uranium hexafluoride, which is a gas, and when forced through a porous plug differentiates very slightly between the heavier and lighter isotopes. Also combined with carbon to form complex compounds, which are very high resistance insulators, e.g., Teflon.

fluorography. The photography of fluoroscopic images.

fluoroscope. Measurement system for examining fluorescence optically. Fluorescent screen assembly used in fluoroscopy.

fluoroscopy. Examination of objects by observing their X-ray shadow shown on a fluorescent screen; used to ascertain contents of packages without unwrapping, quality of welding, etc.

flutter. Rapid fluctuation of frequency or amplitude in radio reception or sound reproduction, especially from disks. See **echo, wow.** (2) Variation in brightness of a reproduced television picture, arising from additional radio reflection from a moving object, e.g., an aircraft.

flux. See **electric-, magnetic-.**

flux density. The number of photons (or particles) passing through unit area normal to the beam, or the energy of the radiation passing through this area.

flux gate. Magnetic reproducing head in which magnetic flux (due to flux leakage from signals recorded on magnetic tape) is modulated by high-frequency saturating magnetic flux in another part of the magnetic circuit.

flux-gate compass. Device in which the balance of currents in windings is affected by the earth's magnetic field.

flux guidance. Directing the electric or magnetic flux in high-frequency heating by shaped electrodes or magnetic materials respectively.

flux link or linkage. Conservative flux across a surface bounded by a conducting turn. For a coil, *flux linkage* is the integration of the flux with individual turns.

flux of a vector. Integral of the product of the area of an element of a surface and component of a field normal thereto.

fluxmeter. Electrical instrument for measuring total quantity of magnetic flux linked with a circuit; it consists of a search coil placed in the magnetic field under investigation, and a ballistic galvanometer, or an

uncontrolled moving-coil element (Grassot) or semiconductor probe generating a Hall voltage. See also **gaussmeter.**

flyback. The return of the beam to its starting point in a radar trace or a line of a TV picture. In the latter the line is blanked out during the process. Also **retrace.**

flying-spot scanner. Device used in television for converting the light variations over the surface of a picture into a train of electric signals, the original picture being scanned by a small but extremely bright spot of light. See **scanner.**

flywheel effect. Maintenance of current in a resonant circuit through electrical inertia of the system during the interval between the exciting impulses. Used in television scanning circuits to control the frame frequency without using synchronizing pulses.

focusing. The convergence to a point of (a) beams of electromagnetic radiation, (b) charged particle beams, (c) sound or ultrasonic beams. See **alternating-gradient-, electrostatic-, magnetic-, phase-.**

focusing coil (or electrode). One used to focus a charged-particle beam by magnetic (or electrostatic) field control.

foetal echo. Detection of foetus by ultrasonic techniques; can be used at earlier stage than can radiography.

folded dipole. An aerial comprising interconnected parallel dipoles, connected together at their outer ends and fed at the centre of one dipole. The dipole separation is a small fraction of the transmitted wavelength.

folded horn. An acoustic horn which is turned back in itself to reduce necessary space.

follow (up) system. A *servo-positioning* system, e.g., on a machine tool, controlled by signals on a tape.

foot. The part of a valve envelope in which the electrodes are mounted.

foot-candle. Unit of illumination of one lumen per square foot. This is the flux density of luminous energy on a surface 1 ft from a uniform source of one candela.

foot-lambert. Unit of luminance of 1 lumen/sq.ft. For reflecting surfaces, the luminance in foot-lamberts is given by the product of the illumination in foot-candles and the reflectivity of the surface.

forbidden. In computing, said of a combination of symbols not in an operating code, and which thereby reveals a fault.

forbidden band. The gap between two bands of allowed energy levels in a crystalline solid. See **energy band, energy gap.**

forbidden transition. Transition of electrons between energy states which, according to Pauli selection rules, have a very low probability in relation to those which have a high probability and are *allowed*.

force. Mechanical force is equal to the rate of change of momentum of a body. Electromotive force, magnetomotive force, magnetizing force, etc., are strictly misnomers.

force on charge. If a charge Q is moving a

velocity V in a magnetic field of intensity B, the normal force is

$$F=B \cdot Q \cdot V \cos \theta,$$

where θ is the angle between the magnetic field and the direction of motion. If the charge is confined to a conductor, the force on a short length l of the latter is

$$F=B \cdot I \cdot l \cos \theta,$$

where I is the current.

forced oscillations. Oscillatory currents whose frequency is determined by factors other than constants of the circuit in which they are flowing, e.g., those flowing in a resonant circuit coupled to a fixed-frequency oscillator. Mechanical forced oscillations occur in a similar way (e.g., in servomechanisms) when controlled by the frequency of a vibrator.

fork frequency. That generated by a self-sustained tuning-fork for signal carrier in facsimile transmission and reception.

fork-tone modulation. A modulation of the picture carrier frequency by the synchronizing (fork) tone, used to ensure that the true fork frequency is being received for comparison with the local fork.

forked circuit. One which divides for either simultaneous or alternative reception from one transmitting system; also called **channel circuit.**

form factor. The ratio of the effective value of an alternating quantity to its average value over a half-period.

formant. Speech spectral pattern, arising in the vocal cords, which determines the distribution of energy in unvoiced sounds and the reinforcement of harmonics in voiced sounds. It is the formant which leads to recognition of speech sounds by aural perception.

formative time. The time interval between the first Townsend discharge in a given gap and the formation there of a self-maintaining glow discharge.

forward current. That which is due to the forward voltage applied to the system.

forward path. The transmission path from the loop activating signal to the loop output signal in a feedback control loop.

forward scatter. (1) Multiple transmission on centimetric waves, using reflection down and forwards from ionization in troposphere; range about 100 miles. (2) Scattering of particles through an angle of less than 90° to the original direction of the beam; cf. *back scatter.*

forward transadmittance. Of an electron beam tube, the complex quotient of the fundamental component of the short-circuit current induced in the second of any two gaps and the fundamental component of the voltage across the first. Also **beam transadmittance.**

forward transfer function. That of the forward path of a feedback control loop.

forward voltage. The voltage of the polarity which produces the larger current in an electrical system.

forward wave. One whose group velocity is in the same direction as the motion of the electron stream in a travelling-wave tube. See **backward wave.**

Foster-Seeley discriminator. One for demodulating FM transmission using a balanced pair of diodes. To these are applied voltages which are sum and difference of limiter signal voltage and half transformer coupled voltage, diode outputs being differenced.

Foster's reactance theorem. Expression for generalized impedance of a number of tuned circuits, in series or parallel, which indicates that such impedances exhibit in order resonant and anti-resonant frequencies.

Foucault currents. See **eddy currents.**

foundation lighting. Adequate general illumination for operations, but lacking artistic or dramatic effect.

four-electrode valve. Any valve with two grids between cathode and anode, e.g., tetrode.

four-terminal resistor. Laboratory standard fitted with current and potential terminals, arranged so that measurement of the potential drop across the resistor is not affected by contact resistances at the terminals.

Fourier analysis. The determination of the harmonic components of a complex waveform either mathematically or by a wave-analyser device.

Fourier principle. Principle which shows that all repeating waveforms can be resolved into sine wave components, consisting of a fundamental and a series of harmonics at multiples of this frequency. It can be extended to prove that non-repeating waveforms occupy a continuous frequency spectrum.

Fourier transform. A mathematical relation between the energy in a transient and that in a continuous energy spectrum of adjacent component frequencies.

Fowler-Nordheim equation. The field emission current (J) from a surface under an applied electric field E_0 ($\sim 10^9$ volt cm^{-1}), i.e., $J \propto E_0{}^2 \exp{}^-C/E_0$, where C is a constant.

Fowler plot. In photoelectronic emission it is a plot of $\ln (JT^{-2})$ vs $\left(\dfrac{h\nu}{kT}\right)$, where J is the photo-current, T is the absolute temperature, ν is the frequency of the radiation, k is Boltzmann's constant and h is Planck's constant.

frame. In U.K., obsolete term for the complete scanning of a TV picture. See **field** (3). In U.S., the total picture, as in sound-film terminology.

frame antenna. One comprising a loop of one or more turns of conductor wound on a frame, its plane being oriented in the direction of the incoming waves, or, in transmission, for the direction of maximum radiation; also called **coil antenna, loop antenna.**

frame direction-finding system. Simple type of direction-finder involving a loop, preferably screened to obviate *antenna effect*, the polar response of which is a figure of eight; the loop is rotated until the received signal vanishes, when the axis of frame is in line with direction of arrival of wave. Also **loop direction-finding system.** See **sense.**

frame frequency. The number of complete picture scannings in television per second. In U.K., 25/sec., in U.S., 30/sec., synchronized with mains frequency to minimize the effect of mains interference. Also **picture frequency.**

frame slip. Lack of exact synchronization of the vertical scanning and the incoming signal, whereby the reproduced picture progresses vertically.

framing. Adjustment of picture-repetition frequency in a television receiver to keep the picture stationary on the screen.

framing oscillator. That which generates frame-scanning voltage or current. Also **field oscillator.**

framing signal. One used in facsimile transmission to adjust the picture to a desired position in the direction of progression.

francium (Fr). The heaviest alkali metal, at. no. 87. No stable isotopes exist. ^{223}Fr of half-life 22 min. is most important. Formerly called **virginium.**

Franklin antenna. Directive antenna comprising a number of radiating elements uniformly spaced along a line at right-angles to the direction of maximum radiation. Each element consists of a vertical wire several half-wavelengths long, the radiation from alternate half-wavelengths being suppressed, to secure maximum radiation along the near horizontal direction. Such a curtain is usually accompanied by a similar, but unexcited, reflector.

Fraunhofer region. See **field** (1).

free. Said of a transducer when it is not *loaded,* e.g., the input has a *free impedance.*

free charge. An electrostatic charge which is not bound by an equal or greater charge of opposite polarity.

free electron theory. Early theory of metallic conduction based on concept that outer valence electrons, which do not form crystal bonds, are free to migrate through crystal, so forming *electron gas.* Now superseded by **energy band** theory.

free field. Non-reflecting space required for measurement of the acoustic response of microphones and loudspeakers. Realized down to low frequencies, i.e., order of a few hundred Hz, in an *anechoic room.*

free-field emission. That from an emitter when the electric gradient at a surface is zero.

free impedance. That at input terminals of a transducer, when its load impedance is zero.

free line signal. A visual signal which is associated with the trunk, toll or junction multiple to indicate which circuit is to be used next.

free oscillations. (1) Oscillatory currents whose frequency is determined by constants of the circuit in which they are flowing; e.g., those resulting from the discharge of a capacitance through an inductance. (2) Mechanical

oscillations governed solely by natural elastic properties of vibrating body.

free-running circuit. Same as **astable circuit.**

free-running frequency. That of an oscillator, otherwise uncontrolled, when not locked to a synchronizing signal, especially a time-base generator in a television receiver.

free-space impedance. For electromagnetic waves the characteristic impedance of a medium is given by the square root of the ratio of permeability to permittivity. (Or 4π times this in unrationalized systems of units.) This gives a free space value of 376·6 ohm.

free-wheel rectifier. One placed in shunt with a highly inductive load to stabilize the main rectifier against sudden changes in the load current.

frequency. Rate of repetition of a periodic disturbance. Measured in *hertz.* Also **periodicity.**

See also:

angular-	Langmuir-
anti-resonance-	Larmor-
assigned-	limiting-
atomic-	line-
audio-	maximum usable-
bass-	medium-
beat-	minimum sampling-
carrier-	natural-
centre-	oscillation-
clock-	pump-
critical-	radio-
cross-over-	reference-
cutoff-	resonant-
cyclotron-	roll-off-
dot-	scanning-
field-	side-
fork-	spark-
frame-	subaudio-
free-running-	superaudio-
fundamental-	supersonic-
high-	threshold-
image-	transition-
instantaneous-	ultra-high-
intermediate-	vertical-
ion cyclotron-	zero-.

frequency allocation. Frequency on which a transmitter has to operate, within specified tolerance. Bands of frequencies for specified services are allocated by international agreement.

frequency band. Interval in the frequency spectrum occupied by a modulated signal. In sinusoidal amplitude modulation, it is twice the maximum modulation frequency, but it is much greater in frequency or pulse modulation.

frequency bridge. An a.c. measuring bridge whose balancing condition is a function of supply frequency, for example, a Robinson bridge.

frequency changer. (1) A machine designed to receive power at one frequency and to deliver it at another frequency. (2) A combination of oscillator and modulator valves used in a superhet receiver to change the incoming signal from its original carrier frequency to a fixed intermediate carrier frequency; also called **conversion mixer, conversion transducer, frequency converter, converter** (in U.S.).

frequency converter. Same as **frequency changer** (2).

frequency demultiplication. See **frequency division**.

frequency departure. Discrepancy between actual and nominal carrier frequencies of a transmitter. Formerly termed **frequency deviation**, a term now applying only to frequency and phase modulation.

frequency-departure meter. Instrument for measuring frequency departure of transmitter, normally forming part of control console.

frequency deviation. (1) In frequency modulation, maximum departure of the radiated frequency from mean quiescent frequency (carrier). (2) Greatest deviation allowable in operation of frequency modulation. In broadcast systems within the range 88 to 108 MHz, the maximum deviation is ±75 kHz. (3) See **frequency departure**.

frequency-discriminating filter. See **electric wave filter**.

frequency discriminator. Circuit, the output from which is proportional to frequency or phase change in a carrier from condition of no frequency or phase modulation.

frequency distortion. In sound reproduction, variation in the response to different notes solely because of frequency discrimination in the circuit or channel. Generally plotted as a decibel response on a logarithmic frequency base.

frequency diversity. See **diversity reception**.

frequency division. Dividing a frequency by harmonic locking oscillators or stepping valve circuits. Specific integral division alone can be obtained; an arbitrary collection of frequencies cannot be divided, except by recording and reproducing at a lower speed; also called **frequency demultiplication**.

frequency doubler. Frequency multiplier in which the output current or voltage has twice the frequency of the input. Achieved by simple push-pull tuned circuits.

frequency doubling. Introduction of marked double-frequency components through lack of polarization in an electromagnetic or electrostatic transducer, in which the operating forces are proportional to the square of the operating currents and voltages respectively.

frequency drift. Change in frequency of oscillation because of internal (ageing, change of characteristic, or emission) or external (variation in supply voltages, or ambient temperature) causes. Also **oscillator drift**.

frequency meter. Simple or complex apparatus for relating an unknown frequency to a standard, ultimately the yearly orbiting of the earth round the sun. The *basic frequency* is that of a quartz-ring oscillator at 10^5 Hz, other frequencies being related to this by dividing, multiplying, or heterodyning. Ultimate standardization is now *via* an *atomic clock*. See **wavemeter**.

frequency-modulated cyclotron. One in which frequency of the voltage applied to the *dees* is varied to keep synchronous orbiting of accelerated particles when their mass increases through relativity effect at the high velocities attained.

frequency modulation. Variation of frequency of a transmitted wave in accordance with impressed modulation, the amplitude remaining constant.

frequency-modulation receiver. One which converts frequency-modulated signals into, e.g., radio signals. Has the advantage of giving a high signal/noise ratio.

frequency monitor. Nationally or internationally operated equipment to ascertain whether or not a transmitter is operating within its assigned bandwidth.

frequency multiplier. Four-terminal device in which output current or voltage has frequency an integral multiple of that of the input. A saturated iron-cored inductance, or a thermionic valve working on the non-linear part of its characteristic, can be used, together with the appropriate frequency-selecting circuits, to produce doubling, which can be repeated.

frequency of collision. That of free electrons with other ions and molecules in ionosphere.

frequency of gyration. That of electrons about a line indicating direction of magnetic field in ionosphere.

frequency of penetration. That of a wave which just fails to be reflected by ionospheric layer.

frequency overlap. Common parts of frequency bands used, e.g., for the regular video signal and the chrominance signal in colour television.

frequency pulling. Change in oscillator frequency resulting from variation of load impedance.

frequency relay. See **relay types**.

frequency response. (1) Ratio of the output (a sound intensity) at a stated point in the field of a sound radiator to the electrical power (*applied power*) causing it, which may also be stated. *Total response* implies the total output acoustic power or the mean-spherical intensity at a stated distance. Inversely, exactly the same considerations apply to a microphone. (2) Relationship between gain and signal frequency in amplifiers.

frequency selectivity. See **selectivity**.

frequency-shift keying. In radio telegraphy, altering the carrier by mark-and-space keying.

frequency-shift transmission. A form of modulation used in communication systems in which the carrier is caused to shift between two frequencies denoting respectively *on* and *off* pulse.

frequency stabilization. Prevention of changes produced in frequency of oscillation of a self-oscillating circuit by changes in supply voltage, load impedance, valve parameters, etc. Achieved by resonating crystals, tuned cavities or transmission lines.

frequency standard. Reference oscillator of very high stability. Usually of quartz-ring type, although atomic beam standards provide the final reference.

frequency swing. Extreme difference between maximum and minimum instantaneous frequencies radiated by a transmitter.

frequency tolerance. Extent to which frequency of the carrier of a transmission is permitted to deviate from its allocation.

frequency translation. Shifting all frequencies in a transmission by the same amount (not through zero).

fresnel. A unit of frequency; 10^{12} Hz.

Fresnel region. See field (1).

Fresnel zones. Zones into which a wavefront is divided according to the phase of the radiation reaching any point from it. The radiation from any one zone will reach the given point half a period out of phase with the adjacent zones.

frictional electricity. Static electricity produced by rubbing bodies together, e.g., an ebonite rod with fur.

fringe area. In television or radio broadcasting, the regions around a transmitter in which good reception is uncertain.

fringing. Incorrect registration of images in colour television, so that edges of colour exhibit departure from correct rendering.

frisking. Searching for radioactive radiation by contamination meter, usually a portable ionization chamber.

front porch. See porch.

front-to-back ratio. The ratio of the effectiveness of a directional antenna or microphone, etc., in the forward and reverse directions.

frontwall cell. Semiconductor cell in which light passes through a conducting layer to the active layer, which is separated from the base metal by a semiconductor.

fry or frying. (1) Noise accidentally added to the sound when recording a record. (2) Noise reproduced when excessive current is passed through the carbon granules in a telephone transmitter.

fuel cell. A chemical source of electricity based on the recombination of elements (e.g., hydrogen and oxygen) normally separated by electrolysis.

fuel element. The smallest individual unit containing fuel material for a reactor.

fuel rating. The ratio of total energy released to initial weight of heavy atoms (U, Th, Pu) for reactor fuel. Usually expressed in megawatts per tonne. In U.S., **specific power.**

fuel rod. A body of nuclear fuel in rod form for use in a reactor. Short rods are termed **slugs.**

full compatibility. Reception by a colour TV receiver of a colour TV signal when the chromaticity information is within the black-and-white video band of frequencies.

full-wave rectification. That in which current flows, during both half-cycles of the alternating voltage, through similar rectifying devices alternately, e.g., in a double diode or bridge rectifier.

function generator. (1) Signal generator with range of alternative non-sinusoidal output waveforms. (2) Element in analogue computer capable of generating voltage wave approximately following any desired single-valued continuous algebraic function of one variable.

function switch. A network with a number of inputs and outputs connected in a computer, so that the output signals give the input information in a different code from that of the input.

function unit. A means of storing a functional relationship and releasing it continuously or in increments.

fundamental crystal. A crystal which is designed to vibrate at the lowest order of a given mode.

fundamental frequency. In a steady periodic oscillation, a frequency which divides into all components in the waveform. See harmonic.

fundamental frequency of antenna. Lowest frequency at which antenna is resonant, when not loaded with terminal inductance.

fundamental mode. Mode of oscillation of antenna at its fundamental frequency. An earthed antenna is characterized by a single node of current at the extreme end of the antenna.

fundamental particle. See elementary particle.

fundamental series. Series of optical spectrum lines observed in the spectra of alkali metals. Energy levels for which the orbital quantum number is 3 are designated **f-levels.**

fundamental units. Units arbitrarily selected to form the basis of a system of measurements. See absolute units.

fundamental wavelength. (1) Wavelength in free space corresponding to the fundamental frequency of an antenna. (2) The main or operating wavelength in a transmitter radiating harmonics.

fused junction. One formed by recrystallization of semiconductor and impurity from a liquid on to pure base crystal.

Fusetron. TN for surge-resisting fuse.

fusion. The process of forming new elements by the fusion of lighter ones; principally the formation of helium by the fusion of hydrogen and its isotopes. The process involves a loss of mass which appears as *fusion energy* (q.v.).

fusion bomb. Same as hydrogen bomb.

fusion energy. Energy released by nuclear fusion, usually the formation of helium from lighter particles. See carbon cycle (1), hydrogen bomb, thermonuclear energy.

fuzz. Extraneous noise introduced in recording, made evident on reproduction.

G

G display. Similar to *F display* (q.v.) but indicating increasing or diminishing range of target by increasing or diminishing lateral extension of the spot.

G-gas. Gaseous mixture (based on helium and *iso*butane) used in low-energy beta counting (e.g., of tritium) by gas flow proportional counter.

G-string. Colloquialism for single wire transmission line loaded with dielectric so that surface wave propagation can be employed.

G-Y signal. Component of colour TV signal which, when combined with the Y (luminance) signal, produces the green chrominance signal.

gadolinium (Gd). A rare element, at. no. 64, at. wt. 157·25. Many stable and unstable isotopes exist.

Gaede diffusion pump. Pump using mercury vapour, which entrains molecules of gas from a low pressure established by a backing pump. Oil of low vapour pressure (apiezon) is a modern alternative.

Gaede molecular pump. Rotary pump which ejects molecules of gas by imparting a drift velocity to their random motion.

gain. (1) In electric systems, generally provided by insertion of an amplifier into a transmission circuit, or by matching impedances by a loss-free transformer. Measured in decibels or nepers, and defined as the increase in power level in the load, i.e., the ratio of the actual power delivered to that which would be delivered if source were correctly matched, without loss, to the load in the absence of the amplifier. (2) In a directional antenna, ratio (expressed in decibels) of voltage produced at the receiver terminals by a signal arriving from the direction of maximum sensitivity of the antenna, to that produced by same signal in an omnidirectional reference antenna (generally a half-wave dipole). In a transmitting antenna, ratio of the field strength produced at a point along the line of maximum radiation by a given power radiated from antenna, to that produced at same point by the same power from an omnidirectional antenna.

See also:

antenna-	loop-
conversion-	power-
current-	reflection-
directive-	transmission-
insertion-	voltage-.

gain amplifier. Thermionic amplifier following the photocell amplifier and preceding the power stages in sound-film projection equipment.

gain bandwidth. Product of maximum amplifier *gain* (between specified impedances) and the bandwidth possible (between *half-power points*).

gain control. Means for varying the degree of amplification of an amplifier, often a simple potentiometer. See automatic-.

galactic noise. That arriving from outer space, similar to electronic circuit noise, but apparently arising from sources in galaxies.

galaxy. A complete stellar system, i.e., the Milky Way or one of numerous others beyond it and apparently containing comparable numbers of stars. See also quasar.

galena. A crystalline form of lead sulphide which behaves like a semiconductor and has been used as a crystal rectifier for radio signals.

Gall alternating-current potentiometer. One of the coordinate type comprising two identical potential measuring circuits, one being in phase and the other in quadrature with the current.

gallium (Ga). A metal with low melting-point (30°C), at. no. 31, at. wt. 69·72. In the form of gallium arsenide, it is an important semiconductor.

galvanic. Obs. for electric.

galvanoluminescence. Feeble light emitted from the anode in some electrolytic cells.

galvanomagnetic effect. See Hall effect.

galvanometer. Current-measuring device depending on forces on the sides of a current-carrying coil normal to magnetic fields in gaps. Developed into Duddell string galvanometer, the *light valve* and *vibrator* for sound-film recording on photographic film. In a moving-magnet instrument, the suspended coil is replaced by an astatic magnet system which is magnetically shielded for very sensitive work. See ballistic-, d'Arsonval-, tangent-, vibration-.

galvanotaxis. Tendency of organisms living in medium to grow or move into particular orientation relative to electric current through medium; also galvanotropism.

gamma. (1) A unit of magnetic field intensity particularly used in magnetic surveys. (2) 10^{-6} gram. (3) A development factor (due to Hurter and Driffield) in photography, given by $\gamma = (D_2 - D_1)/\log(E_2/E_1)$, where D_1 and D_2 are the densities of the photographic images, and E_1 and E_2 are the respective exposure times. Usually deduced from linear part of graph relating D with $\log E$. See also sensitometer. (4) In television, the relationship between the brightness contrast of two points on the transmitted scene and the brightness contrast of the corresponding points on the receiver screen, given by

$$\gamma = \frac{\log R_2/R_1}{\log S_2/S_1}.$$

R and *S* refer to receiver and scene respectively, the subscripts indicating two particular points. (5) See **gamma radiation.**

gamma correction. Introduction of a non-linear output-input characteristic to change effective value of gamma. See **gamma** (4).

gamma detector. A radiation detector specially designed to record or monitor gamma radiation.

gamma radiation. Electromagnetic radiation of high quantum energy emitted after nuclear reactions or by radioactive atoms, when nucleus is left in excited state following emission of α- or β-particle.

gamma-ray energy. This depends on wavelength and controls depth of penetration. Measured by maximum energy of photoelectrons, or diffraction by certain crystal lattices. See **curie, röntgen.**

gamma-ray spectrometer. Instrument for investigation of energy distribution of gamma-ray quanta. Basically a scintillation counter followed by a pulse height analyser.

Gamow-Teller selection rules. See **nuclear selection rules.**

gang capacitor. Assemblage of two or more variable capacitors mechanically coupled to the same control mechanism.

ganging. Mechanical coupling of tuning controls of two or more resonant circuits.

ganging oscillator. One giving a constant output, whose frequency can be rapidly varied over a wide range; used for testing accuracy of adjustment of ganged circuits over their tuning range.

gap. (1) Range of energy levels between the lowest of conduction electrons and the highest of valence electrons. (2) Digits which separate signals for data or programme. (3) Air gap in magnetic circuit to increase inductance and saturation point, e.g., in an audio output transformer. (4) Region between lobes of polar diagram for which response is inadequate. (5) Space between discharge electrodes.

gas amplification. (1) Increase in sensitivity in a gas-filled photocell, as compared with the corresponding high-vacuum cell, due to ionization of the gas caused by the primary photoelectrons. (2) Increase in sensitivity of a Geiger or proportional counter compared with a corresponding ionization chamber.

gas-cooled reactor. Reactor in which cooling medium is gaseous, usually air or carbon dioxide.

gas counter. Geiger counter into which radioactive gases can be introduced.

gas counting. That of radioactive materials in gaseous form. The natural radioactive gases (radon isotopes) and carbon dioxide (^{14}C) are common examples. See also **gas-flow counter.**

gas current. Positive ions discharged at a negative electrode, e.g., grid of a triode, because of ionization by electrons passing between other electrodes, e.g., cathode to anode. Basis of a vacuum gauge.

gas electrode. One which holds gas by adsorption or absorption, so that it becomes effective as an electrode in contact with an electrolyte.

gas factor. Ratio of space current in a gas photoelectric tube to that in a corresponding vacuum one.

gas-filled photocell. One in which anode and photocathode are enclosed in atmosphere of gas at low pressure. It is more sensitive than the corresponding high-vacuum cell because of formation of positive ions by collision of the photoelectrons with the gas molecules.

gas-filled relay. Thermionic tube, usually of the mercury-vapour type, when used as a relay; a thyratron.

gas-flow counter. Counter tube through which gas is passed to measure its radioactivity, e.g., CO_2, which is weakly radioactive because of ^{14}C. The sample is introduced into the interior of the counter, and to prevent the ingress of air the counting gas flows through it at a pressure above atmospheric. Also **flow counter.**

gas focusing. Controlling the beam in a CRT, X-ray tube or other discharge tube by the action of a small amount of residual gas in envelope, which, on becoming ionized by collision, forms core of positive ions along the centre of the beam and provides the necessary focusing field. Also called **ionic focusing.**

gas noise. That arising from ionization in gas-discharge tubes, used for generating 'white' noise as a source for testing, e.g., acoustic devices.

gas tube. Since it is so far impossible to obtain a perfect vacuum, or even one approaching outer space, all tubes and valves are gas tubes; in a so-called gas tube the pressure of residual gas is sufficiently high to influence the operation. See **gas-filled photocell, gas focusing, thyratron.**

gaseous discharge. Flow of charge arising from ionization of low-pressure gas between electrodes, initiation being by electrons of sufficient energy released from hot or cold cathodes. Various gases give characteristic spectral colours, e.g., mercury, sodium vapour, neon, hydrogen, etc.

gash. A ferroelectric compound with an almost square hysteresis loop suitable for constructing a binary cell. (*G*uanidine *a*luminium *s*ulphate *h*exahydrate.)

gate. (1) Electronic circuit which passes impressed signals when permitted by another independent source of similar signals. (2) In reactor engineering, the movable barrier of shielding material used for closing an aperture. See **coincidence-, colour-.**

gate current. The a.c. or pulsed d.c. which saturates core of a reactor or transductor.

gate winding. Winding used to obtain gating action in a magnetic amplifier.

gated-beam tube. Cathode-ray tube in which beam current can be gated by signals applied to a control electrode.

gating. Selection of part of a wave on account of time or magnitude. Operation of a circuit when one wave allows another to pass during specific intervals.

gauge. Instrument for measuring level of some physical parameter, e.g., vacuum gauge, strain gauge, etc.

gauss. CGS electromagnetic unit of magnetic flux density; equal to one line (or maxwell) per sq cm, each unit magnetic pole terminating 4π lines.

Gaussian distribution. Widely encountered spread of values about a nominal mean in systems where statistically large numbers of readings are obtained. Characterized by equal probabilities of values with equal positive and negative deviations from the mean. Also **normal distribution**.

Gaussian noise. That arising with a spectral energy distribution similar to the normal distribution curve of statistics.

Gaussian response. Response, e.g., of an amplifier, for a transient impulse, which, when differentiated, matches the Gaussian distribution curve.

Gaussian units. Formerly widely-used system of electric units where quantities associated with electric field are measured in e.s.u. and those associated with magnetic field in e.m.u. This involves introducing a constant c (the free space velocity of electromagnetic waves) into Maxwell's field equations.

Gaussian well. A particular form of potential energy distribution of a nuclear particle in the field of a nucleus or other nuclear particle.

gaussistor. Valvelike device of bismuth or other magnetoresistive material, which can amplify or oscillate.

gaussmeter. Instrument measuring magnetic flux density. This term is most widely used in the U.S.

gecom. An automatic code to fit all computers, to receive instruction in words. (*General compiler.*)

gee. VHF radio-navigation system depending on pulses sent from two or more transmitting stations, difference of timing determining location on a chart with hyperbolic lattice. (*Ground electronics engineering.*)

Geiger characteristic. Plot of recorded count rate against operating potential for Geiger tube interrupting beam of radiation of constant intensity.

Geiger-Müller counter. One for ionizing radiations, with a valve carrying a high-voltage wire in a halogen or organic atmosphere at low pressure, and a valve (or transistor) circuit which quenches the discharge and passes on an impulse for scaling and counting electronically. Also **Geiger counter**, **GM counter**. See **Townsend avalanche**.

Geiger-Müller tube. The detector of a GM counter, i.e., without associated electronic circuits.

Geiger-Nuttall relationship. An empirical rule for calculating the half-life of some radioactive elements from the range of the alpha-particle emitted. The relation is now found to be more limited in its application than it was initially thought to be.

Geiger region. The part of the characteristic of a *counting tube*, where the charge becomes independent of the nature of the ray intercepted; also called **Geiger plateau** since in this region the characteristic is almost horizontal.

Geiger threshold. Lowest applied potential for which Geiger tube will operate in Geiger region.

Geissler tube. Original gas-discharge tube containing various gases, characterized by a central capillary section for concentrating the glow; useful for spectrometer calibrations.

gel. The apparently solid, often jelly-like, material formed from a colloidal solution on standing.

general instruction. See **macro instruction**.

generation rate. Rate of production of *electron-hole pairs* in semiconductors.

generation time. Average life of fission neutron before absorption by fissile nucleus.

generator. Electrostatic or electromagnetic device for conversion of mechanical into electrical energy. See **current-**, **direct-current-**.

genetics. Study of heredity and mutation including effect of radiation on mutation rates.

Geneva movement. Intermittent movement from a continuous motion by a wheel with slots. Used in film projectors, automation machines, counting mechanisms.

geodesic. A great circle route in navigation.

geomagnetic. Pertaining to the natural magnetization of the earth.

geometric capacitance. That of an isolated conductor in vacuo, uninfluenced by dielectric material, and depending only on shape.

geometric distortion. Any departure from the original perspective in reproduction (apart from definition). See **keystone distortion**.

geometric mean. The nth root of the product of a set of n numbers.

geometrical attenuation. Reduction in intensity of radiation on account of the distribution of energy in space, e.g., due to inverse-square law, or progression area along the axis of a horn.

geometrical cross-section. Area subtended by a particle or nucleus. This does not usually resemble the interaction cross-section.

geometry. In radiation, experimental limitations introduced by the finite geometry of sources and detectors which make measurements, e.g., of energy, ambiguous.

geometry factor. $1/(4\pi)$ of the solid angle subtended by the window or sensitive volume of a radiation detector at the source.

geophysics. Study of physical properties of the earth.

George. Colloquialism for **automatic pilot**.

germanium (Ge). Metallic element, at. no. 32,

at. wt. 72·59, density 5·32 g cm $^{-3}$, energy gap in eV 0·72 at 25°C temperature, coefficient of energy gap -0.0001 eV deg^{-1}C, mobility of electrons 3600 cm^2 V^{-1} sec^{-1} at 25°C, mobility of holes 1700 cm^2 V^{-1} sec^{-1} at 25°C, m.p. 936°C, linear thermal expansion coefficient 6.1×10^{-6} deg^{-1}C at 30°C, specific heat 0·074 cal. g^{-1} deg^{-1}C, dielectric constant 16. Used widely in transistors and crystal diodes.

germanium radiation detector. See **lithium drift germanium detector, semiconductor radiation detector.**

germanium rectifier. One depending on asymmetrical conduction at a Ge interface. Highly efficient as a rectifier, used for railway traction and for power supply on a wide scale, as well as for low-power circuits.

getter. Material (K, Na, Mg, Ca, Sr or Ba), used, when evaporated by high-frequency induction currents, for *cleaning* the vacuum of valves, after sealing on the pump line during manufacture. See **barium-.**

gettering discharge. That used to assist the *getter* in the *clean-up* of vacuum in valves, by the ionization of the remaining gas molecules.

GeV. Abb. for *giga-electron-volt*; unit of particle energy, 10^9 electron-volts. In U.S., sometimes BeV.

ghost. (1) Duplicated image on a television screen arising from additional reception of a delayed, similar signal which has covered a longer path, e.g., through reflection from a tall building or mast. Also **double image.** (2) A specimen for use in radioactivity measurements which behaves similarly to biological tissue. Also **phantom.**

giant source. Large source of radioactivity, e.g., 150,000 curies of ^{60}Co, used for industrial sterilization of packed food or chemical processing (e.g., cross-linking of polymers).

giga-. Prefix indicating 10^9, as in GeV.

giga-electron-volt. See GeV.

gilbert. In the CGS electromagnetic system of units, m.m.f. of an enclosing coil equal to $10/4\pi$ ampere-turn.

Gill-Morrell oscillator. Triode valve with anode and grid fed through a Lecher system, oscillation depending on the *transit time* of the electrons in the valve, but with frequency related to the Lecher line. See **Barkhausen-Kurz oscillator.**

gillion. 10^9 or one thousand million.

Giorgi system. System of units proposed in 1904 and recently adopted by the International Electrotechnical Commission, as the MKSA system; the common *practical units* of ohm, volt, ampere, etc., become identified with *absolute units* of the same entities.

glancing angle. The complement of the *angle of incidence* (q.v.).

glide-path beacon. A directional radio beacon, associated with an ILS, which provides an aircraft, during approach and landing, with indications of its vertical position relative to the desired approach path.

glide-path landing beam. Radio signal pattern from a radio beacon, which aids the landing of an aircraft during bad visibility.

glint. Apparent random motion of the effective centre of reflection of a radar target, leading to noise spread.

Glo-ball. Tiny helium-filled glass ball, for exploring field oscillations in resonant cavities.

glove box. An enclosure in which radioactive or toxic material may be manipulated by the use of gauntlets.

glow discharge. Visible discharge near a cathode when the potential drop is slightly higher than the ionization potential of the gas. Consists of luminous spectral bands.

glow potential. Potential which initiates sufficient ionization to produce gas discharge between two electrodes, but is below sparking potential. It is virtually constant over a wide range of currents.

glow switch. Tube in which a glow discharge thermally closes a contact, starting fluorescent tubes.

glow tube. Cold-cathode gas-filled diode, with no space-current control, the colour of glow depending on contained gas.

glucinum. See beryllium.

glueline. High-frequency heating technique for drying glue films in woodwork construction, by applying electric field in line with the film, with specially shaped electrodes. The film should have a high loss factor compared with medium to be 'glued'.

Golay cell. Pneumatic cell used as detector of heat radiation.

gold (Au). Valuable metallic element of very high density (19·3 gm/cm^3), at. no. 79, at. wt. 196·967, m.p. 1063°C, b.p. 2600°C. Very high absorption cross-section for slow neutrons.

gold-leaf electrometer. *Gold-leaf electroscope* (q.v.) modified for the measurement of very small currents by observing the rate of movement of the gold leaf through a microscope. The *personal dosimeter* (q.v.) is a development of this instrument. See also **Millikan electrometer.**

gold-leaf electroscope. A device for detecting small electric charges which are applied to a piece of thin gold foil usually attached at upper end to a metal electrode. Mutual repulsion between the foil and the similarly charged plate electrode leads to the former being displaced.

Goldschmidt alternator. High-frequency alternator in which the stator and rotor carry a number of windings each tuned to successively higher frequencies; up to 100 kHz can be thus attained.

goniometer. Instrument for the measurement of angles, e.g., that between faces of crystal (*crystal goniometer*), that between heading of vessel and radio-navigation beam (*radio-goniometer*).

governor. Speed regulator on variable speed motor or rotating machine, e.g., d.c. gramophone motor.

gradient meter. A means of measuring the potential gradient of an electrical field at the surface of an electrical conductor.

gradiometer. Magnetometer for measurement of magnetic field.

grain. Individual silver bromide crystal in photographic emulsion, which can be reduced to metallic silver on development if at least one photon of light has been absorbed during exposure. Many developers also cause clumping of individual grains, so producing increased granularity in the image. In nuclear research emulsions individual grain size is about 0·2 micron.

gram-atom. A quantity of an element for which the weight in grams is numerically equal to the atomic weight.

gram-ion. Mass in grams of an ion, numerically equal to that of the molecules or atoms constituting the ion.

gram-molecule. See mol(e).

graphecon. A double-ended storage tube used for the integration and storage of radar information and as a translating medium. The recording of the signals is based on the electrical conductivity induced in the target by a high voltage writing beam.

graphic panel. Master control panel in automation and remote control systems, which shows the relation and functioning of different parts of the control equipment by means of coloured block diagrams.

graphical symbols. Representation of the various types of thermionic valves by diagrams as recommended by BSI.

graphite reactor. One in which fission is produced principally or substantially by slow neutrons moderated in graphite.

grass. Irregular deflection from the time base of a radar display, arising from electrical interference or noise. Also picture noise.

grating. An arrangement of alternate reflecting and non-reflecting elements, e.g., wire screens or closely spaced lines ruled on a flat (or concave) reflecting surface, which, through diffraction of the incident radiation, analyses this into its frequency spectrum. An *optical grating* can contain a thousand lines or more per cm. A *standing-wave* system of high frequency sound waves with their alternate compressive and rarified regions can give rise to a diffraction grating in liquids and solids. With a *criss-cross* system of waves, a 3-dimensional grating is obtainable.

Gratz rectifier. Type of *bridge rectifier* using six rectifying elements for 3-phase supply.

graviton. Hypothetical quantum of gravitational field energy bearing the same relationship to the gravitational field as the photon does to the electromagnetic field. The gravitational interaction between particles is so weak that no discontinuous (quantized) effects have yet been observed experimentally. Hence the existence of the graviton can only be inferred from laws found immutable in other branches of physics.

gravity waves. Liquid surface waves controlled by gravity and not surface tension.

grazing angle. A very small *glancing angle*.

great circle distance. Distance between two points on the surface of a sphere, measured along the arc joining them and centred at the centre of the sphere.

green gun. Electron gun used to excite the green phosphor in a 3-colour TV picture tube.

Greenwich mean time. Mean solar time along the zero meridian of longitude. See also **zulu time.**

grenz rays. X-rays produced by electron beams accelerated through potentials of 25 kV or less. These are generated in many types of electronic equipment but have very low penetrating power.

grey body. One which emits radiation equally at all wavelengths but absorbs more than it emits and is thus not a perfect black body.

grid. Control electrode having an open structure (e.g., mesh) allowing the passage of electrons; in an electron gun it may be a hole in a plate.
See also:
de-ionization-	shield-
injector-	space-charge-
resonator-	suppressor-.
screen-	

grid base. The grid-bias voltage required to produce anode current cut-off for a thermionic valve.

grid bias. D.c. negative voltage applied to control grid of thermionic valve.

grid-bias battery. One providing power for the grid polarization of valves. It usually makes the grid negative with respect to the cathode and so only supplies power if the instantaneous signal swings the grid positive again. U.S. term **C-battery.**

grid characteristics. Graph of grid current against grid potential for a thermionic valve.

grid control. That provided by voltage on grid of a thyratron or mercury-arc rectifier; at a sufficient positive voltage, anode current flows, grid loses control until de-ionization is effected, after loss of anode voltage.

grid-controlled mercury-arc rectifier. One in which initiation of arc in each cycle is determined by phase of voltage on a grid for each anode.

grid-dip meter. Wavemeter in which absorption of energy by tuned circuit at resonance is indicated by decrease of valve grid current.

grid drive. Voltage or power required to drive a valve when delivering a specified load.

grid emission. Thermionic emission of electrons from valve grid consequent upon serious overheating. Secondary grid emission may also occur as a result of electron bombardment.

grid-glow tube. Cold-cathode gas-discharge tube in which glow is triggered by a grid.

grid leak. High value resistor connected between grid and earth, which biases valve when this carries a 'leakage' current.

grid-leak detector. Demodulator circuit in which modulation wave appears across grid-leak resistor. Preferably termed **cumulative-grid detector.**

grid limiter. Series resistor used in grid lead of valve to avoid excess grid-current flow under abnormal operating conditions.

grid modulation. See **suppressor-grid modulation.**

grid neutralization. The method of neutralization of an amplifier through an inverting network in the grid circuit which provides the requisite phase shift of 180°. Also called **phase inverter.**

grid pool tube. A gas-discharge tube with a liquid (or solid) pool-type cathode such as the mercury cathode.

grid resistor. See **grid leak.**

grid return. The external conducting path for the grid current.

grid suppressor resistor. A resistor used to suppress parasitic oscillations by connecting it between the control grid and the tuned circuit of an RF amplifier.

grid swing. The maximum voltage excursion of the control grid potential from the bias potential when a signal is applied to the grid.

groove. Track followed by stylus in disk recording.

gross error. In a measurement, the result of which is expressed in units and a fraction (or decimal), error in the units but not in the fraction (or decimal).

Grotthus-Draper law. Only such energy of electromagnetic radiation as is absorbed by tissue is effective in chemical action following ionization in that tissue.

ground(ed). U.S. for **earth(ed).**

ground absorption. The energy loss in radio-wave propagation due to absorption in the ground.

ground capacitance. See **earth capacitance.**

ground clutter. The effect of unwanted ground return signals on the screen pattern of a radar indicator.

ground-controlled approach. Aircraft landing system in which information is transmitted by a *ground controller* from a ground radar installation at end of runway to a pilot intending to land; also called **talk-down.**

ground-controlled interception. Radar system whereby aircraft are directed on to an interception course by a station on the ground.

ground controller. Person who has control of aircraft over an area, in fog, or during landing, by radar. See **ground-controlled approach.**

ground noise. See **background noise.**

ground ray. Same as **direct ray.**

ground reflection. The wave in radar transmission which strikes the target after reflection from the earth.

ground return. The aggregate sum of the radar echoes received after reflection from the earth's surface.

ground speed. Speed of aircraft or missile relative to the ground and not to the surrounding medium; cf. *air speed.*

ground state. State of nuclear system, atoms etc., when at their lowest energy, i.e., not *excited*; also **normal state.**

ground system. See **earth system.**

ground wave. See **direct ray, wave.**

grounded base. See **common-base connection.**

grounded-cathode amplifier. One with cathode at zero alternating potential, drive on the grid and power taken from the anode; the normal and original use of a triode valve.

grounded collector. See **common-collector connection.**

grounded emitter. See **common-emitter connection.**

grounded-grid amplifier. One with grid at zero alternating voltage, drive between cathode and earth, output being taken from the anode; there is no anode-grid feedback.

group frequency. The frequency which corresponds to the group velocity of a progressive wave in a waveguide or transmission line.

group mixer or **fader.** See **mixer.**

group modulation. Use of one carrier for transmitting a group of telephone or telegraph channels, with demodulation on reception and ultimate separation. Side frequencies of the said carrier may represent different groups.

group operation. The use of one mechanism to operate all the poles of a multiple switch.

group theory. (1) Approximate method for study of neutron diffusion in reactor core, in which neutrons are divided into a number of groups, all members of a group being assumed to have the same velocity, and are maintained thus for a definite number of collisions, before being transferred into the next group. (2) Mathematical methods employed in dealing with special unitary groups. See **Lie algebra.**

group velocity. That of energy propagation for a wave in a dispersive medium. Given by the differential coefficient of the reciprocal of the wavelength, with respect to the frequency, i.e., $d\beta/d\omega$, where β is the delay in radians m^{-1} and $\omega = 2\pi$ frequency in Hz.

grown-diffusion transistor. A junction transistor in which the junctions are formed by the diffusion of impurities.

grown junction. One formed during growth of crystal when desired impurity is added to molten state.

growth. (1) Elongation of fuel rods in reactor under irradiation. (Growth is a change of shape and not volume—it must be distinguished from **S-value.**) (2) Build-up of artificial radioactivity in a material under irradiation, or of activity of a daughter product as a result of decay of the parent.

guard. Signal which prevents accidental operation by spurious signals or avoids possible ambiguity.

guard band. Any additional frequency band on either side of an allocated band (including any *frequency tolerance*), to ensure freedom from interference from other transmissions, e.g., 0·25 MHz between television channels.

guard circle. Inner groove on disk recording

which protects stylus from being carried to centre of turntable.

guard ring. Auxiliary electrode used to avoid distortion of electric (or heat) field pattern in working part of a system as a result of the *edge effect*, or to bypass leakage current through insulator to earth in an ionization chamber.

guard-ring capacitor. A standard capacitor consisting of circular parallel plates with a concentric ring maintained at the same potential as one of the plates to minimize the edge effect.

guidance. System for flight control using electronic means for sensing its own position and that of the target, and adjusting its own velocity, acceleration, orientation, propulsion. Divided into *initial, midcourse* and *terminal.* See **beam riding, homing** (2).

guide field. That component of field in a cyclotron or betatron which maintains particles in their intended path.

guide track. Extra sound-track recorded during the shooting of a film to assist in editing and post-synchronizing, but not used in the final version.

guided missile. Tactical unmanned weapon using *guidance* on course to target.

guided wave. Electromagnetic or acoustic wave which is constrained within certain boundaries as in a waveguide.

Guillemin effect. The tendency of a bent magnetostrictive rod to straighten in a longitudinal magnetic field.

Guillemin line. A network designed to produce a nearly square pulse with a steep rise and fall.

gulp. Several bytes forming part of a word in data processing.

gun. Assemblage of electrodes, comprising cathode, anode, focusing and modulating electrodes, from which the electron beam is emitted before being subjected to deflecting fields.

gun current. Total electronic current flowing to the anode, part of which forms the beam current.

Gunn effect. The production of high-field intensity domains in a semiconductor diode (usually by dipole charges formed across a depletion layer, although other processes, such as charge accumulation, can produce similar effects). These domains can form the basis of negative-resistance microwave semiconductor oscillators.

gutnik. Small encapsulated transistor radio telemeter for swallowing. Self-contained battery can be charged by rectifying an induction current.

guy. Thin tension support for antenna mast or similar structure.

gyration. The motion around a fixed centre or axis.

gyrator. Electronic component which does not obey reciprocity law. Frequently based on the *Faraday effect* in ferrites.

gyro frequency. See **frequency of gyration.**

gyrocompass. Compass which employs a gyroscope element to provide the reference direction.

gyromagnetic effect. See **Barnett effect.**

gyromagnetic ratio. Quantity which expresses numerical value of gyroscopic effects arising in atomic or nuclear physics, as a result of the spin of charged particles producing interaction with a magnetic field. The classical value is given by the ratio of magnetic moment to angular momentum but other analogous definitions are used in systems where the classical value does not apply.

gyroscope. Spinning body in *gimbal* (or similar) *mount* which offers marked resistance to torques tending to alter the alignment of the spin axis, and in which *precession* or *nutation* replaces the direct response of static bodies to such applied torques.

gyrosyn. A gyrocompass using a flux gate. Two of these on the wing tips of an aircraft provide an artificial horizon.

gyrotron. Rotating vibrating fork used as a gyroscope.

H

H-antenna. Two vertical dipoles, one reflecting, $ca.$ $\frac{1}{2}$ to $\frac{1}{3}$ wavelength apart, much used for television reception.

H-bend. A gradual change in the axis of a waveguide to maintain perpendicularity to the plane of polarization.

H-bomb. Same as **hydrogen bomb.**

H display. Modified B $display$ (q.v.) to include angle of elevation. The target appears as two adjacent bright spots and the slope of the line joining these is proportional to the sine of the angle of elevation.

H-network. Symmetrical section of circuit, with 1 shunt branch and 4 series branches.

H-plane. The plane containing the magnetic vector H in electromagnetic waves, and containing the direction of maximum radiation. The electric vector is normal to it.

H-plane bend. That of a waveguide in which its longitudinal axis remains in plane parallel to plane of magnetic field vector.

H-radar. Navigation system in which an aircraft interrogates two ground stations for distance.

H2S. Airborne centimetric radar system using a rotational aerial, radar echoes from the ground being reproduced by intensity modulation on a $plan$-$position$ $indicator$ to 'paint', on a cathode-ray tube screen, a contrast map corrected for slant range.

H-wave. See **TE-wave.**

h-bar. See **Dirac's constant.**

h-parameters. See **transistor parameters.**

Haas effect. The acoustic phenomenon associated with a long-delayed echo which has been applied to reinforcement systems in auditoria. See **reinforcement.**

habit. External form of a perfect crystal, exhibiting all possible facets and the angles between them.

hafnium (Hf). Metallic element, at. no. 72, at. wt. 178·49, which has a high absorption cross-section for neutrons.

halation. In a cathode-ray tube, the glow surrounding spot on phosphor arising from internal reflection within the thickness of the glass; also called **halo.**

half-adder. Unit having two input and two output circuits for binary signals, two such circuits forming one adder.

half-cell and **half-element.** See **single-electrode system.**

half-life. (1) Time in which half the atoms of a given quantity of radioactive nuclide undergo at least one disintegration. Also **half-value period.** (2) See **biological half-life.** (3) See **effective half-life.**

half-power. Condition of a resonant system (electrical, mechanical, acoustical, etc.) when amplitude response is reduced to $1/\sqrt{2}$ of maximum, i.e., by 3 dB. The frequency

of angular difference between symmetrical half-power points is a measure of $selectivity$ of the circuit.

half-residence time. Time in which half the radioactive débris deposited in the stratosphere by a nuclear explosion will be carried down to the troposphere.

half-section. See **section.**

half-silvered. Said of a surface of a transparent material when a deposited metallic film, e.g., formed by anode sputtering, reflects a substantial proportion of incident light.

half-supply voltage principal. If the collector-emitter voltage of a power stage transistor is less than half the supply voltage, the circuit will be inherently safe from thermal runaway as the thermal loop gain will be less than unity.

half-thickness. See **half-value thickness.**

half-tone. Description of image in which there is a continuous range of contrast between black and white.

half-value layer. See **half-value thickness.**

half-value period (or **half-period**). See **half-life** (1).

half-value thickness. The thickness of a specified substance which must be placed in the path of a beam of radiation in order to reduce the transmitted intensity by a half. Also **half-thickness, half-value layer.**

half-wave rectifier. One in which there is conduction for part of one-half of the applied alternating cycle of voltage.

half-width. (1) Half the width of the energy peak in a spectral distribution measured between the two half-amplitude points. It may either be expressed directly or as a percentage of the mean value. (2) Half the width of a response curve between the half-power points.

Hall coefficient. This is defined as

$$R_H = E_y/J_x H_z,$$

where J_x is the current flow in x direction, E_y is electric field developed in transverse direction when a magnetic field H_z is applied in z direction. Also given by

$$R_H = 1/Nce,$$

where N is the number of free electrons of charge e e.s.u. per unit volume and c is velocity of light, deduced from free electron theory of metals. Values for some metals imply positive charges, explained by band theory of solids. In MKSA units,

$$R = E_y/BJ_x \text{ ohm m}^3/\text{weber} = 1/Ne,$$

where B is magnetic induction. See **hole.**

Hall effect. Disturbance of the lines of current flow in a conductor due to the application of a magnetic field, leading to an electric potential gradient transverse to direction of

128

current flow. Also **galvanomagnetic effect.**

Hall mobility. Mobility (mean drift velocity in unit field) of current carriers in a semi-conductor as calculated from the product of the *Hall coefficient* and the conductivity.

Hallwachs effect. Release of electrons from a body when exposed to ultraviolet light. Obs. for **photoelectric effect.**

halo. Same as halation.

halo coupling. *Loop coupling* (q.v.) when the loop is just outside a resonator.

halogen. One of the seventh group of elements in the periodic table, for which there is one vacancy in the outer electron shell, e.g., Cl.

halogen quench Geiger counter. Low voltage tube for which halogen gas (normally bromine) absorbs residual electrons after a current pulse, and so quenches the dis-charge in preparation for a subsequent count.

ham. Colloquialism for holder of radio amateur transmitting licence.

hammer track. Highly characteristic track resembling hammer, formed by decay of ^{8}Li nucleus into two α-particles emitted in opposite directions at right angles to the lithium track.

Hammond organ. One operated by manuals and pedals, in which the sounds are syn-thesized from fundamentals produced by electromagnetic generators and reproduced through loudspeakers.

hand monitor. Radiation monitor designed to measure radioactive contamination on the hands of an operator.

hand reset. Restoration of a magnetic device, e.g., relay or circuit-breaker, by a manual operation.

hard. Adjective, synonymous with *high vacuum*, which differentiates thermionic vacuum valves from gas-discharge tubes and solid-state devices, e.g., transistors, diodes, etc., which can perform similar functions.

hard lighting. High contrast illumination from a small source. Hence *hard* for the type of lamp.

hard shower. One of cosmic-ray particles, several of which are highly penetrating.

hardness. Degree of vacuum in an evacuated space, especially of a thermionic valve or X-ray tube. Also penetrating power of X-rays, which is proportional to frequency.

harmodotron. V-band or higher-frequency (sub-mm) microwave generator, in which resonant cavity similar to rhumbatron is excited to high order mode resonance at a harmonic of the bunching frequency.

harmonic. Sinusoidal component of repetitive complex waveform with frequency which is an exact multiple of basic repetition frequency (the fundamental). The full set of harmonics form a Fourier series which completely represents the original complex wave. In acoustics, harmonics are frequently termed overtones, and these are counted in order of frequency above, but excluding, the lowest of the detectable frequencies in the note; the label of the harmonic is always its frequency divided by the fundamental. Thus the *n*th overtone is the $(n+1)$th harmonic.

harmonic absorber. Arrangement for remov-ing harmonics in current or voltage wave-forms, using tuned circuits or a wave filter.

harmonic analysis. Process of measuring or calculating the relative amplitudes of all the significant harmonic components present in a given complex waveform. The result is frequently presented in the form of a Fourier series, e.g., $A = Ao \sin \omega t + A_1 \sin (2\omega t + \Phi_1) + A_2 \sin (3\omega t + \Phi_2) + \ldots$ etc., where ω = pulsatance and Φ = phase angle.

harmonic antenna. One whose overall length is an integral number (greater than one) of quarter-wavelengths.

harmonic component. Any term, other than the first, in a Fourier series representing a complex wave.

harmonic content. Non-sinusoidal periodic function with the fundamental extracted.

harmonic distortion. Production of harmonic components from a pure sine wave signal as a result of non-linearity in the response of a transducer or amplifier.

harmonic excitation. (1) Excitation of an antenna at one of its harmonic modes. (2) Excitation of a transmitter from a har-monic of the master oscillator; also called **harmonic drive.**

harmonic filter. One which separates harmonics from fundamental in the feed to an antenna; also called **harmonic suppressor.**

harmonic generator. Waveform generator with controlled fundamental frequency, producing a very large number of appreci-able amplitude odd and even harmonic components which provide a series of reference frequencies for measurement or calibration. See **multivibrator.**

harmonic interference. That caused by har-monic radiation from a transmitter and outside the specified channel of radio communication.

harmonic series. One in which each basic frequency of the series is an integral multiple of a fundamental frequency.

harmonic suppressor. Same as **harmonic filter.**

Harris flow. The flow of electrons in a cylindrical beam subjected to a radial electric field to counteract the beam space-charge divergence.

hartley. Computer unit of information, equal to 3·32 *bits*, and conveyed by one decadal code element.

Hartley oscillator. Triode valve oscillator circuit consisting essentially of a parallel resonant circuit connected between grid and anode, cathode being connected to the coil by a tapping point, which becomes *earthy.*

Hartley principle. General statement that amount of information which can be trans-mitted through a channel is the product of frequency bandwidth and time during which it is open, whether time division is used or not. See **information.**

Hartree equation. Equation relating flux density in a travelling-wave magnetron to

E

minimum anode potential required for oscillation in any given mode. Graphs representing this relationship for different mode numbers form a **Hartree diagram**.

hash. Electrical interfering noise arising from vibrators or commutators.

haversine. Half the *versine*. A quantity required in spherical trigonometry.

Hay bridge. A.c. bridge quite widely used for the measurement of inductance.

Hazeltine neutralization. A method of anode neutralization used in single-stage power amplifiers.

head. Recording and reproducing unit for magnetic tape, containing exciting coils and a laminated core, in ring form with a minute gap. Flux leakage across this gap enters the tape and magnetizes it longitudinally. See **flux gate**.

head amplifier. (1) That amplifying output of magnetic or photoelectric reproducer head in a cinema projector, when separate from the main or *gain* amplifier. (2) Pre-amplifier which has to be situated geometrically adjacent to signal source. Used especially in conjunction with radiation detectors.

header. Base of hermetically sealed relay through which conductors pass to make contact, temporary or permanent, with external current.

heading. Horizontal direction in which a vessel or missile points. (This only coincides with the track in the absence of drift and yaw.)

headphones. Two earphones fitted to a headband for retaining them in position.

health physics. Branch of radiology concerned with health hazards associated with ionizing radiations, and protection measures required to minimize these. Personnel employed for this work are *Health Physicists* or *Radiological Safety Officers.*

hearing aid. Small (transistor) amplifier, consisting of microphone, earphone and batteries, used by partially deaf person.

hearing loss. At any frequency the hearing loss of a partially deaf ear is the ratio in decibels of its threshold of hearing to that for a normal ear. (This term is synonymous with *deafness* in scientific work.) 100 times the ratio of the hearing loss to the number of decibels between the normal thresholds of hearing and of feeling is termed the *percentage hearing loss* for the specified frequency.

heart-shaped reception. Same as **cardioid reception**.

heat sink. Usually metal plate especially designed (e.g., with fins) to conduct (and radiate) heat from an electrical component, e.g., a transistor.

heater. Conductor carrying current for heating an equipotential cathode, generally enclosed by the cathode.

heating depth. Thickness of skin of material which is effectively heated by dielectric or eddy-current induction heating, or radiation.

heating inductor. Conductor, usually water-cooled, for inducing eddy currents in a charge, workpiece, or load. Also **work coil**; see also **applicator**.

heating time. Time required after switching on before a valve cathode or other piece of equipment reaches normal operating temperature.

Heaviside-Campbell bridge. An electrical network for comparing mutual and self inductances.

Heaviside layer. See **E-layer**.

Heaviside-Lorentz units. Rationalized CGS Gaussian system of units, for which corresponding electric and magnetic laws are always similar in form. See **rationalized units**.

Heaviside unit function. Step change in a magnitude, with an infinite rate of change, required in pulse analysis and transient response of circuits.

heavy aggregate concrete. Concrete containing very dense aggregate material in place of some or all of the usual gravel, so increasing its gamma-ray absorption coefficient.

heavy hydrogen. Same as **deuterium**.

heavy particle. Fundamental particle heavier than the neutron. See **hyperon**.

heavy water. Deuterium oxide, or water containing a substantial proportion of deuterium atoms (D_2O or HDO).

heavy-water reactor. One using *heavy water* as moderator, achieving sufficiently high neutron flux to manufacture ^{60}Co and other radioisotopes for sources of β- and γ-rays.

hectometric waves. See **medium frequency**.

height control. The means of varying the 'y' deflection of a CRO in TV or in radar.

Heil oscillator. Early form of velocity-modulated transit time oscillator; forerunner of the klystron. Also called **coaxial-line tube, Heil tube**.

heiligtag effect. Error arising in a bearing because of reception of waves from a transmitter taking different paths. Also called **wave-interference error**.

Heisenberg force. Short-range exchange force between nucleons which changes sign if both space or spin coordinates are interchanged. See **short-range forces**.

Heisenberg principle. See **uncertainty principle**.

Heising modulation. Constant-current modulation, arising from one valve driven by signal and another valve driven by carrier, having their anodes fed through the same inductor; the modulated carrier is taken from the anode circuit by capacitive or inductive coupling.

heliatron. Voltage-tuned microwave oscillator in which electron beam follows helical path.

Helipot. TN for precision multi-turn potential divider able to be reset to within 0·1%.

helium (He). Inert gas, at. no. 2, at. wt. 4·0026, with extremely stable nucleus identical to α-particle. Helium only liquefies at temperatures below 4°K, and undergoes a phase change to a form known as *liquid helium II* at 2·2°K. The latter form has many unusual properties believed to be due to a substantial proportion of the molecules existing in the lowest possible quantum energy state. See **superfluid**. Liquid helium is the standard coolant for devices working at cryogenic temperatures.

helix. Spiral path as followed by charged particle in magnetic field.

Helmholtz coils. Pair of identical compact coaxial coils separated by a distance equal to their radius. These give a uniform magnetic field over a relatively large volume at a position midway between them.

Helmholtz resonator. Enclosure with small aperture which forms an acoustic cavity resonator.

Helmholtz's theorem. See **Thevenin's theorem**.

henry. MKSA unit of self and mutual inductance, such that e.m.f. of 1 V is induced in a circuit by current variation of 1 A/s.

heptode. See **pentagrid**.

Hermes. AERE machine, similar to large mass spectrograph, for electromagnetic separation of isotopes with very similar specific charge. (*H*eavy-*e*lement *r*adioactive *m*aterial *e*lectromagnetic *s*eparator.)

hertz (Hz). Formerly Continental, now standard U.K. name for the unit *one cycle per second* (c/s), after Heinrich Rudolf Hertz, who first realized electromagnetic waves of about 1 metre wavelength.

Hertz antenna. Original half-wave dipole, fed at the centre.

Hertzian dipole. Pair of opposite and varying charges, close together, with an electric moment; also **Hertzian doublet**.

Hertzian oscillator. Idealized system envisaged by Hertz, comprising two point charges of opposite sign and separated by an infinitesimal distance, whose electric moment varies harmonically with time.

Hertzian radiator. Original form of radiator used by Hertz, comprising two parallel flat metal plates, connected by straight conductors to a spark gap placed midway between them. See **dipole**.

Hertzian waves. Electromagnetic waves, from e.g., 10^4 to 10^{10} Hz, which have been found useful for communicating information through space.

heterodyne. Combination of two sinusoidal RF waves in a non-linear device with the consequent production of sum and difference frequencies. The latter is the *heterodyne frequency*, and will produce an AF *beat note* when the two original sine waves are sufficiently close in frequency.

heterodyne conversion. Change in the frequency of a modulated carrier wave produced by heterodyning it with a second unmodulated signal. The sum and difference frequencies will carry the original modulation signal and either of these can be isolated for subsequent amplification. The frequency-changing stage of a supersonic heterodyne (*superhet*) radio receiver uses this principle, an oscillator being tuned to a fixed amount above the signal frequency so that

the difference frequency (intermediate frequency) remains constant for all incoming signals.

heterodyne frequency meter. See **heterodyne wavemeter**.

heterodyne interference. That arising from simultaneous reception of two stations the difference between whose carrier frequencies is an audible frequency.

heterodyne oscillator. One which generates a continuously variable frequency by heterodyning two very much higher frequency oscillations—one of which is variable. Also **beat frequency oscillator**.

heterodyne reception. Reception of ICW radio-telegraphy signals by heterodyning the signal with a locally-generated oscillation at a slightly different frequency. The rectified output contains a beat note component which reproduces the incoming signal audibly. Also **beat reception**.

heterodyne wavemeter. One in which a continuously variable oscillator is adjusted to give zero beat with an unknown frequency, the value of which then coincides with the calibration of the oscillator. Also **beat frequency wavemeter, heterodyne frequency meter**.

heterodyne whistle. Variable note heard when tuning an incoming signal with an oscillating receiver and causing interference in reception elsewhere. The constant high-pitched note resulting from two carriers beating.

heterogeneous radiation. Radiation comprising a range of wavelengths or particle energies.

heterogeneous reactor. Reactor in which the fuel and moderator elements in the core form large discontinuous elements arrayed in a lattice structure.

heterostatic operation. See **Dolezalek electrometer**.

Heusler alloys. Alloys of manganese, copper and aluminium (but not containing iron, nickel or cobalt) which are strongly ferromagnetic.

hex. Colloquialism for *uranium hexafluoride*, the compound used in the separation of uranium isotopes by gaseous diffusion.

hexode. Six-electrode electron tube comprising an anode, cathode, control electrode and three grid electrodes.

Hibbert standard. A standard of magnetic-flux linkage suitable for fluxmeter or galvanometer calibration. It comprises a stabilized magnet producing a radial field in an annular gap, through which a cylinder carrying a multi-turn coil can be dropped.

high-fidelity. Said of sound reproduction of exceptionally high quality. By extension, also used of amplifiers and loudspeakers with low distortion. Abb. **hi-fi**.

high-fidelity amplifier. One in which the input signal is reproduced with a very high degree of accuracy.

high-flux reactor. One designed to operate with a greater neutron flux than normal (at present about 10^{15} neutron cm^{-2} $sec.^{-1}$), especially for testing materials. Also **materials-testing reactor.**

high frequency. Any frequency above the audible range, i.e., above 15 kHz, but especially those used for radio communication.

high-frequency alternator. One of several (Alexanderson, Bethenod-Latour, Goldschmidt) used to generate high-frequency currents required by long-wave transmitters early this century.

high-frequency amplification. That at frequencies used for radio transmission. In a receiver, any amplification which takes place before detection, frequency conversion or demodulation.

high-frequency capacitance microphone. One which uses audio variation of capacitance to vary the frequency of an oscillator, or response of a tuned circuit.

high-frequency carrier telegraphy. Carrier telegraphy in which the frequency of the carrier currents exceeds the range which is transmitted over a voice-frequency telephone channel.

high-frequency choke. Inductance coil designed to have high impedance at specified high frequencies, depending on anti-resonance with self-capacitance.

high-frequency heating. Heating (induction or dielectric) in which frequency of current is above mains frequency; from rotary generators up to, e.g., 3000 Hz and from valve generators 1–100 MHz. Also **radio heating.**

high-frequency resistance. That of a conductor or circuit as measured at high frequency, greater than that measured with d.c. because of *skin effect* (increased losses would mean lower resistance for constantly applied p.d.).

high-frequency transformer. One designed to operate at high frequencies, taking into account self-capacitance, usually with bandpass response.

high-frequency welding. Welding by RF heating. See **seam welding.**

high-level diffusion length. The average path length through which a hole diffuses between its generation and recombination in a semiconductor or transistor.

high-level firing time. The time required to establish an RF discharge in a switching-tube after the application of RF power.

high-level (low-level) modulation. Conditions where modulation of a carrier for transmission takes place at high level for direct coupling to the radiating system, or is at lower level than this, with subsequent push-pull or straight amplification. Also **high-(low-) power modulation.**

highlight. Very bright area in a screen or image; the converse of *shadow.*

high-pass (low-pass) filters. Filters which freely pass signals of all frequencies above (or below) a reference value, known as the *cutoff frequency* (f_c). (*N.B.*—Beyond f_c, attenuation only rises slowly and seldom approaches complete cutoff as implied by this name.)

high-power modulation. See **high-level modulation.**

high recombination rate contact. The contact region between a metal and semiconductor (or between semiconductors) in which the densities of charge carriers are maintained effectively independent of the current density.

high-temperature reactor. One designed to attain core temperatures above 660°C. See under **uranium**—*alpha* and *beta.*

high-tension battery. One supplying power for the anode current of valves. Also **anode battery**; in U.S. **B-battery.**

high vacuum. A system so completely evacuated that the effect of ionization on its subsequent operation may be neglected. See also **hard.**

high-velocity scanning. The scanning of a target with electrons of sufficient velocity to produce a secondary emission ratio which is greater than unity.

highway. Defined path for transmission of computer signals. *Trunk* or *bus* (U.S.). If related, several highways can be a *channel.*

HiK capacitor. One in which the dielectric of barium and strontium titanates has permittivities above 1000.

hill-and-dale recording. Disk recording in which cutting stylus and reproducing stylus move at right-angles to the plane of the disk. Now obsolete. Also called **contour recording, vertical recording.** See **stereophonic recording.**

hill-climbing. Continuous or periodic adjustment of self-regulating adaptive control systems to achieve an optimum result.

hindrance. Impedances (0 for zero, and 1 for infinite) used in theoretical manipulation of switching.

Hiperco. RTM for a high-permeability, high-saturation, 35% cobalt-iron alloy with 0·5% chromium, for making magnets, etc.

Hipernik. RTM for isotropic alloy of equal parts of iron and nickel, having its grain orientated for making magnets, etc.

Hipersil. RTM for silicon-iron alloy, highly grain-orientated.

Hittorf tube. Early form of cathode-ray tube. See **cathode dark space.**

Hodectron. Mercury-vapour discharge tube, fired by a magnetic pulse.

hodoscope. Apparatus (e.g., an array of radiation detectors) for tracing paths of charged particles in a magnetic field.

Hoffman binant electrometer. Primitive suspended needle electrometer which, although highly unstable, is the most sensitive direct-current measuring instrument (down to 10^{-19} amp.).

hoghorn. Radar horn in which the waveguide is flared out, so that the electromagnetic wave is reflected from a parabolic surface.

hohlraum. Cavity employed as black-body radiator. A *microwave hohlraum* is a similar device covering microwave frequencies and employed as wideband noise source.

hold. (1) Synchronization control, whereby frequency of the local oscillator is adjusted to that of the incoming synchronizing pulses. (2) The retention in computers of the information contained in one storage device after copying it into another. (3) The maintenance, in charge-storage tubes, of the equilibrium potential by means of electron bombardment.

hold-off voltage. That which just prevents initiation of ionization and discharge by control electrode.

hold-up. The quantity of process material in an isotope separation plant.

holdback. Agent for reducing an effect, e.g., a large quantity of inactive isotope reduces the co-precipitation or absorption of a radioactive isotope of the same element.

holding anode. Auxiliary d.c. anode in a mercury-arc rectifier for maintaining an arc.

holding beam. Widely spread beam of electrons used to regenerate charges retained on the dielectric surface of a storage tube or electrostatic memory.

hole. Vacancy in a normally filled energy band, either as result of an electron being elevated by thermal energy to the conduction band, so producing a hole-electron pair, or as a result of one of the crystal lattice sites being occupied by an acceptor impurity atom. Such vacancies are mobile and contribute to electric current in the same manner as positive carriers, and mathematically are equivalent to positrons. Also **negative ion vacancy.**

hole current. That part of the current in a semiconductor due to the migration of holes.

hole density. The density of the holes in a semiconductor in a band otherwise full.

hole injection. Holes can be *emitted* in *n*-type semiconductor by applying a metallic point to its surface.

hole trap. An impurity in a semiconductor which can release electrons to the conduction or valence bands and so trap a hole.

hollow-cathode tube. A gas-discharge tube in which radiation is emitted from a hollow cathode (closed at one end) in the form of a cathode glow.

holmium (Ho). Rare earth element, at. no. 67, at. wt. 164·930. One stable isotope and several radioactive ones exist.

holohedral. Of a crystal when complete, showing all possible faces and angles.

homer. Any arrangement which provides signals or fields which can be used to guide a vehicle to a specified destination, e.g., a *homing transmitter* for aircraft, a *leader cable* for ships, or *guidance system* for missiles.

homing. (1) Said of a magnetic stepping relay when its wipers return to a datum set of contacts when relay is released. (2) Guided missile guidance system in which a *seeker* is actuated by some influence from target, e.g., infrared rays from engine exhaust, or

radio transmission. (3) Process of directing a vessel straight towards an RDF transmitter.

homocentric. Term applied when rays are either parallel or pass through one focus.

homodyne reception. That using an oscillating valve adjusted to, or locked with, an incoming carrier, to enhance its magnitude and improve demodulation. Also **demodulation of an exalted carrier.**

homogeneous ionization chamber. One in which both walls and gas have similar atomic composition, and hence similar energy absorption per unit mass.

homogeneous radiation. Radiation of constant wavelength (monochromatic) or of constant particle energy.

homogeneous reactor. One in which the fuel and moderator are finely divided and mixed to produce an effectively homogeneous core material. They may both be evenly suspended in a liquid coolant or the moderator itself may be liquid. See, e.g., **slurry reactor.**

homopolar generator. Low voltage d.c. generator based on Faraday disk principle and producing ripple-free output without commutation.

homopolar magnet. One with concentric pole pieces.

homopolar molecule. One without effective electric dipole moment.

honeycomb coil. One in which wire is wound in a zig-zag formation around a circular former. The adjacent layers are staggered, so that the wires cross each other obliquely to reduce capacitance effects between turns.

hook transistor. A junction transistor which uses an extra *p-n* junction to act as a trap for holes, thus increasing the current amplification.

hop. Distance along earth's surface between successive reflections of a radio wave from an ionized region; also **skip.**

horizon sensor. Sensor providing a stable vertical reference level for missiles and depending on the use of a thermistor to detect the thermal discontinuity between earth and space.

horizontal antenna. One comprising a system of one or more horizontal conductors, radiating or responsive to horizontally-polarized waves.

horizontal blanking. The elimination of the horizontal trace in a CRT during flyback.

horizontal component. Component of earth's magnetic field which acts (i.e., exerts a force on a unit pole placed in it) in a horizontal direction.

horizontal hold control. The control in TV which varies the free-running period of the oscillator providing the horizontal deflection.

horizontal polarization. That of an electromagnetic wave when the alternating electric field is horizontal.

horn. Tube of continuously-varying cross-section used in the launching or receiving of radiation (e.g., acoustic horn, electromagnetic horn). Horns are best classified by their geometric shapes which include: compound, conical, cornucopia, exponential (logarithmic), folded, pyramidal, sectoral.

horn feed. The radar feed to a transmitter in the form of a horn.

horn loudspeaker. A loudspeaker in which the radiating device is acoustically coupled to the air by means of a horn.

horn throat and mouth. Beginning and end (flare) of metal horn, between which it may be curved, folded, bifurcated.

horsepower. Measures the rate of doing work. 1 horsepower (HP)$=746$ watt$=3\cdot3\times10^4$ ft lb/min.

horseshoe magnet. Traditional form of an electro- or permanent magnet, as used in many instruments, e.g., meters, magnetrons.

hot. (1) Charged to a dangerously high potential. (2) Colloquial reference to dangerous radioactive substances; hence *hot laboratory*, where extreme precautions against irradiation of humans are taken.

hot cathode. One in which the electrons are produced thermionically.

hot-cathode rectifier. One with emitting cathode heated independently of the rectified current, e.g., a mercury thyratron.

hot spot. A small region of an electrode in a valve (or CRO screen) which has a temperature above the average.

hot-wire instrument. An electrical instrument, e.g., ammeter, which depends for its action on the expansion of a wire or strip which is heated electrically.

hot-wire microphone. One in which a d.c. signal through a hot wire is modulated by resistance variations consequent upon the cooling effect of an incident sound wave.

hour angle. The angle between two meridians in navigation when expressed as a time (24 hours$=360°$, since the earth rotates once per day).

housekeeping. Colloquialism for preparation of memory by computer programme.

housing. Containment of apparatus to prevent damage in handling or operations. Also **wrapround**; see **packaged**.

Howe factor. One expressing the fractional increase in volume of a reactor fuel as a result of the additional atoms produced during fission.

howl. A high-pitched audio tone due to unwanted acoustic (or electrical) feedback.

howler. A device which uses acoustic feedback between a telephone transmitter and a telephone receiver to maintain a continuous oscillation and so provides suitable currents for testing telephonic apparatus.

hue. Physiological attribute of colour perception—the relationship between hue and colour in vision being comparable to that between pitch and frequency in sound.

hum. Objectionable low-frequency components induced from power mains into sound reproduction, either from inadequate smoothing of rectified power supplies, induction into transformers and chokes, unbalanced capacitances or leakages from cathode heaters.

hum-bucking coil. See **bucking coil**.

hum modulation. That which results from inadequate smoothing of anode or grid supplies in transmitters, or in RF amplifier in a receiver.

hummer. A microphone used for balance detection in conductivity measurements.

Hund rule. Transition metals with incomplete inner electron shells have large magnetic moments, as the electron spins in the incomplete shell are self-aligning.

hunting. Desired or undesired fluctuation of a magnitude between limits and at a low rate in a servo system. May arise from instability in part of system; see also **cycling**.

hush-hush. Variation in background noise when a noise-reduction system is in use.

hybrid electromagnetic wave. Wave having longitudinal components of both the electric and magnetic field vectors.

hybrid junction. A waveguide transducer which is connected to four branches of a circuit, designed to render these branches conjugate in pairs. If in the form of a **(hybrid) ring**, it is commonly called a **rat-race**.

hybrid parameters. See **transistor parameters**.

hybrid set. Two or more transformers forming a four-pair terminal network. Four impedances may be connected to these terminals pins so that the branches containing them may be conjugate in pairs.

hybrid-T. Combination of E- and H-junctions for reflection-free radar propagation.

hydraulic amplifier. Hydraulic mechanism used to amplify an applied pressure and to increase the pressure level of a mechanical signal.

hydrogen (H). Lightest element and most simple atom, at. no. 1, at. wt. $1\cdot00797$. Exists in its pure state as a gas (except at very low temperatures) but most widely as a constituent of water (H_2O). Of great importance in the moderation (slowing down) of neutrons, as hydrogen atoms are the only ones of similar mass to a neutron, and are therefore capable of absorbing an appreciable proportion of the neutron energy on collision. Isotopes are, with one proton and one electron,

Name	Atomic symbol	Atomic weight	Mass number	Neutrons
protium	^{1}H	1·007825	1	0
deuterium	D,^{2}H	2·01410	2	1
tritium	^{3}H	3·0221	3	2

hydrogen bomb. An *atomic bomb* (q.v.) surrounded by lithium deuteride which, through the atomic bomb temperature, fuses to helium with very great emission of energy because of loss of mass (1–20 megaton TNT equivalent or more in explosive power, i.e., 1000 times more powerful than

Hiroshima atomic bomb). Also **fusion bomb, H-bomb, thermonuclear bomb.**

hydrogen bond. Type of electromagnetic valence linkage between two atoms through a hydrogen atom, known as a *resonance phenomenon*, in which the bond is alternately attached to each of the atoms, since the stable hydrogen atom cannot be associated with more than two electrons.

hydrogen electrode. For pH measurement, a platinum-black electrode covered with hydrogen bubbles.

hydrogenous. Substance rich in hydrogen and therefore suitable for use as moderator of neutrons.

hydromagnetic. Pertaining to the behaviour of a plasma in a magnetic field. See also **magnetohydrodynamics.**

hydromagnetic instability. See **instability** (1a).

hydrometer. Instrument for measuring liquid density (or specific gravity), consisting of a weighted bulb with a graduated stem. In electrical terminology, often one specifically designed to check the electrolyte of lead-acid accumulators.

hydrophone. Electro-acoustic transducer used to detect sounds or ultrasonic waves transmitted through water. Also **subaqueous microphone.**

hygristor. Resistance element sensitive to ambient humidity.

hygrometer. Instrument for measuring, or giving output signal proportional to, atmospheric humidity. Electric hygrometers make use of *hygristors*. See **electrical-.**

Hymn-88. RTM for a magnetic alloy, mostly nickel, and some molybdenum and iron.

hyp. One-tenth of a *neper.* See **decineper.**

hyperbolic. Said of any system of navigation which depends on difference in cycles and fractions between locked waves from two or more stations, e.g., Decca, gee, loran.

hypercharge. See **strangeness.**

hyperfine structure. Splitting of spectrum lines into two or more very closely spaced components, due to effects such as different isotopic mass (see **isotope structure**); or interaction between the orbital electrons and external fields. See **Stark effect, Zeeman effect.**

hyperon. Cosmic-ray particle having a mass greater than that of a neutron and less than that of a deuteron.

hypersonic. See **Mach number.**

hypertonic, hypotonic. Said of a solution having a respectively higher or lower osmotic pressure than a standard reference solution.

hypsometer. Apparatus for determining the boiling point of liquid.

hysteresis. The retardation or lagging of an effect behind the cause of the effect, e.g., magnetic, dielectric, etc. See **control-.**

hysteresis cycle. See **hysteresis loop.**

hysteresis error. Instruments or control systems may show non-reversibility similar to hysteresis. The maximum difference between the readings or settings obtainable for a given value of the independent variable is the *hysteresis error.*

hysteresis heat. That arising from hysteresis loss, in contrast with that from ohmic loss associated with eddy currents.

hysteresis in a valve oscillator. The anomalous behaviour which can occur, e.g., in reflex klystrons.

hysteresis loop. The relationship between magnetic induction and magnetizing field for a ferromagnetic material is non-reversible —changes in induction during demagnetization lagging on those which occur during magnetization. If readings are continued with reversed field a loop can be plotted, two values of induction being associated with each applied field. The area of this loop measures the energy dissipated during a cycle of magnetization. Similar effects occur with applied mechanical or electrical stresses, and this is by analogy known as *mechanical* or *electric hysteresis.*

hysteresis loss. Energy loss in taking unit quantity of material once round a hysteresis loop. It can arise in a dielectric material subjected to a varying electric field or in a magnetic material in a varying magnetic field, etc.

hysteresis motor. A synchronous motor which starts by reason of the hysteresis losses induced in its steel secondary by the revolving field of the primary.

I

I display. Radar display representing the target as a full circle when the antenna is pointed directly at it, the radius being in proportion to the range.

I neutrons. Those possessing such energy as to undergo resonance absorption by iodine.

I signal. In the NTSC colour TV system, that corresponding to the wideband axis of the chrominance signal.

iconoscope. Electron camera comprising a mosaic of photoemissive material upon which the optical image is focused, and which is scanned by a cathode-ray beam, which restores charge lost through illumination.

Iconotron. RTM for an **image iconoscope.**

ideal crystal. One in which there are no imperfections or alien atoms.

ideal gas. Gas with molecules of negligible size and exerting no intermolecular attractive forces. Such a gas is a theoretical abstraction which would obey the ideal gas law under all conditions. The behaviour of real gases becomes increasingly close to that of an ideal gas as the pressure on them is reduced.

ideal noise diode. One which has an infinite internal impedance and in which complete 'shot-noise' fluctuations are shown by the electric current.

ideal SPF. See **simple process factor.**

ideal transformer. Hypothetical transformer corresponding to one with a coefficient of coupling of unity.

ideal wire. See **Eureka wire.**

identification, friend or **foe.** Military radar device consisting of automatic transponder carried by vessel in order to modify nature of target blip on radar screen when this has been located by a friendly radar station. If blips are not modified in this way, the vessel responsible is assumed hostile.

idiochromatic. Said of the properties of a crystal free from impurity; also **intrinsic.**

idiostatic connection. See **Dolezalek electrometer.**

idle component. Same as **reactive component.**

ignition coil. An induction coil incorporated in the ignition system of an internal combustion engine.

ignition noise or **interference.** The electrical noise arising mainly from ignition sparks, which causes interference in TV or radio systems.

ignition rectifier. Mercury-arc rectifier in which the cathode spot is initiated by a voltage impulse applied to a special electrode dipping into the mercury pool.

ignition temperature. The temperature at which energy balance for a thermonuclear reaction is reached, i.e., energy released in the plasma by fusion equals that lost.

ignition voltage. That required to start discharge in a gas tube.

ignitor. See **pilot electrode.**

ignitron. Mercury-arc rectifier with *ignitor*, which is an electrode which can dip into the cool mercury pool and draw an arc to start the ionization.

ignore. A character in computer terminology which signifies that no action is to be taken.

Ilford C.2. TN for nuclear research emulsion widely used in study of particle tracks.

illinium. See **promethium.**

illumination. Ratio of luminous flux on a surface to normal area of the surface; also **illuminance.**

image admittance. The reciprocal of *image impedance.*

image charge. Hypothetical charge used in electrostatic theory as a substitute for a conducting (equipotential) surface. The change must not modify the field distribution at any point outside this surface.

image converter tube. One in which an optical image applied to a photoemissive surface produces a corresponding image on a luminescent surface.

image dissector. Electron camera in which the optical image is focused on a photoemissive surface, from which electrons are emitted in straight lines in the form of the image. Suitably arranged electric or magnetic fields cause this pattern to sweep across a point anode, which effectively scans the image.

image dissector multiplier. Combination of image dissector and electron multiplier in one tube or unit.

image force. That force on an electric (or magnetic) charge between itself and its image induced in a neighbouring body.

image frequency. That which is as much greater (or less) than the local oscillator frequency as the signal frequency is less (or greater) in a supersonic heterodyne receiver. A transmission on this frequency will be accepted, unless rejected by previous stage tuned to the signal frequency.

image iconoscope. Camera tube in which the image is focused on a photoemissive cathode, electrons from which are focused on to a target, which is front-scanned by a high-energy electron beam.

image impedance. Input impedance of a network or filter section, when it is terminated by the same impedance. If Z_f is open-circuit impedance and Z_g short-circuit impedance, image impedance is given by

$$Z_0{}^2 = Z_f Z_g.$$

image orthicon. Tube in which the electron image pattern from a continuous emissive surface is focused on a storage photomosaic, which is scanned on the reverse side by a low-energy cathode beam. See **orthicon**.

image phase constant. In a filter section, properly terminated in both directions, unreal or imaginary part of the (image) transfer constant; phase delay of the section in radians.

image ratio. The ratio of the image frequency signal input at the aerial to the desired signal input, in a heterodyne receiver, for identical outputs.

image reactor. Hypothetical reactor used in theoretical study of neutron flux distribution as a substitute for a neutron reflector. The treatment is analogous to the *image charge* method in electrostatics.

image response. Unwanted response of a supersonic heterodyne receiver to the image frequency.

image signal. One whose frequency differs from the received signal, in a superheterodyne receiver, by twice the intermediate frequency.

image tube. One in which an optically focused image on a photoemissive plate releases electrons which are focused on a phosphor by electric or magnetic means. Used in X-ray intensifiers, infrared telescopes, electron telescopes and microscopes.

Image Vericon. RTM for TV camera tube, similar to an *image orthicon*.

imaginary number. One represented by a vector parallel to the y axis on an Argand diagram. See **complex number**.

immitance. Combined term covering **impedance** and **admittance**.

impact fluorescence. That which is created by particle bombardment.

impact ionization. The loss of orbital electrons by an atom of a crystal lattice which has experienced a high-energy collision.

impact parameter. The distance at which two particles which collide would have passed if no interaction had occurred between them.

impedance. (1) Complex ratio of sinusoidal voltage to current in an electric circuit or component. Its real part is the *resistance* (dissipative or wattful impedance) and its imaginary part the *reactance* (non-dissipative or wattless), which may be positive or negative according to whether the phase of the current lags or leads on that of the voltage. Resistance, reactance and impedance are all measured in *ohms*. Expressed symbolically the impedance

$$Z = R + jX, \text{ where}$$
$$R = \text{resistance}$$
$$X = \text{reactance}$$
$$j = \sqrt{-1}.$$

(2) Component offering electrical impedance —preferably termed **impedor**.

See also:

acoustic-	bilateral-
anode-	blocked-
anode load-	cathode interface-
antenna-	characteristic-

E*

conjugate-	motional-
driving-point-	open-circuit-
electrode-	output-
free-space-	radiation-
image-	rated-
input-	self-
internal-	short-circuit-
intrinsic-	terminal-
iterative-	transfer-
load-	unilateral-
loaded-	wave-
matching-	waveguide-.
mechanical-	

impedance circle. Locus of the end of the impedance vector in an Argand (R, X) diagram of a system, e.g., drawn to show the variation of input impedance of an improperly terminated line with frequency.

impedance coupling. The coupling of two circuits by means of a tuned circuit or an impedance.

impedance matching. When connection of a load impedance to the output terminals of a system does not result in maximum possible power being transferred to the load, or leads to excessive energy being reflected back towards the source, an impedance matching device (transformer, stub line, etc.) may be connected between the output and the load to minimize or overcome these effects. See **stub tuning**.

impedometer. Device for measuring impedances in waveguides.

impedor. Physical realization of an *impedance*, as also of an inductor, capacitor or resistor.

implant. The radioactive material, in an appropriate container, which is to be imbedded in a tissue for therapeutic use, e.g., needle or seed.

implementation. The various steps involved in installing and operating a data-processing or control system.

implosion. Mechanical collapse of an evacuated device, e.g., a CRT.

importance function. Limiting value, after a long time, of the number of neutrons per generation in the chain reaction initiated by a specified neutron, to which this function applies.

impulse. Obsolete term for **pulse**.

impulse excitation. (1) Excitation of grid of thermionic tube in which the anode current is allowed to flow for only a very short period during each cycle. (2) Maintenance of oscillatory current in a tuned circuit by pulses synchronous with free oscillations, or at a sub-multiple frequency.

impulse generator. Valve circuit providing single, or a continuous series of, pulses, generally by capacitor discharge and shaping, e.g., by the charging of the condensers in parallel and the discharging of them in series.

impulse inertia. That property of an insulator by which the voltage required to cause disruptive discharge varies inversely with its time of application.

impulse transmission. That involving the use of impulses for signalling, thus reducing the effects of low-frequency interference.

impulsive sound. Short sharp sound the energy spectrum of which spreads over a wide frequency range.

impurity. Small proportion (few $p/10^6$) of foreign matter, e.g., indium, added during a melt of a purer (few $p/10^8$) semiconductor, e.g., silicon, germanium, to obtain required conduction for diodes and transistors. The alien substance in the crystal lattice may add to, or subtract from, the average densities of free electrons and holes in the semiconductors. See also **acceptor, donor.**

impurity levels. Abnormal energy levels arising from slight impurities, resulting in conduction in semiconductors.

inactive component. See **reactive component.**

incident beam. Any wave or particle beam the path of which intercepts a surface of discontinuity.

inclination. See **dip.**

increment. The amount by which a variable is increased during a mathematical operation; cf. *decrement* (2).

incremental hysteresis loss. A small pulsation of the magnetic field about a fixed value leading to a small hysteresis loop on the boundary of a full loop.

incremental induction. The difference of the maximum and minimum value of a magnetic induction at a point in a polarized material, when subjected to a small cycle of magnetization; cf. *incremental hysteresis loss.*

incremental permeability. That of magnetic material when measured with small alternating magnetizing forces, but polarized with a steady magnetizing force. For small polarization the incremental permeability equals the initial permeability, which is always less than the differential permeability, but diminishes seriously when the polarization reaches saturation.

incremental resistance. See **differential anode resistance.**

independent drive. System in which the frequency of a transmitter is determined by an oscillator whose output is amplified and subsequently delivered to the antenna.

independent heterodyne. An oscillator, electrically separate from the detector valve, employed for supplying local oscillations used in heterodyne reception.

independent particle model (of a nucleus). Model in which each nucleon is assumed to act quite separately in a common field to which they all contribute.

indeterminacy principle. See **uncertainty principle.**

index value. Pre-set value of a controlled quantity at which an automatic control is required to aim; also **desired value.**

indexing. Regular step-wise (Geneva) motion (linear or rotary) in automatically-controlled mechanisms, timed mechanically, electrically or electronically.

indicator. Obsolete term for **radioactive tracer.**

indicator tube. Miniature CRT in which size or shape of target glow varies with input signal. Variously known as **magic eye, recording level indicator, tuning indicator,** etc.

indicial admittance. Transient current response of a circuit to the application of a *step function* of one volt, using Heaviside operational calculus.

indicial response. Output waveform from a system when a step pulse of unit magnitude is applied to the input.

indirect ray or **wave.** See **ionospheric ray.**

indirectly-heated cathode. One with an internal heater, highly insulated from the cathode on a surrounding ceramic cylinder. Also **unipotential cathode.**

indium (In). Silvery metallic element of soft texture with low m.p. (156°), at. no. 49, at. wt. 114·82, b.p. 2100°C. It has a large cross-section for slow neutrons and so is readily activated. Used in transistor manufacture and as bonding material for acoustic transducers.

induced charge. That produced on a conductor as a result of a charge on a neighbouring conductor.

induced current. That which flows in a circuit as a result of induced e.m.f.

induced dipole moment. Induced moment of an atom or molecule which results from the application of an electric or magnetic field.

induced e.m.f. That which appears in a circuit as a result of changes in the interlinkages of magnetic flux with part of the circuit. The e.m.f. in secondary of a transformer. Discovered by Faraday at Royal Institution in 1831.

induced noise. That which arises in electrodes because of sufficiently high frequencies in the motions in a space charge in a valve.

induced polarization. That which is not permanent in a dielectric, but arises from applied fields.

induced radioactivity. That induced in non-radioactive elements by neutrons in a reactor, or protons or deuterons in a cyclotron or linear accelerator. X-rays or gamma-rays do not induce radioactivity, unless the gamma-ray energy is exceptionally high.

inductance. (1) That property of an element or circuit which, when carrying a current, is characterized by the formation of a magnetic field and the storage of magnetic energy. (2) The magnitude of the capability of an element or a circuit, to store magnetic energy when carrying a current. See **distributed-, mutual-, self-, tuning-.**

inductance-capacitance filter. A circuit arrangement of series inductors and shunt capacitors across the input terminals to eliminate the appropriate ripple frequency from a unidirectional current.

inductance coefficient. Property of a component (inductor) in a circuit whereby back e.m.f. arises because of rate of change of current. It is 1 *henry* when 1 volt is generated by rate of change of 1 amp/sec.

inductance coil. Coil, with or without an iron circuit, for adding inductance to a circuit; also **inductor.**

inductance coupling. That between two circuits by inclusion of inductance common to both.

induction. See **electromagnetic-, electrostatic-, magnetic-, normal-, residual-, self-.**

induction accelerator. See **betatron.**

induction balance. An electrical network, i.e., a bridge, to measure inductance.

induction coil. Transformer for high-voltage pulses, obtained by interrupted d.c. in the primary, as in a petrol engine. The original Ruhmkorff induction coil was magnetically open-circuit and self-interrupting, like a buzzer or relay; used for early discharges in gas tubes.

induction compass. One which indicates the direction of the earth's magnetic field by a rotating coil in which an e.m.f. is induced.

induction field. See **field** (1).

induction furnace. Application of induction heating in which the metal to be melted forms the secondary of a transformer.

induction heating. That arising from eddy currents in conducting material, e.g., solder, profiles of gear-wheels. Generated with a high-frequency source, usually thermionic valves of high power, operating at 10^6–10^7 Hz. Also **eddy-current heating.**

inductive. Said of an electric circuit or piece of apparatus which possesses self or mutual inductance, which tends to prevent current changes.

inductive circuit. One in which effects arising from inductances are not negligible, the back e.m.f. tending to oppose a change in current, leading to sparking or arcing at contacts which attempt to open the circuit.

inductive coupling. That between two circuits by mutual inductance.

inductive interference. That due to the electric supply systems to communication systems.

inductive load. Terminating impedance which is markedly inductive, taking current lagging in phase on the source e.m.f., e.g., electrodynamic loudspeaker or motor. Also **lagging load.**

inductive neutralization. An amplifier in which the feedback susceptance of the self-capacitance of the circuit elements is balanced by the equal and opposite susceptance of an inductor.

inductive pickoff. One in which changes in reluctance of a laminated path alter a current or generate an e.m.f. in a winding.

inductive reactance. Product of the angular frequency ($=2\pi$ frequency) and inductance. Measured in (positive) ohms when the inductance is in henries.

inductive reaction. Same as **electromagnetic reaction.**

inductive resistor. Wirewound resistor having appreciable inductance at frequencies in use.

inductometer. Calibrated variable inductance.

inductor. Same as **inductance coil.**

inductor core. Usually of thin steel laminations for low-frequency use, but for high frequencies iron dust or magnetic ferrites form the cores to reduce eddy-current loss.

Inductosyn. Very highly sensitive angular repeater or resolver.

inelastic collision. A collision in which there is change in the total energies of the particles concerned. This results from excitation or de-excitation of one (or both) of the particles involved.

inelastic scattering. See **scattering.**

inert gases. Elements helium, neon, argon, krypton, xenon and radon, much used (except the last) in gas-discharge tubes. ^{222}Rn has short-lived radioactivity, half-life less than 4 days, and is used for cancer control. Also **noble gases, rare gases.**

inert metal. Alloy (usually Ti-Zr) for which scattering of neutrons by nuclei is negligible.

inertia switch. One operated by an abrupt change in its velocity, as for some meters, to avoid overloading.

inertial damping. That which depends on the acceleration of a system, and not velocity, e.g., a free flywheel coupled to a shaft by eddy currents.

inertial guidance. That based on integration of accelerations of a moving body, e.g., a ship, using very accurate gyros, acting with a stable platform.

infinite-impedance detector. Anode-bend detector, the output being taken from a cathode resistor, which also reduces demodulation distortion.

infinite line. One which is infinitely long, or finite but terminated with its characteristic impedance, and along which there is uniform attenuation and phase delay.

infinite persistence screen. CRT screen on which displayed traces remain visible (up to weeks) unless removed by electric signal applied to clearing electrode. See **Remscope, Storascope.**

information. This is measured by the number of code elements in use multiplied by the logarithm of the number of values which every code element may take. See **bit, Hartley principle.**

information content. Gross IC is the number of *bits* or *hartleys* required to transmit a message via a noiseless system with specified accuracy, regardless of redundancy. Net IC is the minimum number which would transmit the essential information.

information theory. Mathematical analysis of efficiency with which communication channels are employed—the aim being to find the most efficient system of coding for any channel.

infrablack. Amplitude in a television signal beyond the black level of the picture.

infradyne. Supersonic heterodyne receiver, in which the intermediate frequency is higher than that of the incoming signal.

infrared. Spectrum region in which radiation has wavelengths longer than the visible red, e.g., 7500Å to 1 mm, the shortest radio wave. Can penetrate fog or mist.

infrared detection. Rays detected and registered photographically with special dyes; photosensitively with a special Cs—O—Ag surface; by photoconduction of lead sulphide and telluride; and, in absolute terms, by bolometer, thermistor, thermocouple or Golay detector.

infrared maser. One which radiates or detects signals of mm. wavelength. See also **laser**.

infrared radiation. Electromagnetic waves covering a range from the limit of the visible spectrum (the near region) to the shortest microwaves (the far region). A convenient subdivision is as follows: near: $0.75\,\mu$ to $2.5\,\mu$; intermediate: $2.5\,\mu$ to $30\,\mu$; far: $30\,\mu$ to $1000\,\mu$. The absorption of the radiation by crystalline materials results from the excitation of lattice vibrations which occur in a relatively narrow frequency band.

infrared spectrometer. An instrument similar to an optical spectrometer, but employing non-visual detection and designed for use with infrared radiation.

infrasonic. Said of frequencies below the usual audible limit, e.g., 20 Hz.

inherent filtration. That introduced by the wall of the X-ray tube as distinct from added primary or secondary filters.

inherited error. Error in one stage of a multistage computing calculation which is carried over as an initial condition to a subsequent stage.

inhibit. To prevent normal operation of unit by application of pulse to input gate.

inhour. Unit of reactivity equal to reciprocal of period of nuclear reactor in hours, e.g., a reactivity of 1 inhour will result in a period of 1 hour (see **period**, 3).

initial conditions. Values of the parameters of an equation which are used by a computer in calculating a numerical solution.

initial ionizing event. One which starts a pulse in a radiation detector.

initial magnetization curve. A graph showing the relationship between magnetic induction and applied magnetizing field for an initially unmagnetized sample. It extends from the origin (unmagnetized condition) to the tip of the hysteresis loop (saturation).

initial permeability and **susceptibility.** Limiting value of differential permeability and susceptibility respectively when magnetization tends to zero. Applicable to small alternations of magnetization when there is no magnetic bias.

injection efficiency. Fraction of current flowing across emitter junction in a transistor which is due to *minority carriers*.

injector grid. One which injects a modulating voltage into the electron stream of a first detector of a superheterodyne receiver.

inlay. Electronic mixing of television images using masks.

in-line assembly. Assembly by machine, in which a line of insertion heads inserts components one at a time into a wiring board, on

which a circuit has been previously established by printing or etching.

inner bremsstrahlung. A process which sometimes occurs whereby, during beta disintegration, a photon is emitted of energy between zero and the maximum energy available in the transition.

inner ear. Structure encased in bone and filled with fluid, including the cochlea, the semicircular balancing canals and the vestibule.

inner loop. One which operates within a larger loop of feedback.

inner marker beacon. A vertically-directed radio beam which marks the aerodrome boundary in a beam-approach landing system, e.g., ILS.

inner product. See **scalar product**.

in phase. See under **phase**.

in-phase loss. See **ohmic loss**.

in-pile test. One measuring the effects of irradiation while the specimen is subjected to radiation and neutrons in a reactor.

input. (1) That part of the computer which is fed with information in coded form on punched cards or punched tape. (2) Vehicle carrying the information. (3) Information itself, which is read in.

input block. That part of the internal storage of a computer reserved for accepting input data.

input capacitance. That of the control grid of a valve when all other electrodes are earthed except the anode, which has its normal load. See **Miller effect**.

input gap. See **buncher**.

input impedance. That measured between grid and cathode of a valve in the operating condition, comprising conductance and capacitance paths to all other electrodes. See **Miller effect**.

input signal. That connected to the input terminals of any instrument or system (usually electronic).

input transformer. One employed for isolating a circuit from any d.c. voltage in the applied signal and/or to provide a change in voltage.

input unit (or **equipment**). Unit or equipment used to supply coded information to the arithmetic unit of a computer or data-processing system.

input voltage. That applied to the control electrode of a valve, either directly or through a transformer.

inscriber. In computers an *in*put tran*scriber*.

insel bildung. Areas of a cathode surface where the emission is enhanced, because of uneven grid and anode electrostatic forces.

insensitive time. See **dead time** (1).

insert earphone. One for insertion into the meatus, moulded to fit in hearing-aids.

insertion gain or **loss.** That gain or loss in decibels when a transformer or other impedance-matching transducer or network is inserted into a circuit.

instability. (1) In a plasma, sudden deformation of quasi-static distribution, usually due to either (*a*) weakening of magnetic field as a result of pressure exerted by plasma (hydromagnetic or exchange stability), or (*b*) non-Maxwellian velocity distribution in plasma; (*a*) leads to kinks in discharge track, (*b*) to necking or eventual rupture of track. (2) Tendency for a circuit to break into unwanted oscillations.

instantaneous amplitude. Instantaneous vector magnitude of any periodic disturbance.

instantaneous automatic gain control. A rapid action system in radar for reducing the 'clutter'.

instantaneous frequency. That calculated from the instantaneous rate of change of angle of a waveform on a time base.

instantaneous power. For a circuit or component, the product of the instantaneous voltage and the instantaneous current. This may not be zero even for a non-dissipative (wattless) system on account of stored energy, although in this case its time integral must be zero.

instantaneous store. A less precise term for a zero-access store.

instruction. Part of the coded programme which is read in to a computer to set the circuits for processing subsequently inserted data. See address.

instruction code. Unique set of codings, acceptable to computer, to regulate its operation.

instrument landing system. Method of landing aircraft, using master radar beacons and vertical and lateral guidance without visual reference to the ground.

instrument range. The intermediate range of reaction rate in a nuclear reactor, when the neutron flux can be measured by permanently installed control instruments, e.g., ion chambers.

instrument rating. The limits set by the manufacturer for optimum working.

instrument shunt. A resistance of appropriate value connected across the terminals of a current-measuring instrument to extend its range.

instrument transformer. One specially designed to maintain a certain relationship in phase and magnitude between the primary and secondary voltages or currents.

instrumental sensitivity. The ratio of the magnitude of the response to that of the quantity being measured, e.g., divisions per milliamp.

instrumentation. (1) The measuring and control equipment used in the operation of a plant or process. (2) The study of the application of instruments to industrial control processes.

insulant. See insulation.

insulating materials, classification. See class-A, etc.

insulation. Material whose conductivity is so low that it can be neglected in particular applications, e.g., sulphur for electroscopes, PTFE for ionization gauges, rubber, oil or compressed gas for cables, paper or mica for capacitors, porcelain for open transmission lines. More precisely a substance in which the normal energy band and the first excita-

tion band are separated by a forbidden band, which requires several eV to disrupt; also termed **insulant.** See air-.

insulator. Unit of insulating material specifically designed to give insulation *and* mechanical support, e.g., for telegraph lines, overhead traction and transmission lines, plates and terminals of capacitors.

insulator strength. The maximum mechanical and/or the maximum electrical stress which can be applied.

integer. Any number which is an exact multiple of unity.

integral dose. See dose.

integrating ammeter, voltmeter, wattmeter, etc. Instrument which measures the time integral of the named quantity.

integrating circuit or network. One comprising resistors, capacitors and/or inductors, used, e.g., with servomechanisms, the output voltage being proportional to the integral of the input voltage; cf. *differentiating circuit.*

integrating motor. Permanent magnet d.c. motor, angular rotation of which is integration of current in armature.

integrating photometer. One which measures the total luminous radiation emitted in all directions.

integrator. (1) Any device which integrates a signal over a period of time. (2) Unit in a computer which performs the mathematical operation of integration, usually with reference to time.

intelligence. The information contained in a signal.

intensifier electrode. One which provides post-deflection acceleration in the electron beam of a cathode-ray tube.

intensifying screen. (1) Layer or screen of fluorescent material adjacent to a photographic surface, so that registration by incident X-rays is augmented by local fluorescence. (2) Thin layer of lead which performs a similar function for high-energy X-rays or gamma-rays, as a result of ionization produced by secondary electrons.

intensitometer. Instrument for measuring intensities of X-rays during exposures.

intensity level. Level of power as expressed in decibels above an arbitrary zero power level, e.g., sound intensity level on an audition diagram. See phon, volume unit.

intensity modulation. Modulation of the luminosity of the spot on the fluorescent screen of kinescope and/or oscilloscope, by variation of the current in the beam.

intensity of magnetization. Vector of the magnetic moment of an element of a substance divided by the volume of that element.

intensity of radiation. Energy flux, i.e., of photons or particles, per unit area normal to the direction of propagation.

intensity (of wave). The energy carried by any sinusoidal disturbance is proportional to the square of the wave amplitude. See root-mean-square.

interaction. (1) Transfer of energy between

two particles. (2) Interchange of energy between particles and a wave motion. (3) Between waves—see **interference** (2).

interaction factor. That allowing for insertion loss due to network connected between source and load. This vanishes where iterative impedance of network equals that of load or source.

interaction space. That in a vacuum tube where electrons effect transfer of energy between electrodes.

interbase current. The current flowing between the two base connectors in junction-type tetrode transistors.

interceptor missile. A radio-guided missile for the interception of missiles or aircraft.

intercom. Colloq. for *intercom*munication system.

intercut. Same as *cross cut* (q.v.) in the editing of soundfilm and films for television.

interdigital. Said of magnetrons when alternate segments of the anode are strapped at one end, the other segments being strapped at the other end.

inter-electrode capacitance. That of any pair of electrodes in a valve, other electrodes being earthed.

interface. Boundary surface between two media.

interference. (1) Any signal (whether generated naturally, such as atmospherics, or by radio transmitters or electric machinery) other than that to which it is intended that a radio receiver should respond. (2) Interaction between two or more trains of waves of the same frequency emitted from coherent sources. This leads to the establishment of a series of stationary nodes and antinodes known as an **interference pattern.** See **electrical-**.

interference eliminator. Same as **interference trap.**

interference fading. Fading of signals because of interference among the components of the signals which have taken slightly different paths to the receiver.

interference filter. Same as **interference trap.**

interference inverter. A circuit in a television receiver which reverses the drive to the grid of the CRT when there is a large interference pulse.

interference pattern. See **interference.**

interference trap. Means of reducing interference, e.g., a tuned rejector circuit for a single steady transmission, or a band-pass filter to reduce the accepted band of frequencies to the minimum; also called **interference filter, interference eliminator.**

interferometer. Instrument in which an acoustic, optical or microwave interference pattern of 'fringes' is formed and used to make precision measurements, mainly of wavelength. See **acoustic-**.

interlaced scanning. Sequential scanning of alternate lines in a television picture, so that definition is improved and flicker is reduced by allowing time for detailed areas of phosphors to discharge.

interleave. Alternation of parts from different and unrelated messages on a tape.

interleaved transmission. Type of colour TV signal in which luminance and chrominance signals are compressed into the same video band of frequencies.

interlock. (1) Arrangement of controls in which those intended to be operated later are disconnected until the preliminary settings are correct (e.g., it is normally impossible to apply the anode voltage to an X-ray tube before the cooling water is flowing). (2) Adjustment of furniture or scenery in front of a *back projection* so that the proper perspective is preserved. (3) A mechanical and/or electrical device to prevent hazardous operation of a reactor, e.g., to prevent withdrawal of control rods before coolant flow has been established.

intermediate charged vector boson. A hypothetical particle which has been suggested as the agent of weak nuclear interaction forces (as the pion and photon are of strong forces and electromagnetic forces). So far, attempts to detect it have not been successful.

intermediate circuit. Closed tuned circuit for coupling an antenna to a transmitter or receiver.

intermediate coupling. Coupling between spin and orbital angular momenta of valence electrons with characteristics between those of *j.j. coupling* and those of *Russell-Saunders coupling* (qq.v.).

intermediate frequency. Output carrier frequency of a frequency changer (first detector) in a supersonic heterodyne receiver, adjusted to coincide with the centre of the frequency pass band of the intermediate amplifier.

intermediate-frequency amplifier. In a supersonic heretodyne receiver, that which is between the first (frequency changer) and second (final) demodulator, and which provides the main gain and band pass of the receiver.

intermediate-frequency oscillator. That which, in heterodyne reception, is coupled with the output of the intermediate amplifier for demodulation in the second (final) detector.

intermediate-frequency response ratio. The ratio in decibels of an input signal at the intermediate frequency to one at the required signal frequency, which would produce a corresponding output.

intermediate-frequency strip. The chassis of an intermediate-frequency amplifier. Abb. **IF strip.**

intermediate-frequency transformer. One specially designed for coupling valves (and to provide selectivity) in an intermediate frequency amplifier, the pass band being determined by anti-resonance of the windings and coefficient of coupling.

intermediate neutrons. See **neutron.**

intermediate reactor. One designed so that the majority of fissions will be produced by the absorption of *intermediate neutrons.* Also **epithermal reactor.**

intermediate waves. Those whose wavelengths are in the range 50 to 200 metres, bridging the medium and short radio wavelengths.

intermittent control. Control system in which controlled variable is monitored periodically, an intermittent correcting signal thus being supplied to the controller.

intermittent duty. The conditions of use for a component operated at its intermittent rating.

intermittent rating. The specified power-handling capacity of a component or instrument under specified conditions of non-continuous usage.

intermodulation. Undesired modulation of all frequencies with each other in passing through a non-linear element in a transmission path, giving a blur to a desired signal.

intermodulation distortion. Amplitude distortion in which the intermodulation products are of greater importance than the harmonic products, as in AF amplifiers for high-quality speech or music.

internal capacitance. Same as *inter-electrode capacitance*, but specially referring to the anode-cathode capacitance of a valve.

internal conversion. Nuclear transition where energy released is given to orbital electron, which is usually ejected from the atom (conversion electron), instead of appearing in the form of a gamma-ray photon. The *conversion coefficient* for a given transition is given by the ratio of conversion electrons to photons.

internal impedance. (1) That presented by the anode and cathode terminals of a valve under operating conditions, and equal to the differential anode resistance in shunt with the inter-electrode capacitances. (2) That between the output terminals of any signal source or generator.

internal memory. The memory circuits and equipment of a computer which are under its direct control.

internal pair production. Production of electron-positron pair in the coulomb field of a nucleus. For transitions where the excitation energy released exceeds 1·02 MeV, this process will be competitive with both internal conversion and gamma-ray emission. It occurs most readily in nuclei of low atomic number when the excitation energy is several MeV.

internal resistance. That of a valve, battery, photocell, etc., resulting in a drop in terminal voltage when current is taken. Also **source resistance.** See **Thevenin's theorem.**

internal voltages. Those, such as *emission electron energy, contact potential* or *work function*, which add an effect to the external voltages applied to the electrodes of a valve.

international candle. Former unit of luminous intensity stated in terms of a point source, which was expressed in an equivalence of a standard lamp burning under specified conditions; see **candela.**

international common wave or frequency. One which is shared by different countries in a continent.

international electrical units (ohm, watt, amp, volt). Values of the practical units adopted internationally until 1947, since when MKSA units have been employed. The international watt differed by 16 parts in 10^5 from the absolute watt (1 J/sec.).

international temperature scale. A practical scale of temperature which is defined to conform as closely as possible with the thermodynamic Centigrade scale. Various fixed points were defined initially using the gas thermometer, and intermediate temperatures are measured with a stated form of thermometer according to the temperature range involved. The majority of temperature measurements on this scale are now made with platinum resistance thermometers.

interpolation. General procedure for determining values of a function between known or observed values, e.g., entries in a table of logarithms. There are various procedures, depending on the assumption of a curve (line or parabola) which fits localized values. Imitated electrically in some controls for machine-tools.

interpolator. Apparatus for comparing *cards* in separate packs and merging them into one pack.

interpoles. Additional pole pieces used to overcome armature reaction in rotating electric machines.

interpreter. A procedure in digital computers by which a programme stored in an arbitrary code is translated into inactive language and carries out the operations as they are translated.

interrogation. Transmission of a pulse to a *transponder* (pulse repeater).

interrogator responder. An electronic combination of an interrogator with the responder which receives and displays the answering pulses.

interrupted continuous waves. Continuous waves which are interrupted at a constant rate.

interrupter. A device which interrupts periodically the flow of a continuous current, such as the mechanical 'make and break' of an induction coil.

interstage coupling. See **intervalve coupling.**

interstation interference. That from another transmitter on the same or an adjacent wavelength, as distinct from atmospheric interference.

interstation noise suppression. Suppression of radio receiver output when no carrier is received.

interstice. Space between atoms in a lattice where other atoms can be located.

interval. (1) Time which elapses between events. (2) Ratio of the frequencies of two sounds—or in some cases its logarithm. See **semitone.**

intervalve coupling. Those components which

allow one active device (valve or transistor) to drive another (cascade); also **interstage coupling**.

intervalve transformer. One designed to couple the anode circuit of one valve to the grid circuit of another valve, with a specified frequency response.

intranuclear forces. Those operative between nucleons at close range comprising *short-range forces* and *coulomb forces*. According to the hypothesis of charge independence, the former are always the same for two nucleons of corresponding angular momentum and spin regardless of whether these are protons or neutrons. See **isotopic spin**.

intrinsic. See **idiochromatic**.

intrinsic angular momentum. The total spin of atom, nucleus or particle as an idealized point, or arising from orbital motion. When quantized the former and latter are respectively,

$$\frac{1}{2}\frac{h}{2\pi} \quad \text{and} \quad \frac{h}{2\pi},$$

where h=Planck's constant.

intrinsic conduction. Conduction in a semiconductor when electrons are raised from a filled band into the conduction band by thermal energy, so producing hole-electron pairs. It increases rapidly with rising temperature.

intrinsic crystal. One whose photoelectric properties do not depend on impurities.

intrinsic impedance. Wave impedance depending on the medium alone, i.e., $Z_0{}^2=\mu/\epsilon$, where μ is the absolute permeability and ϵ is the absolute permittivity of the medium.

intrinsic induction. Synonymous with **intensity of magnetization**. *N.B.*—In unrationalized units, it is 4π times this quantity.

intrinsic mobility. The mobility of electrons in an intrinsic semiconductor.

intrinsic semiconductor. See **semiconductor**.

intrinsically safe. Said of electrical apparatus in which the open-circuit voltage and/or the short-circuit current are less than specified values under all conditions.

inverse current. See **reverse current**.

inverse networks. A pair of two-terminal networks whose impedances are such that their product is independent of frequency.

inverse square law. Law giving the decrease in intensity of fields or radiation with distance from the source. It applies to any system with a spherical wavefront and negligible energy absorption and is a consequence of Euclidean geometry.

inverse voltage. That generated across a rectifying element during the half-cycle when no current flows. Its maximum value is the *peak inverse voltage*.

inverted-L antenna. One comprising a vertical up-lead joined to one end of a horizontal conductor.

inverted rectifier. (1) Mercury-arc, or other, rectifier arranged to convert d.c. to a.c. See also **inverter** (1). (2) An amplifier which inverts polarity.

inverted speech. Inversion of the order of speech frequencies before modulation for privacy; effected by modulation with a carrier just above the maximum speech frequency, then discarding this carrier and its upper sideband.

inverted-V antenna. Two wires, several quarter-wavelengths long, joined at the top, one lower end being terminated by a resistance, the other end by the transmitter or receiver; the direction of maximum response is horizontal and in the plane of the wires.

inverter. (1) Circuit with gas-discharge tubes, for converting d.c. to a.c. power by a switching operation determined by a local source of a.c. or by an automatic relaxation oscillation, often push-pull. (2) Arrangement of modulators and filters for inverting speech or music for privacy. Also **speech inverter**.

iodine (I). Non-metallic, blackish-grey, crystalline solid element, at. no. 53, at. wt. 126·9044, sp. gr. 4·95, m.p. 114°C, b.p. 184°C. Used in medicine, photography, chemical analysis and for neutron absorption. See I neutrons.

ion. Strictly, any atom or molecule which has resultant electric charge due to loss or gain of valence electrons. Free electrons are sometimes loosely classified as negative ions. Ionic crystals are formed of ionized atoms and in solution exhibit ionic conduction. In gases ions are normally molecular and cases of double or treble ionization may be encountered. When almost completely ionized, gases form a fourth state of matter known as a *plasma* (q.v.). Since matter is electrically neutral, ions are normally produced in pairs. See **aquo-, negative-, positive-**.

ion beam. A beam of ions moving in the same direction with similar speeds, especially when produced by some form of accelerating machine or mass spectrograph. Also **ionic beam**.

ion burn. Damage to the phosphor of a magnetic deflection CRT because of relatively minute deflection of heavy negative ions by the magnetic control which deflects the electron beam. Electric deflection deflects both electrons and heavy negative ions together and there is no burn in an electrostatic CRT.

ion cluster. Group of molecules loosely bound (by electrostatic forces) to a charged ion in a gas.

ion concentration. (1) The number of ions of either sign, or of ion pairs, per unit volume. Also **ionization density**. (2) That expressed in gram-ions per unit volume for the particular ion. See **pH value**.

ion cyclotron frequency. Ions in a magnetic field follow a circular orbit similar to that in a cyclotron. The orbital frequency is a function of the field and the velocity and specific charge of the ion.

ion exchange. Reversible exchange of ions between a solution and an insoluble material containing attracting ions. Used for softening water, separating isotopes, etc.

ion migration. The movement of ions in an electrolyte or semiconductor due to applying a voltage across electrodes.

ion mobility. *Ion velocity* (q.v.) in unit electric field (1 V/cm or 1 V/m).

ion pair. Positive and negative ion produced together by transfer of electron from one atom or molecule to another.

ion source. Device for releasing ions as these are required in an accelerating machine. It usually consists of a minute jet of gas or vapour of the required compound, which is ionized by bombardment with an electron beam.

ion spot. The deformation of target or cathode by ion bombardment, leading in camera tubes to a spurious signal.

ion trap. Means of avoiding ion burn in a CRT, associated with magnetic deflection. The beam can be bent so that the heavy negative ions do not fall on the phosphor. U.S. beam bender.

ion velocity. Velocity of translation of drift of ions under the influence of an electric field in a gas or electrolyte. See also **ion mobility.**

ion yield. The average number of ion pairs produced by each incident particle or photon.

ionic. Appertaining to or associated with gaseous or electrolytic ions. *N.B.—Ion* is frequently used interchangeably with *ionic* as an adjective, e.g., in *ion(ic)* conduction.

ionic beam. See **ion beam.**

ionic bombardment. The impact on the cathode of a gas-filled electron tube of the positive ions created by ionization of the gas. The bombardment may cause electrons to be ejected from the cathode.

ionic bond. Coulomb force between ion pairs in molecule or ionic crystal. These bonds usually disassociate in solution.

ionic concentration. See **ion concentration** (2).

ionic conduction. That which arises from the movement of ions in a gas or electrolytic solution. In solids, it refers to electrical conductivity of an ionic crystal, arising from movement of positive and negative ions under an applied electric field. It is a diffusion process and hence very temperature-dependent.

ionic conductor. One in which conduction is predominantly by ions, rather than by electrons and holes.

ionic crystal. Lattice held together by the electric forces between ions, as in a crystalline chemical compound.

ionic current. Current carried by positively-charged ions in a gas at low pressure. Especially applied to the small current which flows from the filament to the grid of a thermionic tube when the grid is made very negative and vacuum is not perfect.

ionic focusing. Same as **gas focusing.**

ionic formula. Same as **electronic formula.**

ionic-heated cathode. Hot cathode that is heated primarily by ionic bombardment of the emitting surface.

ionic heating. That of a cathode when under ionic bombardment during manufacture.

ionic modulation. That of waves propagated through an ionized gas, the degree of ionization being varied in accordance with the modulating signal, which varies absorption of the beam.

ionic potential. Ratio of an ionic charge to its effective radius, regarded as a capacitor.

ionic radius. Approximate limiting radius of ions in crystals, ranging, for common metals (including carbon), from a fraction to several ångströms.

ionic valve. Cold-cathode rectifier tube comprising two electrodes in the form of a point and a spiral of wire respectively enclosed in a gas at low pressure. Also called **Lodge** valve.

ionium (Io). At. no. 90, the common name of a naturally occurring radioactive nuclide. An isotope of thorium.

ionization. Formation of ions by separating atoms, molecules or radicals, or by adding or subtracting electrons from atoms by strong electric fields in a gas, or by weakening the electric attractions in a liquid, particularly water.

ionization chamber. Instrument used in study of ionized gases and/or ionizing radiations. It comprises an enclosure with parallel plate or coaxial electrodes, between which ionized gas is formed. For fairly large applied voltages the current through the chamber is dependent only upon the rate of ion production. For very large voltages additional ionization by collision enhances the current. The system is then known as a *proportional counter* or *Geiger counter* (qq.v.).

ionization cross-section. Effective geometrical cross-section offered by an atom or molecule to an ionizing collision.

ionization current. (1) That passing through an ionization chamber. (2) Current passed by an ionization gauge, used for measuring low gas pressures.

ionization delay. Time between initiation of ionization in an ionization chamber and the resulting current change in the external circuit.

ionization density. See **ion concentration** (1).

ionization electrometer. One which measures and records radioactive ionization.

ionization gauge. Vacuum gauge formed by small thermionic triode attached to a chamber, in which it is desired to measure residual gas pressure. Current is passed from anode to cathode, the grid being made negative; grid current is measured, giving degree of vacuum.

ionization heat. Increase of sensible heat for complete ionization of a gram-molecule of a substance.

ionization manometer. Ionization gauge in which a grid-current meter is calibrated to read gas pressure directly.

ionization potential. Energy in *electron-volts* (eV) required to detach an electron from a neutral atom, depending on the complexity of the shells; for hydrogen the value is 13·6 eV. Also **electron binding energy, radiation potential.**

ionization time. Delay between the application of ionizing conditions and the onset of ionization, depending on temperature and other factors, e.g., in a mercury-pool rectifier.

ionized atom. One with a resultant charge arising from capture or loss of electrons; an *ion* in gas or liquid.

ionized-gas detector. Early form of detector in which discharge through ionized gas is triggered by the arrival of a signal.

ionizing collision. Interaction in which an ion pair is produced.

ionizing energy. That required to produce an ion pair in a gas under specified conditions. Measured in eV. For air, the value is about 32 eV.

ionizing event. Any interaction which leads to the production of ions.

ionizing particle. Charged particle which produces considerable ionization on passing through a medium. Neutrons, neutrinos and photons are not ionizing particles although they may produce some ions.

ionizing radiation. Any electromagnetic or particulate radiation which produces ion pairs when passing through a medium.

ionizing voltage. See **starting voltage.**

ionogenic. Forming ions, e.g., electrolytes.

ionophone. See **cathodophone.**

ionoscope. A storage camera tube which collects the electric charges of the optical image until removed by the scanning electron beam. See also **emitron, orthicon, Vidicon.**

ionosphere. That part of the earth's atmosphere (the upper stratosphere) in which an appreciable concentration of ions and free electrons normally exist. This shows daily and seasonal variations. See **E-layer, F-layer, hop.**

ionospheric control points. Points in the ionosphere distant 2000 km and 1000 km from each ground terminal, used respectively to control transmission by way of the F2 and E layers.

ionospheric disturbance. An abnormal variation of the ion density in part of the ionosphere, commonly produced by solar flares and having a marked effect on radio communication. See **storm.**

ionospheric forecast (or **prediction**). The forecasting of ionospheric conditions relevant to communication.

ionospheric path error. That in bearing arising from irregularity of ionized layer.

ionospheric ray or wave. Radiation reflected from an upper ionized region between transmitter and receiver. Also called **indirect, reflected** or **sky ray** (or **wave**).

ionospheric regions. These are

D region, between 50 and 100 km,

E region, between 90 and 150 km,

F region, higher than 150 km,

above surface of earth. Internal effective

layers are labelled *E, sporadic E, E2, F, F1, F1½, F2.*

iontophoresis. Migration of ions into body tissue through electric currents. See **electrophoresis.**

iridium (Ir). Metallic element of the platinum family, at. no. 77, at. wt. 192·2, sp. gr. 22·42, m.p. 2454°C. The radioactive isotope ^{192}Ir is a medium-energy γ-emitter used for industrial radiography.

iris. (1) Apertured diaphragm across a waveguide for introducing specific impedances. (2) Adjustable aperture used in conjunction with camera lens for control of exposure.

iron (Fe). Metallic element, at. no. 26, at. wt. 55·847, sp. gr. 7·88, m.p. 1535°C, b.p. 2450°C. It has magnetic properties and high strength.

iron loss. Power loss due to hysteresis and eddy currents in iron or magnetic material transformers or electrical machinery.

irradiance. See **radiant flux density.**

irradiation. The exposure to, and absorption of, ionizing radiation.

isentropic. See **entropy** (2).

island effect. The limitation of emission to small areas of the cathode surface of a valve when the grid voltage is less than a minimum value.

isobaric spin. See **isotopic spin.**

isobars. Nuclides having the same total of protons and neutrons, with the same *mass number* and approx. the same *atomic mass.*

isochromatic. Interference fringe of uniform hue observed with white light source; especially in photoelastic strain analysis where it joins points of equal phase retardation.

isochrone. Hyperbola (or ellipse) on a chart or map, along which there is a constant difference (or sum) in time of arrival of signals from two stations at ends of a baseline.

isochronous modulation (or **restitution**). Modulation in which the time between any two significant constants is theoretically equal to (or a multiple of) the unit interval.

isocline. (1) Contour line on *cyclogram* used in analysis of negative resistance oscillators to determine amplitude and waveform of resulting stable oscillation. (2) Line on a map, joining points where the angle of dip or inclination of the earth's magnetic field is the same.

isocount contour (or **surface**). Line (or surface) drawn so as to join points for which the radiation count rate is the same.

isodiapheres. Nuclides having the same difference between totals of neutrons and protons.

isodose contour (or **surface**). Line (or surface) drawn so as to join points for which the measured radiation dose rate is the same.

isoelectric point. Hydrogen ion concentration in solutions, when the dipolar ions are at a maximum. The point also coincides with minimum viscosity and conductivity.

isoelectronic. Said of similar electron patterns, as in valency electrons of atoms.

isogonic line. Line on a map joining points of equal magnetic declination, i.e., corresponding variations from true north.

Isolantite. A proprietary insulation material for high-frequency applications.

isolation diode. One used to block signals in one direction but to pass them in the other.

isolation transformer. One used to isolate electrical equipment from its power supply.

isolator. Plug or wedge in a waveguide which separates waves, e.g., by absorbing a reflection. See **gyrator**.

isolux. Locus, line, or surface where the light intensity is constant. Also called **isophot**.

isomagnetic lines. Lines connecting places at which a property of the earth's magnetic field is a constant.

isomer separation. The chemical separation of the lower energy member of a pair of nuclear isomers.

isomeric transition. One between two isomeric states, or from one such state to the ground state. The energy is released through *internal pair production, internal conversion* or *specific gamma ray emission*.

isomers. Those nuclides which have the same *atomic number* and the same *mass number*, but are distinguishable by their energy; that having the lowest (*ground*) state is stable, the others are unstable and are said to be in *isomeric states*.

isoperms. Magnetic alloys containing iron, nickel and cobalt, used in high-frequency inductors.

isophot. See **isolux**.

isopleth. See **nomogram**.

isospin. Contraction of **isotopic spin**.

isotones. Nuclei with the same neutron number but different atomic numbers (i.e., those lying in a vertical column of a *Segrè chart*).

isotope. One of a set of chemically identical species of atom which have the same *atomic number* but different *mass numbers*. A few elements have only one natural isotope, but all elements have artificially-produced radioisotopes. Gk. *isos*, same; *topos*, place, suggested to F. Soddy and introduced by him in 1913.

isotope separation. Process of altering the relative abundance of isotopes in a mixture. The separation may be virtually complete as in a mass spectrograph, or may give slight enrichment only as in each stage of a diffusion plant. See **Hermes, Capenhurst plant**.

isotope structure. Hyperfine structure of spectrum lines resulting from mixture of isotopes in source material. The wavelength difference is termed the **isotope shift**.

isotopic abundance. See **abundance**.

isotopic dilution analysis. A method of determining the amount of an element in a specimen by observing the change in isotopic composition produced by the addition of a known amount of radioactive allobar.

isotopic number. See **neutron excess**.

isotopic spin. Misleadingly named quantum number allocated to nuclear particles to differentiate between members of a multiplet (a group of particles differing only in electric charge) in the same way that different quantum numbers in optical spectroscopy represent the multiplet structure of electron energy levels. It has no connection with the nuclear spin of the particle. It was introduced to distinguish the proton and neutron, both being regarded as the same nucleon with isotopic spin either parallel or anti-parallel with some preferred direction. The proton has isotopic spin $+\frac{1}{2}$ and the neutron $-\frac{1}{2}$. The small mass differences existing between members of multiplets is due to the energy difference associated with their differing charges. This form of nomenclature is in use with all baryons and mesons, the number of members of a multiplet of isotopic spin I being $(2I+1)$. Thus singlets have an isotopic spin 0; doublets $\frac{1}{2}$ (isotopic quantum numbers $+\frac{1}{2}$ and $-\frac{1}{2}$); triplets 1 (isotopic spin quantum numbers $+1\frac{1}{2}$, $+\frac{1}{2}$, $-\frac{1}{2}$ and $-1\frac{1}{2}$). The justification for this classification is that all members of a multiplet respond identically to the strong nuclear interactions—their charge difference affecting only electromagnetic interactions. Isotopic spin is conserved for all strong interactions and never changes by more than one unit in a weak one. Also **isobaric spin**.

isotron. Device in which pulses from a source of ions are synchronized with a deflecting field. This separates isotopes, because their acceleration varies with mass.

isotropic. Said of a medium the physical properties of which, e.g., magnetic susceptibility or dielectric constant, do not vary with direction.

isotropic dielectric. One in which the electrical properties are independent of the direction of the applied electric field.

isotropic radiator. Same as **omnidirectional radiator**.

isotropic source. Theoretical source which radiates all its electromagnetic energy equally in all directions.

iterated fission expectation. Limiting value, after a long time, of the number of fissions per generation in the chain reaction initiated by a specified neutron to which this term applies.

iterative impedance. That of a four-terminal network or transducer when the output is terminated with the same impedance or when an infinite series of identical such networks are cascaded.

J

J-antenna. Dipole fed and matched at the end by a quarter-wavelength line.

J-carrier system. A broadband carrier telephony system providing 12 telephone channels and using frequencies up to 140 kHz.

J display. A modified A display with circular time base.

j.j. coupling. An extreme form of coupling between the orbital electrons of atoms. Electrons showing individual spin-orbital coupling also interact with each other.

jack. Socket whose connections are short-circuited until a *jack plug* is inserted. A *break jack* is one which breaks the normal circuit on inserting plug, while a *branch jack* is one which does not.

jack box. One containing switches or connections for changing circuits, especially in aircraft.

jacket. See **can**.

jamming. Deliberate interference of a transmission on one carrier by transmission on another approximately equal carrier, with wobble or noise modulation.

Janus. Transmitting or receiving antenna which can be switched between opposite directions.

jar. Obsolete unit of capacitance, equal to 1/900 μF. See **Leyden jar**.

jet pump. One in which a constricted jet of water entrains and removes gas from an intended vacuum.

jig welding. See **seam welding**.

Jim Creek transmitter. The very powerful radio transmitter (10^2 watts) of the U.S. Navy at Jim Creek, Washington.

jitter. Small rapid irregularities in a waveform arising from fluctuations in supply voltages, components, etc.

Johnson-Lark-Horowitz effect. The resistivity of a metal or degenerate semiconductor, arising from electron scattering by impurity atoms.

Johnson noise. Same as **thermal noise**.

joint. Connection between two lengths of cable, waveguide, etc. Waveguides may be *butt-jointed* (when intermetallic contact is maintained by pressure) or *choke-jointed* (when electrical continuity is maintained by a standing wave with the mechanical joint placed at a current node).

joint access. The means of allowing access to a circuit from both switchboard and selector levels.

joint trunk exchange. An exchange restricted primarily to handling calls over toll and trunk circuits at the same positions.

Joshi effect. Change of current in a gas because of light irradiation.

joule. Unit of work, energy and heat in the MKS system, equal to work done when a force of 1 newton advances its point of application 1 m. One joule$=10^7$ erg$=$ 0·2390 calorie.

Joule effect. (1) Production of heat solely arising from current flow in a conductor. See **Joule's law**. (2) Slight increase in the length of an iron core when longitudinally magnetized; see also **magnetostriction**.

Joule magnetostriction. That for which length increases with increasing longitudinal magnetic field. Also **positive magnetostriction**.

Joule meter. An integrating wattmeter whose scale is in joules.

Joule-Thomson coefficient. Given by the ratio of change in temperature to pressure difference in a porous plug expansion.

Joule-Thomson effect (or Joule-Kelvin effect). Concerned with the change of temperature when a gas expands through a porous plug. The change is proportional to the pressure difference across the plug. There is an inversion temperature below which cooling takes place, otherwise heating occurs on expansion.

Joule's equivalent (J). Defined by $W=JH$, where H denotes the number of units of heat obtained from the complete conversion of W units of work. Also **mechanical equivalent**.

Joule's law. (1) Heat liberated is

$$H=I^2Rt \text{ joule},$$

where
- $I=$current in amperes
- $R=$resistance in ohms
- $t=$time in seconds.

This is the basis of all electrical heating, wanted or unwanted. With a.c., R is an *effective resistance* and I may be confined to a thin *skin* of the conductor. (2) For an ideal gas, that its intrinsic energy at a constant temperature is independent of its volume.

judder. Irregular motion in facsimile transmission or reception, manifested in reproduced picture.

jump. (1) Control of computer so that the next instruction obeyed is not that stored sequentially after the last. Jumps may be *conditional* when they take place only if the result of a previous operation satisfies some specific requirement (e.g., to be negative) or *unconditional* when they represent a rigorous instruction to alter the normal sequence of operation. (2) Deprecated term for transition of electron or nucleus to different energy level.

jumper (or jumper wire). A length of wire used in telephony for the purpose of re-arranging permanent circuit connections.

junction. (1) Area of contact between semiconductor material having different electrical

148

properties, larger than a *point contact*.
(2) Union or division of waveguides in either
H or *E* planes, *tee* or *wye, tapered* to broaden
the frequency response, or *hybrid* to direct
flow of wave energy.

See also:

alloy-	*n-n-*
collector-	*p-n-*
doped-	*p-p-*
emitter-	thermo-
hybrid-	waveguide-.

junction box. Box in which cables are led in
and redistributed, often with facility for
monitoring.

junction circuit. One directly connecting two
exchanges situated at a distance apart less
than that specified for a trunk circuit.

junction coupling. In coaxial-line cavity
resonators, coupling by direct connection to
the coaxial conductor.

junction diode. One formed by the junction
of *n-* and *p-* type semiconductors, which
exhibits rectifying properties as a result of
the potential barrier built up across the
junction by the diffusion of electrons from
the *n*-type material to the *p*-type. Applied
voltages, in the sense that they neutralize this
potential barrier, produce much larger
currents than those that accentuate it.

junction laser. One in which Lossev radiation
from a semiconductor junction is rendered
coherent by a high-current density so that
the junction operates as a solid-state laser.

junction rectifier. One formed by a *p-n* junction by *holes* being carried into the *n*-type
semiconductor. Germanium and silicon
power rectifiers are being increasingly
used for currents higher than about 1 amp.

junction transistor. One in which the base
receives minority carriers through *junctions*
on each side, which may diminish to *points*.
In normal operation the emitter is forward-
biased. Signal currents entering the base
tend to neutralize reverse bias on the
collector junction, thus controlling the
relatively large current flowing in collector
circuit.

K

K-amplifier. Theoretical amplifier with current gain of unity and voltage gain of K, used in circuit analysis.

K-band. Obsolescent frequency band notation, this band corresponding to wavelengths between 1 and 2·7 cm approx.

K-capture. Absorption of an electron from the innermost (K-) shell of an atom into its nucleus. An alternative to the ejection of a *positron* from the nucleus of a radioisotope.

K display. A form of *A display* (q.v.) produced with a lobe-switching antenna. Each lobe produces its own peak and the antenna is directly on target when both parts of the resulting double peak have the same height.

K-electron. One of the two electrons in the K- (innermost) shell of an atom. Its principal quantum number is unity.

K/L ratio. The ratio of the number of internal conversion electrons from the K-shell to the number from the L-shell, which are emitted in a particular decay process.

K-lines. Characteristic X-ray frequencies from atoms due to excitation of electrons in the K-shell. These give information relating to the atomic numbers of the elements bombarded.

K-shell. The innermost shell of electrons in an atom containing $2(=2 \times 1^2)$ electrons.

k-factor. See specific gamma-ray emission.

k-space. Symbol for momentum space or wave-vector space. An important concept in semiconductor energy band theory.

Kaluza theory. A unified field theory of gravitation and electromagnetism which is based on a five-dimensional continuum.

kaon. See under meson.

kayser. Unit for *wave number*, the reciprocal of a wavelength (in cm).

keep-alive arc. Small auxiliary arc in a mercury-arc rectifier to maintain ionization during off-load condition.

keep-alive electrode. See pilot electrode.

keeper. See armature (1).

Keesom relationship. One involving molecular attraction and dipole interaction.

Kelvin ampere-balance. Laboratory instrument for measuring current; in it the calculated force between two coils carrying the current to be measured is balanced by the force of gravity on a weight sliding along a beam.

Kelvin bridge. An electrical network involving two sets of ratio arms used for accurate measurement of low resistances.

Kelvin effect. Skin effect, whereby varying currents in a conductor tend to concentrate near the surface. Important in radio circuits, and in eddy-current (HF) heating.

Kelvin electrometer. One designed to measure high voltages, by balancing the attraction of a disk, surrounded by a guard ring to obviate fringe effects.

Kelvin temperature. Temperature expressed on the Kelvin thermodynamic scale, i.e., that in which measurements are made from absolute zero, by a system independent of the type of thermometer used.

Kelvin-Varley slide. Constant resistance decade voltage divider of the type used in vernier potentiometers. It consists of a resistor of $2r$ ohms shunting two adjacent units in a series array of eleven resistors, each r ohms. The $2r$ resistor may similarly be divided into eleven equal parts and shunted if additional decade(s) are required.

Kennelly-Heaviside layer. See E-layer.

kenotron. Large hot-cathode vacuum diode for industrial rectification and X-ray plants. In U.S. valve tube.

kernel. Mathematical function which defines linear operator. In nucleonics, neutron sources distributed continuously in various configurations may be represented by kernels which are used in calculations of flux and of slowing-down density.

Kerr cell. Device employing *Kerr effect* (1) to modulate a beam of light. It consists of a liquid cell between crossed polaroids, electric fields being applied across the liquid by plane electrodes. Used in recording SOF and formerly in Scophony television system.

Kerr effect. (1) Double refraction produced in certain transparent dielectrics by the application of an electric field. Also **electro-optical effect.** (2) Slight rotation of plane of polarization of light when reflected from the polished surface of magnetized material. Also **magneto-optical effect.**

KeV. Abb. for **kilo-electron-volt**; unit of particle energy, 10^3 electron-volts.

key. Hand-operated spring-loaded switch used for telegraphy transmissions.

key click. Click produced by ringing in key circuit as contact is opened or closed. Normally minimized by use of *key click filter.* Also called **key chirp.**

key light(ing). The main illumination on a cinema or TV set.

keyer. A device for changing the output of a transmitter from one frequency (or amplitude) to another according to intelligence transmitted. Also used for short sound bursts.

keying. Production of signals by operating of key. Classified by the circuit lead in which keying takes place, i.e., anode, cathode, control grid, screen grid, etc., keying.

keying wave. See **marking wave.**

keystone distortion. Distortion arising in a camera tube because the length of the horizontal scan line depends upon its vertical displacement, when the electron beam scans the image plate at an acute angle. It results in distortion of the rectangle into a trapezoidal pattern; it can be obviated by special transmitter circuits.

kicksorter. See **pulse height analyser.**

kilo-. A prefix denoting 10^3.

kilocalorie. Unit of energy equal to 10^3 calories and sometimes denoted by **Calorie.**

kilocurie source. Giant radioactive source—usually in form of ^{60}Co.

kilocycle. One thousand cycles per second (now kHz).

kilo-electron-volt. See **KeV.**

kiloton. Unit of explosive power for nuclear weapons equal to that of 10^3 tons of TNT.

kilovolt-ampere. A unit of apparent electrical power equal to 1000 volt-amperes.

kilowatt. A unit of electrical power equal to 1000 watts.

kilowatt-hour. Unit of work equal to 1000 watts acting for one hour.

kinescope. See **teletube.**

kink instability. See **instability** (1)(*a*).

Kipp relay. A form of trigger or bistable multivibrator circuit.

Kirchhoff's laws. Generalized extensions of Ohm's law employed in network analysis. They may be summarized as:
(1) $\Sigma i = 0$ at any junction.
(2) $\Sigma E = \Sigma iZ$ round any closed path.
(E = e.m.f., i = current, Z = complex impedance.)

Klein-Gordon equation. Describes the motion of a spinless charged particle in an electromagnetic field.

Klein-Nishina formula. Theoretical expression giving cross-section of electrons for scattering of photons. See **Compton effect.**

Klein-Rydberg construction. The point by point construction of the potential energy curve of a diatomic molecule from observed rotational and vibrational levels.

klystron. General name for class of UHF electron tubes (amplifiers, oscillators, frequency multipliers, cascade amplifiers, etc.), in which electrons in a stream have their velocities varied (velocity modulation) by an UHF field; and subsequently impart energy to it or to other UHF fields contained in a resonator. Can be a cm-wavelength generator in which a buncher rhumbatron imparts the modulation which excites a catcher rhumbatron cavity. In the Sutton tube, an electron-stream reflector electrode enables one rhumbatron to perform both functions.

klystron reflection. See **reflex klystron.**

knee. The region of maximum curvature on a characteristic curve, e.g., of a valve.

Knight shift. Shift of nuclear magnetic resonance frequency due to the effect of the magnetic field of conduction electrons.

knock-on electrons. Those which arise from the collision of fast mesons from cosmic rays with the electrons of gaseous atoms in the atmosphere. See **delta-ray.**

knot. Speed of 1 nautical mile per hour (1·15 m.p.h.), used in navigation and meteorology.

Knott's equations. The continuity equations governing the energy redistribution of ultrasonic waves incident at an interface between two media.

Koch resistance. The resistance of a vacuum photocell or phototube when its active surface is irradiated with light.

Kohlrausch's law. The contribution from each ion of an electrolytic solution to the total electrical conductance is independent of the nature of the other ion.

Kolor-rite. TN for daylight colour matching fluorescent lamp tube.

Konel metal. TN of alloy of Ni—Co—Fe—Ti, much used for cathode supports when internally heated.

Kooman's array. A plane array of pinetree aerials, the corresponding dipole elements of the successive pinetrees being substantially co-linear.

Kovar. TN for an alloy of Ni—Co—Fe for glass-to-metal seals over working ranges of temperature, when temperature coefficients of expansion coincide.

Kromayer lamp. Mercury-vapour lamp for generating ultraviolet rays, especially for therapy.

Kronig-Penney model. A mathematical model involving a one-dimensional periodic potential which allows certain properties of metals to be verified by rigorous calculations.

Kruskal limit. Theoretical maximum current for which a plasma discharge can remain stable for any given axial magnetic field.

krypton (Kr). Inert gas in atmosphere, giving off-white discharge. At. no. 36, at. wt. 83·80, m.p. $-157°$C, b.p. $-153°$C.

Kundt effect. Rotation of plane of polarization of light in certain fluids by a magnetic field.

Kundt's tube. Transparent tube in which stationary acoustic waves are established, indicated by lycopodium powder, which congregates in a heap, or even a disk, at the nodes. Used for measuring sound velocities.

Kurie plot. One used for determining the energy limit of a beta-ray spectrum from the intercept of a straight-line graph. Prepared by plotting a function of the observed intensity against energy, the intercept on the axis being the energy limit for the spectrum. Also **Fermi plot.**

kymograph. Instrument for recording physiological muscular waves.

L

L-capture. Absorption of an electron from the L-shell into the nucleus, giving rise to X-rays of characteristic wavelength depending on atomic number of the element. See also **K-capture, X-rays.**

LC coupling. Inductor output load of a thermionic valve is connected through a capacitor to the input circuit of another valve.

L display. A radar display in which the target appears as two horizontal pulses, left and right from a central vertical time base, varying in amplitude according to accuracy of aim.

LD 50. Median *lethal dose* of radiation required to kill, within a given period, 50 per cent. of a large group of animals or organisms.

L-electron. One of up to eight in the L-shell of an atom. Their principal quantum number is two.

L-line. Characteristic X-ray spectrum line produced when vacancy in L-shell is filled by electron. See also **K-lines, M-line.**

L/M ratio. The ratio of the number of internal conversion electrons from the L-shell to the number from the M-shell which are emitted in a particular decay process.

L-network-attenuator or **-filter.** Half an unbalanced T-network. See **L-section.**

L-radiation. A particular X-ray series emitted from an element when the electrons of its L-shell are excited.

L-section. Section or half-section of a wave filter, having one shunt and one series arm.

L-shell. The second shell of electrons in an atom. Contains $8(=2 \times 2^2)$ electrons.

l-s coupling. See **Russell-Saunders coupling.**

labelled. See **radioactive tracer.**

labyrinth. A form of loudspeaker cabinet with partitions to produce a long convoluting path from the back of the drive unit to the outlet opening. By lining the duct with acoustic damping material, the resonances are kept to the lowest frequencies.

ladder network. One which comprises a number of filter sections, all alike, in series, acting as a transmission line with attenuation and delay properties.

lag angle. The phase angle between two waves, or periodic disturbances, of the same frequency. If A reaches peak amplitude before B then the phase angle wave B with respect to A is a *lag* angle; of A with respect to B it is a *lead* angle.

lagging load. See **inductive load.**

lambda limiting process. A method of defining a point electron as the limiting case in which a time-like vector tends to zero.

lambda-particle. Hyperon with hypercharge 0 and isotopic spin 1.

lambda point. (1) Transition temperature of

helium I to helium II. (2) Temperatures characteristic of second-order phase change, e.g., ferromagnetic Curie point. Also λ-**point.**

lambert. Unit of luminance or surface brightness of a source or diffuse reflector, emitting one lumen/cm². The *millilambert* is used for low illuminations, as in cinemas. See **lumen.**

lamination. A thin strip of magnetic material, ideally having high resistivity. Used in magnetic circuits, where unwanted eddy currents are present, to reduce their effect. Also **stamping.**

laminography. See **tomography.**

Landau theory. (1) Theory for calculating diamagnetic susceptibility produced by free conduction electrons. (2) Theory explaining the anomalous properties of liquid helium II in terms of a mixture of normal and superfluids; see **helium, superfluid.**

Landé splitting factor. Constant employed in the calculation of the shift of energy levels produced by a magnetic field, and leading to *hyperfine structure* (q.v.) of spectrum lines; see **Zeeman effect.** It is closely related to the gyromagnetic ratio of the electron.

landing beacon. A radio transmitter used to produce a landing beam.

landing beam. Field from a transmitter along which an aircraft approaches a landing field during blind landing. See **instrument landing system.**

lane. Space between two hyperbolic lines on a navigation pattern, e.g., for Decca navigation.

Langevin theory. (1) A classical expression for diamagnetic susceptibility produced by orbital electrons of atoms. (The quantum mechanical equivalent of this was derived subsequently by Pauli.) (2) An expression for the resultant effect of atomic magnetic moments which enters into explanations of both *paramagnetism* and *ferromagnetism.*

Langmuir dark space. Non-glow region surrounding a negative electrode placed in the luminous positive column of a gas discharge.

Langmuir frequency. The natural frequency of oscillation for electrons in a plasma.

Langmuir law. Law relating current and voltage in a space-charge limited vacuum tube. $i = GV^{3/2}$, where i = current, V = applied voltage, and G = perveance.

Langmuir probe. Electrode(s) introduced into gas-discharge tube in order to study potential distribution along the discharge.

language. In electronic computers, a system consisting of a well-defined set of characters (or symbols) with a code for combining these to form words or expressions.

lanthanum (La). A rare earth element, at. no. 57, at. wt. 138·91.

152

Laplace transform. Mathematical transform, widely used for study of transients in electric networks, which greatly simplifies the form of many differential equations.

Laplace's law. The differential form of Ampère's law for the magnetic field produced by a current-carrying conductor. In vector notation:

$$dH = \frac{I(d l \times r)}{r^3},$$

where dl is an element of the conductor, I the current, and r the vector drawn from the element to the point at which the field component dH is required. *N.B.*—Since action and reaction are equal, this also gives force on element due to unit pole at point.

Laplacian. (1) The vector operator ∇^2 (or div. grad.). If this operator is applied to a potential function $V=f(xyz)$, its value in rectangular coordinates is

$$\frac{\partial^2 V}{\partial x^2} + \frac{\partial^2 V}{\partial y^2} + \frac{\partial^2 V}{\partial z^2}.$$

(2) The negative of the geometrical *buckling* in a reactor.

lapping. The final abrasive polishing of a quartz crystal to adjust its operating frequency. Also smoothing of surface of crystalline semiconductors.

Larmor frequency. The angular frequency of precession for the spin vector of an electron acted on by an external magnetic field.

Larmor precession. The motion experienced in a small uniform magnetic field by a charged particle (or system of charged particles) when subjected to a central force which is directed towards a common point.

Larmor radius. Radius of the circular or helical path followed by a charged particle in a uniform magnetic field.

laryngophone. See **throat microphone**.

laser (*Light Amplication by Stimulated Emission of Radiation*). A coherent light source using a crystalline solid (e.g., a ruby) or gas-discharge tube, in which atoms are pumped simultaneously into excited states by an incoherent light flash. They return to their ground state with the emission of a light pulse for which the energy flux may be 10 mW/cm² (lasting 30 nanosecs.) and the beam divergence $<10^{-3}$ radians. This action is essentially the same as that of a microwave maser. The first continuously operating laser was the gas laser produced by Bell Telephone Laboratory (1960), and it utilized a low-energy discharge in a mixture of neon and helium gases, to produce a sharply defined and intense beam of infra-red radiation.

latching. Arrangement whereby a circuit is held in position, e.g., in read-out equipment, until previous operating circuits are ready to change this circuit; also called **locking**.

latency. Delay, in digital computers, between the directing signal and the availability of the required signals from a memory.

latent neutrons. In reactor theory, the delayed neutrons due from (but not yet emitted by) fission products.

lateral deviation. Error in bearing arising from tilt in the reflecting ionosphere.

lateral inversion. Defect in a reproduced television image, the picture being reversed, the right-hand side appearing on the left, due to a reversal in the connections from the line-scanning generator.

lateral recording. That in which the cutting stylus removes a thread (*swarf*) from the surface of a wax or cellulose blank disk, the modulation being realized as a lateral (radial) deviation as the spiral is traversed. Cf. *hill-and-dale recording, stereophonic recording.*

lateral reproducer. A pickup or other type of reproducer used in lateral recording.

lattice. (1) Regular geometrical pattern of discrete bodies of fissionable and non-fissionable material in a nuclear reactor. The arrangement is subcritical or just critical if it is desired to study the properties of the system. (2) A regular space arrangement of points as for the sites of atoms in a crystal. See **active-, crystal-, space-**.

lattice dynamics. Mechanics of the properties of the thermal vibrations of crystal lattices.

lattice filter. One or more lattice networks acting as a wave filter.

lattice network. One formed by two pairs of identical arms on opposite sides of a square, the input terminals being across one diagonal and the output terminals across the other; also called **bridge network, lattice section**.

lattice spacing (or **parameter**). Length of the edge of a unit cell in a crystal.

lattice vibration. Vibration of atoms or molecules in a crystal due to thermal energy.

Laue pattern. Spot diagram produced on a photographic plate when a heterogeneous X-ray beam is passed through a thin crystal, which acts like an optical *grating*.

launching. Said of the operation of transmitting a signal from a conducting circuit into a waveguide.

Lauritsen electroscope. Sensitive rugged electroscope using a metallized quartz fibre. See **dosemeter**.

lawnmower. Colloquialism for type of radar noise limiter used to reduce *grass* on the display screen.

Lawrence tube. See **tricolour chromatron**.

lawrencium (Lw). Transuranic element, at. no. 103. Its only known isotope has the short half-life of 8 secs. First isolated in the Lawrence radiation laboratories, California.

layer line. One joining a series of spots on X-ray rotating crystal diffraction photograph. Its position enables the crystal lattice spacing parallel to the axis of rotation to be determined.

layers. Ionized regions in space which vary vertically and affect radio propagation. See **E-layer, F-layer**.

lazy arm. Small microphone boom.

lead (Pb). Metallic element, at. no. 82, at. wt.

207·19, m.p. 327·4°C, b.p. 1750°C. The naturally occurring stable element consists of four isotopes:—Pb-204 (1·5%), Pb-206 (23·6%), Pb-207 (22·6%), Pb-208 (52·3%). Used as shielding in X-ray and nuclear work because of its cheapness, high density and nuclear properties; and in accumulator batteries, sound absorbers and solders.

lead accumulator or **cell.** Secondary storage battery with lead-lead peroxide electrodes in sulphuric acid.

lead age. The age of a mineral estimated from the relative number of atoms of the stable radiogenic end product and radioactive parent present. In this case, lead is the end product.

lead angle. See **lag angle.**

lead equivalent. Absorbing power of radiation screen expressed in terms of the thickness of lead which would be equally effective for the same radiation.

lead glass. Dense glass used as transparent radiation shield, especially for X-rays.

lead-in. (1) Unmodulated groove at the start of a recording on a disk, so that the needle falls into the groove correctly before the start of the modulation. The corresponding final groove after modulation ends is the *lead-out.* (2) Cable which connects the transmitter or receiver to the elevated part of an aerial.

lead-lag network. Resistance-capacitance network employed in some circuits for series stabilization.

lead network. One which provides a signal proportional to rate of change, i.e., time differential or derivative, of error signal.

lead-out. See **lead-in.**

lead-screw. In disk recording the screw which feeds the cutting head and its stylus along a radius of the blank disk as it revolves. The rate of feed need not be constant, but can depend on the modulation, or be accelerated for a marker groove.

lead telluride. Thermocouple compound for generating electric power through absorption of heat, e.g., from flame, air, or liquid; this cooling is also the basis of refrigeration units (*Peltier effect*).

lead zirconate titanate. A piezoelectric ceramic with a higher Curie temperature than barium titanate.

leading current. Alternating current which is in advance of the applied electromotive force creating the current. For a pure capacitance in circuit, the current is $\pi/2$ radians in advance of the applied voltage.

leading edge. (1) That of a data card which enters a machine first. (2) See **edges.**

leak detector. (1) Device for indicating points of ingress of gas into a high vacuum system. (2) See **burst-can detector.** (3) Leaky water-mains detector (acoustic).

leakage current. Minute current which is unwanted, especially in high insulated circuits associated with electrometer valves, capacitors and photocells. Such currents can arise from contamination or condensation on what should be clean insulating surfaces of otherwise good insulators.

leakage flux. (1) That which, in any type of electric machine or transformer, does not intercept all the turns of the winding intended to enclose it. (2) That which crosses the air gap of a magnet other than through the intended pole faces. Often called **lost flux.**

leakage neutrons. Those penetrating reflector round core of reactor.

leakage radiation. That escaping from enclosure or shield.

leakage reactance. That, in a transformer, which arises from flux in one winding not entirely linking another winding; measured by the inductance of one winding when the other is short-circuited. Leakage inductance, taken with the effective self-capacitance of a winding, determines the range of frequencies effectively passed by the transformer.

leakage spectrum. The energy distribution of the neutrons escaping from a reactor.

leakance. In electrical circuits, the leakage current expressed by the reciprocal of the insulation resistance of the circuit.

leaky grid. See **demodulation by-.**

leap-frog test. Programme used to test the internal operation of a computer.

least action, principle of. Principle stating that the actual motion of a conservative dynamical system between two points takes place in such a way that the action has a minimum value with reference to all other paths between the points which correspond to the same energy.

least energy principle. That a system is only in stable equilibrium under those conditions for which its potential energy is a minimum.

least squares fitting. Method of combining experimental data to obtain the best relation between two variables.

least time, principle of. Principle stating that the path of a ray, e.g., of light from one point to another (including refractions and reflections) will be that taking the least time. Also known as **Fermat's principle.**

Lecher wires. Two insulated parallel stretched wires tunable by means of sliding short-circuiting copper strip. The wires form a microwave electromagnetic transmission line which may be used as a tuned circuit, as an impedance matching device or for the measurement of wavelengths. Also called **parallel-wire resonator.**

Leclanché cell. A primary cell which has a positive electrode of carbon surrounded by a mixture of manganese dioxide and powdered carbon in a porous pot. The pot and the negative zinc electrode stand in a jar containing ammonium chloride solution. The e.m.f. is approximately 1·4 volt. The *dry cell* is a particular form of Leclanché.

Leduc current. Interrupted direct current in which each pulse is of approximately the same magnitude and in the same sense.

leeway. Sidewards drift of aircraft or vessel resulting in course showing *crab angle* (q.v.).

leg. (1) One side of a loop circuit, i.e., either the *go* or *return* of an electrical circuit. (2) The course between branch points in a computer programme or routine.

Lenard rays. Cathode rays which have passed from the discharge tube through a suitable window.

Lenard tube. Beam tube in which part of the electron beam can be extracted through a part of the wall of the vacuum chamber.

lens. Device for focusing radiation.

lens antenna. (1) Symmetrical device, with centre dielectric or a lattice of metal spheres, capable of concentrating a beam of microwaves over a band of frequencies. (2) System of metal slats, adjusted in dimensions to give such phase delays as to concentrate a beam.

Lenz's law. A current resulting from an induced e.m.f. is in such a direction as to oppose the rate of change of current which generates the e.m.f.

Leo. TN for a large digital computing and data-processing system. (*Lyons electronic office.*)

lepton. Electron, negative muon or neutrino. Positrons, positive muons and anti-neutrinos are *anti-leptons*. (Lepton is also used as a generic title for particles of both groups.) Leptons do not participate in the strong interactions.

lepton number. The number of leptons less the number of corresponding anti-particles taking part in a process. The number appears to be conserved in any process.

lethal mutation. Biological term referring to a mutation leading to the death of the offspring at any stage.

lethargy (of neutrons). The natural logarithm of the ratio of initial and actual energies of a neutron during the moderation process. The lethargy change per collision is defined similarly in terms of the energy values before and after the collision.

leukemia. Disease associated with the overproduction of immature white blood cells. Can be incurred through over-exposure to ionizing radiation.

level. (1) General term for a magnitude, taken as a ratio to a specified datum magnitude, usually in decibels in a power basis. Thus transmission level is in relation to a *zero level* of 1 milliwatt, and sound levels in relation to 10^{-16} watt/cm² (*zero phon level*). (2) Said of a set of contacts in a stepping relay which are operated together. (3) Possible energy value of electron or nuclear particle.

level compensator. A circuit or device to compensate automatically for effects resulting from variations in amplitude of a received signal.

level indicator. Voltage indicator on a transmission line, calibrated to indicate decibels in relation to a zero power level; also called **power-level indicator.** See volume unit.

level setting. Provision for adjusting the base voltage for an irregular waveform, e.g., in

television scanning circuit voltages and signals. See **clamp.**

levitation. Balance of gravity in a keeper of an electromagnet by current in the energizing coils, control being by interrupted beam of light falling on a photocell.

Leyden jar. Original form of capacitor with electrodes of foil on the inner and outer surfaces of a glass jar.

library. Collection of *routines* and *subroutines*, in computers, which can be reliably used without adjustment.

library routine. A ready-made routine available from a library and which can be used immediately; usually on punched cards or a paper tape.

lid. Temperature inversion in the atmosphere which prevents the mixing of the air above and below the inversion region.

Lido. Swimming-pool-type reactor at Harwell, used for research into shielding.

Lie algebra. That dealing with groups of quantities subject to relationships which reduce the number that are independent. These are known as *special unitary* groups, and groups SU (2) and SU (3) have proved very valuable in elucidating relationships between fundamental particles.

Liesegang rings. A banded precipitate formed by ions which react through diffusion in certain gels.

lifetime. The mean period between the birth and death of a charge carrier in a semiconductor. See also **half-life, mean life.**

ligasoid. A colloidal system in which the continuous phase is gaseous and the dispersed phase is liquid.

light-band pattern. See optical pattern.

light current. Current in a photoelectric device because of incident light. Cf. *dark current.*

light efficiency. Measured, e.g., in an electric lamp, by the ratio (total luminous flux/total power input). Expressed in lumen/watt.

light filter. Filter used in photography, etc., to change or control the total (or relative) energy of light distribution from the source. It consists of a homogeneous optical medium (sometimes of a specific thickness as in interference filters) with characteristic light absorption regions.

light integrator. Electronic circuit in which current from a photocell charges a capacitor, and hence triggers a switch at a point governed by the total integrated illuminations.

light-positive. Said of a material whose electrical conductivity increases under the action of light. Most photoconductive substances are light-positive, but some are *light-negative.*

light quanta. The energy of light appears to be concentrated in packets called *photons.* The energy of each photon $= h\nu$, where ν is the frequency and h is Planck's constant.

light-sensitive. Said of thin surfaces of which the electrical resistance, emission of electrons, or generation of a current depends on incidence of light.

light valve. Device for varying the area for light transmission and exposure on a photographic film sound-track. Early types consisted of two ribbons carrying modulating current which opened or closed because of magnetic field normal to their plane. Other types use the Kerr effect with polarized light.

light water. Normal water (H_2O) as distinct from *heavy water* (q.v.).

light-year. An astronomical measure of distance, being the distance travelled by light in space during a year, which is approximately 6×10^{12} miles.

lighthouse tube. Same as **megatron**.

lightning. Luminous discharge of electric charges between clouds, and between cloud and earth (or sea). A path is found by the leader stroke, the main discharge following along this ionized path, with possible repetition.

lightning arrester. A device for the protection of apparatus from damage by a lightning discharge or other accidental electrical surge. A *surge arrester* is effectively a gas-filled diode whose discharge changes from a normal glow discharge to an arc discharge under over-voltage conditions.

lightning conductor. A metal strip connected to earth at its lower end, and its upper end terminated in one or more sharp points where it is attached to the highest part of a building. By electrostatic induction it will tend to neutralize a charged cloud in its neighbourhood and the discharge will pass directly to earth through the conductor.

limen. Smallest difference in pitch (frequency) or intensity of a sound or colour which can be perceived by the senses.

limit. In a line spectrum where the lines are grouped in a series they get closer together towards the shortest wavelength, which is called the series limit.

limited. Pertaining to amplifiers which have reached a condition of *saturation* when the output is constant despite increasing input signal. In servo systems, the control motors will be torque- or velocity-limited outside a particular limiting signal.

limited stability. A property of a servomechanism or communication system which is characterized by stability only when the input signal falls within a particular range.

limiter. Any transducer in which the output, above a threshold or critical value of the input, does not vary, e.g., a shunt-polarized diode between resistors. Particularly applied to the circuit in a frequency-modulation receiver in which all traces of amplitude modulation in the signal have to be removed before final demodulation.

limiter valve. A valve operated as an amplifier, so biased that the output resulting from a large input voltage is substantially the same as that from a small one.

limiting frequency. One for which there is a significant change in response, as contrasted

with a *cutoff frequency*, which (as in a wave filter) may be nominal.

Lindeck potentiometer. One differing from most others in using a fixed resistance but variable current to obtain balance.

Lindemann electrometer. A form of electrometer which uses a metal-coated quartz fibre located between an arrangement of electrodes to which the potentials are applied. The fibre is mounted on, and perpendicular to, a quartz torsion fibre which provides the controlling couple.

Lindemann glass. Low-density glass used for windows of low-voltage X-ray tubes.

line. (1) Transmission line, coaxial, balanced pair, or earth return, for electric power, signals or modulation currents. (2) Single scan in a facsimile or television picture transmission system. (3) Single frequency of radiation as in a luminous, X-ray, or neutron spectrum. (4) Said of a microphone or loudspeaker when it is considerably extended in one direction, for directivity in a normal plane. (5) Direction of an electric or magnetic field, as a line of flow or force.

line amplifier. In broadcasting, that which supplies power to the line, either to a control centre or to a transmitter.

line balance. Matching impedance, equalling the impedance level of the line at all frequencies, for terminating a two-wire line when it divides through a hybrid coil or bridge set into a four-wire line. Used at the ends of trunk lines. Also called **line impedance**.

line bias. The effect of the electrical characteristics of a transmission line on the length of the teletypewriter signals.

line-blanking. The reduction of the amplitude of the television video signal to below the black level at the end of each line period. This allows the transmission of the line-synchronizing pulses.

line coupling. Transfer of energy between resonant (tank) circuits in a transmitter, using a short length of line with small inductive coupling at each end. Also called **link coupling**.

line defect. See defect.

line distortion. That arising in the frequency content or phase distribution in a transmitted signal, as a result of the propagation constant of the line.

line drop. The potential drop between any two points on a transmission line due to resistance, leakage, or reactance.

line equalizer. Device which compensates for attenuation and/or phase delay for transmission of signals along a line over a band of frequencies. Also **lumped loading, phase compensation, phase equalization**.

line flyback. (1) The time interval corresponding to the steeper portion of a sawtooth wave. (2) Return time of an image spot from its deflected position to its starting point.

line focus. CRT in which electron beam meets screen along a line and not at a point. When unintended, due to astigmatism.

line frequency. Number of lines scanned per second in a television image; in the U.S., 525 lines are scanned in 1/30 second. In the older British system there are 405 line periods per complete picture, and 25 complete pictures are transmitted per second, hence the line frequency = 10,125 lines per second. In the new British system there are 625 line periods per complete picture.

line hold. Variable resistance incorporated in the line time-base generator of a TV receiver. Used to control the time constant of the circuit so that the received picture is maintained in step with the transmitted picture.

line impedance. See **line balance**.

line integral. Mathematical concept associated with vector fields. Given by the summation along any path of the product of an element of the path and the component of the field vector parallel to it.

line of flux and **force.** Line drawn in a magnetic (or electric) field so that its direction at every point gives the direction of magnetic (or electric) flux and force respectively at that point.

line of position. One along which a vessel or aircraft is believed to lie. Navigational fixes are obtained from intersecting LOPs.

line-of-sight. Said of a transmission system when there has to be a straight line between transmitting and receiving antennae, as in short waves and radar.

line-of-sight velocity. The velocity at which a celestial body approaches, or recedes from, the earth. It is measured by Doppler shift of the spectral lines emitted by the body as observed on the earth. Also called **radial velocity**.

line oscillator. A valve which has its frequency stabilized either by a resonant low-loss (high-Q) coaxial line, or by a resistance-capacitance ladder which gives the necessary delay (phase shift) in a feedback loop. Also **phase-shift oscillator**.

line-output transformer. The transformer which performs the function of transferring the output of the line time-base generator in a television receiver to the horizontal scanning coils.

line-output valve. In a television receiver, the final amplifying valve in the line time-base generator.

line pad. A resistance-attenuation network which is inserted between the programme amplifier and the transmission line to the broadcasting transmitter. Its purpose is to isolate electrically the amplifier output from the variations of impedance of the line.

line parameters. Those necessary (series impedance and shunt admittance) to specify the electrical characteristics of a transmission line.

line printer. One in which a line of figures is printed mechanically and simultaneously across a roll of paper, the figures being preset while the paper is fed forward one or more lines.

line reflection. The reflection of some signal energy at a discontinuity in a transmission line.

line sequential colour television. System in which successive scanning lines generate images in each of the three primary colours.

line slip. An apparent horizontal movement of part (or all) of a reproduced screen picture due to lack of synchronism between the line frequencies of the signal and the scanning system.

line spectrum. Spectrum in which the radiation is in narrow energy bands, called *lines*, characteristic of atomic state.

line stabilization. Dependence of a valve oscillator on a section of transmission line for stabilization of its frequency of oscillation; e.g., a quarter-wave line acts as a rejector circuit of very high Q, thus giving a highly critical change in phase at the resonant frequency.

line stretcher. A section of an electromagnetic waveguide whose physical length may be varied for adjustment of tuning.

line-synchronizing pulses. Negative pulses introduced into modulation signal applied to cathode-ray tube when an image is built up by scanning successive horizontal lines (e.g., television, radar, etc.). These are filtered out and used to trigger the line time base, thus synchronizing the time base with the transmitted waveform. Since pulses are negative at 'blacker than black' level (tube beam current cutoff), they also eliminate flyback trace on screen.

line-up. Adjustment of a number of circuits in series so that they function in the desired manner when required.

line voltage. That between an electric supply line and earth.

line width. (1) Wavelength spread (or energy spread) of radiation which is normally characterized by a single value. The spread is defined by the separation between the points having half the maximum intensity of the line. See **magnetic resonance-**. (2) In TV, reciprocal of the number of lines per unit length in the direction of line progression.

linear. Said of any device or motion where the effect is exactly proportional to the cause, e.g., rotation and progression of a screw, current and voltage in a wire resistor (Ohm's law) at constant temperature, output versus input of a modulator or demodulator. See **non-linear, ultralinear**.

linear absorption coefficient. See **absorption coefficient**.

linear accelerator. Large device for accelerating electrons or positive ions up to nearly the velocity of light. Particles are accelerated through loaded waveguides by high-frequency pulses or oscillations of the correct phase.

linear amplifier. One for which the output signal level is a constant multiple of the input level.

linear detection. See **detector**.

linear distortion. That which results in the non-linear response of a system, such as an amplifier, to the envelope of a varying signal, such as speech, without distorting (within acoustic perception) the detailed waveform.

linear energy transfer. The linear rate of energy dissipation by particulate or EM radiation while penetrating absorbing media.

linear modulation. Modulation in which the change in the modulated characteristic of the carrier signal is proportional to the value of the modulating signal over the range of the audio-frequency band.

linear motor. A particular form of induction motor in which the starter and rotor are linear and parallel instead of cylindrical and coaxial.

linear network. One with electrical elements which are constant in magnitude with varying current.

linear programming. That which enables a computer to give an optimum result when fed with a number of unrelated variables.

linear rectifier. One in which the output current is strictly proportional to the envelope of the applied alternating voltage.

linear resistor. One which 'obeys' Ohm's law, i.e., under certain conditions the current is always proportional to voltage. Also called **ohmic resistor.**

linear scan. (1) A radar beam which moves with constant angular velocity. Also a radar scan which is projected and fixed in a straight line in order to increase the intensity of echoes in sector scanning. (2) The sweeping of a cathode spot across a screen at constant velocity using a deflecting sawtooth waveform.

linear stopping power. See **stopping power.**

linear sweep. The use of a sawtooth waveform to obtain a linear scan. See **linear scan** (2).

linear time-base oscillator (or **generator**). The electrical apparatus for producing the deflecting voltage (or current) of the linear time base. See **linear scan** (2), **sweep circuit** (1).

linearity control. See **strobe.**

lingering period. The time interval during which an electron remains in its orbit of highest excitation before jumping to the energy level of a lower orbit. The difference in energy will be radiated.

link. (1) A communication circuit or channel which is designed to be in tandem with other circuits or channels. (2) A path between two units of automatic switching apparatus which form part of a central control system.

link coupling. See **line coupling.**

linkage. The product of the total number of lines of magnetic flux and the number of turns in the coil (or circuit) through which they pass.

lin-log receiver. A radio receiver which gives a linear amplitude response for small signals but a logarithmic response for larger signals.

lip microphone. One held in contact with the mouth, to exclude external noise, as at sports events.

lip-synchronized shot. Film or television shot taken so close to a speaker that exact synchronization is essential.

liquid counter. Radiation counter for liquid samples, e.g., liquid scintillation counter.

liquid-drop model. A model of the atomic nucleus using the analogy of a liquid drop in which the various concepts of surface tension, heat of evaporation, etc., are employed.

liquid flow counter. One for continuous monitoring of radioactivity in flowing liquids.

liquid helium II. See **helium.**

liquid-metal reactor. (1) Normally, a reactor designed for liquid-metal (sodium or potassium) cooling. (2) Occasionally, a liquid-metal fuelled reactor.

liquid rheostat. One in which a liquid column is used as the resistive element, the terminals being attached to suitable metal plates, one of which is usually movable, thus providing a continuous variation. Only used where current control is not required to be too precise.

Lissajous figures. Figures formed by combining in space two sinusoidal vibrations executed at right angles to one another. The form of the figure will depend upon the relative frequency and the relative time phase of the vibrations. The system is widely used for comparing frequencies (assumed integral) of two oscillating voltages by applying them respectively to the horizontal and vertical deflecting plates of a cathode-ray oscillograph.

lithium (Li). An element, at. no. 3, at. wt. 6·939, m.p. 186°C, b.p. 1360°C, sp. gr. 0·534. It is the lightest known solid, chemically resembling sodium but less active. It is used in alloys and in the production of tritium.

lithium-drift germanium detector. A semiconductor detector with relatively large sensitive volume which can be used for high resolution gamma-ray spectrometry. The detectors must be stored at the temperature of solid CO_2 and used at that of liquid nitrogen.

Littrow mounting. The mounting of a plane mirror so that it reflects back the emerging light from a prism spectrometer, and the light transverses the prism a second time. By rotating the mirror, the spectrum may be scanned.

litz(endraht) wire. Multiple-stranded wire, each strand of which is separately insulated to reduce the relative weighting of the *skin effect*, i.e., concentration of high-frequency currents in the surface. Much used in compact low-loss coils in filters and in high-frequency tuning circuits.

live. (1) Direct transmission of sound of television without recording. A *live insert* is that part of a transmission, generally from records, which is live. (2) Electrically, connected to a voltage source.

live room. One which has a longer period of reverberation than the optimum for the conditions of performance and listening, with consequent blurring of speech and musical sounds, reduced by introducing additional acoustic damping.

load. (1) Mechanical force applied to a body. (2) Termination of an amplifier or line which absorbs the transmitted power, which is a maximum when this load *matches* the output impedance. (3) The actual power received by a terminating impedance. (4) Material placed between electrodes for the induction of heat through dielectric loss by means of high-frequency electric fields. (5) Concept of circuit current as a load.

load capacitor. That which tunes and maximizes the power to a load in induction or dielectric heating.

load characteristic. Line representing the load impedance, drawn on a set of characteristic curves of a valve, so that the extent of non-linear distortion arising from a stated driving voltage, and the power limitations, can be estimated. Also called **operating characteristic.**

load coil. The coil in an induction heater used to carry the alternating current, which induces the heating current in the specimen or charge.

load compensation. See load stabilization.

load factor. Ratio of average load to peak load over a period.

load impedance. Impedance of the device which accepts power from a source, e.g., amplifiers, loudspeakers, magnetizing coils, etc.

load leads. The connections or transmission lines between the power source of an induction (or dielectric) heater and the load coil or applicator.

load line. On a set of anode characteristic curves for a valve, a line representing the load, straight if entirely resistive, elliptical if reactive.

load matching. Adjusting circuit conditions to meet requirements for maximum energy transfer to load.

load regulator. A device for maintaining the load at a predetermined value, or which varies it in a predetermined manner.

load stabilization. Method of stabilization by variation of apparent load.

loaded concrete. Concrete used for shielding nuclear reactors, loaded with elements of high atomic number, e.g., lead or iron shot.

loaded impedance. That of input of transducer when output load is connected.

loaded push-pull amplifier. Push-pull stage of valve amplification in which the amplitude distortion arising from grid current is minimized by shunting the grids with resistances low in comparison with the effective non-linear grid resistance. Also **low-loading amplifier.**

loading. (1) The addition of inductance to a line for the purpose of improving its transmission characteristics throughout a given frequency band. (2) The introduction of fuel into a reactor.

loading coil. Coil inserted, in series with a line's conductors, at regular intervals. Also called **Pupin coil.**

lobe. Enhanced response of an antenna in the horizontal or vertical plane, as indicated for a lobe or loop in its radiation pattern. The *beam* effect arises from a *major lobe*, generally intended to be along the forward axis. See **back-.**

lobe switching. Scanning downcoming waves and switching into circuit networks which accept a chosen direction for optimum signal/noise ratio, as in *musa*. Also called **beam switching.**

local action. Deterioration of battery due to currents flowing to and from same electrode.

local carrier. Demodulation with an adequate carrier wave inserted before demodulation.

local control. That carried out directly and not from a remote control console.

local hour angle. The angle between a specified meridian and the observer's meridian expressed as a time. See **hour angle.**

local oscillator. That which supplies the frequency for beating in a supersonic heterodyne receiver. It may use electrodes in the first detector or frequency changer, or a separate valve. Also called **beating oscillator.**

local-oscillator valve. The valve in the oscillator of a superheterodyne receiver.

local time. In navigation, that at observer's position. It may be mean, solar or sidereal.

localization. See stereophony.

localizer beacon. A directional radio beacon, associated with the ILS, which provides an aircraft during approach and landing with an indication of its lateral position relative to the runway in use.

location. (1) In computers, a position in a store which holds a word (or part of a word). (2) In radio, radio-location or *radar*.

location shot. One made on location, i.e., away from permanent facilities of a studio.

locator beacon. The 'homing' beacon on an airfield used by the pilot until he picks up the localizer signals of the instrument landing system.

lock. A relay *locks* when, on operation, it trips a ratchet device which holds it in operation after the operating current ceases; or if, on operation, it *makes* a circuit which maintains in a winding of the relay a current sufficient to keep it operated, after the original operating current has ceased. When the *locking circuit* is opened the relay *unlocks, releases, falls off,* or *de-operates.*

lock in. Generally, to synchronize one oscillator with another, as in a homodyne or frequency doubler. One oscillator must be free-running and capable of being pulled.

lock-in amplifier. Synchronous amplifier, sensitive to variation of signal of its own frequency.

locked. Said of an oscillator when it is held to a specific frequency by an external source.

locked (or eccentric) groove. Finishing groove

on the surface of a gramophone record, motion of the needle in this groove operating stopping or record-changing mechanisms.

locking. (1) See **latching.** (2) The control of frequency of an oscillating circuit by means of an applied signal of constant frequency.

locking-on. The automatic following of a target by a radar aerial.

lockover. Circuit which is *bistable.*

lodar. Loran RDF system in which responses to ground and sky waves are registered separately, so avoiding night effect.

Lodge valve. An early form of cold-cathode rectifying valve. See **ionic valve.**

loft aerial. A receiving aerial which may be erected in the loft of a house.

log. *Odometer* (q.v.) for use at sea.

logarithmic amplifier. One with an output which is related logarithmically to the applied signal amplitude, as in decibel meters or recorders.

logarithmic decrement. Logarithm to base ϵ of the ratio of the amplitude of successive oscillations which are diminishing through energy dissipation.

logarithmic horn. The horn of a loudspeaker which is of *exponential* form. Also a metal horn for microwave work.

logarithmic resistance. A variable form of resistor by which the movement of a contact is directly or indirectly proportional to the fractional change of resistance. Such characteristics are also obtainable with potentiometers and capacitors.

logatom. Artificial word without meaning, which has a vowel with an initial and/or final consonant, used in articulation testing.

logger. (1) Arrangement of thermionic valves for obtaining an output indication which is proportional to the logarithm of the input amplitude or intensity. Required in modulation and noise meters. (2) Colloquialism for recorder or print-out device in control system.

logic. Basis of operation as designed and effected in a computer, comprising *logical elements*, which perform specified elementary arithmetical functions, e.g., add, subtract, multiply, divide.

logical comparison. The operation of comparing two numbers in a computer.

logical decision. One which is made on data in accordance with criteria previously inserted into an electronic computer by a programme.

logical design. The basic planning of a data-processing or computer system, and/or the synthesizing of a network of logical elements to carry out a particular function.

logical diagram. Generally a diagram showing logical elements and interconnections without engineering details.

logical elements. The smaller building blocks in a data-processing or computer system, which can be represented by mathematical operators in symbolic logic.

logical operation. Any non-arithmetical operation in a computer.

logical symbol. A graphical representation of a logical element.

logical unit. Section of a computer which performs a stated operation on data presented in the form of digits, e.g., add, subtract, gate, etc.

loktal base. A valve base with a centre pin to lock the base securely in an appropriate socket. It has eight pins which extend directly through the glass envelope of the valve. Also **octal base.**

London forces. Forces arising from the mutual perturbations of the electron clouds of two atoms or molecules. These will be attractive when the molecules are in their ground electronic states.

lone pair. Pair of valency electrons unshared by another atom. Such lone pairs are responsible for the formation of coordination compounds.

long-line effect. Frequency-jumping by an oscillator supplying a load through a long transmission line; due to admittance of line being suitable for oscillation at more than one frequency.

long persistence screen. CRT screen coated with long afterglow phosphor (up to several seconds).

long-range. (1) Of aircraft, ship or missile capable of covering great distances without refuelling. (2) Of radio or navigational system capable of operating over great distances.

long-tail pair. Method of obtaining two antiphase outputs from one input signal (e.g., for use in a push-pull amplifier). Consists of two triodes sharing a common cathode resistor, equal bias being applied to both valves. The input signal is applied to one grid, and the antiphase outputs are taken from the anodes.

long-tail tube. See **remote cutoff.**

long wave. Low-frequency radio signal. (Also used of subaudible and subvisible sound and light waves.)

longitudinal current. See **circulating current.**

longitudinal heating. Dielectric heating in which electrodes apply a high-frequency electric field parallel to lamination. See **glueline.**

longitudinal magnetization. That of magnetic recording medium along an axis parallel to the direction of motion.

longitudinal wave. Normal propagating sound wave in which the motions of the relevant particles are in line with the direction of translation of energy.

looming. A particular form of mirage in which the images of objects below the horizon appear in a distorted form.

loop. (1) Same as *antinode* of displacement in standing waves. See **mesh.** (2) Feedback control system. (3) Closed graphical relationship, e.g., hysteresis loop, etc. (4) Set of instructions used more than once in computer programme.

loop actuating signal. The combined input and feedback signals in closed-loop system.

loop antenna. Same as **frame antenna.**

loop coupling. Small loop connected between conductor and shield of a coaxial transmission line, for collecting energy from one of the series of resonators in a magnetron or a rhumbatron in a klystron valve. See **halo coupling.**

loop difference signal. Output signal at point in feedback loop produced by input signal applied at the same point.

loop direction-finding system. See **frame direction-finding system.**

loop error. The departure of the loop output signal from the desired value. If used as the loop actuating signal it is known as the **loop error signal.**

loop feedback signal. The part of the loop output signal fed back to the input to produce the *loop actuating signal* (q.v.).

loop gain. Total gain of amplifier and feedback circuit; if greater than zero, the system is liable to sustain oscillations. See **Nyquist criterion.**

loop input signal. An external signal applied to a feedback control loop in control systems.

loop output signal. The extraction of the controlling signal from a feedback control.

loop transfer function. The mathematical function expressing the relationship between the output of a properly terminated feedback loop system and the input.

loopstick antenna. See **ferrite-rod antenna.**

loose coupling. See **undercoupling.**

lorac-A. System with a *master* and two *slave* radio stations. (*Long range accuracy.*)

lorac-B. Lorac-A with the addition of a reference radio station.

loran. A tracking system depending on time differences of pulses as received from different but synchronized radio transmitting stations on 2 MHz. The loci of paths with constant time differences are therefore hyperbolic, intersections of two (or more) sets giving a *fix*. (*Long range navigation.*) See **gee.**

loran-C. Loran system on low radio frequencies.

Loschmidt number. (1) Number of molecules at standard temperature and pressure in one millilitre of a perfect gas, i.e., $2 \cdot 687 \cdot 10^{19}$. Used by Stoney in estimating electron charge in 1876. (2) See **Avogadro number.**

loss. Opposite of *gain*, the diminution of power, as in a transformer or line. Loss is realized as a standard in a resistive *pad* (*attenuator*), which introduces a known loss into a measuring circuit.

loss angle. A measure of the loss due to hysteresis in an imperfect dielectric, being the angular difference between its lead angle and 90°.

loss factor. The ratio of the average power loss to the power loss under peak loading.

Lossev radiation. The radiation due to the recombination of charge carriers injected into a *p-i-n* or *p-n* junction biased in the forward direction.

lossless line. See **dissipationless line.**

lossy. Pertaining to a material or apparatus which dissipates energy, e.g., a dielectric material or a transmission line with a high attenuation. The attenuation loss will be expressed in *decibels* (dB) while the rate of loss in the dielectric is proportional to its *loss factor*, i.e., to the product of power factor and dielectric constant.

lost counts. Those not recorded by a counting system due to window absorption or counter dead time (*paralysis time*).

lost flux. See **leakage flux.**

loud hailer. A high-power directional loudspeaker. See **megaphone.**

loudness contour. Line drawn on the audition diagram of the average ear which indicates the intensities of sounds that appear to the ear to be equally loud. In an objective sound-level meter, any component in the applied sound is referred to the *reftone* by these loudness contours.

loudness level. That of a specified sound is the intensity of the *reftone* (1000 Hz, on the *phon* scale), which is adjusted to equal, in apparent loudness, the specified sound. The adjustment of equality is made either subjectively or objectively as in a sound-level meter.

loudspeaker. An electro-acoustic transducer which accepts transmission currents and is particularly designed for radiating sound waves for audition by a number of persons, as contrasted with the telephone receiver which is useful to one person only. See:

biconical horn	magnetic-
Brown-	magnetostriction-
compound horn	monitoring-
crystal-	moving-coil-
double-coil-	mushroom-
double-cone-	open-diaphragm-
duode	piezoelectric-
dynamic-	pneumatic-
electrodynamic-	Riffel-
electromagnetic-	Schlenke-.
electrostatic-	

loudspeaker dividing network. See **dividing network.**

loudspeaker microphone. A microphone and dynamic loudspeaker combined, which is useful for intercommunication systems.

loudspeaker response. Response measured under specified conditions over a frequency range, at a specified direction and distance.

loudspeaker system. Complete reproducing system comprising multiple loudspeakers and dividing networks together with associated horns, baffles, cabinets, etc.

loudspeaking receiver. Driving unit attached to a horn to form a complete loudspeaker. The unit may contain any moving-coil or moving-iron mechanism.

low definition. Facsimile reproduction in which number of picture elements is small.

low fidelity. Opposite of **high fidelity.**

low-flux reactor. Nuclear reactor having a relatively low neutron flux

F

low frequency. Vague term widely used to indicate *audio frequencies* as distinct from *radio frequencies*, but more correctly indicating *radio frequencies* between 30 and 300 kHz.

low-frequency amplifier. One for amplifying audio-frequency signals.

low-frequency compensation. RC network which compensates for signal loss in coupling capacitors.

low-level modulation. See **high-level modulation**.

low-loading amplifier. More recent name for a loaded push-pull amplifier.

low loss. Said of any component or system with little power dissipation, e.g., low-loss line.

low-pass filters. See **high-pass filters**.

low-power modulation. Same as **low-level modulation**.

low tension. Term loosely applied to the currents and voltages associated with the filament or heater circuits of a thermionic tube.

low-tension battery. One supplying power for heating the cathodes of valves. U.S. term A-battery.

low-velocity scanning. Scanning of target by electron beam under conditions such that the secondary emission ratio is less than one.

lower-pitch limit. The minimum frequency for a sinusoidal sound wave which produces a sensation of pitch.

lower sideband. Band of frequencies of modulated signal below the carrier frequency. Cf. *upper sideband*.

lubber line. That on a compass or direction finder against which required heading can be read off.

lumen. Unit of luminous flux, being the amount of light emitted per second in unit solid angle by a small source of one *candela* output. In other words the lumen is the amount of light which falls on unit area per second, when the surface area is at unit distance from a source of one *candela*.

lumerg. A unit of luminous energy, e.g., one erg of radiant energy which has a luminous efficiency of *l* lumen per watt is equal to *l lumerg* of luminous energy.

luminance. Measure of brightness of a surface, e.g., candela/cm² of the surface radiating normally.

luminance channel. In a colour television system, any circuit path intended to carry the luminance signal.

luminance signal. Signal controlling the luminance of a colour television picture, or of a monochrome picture.

luminescence. Emission of light (other than from thermal energy causes), such as *bioluminescence, triboluminescence, galvano-luminescence, photoluminescence, cathodo-luminescence*, etc. Thermal luminescence in excess of that arising from temperature occurs in certain minerals. See above terms, and **afterglow, fluorescence, phosphor, Stokes' law**.

luminescent centres. Activator atoms, excited by free electrons in a crystal lattice and giving rise to electroluminescence.

luminophore. A luminescent material.

luminosity. Apparent brightness of an image. Precisely, density of luminous intensity in a particular direction.

luminosity coefficients. Multipliers of the *tri-stimulus values* of a colour, so that the total represents the luminance of the colour according to its subjective assessment by the eye.

luminosity curve. That which gives the relative effectiveness of perception or of sensitivity of vision in terms of wavelength.

luminous efficiency. Ratio of luminous flux of lamp to total radiated energy flux. Not to be confused with **overall luminous efficiency**.

luminous flux. The total visible light energy emitted per second by a light source. Measured in lumens.

luminous intensity. Ratio of luminous flux in lumens divided by its solid angle of radiation, as if radiating from a point.

luminous sensitivity. In a photoconductive cell, it is the ratio (output current/incident luminous flux) at a constant electrode voltage.

lumped constant. Electrical magnitude may be taken as *lumped* or *concentrated* when its dimensions are small in comparison with the wavelength of propagation of currents in it.

lumped loading. See **line equalizer**.

lumped parameters. In the analysis of electrical circuits, the circuit parameters which, under certain frequency conditions, may be regarded as behaving as localized units of inductance, resistance, capacitance, etc.

lumped voltage. Fictitious voltage formed by adding to the anode voltage of a multi-electrode valve, the sum of the products of the various intermediate electrode voltages and the respective amplification factors associated with these electrodes. The total *space current* is a function of this quantity.

lutecium (Lu). Rare earth element, at. no. 71, at. wt. 174·97. Formerly **cassieopeium**.

lux. Unit of illumination in MKS system, equal to one lumen per square metre.

Luxemburg effect. Cross-modulation of radio transmissions during ionospheric propagation, which limits the power radiated. Caused by non-linearity in the motion of the ions.

luxmeter. An instrument for the measurement of illuminance in lux units.

Lyman series. A spectral line series occurring in the ultraviolet spectrum of hydrogen. The wave numbers are given by the same expression as that for *Brackett lines* (q.v.) but with $N_1 = 1$.

Lysholm grid. A type of grid interposed between the patient and film in diagnostic radiography, in order to minimize the effect of scattered radiation.

M

M display. A radar *A display* (q.v.) in which a pedestal signal is manipulated by a control calibrated in distance along the baseline until it meets the horizontal of the target break.

M-electron. One of 18 in M-shell of atom, having principal quantum number 3.

M-line. Characteristic X-ray spectrum line produced when vacancy in M-shell is filled by electron. See also **K-lines, L-line.**

M/N ratio. (1) In nuclear physics, ratio of conversion fractions for the M- and N-shells; analogous to the K/L ratio. (2) In radiation chemistry, the ion yield.

M-shell. Shell of electrons around an atom after the K- and L-shells, giving characteristic M-lines in X-ray spectra, originated by M-electrons; the complete shell contains 18 electrons ($=2 \times 3^2$).

m-derived network or **filter.** Electric wave-filter element which is derived from a normal (constant-k) element by transformation, the aim being to obtain more desirable impedance characteristics than is possible in the prototype.

Mach number. Ratio of the velocity relative to a medium to the velocity of sound waves in that medium appropriate to the physical conditions existing. At *Mach One*, the speed is *sonic*; below *Mach One*, it is *subsonic*; and above *Mach One* it is *supersonic*, creating a *Mach* (or *shock*) *wave*. *Hypersonic* conditions in air are reached at Mach numbers between five and ten, depending upon altitude and temperature.

Mach principle. That scientific laws are descriptions of nature, are based on observation, and alone can provide deductions which can be tested by experiment and/or observation.

machine equation. That which an analogue computer is actually programmed to solve, if not identical with the equation governing the original problem.

machine language. *Words*, etc., acceptable by a computer as *instructions* for processing data.

machine units. In electric analogue computing, a machine unit is the voltage used to represent one unit of the simulated variable.

machine variable. The varying signal in an analogue computer which reproduces the variations of the simulated variable under investigation.

machine word. The series of *bits* handled in one operation by a digital computer. It may represent a code instruction or a numerical quantity.

McLeod gauge. Vacuum-pressure gauge in which a sample of low-pressure gas is compressed in a known ratio until its pressure can be measured reliably. Used for calibrating direct-reading gauges.

McNally tube. A reflex klystron capable of being tuned to a wide range of frequencies.

macro- (or general) instruction. Basis of operation of an electronic computer, from which a programme is derived for a specific operation.

macroscopic state. One described in terms of the overall statistical behaviour of the discrete elements from which it is formed. Cf. *microscopic state*.

magic eye. See **electric eye, indicator tube.**

magic numbers. Certain numbers (e.g., 2, 8, 20, 28, 50, 82 and 126) of protons (*atomic number*) or neutrons in an atomic nucleus, which result in enhanced stability for a series of isotopes of an element.

magic-T. Combination of an E-plane junction and an H-plane junction, so that it functions as a hybrid, i.e., permitting power to be diverted according to its direction of flow.

magnadur. A ceramic-type material, comprising sintered oxides of iron and barium, used for making permanent magnets and as a good electrical insulator, its eddy-current loss in an a.c. field being very small.

Magnesil. TN for a magnetic alloy (grain-oriented) used for construction of cores of magnetic amplifiers.

magnesium (Mg). At. no. 12, at. wt. 24·312, sp. gr. 1·74, m.p. 651°C, b.p. 1120°C. A light metallic element, often alloyed with aluminium, and used in constructing electronic components.

Magnestat. TN for a type of magnetic amplifier.

magnet. A body which has the property of attracting iron and when freely supported in isolation from other bodies will tend to set in the N-S direction. Occurs naturally in 'stones' containing magnetite. Once known as a *lodestone*. Nowadays permanent magnets are made artificially from hardened steel which has been magnetized by a strong magnetic flux. See **alternating-current-, blow-out-, colour purity-, homopolar-, horseshoe-, permanent-, superconducting-.**

magnet yoke. Sometimes applied to the whole of the magnetic circuit of an electromagnet (or transformer, etc.), but strictly speaking should refer only to the part which does not carry the windings.

magnetic. Said of all phenomena essentially depending on magnetism, especially to the enhanced magnetic effects associated with ferromagnetism. See **diamagnetism, paramagnetism.**

magnetic amplifier. One in which the saturable properties of magnetic material are utilized to modulate an exciting alternating current, using an applied signal as *bias*; the signal output when rectified becomes a magnification of the input signal.

magnetic armature. Ferromagnetic element the position of which is controlled by external magnetic fields.

magnetic bias. Steady magnetic field added to signal field in magnetic recording, in order to improve linearity of relationship between applied field and magnetic remanence in recording medium.

magnetic bottle. The containment of a plasma during thermonuclear experiments by applying a specific pattern of magnetic fields.

magnetic character reading. Scanning and interpretation of characters printed with magnetic ink on documents, e.g., cheques; special lettering is necessary.

magnetic circuit. Complete path, perhaps divided, for magnetic flux, excited by a permanent magnet or an electromagnet. The range of reluctance is not so great as resistance in conductors and insulators, so that leakage of the magnetic flux into adjacent non magnetic material, especially air, is significant.

magnetic compass. Pivoted magnet used to indicate direction of magnetic north in navigation.

magnetic component. That of an electromagnetic wave, which, due to its variation, will give rise to an electric field. Conversely the variation of the electric component of the field will be accompanied by a magnetic field.

magnetic controller. Unit in control system operated by magnetic field applied to magnetic armature.

magnetic core. Unit for *store* or *memory*, consisting of tiny cores or holes punched in a sheet of ferrite, which are magnetized by wires knitted through them.

magnetic cutter. A cutter used in recording in which the motions of the recording stylus are operated by magnetic fields.

magnetic damping. Damping of motion of a conductor by induced eddy currents in it when moving across a magnetic field; particularly applicable to moving parts of instruments and electricity integrating meters.

magnetic declination. See declination.

magnetic deflection. That of an electron beam in a cathode-ray tube, caused by a magnetic field established by current in coils where the beam emerges from the electron gun which forms the beam, the deflection being at right angles to the direction of the field.

magnetic delay line. The use of a magnetostrictive material, e.g., nickel wire, as a delay medium for propagating sound, and thus providing a circulating memory or store in a computer.

magnetic disk. Form of recording medium used in some magnetophones.

magnetic disk memory. One consisting of a pile of rotating disks of magnetic surface material, so spaced that an arm can swing in for recording or reproduction, with low *access time.*

magnetic displacement. An alternative name for **magnetic induction.**

magnetic doublet radiator. A hypothetical radiator consisting of two equal and opposite varying magnetic poles whose distant field is equivalent to that of a small loop aerial.

magnetic drum. Cylinder driven very uniformly at high speed, carrying a layer of magnetic material for registering and reproducing impulses in computers, this being effected by magnetic heads just clearing the rotating surface.

magnetic energy. Product of flux density and field strength for points on the demagnetization curve of a permanent magnetic material, measuring the energy established in the magnetic circuit. Normally required to be a maximum for the amount of magnetic material used.

magnetic ferrites. Those having magnetic properties and, at the same time, possessing good electrical insulating properties by reason of their ceramic structure.

magnetic field. Modification of space, so that forces appear on magnetic poles or magnets. Associated with electric currents and the motions of electrons in atoms.

magnetic field intensity. The magnitude of the field-strength vector in a medium (i.e., the magnetic strain produced by neighbouring magnetic elements or current-carrying conductors). The MKSA unit is the ampere-turn/metre, and the CGS unit is the oersted. Also called magnetic field strength, **magnetic intensity, magnetizing force.**

magnetic-film memory. See thin-film memory.

magnetic flux. The surface integral of the product of the permeability of the medium and the magnetic field intensity normal to the surface. The magnetic flux is conceived, for theoretical purposes, as starting from a positive fictitious north pole and ending on a fictitious south pole, without loss. When associated with electric currents, a complete circuit, the *magnetic circuit*, is envisaged, the quantity of magnetic flux being sustained by a magnetomotive force (co-existent with ampere-turns linked with the said circuit). Permanent magnetism is explained similarly in terms of molecular m.m.fs. (associated with orbiting electrons) acting in the medium. Measured in *lines* (or *megalines*) or *maxwells.*

magnetic focusing. That of an electron beam by applied magnetic fields, e.g., in CRT or electron microscope.

magnetic freezing. Adhesion of relay armature to core due to residual magnetism, after energizing current stops.

magnetic gate. A gate circuit employed in magnetic amplifiers.

magnetic head. Recording, reproducing or erasing head in magnetic recorder.

magnetic hysteresis. Non-definitive value of magnetic induction in ferromagnetic medium for given magnetic field intensity, depending upon past magnetic history of sample.

magnetic hysteresis loop. See hysteresis loop.

magnetic induction. (1) Induced magnetization in magnetic material, either by saturation by coil excitation in a magnetic circuit, or by the primitive method of stroking with another magnet. (2) Magnetic flux density in a medium. It is given by the product of the field intensity and the permeability of the medium. The MKSA unit is the weber/sq.m. and the CGS unit is the gauss.

magnetic intensity. See **magnetic field intensity.**

magnetic lag. The time required for the magnetic induction to adjust to a change in the applied magnetic field.

magnetic leakage. That part of the magnetic flux in a system which is useless for the purpose in hand, and may be a nuisance in affecting nearby apparatus.

magnetic lens. The counterpart of an optical lens, in which a magnet or system of magnets is used to produce a field which acts on a beam of electrons in a similar way to a glass lens on light rays. Comprises current-carrying solenoids of suitable design.

magnetic loudspeaker. Loudspeaker basically dependent on magnetic forces to produce the sound waves.

magnetic memory or **storage.** The use of a film of magnetic material on the surface of a drum or flexible film (tape) for registering and recovering information in the form of *bits* in electronic computing.

magnetic memory plate. A thin plate of a ferrite material in the form of a rectangular grid. Threaded through the holes are wires, which allow pulses of information to be stored by magnetic induction in the material surround of each hole, without mutual interference between holes.

magnetic meridian. The plane through the centre of the earth containing the horizontal component of the earth's magnetic field at the point concerned.

magnetic microphone. A microphone whose action depends upon the induced e.m.f. in a coil (or conductor) when moved relatively to a magnetic field by a sound wave.

magnetic mirror. Device based on the principle that ions moving in a magnetic field tend to be reflected away from higher-than-average magnetic fields. Thus a *magnetic bottle* can be designed for mirror machines.

magnetic modulator. Modulator using a magnetic circuit as the modulating element.

magnetic moment. (1) Vector such that its product with the magnetic induction gives the torque on a magnet in a homogeneous magnetic field. (2) Dipole moment of atom or nucleus associated with electron orbitals and/or electron and nuclear spin. See **Bohr magneton.**

magnetic oxides. The iron oxides which are ferromagnetic and which, suitably fabricated from the powder form, provide efficient permanent magnets.

magnetic pendulum. Suspended magnet executing torsional oscillations in any horizontal magnetic field. It forms the basis of the horizontal component magnetometer.

magnetic pole. Fictitious entity on which magnetic lines of flux terminate, unit pole

terminating 4π lines on the CGS system and 1 line on the MKSA system. *Coulomb's law* for magnetism is based on the force between such poles.

magnetic potential. A continuous mathematical function the value of which at any point is equal to the potential energy (relative to infinity) of a theoretical unit north-seeking magnetic pole placed at that point.

magnetic potentiometer. A flexible solenoid (wound on a non-magnetic base) used with a ballistic galvanometer to explore the distribution of magnetic potential in a field, etc.

magnetic printing. (1) Printing ink containing magnetic elements and capable of being read both visually and by computer. (2) Transfer of recorded signal from one magnetic recording medium or element to another.

magnetic pumping. Use of radio-frequency currents in coils over bulges in the tube of a stellarator, to modulate the steady axial field and provide heat to the plasma. This process is most efficient when there is resonance between the RF signals and vibrations of the molecules of plasma. The process is then called **resonance heating.**

magnetic quantum numbers. Those determining the components of orbital and spin angular momentum in the direction of the applied field.

magnetic reaction. Same as **electromagnetic reaction.**

magnetic recorder. See **magnetophone.**

magnetic recording. (1) The magnetic tape, wire or disk from which the recorded signal may be reproduced. (2) The process of preparing a magnetic recording.

magnetic resonance. See **nuclear magnetic resonance.**

magnetic resonance line width. Case in which the width of the absorption lines depends upon the interaction of the spins with each other and with the crystal lattice. It is measured by the random fluctuating magnetic field (H) which is exerted on a spin by its neighbours, viz, $\Delta H = \mu/d^3$, where μ is the magnetic moment of each spin and d is the interatomic spacing.

magnetic rigidity. A measure of the momentum of a particle. It is given by the product of the magnetic intensity perpendicular to the path of the particle, and the resultant radius of curvature of this path.

magnetic rotation. See **Faraday effect.**

magnetic saturation. The limiting value of the magnetic induction in a medium when its *magnetization* (q.v.) is complete and perfect.

magnetic shell. A magnetized body of dimensions identical to a current-carrying coil, which can conveniently be considered instead of the latter when considering the forces acting in a system.

magnetic shield. Surface of magnetic material which reduces the effect on one side of a magnetic field on the other side. A substantially complete shield is used to protect a.c. indicating instruments from errors arising from external alternating magnetic fields.

magnetic shift register. One in which the pattern of settings of a row of magnetic cores is shifted one step along the row by each fresh pulse.

magnetic spectrometer. One in which the distribution of energies among a beam of charged particles is investigated by means of magnetic focusing techniques.

magnetic storage. See **magnetic memory**.

magnetic storm. Magnetic disturbance in the earth, causing spurious currents in submarine cables; probably arises from variation in particle emission from the sun, which affects the ionosphere.

magnetic susceptibility. The amount by which the relative permeability of a medium differs from unity, positive for a paramagnetic medium, but negative for a diamagnetic one. Equal intensity of magnetization ÷ applied field.

magnetic suspension. Use of a magnet to assist in the support of, e.g., a vertical shaft in a meter, thereby relieving the jewelled bearings of some of the weight.

magnetic tape. Flexible plastic tape, e.g., 0·5 in. wide, on one side of which is a uniform coating of dispersed magnetic material, in which signals are registered for subsequent reproduction. Used for registering television images or sound or computer data.

magnetic-tape reader. One which has a multiple head which transforms the pattern of registered signal into pulse signals in a computer.

magnetic track. Sound-track on a release cinematograph film (*stripe*). In Cinema-Scope there are four such tracks, one on each side of each row of sprocket holes.

magnetic transition temperature. See **Curie point** (1).

magnetic tube. See **trapping region**.

magnetic tuning. Control of very-high-frequency oscillator by varying magnetization of a rod of ferrite in the frequency-determining cavity by an externally applied steady field.

magnetic units. Units for electric and magnetic measurements in which the *permeability* of a vacuum is taken as unity, with no dimensions.

magnetic variation. See **declination**.

magnetic wire. A wire of magnetic material used in recording.

magnetism. Science covering magnetic fields and their effect on materials, due to unbalanced spin of electrons in atoms. See **Coulomb's law**.

magnetization. Orientation from randomness of saturated domains in the body of ferromagnetic material. Denoted by M and related to the field intensity H and the magnetic induction B by $B = \mu_0(H + M)$ webers per sq.m., where μ_0 is the permeability of free space.

magnetization curve. See **hysteresis loop, initial-**.

magnetize. (1) To induce magnetization in ferromagnetic material by direct or impulsive current in a coil. (2) To apply alternating voltage to a transformer or choke having ferromagnetic material, generally laminated, in its core.

magnetizing current. (1) Current (direct or impulsive) in a coil for the magnetization of ferromagnetic material in its core. (2) The a.c. taken by the primary of a transformer, apart from a load current in the secondary.

magnetizing force. See **magnetic field intensity**.

magneto. See **magneto-electric generator**.

magneto bell. A polarized electric bell which is operated by a low-frequency a.c.

magneto system. A local battery telephony system in which the energy required for signalling is obtained from a generator at the subscriber's station.

magnetocaloric effect. The reversible heating and cooling of a medium when the magnetization is changed. Also **thermomagnetic effect**.

magneto-electric generator. A permanent magnet alternator often coupled to an induction coil in the ignition system of a petrol engine. Also called **magneto**.

magnetohydrodynamics. The study of the motion of an electrically conducting fluid in the presence of a magnetic field. The motion of the fluid gives rise to induced electric currents which interact with the magnetic field and this in turn modifies the motion. The phenomenon has applications both to magnetic fields in space and to the possibility of generating electricity. If the free electrons in a plasma or high velocity flame are subjected to a strong magnetic field then the electrons will constitute a current flowing between two electrodes in a flame.

magneto-ionic. Said of components of an EM wave passing through an ionized region and divided into ordinary and extraordinary waves by magnetic field of the earth.

magneto-ionic double refraction. The resolution of a radio wave in the ionosphere into two components, the ordinary and the extraordinary waves, by means of the earth's magnetic field.

magnetometer. Any instrument for measurement either of the absolute value of a magnetic field intensity, or of one component of this, e.g., horizontal component magnetometer for earth's magnetic field. See **proton-, precessional-**.

magnetomotive force. Line integral of the magnetic field intensity round a closed path.

magneton. See **Bohr magneton**.

magneto-optical effect. See **Kerr effect**.

magnetophone. Any recording device involving the magnetization of a medium, e.g., magnetic tape.

magnetoresistance. The resistivity of a magnetic material in a magnetic field, when carrying an electric current, depends on the direction of the current with respect to the field. If parallel to one another the resistivity increases, but if mutually perpendicular it decreases.

Magnetoresistor. TN for device in which resistance is controlled by a magnetic field.

magnetostatics. Study of steady-state magnetic fields.

magnetostriction. Phenomenon of elastic deformation of certain ferromagnetic materials, e.g., nickel, on the application of a magnetizing force. Used in ultrasonic transducers and, in nickel wires, as a *fast memory* in electronic computers.

magnetostriction loudspeaker. Open diaphragm driven at its apex by a nickel rod, in which magnetostriction vibrations are excited by modulation currents in an embracing coil.

magnetostriction microphone. One depending on the generation of an e.m.f. through magnetostriction reaction.

magnetostriction transducer. Any device employing the property of magnetostriction to convert electrical, to mechanical, oscillations, e.g., by using a rod clamped at centre and passing a.c. through a coil wound around the rod. Also called **magnetostrictor.**

magnetostrictive filter. A filter network which utilizes magnetostrictive elements, bars or rods, with their energizing coils.

magnetostrictive oscillation. Oscillation based on the principle of the alteration of dimensions of a bar of magnetic material when the magnetic flux through it is changed—nickel contracts with an increasing applied magnetic field but iron expands in weak fields and contracts in strong fields. The mechanical oscillatory system—e.g., a bar clamped at its centre—can be coupled magnetically to a valve and the latter will supply the power for sustaining the oscillations. For a medium-size bar, the frequency will be between 1 and 25 kHz.

magnetostrictive reaction. The inverse magnetostrictive effect, i.e., a change in magnetization under applied stresses.

magnetostrictor. See magnetostriction transducer.

magnetron. Thermionic valve in which electrons released from a large cathode gyrate in an axial magnetic field before reaching the anode, their energy being collected in a series of slot resonators in the face of the circular anode. The output power, usually impulsive, is taken from a resonator by a small coupling loop.

magnetron effect. Deflection of electrons emitted from a thermionic filament by the magnetic field produced by the filament-heating current.

magnetron modes. Different frequencies of oscillation corresponding to different field configurations and selected by strapping the cavities in various ways to control their phase differences.

magnetron oscillator. One in which radial electrons from a heated cathode are accelerated towards one or more anodes, the electrons being controlled by a near axial magnetic field, and the frequency by resonant cavities in the anodes.

magnetron pulling. Frequency shift produced by long line effect in load.

magnetron rectifier. Thermionic rectifier utilizing the magnetron effect.

magnification factor. See Q.

magnon. Quantum of spin wave energy in magnetic material.

Magnox. Group of magnesium alloys for canning uranium-reactor fuel elements. Best known are Magnox B and Magnox A12. The latter is Mg with 0.8% Al and 0.01% Be (by weight).

main gap. Total gap between the cathode and anode in a gas-discharge tube.

main routine. Coded directions which form the basis of a programme in setting an electronic computer to perform a required operation on data, independently supplied.

main store or memory. Fastest storage device of a computer. See **primary store.**

mains. In U.K., source of power, e.g., power pack, generator, etc. In U.S., **power line, power supply.**

mains frequency. Mainly 50 Hz in U.K., 60 Hz in U.S. In aircraft 400 Hz.

maintained tuning-fork. One associated with a valve so that the latter supplies energy continuously to maintain the fork in steady oscillation. The frequency of the oscillation is substantially that of the free fork, and provides a method for establishing frequencies with great accuracy. Also **tuning-fork oscillator.**

maintaining voltage. That which just maintains ionization and discharge.

Majorana force. Force between nucleons which changes sign if in the wave function of the two particles their space coordinates are interchanged.

majority carrier. In a semiconductor, the electrons or holes, whichever carry most of the measured current. The other is the *minority carrier.*

make impulse. An impulse arising when a circuit is *made,* i.e., closed.

Mallory battery. TN for type of dry cell with mercury zinc electrodes. These have larger capacity than the conventional dry cell and are less subject to polarization. Their e.m.f. is about 1 volt. Used in radiation monitors, miniature electronic devices, etc.

manganese (Mn). Grey-pink, hard, brittle metallic element, at. no. 25, at. wt. 54·9380, sp. gr. 7·20, m.p. 1260°C. Alloyed with other non-ferromagnetic elements of copper and aluminium, it forms a ferromagnetic material. The element is used in some primary batteries; and there are three readily-available isotopes.

manganin. A copper-base alloy, containing 12 per cent. manganese and 4 per cent. nickel, used for making resistor wires, because of its low temperature coefficient and low contact potential.

manipulator. Remote handling device used, e.g., with radioactive materials.

Mansbridge capacitor. See **self-sealing capacitor.**

mantissa. Fractional part of logarithm.

manual control. That carried out by hand.

map comparison unit. A device for superimposing a radar screen display on a map or chart of an area, thus facilitating the identification of radar echoes.

Marconi antenna. Original simple vertical wire, fed between base end and earth.

marconigram. See **radiogram.**

margin. (1) Maximum percentage change in unit of signal before errors arise in decoding. (2) Number of decibels increase in gain necessary to result in oscillation. Applied to feedback amplifiers, telephone repeaters, public-address systems, etc.

mark. Departure (positive or negative) from a neutral or no-signal state (space) in accordance with a code, using equal intervals of time.

mark-and-space. The on-and-off of a signal code, the *mark* corresponding to a depression of a telegraph (Morse) key. The actual current may be positive, zero or negative for mark or space.

mark-space multiplier. Circuit which averages square-wave signals, ratio of mark/space in time being determined by one voltage, amplitude by another voltage. Used in analogue computing.

mark-space ratio. That of the durations of mark and space in a pulse or square waveform.

marker. (1) Pip on radar display for calibration of range. (2) See **sentinel.**

marker antenna. One which gives a beam of radiation for marking air-routes, often vertically for blind or instrument landing.

marker beacon. A radio beacon in aviation which radiates a signal to define an area above the beacon.

marking (or keying) wave. Wave, slightly different in frequency from the spacing wave, which corresponds to the mark of the signal code.

maser (*M*icrowave *A*mplification by *S*timulated *E*mission of *R*adiation). An extremely stable low-noise amplifier or oscillator, operating by the interaction between radiation (photons) and atomic particles (atoms, electrons, molecules). When an electron in a lower energy state is struck by a photon having an energy exactly equal to the next energy level, the photon is absorbed. Conversely, if the electron is in the higher state, an additional photon may be emitted, the electron then falling to the lower energy level. It follows that, if a large number of the high-energy level atoms are irradiated by a small signal of the energy-transition frequency, this signal will be amplified by the additional photons emitted.

mask. (1) In cinema and TV: (*a*) to obstruct light or sound from an object in a significant manner, as a speaker from a microphone, an actor from an actor or camera; (*b*) an aperture in a projector which determines the extent of a projected image on a screen. (2) In computing, a machine word which specifies which part of another word is to be processed, or extracts a selected group of characters from a sequence.

masking. Loss of sensitivity of the ear for specified sounds in the presence of other sounds, one sound *masking* sounds of higher frequency to a marked extent.

mass. In physics, the mass of nuclear particles cannot be regarded as constant, since the mass of any body is increased if it is moving with a velocity comparable with that of light. For such particles the usual practice is to quote 'rest masses' and these are frequently given in terms of energy; see **mass-energy equivalence.**

mass absorption coefficient. See **absorption coefficient.**

mass decrement or **mass defect.** Both terms may be used for (1) the binding energy of a nucleus expressed in mass units, and (2) the measured mass of an isotope less its mass number. The BSI recommend the use of *mass decrement* for (2) above. In the U.S. it is usually used for (1) and *mass excess* for (2). See also **packing fraction.**

mass-energy equivalence. Confirmed deduction from relativity theory, such that

$$E = mc^2,$$

where E = energy, m = mass, and c = velocity of light. Also **Einstein energy.**

mass excess. See **mass decrement.**

mass number. Total of *protons* and *neutrons* in a nucleus, each being taken as a unit of mass. Also **nucleon number.**

mass radiator. Radiator which consists of metal particles suspended in a liquid dielectric to provide a source of broadband electromagnetic radiation up to UHF, when a sufficiently high voltage is applied to the dielectric to produce sparking between the particles.

mass spectrograph. Vacuum system in which positive rays of various charged atoms are deflected through electric and magnetic fields so as to indicate, in order, the charge-to-mass ratios on a photographic plate, thus measuring the atomic masses of isotopes with precision. System used for the first separation for analysis of the isotopes of uranium.

mass spectrometer. A *mass spectrograph* in which the charged particles are detected electrically instead of photographically.

mass spectrum. See **spectrum.**

mass stopping power. See **stopping power.**

mass unit. See **atomic mass unit.**

Massey formula. One giving the probability of secondary electron emission when an excited atom approaches the surface of a metal.

master. In record manufacture, the copper electroplate (negative) obtained by plating the original wax or cellulose surface on which a recording has been cut.

master clock. Oscillator component which generates all the digital impulses required in an electronic computer.

master oscillator. One, often of low power, which establishes the frequency of transmission of a radio transmitter.

master station. Transmitting station from which one or more 'satellite' stations receive a programme for re-broadcasting.

masurium. See **technetium.**

matched load. One with impedance equalling the characteristic impedance of a line or waveguide, so that there is no reflection of power and the received power is the maximum possible. Also **matched termination.**

matching. (1) Adjusting a load impedance to match the source impedance with a transformer or network, so that maximum power is received. Reactance in the load impedance can be neutralized or *tuned out* by an equal reactance of opposite sign. (2) In radar, section in a waveguide for impedance transformation, depending on metal or dielectric therein.

matching impedance. Adjustment of a load to a source so that the maximum power is accepted, i.e., so that there is no reflection loss due to *mismatch*. The principle applies to many physical systems, e.g., non-reflecting optical surfaces, use of a horn in loudspeakers to match impedance of vibrator to that of air, matching load on electrical transmission line.

matching stub. Short- or open-circuited stub line attached to main line to neutralize reactive component of load and so improve matching.

matching transformer. One expressly inserted into a communication circuit to avoid reflection losses because the load impedance differs from the source impedance. In designing for optimum matching, the ratio of the impedances equals the square of the ratio of the turns on the windings.

materialization. Reverse of Einstein energy released with annihilation of mass. A common example is by *pair production* (electron-positron) from gamma-rays.

materials-testing reactor. See **high-flux reactor.**

matrix. (1) Negative obtained by electrodeposition on the waxlike disk on which a sound recording has been made. Also **solid metal negative.** (2) Set of numbers with relationship best displayed by arranging in rows and columns. (3) An electric circuit network or array of magnetic cores analogous to (2). In computing these form a logical network with the matrix elements at the intersections of a rectangular array of input and output leads. Such an array can form a fast store, an encoder or a decoder.

matrix printing. Marking intersections of a matrix of dots so that a recognizable pattern is displayed and registered on a cathode-ray tube.

F*

Matthiessen hypothesis. That the total electrical resistivity of a metal may be equated to the sum of the various resistivities due to the different sources of scattering of free electrons; this also applies to thermal resistivity.

maximum crest. Same as amplitude peak.

maximum permissible concentration. That agreed for radioactive contaminants in air or water.

maximum permissible dose. The recommended upper limit for the dose which may be received during a specified period by a person exposed to ionizing radiation. Also called **permissible dose.**

maximum permissible dose rate or flux. That dose rate or flux which, if continued throughout the exposure time, would lead to the absorption of the *maximum permissible dose.*

maximum permissible level. A phrase used loosely to indicate maximum permissible concentration, dose or dose rate.

maximum usable frequency. That which is effective for long-distance communication, as predicted from diurnal and seasonal ionospheric observation. Varies on an eleven-year cycle. See **Wolf's equation.**

maxwell. The CGS unit of magnetic flux, the MKS unit being the *weber.* One maxwell $= 10^{-8}$ weber.

Maxwell-Boltzmann distribution. The energy distribution among the molecules of a gas in thermal equilibrium.

Maxwell bridge. An early form of a.c. bridge which can be used for the measurement of both inductance and capacitance.

Maxwell relationship. That between the refractive index (n) and the dielectric constant (k) of a medium. For a non-ferromagnetic, $k = n^2$.

Maxwell's circulating current. A mesh or cyclic current inserted in closed loops in a complex network for analytical purposes.

Maxwell's field equations. Mathematical formulations of the laws of Gauss, Faraday and Ampère from which the theory of electromagnetic waves can conveniently be derived.

mayday. Verbal international radio-telephone distress call or signal, corresponding to SOS in telegraphy.

mean free path. (1) Average distance travelled by a sound wave in an enclosure between wall reflections; required for establishing a formula for reverberation calculations. (2) Average distance between intermolecular collision of gas molecules.

mean free time. Average time between collisions of electrons with impurity atoms in semiconductors; also of intermolecular collision of gas molecules.

mean life. (1) The average time during which an atom or other system exists in a particular form, e.g., for a thermal neutron it will be the average time interval between the instant at which it becomes thermal and the instant of its disappearance in the reactor by leakage or absorption. Mean life= $1.443 \times$ *half-life*. Also **average life**. (2) The mean time between the birth and death of a charge carrier in a semiconductor, a particle (e.g., an ion, a pion), etc.

mean power. Average over a period, of power supplied by a transmitter to an antenna.

mean solar day. That calculated by averaging the intervals between the sun's transit across a meridian. *True* or *apparent solar time* differs from this by an amount known as the *equation of time*, which has values up to 17 minutes (positive or negative).

mean spherical response. That of a microphone or loudspeaker taken over a complete sphere the radius of which is large in comparison with the size of the apparatus. For a loudspeaker, this response (total response) determines the total output of sound power, and therefore, in conjunction with the acoustic properties of an enclosure, the average reverberation intensity in the enclosure. For a microphone, this response is substantially equal to the response for reverberant sound.

measuring unit. Device (transducer) which ascertains the magnitude of a quantity to be controlled.

mechanical analogue. That which can be drawn between mechanical and electrical systems obeying corresponding equations, e.g., mechanical and electrical resonators.

mechanical equivalent. See **Joule's equivalent**.

mechanical impedance. Ratio of the total force required to move a body to the velocity resulting, for a specified frequency of motion. It consists of the *real* part, mechanical resistance, which represents the transmission of mechanical power, and the *imaginary* part, which is purely reactive and represents elastic stored energy.

mechanical line. Conception and adjustment of the elements in an acoustic system, such as a sound-box or electrical recorder, so that they form the elements of a wave filter, in analogy with electric wave filters.

mechanical register. An electromagnetic counting device operated by input pulses.

mechanical resonance. Enhanced response to a constant-magnitude disturbing force as the frequency of this force is increased through a *resonant frequency*, at which the reactance of the inertia balances the reactance of the supporting stiffness of the vibrating system.

mechanical scanning. That performed by vibrating or rotating mirrors in simple or primitive systems of television, universally replaced by electronic methods.

mechanomotive force. The rms value of an alternating mechanical force, in dynes or newtons, developed in a transducer.

medium frequency. Between 3×10^5 and 3×10^6 Hz (radio frequencies), sometimes known as **hectometric waves**.

mega-. Prefix denoting one million, e.g., a frequency of one megacycle per second = 10^6 cycles per second (Hz), megawatt =

10^6 watts, megavolt $\equiv 10^6$ volts, megohm $= 10^6$ ohms.

megabit. Unit of a million *bits*, used in large-store calculations.

megaphone. A horn to direct the voice. It can include microphone, amplifier and sound reproducer. Then called a *loud hailer*.

megatron. Name for an electron tube having disk-shaped electrodes in dense parallel layers, the edges of the disk electrodes being fused into and projecting through the glass envelope to serve as contacts; also called **disk-seal tube, lighthouse tube.**

megawatt days per tonne. Unit for heat output from reactor fuel; a measure of burn-up.

megohmmeter. Portable apparatus for indicating values of high resistance, containing a circuit excited by a battery, transistor, converter or d.c. generator. Now used for measuring insulation resistance of all types of capacitor.

Meissner circuit. Oscillating valve circuit in which the resonant circuit is inductively coupled to two coils included in the anode and grid circuits respectively.

Meissner effect. Apparent expulsion of lines of magnetic induction from a superconductor when cooled below superconducting transition temperature in a magnetic field.

mel. Unit of pitch in sound, a pitch of 10^3 mels being associated with a simple tone of frequency 10^3 Hz at an intensity of 40 dB above the threshold of the listener.

melanex. Thin plastic foil used as light shield over scintillation detectors for alpha-particles.

meltback transistor. A junction transistor, in which the junction is formed by allowing the molten doped semiconductor to solidify.

memory. See **store**.

memory capacity. The amount of information, usually expressed by the number of words, which can be retained in a *memory*. It may also be expressed in *bits*. The information is produced, unaltered, on command. Also called **data storage**.

memory location. A unit storage (holding one computer word) position within the main interval memory.

memory register. A register in the storage of a computer.

memory tube. Preferred term **storage tube**.

mendelevium (Mv or Md). Transuranic element, at. no. 101; ^{256}Mv was first produced by bombarding berkelium with high-energy α-particles. Its half-life is only of the order of an hour.

Mercator's projection. That normally used for charts as bearings are always rendered correctly. Great circle routes will not be straight lines.

mercury (Hg). Metallic element, at. no. 80, at. wt. 200·59, sp. gr. 13·6, m.p. $-39°$C, b.p. 357°C. Used in thyratrons, arc rectifiers, switches, etc.

mercury-198. Mercury isotope made from

gold in a reactor. Can be used in a quartz mercury-arc tube, light from which has an exceptionally sharp green line, because of even mass number of a single isotope. This considerably improves comparisons of end gauges in interferometers.

mercury arc. Luminous discharge of wide line spectral content, especially rich in ultraviolet. At one time used for illumination, but now the gas discharge is widely used to excite coloured radiation in fluorescent powders in both hot and cold cathode tubes. Used in factories, restaurants and for advertising. See **mercury-vapour tube.**

mercury-arc converter. A *frequency converter* employing a mercury-arc power converter.

mercury-arc rectifier. One in which rectification arises from the differential migration of electrons and heavy mercury ions in a plasma, formed by evaporation from a *hot spot* on a cathode pool of mercury. This vaporization has to be started by withdrawing an electrode from this pool, thus generating an arc. Used for a.c./d.c. conversion for railways. See **thyratron.**

mercury cell. (1) Electrolytic cell with mercury cathode. (2) Dry cell employing mercury electrode, e.m.f. *ca.* 1·3 volt; see **Mallory battery, Reuben-Mallory cell.**

mercury memory. One depending on the velocity of mechanical impulses in mercury, using transducers to connect to the electronic circuits. The impulses are reshaped for continuous circulation, the train thus establishing a store which can be tapped as required.

mercury-pool cathode. Cold cathode in valve where arc discharge releases electrons from surface of mercury pool, e.g., in *mercury-arc rectifier* (q.v.).

mercury tank. A vessel containing mercury and used to store information by continuous circulation of inpulses through the tank. See **mercury memory.**

mercury-vapour lamp. Quartz tube containing a mercury arc, specially designed to provide ultraviolet rays for therapeutic and cosmetic treatment.

mercury-vapour pump. See **Gaede diffusion pump.**

mercury-vapour tube. Generally, any device in which an electric discharge takes place through mercury vapour. Specifically, a triode valve with mercury vapour, which is ionized by the passage of electrons and reduces the space charge and the anode potential necessary to maintain a given current. The grid is effective in controlling the start of the discharge. See **thyratron.**

merging. Sequencing or interlacing computer data from separate sources, e.g., punched or magnetic tape, so that they are in required order on new tape.

mesa. Type of transistor in which one electrode is made very much smaller than the other to control bulk resistance. Also, by selective etching, the base and emitter are raised above the region of the collector.

mesh. A complete electrical path (including capacitors) in the component branches of a complex network. Also called **loop.** See **absorption.**

mesh current. One assumed to exist throughout the whole section of a mesh.

mesh network. One formed from a number of impedances in series.

mesic atom. Short-lived atom in which negative muon has displaced normal electron.

meson. One of a series of unstable particles with masses intermediate between those of electrons and nucleons, and with positive, negative, or zero charge. Mesons appear to be related to the nuclear field, in a similar way to the relationship between photons and the electromagnetic field (or the hypothetical relationship between gravitons and the gravitational field). They are bosons which, unlike fermions (leptons and baryons), may be created or annihilated freely. They are now classed in three groups: η-mesons, π-mesons (*pions*), and κ-mesons (*kaons*). The *muon*, formerly classified as the μ-meson, is now known to be a lepton. Mesons participate with baryons in strong interactions obeying the law of conservation of parity.

meson field. That which is considered to be concerned with the interchange of proton and neutron in a nucleus, the mesons transferring energy.

mesothorium-I. Symbol MsTh$_I$. The common name for 6·7y^{228}Ra, a member of the thorium series.

mesothorium-II. Symbol MsTh$_{II}$. The common name for 6·13h ^{228}Ac, a member of the thorium series.

mesotron. Obsolete name for **mu-meson.**

message. Collection of *words*, each comprising a number of *bits*, within a pattern in a computer.

metadyne. See **amplidyne.**

metal ceramic. An alloy of a ceramic and a metal which retains its useful properties at very high temperatures. Also **cermet.**

metal insulator. Waveguide or transmission line an odd number of quarter-wavelengths long, which has a high impedance. Used as a support or anchor without normal insulators at very high frequencies.

metallic bond. Bond in which the valence electrons of the constituent atoms are free to move in the periodic lattice.

metallic conduction. That which describes the movement of electrons which are freely moved by an electric field within a body of metal.

metallic-film resistor. One formed by coating a high-temperature insulator, such as mica, ceramic, Pyrex glass or quartz, with a metallic film.

metallic lens. One with slats or louvres which give varying retardation to a passing electromagnetic wave, so that it is controlled or focused. Can also be used for sound waves.

metallic soap. Soaplike wax material which is used for making the original record in disk manufacture. The metallic character arises from the presence of lead or aluminium stearite, which is mixed with carnauba wax, etc.

metallized-paper capacitor. One in which a very thin metallic film is deposited on the paper (or plastic) used as dielectric, in contrast to the older technique of inter-leaving paper and thin metal foil.

metallized (or spray-shielded) valve. One in which the exterior of the envelope is coated with a conducting metallic film which can be connected to cathode or earth, to provide electrostatic shielding of the interior of the valve from external disturbance.

metastable. Of an atomic or nuclear state exhibiting stability for finite time (i.e., about a nanosecond upwards) and then decaying into a different quantum state.

metastasic. Said of electrons which change their shells or are absorbed from a shell into the nucleus.

metre bridge. A Wheatstone bridge in which a normally straight length of resistance wire with a sliding contact is used to form two variable ratio arms in a Wheatstone bridge network.

Metrosil. TN for semiconductor devices containing silicon carbide, and having a non-linear resistance.

mho. Unit of electrical conductance, the reciprocal of the ohm. Also **siemen**.

mica. A naturally occurring mineral which may be sheared into very thin sheets. In the very clear form it is an extremely good insulator even at very high temperatures and is used in the best capacitors. It is also employed to insulate electrodes of valves.

micro-. Prefix for one-millionth; symbol μ.

micro programme. One already registered in a fast store to be called on by main computer programme.

microammeter. A most sensitive form of robust current-measuring instrument, as distinct from a suspended coil galvanometer.

microampere (μA). Equal to 10^{-6} ampere.

microbar. CGS unit of pressure, 1 dyne/cm^2 (10^{-6} bar). Also **barye**.

microcircuits. Those with components formed in one unit of semiconductor crystal.

microelectronics. Technique of solid circuits in which units of semiconductors are formed into several components.

microfarad (μF). Equal to 10^{-6} farad; the unit of capacitance normally used in practice.

microgroove records. Long-playing records with as many as 200–300 grooves per inch, and played at rotational speeds of $33\frac{1}{3}$ or 45 revs. per min.

micromicro-. Prefix for one-million-millionth (10^{-12}); denoted by $\mu\mu$, or *pico-*.

micromicrofarad. Same as **picofarad (pF).**

microminiaturization. Production and use of circuit components of very small dimensions,

involving vacuum-deposited films, e.g., Nichrome on ceramic rods for resistors, oxide layers for capacitor dielectrics.

micromodule. Said of circuits of components of minute but specified dimensions which are formed from the same crystal of material, e.g., germanium. This is fashioned into capacitors, resistors, photocells, diodes, transistors, etc.

micron. Unit for electromagnetic wave-lengths, one-millionth of a metre. Also used for measurement of dust, condensation and suspensions. Abb. μ. See **ångström**.

microphone. An acousto-electric trans-ducer, essential in all sound-reproducing systems. The excess pressure in the sound wave is applied to a mechanical system, such as a ribbon or diaphragm, the motion of which generates an electromotive force or modulates a current or voltage. See also:

anti-noise-	magnetic-
apple-and-biscuit-	magnetostriction-
bidirectional-	moving-coil-
button-	moving-conductor-
capacitor-	Olson-
carbon-	omnidirectional-
cardioid-	parabolic-
crystal-	pressure-
directional-	pressure-gradient-
ear-	push-pull-
electronic-	radio-
high-frequency capacitance-	Reiss-
hot-wire-	stereomicrophone
lip-	Sykes-
loudspeaker-	throat-.

microphone amplifier. One which brings up the low level of the output of a high-quality microphone to a level which can be switched or sent over a line.

microphone boom. A pivoted arm, carrying a microphone, which can be swung into any position in a studio.

microphone response. That measured over the operating frequency range in a particular direction, or averaged over all directions (reverberant response). The characteristic response is usually given by the ratio of the open-circuit voltage generated by the micro-phone to the sound pressure (dyne/cm^{-2}) existing in the free progressive wave before introducing the microphone.

microphonic. Said of a valve which responds to aerial vibrations and/or knocks.

microphonic noise. That in the output of a valve related to mechanical vibration of the electrode system. Also called **micro-phonicity.**

microradiography. Exposure of small thin objects to *soft X-rays*, with registration on a fine-grain emulsion and subsequent enlargement up to 100 times. Also used to signify the optical reproduction of an image formed, e.g., by an electron microscope.

microscopic state. One in which condition of each individual atom has been fully specified. Cf. *macroscopic state*.

microswitch. One operated by very small mechanical movement.

microsyn. An a.c. transducer employed for measuring angular rotation and used as an output device in precision gyroscopes and servomechanisms. Its construction is similar to that of an *E-transformer*.

microwatt. Equal to 10^{-6} watt.

microwave hohlraum. See hohlraum.

microwave resonance. One between microwave signals and atoms or molecules of a medium.

microwave resonator. Effective tuned circuit for microwave signal. Usually a cavity resonator but tuned lines are also employed.

microwave spectrometer. An instrument designed to separate a complex microwave signal into its various components and to measure the frequency of each; analogous to an *optical spectrometer*.

microwave spectroscopy. The study of atomic and/or molecular resonances in the microwave spectrum.

microwave spectrum. The part of the electromagnetic spectrum corresponding to microwave frequencies.

microwaves. Those electromagnetic wavelengths between 1 mm and 30 cm, i.e., from $0.3 \cdot 10^{12}$ to 10^{9} MHz in frequency, thus bridging gap between normal radio waves and heat waves. Also UHF sound waves.

middle marker beacon. A marker beacon, associated with the ILS, used to define the second predetermined point during a beam approach.

migration. Movement of ions under influence of electric field and against the viscous resistance of the solvent. Measured by observation in thin tubes.

migration area. One-sixth of the mean square distance covered by a neutron between creation and capture. Its square root is the *migration length*.

mike. *Microphone* (colloq.).

mile of standard cable. Old unit of attenuation provided by one mile of an arbitrary type of telephone circuit at 800 Hz. Formerly used for estimating possibilities of transmission over circuits. Now displaced by the decibel and neper, which alone are accepted internationally.

Miller bridge. A particular form used to measure amplification factors of valves.

Miller effect. The change of input impedance of a thermionic valve arising from feedback, causing the input capacitance to be larger than the calculated static electrode capacitances. See also **electrostatic reaction, screen grid.**

Miller indices. Integers which determine the orientation of a crystal plane in relation to three crystallographic axes. The reciprocals of the intercepts of the plane on the axes (in terms of lattice constants) are reduced to the smallest integers in ratio. Also called **crystal indices.**

Miller integrating circuit. One which incorpor-

ates a thermionic valve to improve the linearity of pulse-shaping circuits.

milli-. Prefix for one-thousandth; symbol m.

millibar. Unit (10^3 dyne per sq. cm.) used in measurement of atmospheric pressure.

millicurie. Equivalent to 0·001 curie.

Millikan electrometer. An early form of ionization-chamber dosimeter employing a gold-leaf electrometer instead of the modern quartz-fibre type.

Millikan meter. Early form of personal dosimeter.

millimass unit or **mamu.** Equivalent to 0·001 of *atomic mass unit*.

millimicron. Unit for electromagnetic wavelength, one-thousandth of a *micron*, $m\mu$ and $\mu/1000$.

millinile. See nile.

milliradian. Unit for radar or infrared scanning from aircraft, equal to 10^{-3} radian.

milliröntgen. Equivalent to 0·001 röntgen.

Millman tube. A slow-wave travelling-wave tube, i.e., one in which the electron-beam velocity may be less than that of the electromagnetic wave carrying the signal through the periodic structure which constitutes the interaction region of the tube.

miniature (or sub-miniature) valve. One in which all the dimensions are reduced to very small values, to keep down the interelectrode capacitances and the electron transit time. Used in HF, VHF and UHF circuits.

mini-log. Unit transistor encapsulated circuit for universal use in process controllers, computers, etc.

minimum access programme. Computer programme routine involving minimum loss of time for access to store.

minimum clearing. Same as zero clearing.

minimum discernible signal. Smallest input power to any unit which just produces a discernible change in output level.

minimum ionization. The smallest possible value of the specific ionization that a charged particle can produce in passing through a given substance. It occurs for particles having velocities $=0.95c$, where c = velocity of light.

minimum sampling frequency. Lowest frequency pulse system from which continuous modulated carrier signal can be accurately reconstructed.

minitrack. Phase-comparison angle-tracking radio system, used for tracking satellites.

minority carrier. See majority carrier.

mirage. Radio signal for which transmission path includes reflection from layer of rarified air, in analogous manner to optical mirage.

mirror machine. Type of apparatus using magnetic mirror principles for trapping high-energy ions injected into a plasma for fusion studies.

mirror nuclides. Those with the same number of nucleons, but with proton and neutron numbers interchanged.

mirror reflector. Surface or set of metal rods

which reflects a wave geometrically, e.g., *V-reflector* or *dish*.

misfire. Failure to establish, during an intended conducting period, a discharge or arc in a gas-discharge tube or a mercury-pool rectifier.

mismatch. Load impedance of incorrect value for maximum energy transfer (see **impedance matching** for methods of correcting this condition).

mitron. Magnetron which can be tuned over a wide band by voltage control.

mix. See **cross-fade.**

mixer. (1) Collection of variable attenuators, controlled by knobs, which allows the combination of several transmissions to be independently adjusted from zero to maximum. A *group mixer* or *fader* is one which deals with several groups of transmissions. Hence *mixer*, the person who operates mixers. (2) Frequency conversion stage in a supersonic heterodyne receiver, in which there is rectification of the incoming signal, and a larger signal from a local oscillator. Also **conversion detector, first detector.** See **balanced-, crystal-, diode-.**

mixing. (1) General term for combining transmissions in two or more channels into one channel, usually by *fading*. (2) In gaseous isotope separation, the process of reducing the concentration gradient for the lighter isotope, close to the diffusion barrier.

mixing efficiency. A measure of the effectiveness of the mixing process in isotope separation.

mixing point. One in a control system where an output is obtained from two independent inputs. If these are added it is a *summing point*, and if multiplied a *multiplication point*.

mobility. Description of drift of ions (including electrons and holes in semiconductors) under applied electric fields, additional to thermal agitation. Measured in $cm^2 sec^{-1} volt^{-1}$, and related to conductivity and Hall effect.

mode. (1) One of several electromagnetic wave frequencies which a given oscillator may generate, or to which a given resonator may respond, e.g., magnetron modes. In a waveguide the mode gives the number of half-period field variations parallel to the transverse axes of the guide (see **rectangular waveguide mode diagram** in appendix). Similarly for a cavity resonator the half-period variations parallel to all three axes must be specified. In all cases different modes will be characterized by different field configurations. (2) Similarly, one of several frequencies of mechanical vibration which a body may execute or with which it may respond to a forcing signal. See **dominant-.**

mode jump. Switch of an oscillator from one mode to another; also **mode shift.**

moder. See **pulse moder.**

moderating ratio. That of the slowing-down power of a moderator to the macroscopic adsorption cross-section.

moderation. See **degradation.**

moderator. The material in a reactor used for moderating (slowing-down) neutrons. Usually graphite or heavy water. See **lethargy (of neutrons), slowing-down power.**

moderator control. Control of a reactor by varying the position or quantity of the moderator.

modifier. A code element in computing used to alter the address of an operand.

modular. Form of construction in which units, often with differing functions, are built into identical size *modules*, which are therefore quickly interchangeable.

modulated carrier or **wave.** A frequency which can be transmitted or received through space or a transmission circuit, with a superposed information signal which, by itself, could not be effectively transmitted and received.

modulated continuous wave. Transmission in which a carrier is modulated by a tone or interrupted by keying.

modulating electrode. The electrode of a thermionic valve to which a voltage is applied to control the size of the beam current.

modulation. (1) Process of impressing a signal on to a radio-frequency carrier, at *high power* if the modulator is directly connected to a load, e.g., antenna, or at *low power* if there is high-power amplification, e.g., class-A or class-B, between modulator and load. (2) Signals for recording on tape, which can be *dipole* or *phase*, depending on number reversals of magnetization for each signal. (3) Changing from one key to another in music. The continual change from one fundamental frequency to another in speech. See also:

absorption control-
amplitude-
angle-
anode-
arc-
audio-frequency shift-
cathode-
choke-
compound-
conductivity-
cross-
Day-
demodulation
density-
downward-
dual-
frequency-
group-
high-level (low-level)-
hum-
intensity-
ionic-
linear-
over modulation
percentage-
phase-
positive (or **negative**) **light-**
pulse-amplitude-
pulse-code-
pulse-time-
reactance-
screen-
series-
single sideband system
suppressor-grid-
synchronizing-
total-
undermodulation
velocity-
vestigial sideband
z-axis-.

modulation depth. Factor indicating extent of modulation of a wave. *Difference ÷ sum* of peak and trough values of an amplitude-modulated wave, and of extreme deviations of carrier frequency in a frequency-modulated wave. Often expressed as a percentage.

modulation distortion. When a carrier is modulated, any departure from invariability of carrier amplitude and/or addition of side frequencies proportional in amplitude to the corresponding frequencies in the signal, with phases balanced with respect to the initial phase of the carrier.

modulation factor. In an amplitude-modulated wave with maximum and minimum modulated carrier amplitudes, A_{max} and A_{min} respectively, the positive modulation factor is $\dfrac{A_{max}-A}{A}$, the negative modulation factor is $\dfrac{A-A_{min}}{A}$, where A is the amplitude of the unmodulated carrier.

modulation frequency. One impressed upon a carrier wave in a modulator.

modulation hum. Low frequencies which enter a carrier-frequency amplifier and appear as interference after demodulation.

modulation index. For a single modulation frequency, the ratio of the frequency deviation of the carrier to this frequency.

modulation meter. That placed in shunt with a communication channel, giving indication that interprets, in a stated way, instant-to-instant power level in varying modulation currents.

modulation noise. Noise in a modulated carrier, in excess of the signal but varying with it.

modulation pattern. That on a CRO when the modulated wave is connected to the Y deflection system and the modulation signal to the X deflection plates. The result is a trapezoidal pattern which enables the modulation depth to be measured, and distortion introduced during modulation to be readily detected.

modulation suppression. Reduction of modulation in wanted signal in presence of an unwanted signal. Originally but erroneously termed *demodulation*.

modulation transformer. One which applies the modulating signal to the carrier wave amplifier in a transmitter.

modulator. (1) Any device, *light valve*, *galvanometer*, *vibrator*, which regulates the light falling on a photographic sound-track. (2) Any circuit unit which modulates a radio carrier, at *high level* directly for transmission, or at *low level* for amplification of the modulated carrier before transmission. (3) Circuit in an *electronic organ* which changes the pitch of the notes. See **balanced-, capacitor-, ring-**.

module. Geometrical framework in which circuits can be realized, largely from units of specified dimensions.

modulo check. One which anticipates remainder when a processed number is divided by a number which is also carried with data through the process.

modulus. Constant for units conversion between systems. See **amplitude**.

modulus of elasticity. For a substance, the ratio of stress to strain within the elastic range, i.e., where Hooke's law is obeyed.

modulus of rupture. A measure of the ultimate strength, i.e., breaking load, per unit area of a given specimen determined in bending or in torsion.

mol(e). Amount of any defined chemical substance which weighs its molecular weight in grams. Also **gram-molecule**. Symbol **mol**.

molar conductance. The electrical conductance between electrodes 1 cm apart in an electrolyte having 1 mol of solute in 1 litre of solution.

mole-electronics. Technique of growing solid-state crystals so as to form transistors, diodes, resistors in one mass for microminiaturization. Also called **molectronics, molecular electronics**.

mole-fraction. The ratio of the number of atoms of a certain isotope of an element to the total number of the atoms of that element which are present in an isotopic mixture.

molecular beam. Directed stream of un-ionized molecules issuing from a source and depending only on their thermal energy.

molecular beam resonance method. See **Rabi method**.

molecular bond. Bond in which the linkage pair of electrons are provided by one of the bonding atoms ; cf. *atomic bond*.

molecular distillation. An isotope separation in which molecules are evaporated at very low pressures from a surface and are condensed before they encounter collisions.

molecular electronics. See **mole-electronics**.

molecular stopping power. The energy loss per molecule per unit area normal to the motion of the particle in travelling unit distance. It is *approximately* equal to the sum of the atomic stopping powers of the constituent atoms.

molecular streaming. Gas flow through a tube when the mean free path is large compared with the tube diameter.

molecular volume. That occupied by one mole of a substance in gaseous form at standard temperature and pressure (approx. 22·414 litres).

molecular weight. The sum of the atomic weights of the constituent atoms of a molecule.

molecule. Smallest part of an element or compound which (nominally) exhibits all the properties of that specific compound or element.

molybdenum (Mo). Metallic element, at. no. 42, at. wt. 95·94, m.p. 2625°C, b.p. 4800°C, resistivity 4·77 microhm-cm². Its physical properties resemble those of iron. Used in the form of wire for filament heaters in valves, for electrodes of mercury-vapour lamps and for winding electric resistance furnaces. It seals well to Pyrex glass and spot-welds to iron and steel. It is a prominent fission product from nuclear reactors as ^{95}Mo.

monaural. Pertaining to use of one ear

instead of two (cf. *binaural*). In sound recording, the opposite of *stereophonic*. See **monophonic**.

monitor. (1) Apparatus or person whose purpose is to ascertain that equipment, materials or output fulfil prescribed conditions or operate within prescribed limits. (2) Ionization chamber or other radiation detector arranged to give a continuous indication of intensity of radiation, e.g., in radiation laboratories, industrial operations, or X-ray exposure. (3) Unit in large computers which prepares the machine instructions from the source programme, using built-in compiler(s) for one or more programme languages; and feeds these into the processing and output units in sequence, once compiling is completed. It also controls time-sharing procedures. See also:

air-   hand-
automatic-   physiological-
effluent-   picture-
frequency-   water-.

monitoring. Tapping a communication circuit to ascertain that transmission is as required, without interfering with it. Monitoring may be with headphones or an amplifier and loudspeaker or a telegraph recorder.

monitoring amplifier. Same as **bridging amplifier**.

monitoring loudspeaker. One of high quality and exactly matching others, for verifying quality of programme transmission before radiation or recording. Located in a monitoring room adjacent to studio.

monitoring station. National service for verifying the frequencies of emission of radio transmitters allocated within prescribed bands. See **wavemeter**.

monitron. U.S. term for ionization-chamber radiation background monitor of the type used near nuclear reactors.

monkey chatter. Interference between a wanted carrier and the nearer sideband (modulation) of an unwanted transmission in radio reception.

monochromatic radiation. Electromagnetic radiation (originally visible) of one single frequency component. By extension a beam of particulate radiation comprising particles all of the same type and energy. *Homogeneous* or *mono-energetic* is preferable in this sense.

monochromator. Device for converting heterogeneous beam of radiation (electromagnetic or particulate) to homogeneous beam by absorption or refraction of unwanted components.

monochrome. Signal or picture of one colour or hue.

monochrome receiver. TV receiver which reproduces a black-and-white transmission, or a colour transmission in black-and-white. See **compatibility**.

monophonic. Single-channel sound reproduction, recreating the acoustic source, as compared with *stereophonic*, which uses two

or more identical channels for auditory perspective. See **monaural**.

monopulse. Radar using spaced antennae to give phase differences on received signal. This can give precise directional information.

monoscope. An electron-beam tube used to produce a TV picture signal from a fixed object; used for test purposes.

monostable. Of a circuit or system, fully stable in one state only but *metastable* in another to which it can be driven for a fixed period by an input pulse.

Monte Carlo method. Statistical procedure when mathematical operations are performed on random numbers.

Morse code. A system used in signalling or telegraphy, which consists of various combinations of dots and dashes.

Morse equation. An equation which relates the potential energy of a diatomic molecule to the internuclear distance.

mosaic. (1) Photoelectric surface made up of a large number of infinitesimal granules of photoemissive material deposited on an insulating support. Used as emitting electrode in some forms of electron camera, e.g., iconoscope. Mosaics of piezoelectric crystal elements are employed in ultrasonic cameras. (2) The reconstruction of the track of a nuclear particle through a stack of photographic emulsions.

Moseley's law. The frequencies of the characteristic X-rays of the elements show a strict linear relationship with the square of the atomic number. This result of Moseley's researches stressed the importance of atomic number rather than of atomic weight.

Mossbauer effect. Modification of gamma-ray absorption spectrum due to change in frequency of radiation produced by motion of source (analogous to *Doppler effect* in sound). The effect is also shown when the source is stationary and the absorber moves.

mother. Copper electroplate positive which is made from the master in record manufacture. See **matrix (1)**.

motional impedance. In an electromechanical transducer, e.g., a telephone receiver or relay, that part of the input electrical impedance due to the motion of the mechanism; the difference between the input electrical impedance when the mechanical system is allowed to oscillate and the same impedance when the mechanical system is stopped from moving, or *blocked*.

motor-boating. Very-low-frequency relaxation oscillation in an amplifier, arising from inadequate decoupling of common sources of current supply.

Mott scattering formula. Gives the differential cross-section for the scattering of identical particles arising from a coulomb interaction.

mount. That part of a switching tube or cavity which enables it to be connected to a waveguide.

mountain effect. Error or uncertainty produced in radio direction-finding by reflection of signals from mountains.

moving-coil galvanometer. One comprising a coil suspended between the pole pieces of a permanent magnet. In a more robust but less sensitive form in use as ammeters, the coil is pivoted.

moving-coil loudspeaker. The inverse of a *moving-coil microphone* except that the coil is attached to a radiating cone.

moving-coil microphone. One in which an e.m.f. is induced in a coil attached to a diaphragm, the coil being in the radial field of a small pot magnet. Also called **dynamic microphone, electrodynamic microphone.** See **Sykes microphone.**

moving-coil voltmeter. Voltmeter constructed like a galvanometer and used for d.c. measurements.

moving-conductor microphone. One in which the conductor is stretched between the wedge poles of a permanent magnet; also **ribbon microphone.**

moving-iron. Descriptive of one-time drive in microphones and loudspeakers, which depends on the motion of magnetic material which is part of a magnetic circuit.

moving-iron voltmeter. Voltmeter used for a.c. measurements, depending on attraction of a soft iron vane into the magnetic field due to current.

moving-target indicator. One in which the presentation is concerned only with moving objects, reflections from others being balanced out by a register or memory system.

Moviola. See **Acmeola.**

mu. Greek letter used as symbol for **amplification factor; micro-; micron; permeability.** Written μ.

mu-circuit. The part of a feedback amplifier in which the vectorial sum of the input signal and that of the feedback portion of the output is amplified.

mu-factor. *Voltage amplification factor* of grid-to-anode (or another grid) of a valve, current and all other voltages being constant.

mu-meson (or μ-meson). An elementary particle once thought to be a *meson* but now known to be a *lepton.* See **muon.**

multi-address. Technique whereby *words* are directed to temporary locations before being placed in the *main store*; used to reduce the size of the main store.

multicavity magnetron. A magnetron in which the anode comprises several cavities, e.g., the *rising-sun magnetron* in which the resonators are of two different frequencies and are arranged alternately to obtain mode separation.

multichannel. Any system which divides the frequency spectrum of a signal into a number of bands which are separately transmitted, with subsequent recombination.

multigroup theory. Theoretical reactor model in which presence of several energy groups among the neutrons is allowed for. Cf. *two-group theory.*

multimu. Same as **variable mu.**

multinomial distribution. See **binomial distribution.**

multipactor effect. Electron multiplication between adjacent segments of a linear accelerator.

multiple decay or **multiple disintegration.** Branching in a radioactive decay series.

multiple echo. Perception of a number of distinct repetitions of a sound, because of reflections with differential delays, of separate waves following various paths between the source and observer.

multiple-spot scanning. Facsimile transmission in which image is divided into two or more sections which are scanned simultaneously.

multiple unit or **valve.** Envelope containing a number of valves, with or without common cathode.

multiplets. (1) Optical spectrum lines showing fine structure with several components (i.e., triplets or still more complex structures). (2) See under **isotopic spin.**

multiplex. (1) Use of one channel for several messages by *time division multiplex* or *frequency division* (qq.v.). (2) A frequency-modulated stereo radio system.

multiplication. Secondary emission of electrons due to impact of primary electrons on a surface. It varies from zero with Ni and up to 12·3 with Ni-Be alloy. (2) Ratio of neutron flux in sub-critical reactor to that supplied by the neutron source alone. Equal to $1/(1-k)$, where k is the *multiplication constant.*

multiplication constant. The ratio of the average number of neutrons produced by fission in one neutron lifetime to the total number of neutrons absorbed or leaking out in the same interval. Also called **reproduction constant,** and denoted by k.

multiplication-constant infinite (k_∞). The value of k for an infinitely large reactor (i.e., ignoring leakage).

multiplication point. See **mixing point.**

multiplier. (1) See **electron multiplier.** (2) Unit in analogue computer used to obtain the product of two varying signals. See also:

cross-field-	mark-space-
frequency-	photomultiplier
image dissector-	voltage-.

multipole. An assembly of two or more electric or magnetic poles.

multipole moments. These are magnetic and electric, and are measures of the charge, current and magnet (via intrinsic spin) distributions in a given state. These *static* multipole moments determine the interaction of the system with weak external fields. There are also *transition* multipole moments which determine radiative transitions between two states.

multiposition. Said of a system in which output or final control can take on three or more preset values.

multi-stage. Said of a tube in which the electrons are progressively accelerated by anode rings held at increasing potentials.

multitone. Generator, thermionic or mechanical, which produces a spectrum of currents, i.e., a complex current with a large number of components, equally spaced in frequency.

multitron. Power amplifier for very-high-frequency pulse operation.

multivibrator. Original Eccles-Jordan free-running circuit, with two stages of resistance-capacitance valve (and now transistor) amplification, back-coupled. Period of oscillatory waveform is *ca.* sum of time constants of two grid circuits. Easily locked to tuning-fork or sinusoidal drive, producing many harmonics. Normal symmetrical waveform can be markedly altered by varying a grid bias. (Without coupling capacitors circuit becomes bistable and with asymmetrical back coupling it may also be monostable.)

Mumetal. RTM of high-permeability, low-saturation magnetic alloy of about 80 % nickel, requiring special heat treatment to achieve special low-loss properties. Useful in non-polarized transformer cores, a.c. instruments, small relays. Also for shielding devices from external field as in CROs.

Munsell scale. A scale of chromaticity values giving approximately equal magnitude changes in visual hue.

muon. Subatomic particle with rest mass equivalent to 106 MeV (the heaviest known *lepton*). Has unit negative charge (*anti-muons* have positive charge) and half-life of about 2μs. Decays into electron, neutrino and anti-neutrino (*anti-muon* into positron, neutrino and anti-neutrino). Participates only in weak interactions without conservation of parity.

Murray loop test. A bridge test circuit used for detecting earth leakage points in telephone and other similar signal lines.

musa. Antenna which can be rotated by varying the phases of the individual units. (*M*ultiple-*u*nit *s*teerable *a*ntenna). See also **steerable antenna.**

mrush. Fading and distortion in reception in an area through the interaction of waves from two or more synchronized radio transmitters.

mushroom loudspeaker. One in which a vertical horn is fitted with directing baffles, so that the radiated sound power is directed

uniformly over a limited area in its neighbourhood.

musical echo. One in which the interval between the reception of successive echoes is so small that the impulses received appear to have the quality of a musical tone.

muting. Suppression of an output of electronic equipment unless there is adequate signal/noise ratio. Also **desensitizing.**

muting circuit or switch. Arrangement for attenuating amplifier output unless paralysed by a useful incoming carrier; used with automatic tuning devices.

mutual capacitance. An effect which leads to a change of potential of one conductor producing induced electric charges on other conductors. Quantitatively the mutual capacitance between two bodies m and n is given by the coefficient C_{mn} (see under **capacitance coefficients**).

mutual conductance. Transconductance specifically applied to a thermionic valve. Differential change in a space or anode current divided by differential change of grid potential which causes it. Colloquially termed *slope, slope conductance,* or *goodness* of a valve, measuring the effectiveness of the valve as an amplifier in normal circuits. Expressed in milliamperes per volt, and denoted by g_m.

mutual coupling. See **transformer coupling.**

mutual impedance. See **transfer impedance.**

mutual inductance. Generation of e.m.f. in one system of conductors by a variation of current in another system linked to the first by magnetic flux. The unit is the henry when a rate of change of current of 1 amp/sec. induces an e.m.f. of 1 volt.

mutual inductor. A component consisting of two coils designed to have a definite mutual inductance (fixed or variable) between them.

Mycalex. TN for a fused mixture of ground mica and ground glass which forms an insulating material for high temperatures.

Mylar. TN for plastic foil used for base of magnetic tape, and for thin opaque shield over some scintillation detectors.

myriametric waves. Waves of frequency less than 30 kHz, or of length greater than 10 km.

N

N display. A radar *K display* (q.v.) in which the target produces two breaks on the horizontal time base. Direction is proportional to the relative amplitude of the breaks, and range is indicated by a calibrated control which moves a pedestal signal to coincide with the breaks.

N-electron. One from the N-shell, which, when complete, totals 32 electrons.

N-shell. The fourth (after K, L and M) shell of electrons in an atom, containing 32 ($=2 \times 4^2$) electrons when complete. Transitions among electrons in this shell are associated with the optical spectrum of the atom.

N-terminal pair network. One having N-terminal pairs in which one terminal of each pair may coincide with a node.

N.U. tone. *N*umber-*u*nobtainable tone in automatic telephony.

n-n junction. One between crystals of *n*-type semiconductors, having different electrical properties.

n-p junction. One with electron and hole conductivities on respective sides of the semiconductor junction.

n-p-i-n transistor. Similar to *n-p-n* transistor with a layer of intrinsic semiconductor (germanium or silicon of high purity) between the base and the collector to extend the high-frequency range.

n-p-n transistor. A junction transistor with a thin slice of *p*-type forming the base between two pieces of *n*-type semiconductor, which are the collector and emitter, the conduction being electronic.

n-type semiconductor. One in which the electron conduction (negative) exceeds the hole conduction (absence of electrons), the *donor* impurity predominating.

n-unit. That quantity of fast neutrons which will produce in a Victoreen 100r dosimeter the same reading as 1 röntgen of X-rays.

nabla. See del.

nadir. Point opposite the zenith of the celestial sphere.

NaK. Acronym for sodium (Na) and potassium (K) alloy employed as coolant for liquid-metal reactor. It is molten at room temperature and below.

nand. A logical operator corresponding to *not and*.

nano-. Prefix for a millimicro or one-thousand-millionth (10^9). Symbol *n*.

National Physical Laboratory. National authority for establishing basic units of mass, length, time, resistance, frequency, radio-activity, etc. Now government-controlled.

natural abundance. See abundance.

natural background. Radiation due to natural radioactive minerals and gases, cosmic rays, etc. Background measurements also often include the effects of local contamination and of fall-out.

natural frequency. That of free oscillations in a system.

natural modes. See normal modes.

natural period. Time of one cycle of oscillation arising from free oscillation, depending on inertia and elastance of a system. Reciprocal of *natural frequency*.

natural resonance. Response of system to signal with period equal to its own natural period.

natural uranium. That with its natural isotopic abundance.

natural-uranium reactor. One in which natural, i.e., unenriched, uranium is the chief fissionable material.

navarho. A low-frequency long-range radio-navigation system designed to provide bearing and range information for jet aircraft.

navigation. The art of directing a vessel or aircraft by terrestial or stellar observation, or by radio and radar signals.

navigational beacon. One emitting optical or radio signals for navigational purposes.

navigational planets. Those which can be employed in celestial navigation (Venus, Mars, Jupiter and Saturn).

navigational system. Any system of obtaining bearings and/or ranges for navigational purposes by radio techniques.

near-end cross talk. That occurring in a signal channel propagated in reverse direction to signal in interfering channel.

needle counter. Small diameter GM tube.

needle gap. Spark gap between needle points.

needle scratch. Noise emanating from gramophone, specifically due to irregularities in the contact surface of the groove. Also **surface noise.**

needle talk. Direct sound output from transducer of gramophone pickup.

Neelium. See thermoelectric materials.

negater. Type of inverter which interchanges '1' and '0'.

negative. (1) Designation of electric charge, introduced by Franklin, now known to be exhibited by the electron, which, in moving, forms the normal electric current. (2) Image (as in photography or TV) for which luminance values have been reversed.

negative bias. Static potential, negative with reference to earth, applied to electrode of valve or transistor, usually to prevent collection of free electrons.

negative coupling. See under **positive coupling.**

negative crystal. Birefringent material for which the velocity of the extraordinary ray is greater than that of the ordinary ray.

negative electricity. Phenomenon in a body

when it gives rise to effects associated with excess of electrons.

negative electrode. The *anode* of a primary cell (the electrode by which conventional current returns to the cell) but the *cathode* of a valve or voltmeter (connected to the negative side of the power supply).

negative electron. See **negatron** (2).

negative feedback. Reverse of positive feedback, reducing gain of an amplifier by feeding part of output signal back to input out of phase with incoming signal. Gives more uniform performance, greater stability and reduced distortion. Also called **degeneration, reverse coupling.**

negative glow. In a medium-pressure Crookes tube, the glow between the cathode and Faraday dark space.

negative impedance amplifier. One employing negative impedance units to give power gain.

negative impedance converter. Active network for which a positive impedance connected across one pair of terminals produces a negative impedance across the other pair.

negative ion. Radical, molecule or atom which has become negatively charged through the gain of one or more electrons.

negative ion vacancy. Same as **hole.**

negative lens. Diverging lens.

negative light modulation. See under **positive light modulation.**

negative mutual inductance. See under **positive coupling.**

negative pole. The south-seeking pole of a magnet.

negative proton. Anti-proton.

negative resistance. Effective property of most forms of discharge, in that an increase in voltage accompanies decrease of current, as in a carbon arc. This negative resistance can overcome the positive resistance of a tuned circuit and set up continuous oscillations at the resonant frequency, e.g., the *magnetron* and the *dynatron oscillators.*

negative scanning. Scanning a photographic negative, with reversal in the circuits, so that the reproduced image is the normal positive. This saves time in transmitting televised newsfilm.

negative transconductance. Property of certain valves whereby an increase in positive potential on one electrode accompanies a decrease in current flowing to another electrode.

negatron. (1) A four-electrode thermionic tube for obtaining negative resistance, comprising an anode and grid on one side of a cathode, and an anode on the other. (2) A term sometimes used to distinguish the *negative electron* from the positron.

neodymium (Nd). Rare earth element, at. no. 60, at. wt. 144·24. Neodymium glass is used in solid-state lasers and light amplifiers. The beta-active isotope ^{147}Nd can be extracted carrier-free from fission products.

neon (Ne). Light gaseous inert element recovered from atmosphere. Discovered by Ramsay and Travers in 1898. Gk. *neos* =

new. At. no. 10, at. wt. 20·183, m.p. −248·67°C, b.p. −245·9°C. Historically important in that J. J. Thomson, through his parabolas for charge/mass of particles, found two isotopes in neon, the first non-radioactive isotopes to be recognized. Used in many types of lamp, particularly to start up sodium-vapour discharge lamps. Pure neon was the first gas to be used for high-voltage display lighting, being bright orange in colour. Much used in cold-cathode tubes, reference tubes, Dekatrons.

neon time base. Relaxation oscillator produced by charging capacitor through high resistance and allowing periodic discharge through neon glow tube.

neper. Unit of attenuation, adopted by CCI as of equal status with the decibel. If current I_1 is attenuated to I_2 so that

$$I_2/I_1 = \epsilon^{-N},$$

then N is attenuation (the βl of the Continent). In circuits matched in impedance,

1 neper = 8·686 dB.

After John Napier (Lat. Nepero), Scottish scientist, inventor of natural logarithms.

neptunium (Np). Element no. 93, named after planet Neptune, produced artificially by nuclear reaction between uranium and neutrons. It has principal isotopes 237 and 239, and gives its name to the neptunium $(4n+1)$ radioactive decay series. Only the stable end product ^{209}Bi occurs in nature.

Nernst bridge. An a.c. bridge for high-frequency capacitance measurements.

Nernst effect. Analogous to *Hall effect* (q.v.), a difference of potential arising between opposite edges of a metal strip, due to heat flow along the strip when a transverse magnetic field is applied.

Nernst lamp. One depending on the electric heating of a rod of zirconia in air, giving infrared rays for spectroscopy. The rod must be heated separately to start lamp, as the material is insulating at room temperature.

net transport. The difference between the actual *transport* (2) (q.v.) in an isotope separation plant and that which would be obtained by the same plant with raw material of natural isotopic abundance.

network. (1) Any arrangement of electric components, *active* or *passive*, the former implying an internal source of e.m.f. or current. (2) Circuit element described by the following letters, e.g., *L*, having one shunt and one series element; *C*, one shunt followed by one series element in each leg; *T*, two series and one shunt element from junction; *H*, four series and one shunt element; *π*, one shunt, one series, and one shunt element; *O*, one shunt, one series in each leg and one shunt element; *lattice* (*bridge*), element between each input terminal to both output terminals; *bridged-T*, T-network, with series arms bypassed by fourth element; *twin-T*, two T-networks overlaying each other. All networks have 4 terminals (*quadripole*) in 2-wire line, are *balanced* or *unbalanced* to earth potential, and elements or *branches* may be complex.

See also:

all-pass-	lead-
balanced-	linear-
building-out-	m-derived-
constant-*k*-	mesh-
constant-resistance-	non-dissipative-
delay-	O-
dissipative-	passive-
distorting-	pi-
dividing-	reciprocal-
Doba's-	shaping-
equalizing-	star-
equivalent-	symmetrical-
equivalent T-	T-
filter-	twin T-
ladder-	unbalanced-
lattice-	Y-.

network analyser. See network calculator.

network analysis. Process of calculating theoretically the transfer constant of a network.

network calculator. Combination of resistors, inductors, capacitors, and generators used to simulate electrical characteristics of a power generation system, so that the effects of varying different operating conditions can be studied in computers.

network synthesis. Process of formulating a network with specific electrical requirements.

network transfer function. Mathematical expression giving the ratio of the output of a network to its input. The natural logarithm of its value at any one frequency is the *transfer constant*, and gives the attenuation and phase shift for the signal propagated through the network.

Neumann function. A Bessel function of the second kind.

Neumann principle. Physical properties of a crystal are never of lower symmetry than the symmetry of the external form of the crystal. Consequently tensor properties of a cubic crystal, such as elasticity or conductivity, must have cubic symmetry, and the behaviour of the crystal will be isotropic.

neuroelectricity. That generated in the nervous system of an animal or human being.

neutral. (1) Exhibiting no resultant charge or voltage. (2) Return conductor of a balanced power supply, nearly at earth potential, if without a local earth connection. (3) The normal state of atoms and the universe.

neutral conductor. The one nearest in potential to a *neutral point* (3) (q.v.) in a polyphase power system.

neutral filter. One which attenuates all colours uniformly so that the relative spectral distribution of the energy of the transmitted light is unaltered.

neutral point. (1) In magnetism, a point in the field of a magnet where the earth's magnetic field (usually the horizontal component) is exactly neutralized. (2) In meteorology, any point at which the axis of a wedge of high pressure intersects the axis of a trough of low pressure. Sometimes called a **saddle point.** (3) In a polyphase power system a point at the potential of the junction of a set of equal resistors connected to the supply lines.

neutral relay. In U.S. a non-polarized relay.

neutral state. Said of ferromagnetic material when completely demagnetized. Also **virgin state.**

neutral temperature. That for a thermocouple at which the e.m.f. produced has a turning value, e.g., for Cu-Fe 270°C.

neutral-tongue relay. Relay for radio telegraphy in which the tongue is maintained in a central position by springs, being moved to either of the contacts by a current in the appropriate direction.

neutralization. Counteracting a tendency to oscillation through feedback via anode-grid capacitance. Reversed feedback is provided by a balancing capacitance to which is applied voltage equal and in anti-phase to that on the anode. Also **balancing.**

neutralizing capacitance. See balancing capacitance.

neutralizing voltage. That fed back in the process of neutralization.

neutrino. Subatomic particle emitted simultaneously with emission of positron from atomic nuclei. Its existence was first predicted by Pauli in 1933 to avoid beta decay infringing the laws of conservation of energy and angular momentum. It has very weak interactions with matter and was not observed experimentally until 1956. An apparently different neutrino with similar properties is also associated with the decay of a positive muon. Both neutrinos would appear to have approximately zero mass, no charge, and spin of $\frac{1}{2}\left(\frac{h}{2\pi}\right)$. All reactors emit numbers of neutrinos. Neutrino is often used generically to indicate both neutrinos and anti-neutrinos. See also **anti-neutrino.**

neutrodyne. Obsolete name for various feedback circuits for neutralization.

neutrodyning capacitance. See balancing capacitance.

neutron. Uncharged subatomic particle, mass approximately equal to that of a proton, which enters into the structure of atomic nuclei. Interacts with matter primarily by collisions. In a nuclear reactor, slow (thermal) neutrons have energy of 0·05 eV; fast neutrons of about 1 MeV on fission. Spin quantum number of neutron $= \frac{1}{2}$, rest mass $= 1\cdot008665$ atomic mass unit, charge is zero and magnetic moment is $-1\cdot9125$ nuclear Bohr magnetons. Although stable in nuclei, isolated neutrons decay by beta emission into protons, with a half-life of 11·6 min. Neutrons with the lowest energy distribution are termed *thermal* *neutrons,* those with energies between 10^{-2} and 10^2 eV are known as *epithermal neutrons,* between 10^2 and 10^5 eV as *intermediate neutrons,* and the most energetic as *fast neutrons.*

neutron absorption cross-section. The cross-section for a nuclear reaction initiated by neutrons. This is expressed in *barns.* For many materials this rises to a large value at particular neutron energies due to resonance effects, e.g., a thin sheet of cadmium forms an almost impenetrable barrier to thermal neutrons.

neutron age. See **Fermi age.**

neutron balance. For a constant power level in a reactor there must be a balance between the rate of production of both prompt and delayed neutrons, and their rate of loss due to both absorption and leakage. To assist design calculations two constants are widely used. The neutrons per absorption (symbol η) is the average number of neutrons emitted (including delayed neutrons) per neutron absorbed by the fissile material, and the neutrons per fission (symbol γ) is the average number emitted (including delayed neutrons) per atom undergoing fission. Thus $\gamma = \eta$ multiplied by the probability of fission following absorption.

neutron detection. Observation of charged particle recoils following collisions of neutrons with protons; or of charged particles produced by interaction of neutrons with atomic nuclei., e.g., in a boron-trifluoride *counter.*

neutron energy. (1) The binding energy of a neutron in a nucleus, usually several MeV. (2) The energy of a free neutron which in a reactor will be classed in several groups:

Neutrons	*Energy*
high energy	>10 MeV
fast	10 MeV-20 KeV
intermediate	20 KeV-100 eV
epithermal	100 eV-0·025 eV
thermal or slow	approx. 0·025 eV.

See **multigroup theory.**

neutron excess. The difference between the neutron number and the proton number for a nuclide. Also called **isotopic number.**

neutron flux. Number of neutrons passing through 1 sq cm in any direction in 1 second. Power reactors have above 10^{13}, while zero reactors have of the order of 10^8.

neutron gun. Block of moderating material with a channel used for producing a beam of fast neutrons.

neutron hardening. Increasing the average energy of a beam of neutrons by passing them through a medium which shows preferential absorptions of slow neutrons.

neutron leakage. Escape of neutrons in a reactor from the core containing the fissile material; reduced by using a reflector.

neutron number. The number of neutrons in a nucleus. Equal to the difference between the atomic weight (the total number of nucleons) and the atomic number (the number of protons).

neutron-proton exchange forces. Attempt to explain nuclear forces in terms of an energy contribution arising from the exchange of a charged meson between a neutron and a proton.

neutron radiative capture. The capture of a slow-moving neutron by an atomic nucleus leading to the prompt emission of one or more gamma-rays.

neutron shield. Radiation shield erected to protect personnel from neutron irradiation. In contradistinction to gamma-ray shields, these must be constructed of very light hydrogenous materials which will quickly moderate the neutrons.

neutron source. One giving a high neutron flux (e.g., for chemical activation analysis). Apart from reactors these are chemical sources such as a radium-beryllium mixture emitting neutrons as a result of the (αn) reaction, and accelerator sources in which deuterium nuclei are usually accelerated to strike a tritium-impregnated titanium target, thus releasing neutrons by the (DT) reaction. The former are continuously active but the latter have the advantage of becoming inert as soon as the accelerating voltage is switched off.

neutron spectrometer. Instrument for investigation of energy spectrum of neutrons. See **crystal spectrometer, time-of-flight spectrometer.**

neutron velocity selector. In a simple type, the detector is shielded from neutrons of energy less than 0·3 eV by a cadmium shield. (Cadmium is an almost perfect absorber of thermal neutrons.) More advanced systems employ crystal diffraction or time-of-flight methods. See **neutron spectrometer.**

neutrons per absorption, neutrons per fission. See **neutron balance.**

neutrosonic receiver. Supersonic heterodyne receiver employing neutralized triode amplifiers for intermediate frequency.

newton. The unit of force in the MKS system, being the force required to impart, to a mass of one kilogram, an acceleration of one metre per second per second.

Nichrome. TN for a nickel-chromium alloy largely used for heating resistance elements because of its high specific resistance ($\sim 110 \times 10^{-6}$ ohm-cm), and its ability to withstand high temperatures.

nickel (Ni). Silver-white metallic element, at. no. 28, at. wt. 58·71, m.p. 1450°C, b.p. 3000°C, electrical resistivity at 20°C, 10·9 microhm/cm. Used for structural parts of valves. It is magnetostrictive, showing a decrease in length in an applied magnetic field and, in the form of wire, is much used in computers for small stores, the data circulating and being extracted when required.

nickel-iron (NiFe) **secondary cell.** Alkali accumulator using potassium hydroxide electrolyte. It is lighter and more durable than lead cells and has an e.m.f. of 1·2 V. Also known as **Edison cell.**

Nicol prism. Cemented crystal device used to produce polarized light. Now largely superseded by *Polaroid*.

night error or **effect.** See polarization error.

nile. 1 nile corresponds to a reactivity of 0·01. In indicating reactivity changes it is more usual to use the smaller unit, the *millinile*, equal to a change in reactivity of 10^{-5}.

niobium (Nb). Rare metallic element, at. no. 41, at. wt. 92·906, m.p. 2500°C, used for canning fuel elements in fast breeder reactors, using Na as coolant. Formerly **columbium** (Cb).

Nipkow disk. Rotating disk having a series of apertures arranged in the form of a spiral around the circumference; used for mechanical scanning and in the original Baird TV system.

nit. (1) Unit of luminance, one candela/metre². See **candela.** (2) Unit of information equal to 1·44 bits.

niton. See radon.

nitrogen (N). Gaseous element, colourless and odourless. At. no. 7, at. wt. 14·0067, m.p. −209·86°C, b.p. −195°C. Egyptian *ntr*, Assyrian *nitiru*, Gk. *nitron* = saltpetre. Discovered by D. Rutherford 1772. It was the determination by Rayleigh and Ramsay of its atomic weight that led to the discovery of the inert gases in the atmosphere, argon, etc. Approx. $\frac{4}{5}$ of the normal atmosphere is nitrogen, which is also widely spread in minerals, the sea and in all living matter. At one time 'fixed' from the atmosphere with spinning arcs. A neutral filler in filament lamps, in sealed relays, and in Van de Graaff generators; and in high-voltage cables an insulant. Liquid used as a coolant.

nobelium (No). Transuranic element, at. no. 102; first produced by bombarding ²⁴⁴Cm with ¹³C. Known isotopes have extremely short half-life.

noble gases. See inert gases.

nodal point or **node.** Point in a high-frequency circuit where current is a maximum and voltage a minimum, or vice versa.

nodalizer. One which adjusts to minimum disturbing effects in an electrical circuit, such as hum in an amplifier, or interfering current in a bridge.

node. (1) In acoustics, the location where the interference between two or more progressive sound waves results in a standing wave, with either the sound pressure or particle velocity zero or a minimum. (2) Point of minimum disturbance in a system of waves in tubes, plates or rods. The amplitude cannot become zero, otherwise no power could be transmitted beyond the point. (3) In mathematics, a singular point on a curve, which has the property that two branches of the curve, having distinct tangents, pass through the point. (4) In electrical networks, a terminal common to two or more branches of a network, or to a terminal of any branch of a network. (5) See nodal point.

node voltage. That of some point in an electrical network with reference to a node.

nodon rectifier. An electrolytic device in which ammonium phosphate is the electrolyte, aluminium the cathode and lead the anode.

noise. (1) Generally, unwanted sounds. (2) Interference in a communication channel. (3) Clicks arising from radiation due to sudden changes of current. See also:

carrier-	modulation-
contact-	partition-
cosmic-	photon-
flicker effect	random-
fluctuation-	reactor-
galactic-	reference-
gas-	Schottky-
Gaussian-	shot-
grass	stochastic-
hum	thermal-
induced-	valve-
microphonic-	white-.

noise audiogram. That taken in the presence of a specified masking noise.

noise bandwidth. That of an ideal power gain-frequency response curve (i.e., one with a completely flat response over the bandwidth concerned) with the same gain as the maximum of, and subtending the same area as, the actual response curve of an amplifier. This is given by the integral:

$$\int_0^\infty \frac{P_f}{P_{max}} \, df,$$

where P_f is the power gain as a function of frequency f and P_{max} is the maximum power gain.

noise current. That part of a signal current conveying noise power.

noise diode. One operating as noise generator, under temperature-limited conditions. See **shot noise.**

noise factor or **figure.** Ratio of noise in a linear amplification system to thermal noise over the same frequency band and at the same temperature.

noise field. The EM radiation field due to all sources which can interfere with reception of the required signal.

noise generator. Device for producing a controlled noise signal for test purposes. Also **noise source.** See **noise diode.**

noise intensity. That of the noise field in a specific frequency band.

noise level. That measured on a noise meter relative to a specific reference level, and normally expressed in decibels above reference noise (*dbrn*). See **objective noise meter.**

noise limiter. Device for removing the high peaks in a transmission, thus reducing contribution of clicks to the noise level and eliminating acoustic shocks. Effected by biased diodes or other clipping circuits.

noise meter. See objective noise meter.

noise modulation. That of a carrier wave with a noise signal.

noise power. That dissipated in a system by all noise signals present. See **noise level, Nyquist noise theorem.**

noise pressure. That acting on a transducer due to an acoustic noise field.

noise ratio. Noise power in a terminated filter or circuit divided by the signal power; expressed in decibels. Also **signal/noise ratio, speech/noise ratio.**

noise reduction. Procedure in photographic recording whereby the average density in the negative sound-track is kept as low as possible, and hence in the positive print as high as possible, both consistent with linearity, so that dirt in random electro-static scratches produces less noise.

noise resistance. One for which the thermal noise would equal the actual noise signal present, usually in a specific frequency band.

noise source. (1) The object or system from which the noise originates. (2) See **noise generator.**

noise suppressor. Circuit which suppresses noise between usable channels when these are passed through during tuning. An *automatic gain control* (q.v.) which cuts out weak signals and levels out loud signals.

noise temperature. The temperature at which the thermal noise power of a passive system per unit bandwidth is equal to the noise at the actual terminals. The standard reference temperature for noise measurements is 290°K. At this temperature the available noise power per unit bandwidth (see **Nyquist noise theorem**) is 4×10^{-21} joule.

noise transmission impairment. The transmission loss in dB which would impair intellegibility of a telephone system to the same extent as the existing noise signal.

noise voltage. A noise signal measured in rms volts.

noisy blacks. The non-uniformity of the black areas of a picture due to the level changes arising from noise.

Nomag. TN for non-magnetic, high-resistance cast-iron alloy containing nickel (10-12%) and manganese (5%).

nominal line width. In TV, the reciprocal of the number of lines per unit length in the progression direction.

nominal section. Network which is equivalent to section of transmission line, based on the assumption of lumped constants.

nomogram. Diagram for facilitating calcul-ations, based on a formula or empirical data. It contains parallel lines or curves, so that scale points on these can be linked by a straight line or ruler. Also called **align-ment chart, isopleth, nomograph.**

Nomotron. TN for counter tube, with cold-cathode electrodes and multi-electrodes.

non-bridging. See **bridging.**

non-conductor. Under normal conditions, an electrical insulator in which there are very few free electrons.

non-degenerate gas. That which is in-sufficiently concentrated, e.g., hot-cathode electrons or ordinary gases, so that the Maxwell-Boltzmann law is applicable.

non-directional microphone. Same as **omni-directional microphone.**

non-dissipative network. One designed as if the inductances and capacitances are free from dissipation, and as if constructed with components of minimum loss.

Nonex. TN for brand of glass effective in transmitting ultraviolet light.

non-inductive capacitor. One specially designed for impulse work to have its inherent inductance reduced to a minimum.

non-inductive circuit. One in which effects arising from associated inductances are negligible.

non-inductive resistor. Resistor having a negligible inductance, e.g., comprising a ceramic rod, or a special design of winding of resistance wire.

non-leakage probability. For neutron in reactor, the ratio of the actual multiplication constant to the infinite multiplication constant.

non-linear. Not directly proportional, in contrast to *linear* (q.v.).

non-linear distortion. That in which alien tones are introduced by non-linear response of part of a communicating system.

non-linear distortion factor. Square-root of ratio of the powers associated with alien tones to the powers associated with wanted tones in the output of a non-linear distorting device.

non-linear resistance. Non-proportionality between potential difference and current in an electric component.

non-linear resistor. One which does not 'obey' Ohm's law, in that there is departure from proportionality between voltage and current. In semiconductor crystals, ratio between forward and backward resistance may be 1:1000. See **diode, thyrite, transistor.**

non-linearity. Lack of proportionality between output and input currents and voltages of a network, amplifier, or transmission line, resulting in distortion of a passing signal. Marked non-linearity is a property of diodes, triodes, transistors, etc., used in rectification and demodulation.

non-locking relay or **key.** Key or relay which returns to its unoperated condition when the hand is removed, or the current ceases, usually by the action of a spring which is extended on operation. The key is said to *operate* and *restore* and the relay to *make* and *release*, or *operate* and *de-operate* or *fall-off.*

non-magnetic steel. Steel containing about 12% manganese with no magnetic proper-ties. Stainless steel is almost non-magnetic.

nonode. Nine-electrode thermionic valve.

non-ohmic. See **ohmic.**

non-polarized relay. One in which there is no magnetic polarization. Operation depends on the square of the current in the windings, and is therefore independent of direction, as in telephone and a.c. relays.

non-quantized. See **classical.**

non-reactive power. The root-mean-square value of the active power and of the distor-tion power in an electric system.

non-relativistic. Said of any procedure in which effects arising from relativity theory are absent or can be disregarded, e.g., properties of particles moving with low velocity, e.g., 1/20th that of light propagation.

non-resonant antenna. Same as **aperiodic antenna.**

non-spectral colour. That which is outside the range which contributes to white light, but which affects photocells.

non-specular reflection. Wave reflection of light or sound from rough surfaces, resulting in scattering of wave components, depending on relation between wavelength and dimensions of irregularities; also called **diffuse reflection.**

non-storage tube. Camera tube in which output signal depends on incident light intensity, and not on its integration over a defined period of time between scannings. See **iconoscope.**

non-symmetrical. See **asymmetrical.**

non-synchronous. See **asynchronous.**

non-volatile memory. One in computer which holds data even if power has been disconnected. A magnetic memory is nonvolatile, an ultrasonic one is volatile (it does not comply with the above criterion).

nor. Logical computer circuit with an output only if all inputs are zero.

nor element. An *or element* followed by a *negater*, since the output is 1 when none of the inputs are 1.

normal. (1) Perpendicular. (2) Functioning as designed. (3) In its ground state.

normal distribution. See **Gaussian distribution.**

normal induction. That represented by the curve of normal magnetization.

normal magnetization. Locus of the tips of the magnetic hysteresis loops obtained by varying the limits of the range of alternating magnetization.

normal modes. In a linear system, the least number of independent component oscillations which may be regarded as constituting the free, or natural, oscillations of the system. Also **natural modes.**

normal polarization. That of an electromagnetic wave radiated from a vertical antenna, when measured on the ground at not too great a distance from the transmitter. Electric component is vertical, and magnetic component horizontal.

normal pressure. Standard pressure, usually 760 mm Hg at 45° latitude and sea-level, to which experimental data on gases are referred.

normal state. See **ground state.**

north (-seeking) pole. Of a magnet, the one attracted towards the north. Since unlike poles attract, the earth's magnetic pole in the northern hemisphere is a south-seeking pole. It is actually situated in north Canada. In navigation the north-seeking pole is sometimes termed the **positive pole.**

Norton's theorem. That the source behind two accessible terminals can be regarded as a constant-current generator, the current being the short-circuit current, arising in an infinite impedance source in shunt with an admittance which is that measured between the terminals with no source current. See **Thevenin's theorem.**

not and element. An *and element* followed by a *negater*, i.e., the output is 1 when input signals are 0.

note. Musical sound of constant pitch.

note tuning. Obtaining additional freedom from interference in radio-telegraphic receivers for ICW signals, by tuning postdetector circuits for audio-signals.

Novachord. Electronic musical instrument using a single keyboard, sustaining and swell pedals. Lever stops regulate the waveform and envelope of the current applied to loudspeakers, frequencies of the notes being divided by multivibrators from frequencies generated for highest chromatic octave.

noval. A miniature 9-pin valve, the electrode pins of which pass through the glass base.

nozzle. End of a waveguide, which may be contracted in area.

nuclear Bohr magneton. See **Bohr magneton.**

nuclear breeder. A nuclear reactor in which in each generation there is more fissionable material produced than is used up in fission.

nuclear charge. Positive charge arising in the atomic nucleus because of protons, equal in number to the atomic number.

nuclear conversion ratio. That of the fissile atoms produced to the fissile atoms consumed, in a breeder reactor.

nuclear cross-section. See **cross-section.**

nuclear disintegration. Fission, radioactive decay, internal conversion or isomeric transition. Also *nuclear reactions* (q.v.).

nuclear emission. Emission of gamma-ray or particle from nucleus of atom as distinct from emission associated with orbital phenomena.

nuclear emulsion. Thick photographic coating in which the tracks of various fundamental particles are revealed by development as black traces.

nuclear energy. In principle the binding energy of a system of particles forming an atomic nucleus. More usually the energy released during nuclear reactions involving regrouping of such particles (e.g., fission or fusion processes). The term *atomic energy* is deprecated as it implies rearrangement of atoms rather than of nuclear particles.

nuclear field. Postulated short-range field within a nucleus, which holds protons and neutrons together, possibly in shells.

nuclear fission. Spontaneous or induced splitting of atomic nucleus.

nuclear force. That which keeps neutrons and protons together in a nucleus, differing in nature from electric and magnetic forces, gravitational force being negligible. The force is of short range, is practically independent of charge, and is associated with *mesons.* Yukawa has postulated that the force arises in the interchange of mesons between two nucleons. See **meson** and **short-range forces.**

nuclear fusion. Creation of new nucleus by merging of two lighter ones.

nuclear isomer. A nuclide existing in an excited metastable state. It has a finite half-life after which it returns to the ground state with the emission of a gamma quantum or by internal conversion. Metastable isomers are indicated by adding m to the mass number.

nuclear magnetic resonance. The resonance phenomena arising from the transfer of energy between an RF alternating magnetic field and a nucleus situated in a constant magnetic field. This field must be sufficiently strong to decouple the nuclear spin from the influence of the atomic electrons.

nuclear magnetic resonance spectroscopy. Microwave spectroscopy used to study nuclear paramagnetic resonances.

nuclear magneton. See **Bohr magneton.**

nuclear model. One giving some explanation of certain nuclear properties, e.g., independent model, liquid-drop model, one-particle model, optical model, shell model, strong-coupling model, unified model.

nuclear paramagnetic resonance. Precession of nuclear spins of sample molecules in a strong static transverse magnetic field, when resonance occurs with a weak longitudinal RF magnetic field produced by a microwave signal passed through a coil surrounding the sample. This is observed by a sharp increase in the energy absorbed from the coil. Ultrasonic excitation may replace RF.

nuclear photoeffect. See **photodisintegration.**

nuclear potential. The potential energy of a nuclear particle in the field of a nucleus as determined by the short-range forces acting. Plotted as a function of position it will normally represent some sort of *potential well* (q.v.).

nuclear power. That generated as a result of nuclear reactions, i.e., fission or fusion processes.

nuclear radius. The somewhat indefinite radius of a nucleus within which the density of *nucleons* (protons and neutrons) is experimentally found to be nearly constant. The radius in cm is $1 \cdot 2 \cdot 10^{-13}$ times the cube root of the nuclear number (atomic mass). It is not a precise determinable quantity and is dependent on the type of experiment performed.

nuclear reaction. The interaction of photon or particle with nuclear structure.

nuclear reactor. See **reactor** (2).

nuclear selection rules. See **selection rules.**

nuclear stripping. See **stripping.**

nuclear swelling. See **swelling.**

nuclear weapon. One utilizing the release of nuclear energy.

nucleogenesis. Theoretical process(es) by which nuclei could be created from possible fundamental dense plasma. See **ylem.**

nucleon number. See **mass number.**

nucleonics. The science and technology of nuclear physics and nuclear engineering.

nucleons. Protons and neutrons in a nucleus of an atom.

nucleor. The hypothetical core of a nucleon. It is suggested that this is the same for protons and neutrons. See also **quark.**

nucleus. This is composed of protons (positively-charged) and neutrons (no charge), and constitutes practically all the mass of the atom. Its charge equals the atomic number; its diameter is from 10^{-13} to 10^{-12} cm. With protons equal to the atomic number and neutrons to make up the atomic mass number, the positive charge in the protons is balanced by the same number of electrons orbiting in shells at a distance.

nuclide. Isotope of an element as distinguished by the constitution of the nucleus of its atoms. Including natural and man-made radioisotopes, there are about 1400 nuclides among the known elements. Classified as (a) *natural*, long-life parents ($>10^8$ y) and their decay products; (b) *artificial*, created in a reactor or by bombardment; (c) *naturally-induced*, due to naturally-occurring nuclear reactions, e.g., radiocarbon from cosmic ray neutrons in the atmosphere; (d) those which are now extinct but once existed.

null indicator. Any device, such as head telephones, a CRO or a sensitive galvanometer, for determining zero current, or voltage, in a specified part of an electric circuit, e.g., in a bridge or radio goniometer.

null method. Same as **zero method.**

numeric character. Allowable representation of a specific digit in a particular computing system.

numerical analysis. The use of numerical methods to solve mathematical equations of a form which involves more complex processes or relationships, e.g., integration, by means of trial and error.

numerical selector. In telephony, a group selector which is actuated by the numerical part of a subscriber's number.

Nusselt number. The significant non-dimensional parameter in convective heat loss problems, defined by $Qd/k\Delta\theta$, where Q is the rate of heat loss from a solid body, $\Delta\theta$ is the temperature difference between the body and its surroundings, k is the thermal conductivity of the surrounding fluid, and d is the significant linear dimension of the solid.

nutating feed. That to a radar transmitter which produces an oscillation of the beam without change in the plane of polarization. The resulting radiation field is a *nutation field*.

nutation. The periodic variation of the inclination of the axis of a spinning top (or gyroscope) to the vertical.

nutation field. See **nutating feed.**

Nuvistor. TN for subminiature electron tube.

Nyquist criterion. If for a quadripole the complex transfer ratio is plotted for an infinite range of frequencies, and the point $-1 + j0$ is enclosed, the system is unstable; if excluded, the system is stable. Such a plot is a **Nyquist diagram.**

Nyquist limit (or rate). Maximum rate of transmitting pulse signals through a system. If B is the effective bandwidth in Hz, then $2B$ is the maximum number of code elements (*bands*) per sec which can be received with certainty. $1/2B$ is known as the **Nyquist interval.**

Nyquist noise theorem. One by which the thermal noise power P in a resistor at any frequency can be calculated from Boltzmann's and Planck's constants k and h and the temperature. At normal temperatures it reduces to $P = kTdf$, where T is the temperature in °K and df is the frequency interval.

O

O-electron. One occupying place in O-shell of atom.

O-network. One consisting of impedances connected in O formation, with 2 adjacent junction points connected to the input terminals and the other 2 to the output. The balanced form of the π-network is an O-network.

O.P. process. See **Oppenheimer-Phillips process.**

O-shell. The collections of electrons characterized by the principal quantum number 5, starting with the element rubidium which has one electron in its O-shell.

Oak Ridge. U.S. national nuclear research laboratory.

objective lens. Usually the lens of an optical system nearest the object.

objective noise meter. Sound-level meter in which noise level to be measured operates a microphone, amplifier, and detector, the last-named indicating noise level on the phon scale. The apparatus is previously calibrated with known intensities of the *reference tone*, 1 kHz, suitable weighting networks and an integrating circuit being incorporated in the amplifier to simulate relevant properties of the ear in appreciating noise.

oboe. Distance-measuring system using two ground stations and aircraft transponder.

occlusion. The retention of gases at solid surfaces which have to be released and removed in thermionic low-pressure devices before sealing the envelope.

occulting. Flashing with long flashes separated only by short intervals, used in navigation.

occupational exposure. That to radiation arising from a person's employment.

octal base. International version of **loktal base.**

octantal component of error. That sinusoidal component of the error/tone bearing curve taken over the full 360° range of bearings which shows 8 zeros.

octave. The interval between any two frequencies having the ratio 2:1.

octave analyser. A filter in which the upper cutoff frequency is twice the lower cutoff frequency.

octave filter. Bank of filters for analysing the spectral energy content of complex sounds and noises, using adjacent octave bands of frequency over the whole audio range. One-half and one-third octave filters are also used.

octode. Valve containing cathode, anode, and six intermediate electrodes, used as a frequency changer in superheterodyne receivers. The first three electrodes, starting from the cathode, form the local oscillator

system; the remaining five constitute a pentode system whose emission varies at the frequency of the local oscillator.

odd-even check. Verification of correctness of operation by using only even numbers of impulses, so that presence of an odd number reveals a fault.

odd-even nuclei. Nuclei containing an odd number of protons, and an even number of neutrons.

odd-odd nuclei. Those with an odd number of both protons and neutrons. Very few are stable.

odograph. Navigational instrument plotting ship's track direct on chart.

odometer. Instrument measuring distance travelled. On a ship this would be the *log.*

oersted. Unit of field strength in e.m.u. system, such that 2π oersted is field produced at the centre of a circular conductor, 1 cm in radius, carrying 1 abampere (10 amperes). In the MKSA system

$$1 \text{ oersted} = \frac{1000}{4\pi} \text{ ampere-turns/metre.}$$

off-line. Serial usage of units of a computer with data delay (magnetic or punched tape), for subsequent operations under independent control.

off-normal. Said of a stepping relay when not at normal or home position.

offset. Sustained deviation of the control point from the *index* (or *desired*) *value* in a servo system.

offset tone-arm. In a record reproducer, one which is bent or canted to minimize the effect of the arc which the reproducing stylus normally traverses, the path of the original cutting stylus being radial.

ohm. MKSA unit of resistance, such that 1 ampere through it produces a potential difference of 1 volt. See **acoustic-.**

ohm-cm. The CGS unit of resistivity.

ohm-metre. The MKSA unit of resistivity.

Ohm's law. Law, applicable only to electric components carrying direct current, which states that, in metallic conductors, at a constant temperature and zero magnetic field, the resistance is independent of the current. Formulated by G. S. Ohm in 1827. See **resistance** (2).

ohmic, non-ohmic. A resistor is said to be *ohmic* or *non-ohmic*, according to whether or not its resistance is described by Ohm's law. See **linear resistor.**

ohmic contact. One in which part of p.d. is proportional to current across contact. With metals, this arises from diverging current from contact point, which, if clean, has no resistance at all.

ohmic drop. Voltage drop over part of circuit because of current passing through resistance.

ohmic loss. Power dissipation in a circuit arising from resistance, apart from loss from back e.m.f. in electrolytes or motor, or conversion by transformer. Also **in-phase loss, wattfull loss.**

ohmic resistor. See **linear resistor.**

ohmmeter. Portable apparatus, containing battery, adjustable resistors, etc., for measuring electrical resistance.

oil-immersion objective. The filling of the space between the object lens and the object with an oil of same refractive index as glass. This reduces reflection losses, etc.

oil pump. (1) Roughing piston pump using oil as a seal. (2) Vapour pump with oil of low vapour-pressure.

Olson microphone. The original ribbon microphone, using a battery-excited magnet.

omega-particle. The heaviest (1676 MeV) and most recently detected hyperon (Feb. 1964). Its existence provides strong experimental evidence for the classification system developed from *Lie algebra* (q.v.) (group theory).

omegatron. Device for detailed examination of the residual gas in a vacuum tube.

omnibearing. The bearing of an omni-directional transmitter.

omnidirectional antenna. One receiving or transmitting equally in all directions in horizontal plane, with little effect outside.

omnidirectional microphone. One whose response is essentially independent of the direction of sound incidence. Also called **astatic microphone, non-directional microphone.**

omnidirectional radiator. One which radiates energy uniformly in all directions. Also **isotropic radiator, spherical radiator.**

omnidirectional ranging. A system which gives the bearing of an aircraft from a transmitter independently of the direction in which the aircraft is heading.

on-line. Serial usage of units of computer, without delay or storage between units.

on-line operation. Computing procedure such that results for each stage are available immediately.

on-line processing. Automatic data-ordering from data, as and when received.

on-off control. A simple control system which is either on or off, with no intermediate positions. See **continuous control.**

on-off keying. That in which the output from a source is alternately transmitted and suppressed to form signals.

ondoscope. Glow tube operated by strong microwave radiation fields, and used, e.g., in tuning radar transmitters.

one-group theory. Greatly simplified reactor model in which all neutrons are regarded as having the same energy; cf. *multigroup theory.*

one-many function switch. In computers, one in which one input only is excited at a time

and each input gives a combination of outputs.

one-particle model of a nucleus. A form of *shell model* (q.v.) in which nuclear spin and magnetic moment are regarded as associated with one resident nucleon.

one-shot circuit. Same as **single circuit.**

one-shot multivibrator. Same as **flip-flop** as used in U.K.

opaque. Obstructing the passage of radiation (or particle), usually qualified with respect to the radiation involved.

opaque enema. One of barium sulphate, used in radiography to outline colon.

open account. One of a large number of accounts registered on magnetic tape, all of which are re-recorded with up-dating at regular intervals, e.g., daily.

open circuit. Non-delivery of current from a source then said to be not *loaded.* Constant-current sources, e.g., pentode output valves, must be loaded, or destructive high voltages can result even with normal drive.

open-circuit impedance. Input or driving impedance of line or network when the far end is *free, open-circuited,* not *grounded* or *loaded.*

open cycle. Heat engine operation (or in a coolant system) in which the power fluid is *not* recirculated after passing through the power cycle.

open-diaphragm loudspeaker. Most common domestic type, in which a paper or doped-fabric cone diaphragm is driven by a circular coil at its apex. Mounted in a baffle or box, with or without ports or labyrinth.

open-loop control. One in which degenerate feedback is not employed.

open-loop response. That of open-loop control system or of closed-loop system with feed-back path interrupted for test purposes.

open subroutine. Computer subroutine which can not be re-entered by a jump instruction and must therefore be copied each time used. Cf. *closed subroutine.*

open window unit. Same as **sabin.**

open-wire feeder. One supported from insulators on poles, forming a pole route between antenna and transmitter or receiver.

operand. Quantity of a mathematical type on which an operation is to be performed.

operating angle. See **conduction angle.**

operating characteristic. See **load characteristic.**

operating point. The instantaneous operating conditions of a valve or transistor. See also **quiescent operating point.**

operating power. That of reactor under normal operating conditions.

operating ratio. Ratio of effective operating time to total operating time (including test and maintenance). Also **availability ratio.**

operation. (1) Method of use, e.g., class-A operation, push-pull operation, etc. (2) Specific step in mathematical process.

operation time. (1) That required by computer to complete one specific operation. (2) That which elapses between application of electrode voltages to valve and output current reaching specified value.

operator. Mathematical symbol representing a specific operation to be carried out on a particular operand, e.g., Laplacian operator, vector operator. See **del.**

Oppenheimer-Phillips (O.P.) process. A form of *stripping* (q.v.) in which a deuteron surrenders its neutron to a nucleus without entering it.

opposition. Of periodic quantities, a difference of phase of π.

optic axis. (1) The line which passes through the centre of curvature of a lens' surface, so that the rays are neither reflected nor refracted. Also called **principal axis.** (2) Direction(s) in a doubly refracting crystal for which both the ordinary and extra-ordinary rays are propagated with the same velocity. Only one exists in uniaxial crystals, two in biaxial.

optical maser. One in which the stimulating frequency is visible or infrared radiation; see **laser.**

optical model of a nucleus. One explaining interactions by treating nucleus as sphere of constant refractive index and absorption coefficient.

optical pattern. Light thrown back from the side of the groove of a track cut on a disk, or reproduced in a finished gramophone record, the width being proportional to the lateral recorded velocity of the track. Also **Buchmann-Meyer pattern, Christmas-tree pattern, light-band pattern.**

optical quasar. See **quasar.**

optical range. That which a radio transmitter or beacon would have if it were radiating visible light.

optical recording. Technique of recording sound on a track on photographic film, so that variations in average density and in average area along the track modulate a light beam. After conversion by a photo-cell into a current, this is amplified and acoustically reproduced synchronously with the cinematograph picture.

optical reproducer. That part of a cinematograph projector (*head*) concerned with reproducing optically-recorded sound-tracks, by focusing a slit of light on to the film, receiving the fluctuations in a photocell and so providing voltages for amplification.

optical spectrometer. A spectroscope fitted with a graduated circle for measuring the wavelength of the different colour components emitted from a light source. Analogous to an *acoustic spectrometer.*

optical spectrum. The visible radiation emitted from a source separated into its component frequencies.

optical track. Photographic sound-track on film, composed of *variable area* or *variable density* (qq.v.) optical recording of sound signals.

optical transmission (system). Same as **line-of-sight.**

optimal damping. Adjustment of damping

just short of *critical,* which allows a little overshoot. This attains the ultimate reading of an indicating instrument most rapidly while indicating that it is moving freely.

Optimat. TN of transistor system of process control and programming, whereby adjustments are made to detailed controls in a plant to approach best or set performance.

optimum bunching. That which produces the maximum output power at a given frequency in velocity-modulated electron beams.

optimum programming. Coding programme so that *latency* (delays in getting data from register) is minimized. See **minimum access programme.**

optophone. A photoelectric device for training the blind by converting printed words into sounds.

or. Computer term for logical procedure when *either* of two conditions prevails, according to a criterion in the programme.

or element. One for which output is energized if one or more of the inputs are energized, i.e., output signal is 1 when one or more input signals are 1.

orbit-shift coils. These placed on the magnet pole faces of a betatron or synchrotron in the region of the stable orbit so that by passing a current pulse the particles may be momentarily displaced to strike a target placed outside the stable orbit.

orbital. Motion of an electron in a field of force as described by a *wave function,* which expresses the probability of finding the electron in a region, there being one orbital for each quantized orbit.

orbital quantum number. The second quantum number for an orbital electron, indicating the angular momentum associated with the orbit. 0 for *s-states;* 1 for *p-states;* 2 for *d-states* and 3 for *f-states.* See also **principal quantum number.**

Orbitron. TN for large apparatus containing source for the radiotherapy treatment of cancer (surface). The shield, with the requisite window, is carried round a large hollow wheel, the patient being on a stretcher which can be inserted in the wheel. Rotation minimizes the dose elsewhere than in required region.

order. An instruction in a computer programme.

ordinary ray. See **double refraction.**

organic phosphor. Organic chemical used as solid or liquid scintillator in radiation detection.

organic reactor. A reactor in which organic compounds are used as moderator and coolant.

orientation. (1) Alignment (as of nuclear spin axes, etc.). (2) Direction in space. Solids with preferred orientation usually show directional properties.

origin distortion. Distortion of waveform displayed by gas-focused cathode-ray tube employing electrostatic deflection, due to non-linear relation between angular deflection and deflecting voltage at low values of the latter. This results in flattening a waveform where it crosses zero line.

original. Master disk recording.

orthicon. Camera tube in which the external image is focused on to a mosaic, which is scanned from behind by a low-power electron beam; output signal is taken from reaction of this. See **image orthicon**.

ortho-. Prefix indicating parallel alignment of nuclear spins, e.g., in ortho-positronium. Cf. *para-*.

orthogonal network. See matrix (3).

ortho-hydrogen. A hydrogen molecule in which the two nuclear spins are parallel, forming a triplet state.

oscillating capacitor. See **vibrating capacitor**.

oscillation. Sustained and very stable alternation of current in a tuned circuit (inductance and capacitance, or quarter-wave stub). Maintained by synchronous pulses of energy from a valve, to compensate for energy lost through dissipation and output.

oscillation constant. Square root of product of inductance (henry) and capacitance (farad) of a resonant circuit.

oscillation frequency. This is determined by the balance between the inertia reactance and the elastic reactance of a system, e.g., open or short-circuited transmission line, cavity, resonant circuit, quartz crystal.

If

C = capacitance
L = self-inductance of circuit,

then frequency $f = 1/(2\pi\sqrt{LC})$.

In a mechanical oscillating system,

$$f = 2\pi\sqrt{M/S},$$

where M = mass
S = restoring force/unit displacement.

oscillation transformer. Obsolete term for high-frequency transformer for coupling antenna to closed resonant circuit.

oscillator. A source of alternating current of any frequency, which is sustained in a circuit by a valve or transistor, using positive feedback principle. There are two types: (1) *stable-type*, in which frequency is determined by a line or a tuned (LC) circuit, waveform being substantially sinusoidal, and (2) *relaxation-type*, in which frequency is determined by resistors and capacitors, waveform having considerable content of harmonics. Also applies to mechanical systems, velocities being equivalent to currents. See also:

Armstrong-	dynatron-
autodyne-	electron-coupled-
Barkhausen-Kurz-	framing-
blocking-	ganging-
bridge-	Gill-Morrell-
coherent-	Hartley-
Colpitts-	Heil-
constant-frequency-	Hertzian-
coupled-	heterodyne-
crystal-	intermediate-
double-frequency-	frequency-

line-	squegging-
linear time-base-	synchronization of-
local-	thermionic-
magnetron-	transitron-
master-	tuned-
Pierce-	tuned-anode-
preset-	tuned-base-
push-pull-	tuned-collector-
quenching-	tuned-emitter-
reactor-	tuned-grid-
relaxation-	ultra-audion-
resistance-	Van de Pol-
capacitance-	velocity-
ring-	modulated-
self-excited-	Wien-bridge-.

oscillator crystal. A piezoelectric crystal used in an oscillator to control the frequency of oscillation.

oscillator drift. See **frequency drift**.

oscillatory discharge. That of capacitor through inductor when the resistance of circuit is sufficiently low and current persists after the capacitor has completely discharged, so that it charges again in the reverse direction. This process is repeated until all initial energy is dissipated in resistance, including radiation.

oscillatory scanning. That in which the scanning spot moves repeatedly to and fro across the image, so that successive lines are scanned in opposite directions.

oscillatron. Normal CRT for displaying or registering waveforms by deflecting a beam of electrons.

oscillogram. Record of a waveform obtained from any oscillograph. Usually photograph of CRT display.

oscillograph. In U.S. and U.K., an oscilloscope equipment, with addition of a photographic recording system to register waveforms displayed. In France, equipment for direct recording of a trace by the beam on a photographic plate in vacuum.

oscilloscope. (1) Equipment incorporating a cathode-ray tube, time-base generators, triggers, etc., for the delineation of a wide range of waveforms by electron beam. (2) Mechanical or optical equipment with a corresponding function, e.g., Duddell oscilloscope.

osmium (Os). Densest metal, at. no. 76, at. wt. 190·2, m.p. 2700°C. Alloyed with iridium, it forms an extremely hard material.

osophone. Headphone for deaf persons employing bone conduction of sound.

out of phase. See under **phase**.

outer marker beacon. A marker beacon, associated with the ILS or with the standard approach system, which defines the first predetermined point during a beam approach.

outer product. See **vector product**.

outgassing. Removal of maximum amount of residual gas in a valve envelope by baking the whole valve before sealing.

output. (1) Data either in coded or printed form after processing by an electronic computer. (2) Vehicle, punched tape or printed page, by which processed data leaves a computer. (3) Audio, electric or mechanical signal delivered by instrument or system to a load.

output gap. An interaction gap through which an output signal can be withdrawn from an electron beam.

output impedance. That presented by the device to the load, and which determines *regulation* (voltage drop) of source when current is taken. In a linear source, the backward impedance when the e.m.f. is reduced to zero. Also **source impedance**; see **Thevenin's theorem**.

output meter. That which measures output voltage of an oscillator, amplifier or line. Calibrated in *volts* or *power level* in dB in relation to *zero power level* (one milliwatt) when circuit is properly terminated. See **volume unit**.

output noise. See **thermal noise**.

output regulation. Of a power supply, the variation of voltage with load current.

output transformer. One which couples last stage in a valve amplifier to the load, e.g., a loudspeaker or line.

output valve. One designed for delivering power to a load, e.g., line or loudspeakers, voltage gain not being relevant. Final stage of any multivalve amplifier. Also called **power valve**.

output winding. That from which power is withdrawn in a transformer, transductor or magnetic amplifier.

overall efficiency. Ratio of useful output power to total input power.

overall luminous efficiency. Ratio of luminous flux of lamp to total energy input. Not to be confused with **luminous efficiency**.

overall merit. System of rating a communication channel (especially radio broadcasting) on scale 0–5, derived from signal strength, fading, interference, modulation depth, distortion.

overbunching. Bunching condition in a velocity-modulated electron beam when the bunching process is increased beyond the optimum condition.

overcompression. A technique used in the operation of expansion cloud chambers to reduce the waiting time between expansions.

overcoupling. Said of two electrical circuits or mechanical systems tuned to the same frequency when there is sufficient interaction between them for the frequency response curve of the system to show two maxima, displaced to opposite sides of the maximum for either circuit alone. Also **close coupling**.

overcurrent protection. That by relay operated by excessive load current.

overcutting. Too great amplitude of radial motion in cutting the original track on the wax blank or direct recording disk, so that one track cuts into the adjacent track.

overdamping. Damping in excess of *critical damping* (q.v.). An overdamped system

shows a slow non-oscillatory return to equilibrium following any disturbance.

overflow. (1) Coded data which cannot be handled immediately and which is relegated temporarily to a *store* until it can be dealt with. (2) Numbers out of the capacity of a computer, which stop operations and raise an alarm. See **underflow**.

overlag. Electronic mixing of foreground and background images.

overlap changeover. Disk recordings made in such a way that there is about 30 sec of overlap, during which they can be aligned for reproduction, and a changeover effected without interruption.

overload capacity. Excess capacity of a generator over that of its *rating*, generally for a specified time.

overmodulation. Attempted modulation to depth exceeding 100 per cent., i.e., to such a degree that amplitude falls to zero for an appreciable fraction of the modulating cycle, with marked distortion.

override. Said of any electronic controlling equipment, such as autopilot (*George*) or throttle control when, under ordinary or fault conditions, its operation can be superseded by manual control.

overscanning. Deflection of electron beam beyond the phosphor in a television reproducer (kinescope) or a cathode-ray oscilloscope.

overshoot. (1) Extent to which a servo system carries the controlled variable past its final equilibrium position. (2) For a step change in signal amplitude, *undershoot* and *overshoot* are the maximum transient signal excursions outside range from initial to final mean amplitude levels.

overswing diode. See **backwash diode**.

overtone. Of a sound wave, any harmonic other than the first (the *fundamental*).

overtone crystal. A piezoelectric crystal operating at a higher frequency than the fundamental for a given mode of vibration.

overvoltage. The amount by which the operating voltage on a Geiger counter exceeds the threshold value.

Owen bridge. An a.c. bridge of the four-arm (or Wheatstone) type used for the measurement of inductance.

oxide. A binary compound with oxygen.

oxide-coated cathode. One coated with oxides of the alkali and alkaline-earth metals, to produce thermionic emission at relatively low temperatures.

oxide-coated filament. Incandescent filament, coated with oxides of alkali metals, which acts as an oxide-coated cathode.

oxygen (O). Odourless gaseous element, at. no. 8, at. wt. 15·9994, m.p. −218·8°C, b.p. −182·970°C. It is chemically very active and forms one-fifth of the atmosphere.

oxygen effect. That biological tissues are more sensitive to radiation in the presence of oxygen.

P

P display. Display obtained with a *plan-position indicator* unit. A map display produced by intensity modulation of a rotating radial sweep.

P-electron. One in the P-shell surrounding a nucleus. There is 1 such electron in the caesium atom and 10 in thorium, but the shell is not completed in existing elements.

P-gas. One based on argon and used for gas-flow counting.

P-shell. The outermost electron shell for most of the heavy stable elements. Electrons have a principal quantum number six.

p-levels. See **principal series**.

p-state. That of an orbital electron when the orbit has angular momentum of one Bohr unit.

p-i-n diode. A semiconductor *p-n* diode with a layer of an intrinsic semiconductor incorporated between the *p* and *n* junctions.

p-n boundary. The surface on the transition region between *p*-type and *n*-type semiconductor material at which the donor and acceptor concentrations are equal.

p-n junction. Boundary between *p*- and *n*-type semiconductor, having marked rectifying characteristics; used in diodes, photocells, transistors, etc.

p-n-i-p transistor. Similar to *p-n-p* type with a layer of an intrinsic semiconductor (germanium of high purity), between the base and the collector to extend the high-frequency range.

p-n-p transistor. A junction transistor in which a thin slice of *n*-type is sandwiched between slices of *p*-type, and amplification arises from hole conduction.

p-p junction. One between *p*-type crystals having different electrical properties.

p-type conduction. That arising in a semiconductor containing acceptor impurities, with conduction by (positive) holes.

p-type conductivity. That apparently resulting from movement of positive charges; actually *holes* in impurity component in a semiconductor.

p-type semiconductor. One in which the hole conduction (absence of electrons) exceeds the electron conduction, the acceptor impurity predominating.

packaged. Said of an equipment complete for use, particularly of a magnetron with magnet and coupling unit; also of circuit units having their components sealed or encapsulated together; also of a reactor of limited power which can be packaged and erected easily on a remote site.

packed tower. A vertical tower used in distillation, etc., which contains various types of packing to increase the contacting area between the phases.

packing fraction. $(M-A)/A$ for a nuclide, where M is its mass in amu and A is the mass number. $M-A$ is variously known as the *mass defect* and the *mass decrement*.

pad. (1) Small preset adjustable capacitor, to regulate the exact frequency of oscillation of an oscillator, or a tuned circuit in an amplifier or filter; also **padder, trimmer**. (2) Fixed attenuator inserted in a waveguide to obviate reflections.

padder. See **pad** (1).

painting. Colloq., sweeping a narrow beam across a target.

pair. Two electrons forming a non-polar valency bond between atoms.

pair production. Creation of a *positron* and an *electron* when a *photon* passes into the electric field of an atom. See **Compton effect, photoelectric effect**.

pair-production absorption. The absorption of gamma-rays or other photons in the process of pair production.

paired cable. One in which multiple conductors are arranged as twisted pairs but not quadded.

pairing. Deviation from exactness of interlacing of horizontal lines in a reproduced television image, reducing vertical definition. Also **twinning**.

Pal. Colour TV system developed in Germany from the NTSC system, and adopted for general use in Western Europe (except in France).

palladium (Pd). Noble metal, at. no. 46, at. wt. 106·4, sp. gr. 11·40, m.p. 1549°C.

pan. *Panoramic* motion of a camera.

pancake coil. Inductor in which the conductor is wound spirally. Much used in early radio, lately as part of a printed circuit.

panoramic receiver. One in which tuning sweeps over wide ranges, with synchronized display on a CRT of output signals.

paper capacitor. One which has thin paper as the dielectric separating aluminium foil electrodes, these being wound together and waxed. See **self-sealing capacitor**.

paper tape. See **punched tape**.

para-. Prefix indicating anti-parallel alignment of nuclear spins. Cf. *ortho-*.

parabolic microphone. One provided with a parabolic reflector to give enhanced directivity for high audio frequencies and hence greater range for wanted sounds amid ambient noise.

parabolic reflector. One shaped as a paraboloid of revolution. Theoretically produces a perfectly parallel beam of radiation if a source is placed at the focus (or vice versa). Colloq. **dish**.

parafeed coupling. Combination of resistance-capacitance with intervalve-transformer coupling so that anode-feed through a resistance is diverted from the transformer, which can then be designed with reduced dimensions and cost.

para-hydrogen. A hydrogen molecule in which the two nuclear spins are anti-parallel, forming a singlet state.

parallel arithmetic unit. One in which the digits are operated on simultaneously.

parallel circuit. See **shunt circuit.**

parallel computer. One employing parallel arithmetic.

parallel feed. Connection between anode-supply and anode through a high resistance or inductance, while a.c. from the anode is delivered through a capacitance, so that a.c. and d.c. anode currents are separated. **RC** or **LC coupling**, if a resistive or inductive anode load respectively; also **shunt feed.**

parallel memory. One designed to give as nearly as possible uniform access time.

parallel-plate capacitor. Capacitance is

$$C = kA/11.3d \text{ picofarads,}$$

where A = area of plates (cm²)

d = separation (cm)

k = dielectric constant of medium.

If the change in charge is Q in coulombs, and the consequent change in electric potential difference is V (volts), then capacitance C in farads is given by $C = Q/V$.

parallel-plate chamber. Ionization chamber with plane-parallel electrodes.

parallel-plate lens. In aerial systems one constructed of thin parallel conducting plates.

parallel-plate waveguide. One formed by two parallel conducting or dielectric planes. Often realized in atmospheric propagation under suitable meteorological conditions. See **surface duct.**

parallel resonance. See **shunt resonance.**

parallel tracking. Mechanism whereby the pickup arm is kept exactly tangential to the groove during reproduction from a disk record.

parallel-wire line. Form of transmission line widely used at low frequencies where radiation losses will be small. Cf. *coaxial cable.*

parallel-wire resonator. See **Lecher wires.**

paralysis time. The time for which a radiation detector is rendered inoperative by an electronic switch in a control circuit.

paramagnetic resonance. See **nuclear magnetic resonance.**

paramagnetism. Phenomenon involving a material whose relative permeability is slightly greater than unity. Used to attain very low temperatures by adiabatic demagnetization.

parameter. (1) Variable which, for a particular purpose, combines other variables in computers. (2) Arbitrary constant, which has a particular value in specified circumstances in physics. (3) Variable which can take the place of one or more other variables in mathematics.

parametric amplifier. Circuit in which energy is fed from pumping oscillator into signal through a varying reactance (e.g., a germanium diode), made to act as a negative resistance by the pump signal. Also **mavar** (see Abbreviations).

parametric diode. One in which the series capacitance can be varied by a biasing voltage. Can be a valve or solid-state device.

parametric resonance. The condition of a parametric amplifier when energy transfer from the pump circuit through any resonant cavities to the signal is a maximum.

parametron. Device for digital computers, using parametric oscillation.

paraphase amplifier. Push-pull stage incorporating *paraphase coupling.* Also **see-saw amplifier.**

paraphase coupling. A push-pull stage, or series of stages, in which reversed phase is obtained by taking a fraction of the output voltage of the first valve of the amplifier and applying it to a similar balancing first valve, which operates succeeding stages in normal push-pull, the number of transformers being minimized thereby.

parasitic antenna or **aerial.** Unfed dipole element, which acts as a *director* or *reflector.*

parasitic capture. Neutron capture in reactor not followed by fission.

parasitic oscillation. Unwanted oscillation of an amplifier, or oscillation of an oscillator at some frequency other than that of the main resonant circuit. Generally of high frequency, it may occur during a portion of each cycle of the main oscillation. Also **spurious oscillation.**

parasitic stopper. Components which attenuate a feedback path which otherwise maintains unwanted oscillations.

parent. In radioactive particle decay of A into B, A is the parent and B the daughter product.

parent peak. The component of a mass spectrum resulting from the undissociated molecule.

parity. The conservation of parity or space-reflection symmetry (mirror symmetry) states that no fundamental difference can be made between right and left, or in other words the laws of physics are identical in a right- or in a left-handed system of co-ordinates. The law is obeyed for all phenomena described by classical physics, but was recently shown to be violated by the weak interactions between elementary particles.

parity check. Use of a code in which totality of bits must be either even or odd.

parker. Proposed name of unit for **rem** and **rep.**

parscope. Oscilloscope display for weather radars.

parsec. Astronomical unit of length. 3·26 light-years.

partial. A pure tone component of a sound.

partial capacitance. Where there are a number of conductors (including earth), there is partial capacitance between all pairs, the

partial effective capacitance under specified conditions being calculated from this network. See **capacitance coefficients**.

partial compatibility. System of TV transmission for which both monochrome and colour reception are possible, but in which the chrominance information is outside the black-and-white video band of frequencies.

particle. Volume of air or fluid which has dimensions very small in comparison with the wavelength of a propagating sound wave, but large in comparison with molecular dimensions. See also:

alpha-	ionizing-
beta-	recoil-
delta-	relativistic-
elementary-	V-.

particle physics. Study of the properties of elementary particles, some of which have very high energy and very brief existence. See **Summary of atomic physics** in Appendices, also:

anti-baryon	lambda-particle
anti-lepton	lepton
anti-neutrino	meson
anti-neutron	mu-meson
anti-particle	muon
anti-proton	neutrino
baryon	neutron
boson	nucleons
cascade particle	omega-particle
electron	positron
fermion	positronium
graviton	proton
hyperon	quark
intermediate	sigma-particle
charged vector	Xi particle.
boson	

particle scattering. See **scattering**.

particle velocity. In a progressive or standing sound wave, the longitudinal alternating velocity of the air or fluid particles, taken either as the maximum or as rms velocity.

particulate. Having all or some of the properties of a particle or system of particles.

partition noise. That arising when electrons are abstracted from a stream by a number of successive electrodes, as in a travelling-wave tube.

party line. One which connects two or more subscribers' installations to an exchange.

Paschen-Back effect. Extension of *Zeeman effect* (q.v.) obtained with stronger fields which decouple the nuclear and orbital angular momenta. It leads to greater complexity of hyperfine structure.

Paschen series. A series of lines in the infra-red spectrum of hydrogen. Their wave numbers are given by the same expression as that for the Brackett series but with $n_1 = 3$.

Paschen's law. That breakdown voltage, at constant temperature, is a function only of the product of the gas pressure and the distance between parallel-plate electrodes.

pass band. See **filter transmission band**.

passive. Said of transducers or filter sections not containing an effective source e.m.f.

passive network. One of electrical elements in which there is no e.m.f. or other source of energy.

passive radar. That using microwaves or infrared radiation emitted from source and hence not revealing the presence or position of the detecting system. Military use.

passive transducer. One in which output power is obtained solely from received power, as in networks or loudspeaker.

patch. To join together units of apparatus, such as amplifiers, equalizers, etc., by flexible cords terminated on plugs, which are inserted into break jacks, bridged across terminations of each unit.

patch bay. Section of rack-mounted equipment which includes all break jacks which terminate units of equipment.

patch board. One on which problems can be set up and modified as required in analogue computing. See **patch**.

patch in or out. Temporary insertion of spare apparatus, or removal of defective apparatus, in a circuit by patch cords, usually in a patch bay or field.

path. Channel through which signals can be sent, particularly *forward*, *through* and *feedback* paths of servo systems.

Pauli exclusion principle. Fundamental law of quantum mechanics that no two fermions can exist in identical quantum states.

Pawsey stub. Quarter-wavelength coaxial line, added in parallel to end of coaxial line, to match this to a dipole element.

pay-as-you-view. See **toll television**.

peak. (1) The highest instantaneous value of any varying quantity. (2) The magnitude of a local maximum.

peak clipper. Diode or similar device with sharp cutoff at pre-determined level and no effect below this. Used, e.g., in ignition interference suppression and for removing unwanted amplitude modulation from signal employing pulse-width modulation.

peak dose. Maximum absorbed radiation dose at any point in an irradiated body, usually at a small depth below the surface, due to secondary radiation effects.

peak factor. Ratio of peak value of any alternating current or voltage to rms value. Also **crest factor**.

peak inverse voltage. (1) Maximum instantaneous voltage opposite to that which results in conduction, arising across diode rectifiers. (2) See **inverse voltage**.

peak limiter. Circuit for avoiding overload of a system by reducing gain when the peak input signal reaches a certain value.

peak power. Average radio-frequency power at maximum modulation, i.e., of envelope of transmission.

peak programme meter. One bridged across a transmission circuit to indicate the changes in *volume* of the ultimate reproduction of sound, averaging peaks over 1/1000 sec.

peak-to-peak amplitude. See **double amplitude.**

peak voltmeter. One for measuring the peak value of an alternating voltage, e.g., a diode which is biased so that it just conducts. Also called **crest voltmeter.**

peak white. The level in a TV signal corresponding to white.

peaking. Inclusion of series or shunt resonant units in TV circuits, to maintain response up to a maximum frequency.

peaking transformer. One in which the core is highly saturated by current in the primary, thus providing a peaky e.m.f. in the secondary as the flux in the core suddenly changes over.

pebble-bed reactor. One in which the fuel is in the form of pellets stored in a chamber through which the coolant flows. Reactivity is controlled by varying this rate of flow.

PEC (or pec) amplifier. See **photocell amplifier.**

pedestal. Level between peak of the synchronizing signals and minimum of video signal, in the total television signal.

pedestal sit. Colloquialism for separation in TV picture-signal level between black level (in a positive signal) and blanking level.

Peep. TN for electronic air navigation device which projects flight data on to the pilot's windscreen or other screen at eye level. (*Pilot's electronic eye-level presentation.*)

pellicle. A strippable photographic emulsion used to form a *stack* in nuclear emulsion techniques.

Peltier coefficient. Energy absorbed or given out per second, due to the Peltier effect, when unit current is passed through a junction of two dissimilar metals.

Peltier effect. Phenomenon whereby heat is liberated or absorbed at a junction when current passes from one metal to another.

penetrating shower. Cosmic-ray shower containing mesons and/or other penetrating particles. See **cascade shower.**

penetration. Measure of depth of *skin effect* of eddy currents in induction heating, or depth of magnetic field in superconducting metals. Usually to $1/e$ (0·37) of surface value.

penetration depth. That thickness of hollow conductor of the same dimension which, if the current were uniformly distributed throughout the cross-section, would have the same effective resistance as the solid conductor.

penetration factor. (1) The reciprocal of the *amplification factor* is given by the ratio of change of grid voltage to change of anode voltage. (2) Probability of incident particle passing through nuclear potential barrier.

penetration frequency. See **frequency of penetration.**

penetrometer. A device for the measurement of the penetrating power of radiation by the comparison of the transmission through various absorbers.

pentagrid. A frequency-converting tube for supersonic heterodyne receivers; cathode and first two grids form an oscillator and

the modulated electron stream is mixed with incoming signal by the other grids and anode acting as a pentode. Also **heptode.**

pentane lamp. An early photometric standard lamp. Also **Vernon-Harcourt lamp.**

pentode valve. Five-electrode thermionic tube, comprising an emitting cathode, control grid, a screen (or auxiliary grid) maintained at a positive potential with reference to that of the cathode, a suppressor grid maintained at about cathode potential, and an anode. It has characteristics similar to those of a screened-grid valve, except that secondary emission effects are suppressed. See **beam power valve, diode, screened** and **transistor pentodes.**

percentage articulation. That of elementary speech-sounds received correctly when logatoms are called over a telephone circuit in the standard manner.

percentage hearing loss. See **hearing loss.**

percentage modulation. *Depth of modulation* expressed as a percentage, up to 100%.

perfect dielectric. One in which all the energy required to establish an electric field is returned when the field is removed. In practice, only a vacuum conforms, other dielectrics dissipating heat to varying extent.

perhapsatron. Toroidal Pyrex tube for studying pinch instabilities in high-energy plasma. The latter was produced by coupled transformer, excited by one turn taking discharge current from a 38 μF capacitor charged to 15 kV.

perikon detector. Early crystal detector comprising a point contact between crystals of zincite and bornite.

period. (1) Time taken for one complete cycle of an alternating quantity. Reciprocal of *frequency*. (2) The smallest value of the increment of the independent variable in a periodic quantity for which the function repeats itself; see **periodic quantity.** (3) In a reactor, time in which the neutron flux changes by a factor of *e*. (4) Of a radioactive isotope, synonym for U.K. **mean life**, U.S. **half-life.**

period meter. Instrument for measurement of reactor period.

period range. See **start-up procedure.**

periodic antenna. One depending on resonance in its elements, thereby presenting a periodic change in input impedance as the frequency of the drive is varied.

periodic chain. The arrangement of the periodic system in the form of a simple vertical list of elements in order of increasing atomic numbers.

periodic current. An oscillating current whose values recur at equal intervals of time.

periodic damping. That of a system which shows an oscillatory response to any disturbance. See also **underdamping.**

periodic quantity. Regularly varying quantity y whose values recur at equal increments of the independent variable (x), e.g., $y = f(x) = f(x+m) = f(x+2m)$, etc., where m is a constant. Also **pulsating quantity.**

periodic rating. Rating of a component for continuous use with a specified periodically varying load.

periodic system. Classification of chemical elements into periods (corresponding to the filling of successive electron shells) and groups (corresponding to the number of valence electrons).

periodic table. The most common arrangement of the periodic system.

periodicity. (1) Same as **frequency.** (2) Location of element in periodic table.

perisphere. Theoretical sphere surrounding body in which significant electromagnetic forces occur.

peritron. Cathode-ray tube with facility for axial displacement of screen—used when three-dimensional display is necessary.

Permachon. TN for type of storage tube.

permalloy. Group name for class of high permeability nickel-iron alloys. Also another group of molybdenum permalloy materials, e.g., Mo 4%, Ni 79%, Fe 17%.

permanent magnet. Ferromagnetic body which retains an appreciable magnetization after excitation (electric currents, or stroking with another magnet) has been removed. Used on a large scale in loudspeakers, relays, small motors, magnetrons, etc.

permanent memory. Computer storage unit which is not cleared when power is interrupted. Principally magnetic core, drum or tape.

permatron. Hot-cathode gas-discharge diode, gated by applied magnetic field.

permeability. *Absolutely*, ratio of magnetic flux density produced in a medium to magnetizing force producing it. *Relatively*, the ratio of magnetic flux density produced in a medium to that which would be produced in a vacuum by the same magnetizing force. With CGS electromagnetic units, absolute and relative permeabilities are the same. In MKSA units, the former is equal to the product of the latter, and the permeability of free space ($\mu_0 = 4\pi \times 10^{-7}$H/m). Symbol μ. See **differential-, incremental-, initial-.**

permeability bridge. Device for measuring magnetic properties of a sample of magnetic material, fluxes in different branches of a divided magnetic circuit being balanced against each other.

permeability tuning. Adjusting a tuned circuit by varying inductance of a coil (by altering position on axis of a sintered iron core, or by sliding a copper spade over the winding).

permeameter. Instrument for measuring static magnetic properties of ferromagnetic sample, in terms of magnetizing force and consequent magnetic flux.

permeance. Reciprocal of the *reluctance* of a magnetic circuit.

Permendur. TN for alloy of equal parts of iron and cobalt, characterized by high permeability and very high saturation flux density, *ca.* 23,500 gauss.

Perminvar. TN for alloys of ferromagnetic

elements with low hysteresis and high permeability.

permissible dose. See **maximum permissible dose.**

permittivity. *Absolutely*, ratio of electric displacement in a medium to electric field intensity producing it. *Relatively*, ratio of electric displacement in a medium to that which would be produced in free space by the same field. With CGS electrostatic units absolute and relative permittivities are the same. In MKSA units the former is equal to product of the latter and permittivity of free space

$$(8{\cdot}855 \times 10^{-12} \, Fm^{-1}).$$

Symbol k (or ϵ). *Relative permittivity* is also termed **dielectric constant** or **specific inductive capacity.**

persistence. See **afterglow.**

persistence characteristic. Graph showing decay with time of the luminous emission of a phosphor, after excitation is cut off.

persistence of vision. Brief retention of image on retina of eye after optical excitation has ended. An essential factor in cinematography and television.

persistor. Memory element consisting of bimetallic loop operating in cryostat at superconducting temperature.

persistron. Electroluminescent photoconducting device, which gives a steady display on being impulsed.

personal dosimeter. Sensitive tubular electroscope, using a metallized quartz fibre which is viewed against a scale calibrated in röntgens; it is charged by a generator and radiation discharges it.

personnel monitoring. Monitoring for radioactive contamination of any part of an individual, his breath, excretions, etc.

persuador. The electrode in the image orthicon camera tube which deflects electrons returned from the target into the electron multiplier.

perturbation theory. Mathematical method involving solution of simplified equation representing behaviour of system. This is substituted in more exact equation, including second order terms, so as to determine the effect of these second order perturbations. Widely used in quantum mechanics and in allowing for localized neutron flux variations in large reactors.

perveance. The constant G in Langmuir's equation ($i = GV^{3/2}$) for the anode current i of a space charge limited diode operating with anode voltage V. Its unit is the *perv.*

petoscope. Development of *aniseikon* (q.v.) for detecting movement in a wide angle of view, especially in outdoor operations.

Pfund series. A series of lines in the far infrared spectrum of hydrogen. Their wave numbers are given by the same expression as that for the Brackett series but with $n_1 = 5$.

pH value. Concept used as a measure of the acidity of a solution; given by $pH = \log_{10}(1/[H^+])$, where $[H^+]$ is the hydrogen ion concentration. A pH below 7 indicates acidity, and one above 7 indicates alkalinity.

phanatron. Large hot-cathode gas diode for industrial use.

phantastron. Valve circuit for generating pulses which are delayed from trigger pulses.

phantom. See ghost (2).

phantom antenna. Same as **artificial antenna.**

phantom circuit. An additional circuit using a pair of transmission lines in parallel for the *go* and another similar pair for *return*, being fed at the midpoint of bridging transformers.

phantom material. Material which produces absorption and back scatter of radiation very similar to human tissue, and hence used in models to study appropriate doses, radiation scattering, etc. A *phantom* is a reproduction of (part of) the body in this material. Also termed **tissue equivalent material.**

phase. Fraction of a cycle of periodic disturbance which has been completed at a specific reference time, e.g., at the start of a cycle of a second disturbance of the same frequency. Expressed as an angle, one cycle corresponds to 2π radians or 360°. The terms 'in phase', 'in quadrature' and 'out of phase' correspond to phase angles between two periodic disturbances of 0 (or 2π), $\pi/2$ and π respectively. See also **polyphase, single-phase, three-phase, two-phase.**

phase angle. See under **phase.** Also given by $\tan \delta = $ (reactance/resistance) for an alternating-current circuit or an acoustic system.

phase compensation. See **line equalizer.**

phase constant. Imaginary part of propagation constant of a line or of transfer constant of a filter section. It is expressed in radians per unit length or section. Also applied in X-ray analysis of crystal structure to the values assumed for terms of the Fourier synthesis since X-rays give only the absolute magnitudes of the structure factors.

phase converter. On applying a single-phase voltage to one phase of a rotating 3-phase induction motor, it will be found that 3-phase voltages, which are approximately balanced, may be derived from the 3 terminals.

phase corrector. A circuit for correcting phase distortion.

phase delay. Delay, in radians or seconds, for transmission of wave of a single frequency through whole or part of a communication system. Also **phase retardation.**

phase-delay distortion. That of a signal transmitted over a long line, so that difference in time of arrival of components of a complex wave is significant.

phase deviation. In phase modulation, maximum difference between the phase angle of the modulated wave and the angle of the non-modulated carrier.

phase difference. The phase angle by which one periodic disturbance *lags* or *leads* on another.

phase discriminator. Circuit preceding demodulator in phase-modulation receiver. It converts the carrier to an amplitude-modulated form.

phase distortion. That found in the waveform of a transmitted signal on account of the non-linear relation of wavelength constant of a line, or the image phase constant of amplifier or network, with frequency.

phase equalization. See **line equalizer.**

phase focusing. Effect used in electron bunching whereby those lagging in phase tend to gain energy from the field and those leading tend to surrender it. A similar effect takes place in synchrotron-charged particle accelerators.

phase intercept. Phase delay, in radians for zero frequency, for the whole or part of a transmission system; obtained by extrapolating the measured or calculated curve for phase delay with respect to frequency.

phase-intercept distortion. Distortion in a received signal waveform solely because phase delay for zero frequency is not an exact multiple of π.

phase inversion. Any arrangement for obtaining an additional voltage of reverse phase for driving both sides of a push-pull amplifier stage.

phase inverter. See **grid neutralization.**

phase lag, phase lead. If a periodic disturbance A completes a cycle when a second disturbance B has completed $\phi/2\pi$ of a cycle, B is said to *lead A* by a phase angle ϕ. Conversely A *lags* on B by ϕ or *leads B* by $(2\pi - \phi)$.

phase margin. Variation in phase of unity gain from that which would give instability. Readily measured from a Nyquist diagram.

phase meter. Meter for measuring the phase difference between two signals of the same frequency.

phase modulation. Periodic variation in phase of a high-frequency current or voltage in accordance with a lower impressed modulating frequency. Occurs as an unwanted by-product of amplitude modulation, but can be independently produced.

phase resonance. Resonance in which the induced oscillation differs in phase by $\pi/2$ from the forcing disturbance. Also termed **velocity resonance.**

phase retardation. See **phase delay.**

phase-sensitive detector. Detector in which the output signal changes sign when the phase of the input signal is reversed.

phase-shift control. Control of initial discharge of a gas tube, e.g., thyratron, by the phase of voltage applied to a grid.

phase-shift oscillator. Same as **line oscillator.**

phase shifter. Waveguide or coaxial-line component which produces any selected phase delay in signal transmitted.

phase-shifting circuit. One in which the relative phases of components in an a.c. waveform or signal can be continuously adjusted.

phase-shifting transformer. Transformer for which the secondary voltage can be varied continuously in phase. Usually has rotatable secondary winding and polyphase primaries. Also **phasing transformer.**

phase space. Six-dimensional space, three coordinates of position and three of momentum; employed for study of system of particles. See **microscopic state.**

phase splitter. A means of producing two or more waves which differ in phase from a single input wave.

phase stability. The automatic synchronization of the time of rotation of the charged particle in a frequency-modulated cyclotron.

phase swinging. Lack of synchronism throughout individual cycles of frame-frequency generators at the transmitting and receiving ends of a television system, causing received picture to wander over the screen.

phase velocity. Velocity of propagation of any one phase state, such as a point of zero instantaneous field, in a steady train of sinusoidal waves. It may differ from the velocity of propagation of the disturbance, or *group velocity*, and, in transmission through ionized air, may exceed that of light. Also **wave velocity.**

phasing. In facsimile transmission and reception, adjustment of reproduced picture to correspond exactly with the original, achieved by a phasing signal.

phasing transformer. See **phase-shifting transformer.**

phasitron. Thermionic valve with split concentric electrodes, using three-phase voltages and fan beams of electrons. It is used to modulate a high-frequency electron current.

phasmajector. Device for providing standard video signal for testing television circuits, the signal being electronically generated.

phasor. Mathematical term for quantity represented by a complex number. Vectors are conveniently represented by phasors.

phon. Unit of the objective loudness on sound-level scale; the decibel unit of the 1 kHz intensity-level scale which is used for deciding the apparent loudness of an unknown sound or noise, when a measure of loudness is required. This is effected either by subjective comparison by the ear, or by objective comparison with a microphone amplifier and a weighting network. The reference sound pressure level at 1 kHz is 0·0002 microbar.

phonmeter. Apparatus for the estimation of loudness level of a sound on the phon scale by subjective comparison. Also **phonometer.**

phon scale. That for power in sounds. It is logarithmic from 10^{-16} watt/cm^2 (or 1 pico-watt/metre^2 or 0.0002 dyne/cm^2). This is taken as zero, being in the neighbourhood of average threshold of hearing at 1000 Hz.

phonetics. The scientific study of speech and vocal acoustics.

phonic wheel or drum. Elementary synchronous motor capable of being driven with low power from valve oscillators, so that frequency of the latter can be measured by a revolution counter.

phonoelectrocardioscope. An instrument embodying a double beam CRO, whereby the waveforms of two different functions of the heart are shown simultaneously.

phonograph. Any system for the reproduction of sound from a record. Usually a stylus is used to follow the undulations of the record and these are converted into electrical or acoustical vibrations.

phonon. Quantum of thermal energy in an acoustic mode for lattice vibrations of a crystal. In the limit, the interaction of acoustic waves with matter reduces to that between phonons and atoms, although for audible frequencies (v) phonon energies (hv) are negligible. h is Planck's constant.

phonon mean free path (λ). In a dielectric solid, given by $K=\frac{1}{3}Cv\lambda$ approximately, where K is the thermal conductivity of the solid, C is the thermal capacity per unit volume and v is the mean velocity of sound in the solid.

phosphor. (1) Generic name for the class of substances which exhibit *luminescence.* (2) More specifically, the fluorescent coating used on the screen of CRTs or image intensifiers.

phosphoresence. Luminescence which persists for more than 0·1 nanoseconds after excitation. See **afterglow, fluorescence.**

phosphorus (P). Non-metallic element, at. no. 15, at. wt. 30·9738, m.p. 44°C, b.p. 282°C. It occurs in several allotropic forms, white phosphorus being waxy. Red phosphorus by contrast to the white variety is non-poisonous and not very inflammable.

phot. Luminance unit equalling one lumen/cm^2.

Photicon. RTM for a camera tube analogous to the *image iconoscope.*

photocathode. Electrode from which electrons are emitted on the incidence of radiation in a photocell. It is *semitransparent* when there is photoemission on one side, arising from radiation on the other, as in the signal plate of a TV camera tube.

photocell. See **photoelectric cell.**

photocell amplifier. (1) That located in close proximity to the photocell which receives the light beam modulated by the sound-track, when the latter passes through the sound-gate in a projector. (2) Galvanometer current amplifier in which movement of reflected beam of light from galvanometer across boundary between two photovoltaic cells gives greatly amplified output current. Also **PEC** or **pec amplifier, photoelectric galvanometer.**

photocell pickoff. One which is operated by changes in the intensity or magnitude of a light beam.

photocell sensitivity. Ratio of output current to level of illumination. Expressed in mA/lumen.

photochemical cell. Photocell comprising two similar electrodes, e.g., of silver, in an electrolyte; illumination of one electrode results in a voltage between them; also called **Becquerel cell, photoelectrolytic cell.** See also **backwall cell, barrier-layer cell, frontwall cell, photovoltaic cell.**

photoconducting camera tube. One in which the optical image is focused on to a surface, the electrical resistance of which is dependent on illumination.

photoconductive cell. Type of *photoelectric cell* employing photoconductive element (often cadmium sulphide) between electrodes.

photoconductivity. Property possessed by certain materials of varying their electrical conductivity under influence of light.

photocurrent. That released from sensitized surface of photocell on the incidence of light, electrons which form the current being attracted to an anode polarized positively with respect to the surface. The true photocurrent is augmented by the presence of gas through ionization by collision.

photodiode. Combination of photoconducting cell and junction diode into a two-electrode semiconducting device. Widely used as optical sensing device for data processing.

photodisintegration. Nuclear reaction initiated by photon. Also **nuclear photoeffect, photonuclear reaction.**

photoelasticity. Phenomenon whereby strain in certain materials, crystal or foil, results in coloured fringes, when transmitting polarized light between crossed Nicol prisms.

photoelectric cell. Transducer which has an electric output corresponding to the incident light. Also **photocell.**

photoelectric effect. Phenomena resulting from absorption of photons by electrons, resulting in their ejection with kinetic energy equal to the difference between the energy of a photon and the surface *work function* or an atomic binding energy (*Einstein photoelectric equation*). Among these phenomena are photoconductive, photovoltaic, photoemissive and photoelectromagnetic effects. Emission of X-rays on the impact of high-energy electrons on a surface is an inverse photoelectric effect.

photoelectric galvanometer. See **photocell amplifier** (2).

photoelectric multiplier. See **photomultiplier.**

photoelectric threshold. The limiting frequency for which the quantum energy is just sufficient to produce photoelectric emission. Given by equating the quantum energy to the work function of the cathode.

photoelectric tube. Same as **phototube.**

photoelectric work function. The energy required to release a photoelectron from a cathode. It should correlate closely with the thermionic work function.

photoelectric yield. The proportion of incident quanta on a photocathode which liberate electrons.

photoelectricity. Emission of electrons from the surface of certain materials by quanta exceeding certain energy (Einstein).

photoelectrolytic cell. See **photochemical cell.**

photoelectromagnetic effect. See **photomagnetoelectric effect.**

photoelectromotive force. That which apparently arises in a photovoltaic cell, when it is considered as a constant-current source.

photoelectron. One released from a surface by a photon, with or without kinetic energy.

photoemission. Emission of electrons from surface of a body (usually an electropositive metal) by incidence of light.

photoemissive cell. An electronic valve containing a photocathode, the photocurrent flowing from the anode to the cathode. Greater sensitivity, at expense of linearity of response, is obtained by use of gas in the envelope and by operating at the ionization potential of the gas.

photofission. Nuclear fission induced by gamma rays.

photoformer. Function generator using optical projection of shadow mask, representing required waveform.

photoglow tube. One in which the sensitivity is increased by radiation incident on the cathode.

photographic emulsion technique. The study of the tracks of ionizing particles as recorded by their passage through a nuclear emulsion.

photoluminescence. Light emitted by visible, infrared or ultraviolet rays after irradiation.

photomagnetoelectric effect. Generation of electric current by absorption of light on surface of semiconductor placed parallel to magnetic field. Due to transverse forces acting on electrons and holes diffusing into the semiconductor from the surface. Also **photoelectromagnetic effect.**

photomeson. The meson resulting from interaction of photon with nucleus—usually a pion.

photometer. Instrument for measuring intensity of light or illumination in terms of a reference or standard source.

photomosaic. Sheet of material, usually mica, which is covered by a large number of minute photocells formed by depositing a film of silver, breaking it up and activating it by deposition of caesium, which is also evaporated away from intervening areas. Used in *iconoscope* and *image orthicon* camera tubes for television.

photomultiplier. Photocell with series of dynodes, used to amplify emission current by electron multiplication.

photon. A quantum of light or of electromagnetic radiation equal to the value of *Planck's constant* (h) multiplied by the frequency (v) in hertz.

photon noise. That occurring in photocells as a result of the fluctuations in the rate of arrival of light quanta at the photocathode.

photonegative. Material whose electrical conductivity decreases with increasing illumination.

photoneutron. Neutron resulting from interaction of photon with nucleus.

photonuclear reaction. See **photodisintegration.**

photopic vision. Vision based on cones, and therefore sensitive to colour. Possible only with adequate ambient illumination; cf. *scotopic vision.*

photopositive. Material for which conductivity increases with increasing illumination.

photoproton. Proton resulting from interaction of photon with nucleus.

photoradiogram. Still picture transmitted by radio.

photosensitive. Property of being sensitive to action of visible or invisible light, whether subsequent development is required to exhibit sensitivity or not.

phototelegraphy. General term for high-grade facsimile transmission of half-tone pictures.

phototransistor. A 3-electrode photosensitive semiconductor device. The emitter junction forms a photodiode and the signal current induced by the incident light is amplified by transistor action.

phototube. Gas or vacuum emissive cell.

photovaristor. Material, e.g., cadmium sulphide or lead telluride, in which the varistor effect, i.e., non-linearity of current-voltage relation, is dependent on illumination.

photovoltaic cell. A class of *photoelectric cell* which acts as a source of e.m.f. and so does not need a separate battery. See **backwall cell, barrier-layer cell, frontwall cell.**

photovoltaic effect. The production of an e.m.f. across the junction between dissimilar materials when it is exposed to light or ultraviolet radiation.

Photox, Photronic. RTMs for some photochemical cells.

physical mass unit. See **atomic mass unit.**

physiological acoustics. The section of acoustics which deals with the detection and production of sound by living organs.

physiological monitor. Monitor used in operating theatre to record changes in patient's physiological condition.

pi-attenuator. An attenuator network of resistors arranged as in the pi-network.

pi-mode. Operation of a multicavity magnetron, whereby voltages on adjacent segments of the anode differ by π radians (half-cycle).

pi-network. Section of circuit with one series arm preceded and followed by shunt arms to return leg of circuit.

pi-section filter. Unbalanced filter in which one series reactance is preceded and followed by shunt reactances.

Pianotron. Pianoforte in which the normal vibration of the strings is used to modulate the potential applied to electrostatic screw pickups, with amplification and loudspeakers.

picket fence. Arrangement for stabilizing plasma in a *torus*, consisting of bands of coils round the tube, with currents alternating in direction, giving an interior field with periodic cusps.

pickoff. Any device actuated by relative motion and giving an output, electrical or pneumatic, for operating other mechanisms.

pickup. Generally, any transducer used to 'pick up' a signal required for any equipment. Specifically, mechanical-electrical transducer, usually piezo, actuated by a

sapphire or diamond stylus which rests on the sides of the groove on a record, and, by tracking this groove, generates a corresponding voltage for driving a valve amplifier. See also:

alternating-	electromagnetic-
current-	electronic-
capacitor-	piezoelectric
ceramic-	crystal
crystal-	vibration-.

pickup needle. (1) One in which the moving part is a magnetic needle, which by its motion diverts magnetic flux and induces a voltage in coils on the magnetic circuit. (2) Loosely applied to any needle in a gramophone pickup.

pickup plate. One concerned with electrostatic signals from a number of capacitors, as in a television camera mosaic, cathode-ray tube store, electronic musical instrument.

pickup reaction. Nuclear reaction in which an incident particle collects a nucleon from a target atom and proceeds with it.

pickup tube. See **camera tube.**

pico-. See **micromicro-.**

picofarad (pF). Unit for small capacitances, equal to 10^{-12} farad. Once **micromicrofarad ($\mu\mu$F).**

picture. Complete television image, constructed from one or two interlaced fields. In U.S., frame.

picture-chasing. Circuit used in conjunction with intermediate film-scanning system, with continuously-moving film.

picture element. Same as **picture point.**

picture frequency. Same as **frame frequency.**

picture inversion. Conversion of negative to positive image (or vice versa) when carried out electronically. In facsimile transmission it will correspond to the reversal of the black and white shades of the recorded copy.

picture monitor. CRT for exhibiting television picture or related waveform for purposes of control.

picture noise. See **grass.**

picture point. Any one of the large number of minute illuminated areas which go to make up a television image. Also called **picture element.**

picture ratio. Ratio of the width to height of reproduced image, e.g., 5/4. Soundfilm has a number of aspect ratios, original being 4/3. Also **aspect ratio.**

picture signal. That portion of a television signal which carries information relative to the picture itself, as distinct from synchronizing portions. See **pedestal.**

picture telegraphy. Same as **facsimile telegraphy.**

picture traverse. That circuit which controls frame-frequency scanning.

picture tube. See **teletube.**

Pierce oscillator. Original crystal oscillator in which positive feedback from anode to grid in a triode valve is controlled by piezoelectric mechanical resonance of a suitably cut quartz crystal.

G*

piezoelectric crystal. One showing piezo-electric properties which may be shaped and used as a resonant circuit element or trans-ducer (microphone, pickup, loudspeaker, depth-finder, etc.).

piezoelectric effect or **piezoelectricity.** Electric polarization arising in some anisotropic (i.e., not possessing a centre of symmetry) crystals (quartz, Rochelle salt, barium titanate) when subjected to mechanical strain. An applied electric field will likewise produce mechanical deformation.

piezoelectric loudspeaker. A piezoelectric crystal (quartz or Rochelle salt) used to generate mechanical waves (or, conversely, electric potentials due to incident sound waves, i.e., to act as a microphone). Such devices may also be designed to operate under water. Also **subaqueous loud-speaker.**

piezoelectric microphone. See **crystal micro-phone.**

piezoelectric resonator. A crystal used as a standard of frequency, controlling a valve oscillator.

piezoelectricity. See **piezoelectric effect.**

piezoid. Blank of piezo crystal, adjusted to a required resonance with or without relevant electrodes.

pile. (1) The light reflected from a glass plate incident at the Brewster angle will be plane-polarized if transmitted light is partially-polarized normal to the plane of the re-flected polarized light. By using a pile of glass plates the transmitted light becomes increasingly plane-polarized. (2) Abb. for **thermopile**, a close packing of thermocouples in series, so that alternate junctions are exposed for receiving radiant heat, thus add-ing together the e.m.fs. due to pairs of junctions. (3) Original name for pile of graphite blocks which formed the moderator of the first nuclear reactor.

pile oscillator. The means of maintaining a neutron-absorbing body in periodic motion in a nuclear reactor.

pile-up. Set of spring contacts operated in a relay. Also **stack.**

pillowphone. Telephone receiver, concealed in a pillow, for reproducing, without sound radiation, programmes for a listener.

pilot carrier. In a suppressed-carrier system (as in single sideband working), a small por-tion of original carrier wave transmitted to provide a reference frequency with which local oscillator at the receiving end may be synchronized.

pilot electrode. Additional electrode or spark gap which triggers and makes certain main discharge, e.g., in a discharge spark gap for creating ions, or in a TR switch, or a mercury-arc rectifier. Also **ignitor, keep-alive electrode, starter, trigger electrode.**

pilot lamp. Lamp giving visual indication of the closing of a circuit.

pilot spark. See **pilot electrode.**

pilot-wire regulator. An automatic device in transmission circuits, e.g., for compensating

changes in transmission arising from temper-ature variations.

pimpling. Small swellings on the surface of a reactor fuel can—usually caused by swelling of fuel during burn-up.

pinch. An air-tight glass seal through which pass the electrode connections in a thermi-onic valve.

pinch effect. (1) Because of magnetic attraction of parallel currents, tendency for spinning plasma in an evacuated torus, e.g., *zeta* and *stellarator*, to contract, with consequent instability, reduced by a longitudinal magnetic field. Also **rheostriction.** (2) A distortion in disk reproduction which can arise from a slight vertical movement of the pickup stylus at double the recorded lateral frequency.

pinch-off. Cutoff of the channel current by the gate signal in a field-effect transistor.

pincushion distortion. That of a visual image when a square is reproduced with sides curving inwards, as could arise in a CRT. Cf. *barrel distortion.*

pinetree antenna. Vertical array of horizontal dipoles driven by twisted vertical trans-mission lines, so that they all radiate in phase. Unit in *Kooman's array.*

ping. Brief pulse of medium frequency sound, reproduced from the subaqueous reflection of *asdic* ultrasonic signals. Its length in space will be equal to the product of ping duration time and the velocity of sound.

pion. See under **meson.**

pip. Significant deflection or intensification of the spot on a CRT giving a display for identification or calibration. Particularly applied to the peaked pattern of a radar signal. See also **blip.**

pipe work. Collection of pipes in ranks in an organ, simulated in electronic musical instruments.

Pirani gauge. Low-vacuum manometer for measuring gas pressure, depending on heat loss from heated wire, observed by a resist-ance bridge.

piston. Closely fitting sliding short-circuit in a waveguide.

piston attenuator. Device inserted into a waveguide system to regulate transmitted power, or to make measurements.

pitch. (1) Physiological sensation corres-ponding chiefly to frequency of sound wave. (2) Distance between centres of adjacent fuel channels in a reactor.

pitch control. Control of number of lines per unit length in a television image.

pitchblende. A mineral which chiefly consists of uranium oxides but also contains small amounts of radium, hence its importance.

pith-ball electroscope. Primitive detector of charges, whereby two pith-balls, suspended by silk threads, attract or repel each other, according to inverse-square law.

Pitot tube. An open-ended tube which is pointed directly into the flow of a fluid and connected to a manometer, and under appropriate conditions measures the total pressure head. By combining with a static pressure tube, the dynamic pressure may be deduced, and thus velocity of flow.

place. See column.

plan-position indicator. Screen of a CRT with an intensity-modulated and persistent radial display, which rotates in synchronism with a highly directional antenna. The surrounding terrain is thus painted with relevant reflecting objects, such as ships, aircraft and physical features. See **azimuth stabilized-.**

planar diode. Vacuum tube with plane-parallel electrodes. A photoelectric cathode is usually employed.

Planckian colour. The colour or the wavelength-intensity distribution of the light emitted by a black body at a given temperature.

Planckian locus. Line on *chromaticity diagram* joining points with coordinates corresponding to black-body radiators.

Planck's constant. See **Planck's law.**

Planck's law. Basis of quantum theory, that the energy of electromagnetic waves is confined in indivisible packets or quanta, each of which has to be radiated or absorbed as a whole, the magnitude being proportional to frequency. If E is the value of the quantum expressed in energy units, and v is the frequency of the radiation, then $E = hv$, where h is known as *Planck's constant* and has dimensions of energy $\times$ time, i.e., action. Present accepted value is $6.625 . 10^{-27}$ erg. sec. See **photon.**

Planck's radiation formula. Expression for the electromagnetic energy E_v of frequency v radiated per second from unit area of a black body at absolute temperature T is:

$$E_v = \frac{hv^5}{c^3 \left(\epsilon^{\frac{hv}{kT}} - 1 \right)},$$

where c is the velocity of light, h is Planck's constant, k is Boltzmann's constant, and ϵ is the base of natural logarithms.

plane baffle. Plane board, with a hole at or near the centre, for mounting and loading an open-diaphragm loudspeaker unit.

plane earth factor. Electromagnetic wave propagation, the ratio of the electric field strength which would result from propagation over an imperfectly conducting earth to that resulting from propagation over a perfectly conducting plane.

plane of polarization. The plane containing the incident and reflected light rays and the normal to the reflecting surface. The magnetic vector of plane-polarized light lies in this plane. The electric vector lies in the plane of vibration, which is that containing the plane-polarized reflected ray and the normal to the plane of polarization.

plane polarization. That of an electromagnetic wave when the electric (or magnetic) field at any point does not vary in direction (except for reversal) over a cycle. A light beam striking a glass plate at the Brewster angle gives rise to a reflected beam which is plane-polarized.

plane wave. One for which equiphase surfaces are planes.

planigraphy. See tomography.

plasma. Ionized gaseous discharge in which there is no resultant charge, the number of positive and negative ions being equal, in addition to un-ionized molecules or atoms. The term is also used as a synonym for the positive column in a gas discharge.

plasma torch. One in which solids, liquids or gases are forced through an arc within a water-cooled tube, with consequent ionization; de-ionization on impact results in very high temperatures. Used for cutting and depositing carbides.

plasmatron. Discharge tube in which anode current can be regulated by control of plasma, either by a grid or through the electron stream originating the plasma.

plasmoid. Any individual section of a plasma with a characteristic shape.

plastic effect. One arising from incorrect response of the vision receiver, resulting in a defective reproduction of a TV picture, i.e., poor overall tonal gradations.

plate. (1) Each of the two extended conducting electrodes which, with a dielectric between, constitutes a capacitor. (2) See **anode.**

plate battery. *B-battery* or *high-tension battery*, especially for portable equipment.

plate-bend detector. One which applies incoming signal to grid of triode valve with considerable negative bias; the demodulated signal is taken from anode circuit. Also **anode-bend detector.**

plate dissipation. The energy dissipated by the electrons striking the plate electrode of a valve at high velocities.

plate efficiency. Same as **anode efficiency.**

plate load impedance. The total impedance between the plate (anode) and cathode of a thermionic valve, excluding electron stream.

plate modulation. See anode modulation.

plate neutralization. A method of neutralizing an amplifier (i.e., preventing its oscillation) by shifting through π radians a part of the plate-cathode a.c. voltage and applying it to the cathode-grid circuit via a neutralizing capacitor.

plate rectifier. One of large area for large output currents, e.g., for electrolytic bath supply or electric traction.

plate resistance. Strictly the *dynamic plate resistance of a thermionic valve*, defined by

$$R_p = \left(\frac{\Delta E_p}{\Delta I_p} \right) E_G,$$

where ΔE_p and ΔI_p represent very small changes of plate voltage and current for a constant grid voltage E_G.

plateau. See under Geiger region.

plateau length. The voltage range which corresponds to the plateau of a Geiger counter tube.

plateau slope. The ratio of the percentage change in count rate for a constant source to the change of operating voltage. Measured for a median voltage corresponding to the centre of the Geiger plateau.

plated circuit. Printed electrical circuit prepared by electroplating.

Platinotron. TN for a high-power microwave tube for radar, similar to a magnetron.

platinum (Pt). A hard silvery-white metal which is ductile and malleable and highly resistant to acids and heat. At. no. 78, at. wt. 195·09, sp. gr. 21·45, m.p. 1773·5°C, b.p. 3910°C. Used for electrical contacts for high temperatures and for electrodes subjected to possible chemical attack. Also used as a basic metal for resistance thermometry over a wide temperature range.

playback. (1) Immediate reproduction from a wax recording with a very flexible pickup; used for testing the quality of the reproduced sound before actual records are made. (2) Repetition of any form of audio or visual programme.

pleochroic halo. Tiny coloured region around an alpha-emitting particle in a mineral which is somewhat transparent, the size being related to age.

plethysmograph. An apparatus for measuring variations in the size of bodily parts and in the flow of blood through them. See also **electro-arteriograph.**

pliodynatron. Multi-electrode tube for obtaining negative resistance by secondary emission; an oscillator circuit using such.

pliotron. Large hot-cathode vacuum tube for industrial use, with one or more grids. Applied at one time to any vacuum tube.

plotter. Instrument which prepares graph showing relationship between two voltage signals applied to x and y inputs. Widely used as output unit of analogue computer to represent the relationship between two physically different variables.

plug. (1) Termination on wire or cable—used to make electrical connection when inserted into socket. (2) Piece of absorbing material used to close the aperture of a channel through a reactor core, or other source of ionizing radiation.

plug-in. Originally, inductor with contact pins for insertion into sockets, for preselecting a range of tuning in radio circuits. Now applied to most valves and some capacitors, and to many printed circuit boards. See **wired-in.**

plug-in unit. Any panel or component which can be inserted and interchanged in a computing system, particularly a *cross-connection panel,* which can be set up independently.

plugboard. Detachable unit with an area of sockets, which establishes by *jumpers* sets of circuits quickly for specific computer programmes or problems. See also **patch board.**

plumbing. Colloquialism for waveguides and their jointing in establishing microwave systems. Also for piped vacuum systems.

plunger. Device for altering length of a short-circuited coaxial line, taking the effective form of an annular disk. For very high frequencies there need not be a contact, because a small clearance gives sufficient capacitance for an effective short-circuit.

plutonium (Pu). Element, at. no. 94, product of the radioactive decay of neptunium. Has many isotopes in range 232-246. The fissile isotope ^{239}Pu produced from ^{238}U by neutron absorption in a reactor, is the most important.

plutonium reactor. One in which ^{239}Pu is used as the fuel.

pneumatic loudspeaker. One in which a jet of high-pressure air is modulated by a transducer driven by audio currents, e.g., Stentorphone.

Pockel's effect. The *electro-optical effect* (q.v.) in a piezoelectric material.

pocket chamber. A small ionization chamber used by individuals to monitor their exposure to radiation. It depends upon the loss of a given initial electric charge being a measure of the radiation received.

point. Decimal point in numbers using 10 as radix; similarly with 2 as radix. *Fixed* and *floating* points imply omission of point or its movement among the digits by multiplication or division by the radix respectively. Used in computers.

point contact. Condition where current flow to a semiconductor is through a point of metal, e.g., use of end of metal wire as in 'cat's whisker'.

point-contact rectifier. One comprising a metal point pressing on to a crystal of semiconductor and delivering *holes* when made positive.

point-contact transistor. One in which the base electrode is joined by two or more point-contact electrodes.

point counter tube. One employing gas amplification in which the central electrode is a point or a small sphere.

point defect. See **defect.**

point gamma. The contrast *gamma* for a specified level of brightness. In TV reception, the instantaneous slope of the curve connecting log (input voltage) and log (intensity of light output).

point source. One not subtending a measurable angle at the detector or observer. Any source if sufficiently distant.

poison. General term for contaminating materials which, because of their high-absorption cross-section, degrade intended performance, e.g., fission in a nuclear reactor, radiation from a phosphor, emission from a cathode, conduction in a semiconductor.

poison computer. Reactor control instrument consisting of analogue computer operating in real time, and providing continuous record of level of xenon poisoning in a reactor.

Poisson distribution. Statistical distribution, applicable to most nuclear processes such as radioactive decay, and characterized by a small probability of a specific event occurring during observations over a continuous interval (e.g., of time or distance). Cf. *binomial distribution.*

Poisson's equation. A fundamental electrostatic equation which forms one of *Maxwell's field equations.* The differential form of *Gauss's law of electrostatics.*

Poisson's ratio. The ratio of lateral to longitudinal strain, e.g., in a stretched wire.

polar diagram. One showing relative effectiveness of transmission or reception of antenna system in different directions. A contour of equal field strength around a transmitting antenna; or, in a receiving antenna, contour path of a mobile transmitter producing a constant signal at the receiver. Similar diagrams are prepared for sound fields round electro-acoustic transducers, sensitivity curves around photocells, scintillation counters, etc.; and for all other energy detectors or radiators with directional sensitivity.

polar guide. Waveguide filter with circular components.

polar molecule. One with unbalanced electric charges, usually valency electrons, resulting in a dipole moment and orientation.

polar response curve. See **polar diagram.**

polar vector. One which is symmetrical with respect to an axis, or to a plane containing an axis.

polariscope. Instrument for studying the effect of a medium on polarized light. Interference patterns enable elastic strains in doubly refracting materials to be analysed. See **photoelasticity.**

polarity. (1) Distinction between positive and negative electric charges (Franklin). (2) Distinction between positive (north) and negative (south) magnetic poles of electro- or permanent magnet; these poles do not exist, but describe locations where magnetic flux leaves or enters magnetic material. See **electret.** (3) General term for difference between two points in a system which differ in one respect, e.g., potentials of terminals of a cell or electrolytic capacitor, windings of a transformer, video signal, legs of a balanced circuit, phase of an alternating current.

polarization. (1) Change in a dielectric as a result of sustaining a steady electric field, with a similar vector character; measured by density of dipole moment induced. (2) Same as **intensity of magnetization.** (3) Non-random orientation of electric and magnetic fields of electromagnetic wave. (4) For system of particles possessing spin, a measure of the extent of unbalance between parallel and anti-parallel spin vectors. See:

abnormal-	horizontal-
circular-	induced-
dielectric-	normal-
electrolytic-	plane-
elliptic-	vertical-.

polarization current. That causing or caused by polarization (*soakage*) in a dielectric, with possible late discharge.

polarization error. Error in determining the direction of arrival of radio waves by a direction-finder when the desired wave is accompanied by downward components which are out of phase. Formerly termed **night error.**

polarized capacitor. Electrolytic capacitor designed for operation only with fixed polarity. The dielectric film is formed only near one electrode and thus the impedance is not the same for both directions of current flow.

polarized diversity. See **diversity reception.**

polarized plug. One which can be inserted into a socket in only one position.

polarized relay. See **relay types.**

polarizer. Prism of doubly refracting material, or Polaroid plate, which passes only plane-polarized light, or produces it through reflection. See **Nicol prism, pile** (1).

polarizing angle. See **Brewster angle.**

polarograph. Instrument recording current-voltage characteristic for polarized electrode. Used in solution chemical analysis.

Polaroid. TN for an optically neutral filter which transmits only plane-polarized light. It is produced by accurate alignment of minute doubly refracting crystals, e.g., iodine, in a thin sheet of plastic, e.g., cellulose nitrate. The differential absorption of the ordinary and extraordinary rays by the crystals leads to production of the plane-polarized light.

polaron. Electron in substance, trapped in potential well produced by polarization charges on surrounding molecules, analogous to *exciton* in semiconductor.

pole. (1) That part of the anode between adjacent cavities in a multiple-cavity magnetron. (2) The centre of a spherical mirror. (3) See **magnetic pole, poles (magnetic).**

pole-finding paper. Paper prepared with a chemical solution, which, when placed across two poles of an electric circuit, causes a red mark to be made where it touches positive.

pole piece. Specially-shaped magnetic material forming an extension to a magnet, e.g., the salient poles of a generator, motor or relay, for controlling flux.

poles (magnetic). (1) Points on the earth's surface intersected by its magnetic axis (magnetic inclination 90°). *N.B.*—(*a*) Their position drifts continuously, (*b*) the earth's magnetic flux is directed from southern hemisphere towards northern. Consequently, magnetic pole in northern hemisphere is *south-seeking pole* (S pole) and vice versa. (2) Regions at the ends of ferromagnetic materials considered as magnetic sources of opposite polarity (north- or south-seeking), which have the same distant effect as the actual magnet. Made more precise by ferromagnetic balls on the ends of long thin rods of magnetic material (Robison).

poling. See **turnover.**

polonium (Po). At. no. 84, at, wt. 210. Radioactive element discovered in 1895. Important as an α-ray source relatively free from γ-emission, e.g., in α-*n* neutron sources.

polyethylene. See **polythene.**

polymorphic. Capable of existing in more than one crystal form.

polyphase. Set of a.c. power-supply circuits (usually three) carrying currents of equal frequency with uniformly-spaced phase differences. Normally employing common-return conductor.

polyplexer. Radar device acting as duplexer and lobe switcher.

polyrod antenna. One comprising a number of tapered dielectric rods emerging from a waveguide.

polyskop. Swept-frequency display signal generator.

Polystyrene. TN for high grade UHF plastic insulator.

polytetrafluoroethylene. An insulating polymer of very high resistivity and low friction, resistant to moisture and heat. Also **tetrafluorethylene.** TNs **Fluon, Teflon.**

polythene or **polyethylene.** A tough waxy thermoplastic material which is flexible and chemically resistant. A good insulator. Mechanical properties are changed by irradiation. TN **Alkathene.**

ponderable. U.S. term for quantity of radioactive isotope sufficient to be weighed.

pool cathode. An emissive cathode which consists of a liquid conductor, e.g., mercury. See **pool tube.**

pool reactor. See **swimming-pool reactor.**

pool tube. One where a mercury cathode provides electrons by spot impact of the mercury positive ions formed in the vapour. Large sizes are used for power or traction rectification. Also **tank tube.** See **Excitron, ignitron, mercury-arc rectifier.**

porch. In a complete TV picture signal the *front porch* is that between the front of the horizontal blanking signal and the front of the synchronizing signal. The *back porch* is that between the ends of horizontal synchronizing and blanking signals.

port. Place of access to system—used for introduction or removal of energy or material, e.g., *glove-box port.*

portable. Usually applied to equipment which is not dependent upon connection to an electric supply mains.

ported baffle. *Box baffle,* with parts adjacent to the radiating loudspeaker unit which radiate in anti-phase at low frequencies, thus unifying the acoustic response.

ported loudspeaker. Open-diaphragm loudspeaker mounted on a face of a box with adjacent ports.

position. In radio navigation, a set of co-ordinates used to specify location and elevation.

position-finding. Determination of location of transmitting station (e.g., an aeroplane) by taking a number of bearings by direction-finders which receive a signal from the transmitter.

positioner. Electronic equipment for locating tool and work on a machine-tool, either by setting knobs or by a punched tape, up to three dimensions.

positive. Said of point in circuit which is higher in electric potential than earth.

positive amplitude modulation. That in which the amplitude of the video signal, in TV, increases with an increase in luminance of the picture elements.

positive column. Luminous plasma region in gas discharge, adjacent to positive electrode.

positive coupling. When two coils are inductively coupled so that the magnetic flux associated with the mutual inductance is in the same direction as that associated with the self-inductance in each coil, the mutual inductance and the coupling are both termed *positive.* If these fluxes oppose the mutual inductance and coupling they are termed *negative.*

positive electricity. Phenomenon in a body when it gives rise to effects associated with deficiency of electrons, e.g., positive electricity appears on glass rubbed with silk.

positive electrode. (1) That connected to a positive supply line. (2) The *anode* in a voltameter. (3) The *cathode* in a primary cell.

positive electron. See **positron.**

positive feedback. Interconnection between output and input circuits of an amplifier to facilitate the voltage, current, or power drive of input by addition of voltage, current, or power from output. This reduces resistance of the source of amplified power and, if the feedback is sufficient, sets up sustained oscillations independently of input drive. See also **regeneration.**

positive grid oscillator. See **Barkhausen-Kurz** and **Gill-Morrell oscillators.**

positive ion. An atom (or a group of atoms which are molecularly bound) which has lost one or more electrons, e.g., alpha-particle is a helium atom *less* its two electrons. In an electrolyte the positive ions (*cations*) produced by dissolving ionic solids in a polar liquid like water have an independent existence and are attracted to the anode. Negative ions likewise are those which have gained one or more electrons.

positive (or negative) light modulation. Increase from low to high modulation (or decrease from high to low modulation) of radiated carrier with increase of light in transmitted image.

positive magnetostriction. See **Joule magnetostriction.**

positive pole. See **north(-seeking) pole.**

positive rays. Streams of positively-charged atoms or molecules which take part in electrical discharge in a rarefied gas; they are accelerated through a perforated cathode on to a photographic plate, being deflected by magnetic and electrostatic fields. Used in Thomson's parabolic traces and Aston's mass spectrometer; also **anode rays, canal rays.**

positive video signal. One in which increasing

amplitude corresponds to increasing light value in the transmitted image. White is 100 per cent., and the black level makes about 30 per cent. of the maximum amplitude of signal.

positron. *Posi*tive elec*tron*, of the same mass as, and charge of opposite sign to, the normal (negative) electron. Produced in the decay of radioisotopes, and in *pair production* by X-rays of energy much greater than 1 MeV.

positronium. Short-lived particle, combination of a positron and an electron, its mean life being not greater than 10^{-7} sec. The energy levels of positronium are similar to the hydrogen atom but because of the different reduced mass the spectral line frequencies are only approximately half of the corresponding lines of hydrogen.

post-deflection acceleration. In a CRT, acceleration of the beam electrons after deflection, so reducing the voltage required for this.

post-emphasis. Same as de-emphasis.

post-synchronize. To add a sound-recording to an independently shot picture.

postmortem. Programme designed to locate and diagnose in an electronic computer a fault which has caused faulty operation.

pot. A shortened form of *potentiometer* (q.v.), used in sense (2) only.

pot magnet. One embracing a coil or similar space, excited by current in the coil or a permanent magnet in the central core. Main use is with a circular gap at an end of the core for a moving coil. Miniature split-sintered pot magnets are also used to contain high-frequency coils.

potassium (K). Soft, silvery-white, highly-reactive alkali metal. At. no. 19, at. wt. 39·102, sp. gr. 0·86, m.p. 62·3°C, b.p. 760°C. Shows slight natural radioactivity due to ^{40}K (half-life $1·30 \times 10^9$ years). May be used as coolant in liquid-metal reactors, although usually an alloy with sodium is used. See NaK.

potential. (1) Scalar magnitude, negative integration of the electric (or magnetic) field intensity over a distance. Hence all potentials are relative, there being no absolute *zero potential* other than a convention, e.g., earth, or distance from a charge at infinity. (2) The potential energy of a nucleon expressed in terms of its position in the nuclear field, e.g., a central potential is one that is spherically symmetric, i.e., the potential energy V is a function only of the distance r from the centre of the field. See:

accelerating-	ionization-
critical-	separation-
diffusion-	stopping-
driving-	striking-
excitation-	vector-
floating-	Yukawa-
glow-	zero-
ionic-	zeta-.

potential attenuator. Same as *potentiometer*, as contrasted with a normal attenuator which adjusts power.

potential barrier. Maximum in the curve covering two regions of potential energy, e.g., at the surface of a metal, where there are no external nuclei to balance the effect of those just inside the surface. Passage of a charged particle across the boundary should be prevented unless it has energy greater than that corresponding to the barrier. Wave-mechanical considerations, however, indicate that there is a definite probability for a particle with less energy to pass through the barrier, a process known as **tunnelling.** Also **potential hill.**

potential coefficients. Parts of total potential of a conductor produced by charges on other conductors, treated individually.

potential difference. (1) That between two points when maintained by an e.m.f., or by a current flowing through a resistance; unit is one *volt* when one *joule* of work is done (or received) by one *coulomb* moving between the two points. (2) Line integral of a magnetic field intensity between two points by any path.

potential divider. Chain of resistors such that the voltage across one or more is an accurately known fraction of that applied to all; used for calibrating voltmeters. Also **voltage divider, volt-box.** See Kelvin-Varley slide.

potential drop. Difference of potential along a circuit because of current flow.

potential energy. Universal concept of energy stored by virtue of position in a field, without any observable change, e.g., after a mass has been raised against the pull of gravity. In electricity, potential energy is stored in an electric charge when it is taken to a place of higher potential through any route.

potential gradient. Potential difference per unit length along a conductor or through a dielectric; equal to slope of curve relating potential and distance. Also **electromotive intensity.**

potential hill. See potential barrier.

potential scattering. See scattering.

potential trough. Region of an energy diagram between two neighbouring hills, e.g., arising from the inner electron shells of an atom.

potential well. Localized region in which the energy of a particle is appreciably lower than outside. Such a well forms a trap for any incident particle.

potentiometer. (1) Precision-measuring instrument in which an unknown potential difference or e.m.f. is balanced against an adjusted potential provided by a current from a steady source. (2) Three-terminal voltage divider (this usage has been deprecated, rheostat being preferred). The resistance change with shaft rotation or slider position may follow various laws, thus one may have, e.g., linear potentiometer, logarithmic potentiometer, cosine potentiometer, etc. See next entry; and **coordinate-, Gall-, magnetic-.**

potentiometer card generator. Potentiometer wound with length of turns dependent on rate of change of a function, which determines shape of card.

pot(entiometer) cut. Colloquialism for introducing a *cut* (interruption of reproduction) by bringing down a potentiometer to zero for a short time.

potentiometer for alternating voltage. One used for varying unknown voltage by either two variable voltages in quadrature (*coordinate* or *Gall* type) or a single voltage adjustable in magnitude and phase (*Drysdale* or *polar* type).

potentiometer function generator. One in which functional values of voltage are applied to points on a potentiometer, which becomes an interpolator.

potentiometer pickoff. One in which output voltage depends on position of a slide contact on a resistor, thereby transmitting a position.

Poulsen arc converter. An arc oscillator which can oscillate up to a frequency of the order of 100 kHz.

powder core. One of powdered magnetic material with an electrically insulating binding material to minimize the effects of eddy currents, thus permitting use for high-frequency transformers and inductors with low loss in power.

powder pattern. Same as **Bitter pattern.**

powder photograph. X-ray diffraction photograph of powdered crystal sample. Characterized by set of concentric rings produced by rays diffracted at the Bragg angle relative to the incident beam.

power. The rate at which energy is supplied or consumed by a system or device. See **active-.**

power amplification. (1) That provided by valves when delivering power, as contrasted with voltage amplification. (2) Difference between output and input power levels of an amplifier, expressed in *decibels*. See **applied power.**

power amplifier. Stage designed to deliver output power of an amplifier. It may be separate from other parts of the same amplifier, and may contain its own power supply. Intended to give the required power output with specified degree of non-linear distortion, gain not being considered. In some cases the voltage gain is fractional (or in dB negative). Also **power unit.**

power breeder. A nuclear reactor which is designed to produce both useful power and fuel.

power coefficient. The change of reactivity of a reactor with increase in power. In a heterogeneous reactor, due to temperature differences of the fuel, moderator and coolant, it differs from the temperature coefficient.

power-control rod. One used for the control of the power level of a nuclear reactor. May be a full rod or a part of the moderator. In a thermal reactor it is generally a neutron absorber such as cadmium or boron steel.

power density. Energy released per second per unit volume of a reactor core. Expressed normally in watts/cm³.

power detector or **demodulator.** Rectifier

which accepts a relatively large carrier voltage and thereby achieves low non-linear distortion in the demodulation process.

power efficiency. The ratio of power released by a transducer (optical, mechanical, acoustical, etc.) to the electrical power supplied; or vice versa, according to the direction in which the energy conversion takes place.

power factor. Ratio of the total power (in watts) dissipated in an electric circuit to the total equivalent volt-amperes applied to that circuit. In single- and balanced three-phase systems it is equal to $\cos\phi$, where ϕ is the phase angle between the applied voltage and the applied current in a single-phase circuit, or between the phase voltage and phase current in a balanced three-phase circuit.—In normal dielectrics it is exactly equal to $G\ (G^2+\omega^2C^2)^{-\frac{1}{2}}$ (where C=capacitance, G=shunt conductance and $\omega=2\pi$. frequency) and thus nearly equal to $G/\omega C$.

power gain. Power delivered to a load divided by that which would be delivered if the load is matched to the source, expressed in *decibels*. If the ratio is less than unity, there is *loss*, a negative number of decibels.

power-grid detector. A valve which is operated at a fairly high plate voltage with low values of grid leak and grid capacitor. It is a form of *leaky-grid* detector.

power level. See **transmission level.**

power-level diagram. Diagram indicating how maximum power levels vary at different points of a transmission channel, thereby indicating how various losses are neutralized by appropriate amplifier gains.

power-level indicator. See **level indicator.**

power line. U.S. for **mains.**

power loss. (1) The ratio of the power absorbed by a transducer to that delivered to the load. (2) The energy dissipated in a passive network or system.

power output. The total power delivered to the load of a system. Occasionally (according to context) the useful power if an unwanted d.c. component is also supplied.

power pack. Power-supply unit for an amplifier, e.g., in a radio or television receiver, wherein the requisite steady voltages are obtained by rectifiers from mains. Also the last or power stage of an amplifier when this is integral with the power supply proper. See **transistor.**

power range. See **start-up procedure.**

power reactor. One designed to produce useful power.

power supply. (1) Arrangement for delivering available power from a source, e.g., public mains, in a form suitable for valves or transistors, etc., generally involving a transformer, rectifier, smoothing filter, circuit-breaker or other protection, and frequently incorporating electronic regulation. In a full-wave supply use is made of a full-wave rectifier and filter. (2) U.S. for **mains.**

power-supply vibrator. One used to provide an a.c. supply from a d.c. source.

power transformer. In electronics, transformer used to introduce the energizing supply into an instrument or system (distinct from a *signal transformer* (q.v.)).

power transistor. One capable of being used at high power ratings, i.e., of the order of 100 watts, and generally requiring some means of cooling.

power unit. Same as **power amplifier**.

power valve. See **output valve**.

Powerstat. TN for class of adjustable auto-transformers. Similar to **Variac**.

Poynting vector. One whose flux, through a surface, represents the instantaneous electro-magnetic power transmitted through the surface. Equal to the vector product of the electric and magnetic fields at any point. In electromagnetic wave propagation, where these fields have complex amplitudes, half the vector product of the electric field intensity and the complex conjugate of the magnetic field intensity is termed the *complex Poynting vector*, and the real part of this gives the time average of the power flux.

Poynting's theorem. That which shows that the rate of flow of energy through a surface is equal to the surface integral of the Poynting vector formed by the components of field lying in the plane of the surface. Used for calculating the power radiated from an antenna.

practical units. Those adopted for practical use because of the inconvenient size of the CGS units, e.g., volt, ampere, etc., which are defined in terms of the CGS electro-magnetic units multiplied by appropriate powers of ten.

Prandtl number. The ratio of kinematic viscosity to the thermal conductivity of a fluid.

praseodymium (Pr). A rare earth metallic element. At. no. 59, at. wt. 140·907.

pre-amplifier. Amplifier with at least a stage of valve or transistor gain following a high-impedance source from which the level is too low for line transmission and clearance above noise level.

precession. Rotation of spin axis normal to any torque applied to gyro through its bearings on its gimbals; bearing friction results in precession.

pre-conduction current. Low anode current when discharge is not self-sustaining.

pre-emphasis. Marked alteration of response at the beginning of a part of a transmission system (as in magnetic recording and reproduction, frequency modulation and demodulation, disk-recording and repro-duction) for a technical reason, such as noise level or restriction of amplitude. See **de-emphasis**.

pre-fading. Listening to programme material and adjusting its level before it is faded up for transmission or recording.

preferred numbers. Series depending on $\sqrt[5]{10}$ and $\sqrt[10]{10}$ for increasing magnitudes of articles. Another series for small elec-trical components is $\sqrt[6]{10}$, $\sqrt[12]{10}$, $\sqrt[24]{10}$, to suit prevalent tolerances of ±20%, ±10%, ±5%.

pre-formed precipitate. A precipitate which is used for the co-separation of a tracer element and has been formed before mixing with the tracer to be adsorbed.

pre-heating time. Minimum time for heating a cathode before other voltages are applied, thus ensuring full emission. Automatic delays are often incorporated in large amplifiers and radio transmitters.

pre-record. To record programme material before it is required for transmission.

pre-selection. (1) Use of selective circuits in early stages of a supersonic heterodyne receiver, prior to frequency changer, to reduce second-channel interference and cross-modulation and increase the overall sensitivity and selectivity. (2) Type of automatic switching used in telephone exchanges.

preset. Same as **quiescent**.

preset guidance. The guidance of controlled missiles by a mechanism which is set before launching and which is subsequently unalterable.

preset oscillator. One in which frequency is determined by one or more control knobs acting on a resistance-capacitance or a tuned circuit. Slight drifts arise from mains volt-age or temperature variations, but can be made negligible in effect.

pressing. Disk record formed by pressure, with or without heat; the negative of the recording on a stamper is transferred to a large number of pressings for distribution.

pressure-gradient microphone. One which offers so little obstruction to the passage of a sound wave that the diaphragm, in practice a ribbon, is acted on by the difference in the excess pressures on the two sides, and thus tends to move with particle velocity in the wave. Also **velocity microphone**.

pressure microphone. Any type which is operated by the excess pressure in a sound wave, as contrasted with a ribbon or pressure-gradient type. See **capacitor micro-phone**.

pressure-type capacitor. One in which the dielectric is an inert gas under high pressure. Useful for high voltage work.

pressurized-water reactor. Reactor using water cooling at a pressure such that its boiling-point is above the highest temper-ature reached.

pre-storing. The supplying of data to computer store in advance of its requirement for a particular programme.

pre-TR cell. A gas-filled RF switching valve which protects the *TR cell* in a radar receiver from excessive power. Also acts as a block to the receiver frequencies other than the fundamental.

pretzel. Shape of the stellarator, in which a torus is folded into a figure-8, the end loops being flat and in parallel planes.

primary additive colours. Minimum number of spectral colours (red, green, blue) which can be adjusted in intensity and, when mixed, visually make a match with a given colour. This match with real colours cannot be perfect, since one primary may have to be negative in intensity. A typical set of primary additive spectral colours are, in angström units:

red (6400), green (5370), blue (4640).

For colour TV, original colour has to be separated into such arbitrary components.

primary battery. A combination of a number of primary cells.

primary cell. Voltaic cell in which chemical energy of constituents is changed to electrical energy when current is permitted to flow. Such a cell cannot be recharged electrically because of the irreversibility of chemical reaction occurring therein. It may be used as a subgrade standard cell for calibration purposes.

primary constants. Those of capacitance, inductance, resistance, and leakance of a conductor to earth (coaxial or concentric) or to return (balanced) conductor, per unit length of line.

primary current. That formed by primary electrons, as contrasted with secondary electrons, which reduce or even reverse it.

primary electrons. (1) Those incident on a surface whereby secondary electrons are released (see also **primary ionization**). (2) Those released from atoms by internal forces and not by external radiation as with secondary electrons.

primary emission. Electron emission arising from the irradiation (including thermal heating) or by the application of a strong electric field to a surface.

primary flow. That of *carriers* when they determine the main properties of a device.

primary ionization. (1) In collision theory, the ionization produced by the primary particles, in contrast to total ionization, which includes the *secondary ionization* produced by delta-rays. (2) In counter tubes, the total ionization produced by incident radiation without gas amplification.

primary radar. Basic radar in which target reflects some of transmitted energy, which is detected at the transmitter.

primary radiation. Radiation direct from a source.

primary service area. That in which ground-wave reception of a broadcast transmission is sufficient, in relation to interference and ionospheric reflections, to give consistently good reproduction of programmes without fading.

primary standard. A reference component or particular unit of measurement which has worldwide acceptance. See **standard.**

primary store. (1) The main store built into a computer, not necessarily the fast access store. (2) Relatively small immediate or very rapid access store incorporated in some computers for which the main memory is a slower secondary store.

primary subtractive colours. Minimum number of spectral colours (cyan, magenta, yellow) which, when subtracted in the right intensity from a given white, result in a match with a given colour. These are complementary to the primary additive colours and are used in printing colour films, e.g., Technicolor.

primary voltage. That required by the input of a power transformer. Also the voltage of a primary cell.

primary winding. That of a transformer taking the driving current, with or without polarizing current, especially from a valve.

principal axis. (1) The direction of maximum sensitivity or response for a transducer or antenna. (2) See **optic axis.**

principal quantum number. That which specifies the shell the electron occupies.

principal series. Series of optical spectrum lines observed in the spectra of alkali metals. Energy levels for which the orbital quantum number is unity are designated **p-levels.**

print. Positive film used for projection, made from cut and jointed negatives.

print-out. Automatic reproduction by *printer* of data in store or recorded on punched tape or card.

print-through. In magnetic tape recording, the transfer of a recording from one layer to another when the tape is spooled or reeled, giving rise to a form of distortion; also called **transfer.**

printed circuit. Circuit of an equipment, formed into a unit which can be realized by copper conductors laminated on a phenol base. They may be deposited as a powder photographed on a resist and etched, or deposited by electrodeposition. Circuit components are added by hand or dip soldering.

printer. Device for printing results of computation. *Line* (or *page*) prints whole of line or page at a time, *fly* prints from non-stopping typewheel, *wheel* stops and prints, *stylus*, *matrix* or *wire* synthesize characters directly or through inked tape (*Samastronic*), or electric typewriter (*IBM*).

probability density. Quantum mechanics suggests that electrons must not be regarded as being located at a definite point in space, but as forming a cloud of charge surrounding the nucleus, the cloud density being a measure of the probability of the electron being located at this point. See **uncertainty principle.**

probe. (1) Electrode of small dimensions compared with the gas volume, placed in a gas-discharge tube to determine the space potential. (2) Magnetic or conducting device to extract power from a waveguide. (3) Coil or semiconductor sensing element associated with a fluxmeter. (4) Portable radiation detector unit, cable-connected to counting or monitory equipment. (5) Acoustic device using narrow tube to transmit acoustic pressure to measuring microphone.

process control. In a complicated industrial or chemical process, control of various sections of the plant by electronic, hydraulic or pneumatic means, taking rates of flow, accelerations of flow, changes of law, temperatures and pressures into account automatically.

process factor. See **separation factor.**

Prodac. TN for digital process-control system, applicable to large generators, incorporating normal and over-ride facilities control procedure, with data logging and communications.

production reactor. One designed for large-scale production of transmutation products such as plutonium.

programme. Sequence of events to be performed by an electronic computer, as laid down by a programmer, in processing a given class of data, e.g., payroll, insurance instalments, etc. Realized on a punched card or paper tape, which becomes a unit in the *library* for a specific computer.

programme control. System in which the *index* (or *desired*) *value* is changed according to a plan of operations.

programme level. The level, related to zero power level of one milliwatt in 600 ohms, as indicated by a volume-unit (VU) meter, as defined for this purpose.

programme repeater. Amplifier which is of sufficiently high-grade performance for insertion into transmission lines for relaying broadcasting programmes, with or without automatic means for reversing its direction of operation.

programme signal. The electric waves in audio systems which correspond to the various sounds that are to be reproduced.

programme tape. That in computing and data processing which contains the sequence of instructions.

programmed check. A series of tests included in the programme of a particular problem using the appropriate instructions for the computer.

programmer. Person who analyses a problem and sets out a programme for its solution by a specific electronic computer. Also **compiler.**

programming. Working out detailed sequence of steps which determine operations of an electronic digital computer. The programme is realized in holes punched in cards or tapes which *set* the machine prior to the insertion of data to be processed.

progressive heating. Same as **scanning heating.**

progressive interlace. Normal scanning of a television image, whereby all the *odd* lines are scanned (one *field* in U.S.) and then the *even* lines (another *field* in U.S.), and so on.

progressive scanning. Earlier system than above in which all the lines of a television image are scanned sequentially, without *interlace.*

projection television. Reproduction, on a large

open screen, of a television image established on the phosphor of a small CRT.

promethium (Pm). A radioactive element of the rare earth series, at. no. 61, having no known stable isotopes in nature. Its most stable isotope, ^{146}Pm, has a half-life of over 20 years. Formerly called **illinium.**

promoted mixing. See mixing (2).

prompt critical. Condition in which a reactor could become critical solely with prompt neutrons. For a reactor which is super-critical the reaction rate will rise rapidly and in this condition the reactor is difficult to control. See **delayed neutrons.**

prompt gamma. The gamma radiation which is emitted at the time of fission of the nucleus.

prompt neutrons. Those released at a primary fission with practically zero delay.

propagation. Transmission of energy in the form of electromagnetic waves in the direction normal to a wavefront, which is generally spherical, or part of a sphere, or plane. Applies also to acoustic waves.

propagation constant. Measure of diminution in magnitude and retardation in phase experienced by current of specified frequency in passing along unit length of a transmission line or waveguide or through one section of a periodic lattice structure. Given by the natural logarithm of the ratio of output to input current or of acoustic particle velocity. See also **transfer constant.**

propagation loss. The transmission loss for radiated energy traversing a given path. Equal to the sum of the spreading loss (due to increase of the area of the wavefront) and the attenuation loss (due to absorption and scattering).

propagation of light. The propagation of transverse electromagnetic waves through a vacuum, with a velocity of $2 \cdot 9979 \times 10^{10}$cm sec.$^{-1}$ or 186,281 miles sec.$^{-1}$. The ratio *velocity of light in vacuum/velocity of light in medium* is the refractive index of the medium. According to the special theory of relativity, velocity of light is absolute and no body can move at a greater speed.

proper time. The time coordinate in the Lorentz frame for which the total momentum of the body is zero.

proportional control. Feedback signal in a control system which is proportional to the discrepancy between the actual and desired values of the controlled quantity.

proportional counter. One which uses the *proportional region* in a tube characteristic, where the gas amplification in the tube exceeds unity, but the output pulse remains proportional to the initial ionization.

proportional ionization chamber. One in which the initial ionization current is amplified by electron multiplication in a region of high electric field strength, as in a proportional counter. This device is not for counting but measures ionization currents over a period of time.

proportional navigation. Navigation in which

rate of turn is proportional to rate of turn of line of sight.

proportional region. The range of operating voltages for a counter tube (or ionization chamber) in which the gas amplification is greater than unity and is not dependent on the primary ionization.

propulsion reactor. Reactor designed to supply energy for the propulsion of a vehicle—at present invariably a ship.

protection. Use of barriers of radiation-absorbing material for personnel against external radiation, securing remoteness from source and reducing exposure time. For internal sources inhalation must be restricted, also ingestion and other modes of entry of radioactive materials into the body.

protector. Tube in which glow discharge from a cold cathode prevents high voltage across a circuit.

protium. Lightest isotope of hydrogen, of mass unity (^{1}H), most prevalent naturally. The other isotopes are *deuterium* (^{2}H) and *tritium* (^{3}H).

protactinium (Pa). A radioactive element, at. no. 91. One radioactive isotope of the element is ^{233}Pa which is an intermediate in the preparation of the fissionable ^{233}U from thorium.

proton. Elementary particle, of positive charge and unit atomic mass, atom of lightest isotope of hydrogen without its electron. Appears to be joined with neutrons in building up nucleus of atoms, there being Z protons in the nucleus, where Z is the atomic number. (Each chemical element can have a variable number of neutrons to form isotopes.) It is the lightest baryon (rest mass 1.00758 amu), believed to be the only one completely stable in isolation.

proton precessional magnetometer. Precision magnetometer based on measurement of *Larmor frequency* of protons in a sample of water. See **nuclear magnetic resonance** and **proton resonance.**

proton-proton chain. A series of thermonuclear reactions which are initiated by a reaction between two protons. It is thought that the proton-proton cycle is more important than the carbon cycle in the cooler stars.

proton resonance. A special case of *nuclear magnetic resonance.* Since the nuclear magnetic moment of protons is now well known, that of other nuclei is found by comparison of their resonant frequency with that of the proton.

proton-synchrotron. Synchrotron which is modified to allow the acceleration of protons by frequency modulation of the RF acceleration voltage.

prototype filter. Basic type which has the specified nominal cutoff frequencies, but which must be developed into derived forms to obtain further desirable characteristics, such as constancy of image impedance with frequency.

proximity effect. Increase in effective high-frequency resistance of a conductor when it is brought into the proximity of other conductors, owing to eddy currents induced in the latter. It is especially prominent in the adjacent turns of an inductance coil.

proximity fuse. Miniature radar carried in shells or other missiles so that they explode within a preset distance of the target.

psophometer. An instrument, incorporating a suitable weighting network, which gives a visual indication that is equivalent to the aural effect of disturbing voltages of various frequencies.

psophometric voltage. That which measures, by reference of the random spectrum to 800 Hz, the noise in a communication circuit arising from interference of any kind.

public-address system. Sound-reproducing system for large space and outdoor use, usually with high-powered horn radiators, or columns of open diaphragm units which concentrate radiation horizontally.

Puckle time base. The means of generating a sawtooth waveform by charging a capacitor via a pentode and discharging through a multivibrator.

pull-out. Sudden release of an oscillator, which has been pulled off its own frequency of oscillation by a variable-frequency independent oscillator.

pulling. Variation in frequency of an oscillator when the load on it changes.

pulling by crystal. Growing both metal and non-metal crystal by slowly withdrawing a crystal from a molten surface. See **zone refining.**

pulling figure. The stability of an oscillator, measured by the maximum frequency change when the phase angle of the complex reflection coefficient at the load varies through 360° and its modulus is constant and equal to 0·2.

pulsatance. Same as **angular frequency.**

pulsating current. One, usually unidirectional, which takes the form of a succession of isolated pulses.

pulsating quantity. Same as **periodic quantity.**

pulse. One *step* followed by a reverse *step* after a finite interval. A unidirectional flow of current of non-repeated waveform, i.e., consisting of a transient and a zero-frequency component greater than zero. Measured either by peak value, duration, or integration of magnitude over time. Also called **impulse.**

pulse amplifier. Amplifier with very wide frequency response which can amplify pulses without distortion of the very short rise time of the leading edge.

pulse amplitude. That of the crest relative to the quiescent signal level. Usually a mean taken over the pulse duration.

pulse-amplitude modulation. That which is impressed upon a pulse carrier as variations of amplitude. They may be either unidirectional or bidirectional according to the system employed.

pulse carrier. A carrier wave comprising a series of equally spaced pulses.

pulse code. Coding of information by pulses, either in amplitude, length, or absence or presence in a given time interval. See **Baudot code, binary-coded decimal system, Morse code.**

pulse-code modulation. Any form of pulse modulation in which quantized elements of a varying signal form the modulation code. PAM, PTM, PPM & PWM will all be forms of PCM for a continuous signal modulating a pulse carrier.

pulse crest factor. The ratio of the peak and rms amplitudes.

pulse decay time. Time for decay between the arbitrary limits of 0·90 and 0·10 of the maximum amplitude.

pulse delay circuit. One through which the propagation of a pulse signal takes a known time.

pulse demoder. Circuit responding only to a specific *pulse mode.*

pulse duration. (1) Time interval for which the amplitude exceeds a specified proportion (usually $1/\sqrt{2}$) of its maximum value. Also **pulse length, pulse width.** (2) Duration of a rectangular pulse of the same maximum amplitude and carrying the same total energy.

pulse duty factor. The ratio of pulse duration to spacing.

pulse edges. See **edges.**

pulse EHT generator. The means of generating the EHT supply for a TV tube by the rectification of the voltage pulses occurring in the anode circuit of a valve, when the anode current is suddenly 'cut off'.

pulse-forming line. An artificial line which generates short high-voltage pulses for radar.

pulse frequency modulation. Frequency modulation of a pulse carrier wave by an input signal. Also **pulse interval modulation.**

pulse generator. One supplying single or multiple pulses, usually adjustable for PRF, amplitude and width. May be self-contained or require sine wave input signal.

pulse height analyser. (1) A single or multi-channel pulse height selector followed by equipment to count or record the pulses received in each channel. The multichannel units are known as **kicksorters.** (2) One which analyses statistically the magnitudes of pulses in a signal using PCM.

pulse height discriminator. See **discriminator** (2).

pulse height selector. Circuit which accepts pulses with amplitudes between two adjacent levels and rejects all others. An output pulse of constant amplitude and profile is produced for each such pulse accepted. The interval between the two reference amplitudes is termed the *window* or *channel width* and the lower level the *threshold.*

pulse interleaving or **interlacing.** Adding independent pulse trains on the basis of time division multiplex along a common path.

pulse interval modulation. See **pulse frequency modulation.**

pulse ionization chamber. One for detection of individual particles by their primary ionization. It must be followed by a very high-gain stable amplifier, but has a much shorter dead time than a proportional or Geiger counter.

pulse jitter. Irregularities in pulse spacing.

pulse length. See **pulse duration.**

pulse method of measuring sound velocities. The velocity is measured by comparing the time of transit of an acoustic pulse through the medium with that of a radar signal over the same distance.

pulse mode. Coded group of pulses which select a particular communication channel from a common carrier. A *pulse mode multiplex* controls these channels by means of *pulse demoders.*

pulse moder. Circuit generating *pulse mode.*

pulse pile-up. Failure to separate adjacent pulses due to finite resolving time of detection circuit. In nuclear counting it occurs where pulses have random distribution, and in communication circuits when the PRF or pulse duration are variable.

pulse-position modulation. One form of **pulse-time modulation.**

pulse regeneration or **restoration.** Correction of pulse to its original shape after phase or amplitude distortion.

pulse repeater. See **repeater.**

pulse repetition frequency or **rate.** The average number of pulses in unit time.

pulse rise time. Time required for amplitude to rise from 0·10 to 0·90 of its maximum value.

pulse shaping and **re-shaping.** Adjustment of a pulse to square-wave form by electronic means.

pulse spacing. The period (or time interval) between corresponding parts of successive pulses) for a pulse train.

pulse spectrum. The distribution, as a function of frequency, of the magnitudes of the Fourier components of a pulse.

pulse spike. A subsidiary pulse superimposed upon a main pulse.

pulse-time modulation. Modulation by pulses, the duration or relative position of these depending on a signal to be transmitted.

pulse time multiplex. See **time division multiplex.**

pulse train. A series of equally spaced similar pulses.

pulse transformer. A transformer designed to accept the very wide range of frequencies required to transmit pulse signals without serious distortion.

pulse valley. The part of a pulse lying between two specified maxima.

pulse width. See **pulse duration.**

pulse-width modulation. One form of **pulse-time modulation.**

pulsed Doppler. Doppler system in which direction and range can be measured.

pulsed radar system. One transmitting short pulses at regular intervals and displaying the reflected signals from the target on the screen of a CRO.

pulser. A simple *pulse generator* used to drive pulse-controlled circuits.

pulsing. Said of an oscillatory valve when it alternates between, or sustains oscillations of, two frequencies.

pump frequency. Frequency of oscillator used in *maser* or *parametric amplifier* to provide the stored energy released by the input signal.

pumped tube. Transmitting valve, X-ray tube or other electronic device which is continuously evacuated during operation.

pumping speed. Rate at which a pump removes gas in creating a near-vacuum; measured in litres (or cu. ft.) per minute, at s.t.p. against a specified pressure.

punch-through. Collector-emitter voltage breakdown in transistors.

punched card. Card providing stored information or instructions for use in computer or data-processing system.

punched tape. Tape into which coded patterns of holes are punched, constituting the source of input and output data; originated in telegraphy. Also **paper tape.**

punching. Process of preparing punched cards or tapes.

Pupin coil. See **loading coil.**

pure colour. Colour with CIE coordinates lying on the *spectrum locus* or on the *purple boundary.*

pure tone. See **simple tone.**

Purkinje effect. Shift of maximum sensitivity of human eye towards blue end of spectrum at very low illumination levels.

purple boundary. Straight line joining the ends of the *spectrum locus* on a *chromaticity diagram.* The coordinates of all real colours fall within the loop formed by these two lines.

pursuit. Navigation of a missile by electronic means whereby the guided vehicle is always on line of sight.

push-button tuning. Selection by push-button of a number of preset tuned circuits in a receiver, to change wavelengths quickly.

push-pull. (1) General principle of transmission, whereby one leg of a balanced circuit is driven by a periodic waveform, while the other leg is driven by the same waveform with phase reversed. (2) Term applied to sound-tracks which carry sound recordings in anti-phase. They are *class-A* when each carries the whole waveform and *class-B* when each carries half the waveform, both halves being united optically or in a push-pull photocell.

push-pull amplifier. Two thermionic valves so connected that when the grid of one is

positive (with respect to its mean potential) that of the other is negative, and vice versa. Used for the reduction of harmonic distortion in amplifiers, and in short-wave oscillators and amplifiers, etc.

push-pull microphone. Carbon microphone in which two carbon-granule cells are mounted on either side of a stretched diaphragm, so that amplitude distortion arising in one is partially balanced by the opposite phase amplitude distortion in the other.

push-pull oscillator. One depending on a pair of push-pull valves, positive feedback being taken from each anode to the opposite grid, thus maintaining symmetry and balance.

push-push amplifier. One which uses two similar valves (or transistors) with their grids in phase opposition but their anodes parallel-connected to a common load. By this means even-order harmonics are emphasized.

pyramidal horn. One with linear flare-out in both planes; *cf. sectoral horn.*

Pyranon or **Pyranol.** TNs for dielectric impregnator used for paper capacitors.

Pyrex. TN for a heat-resistant glass which has a low dielectric loss and a high insulation resistance.

Pyristor. TN for current-sensitive switch.

pyroelectricity. Polarization developed in some hemihedral crystals by an inequality of temperature.

pyrometer. Instrument similar to thermometer, used particularly for measuring high temperatures. The main types are *resistance, thermoelectric, optical* and *radiation* pyrometers.

pyron detector. Early demodulator formed by a contact between iron and iron pyrites.

pyrophoric metals. Metals liable to spontaneous combustion under conditions which may arise in a nuclear reactor. The nuclear fuels U, Th and Pu are all pyrophoric.

Pyroscan. TN for a detector of infrared rays, *ca.* 10μ wavelength.

Pyrotenax. TN for a modern form of electric power cable consisting of copper conductors embedded in a heat-resistant core of copper sheathing. It is capable of operating at a much higher conductor temperature than traditional cables.

pyrotron. Thermonuclear device based on cylindrical discharge chamber with magnetic mirrors at the ends.

Pythagorean scale. A musical scale in which the frequency intervals are given by the ratios of integral powers of the numbers 2 and 3.

Q

Q. Symbol of merit for an energy-storing device, resonant system or tuned circuit. Parameter of a tuned circuit such that

$$Q = \omega L/R, \quad \text{or} \quad 1/\omega CR,$$

where L =inductance, C =capacitance and R =resistance, considered to be concentrated in either inductor or capacitor. Q is the ratio of shunt voltage to injected e.m.f. at the resonant frequency $\omega/2\pi$. $Q = f_r/(f_1 - f_2)$, where f_r is the resonant frequency, and $(f_1 - f_2)$ is the bandwidth at the half-power points. For a single component forming part of a resonant system, it equals 2π times the ratio of the peak energy to the energy dissipated per cycle. For a dielectric it is given by the ratio of displacement to conduction current. Also called **magnification actor, quality factor, Q-factor, storage factor.**

Q-electron. One of two in the Q-shell for radioactive elements of atomic number 88 and above.

Q-factor. See **Q.**

Q-gas. One based on helium (98·2% He, 1·8% Butane), widely used in gas-flow counting.

Q-meter. Laboratory instrument which measures the Q-factor of a component.

Q-point. Quiescent condition of a valve without excitation, as represented by a point on its characteristic curves.

Q-shell. The outermost electron shell ever normally occupied by electrons.

Q-signal. (1) First of three-letter code for standard messages in international telegraphy. (2) In the NTSC colour system, that corresponding to the narrow-band axis of the chrominance signal. (3) Request for pratique in international flag code.

Q-value. Quantity of energy released in a given nuclear reaction. Normally expressed in MeV, occasionally in amu.

quad. Either four insulated conductors twisted together (*star-quad*), or two twisted pairs (*twin-quad*). Normally a single structural unit of a multiconductor cable.

quadrant electrometer. See **Compton electrometer, Dolezalek electrometer.**

quadraton. A form of high-vacuum tetrode.

quadrature. See under **phase.**

quadrature component. Same as **reactive component.**

quadricorrelator. Abb. for *quadr*ature inform-ation *correlator*. Term applied to certain forms of automatic frequency and phase control systems in television which use the correlation existing between a pair of measurements of a synchronizing signal at quadrature phases to derive additional information.

quadripole. A network with two input and two output terminals. A balanced wave-filter section.

quadripole amplifier. A type of **parametric amplifier.**

quadrivalent. An atom with four electrons in its valency shell.

quadruplex system. A system of Morse telegraphy arranged for simultaneous independent transmission of two messages in each direction over a single circuit.

quadrupole. A collection of charges such that the potential at a distant point from their centre of mass may be expressed by an infinite series of terms in inverse powers of r. The inverse third power term is the **quadrupole potential.**

quadrupole moment. Moment derived from the series expansion (see **quadrupole**) of charges multiplied by space coordinates. The sum of the quadratic terms is the quadrupole moment, which is possessed by most metals.

quality. (1) The timbre or quality of a note, which depends upon the number and magnitude of harmonics of the fundamental. (2) In radiography, etc., it indicates the approximate penetrating power. Higher voltages produce higher quality X-rays of shorter wavelength and greater penetration. (This term dates from a period before the nature of X-rays was completely understood.)

quality control. Maintenance of production specification within a specified tolerance, aided by electronic computer which automatically determines mean or mean-square of deviation of actual measurement from that specified.

quality factor. (1) See **Q.** (2) See **relative biological effectiveness.**

Quanticon. Photoconductive camera tube.

quantity of electricity. Usually denoted by Q, it is the amount of electric charge, the practical unit being the *coulomb.*

quantization. (1) In information theory, division of the amplitude of a wave into a restricted number of finite amplitudes (sub-ranges). (2) In quantum theory, the division of the energy of a system into discrete units (quanta) so that the continuous infinitesimal changes are excluded. (3) In computers, see **digitization.**

quantized computer. See **computer.**

quantum. (1) General term for indivisible unit of any form of physical energy, in particular the *photon,* unit of electromagnetic radiation energy, its magnitude depending only on frequency. See also **graviton, magnon, phonon, roton.** (2) In communications, etc., an interval on a measuring scale, fractions of which are considered insignificant.

quantum efficiency. Number of electrons released in a photocell per photon of incident radiation of specified wavelength.

quantum electronics. That concerned with amplification or generation of microwave power in solid crystals, governed by quantum mechanical laws.

quantum mechanics. See **statistical mechanics.**

quantum number. One of a set, describing possible states of a magnitude when quantized, e.g., nuclear spin. See **azimuthal-, spin-.**

quantum statistics. Those dealing with the distribution of the particles of a given type among the various possible energy values. See **Bose-Einstein statistics, Fermi-Dirac statistics.**

quantum theory. That developed from *Planck's law*. There is a quantum theory for most branches of classical physics.

quantum yield. Ratio of the number of photon-induced reactions occurring, to the total number of incident photons.

quark. It has been shown that all strongly interacting particles could be theoretically accounted for in terms of the manifestations of combinations of three more fundamental particles. These hypothetical entities would have fractional charges. The names **quark** and **ace** have been proposed for them.

quarter-wave antenna. One whose overall length is approximately a quarter of free-space wavelength corresponding to frequency of operations. Under these conditions it is oscillating in its first natural mode, and is half a dipole.

quarter-wave line. Quarter-wavelength section of balanced transmission line (Z_0) designed to operate as a matching device between lines of different impedance levels (Z_1, Z_2), such that $Z_0{}^2 = Z_1 Z_2$. Also called **quarter-wave bar.**

quarter-wavelength stub. Resonating two-wire or coaxial line, approximately one quarter-wavelength long, of high impedance at resonance. Used in antennae, as 'insulating' supports for another line, and as coupling elements.

quartz. Natural silicon oxide in clear crystal form, used in place of glass for transparency to ultraviolet radiation, high-temperature resistance or freedom from radioactive contamination. Also used for piezoelectric resonant element in oscillators and ultrasonic transducers. See **twinning** (1).

quartz-fibre dosimeter. See **personal dosimeter.**

quartz lamp. One which contains a mercury arc under pressure, a powerful source of ultraviolet radiation.

quartz oscillator. One whose oscillation frequency is controlled by a piezoelectric quartz crystal.

quartz thermometer. One with digital read-out obtained by beating a temperature-dependent quartz oscillator against a temperature invariant one.

quasar. Acronym for *quasi*-stell*ar*. Point (starlike) radiation source outside our own galaxy but much more localized than others in observable galaxies. *Radio quasars* (quasi-stellar sources) were discovered first but are now known to be much less common than *optical quasars* (quasi-stellar galaxies.)

quasi-bistable circuit. An astable circuit which is triggered at a high rate as compared with its natural frequency.

quasi-conductor. A medium in which $Q \ll 1$.

quasi-dielectric. A medium in which $Q \gg 1$.

quasi-Fermi levels. Energy levels in a semiconductor from which the number of electrons or holes contributing to conduction under non-equilibrium conditions can be calculated, in the same way as from the true Fermi level under equilibrium conditions.

quasi-optical waves. Electromagnetic waves of such short wavelength that their laws of propagation are similar to those of visible light.

quench. Resistor or resistor-capacitor shunting a contact, to reduce high-frequency sparking when a current is broken in an inductive circuit.

quench time. That required to quench the discharge of a Geiger tube. *Dead time* (q.v.) for internal quenching, *paralysis time* (q.v.) for electronic quenching.

quenched spark converter. A generator which utilizes the oscillatory discharge of a capacitor through an inductor and spark gap as a source of RF power.

quenched spark gap. One in which the discharge takes place between cooled or rapidly moving electrodes.

quenched spark system. An old system in which means are employed to extinguish each spark rapidly, so as to reduce the decrement of the oscillatory currents induced in an antenna or closed oscillatory circuit.

quenching. (1) Process of inhibiting continuous discharge, by choice of gas and/or an external valve circuit, so that discharge can occur again on the incidence of a further photon in a counting tube. Essential in a GM counter. (2) Suppression of oscillation, particularly periodically, as in a super-regenerative receiver.

quenching oscillator. One with a frequency slightly above the audible limit, and which generates the voltage necessary to quench the high-frequency oscillations in a super-regenerative receiver.

queueing. Retardation and waiting of data for acceptance, e.g., for computer processing, or objects in an automation line.

quiescent. General term for a system waiting to be operated, e.g., a valve ready to amplify or a gas-discharge tube to fire. Also **preset.**

quiescent carrier transmission. One for which the carrier is suppressed in the absence of modulation.

quiescent current. Current in a valve or transistor, in the absence of a driving or modulating signal. Also **standing current.**

quiescent operating point. The steady-state operating conditions of a valve or transistor in its working circuit but in the absence of any input signal.

quiescent period. That between pulses in a pulse transmission.

quiescent push-pull amplifier. Thermionic valve or transistor amplifier, in which one side alone passes current for one phase, the other side passing current for the other phase.

quiet automatic volume control. Same as **delayed automatic gain control.** The appli-

cation of this principle is known as *quieting.*

quieting sensitivity. The minimum input signal required by an FM radio receiver to give a specified signal/noise ratio at the output.

Quincke tube. Parallel adjustable sound-transmission tube system forming interference filter.

Quincke's method. One for determining the magnetic susceptibility of a substance in solution by measuring the force acting on it in terms of the change of height of the free surface of the solution when placed in a suitable magnetic field.

R

RC coupling. That between separate stages of an amplifier when due to resistive and capacitive elements only.

R-meter. One calibrated in röntgens, incorporating an ionization chamber and indicating meter for radiation survey.

R-value. Percentage decrease in density of reactor fuel for 1 per cent. burn-up. See also **S-value.**

R-Y signal. Component of colour TV chrominance signal. Combined with luminance (Y) signal it gives primary red component.

rbe dose. See under **dose.**

rms power. The effective average power level of an alternating electric supply. See **root-mean-square value.**

rabbit. See **shuttle, single-state recycle.**

Rabi method. Method for determining nuclear magnetic moments; also known as the **molecular beam resonance method.**

race-track. Discharge tube or ion-beam chamber where particles are constrained to an oval path, as in a calutron.

racemic substance. An optically inactive substance, but containing forms of opposed optical activity.

racon. *Ra*dar bea*con* used by navigators for identifying objects at a distance.

rad. Unit of radiation dose which is absorbed, equal to 100 ergs per gram of the absorbing (often tissue) medium.

radac. (*Ra*pid *d*igital *a*utomatic *c*omputing). A system for the fast and accurate analysis of complex data, e.g., for fire control against missiles and rockets.

radan. (*Ra*dar *d*oppler *a*utomatic *n*avigation). A navigation system independent of any ground-based equipment.

radar. (*Ra*dio *d*etection *a*nd *r*anging). System using pulsed radio waves, in which reflected (*primary radar*) or regenerated (*secondary radar*) pulses lead to measurement of distance and direction of target. In an alternative radar system, continuous RF waves are sent out by the transmitter. See:

Doppler-	passive-
early-warning-	primary-
H-	secondary-.

radar altimeter. See **radio altimeter.**

radar beacon. A fixed radio transmitter whose radiations enable a craft to determine its own direction or position relative to the beacon by means of its own radar equipment.

radar control. Direct control of vehicle (e.g., missile) by radar signals.

radar echo. The returned radar pulse received after reflection.

radar fence. A network of radar warning stations which surround a protected area.

radar homing. Either the use of radar to guide a missile to a target, or homing on a radar transmitter.

radar indicator. Display on a cathode-ray tube of the output of a radar system, either as radial line or as coordinate system for range and direction. The echo signal gives a brightening of the luminous spot, which remains for some seconds because of afterglow. Also **radar screen.** See A-, B-, C-, etc. **display, plan-position indicator.**

radar performance figure. The ratio of peak transmitter power to minimum signal required by receiver.

radar pulse. Repeated pulse on very high frequency, usually a few microseconds in duration, repeated, e.g., 1000 times per second, frequency of the wave being up to 40 GHz.

radar range. Usually given as that at which a specified object can be detected with 50% reliability.

radar relay. Apparatus for relaying the radar video and synchronizing signals to a remote location.

radar resolution. The ability to separate signals received from small closely-spaced targets. Best quoted as an angular separation.

radar scan. The path followed by a radar beam in space.

radar screen. See **radar indicator.**

radar V-beam. One using two fan-shaped beams to produce three-dimensional position data on a number of targets.

radarscope. The display unit in a radar system.

radechon. A storage tube in which the controlling element is a fine mesh grid.

radiac. Class of instruments for radiation survey, in mineral prospecting, working with nuclear reactors, etc.

radial distribution function. For an isotropic liquid, the average number density at a particular distance from a selected molecule.

radial grating. One consisting of radial wires inserted in a circular waveguide to suppress E-modes.

radian. Natural measure of angle, such that one revolution $=2\pi$ radians. Hence 1 radian $=57°\ 17'\ 45''$.

radian frequency. Same as **angular frequency.**

radiant flux. The time rate of flow of radiant electromagnetic energy.

radiant-flux density. A measure of the radiant

power per unit area that flows across a surface. Also **irradiance**.

radiant intensity. The energy emitted per second per unit solid angle about a given direction.

radiated power. The actual power level of the radio signals transmitted by an antenna. See also **effective radiated power**.

radiating circuit. Any circuit capable of sending out power, in the form of electromagnetic waves, into space, especially the antenna circuit of a radio transmitter.

radiation. The dissemination of energy from a source. Its intensity falls off as the inverse square of the distance from the source in the absence of absorption. Applied to electromagnetic waves (radio waves, light, X-rays, γ-rays, etc.), acoustic waves, and emitted particles (α and β, protons, neutrons, mesons, etc.). See:

actinic-	intensity of-
alpha-	ionizing-
annihilation-	resonance-
Cherenkov-	secondary-
electromagnetic-	soft-
external-	spurious-
gamma-	stray-
heterogeneous-	ultraviolet-
homogeneous-	visible-.
infrared-	

radiation chemistry. Study of the chemical effects of high-energy radiation on matter.

radiation counter. One used to detect individual particles or photons in nuclear physics. See:

bubble chamber	lithium-drift german-
Cherenkov	ium detector
counter	nuclear emulsion
cloud chamber	pellicle
film badge	photographic emulsion
gas amplification	technique
gas counter	proportional counter
gas-flow counter	proportional ioniz-
Geiger-Müller	ation chamber
counter	pulse ionization
ionization	chamber
chamber	spark chamber
liquid counter	Sugarman counter
liquid flow counter	track chamber.

radiation damage. The effects of radiation, e.g., γ-rays, neutrons, upon matter. Can involve dissociation, evolution of gases and disruption of crystal structure. In the case of polythene, e.g., continued radiation leads to cross-linking and an increase in the modulus of elasticity with a decrease in the internal mechanical damping.

radiation diagram. See **radiation pattern**.

radiation efficiency. Ratio of actual power radiated by antenna to that provided by the drive. Also **aerial efficiency**.

radiation emittance. Radiated power per unit area of surface. It is expressed either as the total radiation or as the spectral radiant emittance, which is the emittance per unit band in wavelength.

radiation field. See **field** (1).

radiation flux density. Rate of flow of radiated energy through unit area of surface normal to the beam. (For particles this is frequently expressed in number rather than energy.) Also **radiation intensity**.

radiation impedance. Impedance per unit area. Measured, e.g., by the complex ratio of the sound pressure to the velocity at the surface of a vibrating body which is generating sound waves, or by the corresponding electromagnetic quantities.

radiation intensity. (1) Radiation dose rate (deprecated). (2) See **radiation flux density**.

radiation length. The path length in which relativistic charged particles lose e^{-1} of their energy by radiative collisions. See **bremsstrahlung**.

radiation loss. That power which is radiated from a non-shielded radio-frequency transmission line.

radiation pattern. Polar or Cartesian representation of distribution of radiation in space from any source, and, in reverse, effectiveness of reception; also called **radiation diagram**.

radiation potential. See **ionization potential**.

radiation pressure. Minute pressure exerted on a surface normal to the direction of propagation of an electromagnetic wave. In the case of a sound wave in a fluid, the pressure gives rise to 'streaming', i.e., a flow of the fluid medium.

radiation pyrometer. A temperature-measuring instrument which depends on the absorption of radiant heat, using as detector a bolometer, a thermocouple or a thermopile. Photoelectric cells may be used for detecting radiation in and near the visible.

radiation resistance. That part of the impedance of an antenna system related to the power radiated; the power radiated divided by the square of the current at a specified point, e.g., at the junction with the feeder.

radiation sickness. The physiological effects of excessive radiation dose. Nausea, vomiting, skin burn, fall in blood count or death according to amount of exposure.

radiation survey. Evaluation of radiation hazards in a given building or region. This may involve plotting contours of equal radiation intensity.

radiation therapy. The use of any form of radiation, e.g., electromagnetic, electron or neutron beam, or ultrasonic, for the treatment of disease.

radiation trap. (1) Beam trap for absorbing intense radiation beam with a minimum of scatter. (2) Maze or labyrinth formed by entry corridor with several right-angle bends, used for approach to multicurie radiation sources or some accelerating machines.

radiative collision. Collision in which kinetic energy is converted into electromagnetic radiation. See **bremsstrahlung**.

radiator. (1) Element of an antenna which is effective in radiating power as electromagnetic waves; usually a resonating quarter- or half-wavelength wire, or a multiple of this, with suppression or neutralization of elements, or of their radiation in the opposite phase or direction. (2) In radioactivity, the origin of alpha-, beta- and/or gamma-rays; also called a **source**.

radio. Generic term applied to methods of signalling through space, without connecting wires, by means of electromagnetic waves generated by high-frequency alternating currents. Supersedes the obsolete *wireless*.

radio altimeter. Device for determining height, particularly of aircraft in flight, by electronic means. Generally by detecting the delay in reception of reflected signals, or change in frequency, using the Doppler effect; also called **terrain clearance indicator**.

radio astronomy. Study of noise radio waves from celestial objects. Strong, e.g., on 21 cm wavelength, arising from reversal of spin in hydrogen atoms in space. Includes study of meteor streams, solar noise, radio stars and quasars, using radio telescopes.

radio beacon. Stationary radio transmitter which transmits steady beams of radiation along certain directions for the guidance of ships or aircraft, or one which transmits from an omnidirectional antenna and is used for the taking of bearings, using an identifying code. Also **aerophare**. See:

fan marker-	racon
glide-path-	radar-
inner marker-	ramark
localizer-	track guide
marker-	VHF rotating talking-
middle marker-	Z marker-.
outer marker-	

radio beam. Concentration of electromagnetic radiation within narrow angular limits, such as is emitted from a highly directional antenna in the form of a curtain or bowl.

radio bearing. Direction of arrival of a radio signal, as indicated by a loop or goniometer, for navigational purposes.

radio broadcasting. The transmission by means of electromagnetic (i.e., radio) waves, of a programme of sound or TV for general reception. The separation of the frequency channels, usually about 10 kHz, is decided by international agreement.

radio channel. Any channel suitable for the transmission of radio signals.

adio circuit. Communication system including a radio link, comprising a transmitter and antenna, the radio transmission path with possible reflections or scatter from ionized regions, and a receiving antenna and receiver.

radio-colloids. Radioactive atoms in colloidal aggregates.

radio communication. Transmission of information by channels including radio links, fixed or mobile, either coded (telegraph), analysed (video and fascimile), or plain (broadcasting and telephony). Postwar extensions include radar, hyperbolic navigation and telemetry from space vehicles.

radio compass. Originally a rotating loop,

later rendered more sensitive by a goniometer system, and by display on a cathode-ray tube. Any device, depending on radio, which gives a bearing. See **Adcock antenna**, **Bellini-Tosi antenna**.

radio control. The control of a vehicle by radio signals.

radio direction-finding. Formerly used for all methods of direction-finding by radio techniques, now principally for passive reception of direction-finding signals from radio beacons or navigational transmitters, as distinct from active radar.

radio exchange. Radio-receiving station, with multiple power amplifiers, for distributing to subscribers radio programmes on a relay basis, via overhead wires, telephone lines or electric power mains; also termed **rediffusion, relay system**. See **carrier telegraphy**.

radio fix. A position as determined by a radio-navigation system.

radio frequency. One suitable for radio transmission, above 10^4 Hz and below 3×10^{12} Hz approx. Also **radio spectrum**.

radio-frequency amplifier. One designed for operation on radio frequencies, based on tuned circuits.

radio-frequency component (choke, resistor, transformer, etc.). One designed to operate with radio-frequency signals.

radio-frequency heating. See **dielectric heating, induction heating**.

radio-frequency pulse. One containing a number of cycles of radio-frequency oscillation.

radio-frequency spectroscopy. Study of absorption spectrum of nuclei spinning in strong magnetic field by nuclear paramagnetic resonance; or of molecular electron spin resonances.

radio-frequency transformer. One designed for operation on radio frequencies; generally, the primary and secondary windings are tuned with capacitors, thus providing a band-pass filter. Cores, if any, are moulded from iron powders.

radio heating. See **high-frequency heating**.

radio horizon. In the propagation of electromagnetic waves over the earth, the line which includes the part of the earth's surface which is reached by direct rays.

radio link. Self-contained radio circuit capable of working in both directions, for insertion between two land-line circuits.

radio microphone. (1) Microphone which transmits a signal which is demodulated for applying to an audio reproducing system. (2) One with a miniature radio transmitter, which allows entire freedom for the speaker, the transmission being picked up by a receiver nearby.

radio navigation. The use of radio signals for obtaining position fixes or indicating departures from a planned course line.

radio quasar. See **quasar**.

radio range. Specific system of radio homing for aircraft, in which crossed loops are separately modulated with complementary signals, which coalesce on reception when the aircraft is *on course*.

radio receiver. One for receiving and interpreting radio transmissions. Normally *either* a radio-frequency tuned amplifier, demodulator, and a low frequency amplifier, *or* an oscillator (first detector or remodulator), a band-pass intermediate frequency amplifier, a second detector or demodulator, followed by a low-frequency amplifier.

radio relay. See **wire broadcasting.**

radio-relay station. An intermediate station receiving a signal from the primary transmitter and re-radiating it to its destination.

radio silence. 3-minute periods twice every hour when all ship and aircraft transmitters are silent, the operators listening for distress calls on 0·5 MHz.

radio spectrum. See **radio frequency.**

radio station. Complete equipment for the transmission and/or reception of radio telegraphy or telephony, together with the building(s) and power plant.

radio telegraphy. Using a radio channel for telegraph purposes, e.g., by interrupted carrier, change of frequency, modulation with interrupted audio tone; either from fixed or mobile stations.

radio telephone. A combination of radio transmitter and receiver employed in voice communication, sometimes together with a telephone-wire system.

radio telephony. Use of a radio channel for transmission of telephonic speech. Methods include simple modulation, suppressed carrier and one sideband, inverted sidebands and carrier, scrambling before modulation, one in a group modulation, or pulse modulation.

radioactinium (RdAc). Thorium isotope of mass number 227, formed naturally from the β-decay of actinium-227. Radioactinium emits α-particles to produce radium-223.

radioactivation analysis. Method based on measurement of radionuclides formed by neutron irradiation, or less frequently, proton or deuteron irradiation by a particle accelerator.

radioactive atom. Atom which decays into another species by emission of an α- or β-ray (or by electron capture). Activity may be natural or induced.

radioactive chain. Chart of natural radioactive isotopes, showing how one is related to others through radiation and decay, finally to an isotope of lead. Three natural radioactive series exist, their members having mass numbers :—(*a*) $4n$ (thorium series); (*b*) $4n+2$ (uranium series); (*c*) $4n+3$ (actinium series). Members of the $4n+1$ (neptunium series) can be produced artificially. *N.B.*—Radioactive decay leads to a decrease of one in the value of *n* above, whenever an alpha-particle is emitted. Also **radioactive series.**

radioactive decay. See **disintegration constant, half-life, radioactive atom.**

radioactive displacement law. See **displacement law.**

radioactive equilibrium. Eventual stability of products of radioactivity if contained, i.e., rate of formation (quantitative) equals rate of decay. Particularly important between radium and radon.

radioactive isotope. Natural or artificial isotope exhibiting radioactivity, used as a source for medical or industrial purposes. Also **radioisotope.**

radioactive series. See **radioactive chain.**

radioactive standard. A radiation source for calibrating radiation measurement equipment. The source has usually a long half-life and during its decay the number and type of radioactive atoms at a given reference time is known.

radioactive tracer. Small quantity of radioactive preparation added to corresponding non-active material to *label* or *tag* it, so that its movements can be followed by tracing the activity. (The chemical behaviour of radioactive elements and their non-active isotopes is identical.)

radioactive tube. Any electron tube in which the cathode is activated by a radioactive substance. About $1\mu\mathrm{C}$ of radium bromide has been commonly used for such tubes.

radioactivity. Spontaneous disintegration of certain natural heavy elements (radium, actinium, uranium, thorium) accompanied by the emission of α-rays, which are positively-charged helium nuclei; β-rays, which are fast electrons; and γ-rays, which are short X-rays. The ultimate end product of radioactive disintegration is an isotope of lead. Discovered by Becquerel in 1894, classified by Marie Curie, investigated by Rutherford and Soddy early in this century. See **half-life, mean life, radioactive chain, etc.**; also **artificial-, atmospheric-, induced-.**

radiobiology. Branch of science involving study of effect of radiation and radioactive materials on living matter.

radiocarbon age (or dating). See **carbon dating.**

radiochemical purity. The proportion of a given radioactive compound in the stated chemical form. See also **radioisotopic purity.**

radiochemistry. Study of science and techniques of producing and using radioactive isotopes or their compounds to study chemical problems. Not to be confused with **radiation chemistry.**

radioelement. One exhibiting natural radioactivity.

radiogenic. Said of stable or radioactive products arising from radioactive disintegration.

radiogoniometer. See **goniometer.**

radiogram. (1) Message regularly transmitted by radio telegraphy. Once termed **marconigram,** after the inventor, Marconi. (2) See under **radiography.** (3) Combination of gramophone and radio broadcast receiver.

radiography. Registration of images in photographic material by X-rays or gamma-rays, the result being properly termed a *radiogram.* Such images depend on differential absorption of the substances, including human bone or tissue, passed through.

radioisotope. Same as **radioactive isotope.**

radioisotopic purity. The proportion of the activity of a given compound which is due to material in the stated chemical form. See also **radiochemical purity.**

radiolocation. Former term for **radar.** See **radio direction-finding.**

radiology. The science and application of X-rays, gamma-rays, and other penetrating ionizing radiations.

radiolucent. Scattering and partly absorbing radiation (especially X-rays).

radioluminescence. Luminous radiation arising from rays from radioactive elements, particularly in mineral form.

radiolysis. Chemical decomposition induced by ionizing radiation.

radiomesh. French hyperbolic navigational system.

radiometer. Instrument devised for the detection (e.g., *Dicke's radiometer* (q.v.), two-receiver radiometer, subtraction-type radiometer) and measurement (e.g., thermopile, bolometer, microradiometer) of electromagnetic radiant energy. Also for the measurement of acoustic energy. See **Rayleigh disk.**

radiomimetic. Of chemical agents, such as nitrogen mustard and certain mesyloxy compounds, the biological effects of which, in very small doses, closely duplicate those of exposure to ionizing radiations.

radionuclide. Any nuclide (isotope of an element) which is unstable and undergoes natural radioactive decay.

radiopaque. Opaque to radiation (esp. X-rays).

radioparent. Transparent to radiation (esp. X-rays).

radiophare. Same as **radio beacon.**

radiophone. Telephone system employing a radio link, e.g., over oceans or between ships.

radiophotoluminescence. Luminescence produced by irradiation followed by exposure to light.

radiophotoluminescent dosemeter. An integrating radiation dosemeter similar in principle to the thermoluminescent type but in which read-out (emission of light) is obtained by releasing the radiation energy stored in the dielectric by means of irradiation with ultraviolet light. These devices normally use silver activated phosphate glass elements the light output from which is well-matched to the cathode sensitivity of a conventional PM tube.

radioresistant. Able to withstand considerable radiation doses without injury.

radiosensitive. Quickly injured or changed by irradiation. The gonads, the blood-forming organs and the cornea of the eye are regarded as biologically most radiosensitive in man.

radioscope. Electroscope designed for use with radioactive materials.

radiosonde. Sounding balloon which ascends to high altitudes, carrying meteorological equipment which modulates (usually frequency) radio signals transmitted back to equipment on earth.

radiotherapy. Theory and practice of medical treatment of disease, particularly any of the forms of cancer, with large doses of X-rays or other ionizing radiations.

radiothermoluminescence. Luminescence produced by irradiation followed by exposure to heat. Now used for personal dosimetry.

radiothorium (RdTh). A common name for $1 \cdot 90y$ ^{228}Th which is a member of the thorium series.

radiovision. See **closed circuit** (3).

radium (Ra). Rare, extremely-radioactive metal, at. no. 88. A member of the uranium decay series first isolated by Mme. Curie.

radium G (RaG). The common name for ^{206}Pb, which is the stable end product of the uranium series. Natural lead contains $23 \cdot 69 \%$ of this isotope.

radix. Basis of system of numbers, as indicated by position of digits. Normally the radix 10 is used, but many computers use 2, thus reducing the number of entities to be dealt with, because a current *on* or *off*, or the presence or absence of a pulse at a particular time, is a sufficient physical representation of a digit. Also **base** (4).

radix point. That which separates the positive and negative indices of the radix; the decimal point when the radix is 10.

radome. Housing for radar equipment, transparent to the signals, e.g., a plastic shell on aircraft or a balloon on the ground; also **blister, raydome.**

radon (Rn). Gaseous radioactive element, at. no. 86, given off by U, Th and Ac. Extremely hazardous, due to risk of absorption during respiration. Formerly called **niton.** See **emanations.**

raffinate. Liquid layer in solvent extraction system from which required solute has been extracted.

Raman scattering. See **scattering.**

ramark. A non-directional radio beacon, usually pulsed, used as a radar beacon to enable a craft to determine its direction relative to it.

ramp voltage. Steadily rising voltage, as in a sawtooth waveform.

Ramsauer effect. Sharp decrease to zero of scattering cross-section of atoms of inert gases, for electrons of energy below a certain critical value.

random access. Requirement whereby any data in a large store can be located and abstracted without affecting other data.

random coincidence. Simultaneous operation of two or more coincidence counters as a result of their discharge by separate incident particles arriving together (not of common discharge by a single particle, as is normally assumed in interpreting the readings).

random error. One unlikely to be repeated in any series of measurements.

random event. Event which has no effect on the probability of occurrence of subsequent events.

random noise. Noise due to the aggregate of a large number of elementary disturbances with random occurrence in time.

random number. One of a large set in which the digits are equally likely to appear.

range. (1) Maximum distance of radio transmitter at which effective reception is possible (not normally constant). (2) In navigation, the distance from a navigation point. (3) Distance that vehicle can cover without refuelling. (4) Instrument limiting values of a variable for which operation is possible. (5) Length of track along which ionization is produced in a nuclear particle. (6) Distance of effective operation of nuclear forces. (7) Maximum horizontal distance covered by projectile. (8) Distance in medium for which specified reduction of intensity in electromagnetic radiation occurs, e.g., half-value range. (9) Maximum distance for effective acoustical reception. (10) The optical short-wave radio range calculated from straight-line propagation.

range-energy relation. That, usually in form of a graph, between the range of particles (of a given type and initial kinetic energy) and the energy. For mean α-particle ranges between 3 cm and 7 cm, a typical relation for the extrapolated range R_α (in cm of air at 1 atms. and 15°C) versus the energy E (in MeV) is $R_\alpha = 0.318 E^{3/2}$.

range height indicator. A radar display used in conjunction with a plan-position indicator for airport control. It displays a vertical plane on which elevation and bearing of target can be seen.

range-tracking. Radar system in which an electronic device rejects signals other than those from a moving target.

rapid-access memory. One in which time of access to computer is of the order of microseconds.

rare gases. See **inert gases**.

rarefaction. Decrease of density due to expansion of the low pressure wavefront of a sound wave.

raster. The area of the screen of a television picture tube which is illuminated by the pattern of horizontal lines produced, during a complete frame-scanning period. Also **scanning field**.

raster stereoscopy. A form of cinematography in which a three-dimensional image is obtained by using a radial grid of conical plastic lens elements in front of the screen.

rat-race. See **hybrid junction**.

ratchetting. Periodic movement of reactor fuel elements due to thermal cycling and differential thermal expansion effects.

rate action. Same as **booster response**.

rate gyro. Gyroscope with single gimbal, which produces a couple proportional to the rate of rotation.

rate time. Period during which control depends on a rate of change or derivative of position.

rated capacity. General term for the output of an equipment, which can continue in-definitely, in conformity with a criterion, e.g., heating, distortion of signals or of waveform. See **continuous, intermittent, periodic ratings**.

rated impedance. Particularly applied to a loudspeaker, in which impedance rises with frequency, with an added sharp rise at frequency of bass resonance. That resistance, equal in magnitude to the modulus of the minimum impedance above this resonant frequency, which replaces the loudspeaker when measuring the power applied to the loudspeaker during testing.

ratemeter. See **count ratemeter**.

rating. Specified limit to operating conditions.

ratio detector. Detector circuit used for frequency-modulated carriers.

rationalized units. Systems of electrical units for which the factor 4π is introduced in Coulomb's laws in order that it shall be absent from more widely used relationships. In Heaviside-Lorentz units (rationalized Gaussian units), this is done directly, thus modifying values of the unit charge and unit pole. In MKSA units, it is done indirectly by modifying the values of the permittivity and permeability of free space.

rattle echo. Unmusical multiple echo, generally associated with thunder, which is formed by nearby flashes giving rise to a sharp acoustic impulse which is reflected between mountains or strata of air.

ray. General term for the geometrical path of the radiation of wave energy, always in a direction normal to the wavefront, but with possible reflection, refraction, diffraction, divergence, convergence and diffusion. By extension, also the geometrical path followed by a beam of particles in an evacuated chamber. This may be curved in electric or magnetic field. See also:

alpha-	particle
Becquerel-	positive-
cosmic-	reflected wave
direct-	röntgen-
ionospheric-	soft cosmic-.

raydome. Same as **radome**.

Rayescent. RTM for electroluminescent panels.

Rayleigh disk. Small, light, mica disk (in water a lead disk is used), pivoted about a vertical diameter, hung by a fine thread of glass or quartz. If placed at an angle to a progressive sound wave, the disk experiences a torque which depends on the square of the velocity of the molecules in the medium. It provides a useful method of measurement of sound intensity, calculated from the measured torque, using a formula due to König.

Rayleigh distillation. A simple distillation in which the composition of the residue changes continuously during the course of the distillation.

Rayleigh distribution. Spectral distribution of noise energy which has a skew distribution.

Rayleigh scattering See **scattering**.

Rayleigh's theorem. Relation between energy of an impulse function in terms of time and the same energy as the sum of energies of frequency components in a frequency spectrum.

reactance. The imaginary part of the impedance. Reactances are characterized by the storage of energy rather than its dissipation. See also:

capacitive-   inductive-
effective-   leakage-.
equivalent-

reactance chart. A chart of logarithmic scales so arranged that it is possible to read directly the reactance of a given inductor or capacitance at any frequency.

reactance coupling. Coupling between two circuits by means of a reactance common to both, e.g., a capacitor or inductor.

reactance modulation. Use of a variable reactance, e.g., capacitor or inductor, or a reactance valve, to effect frequency modulation.

reactance theorem. See **Foster's reactance theorem.**

reactance valve. One in which, by alteration of a voltage on a grid, anode impedance appears as a changing capacitance or inductance. This can alter resonance frequency of oscillatory circuit and hence effect frequency modulation.

reactatron. A low-noise microwave amplifier with a semiconductor diode.

reaction. Positive feedback by coupling between an anode and a previous grid, whereby a small voltage applied to the latter is reinforced by voltage or current of the former. Original use was to reduce effective loss in a tuned circuit and hence improve its gain and selectivity. Excess of reaction leads to oscillation. Also **retroaction.** See **positive feedback, regeneration.** See also:

armature-   pickup-
capacitance-   reverse-
chain-   slow-
electromagnetic-   thermonuclear-.
fast-

reaction capacitor. Variable capacitor for varying the degree of reaction.

reaction coil. One included in the anode circuit of a thermionic valve and inductively coupled to the grid circuit.

reaction rate. The rate of fission in a nuclear reactor.

reactivation. When a thoriated tungsten filament loses its emission, the raising of the temperature for a time (without anode voltage) to bring fresh thorium to the surface.

reactive component of current (or voltage). Preferred terms for component of vectors representing an alternating current (voltage) which is in quadrature (90°) with the voltage (current) vector; also called **idle, inactive, quadrature,** or **wattless component.**

reactive factor. The ratio of reactive volt-amps to total supply volt-amps.

reactive volt-amperes. Product of the reactive voltage and the amperes in a circuit, or the reactive current (amperes) and the voltage of the circuit; measure of wattless power in circuit. Abb. var (*volt-amperes reactive*).

reactivity. The departure of the multiplication constant of a reactor from unity, measured in different ways. See **cent, dollar, nile.**

reactor. (1) Electric circuit component which stores energy (especially applied to iron-cored chokes). (2) Assembly in which the fission of a greater-than-critical mass of a fissile material is regulated by a moderator. Controlling factor is the neutron flux, regulated by absorbing rods (boron-steel, cadmium). Used for neutron irradiation and for releasing nuclear energy. Classified according to fuel, moderator and coolant (or less frequently according to size, power output or function). Several hundred types of reactor have been tested or suggested, based on the following alternatives: *fuel*: plutonium-239, uranium-233, uranium-235; *moderator*: beryllium, graphite, heavy water, light water, organic liquid (or none with fast reactors); *coolant*: gas, heavy water, light water, liquid metal, organic liquid. Fuels are classed as natural, slightly enriched or heavily enriched, according to the extent to which the proportion of fissile material has been increased beyond its normal isotopic abundance. See also:

advanced gas-   intermediate-
  cooled-   low-flux-
boiling-water-   natural-uranium-
breeder-   organic-
ceramic-   packaged-
circulating-fuel-   pebble-bed-
commutating-   plutonium-
converter-   power-
enriched-   pressurized-water-
  uranium-   propulsion-
exponential-   saturable-
fast-   slow-
gas-cooled-   slurry-
heavy-water-   sodium-cooled-
heterogeneous-   swimming-pool-
high-flux-   tank-
high-tempera-   thermal-
  ture-   thorium-
homogeneous-   uranium-
image-   zero-power-.

reactor noise. Random statistical variations of neutron flux in reactor.

reactor oscillator. Device for producing mechanical oscillation of neutron-absorbing sample in reactor core.

reactor simulator. Analogue computer which simulates variations in reactor neutron flux produced by changes in any operating parameter.

reactor trip. Rapid reduction of reactor power to zero by emergency insertion of control members.

read in. To insert data signals into a computer from punched tape or cards or from magnetic tape.

read-out. (1) Output unit of computer. (2) Data from above in form of mechanical or xerographic printing, registering on magnetic tape, or punching paper tape.

read-out pulse. Pulse applied to binary cells to extract the *bit* of information stored.

readatron. System for reading printed data and converting to digital form.

reading. See **sensing** (2).

reading speed. (1) Rate at which *words* or *bits* can be recovered from *memory*. (2) Rate at which input of computer reads these from programme card or tape.

real absorption coefficient. See **true absorption coefficient.**

real number. See **complex number.**

real-time computer. Process of calculation by analogue electronic computer in which operations are synchronized in time to the effects being studied or calculated, e.g., computer which controls a radio telescope while tracking satellites.

real-time processing. Automatic handling of computer data at the same speed as it is received from transducers or instrumentation of any kind.

rebecca-eureka. Radar system on aircraft carrying low-power interrogator transmitters (*rebecca*), working with fixed beacon responders (*eureka*), sending coded signals when triggered by interrogator pulses.

recalescence. Release of heat in ferromagnetic material as it cools through a temperature at which a change of crystal structure occurs—normally associated with change in magnetic properties.

receiver. Final unit in transmission system where received energy is stored, recorded or converted to preferred form, e.g., Magslip receiver, telephone receiver. See also:

bridge-	monochrome-
broadcast-	neutrosonic-
check-	panoramic-
communications-	radio-
crystal-	regenerative-
crystal gate-	straight-
facsimile-	super-regenerative-
frequency-	supersonic
modulation-	heterodyne-
lin-log-	tuned radio-
loudspeaking-	frequency-.

receiver bandwidth. See **bandwidth.**

receiver noise level. See **noise level.**

receiver sensitivity. See **sensitivity.**

reciprocal networks. Those the product of whose impedances remains a constant at all frequencies; thus an inductance is reciprocal to a capacitance.

reciprocation. Operation of finding a reciprocal network to a given network. Used in the design of electric wave filters.

reciprocity calibration. Absolute calibration of microphone by use of reversible microphoneloudspeaker. See **reciprocity theorem.**

reciprocity constant. See **reciprocity theorem.**

reciprocity principle. That interchange of radiation source and detector will not change the level of radiation at the latter, whatever the shielding arrangement placed between them.

reciprocity theorem. The interchange of electromotive force at one point in a network and the current produced at any other point results in the same current for the same electromotive force. In an *electrical network* comprising two-way passive linear impedances, the so-called transfer impedance is given by the ratio of the e.m.f. introduced in a branch of the network to the current measured in any other branch. By the reciprocity theorem this ratio is equal in phase and magnitude to that observed if the positions of the current and e.m.f. were interchanged. In calibration of *transducers*, the theorem concerns the quotient of the value of the ratio of the open-circuit voltage at the output terminals of the transducer (when used as a sound receiver) to the value of the free-field sound pressure (referred to some arbitrarily selected point of reference near the transducer), divided by the value of the sound pressure at a distance *d* from the point of reference to the current flowing at the input terminals of the transducer (used as a sound transmitter). The value of this quotient, termed the *reciprocity constant*, is independent of the constructional nature of the transducer.

recoil atom. One which experiences a sudden change of direction or reversal, following the emission from it of a particle or radiation. Also **recoil nucleus.**

recoil particle. One which arises through collision or ejection, e.g., Compton recoil electrons.

recombination. (1) Neutralization of ions in gas, by combination of charges or transfer of electrons. Important for ions arising from the passage of high-energy particles. (2) Neutralization of free electron and hole in semiconductor, thus eliminating two current carriers. The energy released in this process must appear as a light photon or, less probably, as several phonons; see **exciton, phonon, photon.**

recombination coefficient. Ratio of the rate of recombination per unit volume to the product of the densities of positive and negative current carriers.

recombination-rate surface (or volume). Rate of recombination/unit volume near surface (or in interior) of a semiconductor.

recombination velocity. See **surface recombination velocity.**

reconditioned carrier. Isolation of a pilot carrier for re-insertion of a carrier adequate for demodulation.

record. (1) (Noun) Data preserved in a suitable store (print, punched card or tape, etc.). (2) (Verb) To enter data in such a store. (3) See **recording** (2).

recorder. (1) Instrument employed for sound recording. (2) Electric measuring instrument which prepares a continuous graphical record of the applied signal. (3) Device for printing or storing output data.

H

recording. (1) The process of making a record of a received signal. (2) A disk, tape or film on which a sound record is stored. Disk recordings thus are commonly known simply as *records*.

recording amplifier. That preceding the recording heads of wax cutters, or any other type of recorder.

recording head. The transducer (magnetic, electric, mechanical or electro-optical) employed to record sound on tape, disk or film.

recording level indicator. See **indicator tube.**

recording systems (*black*). (1) In an AM system, the maximum received power corresponds to the maximum density (blackness) of the recording medium. (2) In an FM system, the lowest frequency corresponds to the maximum density of the recording medium. The converse holds for *white* systems.

recordist. Operator of the controls which determine the amplitude of electric currents which control a sound-recording device.

recovery rate. That at which recovery takes place following radiation injury. It may proceed at different rates for different tissues.

recovery time. (1) Time required for control electrode of gas tube to regain control. (2) Time required by TR switch in radar system to operate. (Usually measured to point where receiving sensitivity is 6 dB below maximum.) (3) For Geiger tube, the period between the end of the dead time and the restoration of full normal sensitivity. (4) For counting system, the minimum time interval between two events recorded separately.

rectangular loop hysteresis. Colloquial expression for hysteresis curve of ferromagnetic or ferroelectric materials suitable for use in bistable or switching circuits. Characterized by very steep slope followed by unusually sharp onset of saturation. Also **square loop hysteresis.**

rectangular pulse. Idealized pulse with infinitely short rise and fall times and constant amplitude. Pulse amplitudes and durations are often specified in terms of those of the nearest equivalent rectangular pulse carrying the same energy (subtending the same area).

rectangular scan. Any scanning system producing a rectangular field.

rectification. The process of converting an alternating into a unidirectional current by means of a rectifier.

rectification by leaky grid. See **demodulation by leaky grid.**

rectification efficiency. Ratio of d.c. output power to a.c. input. Often expressed as percentage.

rectifier. (1) Electric component for converting an alternating current into a direct current by the inversion or suppression of alternate half-waves. (2) Section of a cascade between feed point and withdrawal point in isotope separation.

See also:

bridge-	junction-
cold-cathode-	linear-
contact-	megatron
copper-oxide-	mercury-arc-
crystal-	plate-
dry-	point-contact-
electrolytic-	selenium-
germanium-	silicon-
grid-controlled	silicon-controlled-
mercury-arc-	square-law-
half-wave-	tank-
hot-cathode-	tantalum-
ignition-	thermionic-
ignitron	voltage multiplier
inverted-	wye-.

rectifier leakage current. That passing through a rectifier without rectification (e.g., due to parallel capacitance and finite reverse conduction).

rectifier stack. A pile of rectifying elements (usually semiconductor) series-connected for higher voltage operation. See also **Westeht.**

rectifier voltmeter. Voltmeter in which applied voltage is rectified in a bridge circuit before measurement with a d.c. meter.

rectifying detector. Detector of electromagnetic waves which depends for its action on the rectification of high-frequency currents, as opposed to one using thermal or electrolytic breakdown, or similar effects.

rectifying valve. Any thermionic valve in which direct use is made of unilateral or asymmetrical conductivity effects, as opposed to one used primarily for amplification, e.g., a diode used as a rectifier for the anode voltage supply to a receiver, or a triode used as a detector.

rectigon. Thermionic gas diode formerly used as low-voltage rectifier.

recurrence. See **Regge trajectory.**

re-cycle. Repetition of fixed series of operations, e.g., in isotope separation.

red-conscious. Said of an electron camera which is unduly sensitive to light of long wavelengths. This results in inartistic enhancement of the relative brightness of image areas which are red.

rediffusion. See **radio exchange.**

reduced mass. Quantity $mM/M+m$ used in studying the relative motion of two particles, masses m and M, about their common centre of gravity. This is used in place of the smaller mass when the movement of the larger is ignored.

reduced width. For neutron capture, the width of the nuclear energy level divided by the square root of the resonance energy.

redundancy. (1) Provision of extra equipment, e.g., valves or capacitors in parallel, to retain operation after component failure. (2) The provision of greater-than-minimum number of codings to ensure accuracy of interpretation (decoding) after telegraphic transmission through adverse conditions. (3) Inclusion of symbols in the computer coding, which are not strictly necessary for conveying the desired information, but may be used for verification or safeguarding.

redundancy check. See under **redundancy**.

redundancy rate. Percentage of the maximum *bits* for the message which is above that which is sufficient to convey the essential information.

reed. Vibrating tongue of wood or metal, for generating air vibrations in musical instruments. Metal reeds are used in organ reed-pipes and cane wood for tongue action, as in the clarinet. Can be made plate of a capacitor for electronic amplification.

reed relay. One with contacts on springs in vacuum, operated by external coil.

reference. See regulator.

reference current. One to which a varying current can be referred, especially when current gains are being expressed in decibels.

reference direction. Direction relative to which bearings are given in a navigational system.

reference frequency. Alternation of 1000 Hz, widely used in acoustics and circuit testing, being approximately in the middle of useful ranges of audio frequencies. In radio testing, a modulating frequency of 400 Hz. Once the frequency $5000/2\pi$ was in use for telephony. Also called **reference tone**.

reference level. (1) Used loosely for reference current, noise, power or voltage. (2) Any signal level in a complex waveform which provides a reference point, e.g., the reference black level in a TV signal is the signal level associated with the scanning of a black picture element. For positive modulation, it will be the level at which the synchronizing pulses should be separated from the signal.

reference noise. Circuit noise level corresponding to that produced by 10^{-12} watt of energy at a frequency of 1 kHz.

reference power. Power used as standard level when signal inputs or outputs are expressed in decibels—commonly 1mW in 600 ohms.

reference tone. See reference frequency.

reference tube. See voltage regulator tube.

reference voltage. (1) Closely controlled d.c. signal obtained from regulator tube, Zener diode or standard cell, and used for calibration, e.g., in an analogue computer. (2) A.c. voltage used as phase reference, e.g., in synchro-type servo systems.

reference volume. That transmission voltage which gives zero recording level on the standard *VU meter*.

reflectance. Proportion of incident energy returned by a surface of discontinuity. Also **reflection coefficient, reflectivity**.

reflected wave or ray. (1) See **ionospheric ray**. (2) One turned back from a discontinuity in a continuous medium. (3) One propagated back along a waveguide or transmission-line system as a result of a mismatching at the termination.

reflection. (1) Phenomenon which occurs when a wave meets a surface of discontinuity between two media, and part of it has its direction changed so as not to cross this surface, and in accordance with the *reflection laws*. See also **scattering**. (2) Reduction of power from the maximum possible, because a load is not *matched* to the source, and part of the energy transmitted is returned to the source. Reduction in power transmitted by wave filter because of iterative impedance becoming highly reactive outside the pass bands. In all instances, the loss of power is measured in decibels below maximum possible, i.e., when *properly matched*. (3) Return of neutrons to reactor core after a change of direction experienced in the shield surrounding the core. (4) Change in direction of a beam of charged particles produced by electric and/or magnetic fields, see, e.g., *reflex klystron*, *magnetic mirror*. See also **abnormal-, line-, non-specular-, specular-, total internal-**.

reflection coefficient. (1) See **reflectance**. (2) Ratio of electric voltage or field amplitude of reflected wave to that for incident wave. Given by Z_2-Z_1/Z_2+Z_1, where Z_1 and Z_2 are the impedances of the medium (or line) and the load respectively. For acoustic reflection, Z is the acoustic impedance.

reflection error. That in a RDF bearing due to a component of the signal which has been reflected.

reflection factor. Ratio of current actually delivered to load, to that which would be delivered to a perfectly matched load.

reflection gain. Gain in power received in a load from a source because of the introduction of a matching network, such as a transformer; measured in dB.

reflection laws. For wave propagation :—(1) incident beam, reflected beam and normal to surface are coplanar ; (2) the beams make equal angles with the normal.

reflection layers. Layers of very low-pressure atmosphere progressively ionized by particles from the sun. Each layer progressively refracts electromagnetic waves, so that (below a definite frequency) they are effectively reflected downwards. See **E-layer, F-layer**.

reflection loss. Loss, in dB, of power obtainable from a source into a load because the latter is not matched in impedance to the source. Also **return loss**.

reflection point. Point at which there is a discontinuity in a transmission line, and at which partial reflection of a transmitted electric wave takes place.

reflection tube. CRT in which electron beam is reflected backwards on to a screen.

reflectivity. See **reflectance**.

reflector. (1) Part of an antenna array which reflects energy that would otherwise be radiated in a direction opposite to that intended. (2) Layer of material surrounding core of reactor which turns back some of the neutrons leaving the core. (3) Electrode in reflex klystron connected to negative potential and used to reverse direction of electron beam. U.S. **repeller**. See **corner-**.

reflex amplifier. Early valve circuit in which both high-frequency and low-frequency currents are amplified independently together; also called **dual amplifier.**

reflex baffle. Loudspeaker enclosure in which sound waves propagated from rear of diaphragm are reversed in phase and direction, to be combined with those propagated in forward direction.

reflex klystron. One in which the electron beam, after passing through one rhumbatron, is reversed by a negative reflector and, arriving at the same rhumbatron in anti-phase, returns energy at a possible oscillation frequency; also **klystron reflection.**

refraction. Phenomenon which occurs when a wave crosses a boundary between two media in which its phase velocity differs. This leads to a change in the direction of propagation of the wavefront, in accordance with Snell's law.

refractive index. The absolute refractive index of a transparent medium is the ratio of the phase velocity of electromagnetic waves in free space, to that in the medium. It is given by the square root of the product of complex relative permittivity and complex relative permeability.

reftone. Abb. for **reference tone.** See **reference frequency.**

Regavolt. TN for variable auto-transformer similar to **Variac.**

regeneration. (1) Same as *positive feedback* (q.v.), but particularly applied to a superregenerative receiving valve, which oscillates periodically through self-quenching. (2) Reprocessing of nuclear fuel by removal of fission products. (3) Recovery of reactor from effect of xenon (or other) poisoning. (4) Replacement or reforming of stored data, e.g., in computer register or charge storage tube. See **aperiodic-.**

regenerative detector. One in which high-frequency components in the output are fed back (*reaction, retroaction*) to the input, thus increasing gain and selectivity.

regenerative receiver. One with positive feedback for the carrier, enhancing efficiency of amplification and demodulation.

Regge trajectory. A graph relating spin angular momentum to energy for a nuclear particle. Possible quantized values of spin correspond to large discrete energy increments on the graph. This enables recurrences of nuclear particles to be predicted, the extra energy corresponding to the greater rest mass expected to be associated with such particles. A *recurrence* is a particle identical in all respects, except energy (or mass) and spin momentum, with a known particle, and is regarded as being a higher energy equivalent of the normal particle. See **Summary of atomic physics** in Appendices.

region of limited proportionality. Range of operating voltages for a counter tube in which the gas amplification depends upon the number of ions produced in the initial ionizing event as well as on the voltage. For

larger initial events the counter saturates.

register. (1) Computer device which stores small amount of information, usually one word, with fast access. (2) Mechanical, electrical or electromechanical device which stores and displays data—usually one decimal number, in a measuring system. See **control-.**

regulating rod. Fine control rod of reactor.

regulation. (1) Fractional change in voltage level when load is connected to supply. (2) Control of plant or process at required activity level. (3) Anode voltage range for voltage regulator tube between maximum and minimum anode currents.

regulator. Glow tube which provides a constant potential difference across its electrodes when fed through a resistor; used for supplying small currents at constant voltage at a point in a circuit for control and servo purposes. See **dynamic-.**

re-ignition. (1) Re-establishment of conduction by ionization in a discharge tube, after conduction has been cut off, but during the de-ionization period. (2) A process by which multiple counts are generated in a counting tube by excitation arising from one ionization.

Reinartz circuit. One using reaction controlled by a capacitor, suitable for short-wave reception.

reinforcement. Sound reproduction in which the received enhanced level appears to come from the actual source, as is required, e.g., in theatres. In *public-address* the received level comes from sources not necessarily coalesced with the actual source. See **Haas effect.**

reinsertion. See **direct-current restoration.**

Reiss microphone. Carbon transmitter in which a large quantity of carbon granules between a cloth or mica diaphragm and a solid backing, such as a block of marble, is subjected to the applied sound wave. Characterized by high damping of the applied vibrational forces, and freedom from carbon noise by virtue of packing amongst the granules.

rejection band. Frequency band over which signals are highly attenuated when passed through a wave filter, the iterative impedance becoming highly reactive, causing reflections.

rejector circuit. Parallel combination of inductance and capacitance, tuned to the frequency of an interfering transmission, to which it offers a high impedance when placed in series with the antenna circuit of a receiver.

relative abundance. See **abundance.**

relative address. One which identifies the position of a word in a particular routine; cf. *absolute address.*

relative aperture. Ratio of the minimum vertical (or horizontal) clearance for particle passage in the accelerating chamber of an accelerator, to the particle orbit radius.

relative biological effectiveness. Inverse ratio

of an absorbed dose of ionizing radiation to the absorbed dose of 200 kV X-rays (or sometimes radium γ-rays), which would produce an equivalent biological damage. (*N.B.*—Absorbed doses measure the radiation energy absorbed regardless of physical or biological consequences—see **rad**.) Rounded-off values of rbe used in radiological protection are known as *quality factors*.

relative permeability. See **permeability**.

relative permittivity. See **permittivity**.

relative plateau slope. See **plateau slope**. *N.B.* —The rps is often expressed as percentage change in count rate for 100 volt change in potential.

relative stopping power. See **stopping power**.

relativistic. Said of any deviation from classical physics and mechanics based on relativity theory.

relativistic mass. When a particle is accelerated up to a velocity v which is more than a small fraction of the phase velocity of propagation of light in vacuum *c*, it is said to be *relativistic*, with mass increased according to the formula

$$m = m_0 \sqrt{1 - v^2/c^2},$$

where m_0 =rest mass (at low velocities). Required to be considered in cyclotron, betatron and linear-accelerator design.

relativistic particle. One having a speed comparable to that of light.

relativistic velocity. That of a particle when an appreciable part of imparted energy passes temporarily to accession of mass, e.g., for an electron above 10^4 eV.

relativity. Theory based on the equivalence of observation of the same phenomena from differing frames of reference, having different velocities and accelerations. Most useful special theory announced in 1903 by Einstein, generalized a decade later and verified by line-reddening, bending of light in strong gravitational fields, and the explanation of *precession* (q.v.) of the perihelion of planet Mercury. Two important results of this restricted theory are the *relativistic mass* equation and the principle of *mass energy equivalence* (q.v.). See **field theory**.

relaxation. Exponential return of a system to equilibrium following a sudden disturbance. The time constant of the exponential function is the *relaxation time*.

relaxation inverter. One employing a relaxation oscillator.

relaxation methods. Procedures of fitting together sets of equations so that the residuals in their solutions for unknowns are made as small as possible or tolerable.

relaxation oscillation. One of irregular waveform and frequency. Produced, e.g., by charging and discharging of a capacitor. Other examples are the flapping of a flag in the wind, the heart-beat, the scratching of a knife on a plate, etc. See **multivibrator**.

relaxation oscillator. One, usually electric, which generates relaxation oscillations.

Characterized by peaky or rectangular waveforms, and the possibility of pulling into step (locking) by an independent source of impulses of nearly the same frequency. See **multivibrator, time base**.

relaxation time. (1) See **relaxation**. (2) In isotope separation, the time required in a separating plant under given conditions for the mole fraction of the top stage in the process to attain $(1 - e^{-1})$ of the steady-state value.

relay. (1) An electrically-operated switch employed to effect changes in an independent circuit. Examples are moving-coil, moving-iron, solenoid, balanced-armature, telephone-type, alternating-current, gas-tube, transistor and thyratron relays. All function as single or multiple non-manual switches. (2) Broadcast of a performance, e.g., concert-hall or theatre, not specially staged for broadcasting, taken over a radio link, a music line or by recording. See **relay types**, and also:

armature-	non-polarized-
balanced-beam-	reed-
break-before-	shunt-field-
make	side-stable-
break delay	sticking-
coaxial-	thermionic-
electron-	trigger-
gas-filled-	two-step-.
non-locking-	

relay spring. Flexible part of a relay which keeps it in an unoperated condition. It is stressed on operation, and restores the relay to normal on cessation of the operating current, thereby operating contacts.

relay system. See **radio exchange**.

relay types. *Allström*—a sensitive form using a light beam and photocell. *Control*—one which operates by permitting the next step in a control circuit. *Differential*—one operating on difference between, e.g., two currents. *Frequency*—one operating with a selected change of the supply frequency. *Polarized*—one in which the movement of the armature depends on the current direction on the armature control circuit.

reliability. Probability that an equipment or component will continue to function when required. Expressed as average percentage of failure per 1000 hours of availability. Also **fault rate**.

reluctance. Magnetomotive force applied to the whole or part of a magnetic circuit, divided by the flux in it. It is the reciprocal of **permeance**.

reluctance pickup. Transducer for detecting vibrations or reproducing records, utilizing a change of reluctance in a magnetic circuit, thereby inducing an e.m.f. in a coil-carrying flux.

reluctivity. The reciprocal of **permeability**.

rem. See **röntgen equivalent man**.

remanence. See **residual magnetization**.

Remitron. TN for gas tube employed in counting systems.

remodulation. Transferring modulation from one carrier to another carrier, as in the frequency changer in a supersonic heterodyne radio receiver.

remote control. Control, usually by electric or radio signals, carried out from a distance in response to information provided by monitoring instruments.

remote cutoff tube. U.S. term for **variable mu(tual) conductance valve**, which requires a large increasing negative grid bias to reduce the anode current to zero; also called **extended tube, long-tail tube, supercontrol tube.**

remote-handling equipment. Apparatus developed to enable an operator to manipulate highly radioactive materials from behind a suitable shield, or from a safe distance.

Remscope. TN for an oscilloscope in which a trace on the phosphor is maintained indefinitely.

rep. See röntgen equivalent physical.

repeater. Device which transmits what it receives, but in an amplified or augmented form. Originally applied to telegraph relay, later to one-way or two-way telephonic amplifier, with or without echo suppression. A *pulse repeater* is one which receives pulses from one circuit and transmits corresponding pulses to another circuit. If regenerative, it acts as a pulse regenerator.

repeating coil. Unity-ratio transformer for separating telephonic circuits, with windings balanced to earth.

repeller. U.S. term for **reflector**, as in a klystron.

repetitition rate. Rate at which recurrent signals, usually pulses, are repeated.

representation. System of representing data by numbers, their *position* in relation to the *radix* and their *significance.*

reprocessing. See regeneration (2).

reproducer. (1) Complete sound reproduction system. (2) Loudspeaker.

reproducibility. The precision with which a measured value can be repeated in a process or component.

reproduction. Replay of recorded sounds, or re-creation of sounds from electric waves produced by microphone.

reproduction constant. Deprecated term for **multiplication constant.**

re-radiation. (1) Radiation from resonating elements, e.g., masts, antennae, telephone lines, giving errors in bearings or displaced television images. (2) Phenomenon which occurs when a receiver employing reaction on to the antenna circuit is adjusted to the point of oscillation. The signal strength for nearby receivers is thereby increased.

re-recording. Recording acoustic waveforms immediately upon reproduction from the same, or any other type of, recording medium as that in use.

re-run. Repeat of part of programme for a computer. Most programmes incorporate a re-run point for every few minutes running time. In the event of an error, only the part subsequent to the previous re-run point is repeated.

reservoir. Any volume in an isotope separation plant which is used to store material or to ensure smooth operation.

reset. (1) General term for the preparation of any circuit or apparatus for a fresh performance of its duty. Amplifiers require no resetting, apart from switching on; timers or counters do require resetting. Resetting can be automatic, or initiated by an external signal, arbitrary in time. (2) After (or before) completion of an operation by an electronic computer, restoration of all circuits and mechanisms so that the system is ready to perform its function again, from the beginning of its cycle of operations, in accordance with its *instruction* or *programme.*

reset circuit. One which, when operated, resets a functional circuit, i.e., establishes it in a ready condition for operating.

reset rate. That at which discrete corrections are applied in a control system employing sampling action.

reshaping. Restoration to intended shape, in amplitude and time, of pulses which have become distorted.

residual activity. In a nuclear reactor, the remaining activity after the reactor is shut down following a period of operation.

residual current. That in a thermionic valve when the anode is at the same potential as the cathode (due to the finite velocity of emission of electrons at the cathode).

residual flux density. Flux density remaining after exciting magnetic field has been removed. This depends upon the geometry and nature of the whole magnetic circuit.

residual gas. Small amount of gas which inevitably remains in a 'vacuum' tube after pumping. If present to excess, it causes erratic operation of the tube, which is said to be *soft.*

residual induction. Same as *residual magnetization,* but could be applicable to an entire magnetic circuit.

residual magnetization. Magnetization persisting in ferromagnetic material when the exciting magnetizing force is removed; also termed **remanence.**

residual resistance. That persisting at temperatures near zero on the absolute scale, arising from crystal irregularities and impurities, and in alloys.

resistance. (1) Opposition to motion leading to dissipation of energy. (2) In electrical and acoustic fields, the real part of the impedance, characterized by the dissipation of energy as opposed to storage. *N.B.—* Electrical resistance may vary with strain, temperature, polarity, field illumination, purity of material, etc. See **impedance, ohm, reactance,** and also:

anode load-   base-
antenna-   base-spreading-

Bronson-
contact-
coupling-
dark-
differential-
differential
  anode-
dynamic-
effective-
electrode-

equivalent-
high-frequency-
internal-
logarithmic-
negative-
non-linear-
residual-
spark-
temperature
  coefficient of-.

resistance box. One containing carefully constructed and adjusted resistors, which can be introduced into a circuit by switches or keys. At high frequencies, there are disturbing inductive and capacitive effects which complicate measurements using resistance boxes. These are mitigated by suitable design and the boxes are then described as *non-reactive*.

resistance-capacitance coupling. Coupling in which signal voltages developed across a load resistance are passed to the subsequent stage through a d.c. blocking capacitor.

resistance-capacitance oscillator. One producing a sine waveform of frequency determined by phase rotation in a resistance-capacitance section of an artificial line.

resistance coupling. That between successive stages of an amplifier employing thermionic valves, by which changes in anode potential across a resistance are impressed on the grid of the succeeding valve, the correct grid bias being obtained separately. See **direct coupling.**

resistance noise. Same as **thermal noise.**

resistance strain gauge. Instrument used for stress analysis to measure variation of electric resistance with strain.

resistance thermometer. Thermometer employing resistance changes for temperature measurement. Resistance element may be platinum wire for extreme precision, or semiconductor (thermistor) for high sensitivity.

resistance welding. The technique of welding metals by the passage of an electric current through the area of contact. Heating occurs through the contact electrical resistance which renders the metals plastic, and, after interrupting the current, mechanical pressure is applied at the contact area.

resistance wire. That constructed from a wire of relatively high resistivity.

resistive load. Terminating impedance which is non-reactive, so that the load current is in phase with the source e.m.f. Reactive loads are made entirely resistive by adding inductors or capacitors in series or shunt (tuning).

resisticon. High-velocity camera tube.

resistivity. Specific property of a conductor, which gives the resistance in terms of its dimensions. If R is the resistance in ohms, of a wire l cm long, of uniform cross-section a cm^2, then

$$R = \rho \cdot l/a,$$

where the resistivity ρ is in ohm. cm (*not* ohm cm^{-3}). Also **specific resistance.**

resistor. Electric component designed to introduce known resistance into a circuit and to dissipate accompanying loss of power. Types are wire-wound, composition, ceramic, plates and bars, aqueous solutions, deposited carbon, and carbon-contact plates. See also :

ballast-
bifilar-
bleeder-
boro-carbon-
carbon-
cathode-
Chaperon-
composite-
dropping-

inductive-
linear-
liquid rheostat
non-inductive-
non-linear-
rheostat
silit-
standard-
Thomas-.

resnatron. High-power, high-frequency tetrode in which concentric cylinders form bypass capacitors and cavities from which output energy is taken.

re-solution. Passing back into solution metal previously deposited on an electrode in electrolysis; used for measuring thickness of electrodeposits, recording potential differences.

resolution. (1) The definition of a picture in TV or facsimile (measured by the number of lines used to scan the image of the picture). (2) The smallest measurable difference in wavelength, frequency or energy for a light, sound or particle beam spectrum. (3) The smallest mass difference detectable in a mass spectrograph. (4) The minimum image detail recorded on a photographic emulsion (measured in lines per millimeter). (5) The smallest separation of two points recorded separately by an electron or optical microscope—also the smallest angular separation possible with an optical or radio telescope. (6) The smallest time interval between ionizing particles recorded separately by any detector. Resolution may be limited by imperfections in the detecting system or by physical limitations of the signal, e.g., the wavelength of light sets the ultimate limit to the resolution of an optical microscope. Similarly the ultimate limit for a photographic image depends upon the emulsion grain size, but the actual limit may be set by the resolution of the camera or enlarger lens, or by such causes as camera shake and incorrect focusing.

resolution time. The minimum time between two events recorded separately. The maximum time between two events recorded as coinciding.

resolution-time correction. Correction applied to observed counting rate for random events which allows for those not recorded because of the finite resolution time.

resolver. Apparatus which converts polar coordinates into Cartesian coordinates.

resolving time. See **resolution time.**

resonance. (1) Phenomenon of minimum mechanical or acoustical impedance as the frequency of the applied disturbing force is varied, resulting in a maximum velocity of motion. Rods or plates are potentially vibrating systems with several modes of vibration, the frequencies of resonance generally not being exactly harmonic. The *sharpness* of resonance is measured by the ratio of the dissipation to the inertia of the system, which also measures the rate of decay of the motion of the vibrating system when it is impulsed. See **decay factor**. (2) Balance between positive (inductive) and negative (capacitive) reactance of a circuit, accompanied by large currents and voltages resulting from relatively small applied e.m.f. at resonant frequency. (3) Increased probability of nuclear reaction when energy of incident particle or photon would raise compound nucleus to natural energy level. Shown by increase in effective cross-section to particles of this energy.

resonance absorption. See **resonance (3)**.

resonance bridge. Measuring bridge for which balance depends upon adjustment for resonance.

resonance capture. The capture of an incident particle into a resonance level of the resultant compound nucleus.

resonance curve. One showing variation of current in a resonant circuit in series with an e.m.f., as the ratio of the resonance frequency and the frequency of the generator is varied through unity.

resonance escape probability. In a reactor, the probability of a fission neutron slowing down to thermal energy without experiencing resonance.

resonance heating. See **magnetic pumping**.

resonance integral. One used in reactor theory which is expressed in terms of the logarithm of the resonance escape probability multiplied by the slowing-down power of the absorber.

resonance lamp. One which depends on the absorption and re-radiation of a prominent line from a mercury arc, excited in mercury vapour.

resonance level. An excited level of the compound system which is capable of being formed in collision between two systems, such as between a nucleon and a nucleus.

resonance potential. See **excitation potential**.

resonance radiation. Emission of radiation from gas or vapour when excited by photons of higher frequency.

resonance scattering. See **scattering**.

resonance step-up. Ratio of the voltage appearing across a parallel-tuned circuit to the e.m.f. acting in the circuit (usually induced in the coil) when the circuit is resonant at the applied frequency. See *Q*.

resonances. During nuclear reactions, very unstable mesons or hyperons are frequently created. These decay through the strong interaction with a half-life of the order of 10^{-23} sec. Consequently, such particles are undetectable and their formation as an intermediate step in the reaction can only be inferred from indirect measurements.

These temporary states are known as *resonances* to distinguish them from metastable particles with half-lives of the order of 10^{-10} sec. which are detectable.

resonant cavity or chamber. One in which resonant effects result from the possibility of a modal pattern of electric and magnetic fields, as in magnetrons, klystrons, waveguide couplers. Also applies to acoustics.

resonant circuit. One comprising inductance coil and capacitor in series or parallel. The series circuit has an impedance which falls to a very low value at the resonant frequency; that of the parallel circuit rises to a very high value.

resonant frequency. That at which reactances of a series-resonant circuit, or susceptances of a parallel-resonant circuit, balance out; numerically equal to $1/2\pi\sqrt{LC}$ Hz, where L is inductance in henries and C capacitance in farads.

resonant gap. The interior volume of the resonant structure of a TR cell in which the electric field is concentrated.

resonant line. Parallel wire or coaxial transmission line open- or short-circuited at the ends, and an integral number of quarter-wavelengths long. Used for stabilizing the frequency of short-wave oscillators and in antenna systems.

resonant mode. Field configuration in a tuned cavity. In general, resonance occurs at several related frequencies corresponding to different configurations.

resonant window. A resonant iris which is sealed into the envelope of a switching tube.

resonator. Any device exhibiting a sharply defined electric, mechanical or acoustic resonance effect, e.g., a stub, piezoelectric crystal or Helmholtz resonator. Originally, a circular wire ring containing a small spark gap, used by Hertz for detection of electromagnetic waves. See **coaxial line-, Helmholtz-, piezoelectric-**.

resonator grid. Electrode traversed by an electron beam and which provides a coupling to a resonator.

responder. That part of a transponder which replies automatically to the correct interrogation signal.

response. That of a transmission system at any particular frequency is given by the ratio of the output to input level. If these levels are defined on a logarithmic scale, e.g., in dB, the response of the complete system is the sum of the responses of the separate parts. See also **polar diagram**.

response curve. That which exhibits the trend of the response of a communication system or a part thereof, for the range of frequency over which the system or part is intended to operate. Usually plotted in dB against a logarithmic frequency scale.

response time. Time constant of change in output of a magnetic amplifier after a sudden change in input, or of indication given by any instrument after change in signal level.

responsor or **responser.** Receiver of secondary radar signal from transponder. See **identification, friend or foe.**

rest mass. Newton mass of a particle at zero or low velocities, i.e., not augmented by relativistic mass (Einstein). Associated energy on annihilation is the rest mass multiplied by the square of the velocity of light. See **relativistic mass.**

restore. In computers, return of a variable address or word or cycle index to its initial value.

restorer circuit. See **direct-current restoration.**

retardation coil. Inductor for separating d.c. from a.c., particularly from a rectifier or supply with ripple.

retarded field and **retarded potential.** Those at a point which arise later than at some other point because of finite speed of propagation of waves in the medium.

retarding field. Electric field such as between a positively-charged grid and a lower potential outer grid in a valve, so that electrons entering this region lose energy to the field; also called **brake field.**

retarding-field detector. Use of a retarding-field valve for demodulation of ultra-high-frequency signals.

retarding-field oscillator. One which depends on the electron-transit time of a positive grid oscillator valve.

retention. The fraction of radioactive atoms which are not separable from the target compounds, following the production of these atoms by nuclear reaction or radioactive decay.

retention time. Maximum period for which storage element is able to give required output signal.

retentivity. Residual magnetic induction, after removal of field producing saturation induction.

retina. Light-sensitive surface of eye. Human retina contains two types of sensitive element —*cones* and *rods* (qq.v.).

retrace. See **flyback.**

retroaction. Academic and Continental term for **reaction.**

return. Refers to radar reflections, e.g., land (or ground) return, sea return.

return electrons. Those electrons which, having impinged on the fluorescent screen, are on their way back to the anode.

return interval. Flyback time of time base.

return line flyback. Faint trace formed on the screen of a cathode-ray tube by the beam during the flyback period. Usually suppressed. Also **return trace.**

return loss. Same as **reflection loss.**

return trace. See **return line flyback.**

Reuben-Mallory cell. Small robust primary cell, of very level discharge characteristic, having a zinc anode and a (red) mercuric-oxide cathode, which also depolarizes. Made in minute sizes for hearing-aids, internal radio transmitters, watches.

reverberation absorption coefficient. That of a large plane uniform surface when the incident sound wave is of random intensity and direction, as is the reverberant field in an enclosure.

reverberation bridge. Method of measuring the reverberation time in an enclosure; the rate of decay of the sound intensity is balanced against the adjusted and known decay of the discharge of a capacitance through a resistance.

reverberation chamber. One with the minimum acoustic absorption for acoustic absorption measurements, or for adding echo effects in sound reproduction; also **echo chamber.**

reverberation period. See **reverberation time.**

reverberation response. That of a microphone for reverberant sound, i.e., for the simultaneous arrival of sound waves of random phase, magnitude, and direction. Substantially equal to the mean-spherical response at each frequency of interest.

reverberation response curve. That of a microphone to reverberant sound waves. Plotted with the response in decibels as ordinates on a logarithmic frequency base.

reverberation time. Period, in seconds, required for the decay of the average sound intensity in an auditorium over an amplitude range of one million, or 60 decibels, there being no emission of sound power during this decay; also **reverberation period.**

reverse compatibility. Exhibition of an acceptable black-and-white (monochrome) picture on a television receiver designed to accept and exhibit colour pictures.

reverse coupling. See **negative feedback.**

reverse (or inverse) current. That in an electrode opposite to that intended, e.g., that in a triode grid because of ionization.

reverse emission. That from anode to cathode of thermionic valve.

reverse grid current. That which flows away from the grid of a thermionic tube through the external circuit to the cathode, i.e., in the opposite way to the normal current. It is caused by electronic emission from the grid, due to heating or bombardment, or by the presence of positive ions in the interelectrode space, thus indicating residual gas.

reverse reaction. That which opposes self-oscillation; used to neutralize the effects of interelectrode capacitance coupling.

reverse recovery time. That for the reverse current or voltage in a semiconductor diode to reach a specified value after the instantaneous switching from a steady forward current to a reverse bias in a particular circuit.

reverse voltage. That applied to a valve or semiconductor device in a normally non-conducting direction.

reversible absorption current. That which decreases with time much less rapidly than the 'geometrical' capacitor charging current, normally exponential; returned on short-circuiting the plates.

reversible cell. See **accumulator.**

reversible transducer. One for which the loss

H*

is independent of the direction of transmission.

rewrite. To return data to a store when it has been erased during reading.

Reynolds number. A non-dimensional ratio used for assessing the similarity of motion in viscous fluids. The product of any typical length of a body and its velocity, divided by the kinematic coefficient of viscosity of the fluid.

rhenium (Re). At. no. 75, at. wt. 186·2, sp. gr. 20·53, m.p. 3167°C. Used in high-temperature thermocouples.

Rheostan. TN for electrical resistance wire consisting of an alloy of Ni (25%), Co (52%), Fe (5%) and Zn (18%).

rheostat. Electric component in which resistance introduced into a circuit is readily variable by a knob or handle, or by mechanical means, such as an electric motor.

rheostriction. See pinch effect.

rheotron. Same as betatron.

rho meson. Term formerly used to denote a meson which stopped in a nuclear emulsion without any apparent concomitant decay event or nuclear interaction.

rho-theta. Navigational system which gives distance and bearing from a known point radio source.

rhodium (Rh). At. no. 45, at. wt. 102·905, sp. gr. 12·5, m.p. 2000°C approx. A noble metal, silvery-white and resembling platinum. It is used in catalysts and in alloys for high-temperature thermocouples.

rhombic antenna. Directional antenna comprising an equilateral parallelogram of conductors, each several quarter-wavelengths long, usually arranged in a horizontal plane. The wires are connected together through a resistance at one apex at the end of the longer diagonal, and to the transmitting or receiving apparatus at the other end. The maximum directive effect is along the longer diagonal; used for *musa*.

rhometer. One for the measurement of impurity content of molten metals by means of the variation in electrical conductivity.

rhumb line. Navigational line of constant direction, crossing successive meridians at the same angle.

rhumbatron. Type of cavity resonator used, e.g., in klystron. It acts as a tuned circuit comprising a parallel-disk capacitor surrounded by a single-turn toroidal inductance, and is used to velocity-modulate an electron beam passing through holes in the capacitor disks.

ribbon microphone. Same as moving-conductor microphone.

Rice neutralization. A method of grid neutralization used in single stage power amplifiers.

Richardson-Dushman equation. Original Richardson formula, as modified by Dushman, for the emission of electrons from a heated surface, current density being

$$I = AT^2 e^{-\phi/kT}.$$

T is the absolute temperature, A is a material

constant, k is Boltzmann's constant, with ϕ the *work function* of the surface and e the base of natural logarithms.

Richardson effect. Same as Edison effect.

Rieke diagram. Polar form of load impedance diagram representing the components of the complex reflection coefficient of the oscillator load in a microwave oscillator.

Riffel loudspeaker. One in which the radiating element takes the form of two sheets curved to an edge, on which is located the driving element, the latter being a current-carrying conductor in the longitudinal gap of a magnet. The arrangement provides for radiation in a horizontal plane with restricted vertical radiation.

ring counter. A number of counting circuits in complete series, for sequence operation in counting impulses.

ring modulator. One consisting of four rectifying elements in complete series, which act as a switch, being fed with appropriate currents at the corners.

ring oscillator. One in which a number of valves feed each other in a circle or circus, and in which the frequency is determined by a ring cut from a quartz crystal, suspended at its nodes to minimize damping; used, e.g., as a standard time-keeper (*crystal clock*) at 10^5 Hz.

ring scaler. One in which a ring of counting elements are triggered in turn, the output pulse being from one of them to another ring. See Dekatron.

ringdown. Method of signalling operator in wire telephony.

ringing. (1) Extended oscillation in a tuned circuit, at its natural frequency, continuing after an applied voltage or current has been shut off, dying away according to its *decay constant*, but running into the next oscillation. (2) Multiple displaced black-and-white TV images arising from free oscillations in the system.

riometer. Apparatus for continuously measuring ionospheric absorption.

ripple. (1) The a.c. component in the output of a rectifier delivering d.c., reduced by series choke and shunting (smoothing) capacitor, or Zener diode. Measured as a percentage of the steady (average) current. (2) Small liquid surface wave controlled by surface tension forces.

ripple filter. A low-pass filter which is designed to reduce the ripple current but at the same time permits the free passage of the d.c. current, e.g., from a rectifier. Also smoothing circuit.

ripple frequency. The frequency of the ripple current in rectifiers, etc. It is usually double the supply frequency in a full-wave rectifier.

rise time. Time for pulse signal in an amplifier or filter to rise from 10 to 90 per cent. of the maximum amplitude. Also build-up time.

rising-sun magnetron. One in which the resonances of cavities are of two magnitudes alternatively.

Robinson bridge. An a.c. bridge used for the

measurement and control of frequency. Also **Robinson-Wien bridge**.

Robinson direction-finder. Rotating-loop direction-finding system having an auxiliary loop, which can be reversed in series with the main loop and so permits an audible signal to be balanced by switching, instead of nodalizing signal and noise.

Robinson-Wien bridge. See **Robinson bridge**.

Roboting. TN for system operating driverless trolleys, which follow a cable energized by a route setter.

Rochelle salt. Crystal of sodium potassium tartrate, having strong piezoelectric properties, but high damping. Used as a *bimorph* in microphones, loudspeakers and pickups. Its disadvantage is the limited range of ferroelectric property ($-18°C$ to $23°C$), and high-temperature coefficient.

rocket. Slotted-line VSWR measuring system with end cones to match slotted line to external circuits.

Rocky Point effect. Of obscure origin, occasionally occurring in high-voltage transmitting valves; characterized by a transient but violent discharge between the electrodes, not always damaging the valve. First observed at Rocky Point radio station. Also called **flash arc**.

rods. (1) Non-colour sensitive light-perceptive elements on periphery of human retina. Rods respond to lower illumination levels than *cones* (q.v.). (2) Common geometrical shape for reactor-fuel elements or samples intended for irradiation in reactors.

roentgen. See **röntgen**.

roll-off frequency. Frequency where response of an amplifier or filter is 3 dB below maximum.

roll-out. Method of reading number stored in register of computer, by adding digits to each column in turn until this reverts to zero.

röntgen. Unit of X-ray or gamma dose, for which the resulting ionization liberates a charge of 2.58×10^{-4} coulomb/kg of air.

röntgen equivalent man. Unit of biological dose, given by the product of the absorbed dose in rads and the relative biological effectiveness of the radiation. Abb. **rem**.

röntgen equivalent physical. Obsolete unit of absorbed dose equal to that received by water from an exposure dose of 1 röntgen. This was taken as 93 erg/gm. Hence, 1 rep = 0.93 rad. Abb. **rep**.

röntgen-hour-metre. Unit of γ-ray source strength, producing 1 röntgen/hour at 1 m distance.

röntgen rays. Electromagnetic rays of wavelengths of the order of 1-10 ångström units, between much longer (ultraviolet) and much shorter (γ-rays) wavelengths. Generated by high-voltage electrons hitting a target (usually tungsten) in an evacuated tube.

roof antenna. Same as **flat-top antenna**.

room noise. See **ambient noise**.

root locus. Plot of closed-loop performance, showing *poles* (infinities) and *zeros*, and thereby indicating stability of system.

root-mean-square value. Measure of any alternating waveform, the square-root of the mean of the squares of continuous ordinates (e.g., voltage or current) through one complete cycle. If there are harmonics of the fundamental, the total rms value is the sum of the rms values of the fundamental and the harmonics taken separately. Employed because the *energy* associated with any wave depends upon its *intensity*, i.e., upon the *square* of the *amplitude*. Also **effective value**; see also **rms power**.

rooter. Arrangement of thermionic valves for obtaining an output amplitude which is proportional to the square-root of the input amplitude. Required in compressors for reducing contrast in sound reproduction.

rope. See **window**.

Rossi counter. An important type of spherical tissue-equivalent proportional counter, widely used in radiobiology.

rotary amplifier. Rotary generator, output of which is field-controlled by another generator or amplifier.

rotary converter. Combination of electric motor and dynamo used to change form of electric energy, e.g., d.c. to a.c. or one d.c. voltage to another.

rotary discharger. See **rotary spark gap**.

rotary pump. A pump in which two specially shaped members rotate in contact; suited to large deliveries at low pressure.

rotary spark gap. Rotating disk or wheel having projections on its periphery which pass close to two fixed electrodes. A spark of short duration takes place from one electrode to the disk and thence to the other electrode each time the projections come opposite to the electrodes. Used in early forms of quenched spark system; also called **rotary discharger**.

rotating anode. Tube in which the anode is rotated about an axis normal to, but not through, the target area, so that heat from the impact of electrons is dispersed.

rotating field. One in which the magnitude is constant at a point, but whose direction is rotating about a point in a fixed reference system.

rotating joint. Short length of cylindrical waveguide, constructed so that one end can rotate relative to the other; used to couple two other waveguide systems, normally of rectangular cross-section.

rotator. A device for rotating the plane of polarization of a wave in a waveguide.

rotatrol. Electrodynamic amplifier used as component of some servo systems.

roton. Quantum of rotational energy analogous to the *phonon*.

rotor. See **armature** (3).

rough cut. First assembly of shots forming a sequence in an edited film.

round-off. To limit number of digits in a quantity, generally by adding half the radix and dropping the last digit, successively.

Round valve. Early grid-detector valve (named after inventor), in which the effectiveness of demodulation was increased by heating a pellet which released small quantities of gas.

routine. Prescribed series of operations which can be repeated, in conjunction with others, at any time without special planning. Kept in a *library* in the form of punched paper tape or cards. See **subroutine.**

rubidium (Rb). At. no. 37, at. wt. 85·47, sp. gr. 1·53, m.p. 38·4°C. Highly reactive metallic element, white and resembling sodium.

ruby. A single crystal of aluminium oxide in which a small percentage of aluminium atoms is replaced by atoms of chromium. The latter give rise to the characteristic red fluorescence of ruby when it is irradiated.

ruby laser. An optically-pumped ruby crystal producing a very intense and narrow beam of coherent red light. It is used in light-beam communication and for localized heating.

ruby maser. Maser using ruby crystal in the cavity resonator.

Ruhmkorff coil. Electromagnetic high-voltage generator consisting essentially of a transformer with interrupted d.c. primary supply. Source of energy for early W/T spark transmitters.

rumble. Low-frequency noise produced in disk recording when turntable is not dynamically balanced.

run. Complete set of operations in sequence, e.g., running a magnetic-tape store of open accounts for *up-dating*, changes of address, etc.

runaway electron. One under an applied electric field in an ionized gas which acquires energy from the field at a greater rate than it loses through particle collision.

run-down. That part of a repeated waveform which is a decay, particularly if a large fraction of the cycle, as in a linear time-base generator, e.g., of the Miller type.

Russell-Saunders coupling. An extreme form of coupling between the orbital electrons of atoms. The angular and spin momenta of the electrons combine and the combined momenta then interact. Also **l-s coupling.**

ruthenium (Ru). At. no. 44, at. wt. 101·07, sp. gr. 12·2, m.p. 2450°C. A hard, brittle metallic element.

rutherford. Unit of radioactive decay rate equal to 10^6 disintegrating atoms/sec or alternatively that amount of radioactive material undergoing 10^6 atomic disintegrations/sec. This unit is not officially recognized.

Rutherford atom. Earliest modern concept of atomic structure, in which nearly all the mass of the atom is in the nucleus, the electrons, equal in number to Z, occupying the rest of the atomic volume in a fixed pattern.

Rutherford Laboratory. The high-energy laboratory of the National Institute for Research in Nuclear Science at Harwell.

Rutherford scattering. See **scattering.**

Rydberg constant. The frequencies of spectrum lines in the same series are less than those of the spectrum limit by an amount

$$\frac{R}{(n+k)},$$

where n can have any value, k is a constant for each series, R is a universal constant (termed the *Rydberg constant*). It was first determined empirically from spectrographic data, but has since been shown to be given by

$$R = \frac{2\pi^2 e^4}{ch^3} M_r,$$

where M_r is the reduced mass of the radiating electron, and

e = electronic charge
c = velocity of light
h = Planck's constant.

S

SI units. The modern scientific *international system of units*, identical to the electric MKSA system, but more general and using the six fundamental units listed under *absolute units* (q.v.).

S-meter. A meter used in communication receivers for measuring relative signal strength.

S-type cathode. One which conforms to a rigid specification, with Ni or Pt core heater wire, and Ba and Sr oxides coprecipitated, applied and processed.

S-value. Percentage increase in volume of nuclear fuel after 1 per cent. burn-up. Cf. *R-value*.

s-levels. See **sharp series**.

s-state. State of zero orbital angular momentum.

Sabatier effect. An image reversal phenomenon in photography which appears to be connected with the image development.

sabin. Unit of acoustic absorption; equal to the absorption, considered complete, offered by 1 sq ft of open window to low-frequency reverberant sound waves in an enclosure.

Sabine reverberation formula. Earliest formula (named after investigator) for connecting the period of reverberation of an enclosure, T seconds, with the volume, V in cubic feet, and the total acoustic absorption in the enclosure, ΣaS in sabins, where a is the absorption coefficient of a surface of S sq feet. The formula is $T = \cdot05 V/\Sigma aS$.

saddle point. (1) Point on plot of potential energy against distortion for nucleus at which fission will occur instead of return to equilibrium. (2) Same as **neutral point** (2).

safety circuit. One which either gives warning of faults or abnormalities, or operates a trip on some protective device.

safety factor. Excess load which can be handled without breakdown.

safety rods. Rods of neutron-absorbing material used for shutting down a nuclear reactor in case of emergency.

Saint Elmo's fire. Brush discharge from isolated points above the ground, e.g., ship's mast and aircraft.

samarium (Sm). At. no. 62, at. wt. 150·35, sp. gr. 7·8, m.p. 1052°C, b.p. 1600°C. Feeble naturally radioactive metallic element, can be produced by decay of fission fragments and forms reactor poison.

sample intelligence. The part of a signal which is used to provide evidence of the quality of the whole.

sampling. Selection of an irregular computer signal over stated fractions of time or amplitude (pulse height).

sampling action. Type of control system in which error signal only operates controller intermittently. See **reset rate**.

sampling circuit. One with output of a form suitable for the error signal in a controller employing *sampling action*.

sampling gate. A circuit with an output only when the gate is opened by an activating pulse.

sanatron. Valve circuit for fast time bases.

sandwich. Photographic nuclear research emulsion forming series of thin layers with intervening layers of a material, in which some event or process is to be studied.

sandwich seal. A vacuum-sealing device in disk-seal triodes at high frequencies, consisting of a sandwich of thin copper between two glass cylinders.

Sarah. RTM of a small radar transmitter attachable to an airman's lifejacket. (*Search and-rescue-and-homing*.)

Sargent diagram. Log-log plot of radioactive decay constant against maximum β-ray energy, for various beta emitters.

satellite. (1) Radio transmitting station depending on, or controlled by, another station. (2) Circulating vehicle at a height above the earth to reflect or transmit back radio waves as a means of communication, using computer-controlled terrestrial antennae. See **Goonhilly earth station** in Appendices.

satellite communication. Principle of reflection or regeneration of telegraphic or telephonic signals from earth satellites, using highly directive antennae for transmission and reception, orientated by computer calculation of orbit.

satellite station. (1) One which re-broadcasts a transmission received directly, but on another wavelength. (2) One designed for transmission to, and reception from, earth satellites.

saturable reactor. Inductor in which the core is saturable by turns, carrying d.c. which controls the inductance. Used for modulation and control of lighting, and developed in the magnetic amplifier.

saturation. (1) Condition obtaining when all the electrons emitted from the cathode are swept away to the anode or other electrodes, so that further increase in anode potential produces no corresponding increase in anode current. (2) A similar condition in a screened-grid or pentode valve when all the electrons which pass the screen grid go on to the anode, although the cathode emission is not the limiting factor. (3) Application of a sufficiently intense magnetizing force to result in maximum temporary or permanent magnetization in magnetic material. Usually applied by large currents from an impulse current transformer. (4) Degree to which a colour departs from white and approaches the pure colour of a spectral

line. **Desaturation** is the inverse of this.
(5) Condition where field applied across
ionization chamber is sufficient to collect all
ions produced by incident radiation.

saturation activity. The limiting artificial
radioactivity induced in a given sample by a
specific level of irradiation.

saturation current. (1) The steady current in a
winding of an iron-cored transformer which
causes the inductance of the winding to be
seriously reduced. (2) That passed by a
valve, phototube or ion chamber, when
operated under saturation conditions. Also
total emission.

saturation scale. Minimum visual steps of
saturation, varying with wavelength.

saturation signal. A radar signal whose
amplitude is larger than the dynamic range
of the receiver.

saturation voltage. Voltage applied to a
device in order to operate under saturation
conditions.

sausage antenna. A number of wires connected in parallel, arranged in a parallel
formation around circular spreaders.

Savannah. First nuclear-powered merchant
ship (U.S.).

sawtooth oscillator. See **relaxation oscillator,
time-base generator.**

sawtooth wave. One generated by a *time-base
generator* for scanning in a CRT, for uniform
sweep and high-speed return. See **flyback.**

scalar product. That of two vector quantities
when the result is a *scalar quantity*, e.g.,
work=force × displacement. Known as
the *inner product* and denoted algebraically
by a dot between the vectors (or by a round
bracket enclosing them). Its magnitude is
given by the product of their amplitudes,
and the cosine of the angle between them, i.e.,
$A . B = AB \cos \theta$. See **vector product.**

scalar quantity. That which is completely
represented by a single numerical value
associated with a unit of measurement. It
is *pseudo-scalar* if it carries a sign. See
vector quantity.

scale. (1) Succession of notes, in order of
pitch, covering an octave or frequency
range of two to one. (2) Numerical factor
relating measured quantity to indication of
instrument, e.g., as in weighing.

scale factor. Any numerical which alters a
numeric attached to the magnitude of a
quantity, especially in calculations with an
analogue computer, or that related to the
radix, which shifts the significant figures in a
digital computer.

scale-of-ten. Ring, or other, system of counting elements which divide counts by ten.

scale-of-two. Any bistable circuit which can
divide counts by two, when operated by
pulses. See **flip-flop.**

scaler. Instrument incorporating one or more
scaling circuits, used to register a count.

scaling circuit. One which divides counts of
pulses by an integer, so that they are more
readily indicated to a required degree of
accuracy. If scalers-of-ten are used in

cascade, indications of counts in decimal
numbers are possible, e.g., by *Dekatrons.*

scaling factor. Number of input pulses in a
scaler per output pulse.

scalloping. The axial variation of the focusing
field in a travelling-wave tube which gives
rise to a corresponding variation of the
cross-section of the electron beam.

scallops. Alternating sections with reverse
curvature in a stellarator, to avoid drift in
the plasma.

scan. Systematic variation of a radar beam
direction for search or angle tracking. See
A, B, etc., **display** ; see also **scanning** (2).

scandium (Sc). An un-isolated metal, at. no.
21, at. wt. 44·956. Occurs naturally as an
oxide.

scanner. Mechanical arrangement for covering a solid angle in space, for the transmission or reception of signals, usually by
parallel lines or *scans*. See **disk-, flying-
spot-.**

scanning. (1) Coverage of original or reproduced TV image, now generally by interlaced
horizontal lines controlled electronically, but
originally by mechanically-controlled disks,
drums or mirrors. See also **alternate-.**
(2) Coverage of a prescribed area by a
directional radar antennae or sonar beam.
See also :

circular-	**mechanical-**
coarse-	**negative-**
conical-	**oscillatory-**
electron-	**progressive-**
interlaced-	**variable-speed-**
low-velocity-	**vertical-.**

scanning beam. Any beam of light or electrons
which scans a TV image.

scanning coils. Coils mounted in CRT, and
carrying suitable currents for deflecting
electron beam, to sweep the picture area.

scanning disk. (1) Disk with colour-filter
sectors, used in colour-sequence television.
(2) Rotating disk carrying a series of
apertures, lenses, mirrors or other optical
devices, used for mechanical scanning in
primitive television systems. See **Nipkow
disk.**

scanning field. Same as **raster.**

scanning frequency. Number of times an
image is scanned each second. For TV,
25/sec in U.K., 30/sec in U.S., the scan being
synchronized with the supply mains so that
any interference produced is static.

scanning heating. *Induction heating* where the
workpiece is moved continuously through
the heating region, as in *zone refining* of
germanium; also called **progressive heating.**

scanning line. Trace of single traverse of the
picture by the scanning spot from side to
side in horizontal scanning, or vertically in
vertical scanning.

scanning linearity. Uniformity of scanning
speed for a CRO or TV receiver. This is
necessary to avoid waveform or picture
distortion.

scanning loss. That which arises from relative

motion of a scanning beam across a target, as compared with zero relative motion.

scanning speed. That of a scanning spot across the screen of a CRT. Usually accurately specified for a CRO to facilitate time measurements.

scanning spot. Spot of light formed by scanning beam on the screen of a reproducer, or, in some early forms of transmission, on the object being televised.

scansion. Same as **scanning.**

scattering. General term for irregular reflection or dispersal of waves or particles, e.g., in the acoustic waves in an enclosure, leading to diffuse reverberant sound, *Compton effect* (q.v.) on electrons, light in passing through material, electrons, protons, neutrons in solids, and radio waves by ionization. See **forward scatter, backward scatter.** Particle scattering is termed *elastic* when no energy is surrendered during scattering process—otherwise it is *inelastic.* If due to electrostatic forces, it is termed *Coulomb* scattering but if short-range nuclear forces are involved it becomes *anomalous.* Long-wave electromagnetic wave scattering is *classical* or *Thomson* (q.v.), while for higher frequencies *resonance* or *potential* scattering occurs according to whether the incident photon does or does not penetrate the scattering nucleus. Coulomb scattering of alpha-particles is *Rutherford* scattering. For light, scattering by fine dust or suspensions of particles is *Rayleigh* scattering, while that in which the photon energy has changed slightly due to interaction with vibrational or rotational energy of the molecules is *Raman* scattering. *Shadow* scattering results from interference between scattered and incident waves of the same frequency. See **acoustic-, atomic-.**

scattering amplitude. Ratio of amplitude of scattered wave at unit distance from scattering nucleus, to that of incident wave.

scattering cross-section. Effective impenetrable cross-section of scattering nucleus for incident particles of low energy. The radius of this cross-section is the **scattering length.**

Scenioscope. TN for camera tube which uses a screen of conducting glass instead of mica.

Sceptre. A.E.I. thermonuclear research toroidal discharge apparatus, smaller than but similar to *zeta.*

Schering bridge. A.c. bridge for capacitance and PF measurements. Used with a *Wagner earth* (q.v.), it is capable of high accuracy even with small value capacitors.

Schlenke loudspeaker. One with a large stretched Duralumin diaphragm driven eccentrically by a current-carrying coil located in the circular gap of a pot magnet.

Schlieren pattern. One formed optically to investigate pressure waves in gas or other transparent media through resulting variations of refractive index.

Schmidt lines. Lines in the plot of nuclear magnetic moment against nuclear spin, as a result of the spin of an odd unpaired proton or neutron. Experimental magnetic moments for the majority of such nuclides lie between these lines, which are therefore also known as the **Schmidt limits.** See also **one-particle model of nucleus.**

Schmidt optical system. One used in telescopes and for projection TV, employing a spherical concave mirror in place of the more perfect parabolic type. Resulting spherical aberration is avoided by the use of a moulded correction lens.

Schmitt limiter. A bistable pulse generator which gives a constant amplitude output pulse provided the input voltage is greater than a predetermined value.

Schmitt trigger. Monostable circuit giving accurately-shaped constant-amplitude rectangular-pulse output for any input pulse above the triggering level. Widely used as pulse shaper after radiation counter tubes.

Schottky effect. Increase of saturation with increasing potential gradient near cathode-emitting surface of a triode.

Schottky noise. Strictly, noise in the anode current of a thermionic valve due to random variations in the surface condition of the cathode. Frequently extended to include *shot noise.*

Schrödinger equation. General equation of particle waves forming basic law of wave mechanics. Its solution is a wave function for which the square of the amplitude expresses the probability of the particle being located at the corresponding point.

Schroteffekt. See **shot noise.**

Schuler. Said of 84-minute period in a guidance system which has properties analogous to a *Schuler pendulum.*

Schuler pendulum. Theoretically, an ideal pendulum which will not be affected by the earth's rotation, having a length equal to the earth's radius and hence a period of 84 minutes. Practically, such a pendulum is not physically realisable but a stable-platform servomechanism can be constructed to simulate pendular motion of the above period, and is then said to be *Schuler-tuned.* Conversely a gyroscopically stabilized platform constrained to move parallel to the earth's surface in its motion over the earth will possess a period of 84 minutes and will be conditionally stable. Damping to reduce the maximum error may be introduced by rate feedback.

Schumann plates. Photographic emulsion with low gelatine content for use in ultra-violet light, where gelatine absorption is serious.

Schwartzschild antenna. System of bent plates reflecting a radar wave and achieving a very narrow fan beam pattern of radiation.

Schwinger coupling. Directional coupler which links a longitudinal magnetic field and a transverse magnetic field in separate guides.

scintillation. (1) Undesired transient changes in carrier frequency, arising from the modulation process. (2) Minute light flash

scintillation caused when α-, β- or γ-rays strike certain phosphors, known as **scintillators.** The latter are classed as liquid, inorganic, organic or plastic according to their chemical composition.

scintillation counter. One in which a photomultiplier generates pulses which are used to count nuclear events causing scintillations in a phosphor or crystal.

scintillation crystal. One which scintillates when subject to nuclear radiation, e.g., anthracene crystal.

scintillation spectrometer. Scintillation counter for which all the energy of the incident particle is normally absorbed in the scintillator, so that the amplitude of the resulting light flash (or corresponding signal pulse from photomultiplier) is a measure of the energy of the incident particle or photon. These signal pulses are analysed simultaneously or sequentially by a kicksorter or single-channel pulse height analyser. Different types of events counted can then be distinguished by their differing energies. See **alpha-, beta-,** and **gamma-ray spectrometers.**

scope. (1) A radarscope which produces a mode of information display. See **A, B,** etc. **display.** (2) Colloquial term for **oscilloscope.**

Scophony television. Electromechanical system in which a light beam is modulated by a Kerr cell and scanning is carried out by rotating mirrors. See **skiatron.**

scoring system. One used for sound recording in cinematography, where music has to be synchronized with the film.

scotophor. Material which darkens under electron bombardment, used for screen of CRT in storage oscilloscopes. Recovers upon heating. Usually potassium chloride.

scotopic vision. That which takes place at low illumination levels through the medium of the retinal rods.

scram. General term for emergency shutdown of a plant, especially of a reactor when the safety rods are automatically shot-in to stop the fission process.

scram rod. An emergency safety rod used in a reactor.

scrambler. Multiple modulating and demodulating system which interchanges and/or inverts bands of speech, so that speech in transmission cannot be intelligible, the reverse process restoring normal speech at receiving end.

scrambling circuit. Transmitting circuit which is used to make signals unintelligible unless there is a corresponding unscrambling receiver circuit.

scratch filter. A low-pass filter network to eliminate noise due to traverse of the stylus in the grooves of a record.

screen. (1) End of tube on which is exhibited a display through fluorescent excitation by electron beam; a phosphor. (2) Electrode consisting of a relatively fine mesh network of wires interposed between two other electrodes, to reduce the electrostatic capacitance between them. It is usually maintained at positive potential, and connected to earth through a capacitor. See **electrostatic-, Faraday cage;** see also **aluminized-, intensifying-.**

screen burning. Gradual falling-off in luminosity, sometimes accompanied by discoloration in the fluorescent screen of a CRT, particularly if operated under adverse conditions.

screen factor. Ratio of actual area of grid structure to total area of surface containing the grid.

screen grid. One placed between control grid and anode in a valve, having invariant potential. Originally of slats (Hall), but now of gauze or spirals, with the intention of removing effect of anode circuit on the input control grid circuit, thus eliminating positive feedback and instability through anode/grid capacitance. See **Miller effect.**

screen modulation. That in which the potential of the screen in a multi-electrode valve is varied in accordance with impressed modulating currents.

screened-grid valve. Four-electrode valve, with cathode, control grid, screen and anode. Used as a high-frequency amplifier, where the screen, of unvarying potential, prevents positive feedback, and so greatly enhances stability.

screened pentode. Pentode valve having a fine-mesh auxiliary grid and consequently small grid-to-anode capacitance, for use at high frequencies.

screened wiring. Insulated conductors enclosed in earthed and continuously conducting metal tubes or conduits, mainly for mechanical protection, but frequently for preventing induction to or from conductors.

screening. Use of a screen, in the form of a metal or gauze can, normally earthed, so that electrostatic effects inside are not evident outside, or vice versa. Similarly, nucleus of an atom is screened by its surrounding electrons.

screening constant. Of an element, the atomic number minus the apparent atomic number which is effective for, e.g., X-ray emission.

Scylla. Experimental thermonuclear device using magnetic mirror geometry.

sea cell. Primary electrolytic cell which functions as a source of electric power when immersed in sea-water. A battery of such cells is possible, power being available although cells are partially short-circuited by sea-water. Fitted to lifebelts, etc., so that an indicating light is produced automatically in the event of use at night.

seal, sealing, sealing-in. (1) In a relay, colloquialism for *lock*, whether by holding contact or residual magnetism. (2) In a vacuum tube, the point at which the tube is closed after pumping, and the act of closing off.

sealed tube. Electron tube which is permanently evacuated and not pumped during operation. Cf. *demountable* tube.

seam welding. Uniting sheet plastic by heat

arising from dielectric loss, the electric field being applied by electrodes carrying a high-frequency displacement current. Also **high-frequency welding, jig welding.**

search. Continuous examination and rejection of computer data until a desired value is found.

search coil. (1) Rotating coil in a radiogoniometer or phase adjuster. (2) See **exploring coil.**

search radar. One designed to cover a large volume of space and to give a rapid indication of any target which enters it.

search time. That required to locate a specific item of data for which the address is not known, e.g., in a literature search a reference with a specific keyword might have to be located from an arbitrary position in the memory.

Secam. Colour TV system developed in France and adopted by Russia. For European general use, *Pal* (q.v.) is preferred.

second anode. Second accelerating electrode in an electrostatic electron lens.

second-channel interference. In reception by supersonic heterodyne receivers, the interference from signals which are not desired, but whose frequency differs from local oscillator frequency by the same amount as the wanted signal. Both these signals produce an intermediate frequency output acceptable to the receiver. Discriminated against by pre-oscillator tuning.

second detector. In supersonic heterodyne receivers, detector which demodulates the received signal after passing through the intermediate-frequency amplifier.

secondary battery. A number of secondary cells connected together to give a larger voltage or a larger current than a single cell.

secondary cell. See **accumulator.**

secondary constants. Those for a transmission line which are derived from the *primary constants.* They are the *characteristic impedance* (*impedance level*) as of an infinite line, and the *propagation constant* (*attenuation* and *phase delay constant*).

secondary electrons. Those which are emitted from a surface by electronic bombardment, as distinct from the primary bombarding electrons, or photoelectrons.

secondary emission. Emission of electrons from a surface (usually conducting) by the bombardment of the surface by electrons from another source. The number may greatly exceed that of the primaries, depending on the velocity of the latter and the nature of the surface.

secondary emission factor. Ratio of number of secondary electrons emitted from a surface to the number of primary electrons incident on the surface.

secondary emission multiplier. See **electron multiplier.**

secondary grid emission. Electrons released by bombarding electrons, depending on the surface of the grid material.

secondary memory. (1) Storage unit associated with, but outside, a computer. (2) Part of computer store for data less often required, which has a longer access time than the *primary store* (q.v.). Also **auxiliary, backing,** or **secondary store.**

secondary radar. One in which the received pulses have been transmitted by a responder which has been triggered by a primary radar.

secondary radiation. That produced by the interaction of primary radiation and an absorbing medium.

secondary service area. Area in which reasonably consistent broadcast programmes are received at night via an ionospheric reflected wave.

secondary store. See **secondary memory** (2).

secondary winding. Transformer or transductor winding from which output energy or signal is derived.

section. Unit of a ladder network, derived through design techniques to give specified transmission performance with respect to frequency. If symmetrical (except the *lattice* or *bridge* type), it can be divided into equal *half-sections.*

sector disk. Rotating disk with angular sector removed, interposed in path of beam of radiation. Used to produce known attenuation, or to chop or modulate intensity of transmitted beam.

sectoral horn. Waveguide horn with two surfaces parallel to side of guide, and the other two flared out—classified according to whether the flaring is in the plane of the electric or magnetic field.

secular equilibrium. Radioactive equilibrium where parent element has such long life that activities remain effectively constant for long periods.

Seebeck effect. Phenomenon by which an (thermoelectric) e.m.f. is set up in a circuit in which there are junctions between different bodies, metals or alloys, the junctions being at different temperatures. Also **thermoelectric effect.**

seed. Small, near perfect, single crystal used as nucleus for crystal growing.

see-saw amplifier. Same as **paraphase amplifier.**

Segrè chart. Chart on which all known nuclides are represented by plotting the number of protons vertically against the number of neutrons horizontally. Stable nuclides lie close to a line which rises from the origin at 45° and gradually flattens at high atomic weights. Nuclides below this line tend to be β-emitters, while those above tend to decay by positron emission or electron capture. Data for half-life, cross-section, disintegration energy, etc., are frequently added. See Appendices.

selectance. (1) Ratio of sensitivities of a receiver to two specified channels. (2) With a resonant device, a measure of the fall-off in response on either side of resonance.

selection check. Automatic computer check that correct *address* is consulted in response to programme *instruction.*

selection rules. Those specifying the transitions of electrons or nucleons between different energy levels which may take place (*allowed transitions*). The rules may be derived theoretically through wave mechanics, but are not obeyed rigorously—so-called *forbidden transitions* merely being highly improbable.

selective absorption. Concentration of a compound—usually labelled—in a specific organ, as a result of natural processes.

selective fading. That affecting some parts of a composite signal more than others, e.g., sound and not vision in TV reception.

selective interference. That concentrated into relatively narrow frequency channel(s).

selective network. One for which the loss and/or phase shift are functions of frequency.

selectivity. Ability of a receiver to distinguish by tuning between specified wanted and unwanted signals. Measured by frequency difference for the half-power points of the pass band of the receiver. Often aided by directive reception.

selector circuit. One which selects a specified magnitude in a waveform, e.g., amplitude, phase, frequency or epoch.

selenium (Se). At. no. 34, at. wt. 78·96, sp. gr. 4·81, m.p. 217°C, b.p. 685°C. A non-metallic element which is an important semiconductor and exists in a number of allotropic forms. The 'grey form' becomes electrically conducting when irradiated with light.

selenium cell. (1) Photoconductive cell which depends on change in electrical resistance on incidence of light. (2) Photovoltaic cell which depends on the illumination of a barrier layer between selenium and other materials.

selenium rectifier. One depending on a barrier layer of crystalline selenium on an iron base. Widely used in power supplies for small electronic apparatus.

self-absorption. See self-shielding (1).

self-bias. In a class-A amplifier, grid bias obtained from potential difference across a resistor in cathode circuit taking anode current.

self-capacitance. See capacitance.

self-discharge. (1) Loss of capacity of primary cell or accumulator as a result of internal leakage. (2) Loss of charge from capacitor due to finite insulation resistance between plates.

self-excited oscillator. Normal form of oscillator, in which excitation of the grid circuit is derived from alternating current in the anode circuit.

self-heterodyne. See autodyne receiver.

self-impedance. The ratio of applied e.m.f. to resulting current between any pair of terminals in a network if all other terminals are open.

self-inductance. Realization, in a current-carrying coil, of self-induction. See mutual inductance.

self-inductance coefficient. See inductance coefficient.

self-induction. Property of an electric circuit by which it resists any change in the current flowing (Lenz's law).

self-oscillation. That generated in a valve circuit using positive feedback between the output and input circuits.

self-quenching. Said of counter tubes which do not depend on an external circuit for quenching, the residual gas providing sufficient resetting for the next operation of detecting a further photon or particle.

self-quenching oscillator. Same as squegging oscillator.

self-rectifying. Said of an X-ray tube when an alternating voltage is applied directly between target and cathode.

self-regulating. Said of a system when departures from the required operating level tend to be self-correcting, and in particular of a nuclear reactor where changes of power level produce a compensating change of reactivity (e.g., through negative temperature coefficient of reactivity).

self-scattering. The scattering of radioactive radiations by the body of the material which is emitting the radiation.

self-sealing capacitor. One in which a puncture of dielectric by excessive field strength causes oxidation of the neighbouring metal electrode through heat, with consequent restoration of insulation and negligible loss of capacitance; also called **Mansbridge capacitor.**

self-shielding. (1) In large radioactive sources, the absorption in one part of the radiation arising in another part. (2) A coaxial line is self-shielding in that the return transmission current is in the inside surface of the outer conductor, while the interfering currents, if of sufficiently high frequency, are on the outside surface of the outer conductor.

selsyn (*self-syn*chronous). See synchro- differential-.

semiconductor. Near insulator (at room temperature resistivity is between 10^{-2} and 10^9 ohm cm) showing limited electrical conductivity due to either or both effects below : (a) *intrinsic semiconductor*—crossing of forbidden energy bands through thermal energy. Effect increases with temperature and produces equal number of electron and hole (or n and p) carriers; (b) *extrinsic semiconductor*—introduction of current carriers by donor or acceptor impurities with localized energy levels near top or bottom of forbidden band, so producing n- or p-type conductivity. Important semiconducting materials are Ge, Si, Se, Cu_2O, SiC, PbS, PbTe, etc., being used as rectifiers, photocells, thermistors, etc. See compensated-, degenerate-, n-type-, p-type-.

semiconductor detector. Radio demodulator which employs a semiconductor device.

semiconductor diode. Two-electrode, point contact or junction, semiconducting device with asymmetrical conductivity.

semiconductor generation rate. See **generation rate.**

semiconductor impurity. See **donor, acceptor.**

semiconductor junction. One between donor and acceptor impurity semiconducting regions in a continuous crystal ; produced by one of several techniques, e.g., alloying, diffusing, doping, drifting, fusing, growing, etc.

semiconductor radiation detector. Semiconductor diodes, e.g., silicon junction, are sensitive under reverse voltage conditions to ionization in the junction depletion layer, and can be used as radiation counters or monitors. See **lithium-drift germanium detector.**

semiconductor trap. Lattice defects in a semiconductor crystal that produce potential wells in which electrons or holes can be captured.

semi-empirical mass formula. A formula for the calculation of the mass of an atom of given mass number which is based on the liquid-drop model of the nucleus.

semitone. Difference of pitch between two sounds with a frequency ratio equal to the twelfth root of two.

semitransparent photocathode. One where the electrons are released from the opposite side to the incident radiation.

sender transmitting station. Location of radio transmitters and associated aerials (antennae). It can be (*semi-*)*attended, automatic, booster, master, remotely-controlled, satellite* or *unattended.*

sensation level. Difference in level of a single-frequency sound, as applied to the ear, from the level which is just audible at the same frequency. Number of decibels a single frequency note has to be attenuated before it becomes just inaudible. See **phon, sone.**

sensation unit. Original name of the **decibel**; so called because it was erroneously thought that the subjective loudness scale of the ear is approximately logarithmic.

sense. Relative progression along a line, e.g., up or down, left or right, the line not in itself giving this information. In radio direction-finding, in which the direction of arrival of a radio wave is ascertained by directive antennae, the sense of the direction as determined by a simple loop, frame, or Adcock antenna is indeterminate, but is readily determined by injecting an additional e.m.f. depending on the arriving wave and derived from a simple elevated aerial.

sensing. (1) Removal of 180° ambiguity in bearing as given by simple vertical loop antenna, by adding signal from open aerial. (2) In scanning punched cards or punched tape, process of mechanical, electrical, or photoelectric examination of presence or absence of a hole in a set. Also **reading.**

sensitive time. Period for which conditions of supersaturation in a cloud chamber or bubble chamber are suitable for formation of tracks.

sensitive volume. (1) The portion of an ionization chamber or counter tube across which the electric field is sufficiently intense for incident radiation to be detected. (2) The portion of living cells believed to be susceptible to ionization damage; see **target theory.**

sensitivity. General term for ratio of response (either in time and/or magnitude) to a driving force or stimulus, e.g., galvanometer response to a current, minimum signal required by a radio receiver for a definite output power, ratio of output level to illumination in a camera tube or photocell. See **deviation-.**

sensitometer. Wedge photometer, specially arranged for measuring densities of photographic sound-track on film. See **gamma** (3).

sensitometry. Measurement of *gamma*, e.g., of photographic film used for optical film recording.

sensor. General name for detecting device used to locate (or detect) presence of matter (or energy, e.g., sound, light, radio or radar waves).

sentinel. A symbol used to indicate the end of a specific block of information in a data-processing system. Also **marker, tag.**

separation energy. That required to separate one nucleon from a complete nucleus.

separation factor. The abundance ratio of material taken from the product end of an isotope separation system or unit, divided by that of material taken from the reject end. $(S-1)$ is known in the U.K. as the *separation* or *process factor*; in the U.S. as the *enrichment factor.* $S = (n_1'/n_2') \div (n_1/n_2)$, where n_1 and n_2 are the mol fractions of two isotopes of mass numbers m_1 and m_2 respectively, and n_1' and n_2' are the corresponding quantities after processing.

separation potential. The separative work content per mol.

separative efficiency. Ratio of change in separative work content produced by a separation plant in unit time, to the integrated separative powers of its elements.

separative element. One unit of a cascade forming a complete isotope separation plant.

separative power. Change in separative work content produced by an isolated separative element in unit time.

separative work content. The quantity

$$Q\left\{2(C_2-1)\ln\frac{C_2(1-C_1)}{C_1(1-C_2)}+\frac{(C_2-C_1)(1-2C_1)}{C_1(1-C_1)}\right\}$$

for an isotope separation plant.
Q = molar quantity of the product
C_1 and C_2 = abundances of the required isotope in the raw material and the product. In U.S. termed **value.**

separator. A *flag* used to separate items of data.

separator circuit. One generally employing saturated valves, for separating line and frame synchronizing signals from the picture (video) signal and from each other.

separator valve. Amplifying valve interposed between an oscillator valve and modulated stages, to prevent changes of load conditions of the latter from affecting the frequency of the oscillator.

sequence. The section of a magnetic-tape recording which is free from titles or intervals.

sequence register. A register which is reset following the execution of an instruction and which then holds the memory address of the next instruction.

sequencer. Instrument used for sorting information into the required order for data processing.

sequential colour systems. Colour TV systems in which colour information for each channel is transmitted sequentially. Systems may be *field, line* or *dot sequential.*

sequential control. (1) A teletype system by which a master unit automatically controls the sequence in which data stored on a perforated tape at subsidiary stations is processed. (2) The operation of a computer when instructions are supplied sequentially during the running of a programme.

sequential monitoring. Viewing in a cycle a number of transmissions.

sequential operation. Operation of digital computer in which all instructions are carried out sequentially.

sequential transmission. A technique in TV of transmitting pictures so that the picture elements are selected at regular times and are then delivered to the communication channel in the correct sequence.

Serber force. Close-range force between nucleons arising from both ordinary Wigner force and an exchange force of the Majorana type. See **short-range forces.**

serial arithmetic unit. One in which the digits of a number are operated on sequentially.

serial computer. Computer which operates successively on each *bit* of a word, in contradistinction to a parallel computer where all *bits* are operated on simultaneously.

serial store. A memory for which stored data can only be read out sequentially, with access time depending upon location.

series. (1) Said of electric components when a common current flows through them. (2) Lines in a spectrum described by a formula related to the possible energy levels of the electrons in outer shells of atoms.

series arm. In a wave filter, that which is in series with one leg of the transmission line.

series feed. The application of a direct voltage to the grid or anode via the same impedance through which the signal current flows.

series modulation. Anode modulation in which modulator and modulated amplifier valves are connected directly in series, to eliminate the necessity for a modulation transformer or choke coupling

series-parallel network. One in which the electrical components are composed of branches which are successively connected in series and/or parallel.

series resonance. The condition of a tuned circuit when it offers minimum impedance

to an a.c. voltage supply connected in series with it (due to the circuit reactances neutralizing each other). The term **tunance** is sometimes used in place of *resonance* if this condition is attained by adjustment of a component value and not of frequency.

series stabilization. A technique of stabilization using amplifier feedback in which the feedback and amplifier circuits are in series at each end of the amplifier.

series voltage regulation. That performed by control of a variable impedance in series with the output.

service area. That surrounding a broadcasting station where the signal strength is above a stated minimum and not subject to fading. See also **A-, B-, primary, secondary-.**

service band. That allocated in the frequency spectrum and specified for a definite class of radio service, for which there may be a number of channels.

service quality. Proportion of time during which there is correct functioning on demand of an equipment.

servo. General term for system in which the response is determined by a drive which is actuated by the difference (*error*) between a set target and the actual response. A servo system aids or replaces human action, by force, time of operation or location. Error usually requires valve or transistor amplification.

servo-amplidyne system. One in which an amplidyne, together with a control amplifier, is used to amplify mechanical power.

servo amplifier. One designed to form the part of a servomechanism from which output energy can be drawn.

servo control. Control of any system with a servomechanism.

servo link. A mechanical power amplifier which permits low strength signals to operate control mechanisms that require fairly large powers.

servomechanism. Closed-cycle control system in which a small input power controls a much larger output power in a strictly proportionate manner, e.g., movement of a gun turret may be accurately controlled by movement of a small knob or wheel.

servomechanism types. Type-0 has no integrating term and requires an offset signal. Type-1 has one integrating circuit in its closed loop and a velocity error. Type-2 has two integrating circuits and no errors.

servomotor. The source of power for mechanical movement in a servomechanism.

set. To prepare computer circuits so that they operate when required or triggered. Binary circuits are *set* when in '1' configuration, *reset* or *cleared* when restored to '0'.

set point. See **control point.**

set-up. Ratio between black and white reference levels measured from blanking level for facsimile transmission.

sferics. U.S. for **atmospherics.**

shaded pole. Ringing part of a pole of laminations, so that section of flux has its phase shifted and so the average resultant force across a gap, e.g., to the armature in an a.c. relay, has a longitudinal component, thus increasing sensitivity and reducing noise.

shading. (1) Variations in brightness in a televised image because of local defects in the signal plate of a camera tube, arising from inadequate discharge. Corrected by injecting waveforms in the output signal. See tilt-and-bend-. (2) The technique of varying the directivity of a transducer by controlling the amplitude and phase distribution over the active area of the transducer.

shadow. Ineffectiveness of reception because of an obstacle, e.g., due to the topography of the terrain, between transmitter and receiver.

shadow-mask kinescope. A directly-viewed three-gun cathode-ray tube for colour television display, in which beams from three electron guns converge on holes in a shadow mask placed behind a tricolour phosphor-dot screen.

shadow scattering. See scattering.

Shannon equation. Equation in information theory which gives theoretical limit to rate of transmission of binary digits with a given bandwidth and signal/noise ratio.

shaped-beam tube. One in which the cross-section of the beam of electrons is formed to the shape of various characters.

shaping network. One which determines or restores the shape of a pulse, especially in radar and computing.

shared-channel broadcasting. See common-frequency broadcasting.

sharp series. Series of optical spectrum lines observed in the spectra of alkali metals. Has led to energy levels for which the orbital quantum number is zero being designated s-levels.

sharpness. Equivalent to *selectivity*, but referring more directly to the change in circuit adjustment necessary to alter signal strength from its maximum to a negligible value.

shaving. Machining surface of master disk recording, to give fresh surface for further use.

sheath. (1) Excess of positive or negative ions in a *plasma*, giving a shielding or *space charge* effect. (2) The covering on a cable. (3) The can protecting a nuclear fuel element.

shed. Minute unit of nuclear cross-section, equal to 10^{-24} barn.

shell. (1) Shell-like magnet in which magnetization is always normal to the surface and inversely proportional to the thickness. The *strength of the shell* is the product of magnetization and thickness of shell. (2) Theoretical concept of a double layer of poles, i.e., multitudinous magnetic dipoles, which is, in general, not plane. (3) Pattern of orbital electrons surrounding the nucleus of an atom, characterized by principal quantum numbers. See Bohr atom.

shell model. A model of the nucleus of an atom, with protons and neutrons in shells, by analogy with electrons in shells outside the nucleus.

Shepherd tube. A velocity-modulated microwave oscillator.

shield. Screen used to protect persons or equipment from electric or magnetic fields, X-rays, heat, neutrons, etc. In a nuclear reactor, the shield surrounds it to prevent the escape of neutrons and radiation into a protected area. See biological-, Faraday cage, magnetic-, radiation trap, thermal-.

shield grid. Auxiliary grid which protects control grid from heat and deposition of cathode material.

shielded box. Glove box protected by lead walls, and with facilities for manipulation of contents by remote handling equipment.

shielded line. (1) Line or circuit which is specially shielded from external electric or magnetic induction by shields of highly conducting or magnetic material. (2) Transmission line enclosed within a conducting sheath, so that the transmitted energy is enclosed within the sheath and not radiated.

shielded nuclide. One which, when found among fission fragments, is assumed to have been a direct product, because it is known not to be formed as a result of beta decay.

shielded pair. Balanced pair of transmission lines within a screen, to mitigate interference from outside.

shielding. (1) Prevention of interfering currents in a circuit, due to external electric fields. Any complete metallic shield earthed at one point is adequate. (2) Use of high permeability material, e.g., Mumetal, for shielding devices susceptible to a magnetic field, e.g., cathode-ray beam; the field, direct or alternating, is shunted away from spaces where it would cause interference. (3) Use of dense material to attenuate radiation when this might be harmful to operator or measuring system. (4) Similar use of very light materials to thermalize strong neutron beams.

shielding pond. Deep tank of water used to shield operators from highly radioactive materials stored and manipulated at the bottom.

shielding windows. Dense glass blocks or liquid-filled tanks used as windows for inspecting the interior of shielded boxes.

shift. (1) Movement of a set of figures in a computer, to the right or left, as if multiplied or divided by the *radix*. (2) Movement of a pattern on a CRT phosphor, by imposition of steady voltages, e.g., X-shift, Y-shift. (3) Double use of code, using one code for changing over, as in Telex, teleprinter, teletype, analogous with typewriter keyboard. In teleprinters, one shift is capital letters, the other figures and special signs (*case shift*). (4) Change in wavelength of spectrum line due, e.g., to Doppler or Zeeman effects or to Raman scattering. (5) Change in value of energy level (*level shift*) for a quantum mechanical system arising from interaction or perturbation.

shift pulse. One which initiates and controls a shift in a digital computer.

shift register. A temporary storage or delay register in which the output is released a fixed number of clock pulses after the input.

shim rod. Coarse control rod of reactor. It is usually positioned so that the reactor will be just critical when the rod is near the centre of its travel path.

shimming. Adjustment of magnetic field with soft iron shims or, by extension, small compensating coils.

shock excitation. Excitation of transient currents in an oscillatory circuit (ringing) at its natural resonant frequency by the sudden application or removal of an e.m.f. having some other frequency. The cause of interference by keying clicks from a continuous-wave telegraph transmitter.

shock heating. Heating, especially of a plasma, by the passage of a shock wave.

shock wave. One of high amplitude, in which the group velocity is higher than the phase velocity, leading to a steep wavefront. Strong shocks give luminosity in gases and so are useful for spectroscopic work. Also **blast wave.**

shoran. Short-range navigation, using H-radar. (*Short range.*)

shore effect. Horizontal refraction as a radio wave crosses a shoreline at an angle, because of different retardations of the ground wave. This causes direction-finding errors. Also **coastline effect.**

short-circuit. Reduction of p.d. between terminals to zero by connection of a conductor of zero impedance, in which no power is dissipated. If the short-circuit is not perfect, arcing and damage may result if the circuit is not opened quickly elsewhere. See **dead-short.**

short-circuit impedance. Input impedance of a network when the output is short-circuited, shorted or grounded.

short-circuit transfer admittance. Transfer admittance when all other electrodes are held at constant potentials, i.e., have zero admittance to earth.

short delay line. Any part of an electronic computer which adjusts phases of signals by transmission through coaxial lines, extended coils, or as magnetostriction pulses through nickel wires.

short-range forces. Non-coulomb forces which act between nucleons when very close together, and are responsible for the stability of the nucleons. They are of two kinds: *ordinary forces* and *exchange forces.* See Bartlett force, Heisenberg force, Majorana force, Serber force, Wigner force.

short wave. Rather vague designation of radio transmission with wavelengths between *ca.* 15 and 100 m.

short-wave converter. A combination of a local oscillator and a mixer to produce a beat signal in the appropriate frequency band to convert short-wave signals to the region of standard broadcast frequencies.

shortening capacitor. One inserted in series with an antenna to reduce its natural wavelength.

shot noise. The *Schroteffekt* (small-shot effect) of Schottky, which arises inevitably in the anode circuit of a thermionic valve, because electron emission is not strictly continuous, but consists of a series of random pulses from emitting surfaces.

shoulder. That part of a characteristic curve at which the response tends to fall off, as at the upper limit of thermionic characteristics, the gamma curves of photographic emulsions, or peak response of a transformer.

shower. Result of impact of a high-energy cosmic-ray particle consisting of ionizing particles and photons, directed downwards. See **cascade-, hard-, soft-.**

shower unit. The mean path length for the reduction of 50% of the energy of cosmic rays as they pass through matter.

shrinkage factor. Ratio of thickness of nuclear research emulsion on photoplate before and after processing.

shunt. (1) Addition of a component to divert current in a known way, e.g., from a galvanometer, to reduce temporarily its effective sensitivity. See **Ayrton-Mather-.** (2) Diversion of some flux from the gap in a magnetic circuit by a magnetic slide or screw in a moving-coil indicating instrument.

shunt arm. That which is connected across the loop circuit of a wave filter.

shunt circuit. Electric or magnetic circuit in which current or flux divides into two or more paths before joining to complete the circuit; also **parallel circuit.**

shunt-excited antenna. Antenna consisting of a vertical radiator (frequently the mast itself) directly earthed at the base, and connected to the transmitter through a lead attached to it at a short way above ground.

shunt feed. See **parallel feed.**

shunt-field relay. One with two coils on opposite sides of a closed magnetic circuit, so that a bridging magnetic circuit takes no flux while the currents in the two coils magnetize the circuit in the same direction. Flux passes in this bridging circuit when one current is reversed.

shunt resonance. The condition of a parallel tuned circuit connected across an a.c. voltage supply when maximum impedance is offered to the supply, and the circulating loop current is also a maximum. The term *tunance* is sometimes used in place of *resonance* if this condition is attained by adjustment of a component value and not of frequency. Also **parallel resonance.**

shunt stabilization. A technique in which the feedback and amplifier circuits are in parallel.

shunt voltage regulation. That performed by control of a variable impedance in parallel with the output.

shut down. Reduction of power level in a

nuclear reactor to lowest possible value by maintaining core in subcritical condition.

shut-down amplifier. See trip amplifier.

shutter. Plate used to cover aperture in nuclear reactor core and so prevent egress of beam of radiation.

shuttle. Container for samples to be inserted in, and withdrawn from, nuclear reactors, when they are made radioactive by irradiation with neutrons; also **rabbit**.

side frequency. Any frequency of a sideband.

side lobe. Any lobe of the radiation pattern of an acoustic or radio transmitter other than that containing the direction of maximum radiation.

side-stable relay. One polarized, but without central position, free of *mark* or *space* contacts.

side thrust. Radial force on pickup arm, caused by stylus drag.

side tone. One reaching the receiver of a radio-telephone station from its own transmitter.

side wave. Isolated frequency component in the sideband. Analysis shows that a sinusoidally-modulated wave may be physically resolved into a carrier wave of frequency ω and two side waves of frequencies $\omega + \rho$ and $\omega - \rho$ respectively, where ρ is the modulating frequency.

sidebands. Those added to a carrier in the process of any modulation, the carrier remaining unchanged in the ideal case. Either sideband, without the carrier, contains the transmitted information. See carrier.

siegbahn. Unit for X-ray wavelengths, equal to 10^{-13} cm.

siemen. See mho.

sigma circuit. That operating a scram mechanism in a reactor.

sigma meson. An old term denoting a meson which gives rise to a 'star'. Such mesons are actually *pions* (see **mesons**).

sigma-particle. Hyperon triplet, rest mass equivalent to 1190 MeV, hypercharge 0, isotopic spin 1.

sigma pile. One comprising neutron source and moderating materials without any fissile element. Used in study of neutron properties of moderator.

sigmoid curve. An S-shaped curve which is often obtained in dose-effect curves in radiobiological studies.

sign. Digit representing $+$ or $-$, associated with a number in a digital computer.

signal. General term for the physical conveyor of information, e.g., an audio waveform, a video waveform, series of pulses in a computer. Colloquially, the message itself. In radio, the signal modulates a carrier and is recovered during reception by demodulation.

signal electrode or plate. In a television camera tube, the mosaic from which a signal arises during scanning. See image dissector, image inconoscope, orthicon.

signal element. The portion of a signal occupying the smallest interval of the signal code.

signal elongation. That of the envelope of a modulated signal as a result of delay of components transmitted over longer paths.

signal generator. Oscillator, designed to provide known voltages (usually from 1 volt to less than 1 μ volt) over a wide range of frequencies; used for testing or ascertaining performance of radio-receiving equipment. It may be *amplitude- frequency-* or *pulse-modulated*.

signal level. The level at any point in a transmission system, as measured by a voltmeter (VU meter) across the circuit when properly terminated; expressed in dB or VU in relation to a reference level, now 1 mW in 600 ohms.

signal/noise (S/N) ratio. See noise ratio.

signal output current. The absolute difference between the output current and the dark current of a phototube or a camera tube.

signal shaping. Use of specially designed electric network to correct distortion produced during transmission or propagation of signals.

signal-to-crosstalk ratio. In line telephony, the ratio of the test level in the disturbed circuit to the level of the crosstalk at the same point, which is caused by the disturbing circuit operating at the test level.

signal wave. One which allows intelligence to be conveyed.

signal windings. U.S. term for **control turns** (or **windings**) of a saturable reactor.

silent period. Stated period within each hour during which all marine transmissions must close down and listen on the international distress frequency of 500 kHz.

silica gel. Hygroscopic material much used for desiccation in enclosed electronic equipment. Changes clear to pink with moisture, which can be driven off by heat.

silica valve. One in which the envelope is made of fused silica to withstand high temperatures. Originally developed for use in warships.

silicon (Si). At. no. 14, at. wt. 28·086, sp. gr. 2·42, m.p. 1420°C, energy gap 1·12 electron-volts, temperature coefficient of energy gap -0.0003 electron-volts °C^{-1} at 25°C, mobility of holes 250 cm^2 volt^{-1} sec^{-1} at 25°C, mobility of electrons 1200 cm^2 volt^{-1} sec^{-1} at 25°C, dielectric constant 12. A non-metallic element having semiconducting properties, and occurring in two allotropic forms—dark grey crystals and a brown amorphous powder. Used for transistors and certain crystal diodes.

silicon-controlled rectifier. A three-junction semiconductor device which is the solid-state equivalent of a thyratron.

silicon detector. Stable silicon crystal diode for demodulation.

silicon rectifier. A semiconductor diode rectifier usually based on *p-n* junction in silicon crystal.

silicon resistor. A resistor of special silicon material which has a fairly constant positive temperature coefficient, making it suitable as a temperature-sensing element.

silicone rubbers. An important group of synthetic rubbers (dimethylsiloxene polymers), having both a high and a low temperature resistance which is better than for natural rubbers.

silicone fluids. Range of low vapour pressure oils with widely varying viscosities. Used for waterproofing, as optical coupling fluid, and in vapour diffusion vacuum pumps.

silit resistor. Tubular element made from a mixture of silicon carbide and silicon.

Silsbee rule. That a wire of radius r cannot carry a superconducting current greater than $\frac{1}{2}rH_c$, where H_c is the critical field. The self-magnetic field would then destroy superconductivity.

silver (Ag). Noble metal, at. no. 47, at. wt. 107·870, sp. gr. 10·5, m.p. 960·5°C, b.p. 2180°C. The best electrical conductor and the main constituent of photographic emulsions.

silver mica capacitor. High-stability, low power factor, fixed capacitor prepared by vacuum deposition of silver on thin mica sheets.

Silverstat. TN for control device in which a series of contacts are successively closed by an increasing applied force.

simple harmonic motion. Widely occurring conception of motion described by a sinusoidal function.

simple process. The physical mechanism on which each stage of an isotope separation plant depends.

simple process factor. The separation factor obtainable with a simple process. Its theoretical value is the *ideal SPF*, that attained in a single stage of a cascade is the *effective SPF*, or *stage separation factor*.

simple tone. Single-frequency tone for testing telephone circuits, for voice-frequency telegraphy and for facsimile carrier. Also **pure tone.**

simplex. General term for simple one-way transmission and reception over a channel. *Duplex* is use of the same channel in both directions at the same time, with different transmissions. *Diplex* is use of one channel for two different transmissions in the same direction.

simulated line. Same as **artificial line.**

simulation. Setting up of an electronic circuit, especially with analogue computers, which will obey the same laws as a physical system.

simultaneity. Basic issue in relativity theory. Two outside events are simultaneous to an observer if a clock with him indicates the same time for both. For observers in other frames of reference, this is not so.

simultaneous broadcasting. Transmission of one programme from two or more transmitters on different wavelengths.

sine wave. Waveform of a single frequency, indefinitely repeated in time, the only waveform whose integral and differential has the same waveform as itself. Its displace-

ment can be expressed as the sine (or cosine) of a linear function of time or distance, or both. In practice there must be a transient at the start and finish of such a wave.

singing. Self-oscillation in a transmission system caused by feedback across a source of gain because of unbalance in circuit.

singing arc. See **Duddell arc.**

singing point. Gain of amplifier under specified conditions just below oscillation.

single address. Said of the computer coding accompanying a *word* to guide it to a position in the store.

single-channel pulse height analyser. See **pulse height analyser.**

single circuit. One which accepts a particular signal from a random collection of signals. Also called **one-shot circuit.**

single-electrode system. Electrode of an electrolytic cell and the electrolyte with which it is in contact; also called **half-cell, half-element.**

single-ended. (1) Unit or system designed for use with unbalanced signal, having one input and one output terminal permanently earthed. (2) Valve with all electrodes connected to pins at the same end.

single-parity check. A system for detecting errors in computers similar to a binary code.

single-phase. Pertaining to a.c. power supplies, when one outward and one return conductor are required for transmission.

single pole. Switch, relay, etc., in which connections to only one circuit can be made. A single pole-single way switch is a simple on-off switch.

single-shot multivibrator. A monostable *multivibrator* (q.v.).

single-shot trigger. A circuit which provides a triggering pulse to 'set-off' a complete cycle of events ending with a stable state. Used, e.g., with cathode-ray oscillograph.

single sideband suppressed carrier. See **vestigial sideband transmitter.**

single sideband system. One in which, after modulation, one sideband and carrier are removed by push-pull modulator and filters. This minimizes the bandwidth required for transmission and improves signal/noise ratio. On reception, an identical carrier is inserted before demodulation.

single-state re-cycle. Isotope separation operating on the principle used in gaseous diffusion whereby the diffusing gas is divided into two fractions. One of these portions is recirculated within the first stage while the other proceeds to the next stage. Also **rabbit.**

single-valued. Said of a function for which the dependent variable can have only one value for each set of values of the independent variables.

single-wire circuit. One with a single live wire, including coaxial or concentric, with sheath, earth or frame return.

single-wire feeder. One for an antenna, similar to an ordinary downlead, but connected to the antenna in such a manner that it is terminated in its characteristic impedance, so that no standing waves are formed on it.

sink. Unstable operating region on Rieke diagram.

sintering. Process of bonding powder samples by heating—accompanied by large dimensional changes.

sinusoidal current. One which varies sinusoidally with time, having a *frequency, amplitude* and *phase*.

sinusoidal wave. See **sine wave**.

siphon recorder. A sensitive recorder which uses a fine siphon fed with ink and controlled by galvanometer coil currents. The signals are recorded on a continuously moving paper strip.

siren. A disk, with a ring of similar and equally spaced holes, which is rotated at a uniform speed to interrupt periodically an air jet directed at one side of the disk. A high sound intensity can be obtained at a fundamental frequency given by the product of the number of revolutions/sec and the number of holes in the ring.

site error. That in a radio bearing which depends only on site surroundings and which can be used as a correction to readings.

skew. The deviation from rectangular of a facsimile image.

skiagram, skiagraph. Same as *radiogram*.

skiatron. Cathode-ray tube in which an electron beam varies the transparency of a phosphor, which is illuminated from behind so that its image is projected on to a screen, as in a photographic slide projector.

skin depth. Depth to which high-frequency electromagnetic waves penetrate into a conductor. Measured to the point where the amplitude has been attenuated by a factor $1/e$.

skin dose. Absorbed or exposure radiation dose received by or at the skin of a person exposed to sources of ionization; cf. *tissue dose*.

skin effect. That when metallic conductors are used to carry high-frequency currents, these being limited to a surface layer equal in thickness to the skin depth; it gives a large value for the high-frequency resistance of the conductor. See **litz wire**.

skinner. Length of insulated wire between a laced cable form and the connecting point.

skiograph. Instrument for recording X-ray intensities.

skip. (1) See **hop**. (2) A code instruction to a computer to ignore the following item of data. It may be one or more words or only a byte, according to the coding.

skip distance. Region of no-signal between the limit of reception of the direct (*ground*) wave and the first downcoming reflection from an ionized layer; prominent with short waves.

skirt. Lower side portions of a resonance curve, which should be symmetrical.

skirt dipole. Quarter-wavelength wire, emerging coaxially from the conductor of a coaxial transmission feeder; also **sleeve dipole**.

sky wave or **ray.** See **ionospheric ray, space wave**.

slab resolver. Four contacts on a square, giving $(\pm R \cos \theta)$ and $(1 \pm R \sin \theta)$, concentric with a slab potentiometer fed with $\pm R$ at its ends, θ being angular displacement.

slant range. Radar range to elevated target, as distinct from ground range.

slatted lens. One with shaped metal slats, parallel to E or H vector in wave from waveguide. Used also for low-frequency acoustic waves. Also called egg-box lens.

slave operation (of time base). One operated by an external trigger pulse. Not self-running.

sleeping sickness. Colloquialism for the onset of *slow death* (q.v.) in a transistor.

sleeve. Quarter-wavelength coaxial line for coupling a coaxial line to a dipole at its centre. See **balun**.

sleeve dipole. Same as **skirt dipole**.

sleeving. Tubular flexible insulation for threading over bare conductors.

slicer. Combination of limiters such that only a range of signal between two limiting values is passed on.

slide-back. Original method for measuring high-frequency voltage applied to grid of triode; the voltage peak is measured by increase of negative bias which just cuts off grid current.

slide-back voltmeter. Traditional circuit for alternating-voltage measurement in which an adjusted and known steady voltage just balances the peak of the unknown voltage, as indicated by flow of anode or grid current. See **valve voltmeter**.

slide wire. One for potential division, by sliding a contact along a wire. With a concentric tube with a slot and a probe contact, this becomes a coaxial slotted line for voltage standing-wave ratio (VSWR) measurements at high frequencies.

sliding contact. Tangential movement between contacting metal surfaces, to remove film and establish conduction contact. Wear of contact is proportional to total use.

slip. (1) Vertical shift of television image because of imperfect *field* synchronization. Similarly for *line*, but horizontally. (2) Dislocations in crystals produced by plastic deformations. (3) Reduction of speed of induction motor under load, expressed as percentage of synchronous speed.

slip ring. A conducting ring used with a stationary brush to lead currents into and out of rotating machinery.

slope conductance. See **mutual conductance**.

slope resistance. See **differential anode resistance**.

slot antenna. Radiating element formed by metal surrounding a slot.

slot-fed dipole. One normal to a coaxial line, and coupled to it by adjacent longitudinal slots.

slotted line. Rigid coaxial line, with slot access for contact with central conductor. Used for impedance and VSWR measurements at wavelengths comparable with its length.

slow death. Colloquialism for gradual deterioration of transistor characteristics, due to poisoning by traces of contamination. If the failure arises from moisture collecting on base of a junction transistor, sometimes called **sleeping sickness.**

slow neutron. See **neutron.**

slow reaction. Nuclear reaction associated with weak interactions.

slow reactor. Obsolete term for **thermal reactor.**

slowing-down area. One-sixth of the mean square distance from the neutron source to point where neutrons have required average energy. See also **Fermi age.**

slowing-down density. Rate at which neutrons fall below a given energy level per unit volume.

slowing-down length. The square root of the slowing-down area.

slowing-down power. Increase in neutron lethargy per unit distance travelled in medium. The lethargy is approximately given by the negative of the natural logarithm of the energy of the neutron.

slug. (1) Thick copper band, comparable with a portion of a winding on a telephone-type relay which, through induced eddy currents, retards the operation and fall-off of the relay. (2) Unit of fuel in nuclear reactor, either rod or slab of fissile material encased in a hermetic can of Al, Be, Zr, or stainless steel. Also **cartridge.**

slug tuning. Alteration of inductance in radio-frequency tuning circuits, by inserting a magnetic core (original 'musical' choke) or a copper disk or cylinder (spade tuning).

slurry reactor. One in which fuel exists as a slurry carried by the coolant fluid.

small-signal parameters. See **transistor parameters.**

smear test. A method of estimating the loose, i.e., easily removed, radioactive contamination from a surface. Made by wiping the surface and monitoring the swab. Also **wipe test.**

smearer. A particular circuit employed to eliminate the overshoot of a pulse.

Smith chart. Polar chart with circles for constant resistance and reactance, lines for vector angles, and standing-wave ratio circles. Used for impedance calculations, especially with data from slotted lines for very high frequencies and from waveguides. See **voltage standing-wave ratio;** and Appendices.

smoother. Combination of capacitors and inductors for removing the ripple from rectified power supplies.

smoothing choke. Inductor in a filter circuit which attenuates ripple from a rectifier in supplying d.c. to transistor or valve circuits.

smoothing circuit. See **ripple filter.**

snap. Sudden action in magnetic amplifiers with excessive positive feedback, arising from hysteresis in the core.

sneak circuit. In developing the design of a circuit, a part of the circuit which obviates the required performance and has to be circumvented.

Snell's law. A wave refracted at a surface makes angles relative to the normal to the surface which are related by law $N_1 \sin \theta_1 = N_2 \sin \theta_2$. N_1 and N_2 are the refractive indices on each side of the surface, and θ_1 and θ_2 are the corresponding angles. The two rays and the normal at the point of incidence on the boundary lie in the same plane.

snow. Effect of electrical noise on CRT display of intensity-modulation signals. The effect is random and resembles falling snow and may be seen, e.g., on a TV screen in the absence of a signal.

sodium (Na). At. no. 11, at. wt. 22·9898, sp. gr. 0·971, m.p. 97·5°C, b.p. 883°C. Soft, silvery-white metal, very active, with moderate thermal neutron cross-section. It is used in reactors as a liquid-metal heat-transfer fluid.

sodium-cooled reactor. Reactor in which liquid sodium is used as the primary coolant.

sofar. System of underwater navigation depending on the reception of pulses from distant explosions. (*So*und *f*ixing *a*nd *r*anging.)

soft. Said of valves and tubes when there is appreciable gas-pressure within the envelope, such as gas-discharge tubes and photocells. Particularly said of valves when gas is released from the envelope or electrodes. See **hard, outgassing, Round valve.**

soft cosmic rays. Rays consisting of electrons, positrons, photons and some slow heavy particles which are absorbed by 10 cm of lead.

soft radiation. General term used to describe radiation whose penetrating power is very limited, e.g., X-rays at the low frequency end of the X-ray spectrum.

soft shower. One of soft cosmic rays. See **cascade shower.**

solar cell. Photoelectric cell using silicon, which collects photons from the sun's radiation and converts the radiant energy into electrical power with reasonable efficiency. Used in sputniks and for remote locations lacking power supplies, e.g., for telephone amplifiers in the desert.

solar flare. Bright eruption of sun's surface associated with sun spots, causing intense radio and particle emission.

solder. An alloy used to join metallic surfaces to form a continuous solid medium. The contents of the alloy vary according to the flow temperature which ranges from *ca.* 70°C to 2000°C. Low-temperature (soft) solders are essentially alloys of tin and lead in varying proportions. High-temperature solders are alloys of, e.g., zinc, tin, gold, silver, copper and rhodium. Wood's metal has a melting-point of about 70'C and contains 50% Bi, 25% Pb, 12·5% Sn and 12·5% Cd.

solder-covered wire. Copper wire coated with solder instead of tin, to facilitate connections between components in electrical and electronic apparatus.

solenoid. Current-carrying coil of one or more layers. Usually a spiral of closely-wound insulating wire, in the form of a cylinder, not necessarily circular. Generally used in conjunction with an iron core, which is pulled into the cylinder by the magnetic field set up when current is passed through the coil.

solenoidal field. One in which divergence is zero, and the vector is constant over any section of a tube of force.

solid angle. Angle subtended by a surface at a point—numerically equal to the area intercepted on the surface of a sphere of unit radius. See **steradian**.

solid circuit. (1) Modification of properties of a material, e.g., silicon, so that components can be realized in one mass, e.g., resistors, capacitors, transistors, diodes. (2) Sub-miniature realization of a circuit in three dimensions, e.g., as built up as parts of a semiconductor crystal or by etching or deposition on a substrate.

solid metal negative. See **matrix** (1).

solid solution. A single phase solid containing more than one component.

solid-state capacitor. That presented by a depletion layer at a p-n junction, in which the carriers from each side substantially neutralize each other at low voltages. Effective capacitance depends on applied voltage.

solid-state maser. One made from ruby at a few degrees above absolute zero in intense magnetic fields. A pulse of microwaves raises energy of electrons to a high level so that the crystal acts as an amplifier.

solid-state physics. Branch of physics which covers all properties of solid materials, including electrical conduction in crystals of semiconductors and metals, superconductivity and photoconductivity.

solion. A device for the conversion of the movement of ions in an electrolyte into electrical signals to enable low-frequency sound signals to be detected.

solution. Homogeneous mixture of a gas or solid in a liquid. Simple chemical compounds ionize when dissolved in water, radicals or atoms separating with numbers of electrons which differ from their neutral complement, depending on their valency. Such ions bunch together and regulate the conductivity and acidity. See **pH value**.

Sommerfeld atom. Atomic model developed from Bohr atom but allowing for elliptic orbits with radial, azimuthal, magnetic and spin quantum numbers. Modern theories modify this by regarding the electrons as forming a cloud, the density of which is described in terms of their wave function.

sonar (*sound navigation and ranging*). See **asdic, echo sounding**.

sonde. Small telemetering system in satellite rocket or balloon—used especially in meteorology.

sone. Unit of loudness equal to a tone of 1 kHz at a level of 40 dB above the threshold of the listener.

sonic. See **Mach number**.

sonic bang (or **boom**). The loud and objectionable noise set up by the shock waves created by aircraft or missiles which are travelling at supersonic speeds.

sonics. General term for study of mechanical vibrations in matter.

sonne. German navigational system, forerunner of **consol** and **consolan**.

sonobuoy. Equipment dropped and floated on the sea, to pick up aqueous noise and transmit a bearing of it to aircraft; three of such bearings enable the aircraft to 'fix' the source of underwater noise, e.g., from submarines.

Soret effect. See **thermal diffusion**.

sound. (1) The periodic mechanical vibrations of a medium. (2) The sensation felt when the ear-drum is acted upon by the air vibrations if within a limited frequency range, i.e., 20 Hz to 20 kHz, or less.

sound altimeter. An absolute means of measuring the height of an aircraft above ground using short sound pulses.

sound analyser. One which measures each frequency, or a small band of frequencies in the spectral distribution of energy. Particularly useful for tracing sources of vibration or noise in rotating equipment.

sound articulation. Percentage of all elementary speech-sounds received correctly, when *logatoms* (q.v.) are called over a circuit or in an auditorium, in a standard manner.

sound carrier. The radio-frequency electromagnetic wave which is modulated at a sound frequency, and carries the information which is reconverted to sound vibrations at the receiver.

sound channel. The carrier frequency with its associated sidebands, which are involved in the transmission of the sound in TV.

sound energy density. Density of energy in a diffused sound field, in erg/cm³, averaged over a volume, in reverberation time calculations.

sound field. (1) Region through which sound waves, standing or progressive, propagate from a source. Such fields diverge from point, cylindrical or plane sources. (2) Enclosed space in which the diffused sound waves are random in magnitude, phase and direction, constituting reverberant sound.

sound intensity. Flux of sound power through 1 cm², normal to the direction of propagation. If p=rms excess pressure, ρ=density, c=velocity of propagation, then, in CGS units, the intensity is given by

$$I = p^2/\rho c.$$

sound intensity level. At any audio frequency, the intensity of a sound, expressed in decibels above an arbitrary level, 10^{-16} watt/cm², equivalent in air to a pressure of 0·200 millidyne (rms)/cm²; also **sound level**.

sound interval. The interval between two sounds is the ratio of their fundamental frequencies, or its logarithm.

sound level. Same as **sound intensity level**.

sound-level meter. Microphone-amplifier-indicator assembly which indicates total intensity in decibels above an arbitrary zero. With suitable weighting networks the indicator gives approximately the level on the *phon scale*, at least for the comparison of sounds of a similar class.

sound locator. Apparatus for determining the direction of a sound source.

sound pickup. Used loosely for part of a sound-reproducing system, such as a microphone, the sound-head in a projector, or a reproducer of gramophone recordings.

sound-picture spoilation. The leakage of the sound signal of a TV transmission into the vision circuit of a receiver, giving rise to alternate dark and light horizontal bands in the screen picture.

sound pressure. Rms value of the instantaneous pressure exerted by the sound wave over an integral number of periods. The unit is the dyne/cm², or microbar. Also **excess pressure**.

sound-pressure level. Sound pressure expressed in decibels relative to a reference pressure, which is variously taken as 1 microbar and 2×10^{-4} microbar.

sound probe. Usually a very small microphone to minimize the disturbance of the sound field it is being used to measure. It is often equipped with a fine tube which is alone inserted in the field.

sound ranging. Determination of locality of a source of sound, e.g., from guns, by simultaneously recording through spaced microphones and making deductions from the differences of times of arrival.

sound-recording system. That comprising transducers and associated equipment needed for storing sound in a form suitable for subsequent reproduction.

sound reflection factor. The percentage of energy reflected from a large plane surface of uniform material on the incidence of a sound wave at a specified angle.

sound reinforcement system. One used to increase the uniform sound intensity in large halls by employing public address systems. Also by the control of reverberation time, using an electronic method involving microphones, resonators and loudspeakers.

sound reproducing system. That comprising sound recording, transducers and equipment for sound reproduction. In *monaural* reproduction, the system comprises one or more microphones with pre-amplifiers, mixers, amplifiers and loudspeakers. In *binaural* reproduction, two microphones are used to simulate positions of the ears of a human and reproduction of each component is through a separate audio channel.

sound streaming. The unidirectional flow from a sound source in a fluid medium.

sound-track. Track on magnetic tape or ciné film on which sound signals have been or can be recorded. *Optical tracks* may use variable area or variable density modulation.

sound velocity. In dry air at 0°C, velocity is 332 m/s or 760 miles/h approx. The velocity is greater in liquids, and highest in solids.

sounder. A telegraph receiving instrument in which Morse signals are translated into sound signals which are determined by the time intervals between two sounds.

source. (1) Active pair of terminals, which can deliver power to a load. (2) See **radiator** (2).

source impedance. See **output impedance**.

source range. See **start-up procedure**.

source resistance. See **internal resistance**.

source strength. Activity of radioactive source expressed in dps or curies, or less usually, in the rhm value.

sources of neutrons. Fast neutrons are obtained by (*a*) nuclear transformations and (*b*) fission in nuclear reactors. To obtain slow neutrons a moderator such as paraffin wax is used. For laboratory sources, see **neutron source**.

south (-seeking) pole (of a magnet). That attracted approximately towards geographical south. See under **north (-seeking) pole**. The terrestrial north-seeking magnetic pole lies south of Tasmania. In navigation the south-seeking pole is sometimes known as the **negative pole**.

space charge. Collection or cloud of electrons near a source, e.g., heated cathode, which stabilizes emission by repelling back those not attracted to another electrode, the field in its region being very low.

space-charge grid. One between control grid and cathode of a valve (*bigrid*), to reduce the space charge, so reducing voltages necessary on the anode to operate the valve.

space-charge layer. See **depletion layer**.

space-charge limitation. Condition in a thermionic valve when electron current leaving a cathode is limited by balance between attractive electric forces from other electrodes and repulsion within space charge. See **filament limitation**.

space-charge pentode. Same as **beam pentode**. See **beam-power valve**.

space cloth. A flexible material designed not to reflect radio waves.

space current. See **thermionic current**.

space group. Classification of crystal lattice structures into groups with corresponding symmetry elements.

space lattice. Three-dimensional regular arrangement of atoms characteristic of a particular crystal structure. There are 14 such simple symmetrical arrangements known as *Bravais lattices*. See also **symmetry class**.

space probe. Space vehicle for exploration between the planets, sending back data by radio telemetry.

space-reflection symmetry. See **parity**.

space-time. Normal three-dimensional space

plus dimension of time, modified by gravity in relativity theory.

space wave. Wave from an antenna which is not a ground wave, but which travels rectilinearly in space, apart from reflection (negative *refraction* and *bending*) when it enters an ionized region; also **sky wave**.

spaced antenna. One used with diversity systems, or to enhance directivity.

spaced-carrier operation. A method of operation in mobile radio systems to extend the area effectively covered by the mobile service.

spaced-loop direction-finder. One including two loops spaced sufficiently in terms of the wavelength to enhance their normal directivity, as exhibited by the polar diagram of response.

spacing error. An instrumental one arising from the variation with frequency of the angular spacing of the aerial.

spacing wave. Emitted wave corresponding to *spacing* impulses in a code, e.g., Morse code; also called **back wave**.

spacistor. A form of *depletion-layer transistor* (q.v.). A semiconductor device which uses a single *n-p* junction biased oppositely to a normal transistor junction. In this way the frequency limitations due to the carrier transit times of conventional transistors are avoided.

spaghetti. Colloquial for *insulating sleeving*.

spallation. Any nuclear reaction when several particles result from a collision, e.g., cosmic rays with atoms of the atmosphere; chain-reaction in a nuclear reactor or weapon.

spark. Breakdown of insulation between two conductors, such that the field is sufficient to cause ionization and rapid discharge. See **arc, field discharge, lightning**.

spark absorber. Resistance and/or capacitor placed across a break in an electrical circuit to damp any possible oscillatory circuit which would tend to maintain an arc or spark when a current is interrupted.

spark chamber. Radiation detector in which tracks of ionizing particles can be studied by photographing a spark following the ionization. It consists of a stack of parallel metal plates with the electric field between them raised nearly to the breakdown point.

spark coil. *Induction* or *Ruhmkorff coil* used as the source of high voltage in a spark transmitter. Obsolete for radio, but used in motor cars.

spark frequency. Frequency of repetition of discharge in a spark transmitter.

spark gap. Two or more electrodes between which the spark discharge takes place in a spark transmitter.

spark-gap generator. Radio-frequency generator for induction heating in which a capacitor is charged from a high-tension transformer and discharged through an oscillatory circuit when a spark gap breaks down.

spark-gap modulation. A method of pulse modulation in which a pulse-forming line

discharges across a spark gap in the transmitter circuit.

spark resistance. That between electrodes after the discharge has commenced; if excessive and in the oscillatory circuit, it causes loss of power and a high decrement.

spark spectra. The most important way of exciting spectra is by means of an electric spark. The high temperature reached will generate the spectrum lines of multiply-ionized atoms as well as of uncharged and singly-ionized ones, as distinct from the *arc spectrum* (q.v.). Evaporation of metal from the electrodes leads to additional lines not associated with the gas through which the discharge takes place.

spark system. Oldest form of radio telegraphy, using a *spark transmitter* in which high-frequency currents are generated by charging a capacitor from an induction coil, or other source of high voltage, and then discharging it through an inductance in series with a spark gap. The inductance is coupled to the antenna, which may also form part, or all, of the capacitor.

spark transmitter. See spark system.

sparking potential. Potential difference between the ends of an insulator, sufficient to cause a spark discharge through or over the insulator. Also **sparkover potential**.

speak-back circuit. See talk-back circuit.

special character. Any character recognized by a particular computing system which is neither a letter nor a numeral.

specific activity. See activity.

specific charge. Charge-to-mass ratio of elementary particle, e.g., ratio e/m_e of electronic charge to rest mass of electron ($= 1.759 \times 10^8$ coulomb gm^{-1}).

specific conductivity. The inverse of **specific resistance**.

specific gamma-ray emission. Exposure dose from unit point gamma-ray source at unit distance. Usually expressed in röntgens per hour at 1 cm for a 1 mc source. Formerly known as **k-factor**.

specific inductive capacity. Obsolete name for **permittivity**.

specific ionization. Number of ion pairs formed by ionizing particle per cm of path. Sometimes called the *total specific ionization* to avoid confusion with the *primary specific ionization*, which is defined as the number of ion clusters produced per unit length of track.

specific permeability. Same as **relative permeability**. See permeability.

specific power. U.S. term for **fuel rating**.

specific resistance. See resistivity.

spectral characteristic. Graph of photocell sensitivity, as related to wavelength of radiation.

spectral colour. One with degrees of saturation between no-hue and a pure spectral colour on the rim of the chromaticity diagram.

spectral line. Component consisting of a very narrow band of frequencies isolated in a spectrum. These are due to similar quanta produced by corresponding electron transitions in atoms. The lines are broadened into bands when the equivalent process takes place in molecules.

spectral series. Group of related spectrum lines produced by electron transitions from different initial energy levels to the same final one.

spectrograph. Normally used to describe *spectroscope* designed for use over wide range of frequencies (well beyond visible spectrum) and for recording the spectrum photographically. See **mass-, X-ray-**.

spectrometer. Instrument used for measurements of wavelength or energy distribution in a heterogeneous beam of radiation. See also:

acoustic-	microwave-
alpha-ray-	neutron-
beta-ray-	optical-
crystal-	scintillation-
gamma-ray-	time-of-flight-
infrared-	ultraviolet-
mass-	X-ray-.

spectrophotometer. Instrument for measuring photometric intensity of each colour or wavelength present in an optical spectrum.

spectroradiometer. A spectrometer for measurements in the infrared.

spectroscope. General term for instrument (spectrograph, spectrometer, etc.) used in spectroscopy. The basic features are a slit and collimator for producing a parallel beam of radiation, a prism or grating for dispersing different wavelengths through differing angles of deviation, and a telescope, camera or counter tube for observing the dispersed radiation.

spectrum. Arrangement of components of a complex colour or sound in order of frequency or energy, thereby showing distribution of energy or stimulus among the components. A *mass spectrum* is one showing the distribution in mass, or in mass-to-charge ratio of ionized atoms or molecules. The mass spectrum of an element will show the relative abundances of the isotopes of the element. See also:

absorption-	electromagnetic wave
acoustic-	line-
band-	optical-
continuous-	X-ray-.

spectrum analyser. (1) Electronic spectrometer usually working at microwave frequencies and displaying energy distribution in spectrum visually on a cathode-ray tube. (2) Pulse height analyser for use with radiation detector.

spectrum line. Isolated component of a spectrum formed by radiation of almost uniform frequency. Due to photons of fixed energy radiated as the result of a definite electron transition in an atom of a particular element.

spectrum locus. Locus of points on chromaticity diagram representing visual stimulus

produced by light beams of uniform wavelength (*monochromatic radiations*).

spectrum stripping. Process of separating a required gamma-ray spectrum from background, scattered radiation and other spectra. Can be carried out analytically or automatically, e.g., by analogue technique.

specular reflectance. Quotient of reflected to incident luminous flux for a polished surface.

specular reflection. General conception of wave motion in which the wavefront is diverted from a polished surface, so that the angle of the incident wave to the normal at the point of reflection is the same as that of the reflected wave. Applicable to heat, light, radio and acoustic waves. See **reflection laws**.

specular transmittance. See **transmittance**.

speculum. Alloy of 1-tin to 2-copper, providing wide spectral reflection from a grating ruled on a highly polished surface.

speech clipping. Removal of high peaks in speech, to get higher loading of transmitting valves, with some change in intelligibility.

speech equalizer. Circuit for correcting the excessive low-frequency gain of high-fidelity amplifiers designed for reproduction of records.

speech frequency. See **voice frequency**.

speech inverter. See **inverter** (2).

speech/noise ratio. See **noise ratio**.

speech scrambler. See **scrambler**.

speed. Number of standard words/min or *bands*.

speed of light. See **velocity of light**.

spelter. Zinc of about 97 per cent. purity, containing lead and other impurities.

spent fuel. Reactor fuel element which must be replaced due to (*a*) swelling and/or bursting, (*b*) burn-up or depletion, (*c*) poisoning by fission fragments. The fissile material is not exhausted, and so-called spent fuel is normally subsequently reprocessed.

sphere gap. Spark gap between spherical electrodes.

spherical aberration. Loss of image definition arising from the geometry of a spherical surface. Parabolic mirrors are normally used in astronomical telescopes to avoid this defect. See also **Schmidt optical system**.

spherical radiator. Same as **omnidirectional radiator**.

spherics. Same as **atmospherics**.

spigot. A projection in the base of a thermionic valve which serves to locate this in its correct position in the holder.

spike. (1) Sharp isolated waveform in currents for medical use. (2) Initial rise in excess of the main pulse in radar transmission. (3) Zone surrounding track of charged particles in which atoms have been displaced (or heating has occurred, i.e., **thermal spike**). (4) See **pulse**.

spill. Spread of charge from one element of charge storage tube to surrounding ones.

spin. (1) Of an electron, the quantized angular momentum which adds to the orbital angular momentum, thus producing fine structure in line spectra. (2) Of an elementary particle, its quantized angular momentum in the absence of orbital motion. (3) Of a nucleus, its quantized angular momentum, including contributions from the orbital motion of nucleons. (*N.B.*—Quantized spin is always an integral multiple of half the *Dirac unit ħ*.)

spin quantum number. Contribution to the total angular momentum of the electron of that due to the rotation of the electron about its own axis.

spinthariscope. Eyepiece system through which individual scintillations arising from impact of alpha-particles on zinc sulphide can be seen.

spiral lead-in: spiral throw-out. Grooves in disk recording used to guide the stylus into and out of the modulated track.

spiral time base. Arrangement for causing the fluorescent spot to rotate in a spiral path at a constant angular velocity, to obtain a much longer baseline than is possible with linear deflection. Used for detailed delineation of events relatively widely spaced in time, with or without a memory through long-glow or photography.

spiratron. Travelling-wave tube with radial electrostatic beam focusing.

splash baffle. See **arc baffle.**

split-anode magnetron. Early type with split anode, to give a push-pull output from electrons from the filament cathode, when gyrating in a (nearly) coaxial magnetic field.

split-beam CRT. A tube containing only one electron gun, but with the beam subdivided so that two traces are obtained on the screen.

split-flow reactor. One in which the coolant enters at the central section and flows outwards at both ends (or vice versa).

splitting ratio. See **cut** (4).

spontaneous fission. Nuclear fission occurring without absorption of energy. The probability of this increases with increasing values of the fission parameter Z^2/A (Z=at. no., A=at. wt.) for the fissile nucleus.

sporadic E-layer. Ionization in *E-layer* of ionosphere occurring erratically during the day as a result of particulate emission from sun, and leading to *sporadic reflections*.

spot. Point on a phosphor which becomes visible through impact of electrons in a beam. See **ion burn.**

spot-knocking. Removal of local imperfections on an electrode by operating the tube concerned at very high voltages so that spark erosion of discharge takes place.

spot speed. In facsimile recording, the speed of the recording or scanning spot within the allotted time. In TV, the product of the number of spots in a scanning line multiplied by the number of scanning lines/sec.

spot wobble. Oscillatory movement normal to TV scanning lines, to smooth them and make them less obvious. The amplitude of the wobble is slightly greater than the width of a line and about 200–400 vertical cycles occur during each line period.

spotlight. Lamp with a highly-concentrated light beam.

spray points. The collection of sharp points charged to high d.c. potential which are used to transfer charge in electrostatic machines such as the Van de Graaff, Wimhurst, etc.

spray-shielded valve. See **metallized valve.**

spread. Angle, in degrees, within which fall a number of bearings, ostensibly of a distant radio transmitter, when corrected for site and other errors.

spreader. Wooden or metal spar for keeping the wires of a multiwire antenna spaced apart.

spurious coincidences. Those recorded by a coincidence-counting system, when a single particle has not passed through both or all the counters in a system. They usually result from the almost simultaneous discharge of two counters by different particles.

spurious counts. Those arising in counter tubes from leakages and defects in external quenching circuits.

spurious oscillation. See **parasitic oscillation.**

spurious pulse. One arising from self-discharge of particle counter leading to erroneous counting.

spurious radiation. Undesired transmission, e.g., harmonics of carrier or modulation, outside specified band, causing interference with reception of other transmissions.

spurious response ratio. Ratio of field strengths of signals producing spurious and required response in telecommunication receiving equipment, e.g., of image frequency or intermediate frequency relative to required signal frequency.

sputtering. In a gas discharge, the removal of atoms from the cathode by positive ion bombardment, like a cold evaporation. These unchanged atoms deposit on any surface and are used to coat dielectrics with thin films of various metals.

square-law. Said of any device, such as a rectifier or (de)modulator, in which the output is proportional to the square of the input amplitude.

square-law capacitor. Variable vane capacitor, used for tuning, in which capacitance is proportional to the square of the scale reading, so that wavelength of circuit which it tunes becomes directly proportional thereto.

square-law demodulator. See **detector** (2).

square-law rectifier. One in which the rectified output current is proportional to the square of the applied alternating voltage.

square loop hysteresis. See **rectangular loop hysteresis.**

square wave. Pulse wave with very rapid (theoretically infinite) rise and fall times, and pulse duration equal to half period of repetition. Mark-space ratio of unity.

square-wave generator. Variable frequency oscillator with square-wave output.

square-wave testing. That of AF amplifiers using a square-wave input and an oscilloscope to study output waveform. Since square waves have extended harmonic components this enables a rapid check to be made of the amplifier performance over a wide frequency range.

squareness ratio. Ratio of retentivity to saturation flux density in square loop hysteresis material.

squegging. Mode of oscillation of an oscillator when operated under certain conditions, e.g., with excessive resistance in the grid circuit. The oscillations build up to a certain value and then abruptly stop, the process being repeated at a rate determined by the time constant of the capacitance and resistance of the grid circuit. Used also in a CRT time-base generator.

squegging oscillator. One in which normal oscillation is periodically shut off because of grid current blocking and release. Used in super-regenerative receivers. Also **self-quenching oscillator.**

squelch. Colloquialism for automatically reducing the gain of a receiver to cope with a strong signal.

squint. Difference between the geometrical axis of an aerial array and the axis of the radiation pattern; particularly applicable to very-high-frequency Yagi arrays.

stability factor. Ratio $\dfrac{\Delta I_c}{\Delta I_c{}^0}$ for transistor, where ΔI_c is change in collector current for a particular circuit which results from a change in temperature, and $\Delta I_c{}^0$ is the corresponding change measured under conditions of zero emitter current. The larger this factor the greater the sensitivity of the circuit to temperature changes.

stabilization. (1) Maintenance of voltage at a point in circuit, against variations of load or supply voltage. Series pentode valves or shunt-discharge tubes are usual. (2) Maintenance of frequency of a transmitter within limits prescribed internationally for its type of service. Often effected by quartz-crystal drive in a thermostatically-controlled oven.

stabilized-feedback amplifier. One in which amplification is stabilized against changes in supply voltages, etc., by the application of negative feedback.

stabilized glass. That which discolours less than usual when irradiated.

stabilizer tube. Gas-discharge tube, the voltage across which is much more stable than a voltage applied to it in series with a resistor.

stabilivolt. Tube containing a gaseous discharge through a series of electrodes, potentials between which tend to a greater constancy than the voltage applied through a resistance to end electrodes.

stable. (1) Said of systems not exhibiting sudden changes, particularly atoms which are not radioactive. (2) Used to indicate the incapability of following a stated mode of spontaneous change, e.g., *beta-stable* means incapable of ordinary beta disintegration but could be capable of isomeric transition or alpha disintegration, etc. (3) State

of an amplification system when it satisfies the Nyquist criterion, either conditionally or unconditionally.

stable orbit. The circle of constant radius described by accelerated particles; cf. *betatron, synchrotron.*

stable oscillation. One for which amplitude and/or frequency will remain constant indefinitely. *N.B.*—A statically stable system may be dynamically unstable and follow a divergent oscillation when subjected to a disturbance. In this sense the term *dynamically-stable* means that an induced oscillation will be convergent, i.e., of decreasing amplitude.

stable platform. Structure which can be controlled in position with great precision, e.g., by gyroscopes, and which forms base for other information to be measured and transmitted by telemetry, e.g., from satellites.

stack. (1) Same as **pile-up.** (2) Pile of photographic plates exposed to radiation together, and used to study tracks of ionizing particles.

stacked array. See **tiered array.**

stage. Unit of cascade in isotope separation plant, consisting of single separative element or group of these elements, operating in parallel on material of same concentration.

stage efficiency. Ratio of a.c. output power to d.c. input power for any stage of an electronic amplifier.

stage separation factor. See **simple process factor.**

stagger-tuned amplifier. One with couplings tuned to different high frequencies to give a band-pass response.

staggered tuning. Attempt to get a wideband response by a number of tuned circuits, having slightly different frequencies of resonance.

stalloy. Steel with 3·5 per cent. silicon content which has low hysteresis loss, and is used for transformer stampings.

stamper. Metal negative used in producing disk recordings by pressing.

stamping. See **lamination.**

standard. Established unit of measurement, or reference instrument or component, suitable for use in calibration of other instruments. Basic standards are those possessed or laid down by national laboratories or institutes, e.g., BSI, NPL, etc.

standard atmosphere. That corresponding in pressure to a column of mercury 76 cm high, of density $13·59518$ gm cm^{-3}, with gravity $980·665$ cm sec^{-2}, which is a barometric pressure of $1·013250 . 10^6$ dyn. cm^{-2}.

standard cell. See **Weston standard cadmium cell.**

standard chamber. Ionization chamber used for calibration of radioactive sources, or of absolute values of exposure doses.

standard deviation. Applicable to all observations; the rms deviations of a series of random measurements from their mean. If the distribution of the errors is Gaussian, this leads also to the *probable error.* If $\bar{\gamma}$ is the mean value of a series of n-like quantities γ_p, then the standard deviation is given by

$$\sigma = \left[\frac{\sum\limits_{p=1}^{n} (\gamma_p - \bar{\gamma})^2}{n} \right]^{\frac{1}{2}}.$$

See **Gaussian distribution.**

standard illuminant. Illuminant used for accurate colour measurements. The CIE specify three alternative standards: A, B, C, with corresponding colour temperatures: 2848°K, 4800°K, 6500°K (the latter sources being realized by use of specified liquid filters).

standard M-gradient. The constant rate of change of excess modified refractive index with height, regarded as equivalent to normal propagation conditions at a given place on the earth. The excess modified refractive index is the amount by which the modified refractive index exceeds unity.

standard man. Averaged characteristics of adult human body as specified in the recommendations of the ICRP.

standard radio atmosphere. Radio atmosphere having the standard M-gradient.

standard refraction. The refraction arising in a standard radio atmosphere.

standard resistor. One of high accuracy and stability intended for laboratory reference purposes and usually of four-terminal construction.

standard source. One of long life, which can be used for calibrating *radiac meters,* and for comparison with other sources.

standard time. See **zone time.**

standard volume indicator. One specified for indicating power levels on a transmission line, particularly when the level is fluctuating, as with speech.

standing current. See **quiescent current.**

standing-off dose. Absorbed dose after which occupationally exposed radiation workers must be temporarily or permanently transferred to duties not involving further exposure. (Doses are normally averaged over 13-week periods, and standing-off would then continue for the remainder of the corresponding period.)

standing wave. Pattern of maxima and minima when two sets of oppositely travelling waves of the same frequency interfere with each other. They can be produced, e.g., by a source of sound held over the open end of a pipe of suitable length closed with a rigid bung; in this case it is the incident and reflected waves which interfere.

standing-wave indicator. Voltage detector, germanium or silicon diode, for sliding along a *slotted line,* detecting maxima and minima voltages in an electromagnetic standing-wave system. Shunt for sliding along open transmission wire, to detect maxima and minima currents therein. In acoustic stand-

ing wave, the maxima and minima may be detected by means of a probe microphone.

standing-wave meter. One designed to measure VSWR in waveguide.

standing-wave ratio. Where standing and progressive waves are superimposed, the SWR is the ratio of the amplitudes at nodes and antinodes. For a transmission line or waveguide, it is equal to $\dfrac{1-r}{1+r}$ where r is the coefficient of reflection at the termination. It may alternatively be defined by the reciprocal of this value as shown by its value being numerically greater than unity.

star. (1) Radiating tracks in photographic emulsion, arising from particle disintegration on collision, and dispersal of energy to other particles. (2) Celestial body similar to the sun, an incandescent mass undergoing thermonuclear reactions.

star-mesh transformation. Technique for simplifying network, whereby any number of branches meeting at a point can be replaced by an equivalent mesh, thereby reducing the number of connections.

star network. One with many branches connected at a point; a *T-* or *Y-network* has three branches.

star-quad. See **quad.**

Stark effect. Splitting of atomic energy levels, and of corresponding emission spectrum lines, by placing source in region of strong electric field; cf. *Zeeman effect.*

start-stop control. An automatic control system in which the controlling element exercises control by allowing the given process to either proceed or cease.

start-up procedure. Procedure followed when bringing a nuclear reactor into operation. It involves four successive stages:

(*a*) **source range**—a neutron source is introduced to generate the required neutron flux.

(*b*) **counter range**—the reactor is just critical, but counters are required to monitor neutron flux changes.

(*c*) **period range**—changes in reactivity are monitored on period meter.

(*d*) **power range**—reactor is operating within its designed power ratings.

starter. See **pilot electrode.**

starter gap. The conducting path between the pilot electrode and the electrode to which the starting voltage is applied in a glow-discharge tube.

starter voltage. That applied to the pilot electrode of a cold-cathode discharge tube or mercury-arc rectifier; cf. *starting voltage.*

starting anode. The one used first to strike an arc in a rectifier with a mercury-pool cathode.

starting-up time. That required by instrument or system (e.g., nuclear reactor, chemical plant, etc.) to reach equilibrium operating conditions.

starting voltage. That which initiates current

I

passing in a gas-discharge tube after non-conduction; much greater than that required to maintain conduction. Also **ionizing voltage, striking voltage.** See **threshold voltage.**

stat-. Prefix to name of unit, indicating derivation in obsolete electrostatic system of units, e.g., statampere, statohm, etc.

state. The energy level of a particle as specified by the appropriate quantum numbers.

static. (1) Non-movable or non-rotating, e.g., a transformer or rectifier is a static converter. (2) Electrostatic. (3) Said of all electrical disturbances to a radio system which arise through electrostatic induction, particularly from lightning flashes. See also **atmospherics.**

static characteristic. Curve, or set of curves, which describes relation between specified voltages and currents of electrodes under unvarying conditions, as compared with *dynamic characteristic*, which implies operation under normal load conditions.

static frequency changer. Transformer having a magnetically saturated iron core arranged to accentuate the harmonic content of the secondary current; formerly used for production of high-frequency currents for radio transmission.

static memory. One in which information is fixed in space and is available at any time.

static multipole moments. The electric and magnetic multipole moments of a system which are measures of the charge and magnetic distributions in a given state. These determine the interaction of the system with weak external fields by contrast with the *transition multipole moments* which determine the radiative transitions between two states.

station. (1) Location of radio transmitters and/or receivers with antennae, for sending or receiving radio signals on one or more wavelengths. (2) In an automatic production line, a location where a specific operation is performed, e.g., heating by induction or radiation.

station frequency. The frequency of the carrier wave on which a station is transmitting.

stationary gap. A device used in conjunction with a pulse-forming line to form short-period high-voltage pulses for modulating microwave oscillators.

stationary wave. Earlier alternative to *standing wave.*

statistical error. That arising in measurements of average count rate for random events, as a result of statistical fluctuations in the rate.

statistical mechanics. Theoretical predictions of the behaviour of a macroscopic system by applying statistical laws to the behaviour of component particles. *Quantum mechanics* is an extension of classical statistical mechanics, introducing the concepts of the quantum theory, especially the Pauli exclusion principle. *Wave mechanics* (q.v.) is a further extension based on the Schrö-dinger equation, and the concept of particle waves.

statistical weight. (1) In statistical mechanics, the number of possible microscopic states for the system. (2) In nuclear reactors, the effect on reactivity of introducing an absorber at a given site. Also called **weighting factor.**

stator. Stationary part of machine, especially a dynamo or motor; cf. *armature* (3).

steady-state. Said of any oscillation system which continues unchanged indefinitely.

steel. An alloy of iron with carbon and other elements. If used as a structural material in reactors as a neutron or thermal shield, it must be outside the active section. This restriction is due to some of its constituents becoming strongly radioactive, with long lives, on capturing neutrons.

steerable antenna. Fixed multiple antenna, major lobe of sensitivity being adjustable by altering the phase of the separate contributions of the elements. See **musa.**

steering. Alteration by mechanical or electrical means of the direction of maximum sensitivity of a directional antenna, e.g., a radar or radio telescope.

Stefan-Boltzmann law. The total radiated energy from a black body per unit area per unit time is proportional to the 4th power of its absolute temperature, i.e., $E = \sigma T^4$, where σ (the *Stefan-Boltzmann constant*) is equal to $5 \cdot 672 \times 10^{-5}$ erg cm^{-2} sec^{-1} deg^{-4}.

Steinmetz coefficient. Constant β in Steinmetz's relationship for calculating hysteresis losses. Loss $= \beta(\text{flux})^{1 \cdot 6}$.

stellarator (*stellar generator*). Twisted torus for nuclear fusion experiments. The reverse curvatures are to balance out first order drift, which is serious in the simple torus because of variation of magnetic field across the section of the plasma.

stenode. Supersonic heterodyne receiver in which there is very sharp tuning in intermediate-frequency circuits, using piezo-electric quartz crystals, with frequency correction of audio signal after demodulation.

step. (1) Off-setting of a hole in a reactor shield to form a zig-zag joint and so avoid leaving a straight path for the possible escape of radiation. (2) A single specific instruction in a computer routine.

step-down amplifier. One in which a high negative voltage is applied to the anode, current in a positive grid being taken as a measurement of the voltage.

step-down transformer. The reverse of the *step-up transformer*, i.e., in an electrical transformer the transfer of energy from a high to a low voltage.

step function. A function which is zero for all time preceding a certain instant, and has a constant finite value thereafter.

step-up transformer. Audio-frequency transformer with more secondary than primary turns. It thus transforms low to high voltages, and matches low to high impedances.

stepping tube. One in which an electron beam can be moved between a few stable positions by applied voltages and so act as a switch.

steradian. Unit of spherical or solid angle, cone of a whole sphere when divided by 4π.

sterba antenna. Vertical panel, with reflector, formed by folding a single wire so that normal radiation arises from current of one phase in the vertical sections, while the other phase current balances out in space in the crossovers.

stereo. Abb. for *stereophony, stereophonic* or *stereoscopy.*

stereochemistry. Section of chemistry involving arrangement in space of the atoms within a molecule.

stereomicrophone. Dual microphone with, e.g., interlocking figure-of-eight polar diagrams; used to provide signals for both channels of a stereophonic sound reproduction system.

stereophonic recording. (1) Use of adjacent tracks on magnetic tape, with multiple recording and reproducing channels. (2) Use of spiral cut on disks, two channels being represented by stylus motions at right-angles, each at 45° to the surface; reproduction is by one stylus operating a double crystal.

stereophony. Method of sound reproduction which establishes an acoustic pattern similar to that in the originating enclosure. This requires, after trials, at least three matched channels, as compared with two, when the outputs of the channels are applied to the ears. Also **auditory perspective, localization.**

stereoscope. Device for producing apparently binocular (3D) image by presenting differing plane images to the two eyes.

stereoscopy. Sensation of depth obtainable with binocular vision due to small differences in parallax, producing slightly-differing images on the two retinas.

Stern-Gerlach experiment. Atomic beam experiment which provided fundamental proof of quantum theory prediction that magnetic moment of atoms can only be orientated in certain fixed directions relative to an external magnetic field.

sticking probability. Probability of an incident particle, which reaches the surface of a nucleus, being absorbed and forming a compound nucleus.

sticking relay. One which *makes* when d.c. current passes through the operating coil, remains operated on cessation of current (thereby saving current), and falls off or *releases* on a reversed current.

sticking voltage. Potential in electron beam tube above which electrons collected at screen cannot all be dispersed, leading to negative charge accumulating and neutralizing the excess voltage. In a CRT, that accelerating voltage which fails to increase brightness of spot on a phosphor because of insufficient secondary electron emission or conduction for dispersal of incident electrons.

stiction. *St*atic f*riction* in a mechanism.

stilb. Unit of intrinsic brightness, equal to 1 candle/cm² of a surface.

stimulated emission. See laser, maser.

stitch welding. *Seam welding* (q.v.), using small mechanically-operated electrodes, similar to a sewing machine.

stochastic. Said of any system operation in which an element of chance cannot be excluded.

stochastic noise. That which maintains a statistically random distribution.

stoichiometry. Determination of exact proportion of elements to make pure chemical compounds, alloys such as semiconductors, or ceramic crystals.

stoke. The CGS unit of kinematic viscosity (ratio of viscosity/density).

Stokes' law. (1) Incident radiation is at a higher frequency and shorter wavelength than the re-radiation emitted by an absorber of that incident radiation. (2) Law governing the motion of a sphere under streamline conditions in a fluid; the force opposing the motion being given by $F = 6\pi a \eta v$, where a = sphere radius, η = coefficient of shear viscosity, v = velocity.

Stokes' theorem. That the surface integral of the curl of a vector function equals the line integral of that function around a closed curve bounding the surface, i.e.,

$$\int_s \Delta \times v.dS = \int v.dS.$$

stop. Hole in diaphragm (aperture) which sets its location limits to a beam of light from a source. In conjunction with a lens, the *stop* is the ratio of diameter of hole to focal length, determining, with time, *exposure.*

stop-go control. See start-stop control.

stopper. (1) Resistance next to the grid of a valve to reduce high-frequency potentials on the grid and consequent build-up of parasitic oscillations. (2) Resistor-capacitor combination for decoupling anode or grid supply circuits, in order to obviate oscillation or motor-boating in thermionic amplifiers. See **anode-.**

stopping equivalent. Thickness of a standard substance which would produce the same energy loss as the absorber under consideration. The standard substance is usually air at NTP, but can be Al, Pb, H_2O, etc. See **air equivalent.**

stopping potential. Reverse difference of potential required to bring electrons to rest against their initial velocity from either thermal or photoelectric emission.

stopping power. Energy loss resulting from a particle traversing a material. The *linear stopping power* S_L is the energy loss per unit distance and is given by $S_L = -dE/dx$, where x is path distance and E is the kinetic energy of the particle. The *mass stopping power* S_M is the energy lost per unit surface density traversed and is given by $S_M = S_L/\rho$, where ρ is the density of the substance. If A is the atomic weight of an element and n the number of atoms per unit volume, then the *atomic stopping power* S_A of the element is defined as the energy loss per atom per unit area normal to the motion of the particle, and is given by $S_A = S_L/n = S_M A/N$, where N is Avogadro number. The *relative stopping power* is the ratio of the stopping power of a given substance to that of a standard substance, e.g., air or aluminium.

storage battery. See **accumulator.**

storage capacity. Maximum number of *bits* which can be stored, located, and recovered for a *main* or *fast* store of a computer.

storage element. One unit in a memory, capable of retaining one *bit* of information. Also the smallest area of the surface of a charge storage tube which retains information different from that of neighbouring areas.

storage factor. See Q.

storage oscilloscope. One in which trace is retained indefinitely or unless deliberately wiped off. See **Remscope, scotophor, Storascope.**

storage tube. (1) Camera tube in which charge produced by an image can be retained in an element of capacitor. See **iconoscope.** (2) Tube which stores charges deposited on a plate or screen in a CRT, a subsequent scanning by the electron beam detecting, reinforcing or abolishing the charge. Also **memory tube, store.**

Storascope. TN for infinite persistence cathode-ray oscilloscope.

store. (1) In an electronic digital computer, an essential unit, in which the *programme* and *data* are stored until dealt with by the *logical* units. Also used to store results until they can be *read out* by some printing mechanism, e.g., tabulator, typewriter, or xerographic printer. Stores can be magnetic drums, magnetic tape, electrostatic storage tubes, mercury or nickel lines, or saturable cores, or superconducting elements. Also called **memory.** (2) Another name for **storage tube.**

See also:

circulating memory	mercury memory
cold-	non-volatile memory
delay-line (1)	parallel memory
electrostatic memory	permanent memory
fast-	persistor
ferrite-bead memory	primary-
internal memory	random access
magnetic drum	secondary memory
magnetic memory	serial-
magnetic tape	static memory
magnetotriction	temporary memory
main-	twistor
	volatile
	Williams tube
	zero-access-.

storm. Large persistent ionospheric disturbance to transmission on band 3 to 30 MHz, or one coincident with magnetic disturbances lasting several days. Caused by particle emission from the sun and hence delayed after SID and SPA.

straggling. Variation of range or energy of particles in a beam passed through absorbing material, arising from random nature of interactions experienced. Additional straggling may arise from instrumental effects such as noise, source thickness and gain instability.

straight-line capacitor. One in which capacitance varies linearly with scale reading.

straight-line frequency capacitor. One in which capacitance is inversely proportional to the square of the scale reading, so that the frequency of the circuit which it tunes is directly proportional thereto.

straight-line wavelength capacitor. One in which capacitance is proportional to square of scale reading, so that it can tune a circuit with a linear relationship between scale reading and wavelength.

straight receiver. One in which all high-frequency amplification, if any, is at the same frequency as that of the original signal. See **supersonic heterodyne receiver.**

strain. Relates to the dimensional change in a medium when subject to a mechanical, or other form of, stress. It is specifically defined as the ratio of the dimensional change (in length, area or volume) to the original (or unstrained) dimension.

strain-gauge element. (1) Nickel wire or printed circuit resistance mat cemented to a surface under stress and used either (*a*) to measure the static strain in terms of the change in resistance of the element, or (*b*) to measure the dynamic strain resulting from vibration, in terms of the consequent modulation of a steady current passed through the element. (2) An inductance type in which inductance variations are produced by the movement of an armature within a coil (or coils) which is produced by the strain in the structure under test.

strain viewer. Eyepiece or projection unit of polariscope.

Stranducer. RTM for transducer incorporating strain gauges.

strangeness. Quantum number which represents unexplained delay found in strong interactions between certain elementary particles. The strangeness number is zero or integral (positive or negative), and is conserved during interactions but not during decay. K-mesons and hyperons which have a non-zero strangeness number are termed *strange particles*. Strangeness is associated with the displacement of the charge centre of particle multiplets from the expected value. An alternative quantity expressing the same concept and more widely used today is *hypercharge*. The hypercharge of a particle is equal to its strangeness plus its baryon number (1 for baryon, -1 for anti-baryon, 0 for meson). Since leptons do not participate in strong interactions they are not assigned strangeness or hypercharge numbers.

strapping. Alternate connection of segments

in a magnetron, to stabilize phases and mode of resonance in the cavities.

stratosphere. The earth's atmosphere, above the troposphere and below the ionosphere, having little moisture, and isothermal in character.

stray capacitance. Any occurring within a radio circuit other than that intentionally inserted by capacitors, e.g., capacitance of connecting wires, giving rise to *parasitic oscillation* (q.v.).

stray flux. The leakage magnetic flux from an iron-cored device such as a transformer.

stray radiation. Direct and secondary radiation from irradiated objects, which is not serving a useful purpose.

stray resonance. That arising from unwanted inductance and capacitance, e.g., in leads between conductors, in leads inside canned capacitors, between turns of inductors.

strays. Same as **atmospherics.**

streaking. Extension of image element because of relative phase delay of frequency components in the video signal.

streaming effect. See **channelling effect.**

streamline. A line in a fluid such that the tangent at any point follows the direction of the velocity of the fluid particle at the point at a given instant. When the streamlines follow closely the contours of a solid object in a moving fluid, the object is said to be of streamline form.

stress. The force acting on unit area of a medium as in theory of elasticity. Also arises from electrical or thermal fields in a medium.

Stretch. IBM digital computer which can deal with one million logic operations/sec.

striation. Phenomenon in low-pressure gas discharge, which forms luminous bands across the line between electrodes.

striking potential. That sufficiently large to break down a gap and cause an arc, or start discharge in a cold-cathode tube.

striking voltage. See **starting voltage.**

string. Series of computer numbers in order but not related, i.e., at random.

string electrometer. An electrometer consisting of two metal plates oppositely charged, between which a conducting fibre is displaced from a middle position in proportion to the voltage between the plates.

stringer. Group of reactor fuel elements strung together for insertion into one channel of the core.

stripe. Magnetic sound-track(s) on cinematograph film for soundfilm reproduction, adjacent to the printed picture track. Can be added to final print.

stripline. Waveguide formed of strips of copper on dielectrics, formed by etching a printed circuit.

stripper. The section of an isotope separation system which strips the selected isotope from the waste stream.

stripping. A phenomenon observed in deuteron (or heavier nuclei) bombardment in which only a portion of the incident particle merges with the target nucleus, the remainder proceeding with most of its original momentum practically unchanged in direction. See **Oppenheimer-Phillips process.**

strobe. (1) General term for detailed examination of a designated phase or epoch of a recurring waveform or phenomenon. (2) Enlargement or intensification of a part of a waveform as exhibited on a CRT. Also called **linearity control.** (3) Process of viewing vibrations with a stroboscope ; colloquially the **stroboscope** itself.

strobe pulse. Pulse much shorter than a repeated waveform, for examination of a display.

stroboscope. Apparatus producing appearance of slow or zero motion when a vibrating or rotating object is intermittently illuminated by short flashes, e.g., from a xenon or neon gas-discharge tube, at a suitable integral or aliquot frequency.

strobotron. Triggered gas-discharge tube used as pulsed light source in stroboscope.

stroke. The time-base sweep moving with uniform velocity across the screen of a CRO. See **flyback.**

strong interaction. One produced by short-range nuclear forces associated with mesons and completed in a time of the order of 10^{-23} sec. Baryons and mesons participate in strong interactions.

strontium (Sr). At. no. 38, at. wt. 87·62, sp. gr. 2·6, m.p. 757°C, b.p. 1300°C. Has similar chemical properties to calcium. Compounds give crimson colour to flame and are used in fireworks. The radioactive isotope of strontium (*strontium-90*) is produced in fission of uranium and has a long life, hence its presence in 'fall-out' after a nuclear explosion.

strontium age. The interval of time between some geological event, such as the formation of a mineral, and the present time as determined by radioactivity. When the age is estimated from the relative numbers of atoms of a stable radiogenic end product and radioactive parent present, the method is designated by the name of the end product, in this case strontium.

strontium unit. One used to measure the concentration of radioactive strontium-90 in calcium, an SU being 10^{-12} Ci/gm.

strophotron. Multi-reflection oscillator for microwave frequencies, capable of a small range of tuning. See **Barkhausen-Kurz oscillator.**

stub. An auxiliary section of a waveguide or transmission line connected at some angle with the main section. See **coaxial-, quarter-wavelength-.**

stub aerial. Straight wire which resonates at a wavelength *ca.* four times its length.

stub tuning. Use of one or more shunt stubs (quarter-wave resonating lines) for adjusting transmission lines (open or coaxial) for maximum power transfer.

stylus. Needle for cutting or replaying a disk recording.

subaqueous loudspeaker. See **piezoelectric loudspeaker.**

subaqueous microphone. See **hydrophone.**

subatomic. Said of particles or processes at less than atomic level, e.g., radioactivity, production of X-rays, electron shells, nuclear shells.

subaudio requency. One below those usefully reproduced through a sound-reproducing system or part of such system.

subcarrier. One frequency which is modulated over a narrow range by a measured quantity, and then used to modulate (with others) a carrier that will be finally demodulated on reception.

subcritical. Of fissile material for which the multiplication factor is less than unity.

subharmonic. Having a frequency which is a fraction of a fundamental. Subharmonics appear in some forms of non-linear distortion.

submerged heating. Induction heating in a workpiece which is submerged in a quenching liquid.

submerged repeater. One used at intervals along a submarine cable to compensate losses. Built into the cable, or in a steel enclosure to withstand pressure, it has components of long life, e.g., 20 years.

submerged resonance. See **tailing hangover.**

submicron. A particle whose diameter is less than a micron.

subminiature valve. See **miniature valve.**

submodulator. Low-frequency amplifier which immediately precedes the modulator in a radio-telephony transmitter.

subroutine. Short routine, ready-made for a given computer, which is used as a unit to shorten the total programme peculiar to a desired calculation.

subscription television. See **toll television.**

subsonic. Deprecated term for **infrasonic.** See **Mach number.**

substrate. Material of panel on which subminiaturized components are mounted or deposited, e.g., glass or ceramic.

sudden death. Colloquialism for abrupt failure of transistor. (This term originally applied to point contact types where displacement of the contact frequently caused unexplained failures. Comparable breakdown of junction types may occur, e.g., as a result of thermal runaway or reverse voltages.)

Sugarman counter. Form of gas-flow proportional counter.

Suhl effect. The reduction in lifetime of holes injected into an n-type semiconducting filament by deflecting them to the surface, using a powerful transverse magnetic field. Reverse of *Hall effect* (q.v.).

sulphur (S). At. no. 16, at. wt. 32·064, sp. gr. 2·07, m.p. 112·8°C, b.p. 444·6°C. Used for vulcanizing rubber. Vapour bath at boiling point is also used as fixed point in platinum resistance thermometry.

summation check. Figure added to a *word*, indicating a summation of the digits so that accuracy of processing can be verified.

Totals arrived at by two methods for verifying processing of data.

summing point. See **mixing point.**

sunshine unit. U.S. term for **strontium unit.**

sup. See **suppressor grid.**

superaudio frequency. One above those usefully transmitted through an audio-frequency reproducing system, or part of such.

superconducting amplifier. One using superconductivity to give noise-free amplification.

superconducting gyroscope. Frictionless gyroscope supported in vacuum through magnetic field produced by currents in superconductor.

superconducting magnet. Electromagnet in which field is produced by currents in conductors at, or near, superconducting temperatures, thus avoiding problems of heat dissipation.

superconductivity. Property of many pure metals within a few degrees of absolute zero (°K) of having no resistance (dissipation) to a flow of electric current. When the current is established it continues nearly indefinitely. Property removed by a magnetic field; hence the phenomenon can be used for memory or store for computers. Such a device is known as a *cryotron*. The superconducting characteristics of a material are the critical transition temperature and the critical magnetic field curve.

superconductor. A metal showing superconducting properties, e.g., lead.

supercontrol tube. See **remote cutoff.**

supercritical. Assembly of fissile material for which the multiplication factor is greater than unity.

superfluid. Condensed degenerate gas in which a significant proportion of the atoms are in their lowest permitted energy state. In practice, this affects only liquid helium II.

superhet. Acronym for **supersonic heterodyne.**

superheterodyne receiver. See **supersonic heterodyne receiver.**

Supermalloy. TN for magnetic material similar to permalloy but with higher permeability.

Supermendur. TN for rectangular hysteresis loop magnetic material containing iron, cobalt and vanadium.

supermultiplets. Group of particle multiplets with masses differing by equal amounts between the groups. *Lie algebra* (q.v.) explains these mass differences in terms of differences in the *hypercharge*, in the same way that the smaller mass differences within a multiplet are explained in terms of differences in *isotopic spin*. Most supermultiplets contain eight particles and this theory is colloquially known as the **eightfold way.**

superposed, superposition. Said of that added to a normal circuit or circuits, e.g., a phantom on telephone circuits or d.c. telegraphy on a telephone circuit.

superposition theorem. That any voltage/current pattern in a linear network is additive to any other voltage/current pattern.

super-regeneration. Reaction in a receiver to

a degree that normally causes self-oscillation. This is prevented by the application of a *quenching* voltage to the reacting valve, which is thereby intermittently paralysed at a frequency which is high enough to be inaudible. Also **super-reaction.**

super-regenerative receiver. One with sufficient positive feedback to result in a quenched supersonic oscillation (squegging), with consequent increase in sensitivity, but also increase in distortion of demodulated signals.

supersaturation. Water-vapour in the air in excess of 100 per cent. relative humidity. Condensation can take place on nuclei, particularly ions produced by high-speed charged particles, exhibiting a track of minute but visible water drops, as in a (*Wilson*) *cloud chamber.*

supersonic. Deprecated term for **ultrasonic.** See **Mach number.**

supersonic amplification. That at a supersonic frequency, i.e., following the frequency changer in a supersonic heterodyne receiver.

supersonic frequency. The frequency used for the post-frequency-changer amplification in a supersonic heterodyne receiver, viz., from 100 to 450 kHz. See **Mach number.**

supersonic heterodyne receiver. One in which received signal has the frequency of its carrier wave changed, by means of the heterodyne principle, to some predetermined frequency above the audible limit, after which it is amplified and finally rectified and demodulated. Also **beat receiver, double detection receiver.**

super-voltage. Voltage, over a million volts, applied to X-ray tubes or accelerators in therapy.

supervisor. A computer routine that controls sequencing, compiling and execution of all programme runs.

supervisory control. That of a number of separate units from a single control consol using one or more common transmission channels.

suppressed carrier system. One in which the carrier wave is not radiated but is supplied by an oscillator at the receiving end; used with single sideband working because of the phase distortion which arises when a carrier is inserted between two sidebands.

suppressed zero. Said of meter when the zero position is off-scale, i.e., beyond the range of travel of the pointer.

suppression. Elimination of specified data or digits, e.g., initial zeros.

suppressor. (1) Component, usually a low resistance adjacent to an electrode of a valve, to obviate conditions which promote parasitic oscillations. (2) Component, such as a capacitor or resistor, or both, which damps high-frequency oscillations liable to arise on breaking a current at a contact, causing radio interference.

suppressor grid. That between anode and screen in pentode valves to repel secondary electrons back to the anode. Abb. **sup.**

suppressor-grid modulation. Insertion of the signal voltage into the (suppressor) grid circuit of a valve which rectifies and is driven by the carrier; also **grid modulation.**

surface absorption coefficient. See **absorption coefficient.**

surface barrier. Potential barrier across surface of semiconductor junction due to diffusion of charge carriers.

surface-barrier transistor. One in which very thin barriers (by means of etching techniques) are used, so permitting the frequency range to be extended to 100 MHz.

surface charge. See **bound charge.**

surface density. (1) The amount of electric charge per unit surface area. (2) In nuclear physics, mass per unit area, a common method of indicating the thickness of an absorber, etc.

surface duct. Atmospheric propagation duct for which the earth's surface forms the lower boundary.

surface leakage. That along the surface of a non-conducting material or device. May vary widely with contamination, humidity, etc. It sets a practical limit to the value of high resistors for use with electrometers, etc.

surface lifetime. The lifetime of current carriers in the surface layer of a semiconductor (where recombination takes place most readily). Cf. *volume lifetime.*

surface noise. Same as **needle scratch.**

surface recombination velocity. Electron-hole recombination on surface of semiconductor occurs more readily than in the interior, hence the carriers in the interior drift towards the surface with a mean speed termed the surface recombination velocity (*SRV*). Defined as the ratio of the normal component of the impurity current to the volume charge density near the surface.

surface sterilization. Radiation with low-energy rays which penetrate thin surface layers only, e.g., with ultraviolet rays.

surface tension. The force in the surface of a liquid arising from unbalanced intermolecular attraction. Units of measurement are dyne cm^{-1}. See also **liquid-drop model.**

surface wave. See **wave.**

surge. A large, but momentary, increase in the voltage of an electric circuit.

surge arrester. See **lightning arrester.**

surge generator. See **impulse generator.**

surge impedance. See **characteristic impedance** (1).

survival curve. One showing the percentage of organisms surviving at different times after they have been subjected to large radiation dose. Less often, one showing percentage of survivals at given time against magnitude of dose.

susceptance. IP of admittance equal to $\dfrac{-X}{R^2+X^2}$ for a circuit of impedance $R+jx$.

susceptibility. See **magnetic** or **electric susceptibility.**

swarf. In cutting original gramophone disks the thin thread of waste which is removed from the cellulose surface by a cutting stylus. It is often electrified and inflammable. Also **chip.**

sweep circuit. (1) That which supplies deflecting voltage to one pair of plates or coils of a cathode-ray tube, the other pair being connected to the source of current or voltage under examination. See **linear scan** (2). (2) In TV and facsimile transmission the circuits which produce the required *rectangular scan* (q.v.) by the electron beam.

sweep frequency. That of a *sweep circuit.*

swell. Volume control on an electronic organ, operated by potentiometer.

swelling. Of fissile materials, a change in volume without necessarily any change in shape, which may occur during radiation. See **R-value, S-value.**

swimming-pool reactor. One which uses ^{235}U in aluminium cans immersed in water. Also **aquarium reactor, pool reactor.**

swing. (1) Extreme excursion from positive peak to negative peak in an alternating voltage or current waveform. (2) Angle, expressed as plus or minus half the excursion or spread, over which the dial of a goniometer or rotating aerial must be swung to estimate the reading of a bearing.

swinging choke. Iron-cored inductor with saturable core, used in smoothing circuits where decreasing impedance with increasing current improves regulation.

switch. (1) Device for opening and closing an electric circuit. (2) Electronic circuit for switching between two independent inputs, e.g., by valves, tubes, or transistors. (3) Reversal by saturation of the residual flux in a ferrite core. (4) Point in a computer programme at which a branch is possible. See **anti-capacitance-, waveguide-.**

switching. Alternative to *clamp* (q.v.); connection of a circuit point to a known potential for a definite period of time.

switching constant. The ratio of *switching time* and *magnetizing force.*

switching time. Time required for complete reversal of flux in ferroelectric or ferromagnetic core.

Sykes microphone. One in which a limp coil in the radial field of a pot magnet generates an e.m.f. when subjected to an acoustic pressure.

symbolic address. See **floating address.**

symbolic logic. Logic, based on symbolic non-verbal language, used in computers. Used in Boolean algebra.

symmetrical deflection. The application of a voltage to a pair of deflection plates in a CRO such that it varies symmetrically above and below an average value, which is equal to the final anode potential of the tube. This procedure minimizes the possibility of trapezoidal distortion of the screen image.

symmetrical network. One which can be divided into two mirror half-sections.

symmetry class. Crystal lattice structures can show 32 combinations of symmetry elements—each combination forming a possible symmetry class.

sync. separator. Circuit which separates video signal from synchronizing impulses in TV reception.

synchro. A general term used for a family of self-synchronous angle data transmitters and receivers. Also **selsyn.**

synchrocyclotron. Cyclotron in which frequency-modulated voltage is used to accelerate particles to relativistic energies at the expense of a pulsed output instead of a continuous output.

synchronism. Said of two oscillations of the same frequency when the phase angle between them is zero.

synchronization. Adjustment of line and frame frequencies in a TV receiver to coincide with those at the transmitter, and the keeping of them so adjusted and locked.

synchronization of oscillators. Phenomenon when two oscillators, having nearly equal frequencies, are coupled together. When the degree of coupling reaches a certain point, the two suddenly pull into step.

synchronizing modulation. The range of modulation depth reserved for the synchronizing pulses, as distinct from that for picture (video) signals.

synchronizing pulse. That transmitted at the beginning and/or end of each frame and scanning line to regulate synchronization.

synchronizing signals. Those added between the video scanning signals to trigger line and field scanning pulses in the receiver.

synchronizing valve. One used for injecting the synchronizing pulses into a time-base circuit.

synchronizing wheel. Rotating wheel, having a large moment of inertia, used for maintaining synchronism over the individual cycles of frame and line frequency oscillators.

synchronous carrier system. Simultaneous broadcasting by two or more transmitters having the same carrier frequency, the various drive circuits being interlocked to avoid heterodyne beats between them.

synchronous clock. One for which time-keeping is controlled by frequency of a.c. (mains) supply.

synchronous computer. One in which all operations are timed by a master clock, which may be a separate pulse source or timed pulses reproduced from a track on a magnetic drum.

synchronous gate. Time gate controlled by clock pulses and used to synchronize various operations in computer or data-processing system.

synchronous homodyne. Reception in which the incoming modulated carrier has added to it a local oscillation of correct phase, with possible locking.

synchronous motor. A.c. electric motor designed to run in synchronism with supply voltage.

synchronous spark gap. *Rotary spark gap* (q.v.).

driven by a synchronous motor running from the same supply as that for the transformer furnishing the high voltage.

synchronous vibrator. See **vibrator**.

synchronous voltage. That required to accelerate electrons from rest to a velocity equal to the phase velocity of the EM wave being propagated down a travelling-wave tube.

synchrotron. Machine for accelerating charged particles in a vacuum; they are guided by an alternating magnetic field as in a betatron, with acceleration by a high-frequency electric field at one point in their orbit.

syntonic jars. Two similar Leyden jars each fitted with spark gap and equal lengths of conductor for connecting thereto. When one jar is charged and then discharged across the gap, a spark appears at the gap connected to the other jar. Used by Lodge to demonstrate the principle of *syntony*.

syntony. See **current (or voltage) resonance**.

system engineering. The approach to auto-mation in which the design is based on general consideration of the required process and the available control elements, rather than on replacing manual operators by automatic devices which function in a similar manner.

systems analysis. Complete analysis of all phases of activity of an organization, and development of a detailed procedure for all collection, manipulation and evaluation of data associated with operation of all parts of it.

systematic errors. Those arising from faulty calibration or other regular causes, which can be evaluated and corrected.

Szilard-Chalmers process. One in which a nuclear transformation occurs with no change of atomic number, but with breakdown of chemical bond. This leads to formation of free active radicals from which material of high specific activity can be separated chemically.

I*

T

T-antenna. One comprising a top conductor with a vertical downlead attached at the centre; much used for long waves.

T-attenuator. One comprising three resistors, one end of each being connected together. The free ends of two are connected respectively to an input and an output terminal while the free end of the third is connected to both the remaining input and output terminals.

T-core. Type of iron stamping which forms a closed magnetic circuit.

TE-wave. Abb. for *transverse electric* wave, having no component of electric force in the direction of transmission of electromagnetic waves along a waveguide. Also **H-wave** (since it must have magnetic field component in direction of transmission).

TM-wave. Abb. for *transverse magnetic* wave, having no component of magnetic force in the direction of transmission of electromagnetic waves along a waveguide. Also **E-wave** (since it must have electric field component in direction of transmission).

T-network. Network formed of two equal series arms with a shunt arm between them.

TR tube. Switch which does or does not (*ATR tube*) permit flow of high-energy radar pulses. It is a vacuum tube containing argon for low striking, and water-vapour to assist recovery after the passage of a pulse. Employed to protect a radar receiver from direct connection to the output of the transmitter when both are used with the same scanning aerial through a common waveguide system.

T-section filter. T-network ideally formed of non-dissipative reactances, having frequency pass band over which attenuation is theoretically zero or very low. See **Butterworth filter, Chebyshev filter.**

TW antenna. See **travelling-wave antenna.**

tabulator. Machine which prints a line up to 100 characters at a time, on continuous paper, at a speed of, e.g., 900 lines a minute, controlled by punched cards or the output signals of an electronic computer.

tacan. U.S. navigation system using ultra-high radio frequencies. The airborne equipment directly indicates both bearing and range for any station to which it is tuned. (*Ta*ctical *a*ir *na*vigation.)

tachometer. Instrument measuring rate of rotation.

tachometer generator. Unit of servo system which generates a feedback signal proportional to the rate of rotation of the servo motor.

tacitron. Thyratron in which extinction is effected by small negative bias.

tag. See **sentinel.**

tag block. Terminal block, holding varying numbers of double-ended solder tags, which is fitted to every panel of apparatus supported on standard apparatus racks. External wiring to a unit can then be connected without interference with the internal wiring, which is completed during manufacture. External connections to bays of apparatus are also made to tag blocks mounted at the top of the racks by cable forms.

tagged. Synonymous with **labelled.**

tailing. Lag in response from black to white in facsimile reproduction.

tailing hangover. Blurring of reproduced picture because of slow decay in electronic circuits. Also **submerged resonance.**

talbe. Air sea rescue system using VHF continuous waves. (*Talk* and *listen* *beacon*.)

talbot. Unit of luminous energy, such that 1 lumen is a flux of 1 talbot per second.

talk-back (or **speak-back**) **circuit.** One which enables the controller of a programme to give directions to those originating a performance or rehearsal in a studio or location.

talk-down. See **ground-controlled approach.**

tamper. Heavy casing placed round core of nuclear weapon to delay expansion and act as a neutron reflector.

tandem. Connection of output of one four-terminal network to input of a second.

tandem generator. One equivalent to two Van de Graaff generators in series so that twice the normal voltage is obtained.

tangent galvanometer. A vertical circular coil with its plane parallel to the meridian. If I is the current flowing, and r the effective coil radius in cm, then

$$I = \frac{5rH}{\pi n} \tan \theta \text{ amps,}$$

where θ is the angle of deflection of a magnetometer needle placed at the centre of the coil, n the number of turns in the coil, and H the horizontal component of the earth's magnetic field in oersteds.

tank. Said of any storage, e.g., signals in mercury delay line, oscillations in tuned circuit.

tank circuit. Section of a resonating coaxial transmission line or a tuned circuit, which accepts power from an oscillating valve and delivers it, harmonic-free, to a load.

tank line. Quarter-wavelength line used as a frequency stabilizer in an ultra-short-wave radio transmitter.

tank reactor. Covered swimming-pool reactor.

tank rectifier. Mercury-arc rectifier enclosed in a metal tank with vitreous seals for the conductors.

tank tube. Same as **pool tube.**

tantalum (Ta). At. no. 73, at. wt. 180·948, sp. gr. 16·6, m.p. 2850°C. Used for lamp filaments, and in electrolytic rectifiers and capacitors.

tantalum capacitor. Miniature electrolytic capacitor employing tantalum foil.

tantalum rectifier. One employing a tantalum electrode.

tape. In electronics, tape is used : (1) as insulator, e.g., round conductors in multiple cables ; (2) for instruction and control, e.g., perforated paper tape as input for control system or computer; (3) see **magnetic tape**.

tape deck. (1) Platform incorporating essentials for magnetic recording—motor(s), spooling, recording and erasing heads—for adding to amplifier, microphone, loudspeaker, to form a complete recording and reproducing equipment. See **capstan, cassette.** (2) A main store in the form of a long magnetic tape on which computer data is stored as in a file, with facilities for rewinding and searching (*random access*) and replacement of data.

tape reader. Apparatus for providing signals related to holes punched in paper tape as it is moved by a tape transport.

tape recording. Longitudinal recording on magnetic particles dispersed in a medium carried on plastic tape. There is a residual magnetization of a high-frequency biasing current, modulated by the signal current. The residual m.m.f. in the particles allows the modulation to be reproduced by induction in a magnetic circuit.

tapper. Electromechanically-operated hammer for decohering a coherer after actuation by high-frequency signals.

tapping. An intermediate connection on a circuit element such as a resistor, often used to vary the potential applied to another electrical system.

tared filter. One intended for measurement of weight of deposited precipitate.

target. (1) Any electrode or surface upon which electrons impinge at high velocity, e.g., fluorescent screen of a CRT or any intermediate electrode in an electron multiplier, or anode or anti-cathode in an X-ray tube. (2) Plate in a TV camera tube on which external scenes are focused and scanned by an electron beam. (3) Reflecting object which returns a minute portion of radiated pulse energy to the receiver of a radar system. (4) Material irradiated by beam from accelerator.

target strength (*T*). Defined in dB by $T = E - S + 2H$, where E = echo level, S = source level and $2H$ = transmission loss, all measured in dB.

target theory. Proposed explanation of radiobiological effects, in which only a small sensitive region of each cell is susceptible to ionization damage.

Tchebycheff filter. See **Chebyshev filter.**

tearing. Break-up of image in facsimile transmission due to faulty synchronization.

technetium (Tc). At. no. 43, radioactive element

not found in nature, first produced as a result of deuteron and neutron bombardments of molybdenum. The most common isotope ^{99}Tc has a half-life of $2·1 \times 10^6$ years. Found among fission products of uranium, and present (unexplained) in the spectra of some stars. Formerly called **masurium.**

Technetron. TN for a French transistor which employs centripetal striction to vary the conductance of the semiconductor.

Technitron. TN for a high-frequency triode.

Teflon. TN for **polytetrafluoroethylene.**

tele-. Prefix indicating instrument operating by telemetry, e.g., televoltmeter, etc.

telearchics. See **telecontrol.**

Telebit. Digital system for data transmission through interplanetary space.

telechrome. Early design of colour TV tube.

teleciné. Projector of cinematograph films, specially designed for television scanning and video transmission.

telecommunication. Any communication of information in verbal, written, coded or pictorial form by electric means, whether by wire or by radio.

telecontrol. Control of mechanical devices remotely, either by radio (as ships and aircraft), by sound waves or by beams of light. Also called **telearchics.**

teledeltos. Sensitive paper used in direct electric recording (electrography).

Telediphone. RTM of device for monitoring, recording and reproducing for written record any programme material.

telefilm. Regular soundfilm made specially for subsequent transmission for television.

Telegon. TN for synchro with no commutator or slip rings.

telegraph. Combination of apparatus for conveying messages over a distance by electrical pulses sent along special overhead wires or underground cables. Such pulses are often interrupted audio-frequency currents, except for railways.

telegraph word. Standard word for calculations is five letters and letter space, e.g., PARIS .

telegraphy receiver. Radio receiver designed for use in radio telegraphy.

telemetry. Transmission to a distance of measured magnitudes by radio or telephony, with suitably coded modulation (e.g., amplitude, frequency, phase, pulse).

telepantoscope. Device similar to an iconoscope, except that scanning motion of the beam is in one direction only (i.e., line-scanning direction), the frame scanning being accomplished by mechanical means.

telephone. Combination of apparatus for conveying speech over a distance by audio-frequency current sent along special overhead wires or underground cables. Dependent on valve amplifiers (*repeaters*) for compensating circuit loss in long-distance telephony.

telephone capacitor. Fixed capacitor, having a capacitance *ca*. 0·001 μF in parallel with telephones in a crystal receiver to by-pass radio-frequency currents.

telephone transmitter. (1) That of a radio telephone system. (2) A telephone microphone.

telephony. The conversion of a sound signal into corresponding variations of electric current (or potential), which is then transmitted by wire or radio to a distant point where it is reconverted into sound.

telephoto-lens. A combination of a convex and concave lens to increase the effective focal length (and so magnify the image) without altering distance between the lens and film of the camera.

teleprinter. Telegraph transmitter, having a typewriter keyboard and a type-printing telegraph receiver; widely used in commercial offices and for public and news services. U.S. **teletypewriter.**

teleradiography. A technique in medicine to minimize distortion in taking X-ray photographs, by placing X-ray tube at some distance from the body.

telerecording. Recording television programme material on film, for editing and subsequent transmission. See **Ampex.**

telescope tube. One in which an infrared object is focused on a photoemitting surface, electrons from which form an enlarged image, exhibited on a fluorescent screen. See **image tube.**

teletherapy. Treatment by X-rays from a powerful source at a distance, i.e., by high-voltage X-ray tubes, or radioactive sources, such as cobalt-60 or caesium, up to 2000 curies.

teletorium, telestudio. Enclosure, sound-proofed and treated acoustically, used for live TV or broadcasting programmes.

teletron. Cathode-ray tube specially designed for synthesizing television test images, either for direct viewing or for projection.

teletube (*tele*vision *tube*). A cathode-ray tube specifically designed for the reproduction of television images. Also **picture tube,** in U.S. **kinescope.**

Teletype. TN for output device used in data transmission systems, capable of producing both printed and punched tape outputs or of printing from a punched tape.

teletypewriter. U.S. *teleprinter* (q.v.), start-stop telegraph system using normal keyboards.

television. Electric transmission of visual scenes and images by wire or radio, in such rapid succession as to produce, in the observer at the receiving end, illusion of being able to witness events as they occur at the transmitting end. Effected by scanning a *raster* which covers the direct image or a picture on a standard film. In *broadcasting* the transmission is by radio waves. In *closed circuit,* the transmission is by line. In the British system the picture is scanned in 405 (VHF) or 625 (UHF) lines, both at 25 times/sec., and with interlaced scanning, whereby the odd number lines are scanned first, followed by the even lines,

so giving 2 frames for a complete picture. See **Ampex.**

television amplifier. One operating uniformly in gain and phase delay, between zero and several MHz, depending on the definition in the system.

television cable. One capable of transmitting frequencies sufficiently high to accommodate television signals without undue attenuation or relative phase delay; usually coaxial, with as much air insulation as possible.

television camera. Converter of an external scene into a video signal for transmission. Prominent are *emitron, iconoscope, image orthicon, photoconducting orthicon, Vidicon.*

television channel. One with a sufficiently wide frequency band ($\sim$4 MHz) to be used for TV transmission.

television field frame. In interlaced scanning, a *frame* consists of the full sequence of scanning lines, and is divided into two or more equal fields scanned alternately.

television microscope. A device which gives a much enlarged image of a very small object using TV techniques.

televisor. Television receiver.

telex. An audio-frequency teleprinter system for use over telephone lines (provided by the GPO). (Automatic *Tele*typewriter *Exchange Service*.)

telluric current. Current in, or put into, the earth for strata exploration.

tellurium (Te). At. no. 52, at. wt. 127·60, sp. gr. 6·24, m.p. 452°C. A semi-metallic element similar to sulphur.

tellurometer. Radar instrument used for distance measurements in surveying.

Telstar. First international telecommunications satellite, sponsored by American Telephone and Telegraph Company, and designed in Bell Laboratories.

temperature coefficient of resistance. In any conductor, if

$$R = R_0[1 + \alpha(T - T_0)],$$

R is the resistance at temperature T, compared with the resistance R_0 at temperature T_0, the coefficient being α (at temperature T_0). Useful only for *linear* resistances, i.e., pure metals.

temperature cycle. A method of processing used for thick photographic nuclear research emulsions to ensure uniform development. These must be soaked in the solutions at refrigerated temperatures, and then warmed for the required processing period.

temperature-limited. Said of a thermionic device operated under saturation conditions, i.e., with the electrode currents limited by the cathode temperature.

temporary memory. One for holding data during a stage in processing. Also called **buffer memory.**

tension. A term often used to designate a potential difference, e.g., low-tension (such as accumulator) or high-tension supplies (such as power-distribution cable).

tensor. (1) A type of mathematical parameter with components most easily represented by matrix arrangements (e.g., the elastic constants of a substance can be represented by a 6×6 matrix array of 36 tensor components, 21 of which are independent in completely anisotropic materials). A zero rank tensor is a *scalar quantity*, a first rank one is a *vector quantity*. (2) See **amplitude**.

tensor force. A non-central force in nuclear physics whose direction depends in part on the spin orientation of the nucleons.

tenth-value thickness. Thickness of absorbing sheet which attenuates intensity of beam of radiation by a factor of ten.

tera-. A prefix denoting 10^{12}.

terbium (Tb). At. no. 65, at. wt. 158·924, a rare earth metallic element.

term diagram. Energy level diagram for isolated atom in which levels are usually represented by corresponding quantum numbers.

terminal. (1) Point in an electrical circuit at which any electrical element may be connected. See **air-**. (2) A station at which data may enter or leave a communication or computing network.

terminal impedance. End or load impedance.

termination. Any device placed at the end of a transmission system. Approximately equivalent to load but implies greater interest in the physical nature of the device, i.e., the load is often 3 ohms when the termination is a loudspeaker. See **matched load.**

ternary fission. The splitting of the nucleus into three nuclear fragments. Not of universal acceptance in its general application.

terrain clearance indicator. Same as **radio altimeter.**

tertiary winding. (1) Extra winding on an audio-frequency transformer for monitoring purposes. (2) Third winding of a hybrid coil, the other two being exactly equal, inserted in series with two legs of the circuit.

tesla. A unit of magnetic induction in the MKSA electromagnetic system, equal to 1 Wb/sq. m.

Tesla coil. Simple source of high-voltage oscillations for rough testing of vacua and gas (by discharge colour) in vacuum systems.

test chart. See **test pattern.**

test jack. One with contacts in series with a circuit, so that a testing device can be immediately introduced to locate faults.

test pattern. A transmitted chart with lines and details to indicate particular characteristics of transmission system. Used in television for general testing purposes. Also **test chart.**

test point. Designated junction between components in an equipment, where the voltage can be stated (as a minimum or with tolerance) for quick verification of correct operation.

test programme. See **check programme.**

test routine. A check routine in computers.

tête-beche. Single sideband transmitters placed far apart but operating over the same frequency range. One uses the upper, the other the lower, sideband for transmission.

tetrafluoroethylene. See **polytetrafluoroethylene.**

tetravalent. Of an atom which has four electrons in the outer or valency shell.

tetrode. Four-electrode electron valve, e.g., *screened-grid valve* (q.v.). See **beam-.**

tetrode transistor. Transistor with additional base contact to improve HF performance.

thallium (Tl). At. no. 81, at. wt. 204·37, sp. gr. 11·85, m.p. 303·5°C. White malleable metal like lead. Several thallium isotopes are members of the uranium, actinium, neptunium and thorium radioactive series. Thallium isotopes are used in scintillation crystals.

theories of light. Interference and diffraction phenomena are explained by the *wave theory*, but when light interacts with matter the energy of the light appears to be concentrated in *quanta* called photons. The quantum and wave theories are supplementary to each other.

thermal analysis. The presence of discontinuities in cooling (or heating) curves of a body to reveal the presence of phase transformations.

thermal catastrophe. See **thermal runaway.**

thermal column. Column or block of moderator in reactor which guides large thermal neutron flux to given experimental region.

thermal conductivity. See **conductivity** (3).

thermal converter. The combination of a thermoelectric device (e.g., thermocouple) and an electrical heater, thus converting an electrical quantity into heat, and into a voltage. Used in telemetering systems; also **thermo-element.** See **thermocouple meter.**

thermal cross-section. Effective nuclear cross-section for neutrons of thermal energy.

thermal cycle. An operating cycle by which heat is transferred from one part to another. In reactors, separate heat transfer and power circuits are usual to prevent the fluid flowing through the former (which becomes radioactive) from contaminating the power circuit.

thermal diffusion. The process in which a temperature gradient in a mixture of fluids tends to establish a concentration gradient. Has been used for isotope separation. Also known as **Soret effect.**

thermal electromotive force. That which arises at the junction of different metals because of a temperature different from the rest of the circuit. Widely used for measuring temperatures relative to that of the cold junction, e.g., ice in a vacuum flask.

thermal excitation. Collision processes between particles by which atoms and molecules can acquire extra energy.

thermal fission factor. In reactors, the ratio of the fast neutrons emitted from, to the thermal neutrons absorbed by, the fuel elements.

thermal inertia. See **thermal response.**

thermal leakage factor. Ratio of number of thermal neutrons lost from, and absorbed in, reactor core.

thermal neutrons. See **neutron.**

thermal noise. That arising from random (Brownian) movements of electrons in a resistor, and which limits the sensitivity of electronic apparatus dealing with transmission frequency bands. The noise voltage V is given by

$$V = \sqrt{4RkT} \cdot \delta f,$$

where
- δf = frequency bandwidth
- R = resistance of source
- k = Boltzmann's constant
- T = temperature, degree Kelvin.

Also known as **circuit noise, Johnson noise, output noise, resistance noise.** See **Nyquist noise theorem.**

thermal reactor. One for which fission is principally due to thermal neutrons. Formerly called **slow reactor.**

thermal response. Of a reactor, its rate of temperature rise if no heat is withdrawn by cooling. The reciprocal of this is **thermal inertia.**

thermal runaway. The effect arising when the current through a semiconductor creates sufficient heat for its temperature to rise above a critical value. The semiconductor has a negative temperature coefficient of resistance, so the current increases and the temperature increases again, resulting in ultimate destruction of the device. Also **thermal catastrophe.**

thermal shield. Inner shield of reactor, used to protect biological shield from excess heating.

thermal siphon. Siphon used in isotope separation to maintain a circulation in a closed liquid loop by keeping a temperature difference between the two vertical sides of the loop.

thermal spike. (1) Surface (hot spots) due to friction. (2) See **spike** (3).

thermal tuning. Change of resonant frequency in oscillator or amplifier produced by controlled temperature change. Employed, e.g., with crystal (see **quartz thermometer**) or with resonant cavity in microwave tube.

thermal utilization factor. Probability of thermal neutron being absorbed by fissile material (whether causing fission or not) in infinite reactor core.

thermalization. Process of slowing fast neutrons to thermal energies. In reactors, normally the function of the moderator.

thermion. A positive or negative ion emitted from incandescent material.

thermionic amplifier. Any device employing thermionic vacuum tubes for amplification of electric currents and/or voltages.

thermionic cathode. One from which electrons are liberated as a result of thermal energy, due to high temperature.

thermionic conduction. That which arises through electrons being liberated from hot bodies.

thermionic current. One represented by electrons leaving a heated cathode and flowing to other electrodes, as distinguished from current which flows through the cathode for heating it. Also called **space current.**

thermionic detector. Valve used for detection of radio-frequency alternating currents. See **Fleming diode.**

thermionic emission. That from a thermionic cathode in accordance with the Richardson-Dushman equation. See **Edison effect.**

thermionic generator. Producer of HF a.c. using thermionic valves for converting d.c. to a.c. in an oscillator.

thermionic oscillator. A generator of electrical oscillations using high-vacuum thermionic valves.

thermionic rectifier. A thermionic valve used for rectification or demodulation.

thermionic relay. Three (or more)-electrode valve in which the on-off potential applied to one electrode controls the current flowing to another, usually without expenditure of energy at the control electrode.

thermionic valve. One containing a heated cathode from which electrons are emitted, an anode for collecting some or all of these electrons, and, generally, additional electrodes for controlling flow to the anode. Normally, the glass or metal envelope is evacuated but a gas at low pressure is introduced for special purposes. Also called **thermionic tube, thermionic vacuum tube.**

thermionic work function. Thermal energy surrendered by electron liberated through thermionic emission from a hot surface.

thermionics. Strictly, science dealing with the emission of electrons from hot bodies. Applied to the broader subject of subsequent behaviour and control of such electrons, especially in vacuo.

thermistor (*therm*al res*istor*). Semiconductor, a mixture of cobalt, nickel and manganese oxides with finely divided copper, of which the resistance is very sensitive to temperature. Sometimes incorporated in a waveguide system to absorb and measure all the transmitted power. Also used for temperature compensation and measurement.

thermistor bridge. One used for measuring microwave power absorbed by a thermistor, in terms of the resulting change of resistance.

thermite. A mixture of aluminium power and half the equivalent amount of iron oxide (or other metal oxides) which gives out a large amount of heat on igniting with magnesium ribbon. The molten metal forms the medium for welding iron and steel.

thermoammeter. A.c. type in which current is measured in terms of its heating effect.

thermocouple. A bimetallic junction the potential difference across which varies with temperature when heated, and forms basis for temperature-measuring and for control devices. See **thermoelectricity.**

thermocouple ammeter. See **thermocouple meter.**

thermocouple meter. A combination of a thermocouple and an ammeter or voltmeter.

The current in the external circuit passes through a coil of suitable gauge wire which is electrically insulated from the thermo-junction but in very close thermal contact; sometimes couple and heater are enclosed in an evacuated quartz bulb. The rise in temperature of the junction causes the thermoelectric current to flow in the con-nected meter. If the heating current is alternating (such devices may be used for radio-frequency currents), then the meter indications are *root-mean-square* values.

thermocouple voltmeter. See **thermocouple meter.**

thermocouple wattmeter. One which uses thermoelements in suitable bridge to measure average a.c. power.

thermoelectric cooling. Abstraction of heat from electronic components by Peltier effect, greatly improved and made practicable with solid-state materials, e.g., $Bi_2 Te_3$. Devices utilizing this effect, e.g., frigistors, are used for automatic temperature control, etc., and are energized by d.c.

thermoelectric effect. See **Seebeck effect.**

thermoelectric materials. Any set of materials (metals) which constitute a thermoelectric system, e.g., *binary* (bismuth or lead tellu-ride), *ternary* (silver, antimony, telluride), and *quaternary* (bismuth, tellurium, selenium and antimony, termed *Neelium*).

thermoelectric module. Large-area thermo-junction formed from semiconductors and used to cool small electronic assemblies by Peltier heat absorption.

thermoelectric power. Defined as dE/dT, i.e., the rate of change with temperature of the thermo e.m.f. of the hot junction of a thermocouple.

thermoelectricity. Interchange of heat and electric energy. See **Peltier effect, Seebeck effect.**

thermoelement. See **thermal converter.**

thermojunction. See **thermocouple.**

thermoluminescence. Release of light by previously irradiated phosphors upon sub-sequent heating.

thermomagnetic effect. See **magneto-caloric effect.**

thermomolecular pressure. The pressure difference arising between two vessels con-taining a gas at different absolute temper-atures (T_1 and T_2), which are separated by a thermally-insulating partition containing an orifice whose dimensions are small compared with the mean free path of the gas molecules. Elementary kinetic theory gives $P_1/P_2 = \sqrt{T_1/T_2}$ as the pressures in the respective vessels. The effect is important in low-temperature gas systems.

thermonuclear bomb. See **hydrogen bomb.**

thermonuclear energy. Energy released by fusion process, which occurs between particles as a result of their thermal energy and not because of electric acceleration. The rate of reaction increases rapidly with temperature and, in the hydrogen bomb, a fission bomb is used initially to obtain the required temperature ($>20 \times 10^6$ °C) to make use of thermonuclear reactions. The energy of most stars is believed to be acquired from exothermic thermonuclear reactions. For laboratory experiments designed to release thermo-nuclear energy, see **DCX, mirror machine, Perhapsatron, Sceptre, stellarator, zeta.**

thermonuclear reaction. One involving the release of thermonuclear energy.

thermophone. Electroacoustic transducer in which heating effect of current produces calculable sound-pressure wave that can be used for calibration of microphones.

thermopile. See **pile** (2).

thermostat. An apparatus which maintains a system at a constant temperature which may be pre-selected.

Thevenin's theorem. The current produced in a load, when connected to a source, is that produced by an e.m.f. equal to the open-circuit voltage of the source, divided by the load impedance plus the apparent internal impedance of the source, i.e., the conjugate impedance to that which extracts maximum power from the source. Also known as **Helmholtz's theorem.** See **Norton's theorem.**

thick source. Radioactive source with appreci-able self-absorption.

thick target. One which is not penetrated by primary or secondary radiation beam.

thick-wall chamber. Ionization chamber in which build-up of ion current is produced by contribution of knock-on particles arising from wall material.

thimble ionization chamber. A small cylind-rical, spherical, or thimble-shaped ionization chamber, volume less than 5 cc with air-wall construction. Used in radiobiology.

thin-film memory. The use of an evaporated thin film of magnetic material on glass as an element of a computer memory, when a d.c. magnetic field is applied parallel to the sur-face. A large capacity memory will contain thousands of these elements which can be produced in one operation. Also **magnetic-film memory.** See **magnetic memory.**

thin source. Radioactive source with negli-gible self-absorption.

thin target. One penetrated by primary radiation beam so that detecting instru-ment(s) may be used on opposite side of target to source.

thin-wall chamber. Ionization chamber in which the number of knock-on particles arising from wall material is negligible.

thin-window counter tube. A counter tube with a defined portion of the enclosure having a low absorption for the radiation to be measured.

thistle microphone. See **apple-and-biscuit microphone.**

thixotropy. Some liquids show a time-dependent viscosity, which can increase with time when the liquid is undisturbed, but the viscosity returns to its original value on shaking.

Thomas resistor. A standard manganin resistor which has been annealed in an inert atmosphere and sealed into a suitable envelope.

Thomson effect. Phenomenon of production of an e.m.f. in a uniform conductor, when parts of the body are at different temperatures.

Thomson scattering. The scattering of electromagnetic radiation by electrons in which, on a *classical* interpretation, some of the energy of the primary radiation is lost because the electrons radiate when accelerated in the transverse electric field of the radiation. The scattering cross-section according to Thomson is given by

$$b = \frac{8}{3}\pi \left(\frac{e^2}{mc^2}\right)^2 = 0.66 \text{ barn/electron,}$$

where e is electronic charge, m is mass of electron and c is the velocity of light.

Thoraeus filter. A combination of tin, copper and aluminium metal filters, used to harden radiation in the 200–400 kV region.

thorianite. A mineral which largely comprises thorium oxides with the oxides of uranium and the cerium metals, etc. It is an important source of thorium and uranium.

thoriated filament. Tungsten filament containing a small proportion of thorium to reduce the temperature at which copious electronic emission takes place. With heat, the thorium diffuses to the surface, forming a tenuous emitting layer.

thorides. Naturally-occurring radioactive isotopes in the radioactive series containing thorium.

thorite. A mineral mainly consisting of thorium silicate (Th SiO_4).

thorium (Th). At. no. 90, at. wt. 232·038, sp. gr. 11·2, m.p. 1845°C. A metallic radioactive element, dark-grey in colour. It occurs widely in beach sands. The thorium radioactive series starts with thorium of mass 232. It is fissile on capture of fast neutrons, and is a fertile material, ^{233}U (fissile with slow neutrons) being formed from ^{232}Th by neutron capture and subsequent beta decay.

thorium reactor. Breeder reactor in which ^{233}U is bred in a blanket of ^{232}Th.

thorium series. The series of nuclides which result from the decays of ^{232}Th. The mass numbers of all the members of the series are given by $4n$, n being an integer.

thoron (Tn). 54·5s ^{220}Rn. Thorium emanation and an isotope of radon.

three-electrode valve. See **triode valve**.

three-level maser. Solid-state maser involving three energy levels.

three-phase. An electric supply system in which the alternating potentials on the three wires differ in phase from each other by 120°.

three-wire system. A supply system in which, e.g., 220 volt is the p.d. between two of the wires, while approximately 110 volt exists between the other 2-wire combinations.

threshold. See **pulse height selector**.

threshold amplitude. The lowest amplitude level which a pulse height selector or window discriminator will accept.

threshold current. That at which a gas discharge becomes self-sustaining.

threshold effect. The marked increase in background noise which occurs in a valve circuit when on the verge of oscillation.

threshold energy. Minimum energy for a particle which can just initiate a given endoergic reaction. Exoergic reactions may also have threshold energies.

threshold frequency. Minimum frequency in a photon which can just release an electron from a surface.

threshold of audibility. See **threshold of sound**.

threshold of feeling. Minimum intensity or pressure of sound wave which causes sensation of discomfort or pain in average normal human listener.

threshold of sound. Minimum intensity or pressure of sound wave which average normal human listener can just detect at any given frequency. Commonly expressed in dB relative to 2×10^{-4} microbar or 1 microbar. Also **threshold of audibility**.

threshold signal. In navigation, smallest signal for a change in positional information.

threshold value. The minimum input which produces a corrective action in a control system.

threshold voltage. That which must be applied to a radiation counter before pulses can be observed. Also **starting voltage**, especially in U.S.

throat microphone. One worn against the throat and actuated by contact pressure against the larynx. Also **laryngophone**.

thulium (Tm). Element, at. no. 69, at. wt. 168·934, with radioactive isotope emitting 84 KeV gamma-rays, frequently used in radiography.

thyratron. Originally an RTM for a gas-filled triode operating in an atmosphere of mercury vapour. Now applied to any gas-filled triode, other common gas fillings being argon, helium, hydrogen and neon. Ionization starts with sufficient positive swing of the negative grid potential, and anode and grid potentials lose control.

thyratron characteristic. Graph relating grid-striking voltage to anode voltage for a thyratron.

thyratron firing angle. Phase angle of a.c. anode voltage supply to thyratron (measured relative to zero) at instant when it strikes.

thyristor. Thyratron-like semiconductor device for bistable switching between high conductivity and high-frequency modes.

thyrite. Device having an inverse exponential resistance/voltage characteristic, which limits rise of voltage in circuits.

ticker. A printing tape machine, operated by start-stop signals.

tiered array. Antenna comprising a number of radiating elements, one above the other. Also called **stacked array**.

tight coupling. That between two circuits which causes alteration of the current in either to affect materially the current in the other. In mutual reactance coupling, coupling is said to be *tight* when the ratio of mutual reactance to the geometric mean of the individual reactances (of the same sign) of the two circuits approaches unity.

tilt-and-bend shading. Adjusted compensation for inequality of effectiveness of a mosaic in translating an image of an external screen into a video signal.

timbre. The characteristic tone or quality of a sound, which arises from the presence of various harmonics or overtones of the fundamental frequency.

time base. (1) In a graph of the variation of a parameter with time, the horizontal scale is the time base. (2) In a cathode-ray tube, it is the voltages (or currents) producing a deflection of the cathode beam which is uniform with time. See **circular-, spiral-.**

time-base generator. Any circuit for deflecting the spot in the horizontal or vertical direction (or in some cases in a circular path) in a known manner (usually linearly) with time and at adjustable frequency. Same as *scanner* and *scanning* in television. Two time bases of sawtooth waveform are required in a TV receiver, one to operate at line frequency, the other to deflect the beam vertically at frame frequency, thus ensuring that each line is traced below the previous one.

time constant. When any quantity varies exponentially with time, the time required for a fractional change of amplitude, equal to :

$$\left(1 - \frac{1}{e}\right) = 63\%,$$

where e is the exponential constant (the base of natural logarithms). For a capacitance C to be charged from a constant voltage through a resistance R, the time constant is RC. For current in an inductance L being fed from a constant voltage through a resistance R, the time constant is L/R. See **response time.**

time-delay relay. One which closes contacts in one circuit a specified time after those in a second circuit have been closed. Widely used to delay application of high-tension voltage until valve cathodes are hot.

time discriminator. Circuit which gives an output proportional to the time difference between two pulses, its polarity reversing if the pulses are interchanged.

time division multiplex. System of multiplex transmission which allocates a physical channel in sequence to a number of communication channels, each using timed pulses.

time gate. Gate circuit which is open only at specified times.

time interval meter. One recording elapsed time between events.

time-of-flight spectrometer. Spectrometer used with beams of particles, especially neutrons. A chopper admits the particles to a flight tube in short bursts, and another at the far end of the tube allows through only those for which the time of flight corresponds to the interval between the choppers opening. See **sector disk.**

time of operation. (1) In relays, time between application of current or voltage and occurrence of a definite change in circuits controlled by its contact. (2) Time between the occurrence of a primary ionizing event and the occurrence of the count.

time response. See **booster response.**

time scale. When time is a variable in analogue computing, time scale can be *real* (tracking satellites or aircraft), *fast, slow* or *extended.*

time-shared amplifier. Form of multiplex system in which one amplifier handles several signals simultaneously, using successive short intervals of time for each.

time signal. One indicating standardized time, radiated by radio or over telegraph lines for exact calibration. Now determined by atomic clock.

timer. Device, operated by electric motor, clockwork, or an electronic or resistor-capacitor circuit, which opens or closes at specified times, with or without delay, control circuits for lighting lamps, operating motors or valves, etc., in a process controller.

tin (Sn). At. no. 50, at. wt. 118·69, sp. gr. 7·31, m.p. 231·85°C. Exists in three allotropic forms, and is a soft, silvery-white metal, ductile and malleable. Used with lead in low melting-point solders for electrical connections.

tin-nickel. Metal finish which results from the simultaneous electrodeposition of tin and nickel on a polished surface of brass in a carefully controlled bath, resulting in a non-tarnishable and non-corrodible polished surface of low friction.

tin-zinc. Metal finish which results from the simultaneous electrodeposition of tin and zinc on a clean steel surface, giving a non-corrodible finish to chassis for electronic apparatus.

Tinkertoy. NBS project for stacking wafers carrying printed circuits (*micromodules*) forming components.

Tintometer. TN for type of visual colorimeter.

tissue dose. Absorbed depth dose (cf. *skin dose*) of radiation received by specified tissue.

tissue equivalent material. See **phantom material.**

titanium (Ti). At. no. 22, at. wt. 47·90, sp. gr. 4·5, m.p. 1850°C. A metallic element resembling iron, sometimes used for the solid horns of magnetostrictive generators.

toggle. Bistable trigger circuit, a multivibrator with coupling capacitors omitted, which switches between two stable states depending on which valve (or transistor) is triggered. Once **flip-flop** in U.K.

tolerance dose. The maximum dose which can be permitted to a specific tissue during radiotherapy involving irradiation of any other adjacent tissue.

toll television. Programme service which, through technical scrambling devices, is available only by *ad hoc* payment. Also called **pay as you view, subscription television.**

tomography. Movement of source and photoplate in diagnostic radiography so that the ratio of their distances from a chosen plane in the body remains constant, this plane only giving rise to a defined image. Also **body-section radiography, laminography, planigraphy.**

tone. (1) Brightness of an area in a television picture, described as *high-key* or *low-key*. (2) That frequency of interruption of the scanning light in facsimile which becomes the carrier in transmission. (3) Single frequency for testing, but a tone is now often used to specify a complex note, having a constant fundamental frequency. In acoustics a sinusoidal sound wave is termed a pure tone.

tone control. Control for altering characteristic response of audio-frequency amplifier in order to obtain more pleasing output quality.

tone-control circuit. A circuit network or element to vary the frequency response of an audio-frequency circuit, thus varying the quality of the sound reproduction.

tone-control transformer. One in which the leakage and/or self-capacitance can be altered in such a way as to regulate its response over the operating frequency range.

tonometer. An electronic instrument for measuring hydrostatic pressure within the eye.

top shot. One taken with axis of camera nearly vertical.

topping. Operation such that the anode voltage remains constant during a cycle, achieved by a catching diode leading to a definite potential.

toroid or **torus.** Said of a coil or transformer which corresponds in shape to an anchor ring. Adopted because of the ease of making windings exactly balanced to the circuit in which they are inserted and to earth. Also, on account of the entire enclosure of magnetic field, there is no interaction with similar adjacent coils; thus, loading coils for transmission lines are threaded on rods in a pot. Also **doughnut, donut** in U.S.

torque amplifier. (1) Mechanical amplifier in which torque (not angle) is varied by differential friction of belts on drums. (2) Electrical servo system performing the same function.

torque motor. One exerting high torques at low speeds.

torr. Unit of low pressure equal to pressure head of 1 mm of mercury.

torus. See **toroid.**

total absorption coefficient. The coefficient which expresses energy losses due to both absorption and scattering. Relevant to narrow beam conditions. Preferred term **attenuation coefficient.**

total body burden. (1) The summation of all radioactive materials contained in any person. (2) The maximum total amount of radioactive material any person may be permitted to contain.

total capacitance. That between any one conductor and all of a series of other conductors forming a complete system. See **capacitance coefficients.**

total cross-section. The sum of the separate cross-sections for all processes by which an incident particle can be removed from a beam. If all the atoms of an absorber have the same total cross-section, then this is identical with the *atomic absorption coefficient* (q.v.)

total electron binding energy. That required to remove all the electrons surrounding a nucleus to an infinite distance from it and from one another.

total emission. See **saturation current.**

total internal reflection. Complete reflection of incident wave at boundary with medium in which it travels faster, under conditions where Snell's law of refraction cannot be satisfied. The angle of incidence at which this occurs (corresponding to an angle of refraction of 90°) is known as the **critical angle.**

total modulation. Amplitude modulation to a depth of 100 per cent.

Townsend avalanche. Multiplication process whereby a single charged particle, accelerated by a strong field, causes, through collision, a considerable increase in ionized particles.

Townsend coefficient. Number of ionizing collisions per cm of path in the direction of an applied electric field.

Townsend discharge. A *Townsend avalanche* initiated by an external ionizing agent.

trace. Image on phosphor on electron beam impact, forming the display.

trace chemistry. Small traces of chemical reagents, e.g., radioactive materials which are free of non-radioactive isotopes, behave quite differently from the same material in ordinary amounts and concentrations. A knowledge of trace chemistry behaviour is of considerable importance in the handling of radioactive materials. Cf. *tracer chemistry.*

trace routine. One which verifies operations by a *programme*, or reveals faults.

tracer atom. Labelled atoms introduced into a system to study structure or progress of a process. Traced by variation in isotopic mass or as a source of weak radioactivity which can be detected, there being no change in the chemistry. See **radioactive isotope.**

tracer chemistry. The use of isotopic tracers in chemical studies.

tracer compound. One in which a small proportion of its molecules is labelled with a radioactive isotope.

track. (1) To move a camera on a dolly in a defined path while taking a shot. (2) Single circumferential area on a magnetic drum, or longitudinal area on a magnetic tape, alongside other tracks, allocated to specific recording and reproduction channel. (3) Space on a disk or soundfilm allocated to one channel of sound-recording and hence reproduction. See **stripe.** (4) Path followed by particle especially when rendered visible in photographic emulsion by cloud chamber, bubble chamber or spark chamber. (5) Course followed by ship or aircraft.

track chamber. A bubble, cloud or spark chamber which can be used to render the tracks of ionizing particles visible.

track guide. A radio-beacon system providing one or more tracks for the guidance of craft.

tracking. (1) Automatic holding of radar beam onto target through operation of return signals. (2) Following, by an inverse feedback or servo loop, variation of a quantity. (3) The accuracy of a gramophone stylus pickup in following a given path. (4) Excessive leakage current between two insulated points, e.g., due to moisture. (5) The accuracy of adjustment of a set of ganged tuned circuits.

trailing edge. See **edges.**

Trancor. TN for grain-orientated silicon-iron magnetic alloy.

Transactor. TN for interrogator/reply device for inputs to a computer.

transadmittance. Output a.c. current divided by input a.c. voltage for electronic device when other electrode potentials are constant.

transceiver. Equipment, e.g., 'walkie-talkie', in which circuitry is common to transmission or reception.

transconductance. Reciprocal of *transfer impedance.* Ratio of current in one part of a circuit to e.m.f. or p.d. in another part, not necessarily with the same frequency, e.g., in a demodulator. U.S. term for **mutual conductance.** See **conversion-, negative-.**

transcriber. In a computer, the apparatus for the purpose of transferring input or output data from a register of information in one programme language to the medium and language employed in a different computer, or the reverse process.

transducer. General term for any device which converts a physical magnitude of one form of energy into another form, generally on a one-to-one correspondence, or according to a specified formula, e.g., an electroacoustic transducer is one receiving an alternating current (or e.m.f.), and supplying an output of acoustic waves, as in a loudspeaker. The microphone operates in the reverse way. An electric motor is an electromechanical transducer.

See:

active-	frequency
feedback-	changer

magnetostriction-	piezoelectric crystal
passive-	reversible-.

transducer translating device. One for converting error of the controlled member of a servomechanism into an electrical signal that can be used for correcting the error.

transductor. Arrangement of windings on a laminated core, which, when excited, permits current amplification. Part of a *magnetic amplifier.* See **auto-.**

transfer. (1) In computers, transmission of information from one device to another. (2) See **print-through.**

transfer admittance. Ratio of current in any part of a mesh to e.m.f. producing it and applied in another part of the mesh, all other e.m.fs. being reduced to zero.

transfer characteristic. (1) Relation between voltage on one electrode and current passing through another under specified conditions. (2) Relationship between TV camera tube illumination and corresponding signal current.

transfer constant. The natural logarithm of the ratio of the output and input currents for a network.

transfer current. Current in a control electrode to initiate ionization and gas discharge.

transfer function. A mathematical expression relating the output of a closed loop servo system to its input.

transfer impedance. In any network or transducer, complex ratio of rms voltage applied at any pair of terminals to the rms current at some other pair. Also **mutual impedance.**

transfer instrument. Instrument which gives d.c. indication independent of frequency, including zero frequency, so that when calibrated for d.c., it can be used for calibrating a.c. instruments, as with electrostatic wattmeters.

transfer port. The aperture through which items are inserted into or removed from a dry box, glove box or shielded box (usually by sealing into plastic sac attached to the rim of the port).

Transfluxor. TN for arrangement with apertured magnetic cores for memory or switching.

transform. (1) A transform is essentially an operator which enables a mathematical expression to be changed from one form to another, which may be more appropriate to the solution of the problem in view. For example, a differential equation may be transformed from a time variation to a frequency spectrum. In control systems a set of transforms is involved which largely concern the transfer function. The operator $\left(\dfrac{d}{dt}\right)$ where t is time, is converted into an operator p, the Heaviside operator, so that the differential equations become algebraic polynomials in p and may be solved algebraically. To the solution thus found is applied the inverse transform to obtain the time solution. (2) In digital computers, a transform changes the form of the information without a significant alteration in meaning.

transformation. (1) Change of phase, e.g., by reactor fuel. (2) Change of variable or co-ordinates in mathematics. (3) See **atomic transmutation.**

transformation constant. See **disintegration constant.**

transformation ratio. That of turns of primary and secondary windings of a transformer; more precisely, that of e.m.fs. in the windings. See **turns-ratio.**

transformer. An electrical device without any moving parts which transfers energy of an alternating current in the primary winding to that in one or more secondary windings through electromagnetic induction. Except in the case of the auto-transformer there is no electrical connection between the two windings, and usually (excepting the isolating transformer) a change of voltage is involved in the transformation. Mechanical transformers involve lever arms of different lengths.
See:

audio-frequency-	line-output-
auto-	matching-
buck-	oscillation-
constant-current-	output-
constant-voltage-	peaking-
coupling-	phase-shifting-
crossbar-	power-
current-	pulse-
delta-matching-	radio-frequency-
direct-current-	step-down-
doorknob-	step-up-
E-	tone-control-
high-frequency-	tuned-
induction coil	Variac
intermediate-	waveguide-
frequency-	workhead-.
intervalve-	

transformer coupling. Transference, in both directions, of electrical energy from one circuit to another by a transformer, of any degree of coupling, the primary being in one circuit, the secondary in the other. Also **mutual coupling.**

transformer ratio. See **turns-ratio.**

transformer-ratio bridge. An a.c. bridge similar to a *Wheatstone bridge*, but with two transformer windings used for the two ratio arms.

transformer tapping. A means of varying the voltage ratio of a transformer by making a connection to a point on the winding intermediate between the ends.

transient. Any non-cyclic change in a part of a communication system. The most general transient is the *step*, while the steady state is represented by any number of sinusoidal variations. See **Heaviside unit function.**

transient analyser. Test instrument which generates repeating transients and displays

their waveform at different points in the system under investigation, on a CRT screen.

transient distortion. That which arises only when there is a rapid fluctuation in frequency and/or amplitude of the stimulus. It may be measured as the decay rate of the response following the sudden withdrawal of a steady stimulus.

transient equilibrium. Radioactive equilibrium between daughter product(s) and parent element of which activity is decaying at an appreciable rate. Characterized by ratios of activity, but not magnitudes, being constant.

transient state. Transition period and associated phenomena between steady states in the repetition of a waveform.

transient wave. One set up in transmission circuits or filters because of changes in the current amplitude and/or frequency, the effects dying out in local circuits but propagated along lines.

transistor. Three-electrode semiconductor device with thin layer of n- (or p-) type semiconductor sandwiched between two regions of p- (or n-) type, thus forming two p-n junctions back to back. The emitter junction is given a forward bias and the collector junction a reverse bias. Current carriers entering the emitter diffuse through the base to the collector so i_c is of the order of $0.98\ i_c$. Due to the low forward resistance of the emitter junction and the high reverse resistance of the collector junction, considerable power gain is possible for signals in the emitter or base leads. The latter arrangement also gives current gain. Amplification in the p-n-p transistor is due to *hole conduction*, that in an n-p-n transistor to *electron conduction*; see also **transistor construction.**
See :

bipolar-	n-p-n-
complementary-	p-n-i-p-
depletion-layer-	p-n-p-
drift-	photo-
filamentary-	point-contact-
hook-	surface-barrier-
junction-	tetrode-
meltback-	unipolar-.
n-p-i-n-	

transistor amplifier. One which uses transistors as the source of current amplification. Depending on impedance considerations, there are three types, with base, emitter or collector grounded.

transistor characteristics. General name for graphs relating d.c. electrode currents and/or voltages in a manner similar to that adopted for thermionic valves. Many such characteristics are employed, especially: (1) *feedback characteristic*—input voltage v. output voltage for fixed input current; (2) *input characteristic*—input voltage v. input current for fixed output voltage; (3) *output characteristic*—output current v. output voltage for fixed input current; (4) *transfer characteristic*—output current *v.* input current for fixed output voltage.

transistor constructions. The sandwich forms described under *transistor* (q.v.) are known as *junction transistors*, e.g., *n-p-n, p-n-p*, etc., and they can be grown by carefully controlled addition of impurities to a single crystal of the base material, or they may be 'fabricated' by fusing gallium or indium to either side of a thin slice of the base crystal (the *dice*). In the modern so-called *mesa* technique, the dice are vacuum-coated by evaporating the desired metal on to the surfaces. The earliest type of semiconductor detector was the *cat's whisker and crystal* of crystal broadcast receivers; *point-contact transistors* are still used.

transistor equivalent circuit. For the purpose of circuit analysis, a transistor may be represented by a four-terminal network having two common terminals. In this way the transistor characteristics may be expressed in terms of four independent variables, the input and output voltages and currents of the equivalent circuit. Such possible circuits are called the *common base*, the *common emitter* and the *common collector*.

transistor parameters. In circuit analysis, the performance of transistors is calculated from parameters obtained from the slope of the various characteristic curves (as with thermionic valves). Many such sets of parameters have been used, the most widely adopted probably being the hybrid parameters h_{11}, h_{12}, h_{21}, h_{22} (also **h-parameters, small-signal parameters**). These are given by the slopes of the input, feedback, transfer, and output characteristics respectively at the selected working points.

transistor pentode. A point-contact transistor having one collector but three emitters, used for purposes of switching, mixing or modulation.

transistor power-pack. One in which high-tension supply of low power is obtained by rectifying high-voltage current from a transistor oscillator fed at low voltage.

transistor tetrode. Transistor which operates at higher frequencies than the pentode, having an emitter, collector, and two base connections.

transistorized. Said of equipment in which all circuits employ transistors and not thermionic valves.

transit angle. Product of delay or transit time and angular frequency of operation. In a velocity-modulated valve the transient time corresponds to the time taken for an electron to pass through a drift space.

transit time. (1) Time taken by an electron to go from cathode to anode of a thermionic valve; important factor in the operation of valves at very high frequencies. (2) In a transistor, the time necessary for injected charge carriers to diffuse across the barrier region.

transition. (1) Change in nuclear configuration associated with α-, β- or γ-emission, and having probability given by relevant selection rule. (2) In electrons, change of energy level associated with emission of quantum of radiation.

transition energy. That at which phase focusing changes to defocusing in a synchrotron accelerator. This necessitates a sharp arbitrary change of phase in the radio-frequency field.

transition frequency. That at which frequency response of disk recording system changes from constant amplitude to constant velocity characteristic.

transition metal. One of the group which have an incomplete inner electron shell. They are characterized by large atomic magnetic moments.

transition multipole moments. See **static multipole moments.**

transition probability. The probability per unit time that a system in state k will undergo a transition to state l. In the case of radio-active transitions, the total transition probability is termed the *disintegration constant.*

transition region. That over which the impurity concentration in a doped semiconductor varies.

transition temperature. (1) That corresponding to a change of phase. (2) That at which a metal becomes superconducting.

transitron. Pentode in which suppressor grid acts as control grid, characterized by negative mutual conductance between suppressor and screen grids.

transitron oscillator. One in which a retarding field is established by holding a screen grid negative in a screened-grid valve.

translator device. Device for converting information from one form into another. In computers, a network such that input signals in a certain code are changed to output signals in another code without altering the significance of the information.

transmission. Conveying electrical energy over a distance by wires, either to operate controls or indications (*telemetering*), acoustic information (*telephony, broadcasting*) or pictorial information (*facsimile, television*). Also used to denote radio, optical or acoustic wave propagation. See **direct-current picture-, transmittance.**

transmission band. Section of a frequency spectrum over which minimum attenuation is desired, depending on the type and speed of transmission of desired signals.

transmission bridge. A device in telephony which separates a connection into incoming and outgoing sections for the purpose of signalling, at the same time allowing the through transmission of voice frequencies.

transmission coefficient. (1) Probability of penetration of a nucleus by a particle striking it. (2) Ratio of intensities of transmitted (refracted) and incident waves at the interface between two media. Also **transmissivity.**

transmission density. The common logarithm of the transmittance of a processed photographic emulsion.

transmission experiment. One in which radiation transmitted by a thin target is measured in order to investigate the interaction which takes place. Such experiments are used in the measurement of total cross-sections for neutrons.

transmission gain. The increase of power (usually expressed in dB) in a transmission from one point to another.

transmission level. Electric power in a transmission circuit, stated as the decibels or nepers by which it exceeds a reference level. Also called **power level.**

transmission line. General name for conductors used to transmit electric or electromagnetic energy, e.g., power line, telephone line, coaxial feeder, G-string, waveguide, etc. Also for the acoustic equivalent.

transmission-line amplifier. One in which grids of valves are driven by, and anodes feed, appropriate points on an artificial line, giving wideband amplification. Also called *distributed amplifier.*

transmission-line control. Control of frequency of an oscillator by means of a resonant line in the form of a tapped quarter-wavelength stub.

transmission loss. Difference between the output power level and the input power level of the whole, or part of, a transmission system in decibels or nepers.

transmission measuring set. Calibrated set of equipment for the measurement of transmission loss in a communication system.

transmission modes. Field configurations by which electromagnetic or acoustic energy may be propagated by transmission lines, especially waveguides.

transmission primaries. In colour TV, the set of three primaries chosen so that each one corresponds in magnitude to one of the independent signals which comprise the colour signal.

transmission speed. The number of bits or elements of information transmitted in unit time.

transmissivity. See **transmission coefficient** (2).

transmit-receive tube. See **TR tube.**

transmittance. Ratio of energy transmitted by a body to that incident on it. If scattered emergent energy is included in the ratio it is *diffuse transmittance*—otherwise *specular transmittance*. Also **transmission.**

transmitted carrier system. One in which the carrier wave is radiated. Cf. *suppressed carrier system.*

transmitted power. That propagated through any type of communication channel.

transmitter. Strictly, complete assemblage of apparatus necessary for production and modulation of radio-frequency current, together with associated antenna system, but frequently restricted to that part concerned with the conversion of d.c. or mains a.c. into modulated RF current.

See :

arc-	single sideband
Chireix-	system
crystal-controlled-	suppressed carrier
Doherty-	system
facsimile	telephone-
telegraphy	television
radar	vestigial-sideband-
radio	walkie-talkie.

transmitter frequency tolerance. Maximum permitted **frequency departure.**

transmitting valve. One which handles output power of a radio transmitter; may be in parallel or push-pull with others.

transmittivity. Transmittance of unit thickness of non-scattering medium.

transmutation. See **atomic transmutation.**

transparency. (1) Ratio of free space (window) between grid or screen wires in a thermionic valve, to their total area. (2) Proportion of energy, or number of incident photons or particles, which pass through the window of an ionization chamber or Geiger counter.

transpolarizer. Ferroelectric dielectric impedance, controlled electrostatically.

transponder. A form of transmitter-receiver which transmits signals automatically when the correct interrogation is received, e.g., a radar beacon mounted on a flight vehicle (or missile), which comprises a receiver tuned to the radar frequency and a transmitter which radiates the received signal at an intensity appreciably higher than that of the reflected signal. The radiated signal may be coded for identification.

transport. (1) Mechanism for moving tape during recording or reproducing. (2) Rate at which desired material is carried through any section of processing plant, e.g., isotopes in isotope separation.

transport cross-section. The reciprocal of the *transport mean free path.*

transport mean free path. If Fick's law of diffusion is applicable to the conditions in a nuclear reactor, then the mean free path is three times the diffusion coefficient of neutron flux. In practice, the theory usually has to be modified to account for anisotropy of scattering and persistence of velocities.

transport number. The fraction of the total current flowing in an electrolyte which is carried by a particular ion.

transport theory. Rigorous theoretical treatment of neutron migration which must be used under conditions where Fick's law does not apply. See **diffusion theory.**

transposition. Ordered interchange of position of the lines on a pole route, and also of phases in an open power line, so that effects of mutual capacitance and inductance, with consequent interference, are minimized. See **barrel.**

transreceiver. Equipment for both transmission and reception. See **transceiver.**

transrectification factor. Ratio of change in average output current to change in alternating voltage applied to a rectifier.

transuranic elements. The artificial elements 93 and upwards which possess heavier and more complex nuclei than uranium, and which can be produced by the neutron bombardment of uranium. More than ten of these have been produced, e.g., neptunium, plutonium, americium, curium, berkelium, etc.

transvar. Variable directional coupler for transferring power between waveguides.

transverse-beam travelling-wave tube. One in which the directions of propagation of the electron beam and the electromagnetic wave carrying the signal are mutually perpendicular. The *cavity magnetron* is an example of such a tube.

transverse electric wave. See TE wave.

transverse-field travelling-wave tube. One in which the electric fields associated with the signal wave are normal to the direction of motion of the electron beam.

transverse heating. Dielectric heating in which electrodes impose a high-frequency electric field normal to layers of laminations.

transverse magnetic wave. See TM wave.

transverse magnetization. In magnetic recording, that which is perpendicular to the direction of motion of the recording medium.

transverse wave. A wave motion in which the disturbance of the medium occurs at right angles to the direction of wave propagation, e.g., as for the vibration of a violin string.

transwitch. *P-n-p-n* silicon device for on-off switching in computers.

trap. Crystal lattice defect at which current carriers may be trapped in a semiconductor. These traps can increase recombination and generation or they may reduce the mobility of the charge carriers.

trap amplifier. Parallel amplifier with grid of the amplifying valve connected to grid of an amplifying valve in a main amplifier; used to ensure that a short-circuit on any part of the monitoring equipment, such as headtelephones, does not affect transmission in the main current.

trapezium diagram. Pattern on the screen of a cathode-ray oscillograph when an amplitude-modulated radio-frequency voltage is applied to one pair of plates and the modulating voltage is applied to the other pair.

trapezium distortion. That associated with *trapezium effect.*

trapezium effect. Phenomenon in which deflecting voltage applied to the deflector plates of a cathode-ray tube is unbalanced with respect to the anode. If equal alternating voltages, of different frequencies, are applied to the two sets of plates, resulting pattern on the screen is trapezoidal instead of square.

trapping region. Three-dimensional space in which particles from the sun are guided into paths towards the magnetic poles, giving rise to *aurora*, and otherwise forming ionized shells high above the ionosphere. Also **magnetic tube.**

travelling wave. A wave carrying energy continuously away from the source.

travelling-wave amplifier. One using a travelling-wave tube, or one in which simple valves inject gain at repeated points in a wave filter.

travelling-wave antenna. One in which radiating and non-radiating elements are formed into the edges and diameters of a series of adjacent square boxes. Also **box antenna, TW antenna.**

travelling-wave magnetron. Multiple-cavity magnetron in which cavities are coupled by travelling-wave system.

travelling-wave maser. One in which the signal interacts with the electrons in paramagnetic material while propagated through a travelling-wave system.

travelling-wave tube. One in which energy is interchanged between a helix delay line and an electron beam, which can be at an angle. Used to amplify ultra-high and microwave frequencies. See also **transverse-beam-, transverse-field-.**

tray. See card box.

tree. A number of connected circuit branches which do not include meshes.

triad. (1) A colour cell of a screen, with three-colour dots on the phosphor. (2) Group of three neighbouring isobaric nuclides. (3) Group of three transmitters used to obtain navigational position fix. (4) A trivalent atom.

tribo-electrification. Separation of charges through surface friction. If glass is rubbed with silk, the glass becomes *positive* and the silk *negative*, i.e., the silk takes electrons. Phenomenon is that of contact potential, made more evident in insulators by rubbing.

triboluminescence. Luminescence generated by friction.

trickle charger. A charging system which maintains a storage battery in a fully-charged state.

tricolour chromatron. Directly-viewed single-gun cathode-ray tube for colour television display, in which the screen is composed of successive horizontal stripes of the three-colour phosphors, the electron beams being deflected onto the correct phosphors by an electrostatic field produced by a grid of horizontal wire electrodes. Also **Lawrence tube.**

tricon. Navigational system based on pulsed signals from a transmitting triad, which arrive simultaneously when the aircraft is on course.

trigatron. An envelope with an anode, cathode and trigger electrode, containing a mixture of argon and oxygen. The device operates as an electronic switch in which a low-energy pulse ionizes the gas in the switch, and permits discharge of a much higher energy pulse across the main electrodes.

trigger. Manual or automatic signal for an operation to start.

trigger circuit. A circuit having a number of states of electrical condition which are either stable (or quasi-stable) or unstable with at least one stable state, and so designed that desired transition can be initiated by the application of suitable trigger excitation.

trigger electrode. See **pilot electrode.**

trigger level. The minimum input level at which a trigger circuit will respond.

trigger pulse. One which operates a trigger circuit.

trigger relay. (1) Relay which, when operated, remains in its operated condition when the operating current or other control is removed, because of residual magnetism or a mechanical latch. (2) Gas-filled triode or certain combinations of high-vacuum thermionic valves, in which a disturbance of sufficient magnitude can initiate or terminate a discharge but has no subsequent control thereof.

trimer. A substance in which the molecules are formed from three molecules of a monomer.

trimmer. See **pad, trimming capacitor.**

trimming capacitor. Variable capacitor of small capacitance used in conjunction with ganging for taking up the discrepancies between self and stray capacitances of individual ganged circuits, so that they remain in step for all settings of the main tuning control. Also called **trimmer.**

trinoscope. Cathode-ray tube with three electron guns, designed for the reproduction of television images in colour.

triode. See **triode valve.**

triode-hexode. Combination of triode and hexode in the same envelope, used as a frequency converter in a supersonic heterodyne receiver. The triode section is used as oscillator and the hexode as demodulator.

triode valve. Thermionic vacuum tube containing an emitting cathode, an anode, and a control electrode or grid, whose potential controls the flow of electrons from cathode to anode. Also called **three-electrode valve.**

trip. Automatic reduction of reactor power initiated by signal from one of the safety circuits. This may be a gradual power setback or an emergency trip.

trip action. Instability in an incorrectly operated magnetic amplifier.

trip amplifier. Amplifier operating the trip mechanism of a nuclear reactor; also termed **shut-down amplifier.**

trip value. The current or voltage required to operate a relay.

triple detection. The use of two frequency converters before the final detection of incoming signals.

tri-stimulus values. Amounts of each of three colour primaries that must be combined to form an objective colour match with a sample.

tritium. The radioactive isotope of hydrogen of half-life 12·5 years, mass no. 3. It is very rare, the abundance in natural hydrogen being one atom in 10^{17}, but tritium can be produced artificially by neutron absorption in lithium. Can be used to label any aqueous compound, and consequently is of great importance in radiobiology.

tritium unit. A proportion of tritium in hydrogen of one part in 10^{18}. This represents 7 d.p.m. in 1 litre of water.

triton. The tritium nucleus consisting of one proton combined with two neutrons.

trochoid. The path traced out by a point on a line drawn through the centre of a circle, when it rolls along a straight line.

trochoidal mass analyser. A form of mass spectrometer in which the ion beams traverse trochoidal paths within electric and magnetic fields mutually perpendicular.

trochotron (*trocho*idal magne*tron*). High-frequency counting tube, using crossed electric and magnetic fields to deflect a beam on to radially disposed electrodes.

troland. A unit of illuminance used in retinal work. If the apparent area of the entrance pupil of the eye is 1 sq mm, then the troland is the visual stimulation obtained from an illuminance of 1 candle/sq m.

trombone. A U-shaped length of waveguide which is of adjustable length for use in a waveguide circuit.

troposphere. Atmosphere up to region where temperature ceases to decrease with height. This upper boundary is termed *tropopause.*

tropospheric. Said of reflection, absorption or scattering of a radio wave when encountering variations in the troposphere.

tropospheric wave. A radio wave whose path between two points at or near the earth's surface lies wholly within the troposphere and will be governed by meteorological conditions.

tropotron. A form of magnetron valve.

trouble-shooting. U.S. for **fault-finding.**

true (or real) absorption coefficient. The absorption coefficient applicable when scattered energy is not regarded as absorbed. Applicable to broad-beam conditions.

true azimuth. That measured relative to true geographical north.

true bearing. That measured clockwise from true geographical north.

true coincidences. Those produced by a single particle discharging both or all counters; cf. *spurious coincidences.*

true north. The direction of the geographical north pole.

true watts. Power dissipated in an a.c. circuit.

trunk. (1) A circuit in automatic telephony connecting selectors of different rank in an automatic switching network (or between one selector bank and a manual position). (2) See **highway.**

Tschebycheff filter. See **Chebyshev filter.**

tube. (1) Enclosed device with gas at low pressure, depending for its operation on ionization originated by electrons accelerated from a cathode by a field applied by an anode. (2) U.S. for all vacuum and gas-discharge devices. The term is becoming more widely used in Britain, where formerly *valve* was almost universally employed. See also **valve.**

See :

ATR-	Hittorf-
accelerator-	iconoscope
alpha counter	image-
anti-cyclotron	indicator-
Apple-	kenotron
backward wave	klystron
Banana-	Kundt's-
band ignitor-	Lenard-
beam-forming	line focus
electrode	magnetron
beam-indexing	megatron
colour TV-	mercury-vapour-
boron chamber	negatron
camera-	non-storage-
Carcinotron	octode-
cathode-ray-	orthicon
Charactron	Photicon
charge-storage	photoconducting
Chromaton	camera-
chromatron-	photoglow-
Coolidge-	plasmatron
corona-	pool-
counter-	pumped-
Crookes-	resnatron
diode	rotating anode
discharge-	sealed-
dissector-	self-rectifying
double-beam	skiatron
cathode-ray-	stabilizer-
drift-	stepping-
electrometer-	storage-
electron-	TR-
electron-	telescope-
discharge-	thin-window counter-
electron	thyratron
multiplier	transverse-beam
excitron	travelling-wave-
Faraday-	transverse-field
gas-	travelling-wave-
gated-beam-	travelling-wave-
Geissler-	trigatron
glow-	velocity-modulated
grid-glow-	oscillator
halogen quench	Vidicon
Geiger-	voltage regulator-
Heil oscillator	Williams-
hexode	X-ray-.

tube of force. Space enclosed by all the lines of force passing through a closed contour. It is of unit magnitude when it contains unit flux.

tube ring. Undesired ringing noise, sustained in an amplifying system because of continual mechanical impulsing of a microphonic valve from some external source.

tune. To adjust for resonance or syntony, especially musical instruments or radio receivers.

tuned amplifier. One containing tuned circuits, and therefore sharply responsive to particular frequencies.

tuned anode. Inductance coil shunted by a capacitor (either or both of which may be variable) in series with the lead to the anode of a thermionic valve.

tuned-anode coupling. That between stages of

a high-frequency thermionic valve amplifier, in which coupling impedance is a tuned-anode circuit.

tuned-anode oscillator. One in which the tuning circuits are in series with the anode of a valve.

tuned-anode tuned-grid or **tuned-plate tuned-grid oscillator.** A valve oscillator with correspondingly tuned circuits in series with both the anode and the grid leads. In most cases no additional external feedback path is required.

tuned antenna. One operating at its natural resonant frequency.

tuned-base oscillator. One in which the tuning circuits are in series with the base of a transistor (equivalent to *tuned-grid oscillator*).

tuned cell. Adjustable cavity in a waveguide structure, particularly in a wave-filter section.

tuned circuit. One comprising an inductance coil and a capacitor in series or in parallel, and offering a low or high impedance respectively to alternating current at the resonant frequency.

tuned-collector oscillator. One in which the tuning circuits are in series with the collector of a transistor.

tuned-emitter oscillator. One in which the tuning circuits are in series with the emitter of a transistor.

tuned-grid circuit. Parallel-tuned circuit included between grid and cathode of a thermionic valve, to provide high step-up at the resonant frequency. See *Q*.

tuned-grid oscillator. One in which the tuning circuits are in series with the grid of a valve.

tuned magnetron. One capable of tuning to a range of frequencies.

tuned oscillator. Valve or transistor oscillator with frequency controlled by tuned circuit. Classified according to the electrode tuned, e.g., *tuned anode (tuned plate)*, *tuned anode tuned grid (tuned plate tuned grid)*, *tuned base*, *tuned collector*, etc.

tuned-plate tuned-grid oscillator. See **tuned-anode tuned-grid oscillator.**

tuned radio-frequency receiver. Radio receiver which does not use frequency changing before detection.

tuned relay. One which responds only at a resonant frequency.

tuned transformer. Interstage coupling transformer in which one, or more usually both, winding(s) are tuned to resonate with the signal frequency. A higher secondary voltage can be built up than would be the case without resonance.

tuner. Assemblage of one or more resonant circuits, used for accepting a wanted signal and rejecting others. See **turret tuner.**

tungsten (W). At. no. 74, at. wt. 183·85, sp. gr. 19·3, m.p. 3370′C. A hard grey metal, resistant to corrosion, used in cemented carbides for grinding tools, and as wire in incandescent electric lamps. Also **wolfram.**

tungsten carbide. A compound of tungsten

and carbon with a m.p. ~2800°C. Proposed use in high-temperature reactors.

tuning. (1) Operation of adjusting circuit settings of a radio receiver to produce maximum response to a particular signal, generally by varying one or more capacitors and/or inductors. Also **tuning-in.** (2) Carrying out a similar process by electronic or thermal means. See **'cookie-cutter-'**, **'crown-of-thorns-'**, **electronic-**. (3) See **current (or voltage) resonance.**

tuning capacitor. Variable capacitor for tuning purposes, generally consisting of air-spaced vanes; several can be ganged.

tuning coil. See **tuning inductance.**

tuning control. Mechanical means for tuning a resonant circuit, e.g., a knob or slider.

tuning curve. That relating the resonant frequency of a tuned circuit to the setting of the variable element, e.g., a capacitor.

tuning-fork control. Control of the frequency of the wave emitted from a radio transmitter by a tuning-fork oscillator, fundamental of which can be multiplied by valve circuits.

tuning-fork oscillator. See **maintained tuning fork.**

tuning-in. See **tuning.**

tuning indicator. A simple cathode-ray tube in which a metal fin is used as a control electrode. The opening or closing of a fluorescent pattern is the indication of balance. Used as detector for a.c. bridges, and as a tuning indicator in radio receivers. Also **indicator tube.**

tuning inductance. Fixed or variable inductor used for tuning. Also **tuning coil.**

tuning note. Steady musical note radiated from a broadcasting transmitter before a programme, to facilitate the tuning of receivers and for lining-up studio and landline circuits.

tuning-out. Opposite of *tuning-in*, i.e., adjustment for minimum response to a signal, consistent with acceptance of another.

tunnel diode. Junction diode with such a thin depletion layer that electrons bypass the the potential barrier. Negative differential resistance is exhibited. It may be used for low noise amplification up to 1000 MHz. Also **Esaki diode.**

tunnel effect. Piercing of a narrow potential barrier by a current carrier which cannot do so classically, but according to wave mechanics has a small probability of penetrating.

tunnelling. See **potential barrier.**

turbo-generator. The arrangement of a steam turbine coupled to an electric generator for electric power production.

turbulent flow. Fluid flow in which the particle motion at any point varies rapidly in magnitude and direction. This irregular eddying motion is characteristic of fluid motion at high Reynolds numbers.

turnover. (1) Reversing the legs of a balanced transmission circuit. This test is very important in all transmission measurements with balanced circuits, because if the same results are not obtained when any legs of

the balanced system are interchanged, the presence of longitudinal currents is indicated, and no measurement can be accurate unless such currents are eliminated. Also called **poling.** (2) In radiobiology, the rate of renewal of a particular chemical substance in a given tissue. (3) In isotope separation, the total flow of material entering a given stage in a cascade.

turns-ratio. Ratio N of the turns in any pair of windings on a transformer. Power passing between windings changes its impedance level inversely as N^2, because the e.m.f. is proportional to the number of turns.

turnstile antenna. Two normal dipoles, crossed over at their centre, driven with equal currents in quadrature.

turret tuner. As used in television receivers, a drum carrying preset tuned circuits which can be rotated for selection.

tweeter. A loudspeaker used in high-fidelity sound reproduction for the higher frequencies (>5 kHz). See **cross-over frequency, woofer.**

twin crystal. Imperfect growth of crystals, whereby two lattices have a common face, leading to double resonance and unsuitability for radio oscillator or filter use.

twin feeder. Twin-wire transmission line, balanced to earth, on a pole route.

twin-quad. See under **quad.**

twin T-network. One consisting of two T-networks which have both their input terminal pairs and their output terminal pairs connected in parallel.

twin-triode. A combination of two triode valves within the same envelope.

twinning. (1) Intergrowth of crystals of near symmetry, such that (in quartz) the piezo-electric effect is not sufficiently determinate. See **twin crystal.** (2) See **pairing.**

twisted-pair cable. A cable formed by twisting together two thin conductors which are each separately insulated. This arrangement can reduce their intercapacitance.

twister. Plate with slats giving double reflection of a radar wave, one being half-wave retarded, to give a twist in direction of polarization of electric component of wave.

twistor. Unit of a computer store, consisting of a set of copper wires crossing magnetic wires. At the point where pulses in dissimilar wires are coincident, helical magnetization is locally produced. For detection, a single pulse in a copper wire which over-rides reverse magnetic field if present as a stored *bit*.

two-address programme. That which is used in non-sequential computers where each instruction must include the address of two registers—one for the operand and one for the result of the operation.

two-body force. A type of interaction between two particles which is unmodified by the presence of other particles.

two-circuit tuner. One, formerly much used for receivers, in which the antenna circuit is tuned by an inductance and variable capacitor inductively coupled to a closed resonant circuit to which the detector is connected.

two-group theory. Simplified theoretical treatment of neutron diffusion in which only two energy groups are considered, i.e., partly-thermalized neutrons are neglected.

two-phase. Of a.c. systems using two phases, whose voltages are displaced from one another by 90 electrical degrees.

two-position action. Control action of bang-bang type in which control element always takes one of two fixed positions.

two-step relay. Telephone relay which is partially operated by a weak current, and so makes an *x-contact* or *fly contact*, thereby closing a winding in a local circuit. This passes sufficient current for full operation of the remaining contacts of the relay, which *locks*.

two-terminal pair network. See **quadripole**.

two-tone keying. Keying of modulated continuous wave through a circuit which changes the modulation frequency only.

two-way circuit. Bidirectional channel, which operates stably in both directions; particularly used when sources at distances are contributing a montage to a programme.

Tyndall effect. The scattering of light by particles of matter in the path of the light.

type-0, -1 and -2. See **servomechanism types.**

U

U-centre. See colour centre (1).

UT chart. A contour chart giving, for a stated time, the worldwide values of an ionospheric characteristic such as the critical frequency.

Uehling force. Interaction arising from vacuum polarization.

Ulbricht sphere. A sphere used for the integration of the incident light used in a sphere photometer. It depends on the Sumptner principle that if a light source is placed at any point inside a sphere with perfectly diffusing walls, then every portion of the interior is equally illuminated.

ultor. Anode, especially in a cathode-ray tube, which has highest potential with respect to cathode.

ultra-audion oscillator. A Colpitts oscillator in which the resonant circuit is a section of transmission line.

Ultrafax. TN system for very high speed transmission of printed information, which utilizes radio, TV, facsimile and film recording.

ultra-high frequencies. Those between 3×10^8 Hz and 3×10^9 Hz.

ultralinear. Said of a power amplifier when non-linear distortion is reduced to a very low value, e.g., when a pentode valve has negative feedback applied to its screen.

ultramicroscope. An instrument for viewing fog or smoke particles by scattered light against a dark background.

ultramicrowaves. Submillimetre waves having frequencies between 300 GHz and 3000GHz.

ultrasonic. Said of frequencies above the upper limit of the normal range of hearing at or about 20 kHz.

ultrasonic cleaning. The separation of dirt and other foreign particles from a substance by subjecting it to ultrasonic irradiation. Solid bodies such as glass lenses can be polished in this way.

ultrasonic coagulation. Under suitable conditions, coalescence of particles into large aggregates by ultrasonic irradiation.

ultrasonic delay line. A delay line which utilizes the finite time for the propagation of sound in liquids or solids to produce variable time delays. Such systems may also be used for storage in digital computers. Mercury or quartz are used as transmitting media, and nickel wire for magnetostrictive delay lines.

ultrasonic depth finder. Instrument for measuring or displaying the depth of water under a ship, by measuring the time of propagation of a pulse of ultrasonic waves to the sea bed and back.

ultrasonic detector. Electroacoustic transducer for the detection of ultrasonic radiation.

ultrasonic dispersion. High-intensity ultrasonic waves can produce a dispersion of one medium in another, e.g., mercury in water.

ultrasonic flaw detection. Use of ultrasonic waves to detect cracks or flaws in castings, etc.

ultrasonic generator. One for the generation of ultrasonic waves, e.g., quartz crystal, ceramic transducer, supersonic air jet, magnetostrictive vibrator.

ultrasonic grating. The presence of acoustic waves in a medium leads to a periodic spatial variation of density, which produces a corresponding variation of refractive index. Diffraction spectra can be obtained in passing a light beam through such a sound field.

ultrasonic machining. Machining, usually with the lower part of a velocity transformer, which is rigidly attached to a magnetostrictive generator driven by an oscillator. Since the motion is non-rotary, it is possible to drill square section holes.

ultrasonic soldering. That in which specially designed soldering bit replaces the machine tool. Soldering is particularly difficult with aluminium, and the application of ultrasonics is supposed to break up the aluminium oxide layer.

ultrasonic stroboscope. That in which an ultrasonic field is applied to the modulation of a light beam to obtain stroboscopic illumination.

ultrasonic underwater communication. Communication carried out underwater using a modified sonar system.

ultrasonics. General term for the study of ultrasonic sound and vibrations.

ultraudion. Lee de Forest's name for the earliest form of detector circuit employing reaction from three-electrode valve.

ultraviolet microscope. Instrument using ultraviolet light for illuminating an object. Its resolving power is considerably increased, as the resolution varies inversely with the wavelength of the radiation. Still higher resolution is obtained in the *electron microscope*.

ultraviolet radiation. The electromagnetic radiation in the wavelength range 4×10^{-5} cm to 5×10^{-7} cm approximately, covering the gap between X-rays and visible light rays.

ultraviolet spectrometer. An instrument similar to an *optical spectrometer*, but employing non-visual detection and designed for use with ultraviolet radiation.

umbilical cord. A cable which can be cut or disconnected quickly to permit the feeding in of last minute information immediately prior to the launching of a missile.

umbra. Region of complete shadow of an illuminated object.

Umklapp process. A type of collision between phonons, or between phonons and electrons, in which crystal momentum is not conserved. Such processes provide the greater part of the thermal resistance in solid dielectrics.

unbalanced. (1) Bridge circuit in which detector signal is not zero. (2) Pair of conductors in which magnitudes of voltage or current are not symmetrical with reference to earth.

unbalanced circuit. Circuit whose two sides are inherently unlike.

unbalanced network. One arranged for insertion into an unbalanced circuit, the earthy terminal of the input being directly connected to the earthy terminal of the output.

uncertainty principle. Pairs of quantities with the dimensions of action (e.g., time and energy) cannot be determined to an accuracy such that the product of the errors is less than Dirac's constant. Also known as **Heisenberg principle, indeterminacy principle.**

unconditional jump. See **jump** (1).

unconditionally stable. Said of amplification in a system which continues to satisfy the Nyquist criterion when the gain is reduced.

undamped oscillations. Same as **continuous oscillations.**

underbunching. Less than optimum efficiency in a velocity-modulation system.

undercoupling. Said of two electrical circuits or mechanical systems tuned to the same frequency when the resonant frequency of the system remains the same as that of the separate circuits. Also **loose coupling** ; cf. *overcoupling.*

underdamping. Sometimes synonymous with **periodic damping**, but often restricted to cases where a critically-damped response would be preferable.

underflow. In computers, a number below the accepted minimum which cannot be stored (in its existing form) with adequate accuracy by a particular computer register. See **overflow.**

undermodulation. (1) That which is unnecessarily low in relation to the possible level of modulation that can be accommodated on a cinema sound-track. (2) That state of adjustment of a radio-telephone transmitter at which the peaks of speech or music do not produce 100 per cent. modulation, so that carrier power is not used to full advantage.

undershoot. See **overshoot** (2).

undistorted output. That of a valve amplifier free from non-linear distortion. The *maximum undistorted output* is defined as that obtained with a specified degree of non-linear distortion.

undistorted transmission. That of any type of line for which the velocity of propagation and coefficient of attenuation are both independent of frequency.

unexcited. Said of an atom in its ground state. See **excitation** (2).

unfired. Said of any gas-discharge device when in an un-ionized state.

uniaxial. Anisotropic medium which has only one optic axis.

uniaxial crystal. A doubly refracting crystal with only one optic axis.

unidirectional antenna. One in which the radiating or receiving properties are largely concentrated in one direction.

unidirectional current. One which, on average, maintains the same direction in a circuit. It may fluctuate or go negative.

unified field theory. Combined gravitational, electromagnetic and relativity theory postulated on the basis that no laboratory experiment can detect any motion of the laboratory. Best known is probably that of Einstein, but so far no such theory leads to new predictions which could be used as a test of validity.

unified model of nucleus. Modern model of nucleus combining as many valuable features from older models as possible.

unified scale. The scale of atomic and molecular weights which is based on the mass of the ^{12}C isotope of carbon being taken as 12 exactly, and giving the amu as equal to $1·66 \times 10^{-24}$ gm. This scale was adopted in 1960 by the International Unions of both Pure and Applied Physics and Pure and Applied Chemistry, and explains the designation unified scale.

uniform field. One which is described by the same vector at all points.

uniform line. One for which the electric properties of any short element of length are identical.

unilateral impedance. Any electrical or electromechanical device in which power can be transmitted in one direction only, e.g., a thermionic valve or carbon microphone.

unilateral transducer. One for which energy can be transmitted only in the forward direction.

unilateralization. Neutralization of feedback so that transducer or circuit has unilateral response (i.e., there is no response at the input if the signal is applied to the output terminals). While many valve circuits are inherently unilateral equivalent, transistor ones require external neutralization.

unipolar transistor. One with one polarity of carrier.

unipole antenna. Isotropic antenna conceived as radiating uniformly in phase in all directions. Theoretically useful, but not realizable in practice.

unipotential cathode. Same as **indirectly-heated cathode.**

uniselector. An electromechanical selector in automatic telephony which has only rotary motion.

unit. A dimension or quantity which is taken as a standard of measurement.

unit conversion. Expressing a quantity, specified in one system of units, in a different one; See **Unit conversion table** in Appendices.

unit quantity of electricity (*CGS system*). That

which, in a vacuum, when at a distance of 1 cm from an equal charge, repels it with a mechanical force of 1 dyne.

unit sampling. Gating in *pulse code modulation.*

unit step function. One of unit amplitude.

unitor. Multiple plug and socket device, suitable for temporary multichannel connections, especially of interchangeable chassis.

universe. The *galaxy* containing our sun, i.e., all the stars forming the Milky Way.

univibrator. Term for monostable multivibrator circuit. See **flip-flop.**

unloaded antenna. One containing no inductance coils to increase its natural wavelength.

unloaded wavelength. Natural wavelength of an unloaded antenna.

unmarried print. Pair of *reels*, one carrying the optical, the other a photographic or magnetic sound-track, of a film, the reels being synchronized during projection.

unstable system. One capable of undergoing a spontaneous change.

untuned antenna. One not separately tuned to the operating frequency, although effectively tuned by coupling to one or more resonant circuits.

untuned circuit. One not sharply resonant to any particular frequency.

up-time. Period during which a computer is available for intended operation. See **downtime.**

upper sideband. Band of frequencies of modulated signal above carrier frequency. Cf. *lower sideband.*

uptake. In radiobiology, the quantity (or proportion) of an administered substance subsequently to be found in a particular organ or tissue.

uranides. Name for the elements beyond protactinium in the periodic system.

uraninite. See **pitchblende.**

uranium (U). At. no. 92, at. wt. 238·03, sp. gr. 18·68, m.p. 1150°C. A hard grey metal with a number of isotopes. ^{235}U is the only naturally occurring readily fissile isotope, and exists as one part in 140 of natural uranium. ^{238}U is a fertile material with a small fission cross-section for fast neutrons. ^{233}U is a fissile material produced by the neutron irradiation of ^{232}Th. ^{235}U has great value in nuclear reactors and weapons. Metallic uranium exists in three phases known respectively as *alpha*, *beta* and *gamma* uranium (all of which show normal radioactive alpha emission) with transition temperatures of 660°C and 770°C. The large expansion associated with the transition from alpha to beta uranium precludes the operation of nuclear reactors with metallic uranium fuel elements at temperatures above 660°C.

uranium hexafluoride. A gaseous compound of uranium with fluorine which is used in the gaseous diffusion process for separating the uranium isotopes. Very corrosive.

uranium reactor. One in which the fissile fuel material is uranium. Such reactors are classified according to whether they use natural or enriched uranium fuel elements— the proportion of the fissile ^{235}U isotope in the latter having been artificially increased by gaseous diffusion.

utilization factor. A transformer feeding a rectifier system cannot supply the same maximum current as one feeding an a.c. circuit. The ratio of the maximum permitted currents in these two cases is the utilization factor.

V

V-band. Frequency band from 4·6 Hz to 5·6 × 10¹⁰ Hz, used in radar.

V-beam. Scanning by 'fan' beams, one vertical, the other inclined. Interval between reflections depends on target elevation.

VHF rotating talking beacon. An automatic very-high-frequency radio-telephone beacon system, having a radiation pattern which is continuously rotated.

V-particle. Particle exhibiting V-shaped decay track, now identified as neutral meson or hyperon.

V reflector. Two intersecting plane surfaces used as a reflector.

VR tube. See **voltage regulator tube.**

VU meter. Instrument calibrated to read intensity of electro-acoustic signals directly in volume units.

vacancy. Unoccupied site for ion or atom in crystal.

vacuum arc furnace. One in which a small specimen is heated by a high-voltage arc in an inert gas, e.g., argon at low pressure.

vacuum photocell. High-vacuum photoemissive cell in which anode current equals total photoemission currents, so that strict proportionality between current and incident illumination is obtained.

vacuum pump. General term for apparatus which displaces gas against a pressure. Classed as *backing, cryosorption, diffusion, Gaede, ion getter, jet, mercury-vapour, oil, rotary, water jet.*

vacuum switch. One operating in a vacuum to avoid arcing.

vacuum tube. See **valve.**

vacuum tube voltmeter. U.S. for **valve voltmeter.**

valence band. Range of energy levels of electrons which bind atoms of a crystal together.

valence electrons. Those in the outer shell of an *atom*, which, by gaining, losing or sharing such electrons, may combine with other atoms to form *molecules.*

value. U.S. term for **separative work content.**

valve. (1) Simple vacuum device for amplification by an electron stream, covering *bigrid, diode, hexode, pentode, screened-grid, triode,* etc. Also *bottle.* See **tube** (2). (2) Regulating element in fluid-flow line, e.g., servo-controlled pressure valve.

See also:

absorber-	electron-beam-
acorn	four-electrode-
beam-deflection-	ionic-
beam-power-	light-
control-	limiter-
cooled-anode-	line-output-
coplanar-grid-	local-oscillator-
doorknob-	Lodge-
miniature-	separator-
multiple unit	silica-
output-	synchronizing-
reactance-	thermionic-
rectifying-	transmitting-
Round-	water-cooled-.

valve adaptor. Device which enables a valve to be fitted to a socket for which it was not originally designed.

valve admittance. Control-grid admittance under working or specified conditions, including that arising from base sockets, but not from leads thereto.

valve base. Insulating cap cemented to the envelope of a valve and fitted with contacts connected to the electrodes. It enables the valve to be plugged readily into the circuit.

valve characteristic. Graphical relation between voltage and current for specified electrodes, all other potentials being maintained constant.

valve coupling. Use of a valve which permits power to pass in only one direction in a transmission channel.

valve noise (or hiss). That present in telephones connected in the anode circuit of a valve, in the absence of signals applied to the grid, due to shot effect, microphonic action of electrodes, thermal agitation voltage, etc.

valve nomenclature. Description of valves in terms of the number of electrodes therein, e.g., *diode, triode, tetrode, pentode, hexode, heptode, octode,* etc.

valve parameters. Numerical quantities obtained from the characteristic curves and used in circuit analysis. See **amplification factor, differential anode resistance, mutual conductance, transistor parameters.**

valve rustle. Rustling noise accompanying soundfilm reproduction; caused by the clashing of the ribbons in the light valve in the recorder through over-modulation or incorrect adjustment.

valve socket. Arrangement of contact springs or pins into which a valve can be plugged for connection to the rest of the circuit; moulded into low-loss materials.

valve tube. See **kenotron.**

valve voltmeter. Valve used for measuring voltages, rectified output current being dependent on voltage applied to the input. It uses practically no power for operation, and can be calibrated at low frequencies for use at very high frequencies. See **diode voltmeter, slide-back voltmeter.** U.S. term **vacuum tube voltmeter.**

Van Allen belts. Electron belts at 1500–20,000 miles above the surface of the earth, the electrons travelling great distances along the lines of the earth's magnetic field.

Van de Graaff generator. Very high voltage electrostatic machine, using a high-speed belt to accumulate charge in a large Faraday cage, which takes the form of a metal globe. Recent models use Freon or nitrogen gas under high pressure. Employed as voltage source for accelerator tubes, e.g., in neutron sources.

Van de Pol oscillator. Class of pentode relaxation oscillator.

Van der Waals force. Interaction force between neutral atoms (or molecules with no permanent dipoles) due to resultant of instantaneous dipole moments. It is a weak attraction which decreases very rapidly with distance.

vanadium (V). A very hard whitish metal, at. no. 23, at. wt. 50·942, sp. gr. 5·866, m.p. 1715°C. It is used in alloys.

vane wattmeter. Unit for measuring power transmitted in waveguide, depending on mechanical forces induced in a vane.

Vapotron. TN for high-frequency transmitting valve, the anode of which is cooled by boiling water.

var. (1) Navigational system giving mutually perpendicular courses, one displayed visually, the other aurally. (*V*isual-*a*ural *r*ange.) (2) Abb. for *volt-amperes reactive*. See **reactive volt-amperes.**

varactor. Two-electrode semiconductor with non-linear capacitance instantaneously dependent on voltage.

variable area. Pertaining to a sound-track, which is divided laterally into transparent and opaque areas. The line which divides these areas corresponds to the waveform of the recorded signal.

variable density. Pertaining to a sound-track, whose average light transmission varies in proportion to some characteristic of the applied signal.

variable elevation beam antenna. One comprising a large number of dipoles, the maximum lobe of radiation being controlled by adjusting the phases of separate contributions.

variable-focus lens. See zoom lens.

variable mu(tual) conductance valve. See remote cutoff tube.

variable reluctance pickup. One in which the reluctance of a magnetic circuit is varied, the consequent modulation of flux from a permanent magnet generating an e.m.f., as in a transducer or gramophone record reproducer.

variable resistance. See rheostat.

variable-speed scanning. System of scanning employed in a velocity-modulation system, a constant intensity beam going slow for highlights and fast for dark areas.

Variac. TN for auto-transformer in form of a toroid winding on ring laminations, the output voltage being varied by a rotating brush contact on the turns.

variate. Measurement of machined part, as compared with that intended, in machine-tool control.

varimeter, varmeter or **varometer.** Equivalent terms for instrument measuring reactive volt-amps in circuit.

variocoupler. Device comprising two inductance coils whose mutual inductance can be varied for variable inductive coupling between two circuits.

variometer. Variable inductance comprising two coils connected in series and arranged one inside the other, the inner coil being rotated to vary the mutual inductance between them and hence the total inductance.

varistor. Two-electrode semiconductor with a non-linear resistance dependent on instantaneous voltage. Cf. *varactor*.

Varley loop test. A bridge test circuit used for detecting and locating faults in signal lines.

vector addition. Compounding of two vector quantities according to parallelogram law.

vector algebra. Manipulation of symbols representing vector quantities according to laws of addition, subtraction, multiplication and division, which these quantities obey.

vector potential. Vector quantity in electromagnetic field theory, whose component along any axis at any point is equal to $\Sigma i . dl/r$, where $i . dl$ is a current element parallel to the axis at a distance r from a point, the summation extending throughout all space.

vector product. That of two vector quantities when the result is a *vector quantity*, e.g., the product of E and H is a Poynting vector, which represents an energy flux. Known as the *cross* (or *outer*) *product* and denoted algebraically by a full multiplication sign between the vectors (or by a square bracket enclosing them). It represents a vector with amplitude equal to the product of the amplitudes of the component vectors, and the sine of the angle between them; and direction perpendicular to the plane containing the component vectors. See **scalar product.**

vector quantity. That which has direction as well as numerical value. A symbol representing a vector quantity is distinguished by being printed in a heavy type, or by having a bar printed above or below it. See **scalar quantity.**

vector ratio. That between two alternating quantities, in which both relative amplitudes and phases are expressed as vectors.

vectorscope. Instrument which displays phase and amplitude of an applied signal, e.g., of chrominance signal in a colour television system.

vee antenna. Line radiator folded in a V in the horizontal plane.

Velocitron. TN for U.S. external cavity type of reflex klystron.

velocity. In a wave, distance travelled by a given phase of a wave divided by time taken. It is a vector quantity (cf. *speed*, which is scalar) so that it has a magnitude and a direction expressed relative to some frame of reference.

velocity microphone. See **pressure-gradient microphone.**

velocity-modulated oscillator. One in which an electron beam is velocity-modulated (*bunched*) by passing through a toroidal cavity resonator (*rhumbatron*), the energy exciting a further cavity (*collecting*) and feeding back into the first.

velocity modulation. (1) System in which gradation of light and shade in the reproduced television picture is effected by variation of velocity of the scanning spot across the phosphor, the intensity of illumination of the spot being constant. (2) Modulation in a klystron in which the velocities of the electrons, and hence their *bunching*, is related to radio signals to be transmitted.

velocity of energy. Energy flux per unit area divided by the energy density.

velocity of light. $2·997923 \times 10^{10}$ cm/s *in vacuo*. This is a natural constant equal to the maximum speed with which energy can be propagated and normally represented by *c*. In a dispersive medium the phase and group velocities are not the same, both normally being less than *c*. In waveguides and ionized media the phase velocity may be greater—the geometric mean of the phase and group velocities then being equal to *c*.

velocity of sound. In dry air at s.t.p. this is 331·4 m/s (750 m.p.h.). In fresh water 1410 m/s and in sea water 1540 m/s. These values are used for sonar ranging, but do not apply to explosive shock waves. They must be corrected for variations of temperature, humidity, etc.

velocity resonance. See **phase resonance.**

velocity sorting. Any procedure which selects electrons according to their velocities.

velodyne. Tachogenerator in which rotational speed of output shaft is proportional to applied voltage through feedback.

venturi meter. One in which flow rate is measured in terms of pressure drop across a venturi (or tapered throat) in a pipe.

Verdet's constant. The rotation (of the plane of polarization) per cm per gauss in the Faraday effect, which relates to the rotation of the plane of polarization of light by transparent substances in a magnetic field.

verifier. Apparatus which verifies the accuracy of punched cards or tapes and indicates errors, or which compares two punched cards or tapes.

veristron. Device based on electron spin, analogous to *maser*.

vernier potentiometer. Precision pattern based on the *Kelvin-Varley slide* (q.v.). Balance can be attained purely by the operation of switches, so that the possibility of wear associated with sliding contacts is avoided.

Vernon-Harcourt lamp. See **pentane lamp.**

versine. A trigonometrical function of an angle, required for the solution of spherical triangles. It is given by

$$\text{vers(ine)} \; \theta = 1 - \cos(\text{ine}) \; \theta.$$

vertical effect. Early name for **antenna effect** (1).

vertical frequency. That of frame-scanning voltage in a normal television system in which the line scanning is horizontal.

vertical polarization. State of electromagnetic wave when electric component lies in the vertical plane and magnetic component in the horizontal plane, as in the wave emitted from a vertical aerial.

vertical recording. Same as **hill-and-dale recording.**

vertical scanning. That in which individual lines are vertical, not, as normal, horizontal.

very high frequencies. Those between 30 and 300 MHz.

very low frequencies. Those between 10 and 30 kHz.

vestigial sideband. Reduction of total bandwidth associated with normal amplitude modulation by markedly reducing one sideband, especially for television transmission.

vestigial-sideband transmitter. Station which radiates either a carrier with only one full sideband (for bandwidth reduction) ; or the one full sideband with no carrier (for maximum economy of radiated power), i.e., **single sideband suppressed carrier.**

vibrating capacitor. Basically an electrometer in which the potential on the electrode of a capacitor is varied by mechanical oscillation, so that the steady applied potential is converted to an alternating potential, which can be more easily amplified. Also **oscillating capacitor.** See **Vibron.**

vibrating-reed amplifier. D.c. amplifier in which vibrating-reed electrometer converts signal to a.c. This is amplified and fed to phase-sensitive detector to convert back to d.c. output. Also known as **chopper amplifier, contact-modulated amplifier.**

vibrating-reed electrometer. One which, through a driven reed varying a capacitance, converts a steady voltage into an alternating voltage, which is more readily amplified. Also **dynamic-capacity electrometer, vibrating-capacitor electrometer.** See **Vibron.**

vibration galvanometer. Moving-coil taut-suspension galvanometer with natural frequency of vibration of coil tunable, usually over range 40 Hz–1000 Hz. Small a.c. currents at the resonant frequency excite a large response, hence these instruments form sensitive detectors for circuits such as a.c. bridges or potentiometers.

vibration pickup. One which uses some form of microphone or transducer (e.g., crystal, capacitance, electromagnetic) to transform the oscillatory motion of a surface, e.g., of machinery, into an electrical voltage or current.

vibration-rotation spectrum. The infrared end of the electromagnetic spectrum which arises from vibrational and rotational transitions within a molecule.

vibrator. Electromechanical interrupter used to supply intermittent d.c. current to primary of transformer when voltage conversion is

K

required. Formerly widely used for radio equipment in vehicles. The *synchronous vibrator* also uses contacts operating simultaneously to rectify the a.c. voltage from the transformer secondary.

vibrometer. An instrument used for the measurement of displacement, velocity, or acceleration of a vibrating body.

Vibron. RTM for a vibrating-reed capacitor, which can modulate an applied potential and so facilitate its amplification and measurement.

vibrotron. A special form of triode valve in which the anode can be vibrated by a force external to the envelope.

video. Relates to the bandwidth (~megahertz) and spectrum position of the signal arising from TV scanning.

video amplifier. Wideband amplifier which, in a TV system, passes the picture signal. Bandwidth extends from zero frequency to one which maintains the required definition.

video carrier. The carrier wave which is modulated with a video signal.

video pulse. Colloquialism for fast rise-time pulse, i.e., one covering a wide frequency spectrum.

video signal. That part of a TV signal which conveys all the information (intensity, colour, and synchronization) required for establishing the visual image in monochrome or colour TV.

video stretching. A method of increasing the duration time of a video pulse.

video tape recording. Registering television signals on a high-speed magnetic tape (transverse or longitudinal) for subsequent transmission. See Ampex.

Vidicon. RTM for a camera tube operating on the photoconducting principle.

Villari effect. Temporary change in magnetization, arising from longitudinal stretching.

virgin neutrons. Neutrons which have not yet experienced a collision, and therefore retain their energy at birth.

virgin state. Same as **neutral state.**

virginium. See francium.

virtual cathode. Region in a space charge where a potential minimum gives the effect of a source of electrons.

virtual height. The apparent height of an ionized layer as deduced from the time interval between the transmitted signal and the resulting ionospheric echo at normal incidence.

virtual quantum. In higher order perturbation theory, a matrix element which connects an initial state with a final state involves intermediate states in which energy is not conserved. A photon or quantum in one of these states is designated a virtual quantum. This concept enables the coulomb energy between two electrons to be regarded as arising from the emission of virtual quanta by one of the electrons and their absorption by the other.

virtual state (or level). (1) Nuclear state which can only be deduced indirectly, or that

appears as an intermediate stage when the transition between two other states is computed. See **resonance.** (2) Sometimes the so-called unbound singlet state of the deuteron. This state was 'invented' to calculate a theoretical binding energy of the singlet neutron-proton system that would give the same scattering length as that observed in neutron-proton scattering.

viscosity. (1) Damping forces opposing relative motion in fluids. (2) Colloquially, by analogy, delay or creep in the magnetization, following changes in the magnetizing force applied to a ferromagnetic material.

viscous damping. Opposing force or torque proportional to velocity, e.g., resulting from viscosity of oil or from eddy currents.

visible radiation. Electromagnetic radiation which falls within wavelength range of 7800Å to 3800Å, over which the normal eye is sensitive.

visible speech. Display of oscillogram patterns corresponding to characteristic speech sounds—used as an aid to the speech training of the totally deaf.

visiogenic. Artistically suitable for visual reproduction by television.

visual range. (1) Optical range of transmitter. (2) Observable range of ionizing particle in bubble chamber, cloud chamber or photographic emulsion.

visual record. One which can be read directly without decoding.

vitreosil. Translucent form of silica with a low expansion coefficient, used for apparatus required to withstand sudden temperature changes.

vocabulary. Set of *words* for data processing.

vocoder. System for synthetic speech using recorded speech elements.

vodas. Device used for the suppression of echoes in transoceanic radio telephone. (*V*oice-*o*perated *d*evice *a*nti-*s*ing.)

voder. System for producing synthetic speech through keyboard control of electronic oscillators. (*V*oice *o*peration *dem*onstrator.)

vogad. A device used in telephone systems, which is voice-operated, and gives an approximately constant volume output for a wide range of input signals. (*V*oice-*o*perated *g*ain-*a*djusting *d*evice.)

voice coil. The coil attached to the cone of a loudspeaker. The coil currents react with the magnetic field to drive the cone. Also used in microphones to generate signal.

voice filter. Device which deliberately distorts speech for specific purpose, e.g., telephonic imitation.

voice frequency. One in the approximate range 200 Hz–3500 Hz (that required for the normal human voice). Also **speech frequency.**

voice-frequency telegraphy. See **carrier telegraphy.**

Voigt effect. Double refraction of electromagnetic waves passing through a vapour, when an external magnetic field is applied.

Voigt notation. The labelling of elastic constants and elastic moduli in which numbers

1, 2, 3, 4, 5 and 6 replace the pairs of letters *xx*, *yy*, *zz*, *yz*, *zx*, and *xy* respectively.

volatile. Said of stored data which is lost if power supply should fail.

Volscan. Analogue computer which can handle all landing approach problems.

volt. MKSA unit of p.d. or e.m.f., such that the p.d. across a conductor is 1 volt when 1 amp in it dissipates 1 W of power. This is 1 J/s, or 10^7 erg/s, a mechanical unit.

volt-ampere-hour. MKSA unit of apparent power, equivalent to the watt-hour. See **power factor.**

volt-amperes. Product of actual voltage (in volts) and actual current (in amperes), both rms, in a circuit. See **active-, reactive-.**

volt-box. Same as **potential divider.** Also called **volt ratio-box.**

volt-ohm milliammeter. An electrical d.c. test instrument measuring voltage, resistance and current. Usually a.c. volts can also be measured.

Volta effect. Potential difference which results when two dissimilar metals are brought into contact; basis of voltaic cells and corrosion.

voltage amplification. See **voltage gain.**

voltage amplification factor. See **amplification factor.**

voltage divider. See **potential divider.**

voltage doubler. Power-supply circuit in which both half-cycles of a.c. supply are rectified, and the resulting d.c. voltages are added in series.

voltage drop. (1) Diminution of potential along a conductor, or over an apparatus, through which a current is passing. (2) Possible diminution of voltage between two terminals when current is taken from them.

voltage efficiency (of secondary cell). The ratio of average terminal p.d. during discharge to that during charge.

voltage-fed antenna. One which is fed with power from a line at a point of high impedance, where, through resonance, there is a voltage loop in the standing-wave system.

voltage feedback. See **current feedback.**

voltage gain. Ratio of output to input voltages for amplifier, attenuator, etc. If the impedances are the same this is often expressed in dB.

voltage generator. Concept of ideal signal source, often with no internal impedance.

voltage jump. Abrupt discontinuity in voltage drop across discharge tube, normally associated with a marked change in the geometry of the discharge.

voltage level. *P-p* value at any point in a network expressed relative to a specified reference level. When this is 1 volt the symbol *dBv* is used.

voltage multiplier. Circuit for obtaining high d.c. potential from low voltage a.c. supply, effective only when load current is small, e.g., for anode supply to CRT. A ladder of half-wave rectifiers charges successive capacitors connected in series on alternate half-cycles. See **Westeht.**

voltage reference tube. A glow-discharge tube designed to operate with anode-cathode voltage as nearly as possibly constant, regardless of the anode current, and hence suitable for use as a standard of p.d.

voltage regulation. The percentage variation in the output voltage of a power supply for either a specified variation in supply voltage or a specified change of load current.

voltage regulator tube. One in which, over a practical range of current, voltage between electrodes in a glow discharge remains substantially constant; also called **reference tube, VR tube.** See **Stabilovolt.**

voltage resonance. See **current resonance.**

voltage ripple. The peak-to-peak a.c. component of a nominally d.c. supply voltage.

voltage stabilizer or **voltage stabilizing tube.** See **voltage regulator tube.**

voltage standing-wave ratio. Ratio between a maximum and a minimum in a standing wave, particularly on a transmission line or in a waveguide, arising from inexact impedance terminations.

voltaic cell. Any device with electrolyte (ionized chemical compound in water) and two differing electrodes, which establish a difference of potential.

voltaic pile. Earliest form of battery, invented by Volta of Como (1745-1827). Each cell consisted of a thin sheet of zinc and of copper separated by cloth moistened with dilute sulphuric acid. A large number of cells of this type connected in series form a compact high-voltage source. Such a battery, known as the *Zamboni pile*, has been used with infrared sighting systems.

voltameter. Electrolytic cell employed for electrolysis ; cf. *coulometer.*

voltmeter. An instrument for measuring electric potential differences. See :

cathode-ray-	moving-coil-
corona-	moving-iron-
corrosion-	peak-
digital-	rectifier-
diode-	slide-back-
electronic-	thermocouple meter
electrostatic-	valve-.

volume. In electronics, the magnitude of any electric AF signal as measured on a standard volume indicator and expressed in VU. More generally, the loudness of any sounds produced by electronic (or other) means.

volume compression and expansion. Automatic compression of the volume range in any transmission, particularly in speech for radio-telephone transmission, so that the envelope of the waveform is transmitted at a higher average level with reference to interfering noise levels. After expansion at receiving end, resulting transmission is freer from noise. See **compander.**

volume compressor. In communication systems depending on amplitude modulation, intelligence transmitted as a modulation is limited to 100 per cent. So that this is not exceeded with very loud sounds in the modulation, original transmission has to be compressed into relatively small dynamic range to maintain high signal/noise ratio.

volume control. The gain control on an audio amplifier.

volume indicator. Voltage-measuring device which, when placed across a communication channel carrying current for later conversion into sounds, gives a relative estimation of the apparent loudness of these reproduced sounds. The original type of volume indicator used a bridging transformer and an anode-bend rectifying thermionic valve, but indicators having a metal-oxide rectifier are now used. The indicating meter and associated circuit have in each case a time constant which results in the indication being an integration over a short period, e.g., 0·2 sec. of the varying voltage waveform applied to it. In these volume indicators frequency-weighting networks are not incorporated for programmes.

volume ionization. The mean ionization density in any given volume without reference to the specific ionization of the ionizing particles.

volume lifetime. That of current carriers in the bulk of a semiconductor ; cf. *surface lifetime*.

volume limiter. Circuit which automatically restricts volume of signal to specified maximum level.

volume range. Ratio, in dB, of maximum and minimum volumes at which a communication channel can satisfactorily operate.

volume sterilization. Complete trans-irradiation of material, e.g., foodstuffs, with penetrating radiation, e.g., high-energy gamma-rays from large radioactive sources.

volume unit. One used to measure variations of modulation in a communication circuit, e.g., telephone or broadcasting. The unit is the decibel expressed relative to a reference level of 1 mW in 600 ohms, and standard volume indicators are calibrated in these units.

vor. Navigational ranging system in which any bearing can be obtained from the phase difference between an audio reference phase and one which is a function of azimuth, termed the variable phase. (*V*ery-high-frequency *o*mnidirectional *r*adio range.)

W

Wagner earth. A pair of impedances, with their common point earthed, connected across an a.c. bridge network in order to neutralize the effect of stray capacitances. This is done by simultaneously balancing the normal bridge and that formed by the Wagner earth with the ratio arms.

Walden empirical rule. $\eta_0 \Lambda_0 =$ constant, where η_0 is the viscosity of solvent and Λ_0 is the equivalent conductance of electrolyte at infinite dilution. The relationship has only a limited application.

walkie-talkie. A sender and receiver radio communication set which is carried by a single person, and may be operated while moving.

wall effect. (1) The contribution of electrons liberated in the walls of an ionization chamber to the recorded current. (2) The reduction in the count rate recorded with a Geiger tube due to ionizing particles not having the energy to penetrate the walls of the tube. (3) Reduction of free-fall velocity of a body in a tube containing liquid, due to nearness of wall.

wall energy. The energy per unit area stored in the domain wall bounding two oppositely magnetized regions of a ferromagnetic material.

wall-less ionization chamber. One in which (by the use of a guard ring) the collecting volume is defined by the applied field and the contribution of knock-on particles to the ionization is avoided.

Wallman amplifier. See **cascode amplifier**.

wamoscope. Wideband display tube incorporating a travelling-wave tube. (*Wave*-mo*d*ulated oscillo*scope*.)

warble tone. Single-frequency current for testing microphones and loudspeakers, to mitigate errors arising from standing waves in the acoustic space. By cyclic variation of frequency, the standing-wave pattern is constantly changing and errors balanced out.

warbling carrier system. Method of increasing degree of secrecy obtainable with a radio-telephone system using inversion. It consists of rapid variation of the carrier frequency of the transmitter over a range of a few hundred Hz, so that the inverted speech transmitted through space cannot be readily re-inverted by heterodyning.

waste. (1) Depleted material rejected by an isotope separation plant. (2) Unwanted radioactive material for disposal.

water-cooled valve. Large thermionic vacuum tube in which the heat generated by the electronic bombardment of the anode is carried away by water circulating around or through it. In the former the anode is made an integral part of the envelope. Cf. *cooled-anode valve.*

water jet pump. See **jet pump**.

water monitor. One for measuring the level of radioactivity in a water supply. Similar to, but much more sensitive than, an effluent monitor.

watt. MKSA unit of electric power, equal to 1 J/s or 10^7 erg/s. Thus 1 horsepower (HP) equals 746 W.

watt-hour. MKSA unit of electrical energy, being the work done by 1 W acting for 1 hour, and thus equal to 3600 J or $3 \cdot 6 \times 10^{10}$ erg.

watt-hour meter. Integrating meter for measurement of total electric energy consumed in a circuit. The conventional domestic electricity meter is of this type.

wattfull loss. See **ohmic loss**.

wattless component. See **reactive component**.

wattmeter. An instrument for the measurement of the active power (active volt-amperes) in a circuit. See **electrodynamic-, electronic-, electrostatic-, thermocouple-, vane-**.

wave. General term for a time-varying quantity which is also a function of position. A wave proceeding from a radiator will carry energy into the medium. Mechanical waves will involve displacement of the actual medium, while in electromagnetic waves, it is the change in the electric and magnetic field intensities from their equilibrium values which represents the wave disturbance. In the case of electromagnetic waves radiated from a relatively low antenna, one component is the *ground wave*. A component dependent on the nature of the surface is a *surface wave*. See :

alpha-	modulated
beta-	continuous-
carrier-	plane-
continuous-	quasi-optical-
cylindrical-	reflected-
damped-	sawtooth-
decametric-	shock-
decimetric-	side-
dwarf-	sine-
electromagnetic-	space-
floating-carrier-	spacing-
gravity-	square-
Hertzian-	standing-
hybrid electro-	stationary-
magnetic-	TE-
intermediate-	TM-
longitudinal-	transient-
marking-	transverse-
modulated carrier	travelling-.

wave analyser. One which separates frequency components in a continuously repeated waveform, usually by a quartz-crystal filter or gate. In the heterodyne type, the heterodyne principle is applied to measure the various frequency components of a wave.

wave angle. Either angle of elevation or azimuth, of arrival or departure of a radio wave with reference to the axis of an antenna array. Applied similarly to sonar.

wave antenna. Directional receiving antenna comprising a long wire running horizontally to the direction of arrival of the incoming waves, at a small distance above the ground. The receiver is connected to one end, and the other end is connected to earth through a terminating resistance. The induced voltage depends on the forward tilt of the ground wave. Also called **Beverage antenna.**

wave clutter. Spurious marine radar echoes from waves on the sea.

wave duct. Atmospheric or other duct bounded by discontinuities which enable it to support waveguide-type propagation.

wave filter. Combination of sections and half-sections, each comprising inductances and capacitances, so proportioned to give frequency cutoff effects because of sudden changes of reactance. See **filter.**

wave function. Mathematical equation representing the space and time variations in amplitude for a wave system. The term is used particularly in connection with the Schrödinger equation for particle waves.

wave impedance. Complex ratio of transverse electric field to transverse magnetic field at a location in a waveguide. For an acoustic wave, it is pressure/particle velocity.

wave interference. Relatively or completely stationary patterns of amplitude variation over a region in which waves from the same source (or two different coherent sources) arrive by different paths of propagation; *constructive interference* arises when the two waves are in phase and their amplitudes add; *destructive interference* arises when they are out of phase and their amplitudes partly or totally neutralize each other.

wave-interference error. See **heiligtag effect.**

wave mechanics. The modern form of the quantum theory in which all events on an atomic or nuclear scale are explained in terms of interaction between the associated wave systems as expressed by the Schrödinger equation. See also **statistical mechanics.**

wave number. The reciprocal of *wavelength*, i.e., the number of waves in unit distance.

wave parameter. See **wavelength constant.**

wave scattering. See **scattering.**

wave theory. Macroscopic explanation of diffraction, interference and optical phenomena as an electromagnetic wave, predicted by Maxwell and verified by Hertz for radio waves.

wave tilt. The angle between the normal to the ground and the electric vector, in a ground wave polarized in the plane of propagation.

wave train. Group of waves of limited duration, such as those resulting from a single spark discharge in an oscillatory circuit.

wave trap. A circuit tuned to parallel resonance connected in series with the signal source to reject an unwanted signal, e.g., between a radio receiver and the aerial. See also **wave filter.**

wave velocity. See **phase velocity.**

waveband. Range of wavelengths occupied by transmissions of a particular type, e.g., the *medium waveband* (from 200 to 550 metres) used for broadcasting.

waveform, waveshape. Graph showing variation of amplitude of electrical signal, or other wave, against time. If repeating periodically, the waveform is referred to as *steady-state*; and, if varying without cyclic repetition, as *transient.*

wavefront. (1) Leading portion of an advancing wave, e.g., a voltage surge propagated along a transmission line. (2) Imaginary surface joining points of constant phase in a wave propagated through a medium. The propagation of waves may conveniently be considered in terms of the advancing wavefront, which is often of simple shape, such as a plane, sphere or cylinder.

waveguide. (1) Hollow metal conductor within which very-high-frequency energy can be transmitted efficiently in one of a number of modes of electromagnetic oscillation. Dielectric guides, consisting of rods or slabs of dielectric, operate similarly, but normally have higher losses. (2) Also exists around earth for electromagnetic waves, the ionized layers (ionosphere) acting as the upper reflecting 'surface'. (3) Also exists for acoustic waves, the temperature reversal region in the atmosphere corresponding to the electromagnetic ionized layer. Similar conditions obtain under sea. The acoustic waveguide corresponding to (1) has very rigid walls.

waveguide apparatus. See **converter, hybrid junction, iris (1), pad, plunger, resonator, termination, transformer, tuner.**

waveguide attenuator. Conducting film placed transversely to the axis of the guide.

waveguide bend. One in a length of rigid or flexible waveguide. See **E-bend, H-bend.**

waveguide-choke flange. Coupling flange between waveguide sections which offers zero impedance to signal without requiring metallic continuity.

waveguide coupler. Arrangement for transferring part of the signal energy from one waveguide into a second crossing or branching off from the first, e.g., in a *directional coupler* the direction of the flow of the energy transferred to the second guide reverses when the direction of propagation in the first guide is reversed.

waveguide filter. One having distributed properties, giving frequency discrimination in a waveguide wherein it is inserted.

waveguide impedance. Ratio Z derived from

$$W = V^2/Z, \text{ or } I^2Z,$$

where W is power, V a voltage and I a current, the last two being defined in relation to type of wave and shape of waveguide.

waveguide junction. Unit joining three or more waveguide branches, e.g., hybrid junction.

waveguide modes (for electromagnetic waves). (1) *Rectangular section:* TE$_{mn}$—transverse electric mode in which m, n are the number of half-period variations in the electric field respectively parallel to the narrow and broad sides of the guide. TM$_{mn}$—transverse magnetic mode ; substitute magnetic for electric in definition of TE$_{mn}$. (2) *Circular section :* TE$_{mn}$—the mode which has m diametral planes in which longitudinal component of magnetic field (H) is zero, and n cylindrical surfaces at which tangential component of the electric field is zero. T$_{mn}$ —this mode has m diametral planes and n cylindrical surfaces (of non-zero radius) at which the longitudinal component of the electric field is zero.

waveguide stub. One consisting of a piston (*waveguide plunger*) which moves in a short length of pipe connected to the waveguide; used for tuning or detuning the guide.

waveguide switch. One which switches power from waveguide A to B or C, with considerable loss between A and C or B, and between B and C.

waveguide transformer. Unit placed between waveguide sections of different dimensions for impedance matching.

wavelength. (1) Distance between two similar and successive points on an alternating wave, e.g., between successive maxima or minima; equal to the velocity of propagation divided by the frequency of the alternations when travelling along wires. (2) Distance, measured radially from the source, between two successive points in free space at which an EM or acoustic wave has the same phase; for an EM wave, it is equal, in metres, to $300,000 \div$ frequency (kHz). (3) That associated with electrons in motion when considered as a wave train. It is

$$\lambda = h/p,$$

h being Planck's constant and p the momentum of the electron. Similar for neutrons and other particles.
See :

Compton-	dominant-
critical-	fundamental-
de Broglie-	unloaded-.

wavelength constant. The imaginary part of the propagation coefficient.

wavelength minimum. For an X-ray spectrum, given by :

$$\lambda_{min} = \frac{hc}{V_{max}e},$$

where V_{max} =highest voltage applied to tube
λ_{min} =wavelength
h =Planck's constant

e =electronic charge
c =light velocity.

wavemeter. Instrument for measuring or indicating frequency or wavelength of electromagnetic radiation, either for laboratory use or for monitoring in a radio transmitting station. See **absolute-, absorption-, cymometer, frequency meter, heterodyne-.**

wavicle. Quantum mechanical entity which shows both particle and wave properties— related by principle of complementarity.

wax master. Original recording cut on wax disk.

weak interaction. One due to forces of the order of 10^{-13} times as great as in strong interactions. It appears to be due to some unknown coupling between fermions (baryons and leptons) and may be due to an as yet undetected boson (the intermediate charged vector boson) in the same way that strong interactions are believed to be due to meson coupling. Decay produced by weak interactions is not subject to conservation of strangeness, parity or hypercharge.

weber. MKSA unit of magnetic flux. An e.m.f. of 1 volt is induced in a circuit through which the flux is changing at a rate of 1 Wb/s. The weber is equal to 10^8 maxwell. It is also used as the MKSA unit of magnetic pole strength.

Weber-Fechner law. That the physiological sensation produced by a stimulus is proportional to the logarithm of the stimulus.

wedge. (1) Total attenuator, in the form of a wedge of absorbing material, for terminating a waveguide. (2) Insertion of various lossy materials, put into a section of waveguide, to add fixed or variable attenuation in the circuit.

weighting factor. See **statistical weight** (2).

weighting network. One designed to produce unequal attenuation for different frequency components of a signal, thereby weighting these differently in the final output.

Weiss theory. Early theory of ferromagnetism based on the concept of independent molecular magnets.

Weissenberg method. A technique of X-ray analysis in which the crystal and photoplate are rotated in a beam of X-rays while the plate is moved parallel to the axis of rotation.

welding. The process of joining metal by applying pressure or heat (or both). In resistance welding the metals to be joined are heated to a plastic or fused condition by means of a localized electric current.

well. Pictorial term used to depict variation of potential energy of nucleon with distance from nucleus.

well counter. One used for measurements of radioactive fluids placed in a cylindrical container surrounded by the detecting element (scintillation crystal or sensitive volume of special Geiger tube).

Wenner winding. Form of winding used in WW resistors to construct standard resistances of low residual reactance for use at relatively high frequencies.

Westeht. TN for copper-oxide rectifier voltage multiplier unit used for EHT supply to CRT.

Weston standard cadmium cell. Practical portable standard of e.m.f. in which the cathode is $12\frac{1}{2}\%$ Cd and $87\frac{1}{2}\%$ Hg by weight, with anode of amalgamated Pt or highly purified Hg. Saturated sol. aq. Cd_2SO_4 as electrolyte. E.m.f. of cell at 20°C is 1·018636 V, temperature coefficient only 0·00004 V/°C. Used for calibrating potentiometers and hence all other voltage-measuring devices. Cells with unsaturated solutions have a lower temperature coefficient of e.m.f., but do not give an equally high absolute standard of reproducibility. Also **standard cell.**

wet cell. Primary cell with liquid rather than paste electrolyte.

wet electrolytic capacitor. One in which the negative electrode is a solution of a salt, e.g., aluminium borate, which is suitable for maintaining the aluminium oxide film without spurious corrosion.

Wheatstone bridge. An apparatus for measuring electrical resistance by the null method, comprising two parallel resistance branches, each branch consisting of two resistances in series. Prototype of most other bridge circuits.

Whirlwind. High-speed computer at MIT.

whisker. See cat's whisker.

whisker resistance. The resistance of a whisker element in a semiconductor.

whistlers. Atmospheric electric noises which produce relatively musical notes in a communication system.

white noise. Noise currents or acoustic waves which contain a wide range of adjacent random frequencies of uniform energy; used for audio testing. Obtained from amplification of noise generated by gas discharge or a saturated diode.

white radiation. Strictly, radiation of a continuous spectrum of visible light by a hot body. It is seen by the eye as white. Applied by extension to continuous spectra in other wavelength bands, as in *white noise, white spectrum.*

white reference level. Signal modulation level corresponding to maximum brightness (white) in monochrome facsimile transmission.

white spectrum. See X-ray spectrum.

wideband amplifier. One which amplifies over a wide range of frequencies, with corresponding low gain.

width. The spread or uncertainty in a specified energy level, which arises as a result of Heisenberg's indeterminacy principle, and is proportional to the instability of the state concerned.

Wiedemann effect. Tendency to twist in a rod carrying a current when subject to a magnetizing field.

Wiedemann store. One employing the Wiedemann effect in magnetostrictive elements.

Wien bridge. A four-arm a.c. bridge circuit used for measurement of capacitance and power factor.

Wien-bridge oscillator. One in which positive feedback is obtained from a Wien bridge, the variable frequency being determined by a resistance in an arm of the bridge.

Wien's laws (for radiation from a black body). (1) *Displacement law* : $\lambda_m T =$ constant (σ), $\sigma = 0\cdot290$ cm deg. (2) *Emissive power* (E_λ) : within the maximum intensity wavelength interval $d\lambda$, $E_\lambda = CT^5 d\lambda$, where $C = 1\cdot288 \times 10^{-4}$ erg/cm³ sec. deg⁵, T being absolute temperature. (3) *Emissive power (dE)*: in interval $d\lambda$, $dE = A\lambda^{-5}$ exp. $(-B/\lambda T)d\lambda$. λ_m is wavelength at E max.

Wigner effect. The process by which an atom is knocked out of its position in a crystal lattice by direct nuclear impact, and the displaced atom may come to rest at an interstitial position or at a lattice edge. A change of chemical or physical properties may result, e.g., graphite bombarded by neutrons changes in dimensions, and the shape of the crystal lattice is altered. Also **discomposition effect.**

Wigner energy. The energy stored within a crystalline substance resulting from the Wigner effect.

Wigner force. Ordinary (non-exchange) short-range force between nucleons.

Wigner theorem. A deduction from the quantum theory which states that in a collision of the second kind the angular momentum of the electron spin is conserved.

Williams tube. TN for CRT, used as electrostatic store, e.g., in computing.

Wilson chamber. Cloud chamber of expansion type.

Wilson effect. Production of electric polarization when dielectric material is moved through region of magnetic field.

Wilson electroscope. Single-leaf electroscope, with the leaf hanging vertically in its most sensitive position.

Wimshurst machine. Early type of electrostatic induction generator.

winding. See coil (3) and also:

bias-	primary-
bifilar-	secondary-
gate-	tertiary-
non-inductive	Wenner-.
resistor	

window. (1) Strips of metallic foil, of dimensions calculated to give radar reflections and hence confuse locations derived therefrom. Also **chaff, rope.** (2) Thin portion of wall of radiation counter through which low-energy particles can penetrate. (3) See **pulse height selector.** (4) Conducting diaphragm(s) inserted into waveguide which act inductively or capacitively according to how they are positioned.

wipe-out. Interference of such intensity as to render impossible the reception of desired signals.

wipe-out area. Area surrounding a radio transmitter where wipe-out occurs.

wipe test. See **smear test.**

wiper. See **brush.**

wire broadcasting. Distribution of broadcast programmes to multiple receivers on direct wired circuits ; also **radio relay.**

wire recorder. Early type of magnetic recorder with recording medium in form of iron wire.

wired-in. Said of components connected in circuits, particularly subminiature valves and semiconductor devices, which are too small to be plugged safely into a holder.

wired wireless. See **carrier telegraphy.**

wireless. Obsolete name for *radio.*

wiresonde. Meteorological data-transmission system operated by cable from captive balloon.

wirewound resistor. One with metallic wire element.

wiring diagram. Form of circuit diagram modified to indicate the physical layout of the connections.

wobbulator. Colloquialism for a signal generator, whose frequency is automatically varied periodically over a definite range; used to test frequency response of systems.

wolfram. See **tungsten.**

Wolf's equation. $R = k(10G + N)$, where
R = relative sunspot number
k = a constant, depending on the type of telescope used
G = observed number of sunspot groups and
N = observed total number of sunspots, either singly or in groups.
Used in forecasting *maximum usable frequency.*

woofer. Large loudspeaker used to reproduce lower part of audio-frequency spectrum only; see **cross-over frequency, tweeter.**

word. Set of digits, having definite meaning, often of a given number, called (*machine*) *word length.*

word time. Time required to read and rewrite one word in a computer. (The time quoted is normally that for the immediate access store.)

work. Manifestation and interchange of energy. Defined by the advancement of the point of application of a force. Unit is the *erg,* whereby a force of one dyne advances its point of application 1 cm.

work coil. Same as **heating inductor.**

work function. The minimum energy required by an electron (a few electron-volts) for it to pass through a potential barrier. Important for electrode metals in valves. Also **electron affinity.** See **photoelectric-, thermionic-.**

workhead transformer. One associated with the workpiece in induction heating when the generator is at a distance and feeds power through a cable.

working memory. That part of internal storage of a computer directly reserved for operations currently being conducted.

working storage. That part of a computer store reserved for intermediate results during processing.

wow. Low-frequency modulation introduced in sound reproduction system as a result of speed variation. Similar to, but lower in frequency than, *flutter.*

wrap-round. See **housing.**

wrinkling. Uneven texture developing on surface of metallic uranium.

write. (1) To make a permanent record of coded data on, e.g., punched card or tape, magnetic tape, or print by electric typewriter, Xerox. (2) To feed data into a computer store.

writing speed. Speed of deflection of trace on phosphor, or rate of registering signals on charge storage device.

Wulf string electrometer. Type of string electrometer used with an ionization chamber to measure short current pulses.

wye rectifier. Full-wave rectifier system for a three-phase supply.

K *

X

X-band. Frequency band widely used for 3 cm radar. Now correctly designated **Cx-band.**

X-cut. Special cut from a quartz crystal, normal to the electric (x) axis.

X-guide. A transmission line with an X-shaped cross-section dielectric, used for guiding surface waves.

X-plates. Pair of electrodes in a CRT to which horizontal deflecting voltage is applied in accordance with Cartesian coordinate system.

X-ray crystallography. The study of crystal structures by their use for diffracting X-rays.

X-ray diffractometer. See **X-ray spectrograph** (1).

X-ray fluorescent spectrometry. A method of chemical analysis in which the sample is bombarded by very hard X-rays or gamma-rays, and secondary radiations, characteristic of the elements present, are studied spectroscopically.

X-ray focal spot. That small area of the target (anode) of an X-ray tube on which the electron beam is incident, and from which emitted X-rays emerge. High-power tubes frequently have a line focus to minimize localization of the heat dissipated at the anode.

X-ray hardness. Degree of penetration of X-rays, which is greater the shorter the wavelength.

X-ray microscope. A modification of the *electron microscope* in which a transmission type of X-ray target is also the seal of a vacuum system. Hence, specimen can be examined and photographed in air.

X-ray spectrograph. (1) Instrument for recording an X-ray diffraction pattern. Also **X-ray diffractometer.** (2) Recording *X-ray spectrometer.*

X-ray spectrometer. Instrument for investigation of X-ray spectra by crystal diffraction.

X-ray spectrum. The wavelength distribution of the energy radiated from an X-ray tube. It consists of a line spectrum characteristic of target metal, superimposed on a continuous background. The former is known as the *characteristic spectrum* (q.v.), the latter as the *white spectrum.*

X-ray television. A closed-circuit system used for non-destructive industrial testing of materials, joints, etc., avoiding use of photographic film.

X-ray therapy. Use of X-rays for medical treatment.

X-ray tube. The vacuum tube in which X-rays are produced by a cathode-ray beam incident on an anode (or anti-cathode). Such tubes may be sealed high vacuum or continuously pumped. Older designs had small residual gas pressure and cold cathodes.

X-rays. Electromagnetic waves of short wavelength (0·01Å to 50Å) produced when cathode rays impinge on matter. They may be detected photographically by fluorescence and by ionization produced in gases. They penetrate matter which is opaque to light; this makes them valuable for examining inaccessible regions of the body. When X-rays fall on matter, secondary X-rays (*characteristic X-rays*) are emitted which contain monochromatic radiations which vary in wavelength according to the atoms from which they are scattered. See **Compton effect, K-capture, L-capture, Moseley's law.**

X's. Same as **atmospherics.**

X-unit. A unit expressing the wavelength of X-rays or gamma-rays, equivalent to approximately 10^{-11} cm or 10^{-3}Å. (1 XU = $1 \cdot 00202 \pm 0 \cdot 00003 \times 10^{-3}$Å.)

X-wave. The extraordinary component of electromagnetic waves.

x-y recorder. One which traces on a chart the relation between two variables, not including time. Time may be introduced by moving the chart linearly with time and controlling one of the variables so that it changes proportionally with time.

xenon (Xe). Inert gas, at. no. 54, at. wt. 131·3, m.p. $-112°C$, b.p. $-108°C$, with isotope having highest known capture cross-section for thermal neutrons ($2 \cdot 7 \times 10^6$ barn). Formed as a result of radioactive decay of uranium fission fragments, this isotope ^{135}Xe is the most serious reactor poison, and may delay restart of a reactor after a period of shutdown.

xerographic printer. One using photoelectric principles adapted to high-speed read-out and printing of data on forms. Data is presented by cathode-ray tube and formed by an image from a negative, both being printed together.

xerography. Non-chemical photographic process in which light discharges a charged dielectric surface. This is dusted with a dielectric powder, which adheres to the charged areas, rendering the image visible. Permanent images can be obtained by transferring particles to suitable backing surface (e.g., paper or plastic) and fixing, usually by heat.

xeroradiography. Radiography in which a xerographic, and not photographic, image is produced. This enables reduced exposure times to be used for low kV radiography.

Xerox. TN for a system of xerographic printing suitable for computer output printing on forms, which are printed at the same time.

Xi particle. Hyperon sometimes known as the *cascade particle* since it decays to a nucleon through a two-step process apparent as a cascade in a photographic emulsion and bubble chamber. It has hypercharge -1 and isotopic spin $\frac{1}{2}$.

Xylonite. TN for plastic of cellulose-nitrate type.

Y

Y-aerial. A delta-matched antenna.

Y-cut. Special cut of a quartz crystal, normal to the mechanical (y) axis.

YIG (*yttrium iron garnet*). A material which has a lower acoustic attenuation loss than quartz and has been considered for delay lines.

YIG filter. One using a yttrium-iron-garnet crystal which is tuned by varying the current by a surrounding solenoid, a permanent magnet being used to provide the main field strength.

YIG-tuned parametric amplifier. One in which tuning is achieved with a YIG filter.

Y-network. Same as T-network.

Y-parameter. The short-circuit admittance parameter of a transistor.

Y-plates. Pair of electrodes to which voltage producing vertical deflection of spot is applied in accordance with Cartesian coordinate system.

Y-rectifier. See **wye rectifier**.

Y-signal. The monochromatic signal in colour television which conveys the intelligence of brightness. It is combined with the three chrominance components to produce the three colour primary signals. See **B-Y, G-Y, R-Y signals**.

Y12. Code name for electromagnetic separation plant at Oak Ridge.

Yagi antenna (or **array**). System of end-fire radiators or receivers, characterized by directors in front of the normal dipole radiator and rear reflector.

yield. (1) Ion pairs produced per quantum absorbed or per ionizing particle. (2) See **fission yield**.

yield-G. A term in radiation chemistry, being the number of molecules produced or converted per 100 eV of energy absorbed.

yield point. The least stress at which a solid will deform without adding to the load. It has not a well-defined value with some materials, and with others it does not exist. In such cases the *yield strength* is used.

yield strength. The stress corresponding to a specific amount of permanent deformation.

ylem. The basic substance from which it has been suggested that all known elements may have been derived through nucleogenesis, i.e., fusion of fundamental particles to form nuclei. It would have a specific gravity of 10^{13} g/cm³, and would consist chiefly of neutrons.

yoke. (1) Combination of current coils for deflecting the electron beam in a CRT. (2) Part or parts of a magnetic circuit not embraced by a current-carrying coil, especially in a generator or motor, or relay.

Young's modulus. A modulus of elasticity applicable to the stretching of a wire (or thin rod) or to the bending of a beam. It is defined as the ratio

$$\frac{\text{longitudinal stress}}{\text{longitudinal strain}}.$$

yrneh. Unit of reciprocal inductance (*henry* backwards).

ytterbium (Yb). At. no. 70, at. wt. 173·04. A rare-earth metallic element.

yttrium (Y). At. no. 39, at. wt. 88·905, m.p. 1490°C. A rare-earth metallic element.

Yukawa potential. A potential function of the form $V = \dfrac{V_0 \exp. (-kr)}{r}$, r being distance.

Characterizes the meson field surrounding a nucleon. The exponential tail of the Yukawa potential extends with appreciable strength to larger values of r than does the coulomb potential.

Z

z-axis. The optical axis of a quartz crystal which is perpendicular to both x- and y-axes.

z-axis modulation. Variation of intensity of beam, producing varying intensity of brightness of trace.

Z marker beacon. A form of marker beacon radiating a narrow conical beam along the vertical axis of the cone of silence of a radio range.

z-parameters. The open-circuit impedance parameters of transistors.

Zamboni pile. See **voltaic pile.**

Zebra. (1) TN for single-gun colour TV tube. (2) TN for small digital computer.

zebra time. See **zulu time.**

Zeeman effect. The splitting-up of the individual components of the line spectrum of a substance when placed in a strong magnetic field.

Zener breakdown. Temporary and nondestructive increase of current in diode because of critical field emission of holes and electrons in depletion layer at definite voltage.

Zener current. Current produced in an insulator by electrons which have been raised in energy from the valence bond to the conduction bond through the agency of a strong electric field.

Zener diode. One with characteristic showing sharp increase of reverse current at a certain negative potential, and suitable for use as voltage reference. The effect is now believed due to Townsend discharge and not to Zener current, and the device is therefore also known as a **breakdown diode** or **avalanche diode.** See **Townsend avalanche, Zener breakdown.**

Zener effect. Pronounced and stable curvature in the reverse voltage/current characteristic of a semiconductor point-contact diode; predicted by Zener, and widely used in bridges as a control element.

Zener voltage. The electric field necessary to excite the Zener current, being of the order of 10^7 volt cm^{-1}. It is also used to denote the negative voltage which remains approximately constant over a range of current values in the reverse voltage-ampere characteristic of a semiconductor. See **Zener diode.**

zenith. The point on the celestial sphere which is directly overhead.

zenith distance. The angular distance subtended at the observer by his zenith and a celestial body. It may be expressed as an angle (when it is also known as *co-altitude*) or as the length of the arc on the earth's surface subtended by the angle.

zepp antenna. Horizontal half-wavelength antenna fed from a resonant transmission line. It is connected at one end to one wire of the transmission line and the transmitter or receiver is connected between the two wires, the length of the line being critical.

Zernike-Prins formula. A formula which modifies the simpler scattering expression for use with suspensions of high concentrations.

zero-access store. Fast store from which information is immediately available, e.g., delay line containing one word.

zero-address instruction. Operation in computing where the location of the operands is defined by the order code and not specified independently.

zero adjust. Control for setting the reading of a device to the zero mark in the absence of any signal.

zero beat. Heterodyne reception with no difference frequency between oscillator or carrier; homodyne.

zero-beat reception. Of radio telephony, reception in which a locally-generated oscillation, having the same frequency as the incoming carrier, is impressed simultaneously on the detector. See **homodyne reception.**

zero clearing. Operation of any arrangement for balancing antenna effect to obtain the sharpest minimum signal in determining a radio bearing with a rotating antenna or a radiogoniometer. Also called **minimum clearing.**

zero compression. A technique of data processing in a computer used to eliminate the storing of non-significant leading zeros.

zero-cut crystal. Quartz crystal cut at such an angle to the axes as to have a zero frequency/temperature coefficient. Used for accurate frequency standards.

zero-energy reactor. See **zero-power reactor.**

zero error. (1) Residual time delay which has to be compensated in determining readings of range. (2) Error of any instrument when indicating zero, either by pointer, angle, or display.

zero frequency. Implication that a complex signal, such as video, has a reference value which must be transmitted without drift.

zero gravity. The condition, as in an orbiting satellite, when centrifugal force exactly counterbalances gravitational attraction.

zero level. Any voltage, current or power reference level when other levels are expressed in dB relative to this.

zero-level address. An instruction address in which the address part of the instruction is the *operand*.

zero method. Measuring system in which an unknown value can be deduced from other values when a sensitive, but not necessarily calibrated, instrument indicates zero deflection, as in a Wheatstone bridge or potentiometer; also called **null method**.

zero-point energy. Total energy at the absolute zero of temperature. The uncertainty principle does not permit a simple harmonic oscillator particle to be at rest exactly at the origin, and by the quantum theory the ground state still has one half-quantum of energy, i.e., $h\nu/2$, and the corresponding kinetic energy.

zero-point entropy. As follows from the third law of thermodynamics, the entropy of a system in equilibrium at the absolute zero must be zero.

zero potential. Theoretically, that of a point at infinite distance, used for defining capacitance. Practically, the earth is taken as being of invariant potential. That of any large mass of metal, e.g., equipment chassis.

zero power level. Arbitrary power level for referring other power levels, either in decibels or nepers. Zero power level was formerly 5·8 mW in the U.S., but is now 1 mW, both at the standard 600 ohm impedance level.

zero-power reactor. An experimental reactor with an extremely low neutron flux so that the small power level involved means that no forced cooling is required.

zero stability. Drift in no-signal output level of amplifier or indicator, either with time or with operating conditions (e.g., mains voltage supply).

zero suppression. The elimination, in computing and data processing, of non-significant zeros to the left of the integral part of a quantity, especially before the start of printing.

zerograph. An early form of *telegraph*.

zeta. An apparatus, toroid in shape, used at Harwell for the study of controlled thermonuclear reactions. (*Zero energy thermonuclear apparatus.*).

zeta potential. Potential difference between a liquid phase and liquid adsorbed on a solid surface.

zeus. Zero-energy research reactor used to obtain design data for Dounreay fast reactor.

zinc (Zn). At. no. 30, at. wt. 65·37, sp. gr. 7·14, m.p. 419°C. A hard bluish-white metal, constituent of alloys such as brass. Used as an electrode in a Daniell cell and in dry batteries.

zinc orthosilicate. The mineral willemite which is used for fluorescent screens of CRTs. On bombardment with the rays they give a green glow.

zinc telluride. A semiconductor capable of high-temperature operation (up to about 750°C) without excessive intrinsic conductivity.

Zircaloy. Generic name of a class of zirconium alloys developed for canning reactor fuel rods, especially in the U.S. *Zircaloy AJR* is a U.K. alloy, with 1% Cu and 1·5% Mo, used in gas-cooled reactors.

zirconium (Zr). At. no. 40, at. wt. 91·22, sp. gr. 6·4, m.p. 1900°C. Metallic element used in alloys. The zirconates are finding application as acoustic transducer materials. Tritium adsorbed in zirconium is a possible target in accelerator neutron sources.

zirconium lamp. One having a zirconium oxide cathode in an argon-filled bulb. It provides a high-intensity point source with only a small emission of the longer visible wavelengths.

zitterbewegung. The oscillatory motion of a Dirac electron as given by the Heisenberg equation of motion.

zone. Region between two bounding surfaces which are approximately parallel.

zone levelling. An analogous process to *zone refining*, carried out during processing of semiconductors in order to distribute impurities evenly through sample.

zone of a crystal. A set of crystal faces meeting (or doing so if extended) in a series of edges which are all parallel.

zone of audibility. The hearing of explosions at great distances from the source, although at smaller distances there is a *zone of silence*.

zone of silence. Local region where sound or electromagnetic waves from a given source cannot be received at a useful intensity level.

zone plate. A transparent plate on which there are alternate transparent and opaque concentric rings such that only every other half-period of the incident wavefront is transmitted, so that the plate will act like a lens and form an image.

zone refining (or purification). Passage of a melted zone along a trough of metal, particularly germanium. Impurities dissolve into this and do not freeze out. Heating is performed by high-frequency eddy currents from a coil which progresses along the trough.

zone television. System in which different parts or zones of the image are scanned by separate devices and separately transmitted to the receiver, where they are recombined.

zone time. Internationally accepted system by which surface of earth is divided into 24 zones over which a constant time scale, differing from GMT by an exact number of hours, operates.

zoom. The process of rapidly enlarging, by optical or electronic means, a part of a TV picture at the transmitter.

zoom lens. A lens system, used with a television camera, which avoids the interruptions if employing a lens turret, by permitting the continuous variation of the focal length of the camera over a considerable range. Also **variable-focus lens**.

zulu time. Used in telecommunications for *GMT*; formerly termed **zebra time**.

zwitterion. A dipolar ion, i.e., an electrically neutral molecule but having a dipole moment. The majority of aliphatic amino-acids form such dipolar ions.

APPENDICES

SYMBOLS, PREFIXES AND EQUIVALENTS

Greek Alphabet and Usual Meanings in Electronics, Telecommunications and Nuclear Science

N.B.—Lower case letter is employed unless U (upper case), or U and L (either) are added after entry.

Alpha	A	α	Attenuation coefficient, absorption coefficient, angles. Helium nucleus. Current gain of transistor, fine structure constant.
Beta	B	β	Phase constant, angles, fast moving electron, ratio of velocity to velocity of light.
Gamma	Γ	γ	Complex propagation constant (U or L), angles, quantum, parameter in crystal design, gyromagnetic ratio, polarizability of molecule, unit of magnetic field (10^{-5} oersted), contrast of photographic emulsion.
Delta	Δ	δ	Increment or decrement (U or L), decay factor, piezoelectric strain constant, recoil electron, baryon (U) (hypercharge $+1$, isotopic spin 3/2), dielectric loss angle.
Epsilon	E	ϵ	Permittivity, exponential constant, fast fission factor, electric susceptibility (CGS only), piezoelectric stress constant.
Zeta	Z	ζ	Non-cartesian coordinates, electrokinetic potential.
Eta	H	η	Efficiency, intrinsic impedance, electric susceptibility, thermal fission factor, Steinmetz constant, meson (hypercharge 0, isotopic spin 0).
Theta	Θ	θ	Angles, phase displacements, polar coordinate, Curie temperature, angle of declination of earth's magnetic field.
Iota	I	ι	Instantaneous current.
Kappa	K	κ	Magnetic susceptibility, coefficient of coupling, kaon or kappa meson (hypercharge ∓ 1 isotopic spin 1).
Lambda	Λ	λ	Wavelength, mean free path, magnetostrictive coefficient, baryon (U) (hypercharge 0, isotopic spin 1), permeance (U).
Mu	M	μ	Permeability, amplification factor, refractive index, prefix for micro, dipole moment, micron, muon, Bohr magneton.
Nu	N	ν	Frequency, neutrino, average number of neutrons in fission.
Xi	Ξ	ξ	Non-cartesian coordinate, change in lethargy, baryon (U) (hypercharge -1, isotopic spin $\frac{1}{2}$).
Omicron	O	o	
Pi	Π	π	Ratio-circumference to diameter, electron distribution, quantum transition with $\Delta m = 0$, Peltier coefficient (U), Poynting vector (U), pion or pi meson (hypercharge 0, isotopic spin 1).
Rho	P	ρ	Resistivity, volume charge density.
Sigma	Σ	σ	Surface charge density, conductivity, quantum transition with $\Delta m = \mp 1$, nuclear cross-section, summation (U), baryon (U) (hypercharge 0, isotopic spin 1), Thomson coefficient.
Tau	T	τ	Period, standard deviation, time constant, relaxation time, transmission factor, half-life, Fermi age.
Upsilon	Υ	υ	First point of Aries (reference point in celestial navigation) (U).
Phi	Φ	ϕ	Magnetic flux (U or L), angles, polar coordinate, work function, angle of dip of earth's magnetic field, velocity potential.
Chi	X	χ	Susceptibility, (electric or magnetic) mass susceptibility.
Psi	Ψ	ψ	Dielectric flux (U or L), angles, non-Cartesian coordinate, wave function.
Omega	Ω	ω	Resistance (U or L), solid angle (U or L), pulsatance, angular velocity, Verdet constant, baryon (U) (hypercharge -2, isotopic spin 0).

NOTE.—The symbols ϕ ω ρ τ θ ψ have all been used at some time to indicate special classes of heavy meson (usually those which have followed one specific decay scheme). This form of nomenclature is now obsolescent.

When a Greek letter is used to denote an elementary particle the corresponding anti-particle is commonly denoted by the same letter with a tilde, e.g., ν neutrino, $\tilde{\nu}$ anti-neutrino.

Electricity and Magnetism
(See also pages 404-8)

quantity of electricity	Q
charge density	ρ
surface charge density	σ
electric potential	V, ϕ
potential difference, tension	U, V
electromotive force	E
electric field strength	E
electric flux	Ψ
electric displacement	D
capacitance	C
permittivity $D = \varepsilon E$	ε
permittivity of a vacuum	ε_0
relative permittivity $\varepsilon_r = \varepsilon/\varepsilon_0$	ε_r
dielectric polarization $D = \varepsilon_0 E + P$	P
electric susceptibility	χ_e
polarizability	α, γ
electric dipole moment	p
electric current	I
electric current density	j
magnetic field strength	H
magnetomotive force $F_m = \oint H_s ds$	F_m
magnetic induction, magnetic flux density	B
magnetic flux	Φ
permeability $B = \mu H$	μ
permeability of a vacuum	μ_0
relative permeability $\mu_r = \mu/\mu_0$	μ_r
magnetization $B = \mu_0(H + M)$	M
magnetic susceptibility	χ_m
electromagnetic moment $E_p = -m.B$	μ, m
magnetic polarization $B = \mu_0 H + J$	J
resistance	R
reactance	X
quality factor $Q = |X|/R$	Q
impedance $Z = R + iX$	Z
admittance $Y = 1/Z = G + iB$	Y
conductance	G
susceptance	B
resistivity	ρ
conductivity $\gamma = 1/\rho$	γ, σ
self inductance	L
mutual inductance	M, L_{12}
coupling coefficient $k = L_{12}/(L_1 L_2)^{1/2}$	k
phase number	m
loss angle	δ
number of turns	N
power	P
electromagnetic energy density	w
Poynting vector	S
magnetic vector potential	A

Atomic and Nuclear Physics

mass number, nucleon number	A
atomic number, proton number	Z
neutron number $N = A - Z$	N
elementary charge (of positron)	e
electron mass	m, m_e
proton mass	m_p
neutron mass	m_n
meson mass	m_π
nuclear mass (of nucleus: $^A X$)	$m_N, m_N (^A X)$
atomic mass (of nuclide: $^A X$)	$m_a, m_a (^A X)$
(unified) atomic mass constant $m_u = m_a(^{12}C)/12$	m_u
relative atomic mass m_a/m_u	A_r
Planck constant $(\hbar = h/2\pi)$	h
principal quantum number	n, n_l
orbital angular momentum quantum number	L, l_i
spin quantum number	S, s_l
angular momentum quantum number (including electron spin)	J, j_i
magnetic quantum number	M, m_l
nuclear spin quantum number	I (atomic physics) J (nuclear physics)
hyperfine quantum number	F
rotational quantum number	J, K
vibrational quantum number	υ
quadrupole moment	Q
Rydberg constant	R_∞
Bohr radius	a_0
fine structure constant	α
mass excess $m_a - Am_u$	Δ
packing fraction Δ/Am_u	f
nuclear radius $R = r_0 A^{1/3}$	R
magnetic moment of particle	μ
magnetic moment of proton	μ_p
magnetic moment of neutron	μ_n
magnetic moment of electron	μ_e
Bohr magneton	μ_B
nuclear magneton	μ_N
g factor, e.g., $g = \mu/I\mu_N$	g
gyromagnetic ratio	γ
Larmor (angular frequency)	ω_L
level width	Γ
mean life	τ
reaction energy	Q
cross section	σ
macroscopic cross section $\Sigma = n\sigma$	Σ
impact parameter	b
scattering angle	θ, ϕ
internal conversion coefficient	α
disintegration energy	Q
half-life	$T_{1/2}$
reduced half-life	$fT_{1/2}$
decay constant, disintegration constant	λ
activity	A
Compton wavelength $\lambda_C = h/mc$	λ_C
electron radius	r_e
linear attenuation coefficient	μ, μ_l
atomic attenuation coefficient	μ_a
mass attenuation coefficient	μ_m
linear stopping power	S, S_l
atomic stopping power	S_a
linear range	R, R_l
recombination coefficient	α

Common U.K. and U.S. Abbreviations for Electrical Units

U.S.	U.K.	Unit
a, amp	A	ampere
f	F	farad
h	H	henry
Ω	Ω	ohm
v	V	volt
va	VA	volt-ampere
w	W	watt
wh	Wh	watt-hour
MΩ	MΩ	megohm
db	dB	decibel
db	db	decibel (in telecomm.)
mh	mH	millihenry
μf	μF	microfarad
$\mu\mu$f	pF	picofarad
kwh	kWh	kilowatt-hour
cps	c/s	cycle per second
kc, KC	kc/s	kilocycle per second
mc, MC	Mc/s	megacycle per second
kmc, KMC	Gc/s	gigacycle per second
mev	MeV	mega-electron-volt
bev	GeV	giga-electron-volt

Abbreviation is same for singular and plural, and no dot (point, period) is to be used; K and M can be used on circuits, also p for puffs; m (U.S.) = M (U.K.); billion (U.S.) = 10^9, (U.K.) = 10^{12}, hence avoided.

Verbal Interpretation of Letters

U.S.	ITU, NATO	U.S.	ITU, NATO
Able	Alpha	Nan	November
Baker	Bravo	Oboe	Oscar
Charlie	Charlie	Peter	Papa
Dog	Delta	Queen	Quebec
Easy	Echo	Roger	Romeo
Fox	Foxtrot	Sugar	Sierra
George	Golf	Tare	Tango
How	Hotel	Uncle	Uniform
Item	India	Victor	Victor
Jig	Juliet	William	Whiskey
King	Kilo	X-ray	X-ray
Love	Lima	Yoke	Yankee
Mike	Mike	Zebra	Zulu

Expressions in Radio Communication

out	finished, no reply expected
over	finished, reply expected
roger	message received
wilco	message received, will be complied with

Electronic Term Equivalents

U.S.	U.K.
A-battery	low-tension, LT
audio-frequency, AF	low-frequency, LF
B-battery	high-tension, HT
C-battery	grid-bias, GB
conversion trans-conductance	conversion conduct-ance
converter	frequency changer
field-frequency	frame-frequency
kinescope	CRT television
long-tail tube	
remote cutoff tube	variable mutual
super-control amplifier	conductance valve
pix	picture
terminal	binding post
tickler coil	feedback coil
transconductance	mutual conductance
trouble-shooting	fault-finding
tube	valve
ultor	CRT anode, max
VTVM, vacuum tube voltmeter	VV, valve voltmeter

Contact Nomenclature Equivalents

U.K.	U.S.
spring	pole
front, make	form A, NO, normally-open
back, break	form B, NC, normally-closed
changeover	form C, DT, transfer double-throw
	single-throw, ST
make-before-break	double-break, DB
centre	double-make, DM
step	throw

306

International Morse Code

A · —	N — ·	1 · — — — —
B — · · ·	O — — —	2 · · — — —
C — · — ·	P · — — ·	3 · · · — —
D — · ·	Q — — · —	4 · · · · —
E ·	R · — ·	5 · · · · ·
F · · — ·	S · · ·	6 — · · · ·
G — — ·	T —	7 — — · · ·
H · · · ·	U · · —	8 — — — · ·
I · ·	V · · · —	9 — — — — ·
J · — — —	W · — —	0 — — — — —
K — · —	X — · · —	Period · · · · · ·
L · — · ·	Y — · — —	Comma · — · — · —
M — —	Z — — · ·	

Electronic Gates performing Logical Operations
Output of four input gates for various input conditions.

No. of inputs of value 1	0	1	2	3	4
And operation	0	0	0	0	1
Or ,,	0	1	1	1	1
Not-and ,,	1	1	1	1	0
Nor ,,	1	0	0	0	0

Electrical Supply Voltages and Frequencies in Various Countries

Country	Hz	Voltage	Country	Hz	Voltage
Algeria	50	127, 220	Malaya	50	230
Argentina	50	220	Mexico	60	110, 120, 125
Australia	50	230, 240	Morocco	50	115, 200
Austria	50	110, 115, 200, 220	Netherlands	50	125, 127, 150, 220
Belgium	50	110, 220	New Zealand	50	230
Brazil	50/60	110, 115, 120, 125, 127, 220, 230	Nigeria	50	230
			Norway	50	230
			Pakistan	50	230, 400
Bulgaria	50	220, 380	Peru	50/60	110, 200
Canada	60	110, 115, 120	Poland	50	220
Chile	50	110, 220	Portugal	50	110, 220
China	50/60	110, 220	Rumania	50	120, 220
Colombia	60	100, 115, 120, 220	Sierra Leone	50	230
			South Africa	50	220, 380
Czechoslovakia	50	110, 220	Spain	50	127, 150, 220
Denmark	50	220	Sweden	50	110, 127, 150, 220
Eire	50	220, 380			
Finland	50	220	Switzerland	50	120, 125, 127, 135, 220
France	50	120, 220			
Germany	50	110, 120, 125, 127, 210, 220	Syria	50	115, 200
			Tunisia	50	220
Ghana	50	240, 415	Turkey	50	220
Greece	50	220, 380	U.A.R.	50/60	110, 115, 200, 220
Hungary	50	220			
India	50	220, 230	England	50	230, 240
Indonesia	50	110, 127	U.K. Scotland	50	240
Iran	50	220	N. Ireland	50	230, 400
Iraq	50	220	(outside Belfast and Londonderry)		
Israel	50	230, 400	Uruguay	50	220
Italy	50	110, 120, 127, 150, 220	U.S.A.	60	115, 120
			U.S.S.R.	50	110, 120, 220
Japan	50	100	Venezuela	50/60	120
Kenya	50	250	Yugoslavia	50	220
Korea	60	105			

SOME TERMS ENDING WITH 'TRON'

Those defined in the main text are distinguished
by an asterisk.

Accutron. U.S. trade mark for a 'transistorized' watch.

Acetron. U.K. trade mark for mercury-vapour rectifiers.

Additron. *

aeratron. Self-balancing electronic a.c. potentiometer.

Aeriotron. U.S. trade mark for a group of early receiving valves.

Air-tron. U.S. trade mark for an electric deodorizer.

alcatron. A field-effect semiconductor tetrode developed in France.

Alphatron. *

Amplitron. *

Anotron. *

anticyclotron. *

Aquatron. U.S. trade mark for an electrolytic water purifier.

arcatron. A Swiss cold cathode tube for power control.

arcotron. German high-vacuum valve with an external control electrode.

Arditron. U.K. trade mark for a photographic flash lamp.

Argostron. U.K. trade mark for a stroboscopic discharge tube.

aspatron. Portable British atomic pile.

astron. *

atomotron. A form of high-voltage generator.

Audiotron. U.S. trade mark for an early type of receiving valve.

Auditron. U.K. trade mark for sound equipment.

augetron. British electron multiplier tube.

Autotron. U.K. trade mark for a photocell control system.

axiotron. *

Bailectron. U.K. trade mark for measuring apparatus.

balitron. An electron tube having a stable negative resistance characteristic.

Ballastron. U.S. trade mark for: (1) barretter lamp; (2) power factor improvement capacitor for use with fluorescent lamps.

barytron. Same as **mesotron**.

betatron. *

Bevatron. *

biosteritron. A device to produce ultraviolet radiation.

biotron. *

Calortron. U.K. trade mark for temperature-measuring equipment.

calutron. *

capacitron. *

Carcinotron. *

Cardatron. U.S. and U.K. trade marks for electronic data processing machines.

cardiotron. *

Carmatron. U.K. trade mark for electron tubes, particularly a tube consisting of a combination of a magnetron and a backward wave oscillator.

cartographatron. An electronic device for processing transportation data and displaying it in map form.

Cathetron. Same as **Kathetron**.

cavitron. An ultrasonic hand welding and soldering unit, particularly for wires and thin sheets.

Celatron. U.K. trade mark for thermoplastic material.

Cetron. U.S. trade mark for a range of valves.

Charactron. *

chromatron. *

Chronotron. Trade mark for a device giving a known time delay.

chronotron. *

circlotron. A one-port non-linear crossed-field microwave amplifier.

Clarotron. Trade mark for an early range of receiving valves.

Cleartron. Trade mark for an early triode valve.

Climatron. An electronically controlled greenhouse.

clinitron. Apparatus devised to diagnose diabetes.

Cobaltron. U.K. trade mark for gamma-ray apparatus.

Colortron. Three-colour CRT for colour TV.

Colotron. U.K. trade mark for television apparatus and cathode-ray tubes.

combustron. Trade mark for combustion analysis equipment.

compactron. (1) A solid-state radio device. (2) A form of multi-electrode valve. (3) A type of photoconductive cell.

Comparatron. U.K. trade mark for electronic test equipment.

Composertron. *

Con-electron. U.K. trade mark for photocell control equipment.

Connectron. Trade mark for a discharge flash lamp.

convertron. A d.c. voltage regulating device based on silicon-controlled rectifiers.

cosmotron. *

Crestatron. *

croystron. Name suggested for solid-state devices.

cryotron. *

cybertron. An electronic trial-and-error learning machine.

cycletron. British name for a type of cyclotron.

cyclosynchrotron. A particle accelerator.

cyclotron. *

deflectron. Type of cathode-ray tube with electrostatic deflection.

Dekatron.*

Derritron. Name of a British radio firm.

Detectron. (1) Trade mark for a range of early radio valves. (2) Japanese name for certain valves.

Diatron. U.S. trade mark for a mass spectrometer.

digitron.*

diocotron. An effect utilized in some microwave valves where amplification in an electron stream in crossed electric and magnetic fields is due to the relative speeds of the electrons.

dioctron. A microwave amplifier utilizing space-charge waves in an electron beam moving in crossed electric and magnetic fields.

Diotron. (1) Trade mark. (2) A computer circuit. (3) An instrument using a temperature-limited diode.

donutron. An all-metal tunable magnetron.

duodynatron.*

Duratron. Trade mark for a hearing aid.

dynatron.*

dynectron. Mercury commutator in a vacuum envelope.

Dyotron.*

easitron. A form of travelling-wave tube in which the electron beam is in close contact with a periodic structure.

Ectron. U.K. trade mark for electroconvulsant apparatus.

Eittron. U.K. trade mark for radiation-sensitive elements.

Elcotron. U.K. trade mark for machine tools, vehicles, engines, electric motors.

electron.*

Electrothanotron. U.K. trade mark for an electrical humane killer.

Elliotron. U.K. trade mark for electrical apparatus.

Eltron. U.K. trade mark for medical instruments.

emitron.*

empretron. A mercury-pool arc tube allowing operation at rates in the kilocycle range.

Entron. Name of a U.S. firm operating a community TV system.

Eratron. (1) U.K. trade mark for electrical goods. (2) U.K. trade mark for radio and television apparatus.

Estiatron.*

evatron. A thermionic rheostat for automatic control.

excitron.*

faratron. Device for controlling liquid levels.

farvitron. A partial-pressure indicator that can be used for the continuous analysis of gases present in a vacuum system.

fawshmotron. Fast wave simple harmonic motion microwave amplifier.

Ferrotron. U.K. trade mark for a magnetic plastic material.

flashtron. Type of gas discharge relay.

Flextron. Trade mark for a TV enlarging lens for receivers.

frenotron. A type of diode-triode.

furnatron. A type of electronic furnace control.

Galvatron. U.S. trade mark for a recorder.

Gammatron. Trade mark for a range of valves.

gasomagnetron. Russian gas-filled magnetron.

Gastron. Trade mark for a Russian gas content tube.

gausitron. Same as gusetron.

Genotron. Trade mark for high-voltage rectifier valves.

Germitron. Trade mark for an ultraviolet lamp unit.

golfotron. An electronic device to simulate the results of golf shots played indoors.

Gravitron. U.K. trade mark for electronic discharge tubes.

gusetron. Mercury-arc rectifier with a high-voltage starting anode.

gyrotron.*

Halltron. U.S. trade mark for a Hall-effect device to produce a direct current proportional to true power.

Haltron. U.S. trade mark for a range of valves.

harmodotron.*

hartron. Sound recording device using grooves embossed on a tape.

Heenatron. U.K. trade mark for electronically controlled motors.

heliatron.*

Hetron. U.S. trade mark for polyester resins.

Hodectron.*

Hydrotron. U.K. trade mark for electrical apparatus to counteract scale formation in heat exchangers.

hyrotron. A Russian mercury-arc rectifier in which the arc is controlled by a rotating magnetic field.

Hytron. U.S. trade mark for a range of valves.

iatron. A storage display tube.

Iconotron.*

Idotron. U.K. trade mark for photocell measuring devices.

ignitron.*

Illitron. Trade mark for HF heating equipment.

Infratron. U.S. trade mark for an electric space heater.

injectron. A high-voltage switching tube.

Inkatron. U.K. trade mark for ink measuring and controlling equipment in printing.

instron. Machine for determining tensile strength.

ionotron. Device to dissipate electrostatic charges by means of radioactively ionized air.

Isatron. U.S. trade mark for a mass spectrometer.

isotron.*

jet-tron. A pneumatic switch designed to produce an electric signal when an object interrupts the flow of air across an opening.

Kalistron. Trade mark for a plastic material.

kallitron. A periodic combination of two triode valves to obtain negative resistance.

Kartron. Trade mark for a short-circuited turn tester.

Kathetron. Trade mark for a glow discharge tube.

kenopliotron. Diode-triode in which the triode cathode is also the diode anode. This common electrode is heated by electron bombardment.

kenotron.*

kevatron. High-voltage particle accelerator.

Kilotron. Trade mark for an Australian rectifier valve.

Kinetron. Trade mark for a type of cathode-ray tube.

klystron.*

Kni-tron. U.S. trade mark for a form of rectifier.

Kodatron. Trade mark for gas-discharge flash lamp.

koremetron. An optical instrument for measuring eye ɼupil diameter.

Kotron. Trade mark for a selenium rectifier.

larmatron. A quasi-parametric electron beam amplifier.

larmotron. A d.c. pumped quadrupole amplifier.

Leartron. U.S. trade mark for a gramophone pickup.

Lectron. (1) Trade mark for a solder. (2) An electron tube for reading magnetic tape.

Leveltron. U.K. trade mark for a capacitance-operated proximity switch.

Lightron. U.K. trade mark for electrical apparatus.

limitron. An electronic comparator.

loprotron. A beam-switching tube.

Lumatron. U.K. trade mark for a range of cathode-ray tubes.

Lumetron. U.S. trade mark for a colorimeter.

Luxtron. U.S. trade mark for a photovoltaic cell.

Lytron. Trade mark for polyelectrolyte resins.

magnetron.*

martenetron. Name for materials wherein a discontinuous change of resistance occurs at different temperatures, depending on whether the temperature is being increased or decreased.

Maxitron. Trade mark for an X-ray tube.

Mectron. U.K. trade mark for electronic apparatus.

megatron.*

Mellotron. Type of electronic organ.

Meltron. U.K. trade mark for metal goods.

memotron. CRT in which a trace on a screen is reinforced and exhibited continuously.

mesotron.*

meteotron. Device for making artificial rain-clouds.

Metron. (1) Trade mark for a device to measure beverages. (2) Marine depthmeter.

Microtron. (1) Russian electron accelerator. (2) U.K. trade mark for thermionic valves. (3) A type of low-power magnetron made in the U.S.

Miniatron. U.K. trade mark for electronic discharge tubes.

Minitron. U.S. trade mark for a range of small receiving valves.

mitron.*

mnemotron. A polycathode glow tube for counting.

Mobaltron. U.K. trade mark for gamma-ray apparatus.

monitron.*

Monotron. (1) Trade mark for monoscope tube; also called **Videotron.** (2) A form of klystron.

Moritron. U.K. trade mark for measuring apparatus.

Mortron. (1) U.K. trade mark for air-conditioning equipment. (2) U.S. trade mark for an electronically-operated insect killer.

Motron. U.S. trade mark for a servomechanism.

multitron.*

myriatron. An image converter tube.

negatron.*

neostron. A stroboscopic tube.

Neotron. (1) A gas-filled pulse generator valve. (2) A French valve firm and its products.

Neptron. (1) U.S. trade mark for a range of thermionic valves. (2) U.K. trade mark for diathermy instruments.

neutron.*

Nevitron. U.K. trade mark for electric rectifiers and discharge apparatus.

Nimatron. Automatic equipment to play the Chinese game of Nim.

Nobatron. U.S. trade mark for a d.c. power pack.

Nodistron. U.S. trade mark for electric discharge tubes.

Nomotron.*

Nudistron. U.K. trade mark for electric discharge tubes.

Nutron. Trade mark for an early range of receiving valves.

Oaktron. U.S. trade mark for a form of switch.

omegatron.*

Omitron. U.K. trade mark for machines for metals and plastics.

Ophitron. TN for microwave oscillator tube.

Orbitron.*

orgatron. An electronic organ.

Orthotron. (1) Trade mark for an electronic linear accelerator. (2) A cross-field travelling-wave tube.

ovitron. Liquid valve in which a tantalum film is broken by d.c. control on a platinum electrode.

Oxytron. Trade mark for a Danish valve.

palletron. An electron resonator for the production of high voltages.

Pamotron. U.K. trade mark for electronic apparatus for detecting flaws in paper.

Pantatron. Trade mark for gamma radiographic equipment.

parametron.*

Paxitron. U.K. trade mark for lighting, heating, cooking, ventilating equipment, etc.

Penatron. U.S. trade mark for machine tools.

Penetron. (1) Same as **mesotron.** (2) A thickness-measuring device.

peniotron. A Japanese type of fawshmotron amplifier.

310

pentatron. A double triode with a common cathode.

Pentron. Trade mark for a magnetic wire recorder.

perceptron. A machine for character recognition.

perhapsatron. *

peritron. *

permactron. A double-beam electron wave tube.

permatron. *

persistron. *

phanotron. *

phantastron. *

phasitron. *

Phen-O-tron. Name of a firm making printed circuit devices.

Philcotron. Trade mark for an electrolytic rectifier.

Phonotron. Trade mark for an early range of receiving valves.

Phosphotron. U.K. trade mark for electroluminescent lighting units.

photo-augetron. A photomultiplier tube.

photoelectron. *

photoneutron. *

Pianotron. *

Picturetron. U.K. trade mark for CRTs.

plasmatron. *

Platinotron. *

pliodynatron. *

pliotron. *

plomatron. A grid-controlled mercury arc valve.

Polatron. Trade mark for a TV tube incorporating a neutral filter.

Polytron. (1) A name once proposed for a suspected elementary particle. (2) U.K. trade mark for mixing machines. (3) U.K. trade mark for surgical, medical, veterinary and dental products.

positron. *

Powertron. U.K. trade mark for filters, sparking plugs, electric lamps and lighting apparatus.

Preceptron. Trade mark for a deaf-aid.

Precipitron. U.S. and U.K. trade marks for an electrical precipitating device.

prionotron. U.S. term for velocity-modulated electron tube.

Pulsatron. (1) A gas-filled pulse-generating tube. (2) U.K. trade mark for ultrasonic cleaning machines.

Puritron. U.S. trade mark for a form of air-conditioning unit.

pyrotron. *

quadratron. *

Quickitron. Trade mark for storage tube with display target.

Quotron. U.S. trade mark for a device for displaying Stock Market prices.

Radiotron. Trade mark for a range of thermionic valves.

raytron. A device for locating earth faults.

reactatron. *

readatron. *

rebatron. A millimetre wave generator using the 'bunching' principle.

rebatron-harmodotron. A rebatron where high harmonics of a 'bunching' frequency are extracted from a tightly packed electron beam.

Rectron. Trade mark for a range of German thermionic valves.

Remitron. *

Resistron. U.K. trade mark for measuring apparatus.

resnatron. *

Resotron. U.K. trade mark for X-ray apparatus.

rheotron. *

rhumbatron. *

Robotron. (1) Name of a firm and its flash-tube products. (2) U.K. trade mark for information-processing machines.

Rotametron. Trade mark for an industrial electronic weighing system.

Rototron. Trade mark for an electronic telephone ringing device.

ryotron. A cryogenic electronic variable inductor.

sanatron. *

Sanitron. U.K. trade mark for medical instruments.

Saxtron. U.K. trade mark for aerial wire.

Scanatron. U.K. trade mark for colour-correcting equipment used in printing.

Scriptron. Trade mark for a cathode-ray tube used to display letters and numerals.

Sealtron. Name of a U.S. firm making glass-to-metal hermetic seals.

Secatron. Trade mark for register control equipment used in printing.

Securitron. U.S. trade mark for an electronic protection system.

Selectron. (1) U.S. trade mark for a polystyrene moulding compound. (2) Trade mark for a storage tube.

Seletron. Trade mark for a range of selenium rectifiers.

sendytron. A Japanese mercury-arc tube with a high-voltage starting electrode.

sentron. A Japanese short-wave valve.

Servotron. (1) Mercury-arc rectifier with a high-voltage starter. (2) U.S. trade mark for an electronic motor control.

shucktron. An American machine to separate the viscera and muscle of the calico scallop from the shell.

sigmatron. A cyclotron and a betatron operating in tandem to produce very high energy X-rays.

Silectron. U.S. trade mark for a magnetic core material.

skiatron. *

skirtron. A form of wideband klystron.

Soldetron. Trade mark for a soldering iron.

Sortron. Trade mark for a gauging device.

Sparcatron. (1) U.K. trade mark for industrial oils. (2) U.S. trade mark for spark machining equipment and technique.

spiratron. *

spirotron. A device for decelerating high-speed particles.

spraytron. Electrostatic spraying equipment made in Switzerland.

Stabilatron. U.K. trade mark for gamma-ray apparatus.

Stabilotron. U.S. trade mark for electrical apparatus, including a type of magnetron.

statitron. A form of Van de Graaff generator.

Statron. Trade mark for electrostatic spray-gun equipment.

Stemmatron. U.K. trade mark for electronic tubes.

stenotron. A Russian gas-filled transmitting valve.

Stethetron. U.K. and U.S. trade marks for electronic stethoscopes.

stomertron. A device for simulating solar proton streams.

strobotron.*

strophotron.*

Subnitron. U.K. trade mark for electric discharge tubes.

Supertron. Trade mark for an early range of receiving valves.

symetron. A multiple-valve ring-type amplifier.

synchrocyclotron.

synchrotron.*

Syntomatron. U.K. trade mark for electronic tubes.

syntron. A type of electric hammer in which half-wave rectified a.c. is fed to a solenoid.

syrotron. Ion-cyclotron resonance-mass spectrometer.

tacitron.*

takktron. High-voltage glow-discharge rectifier.

Tapestron. Trade mark for a plastic wall screen.

Technetron.*

Technitron.*

Telemitron. U.K. trade mark for electronic discharge devices and television apparatus.

teletron.*

tellertron. An American digital-computing system.

Temtron. U.S. trade mark for an air-conditioning unit.

Tepitron. U.K. trade mark for heating and drying apparatus.

Teratron. U.K. trade mark for electronic tubes.

Theratron. U.K. trade mark for medical and veterinary instruments.

Thermatron. U.S. trade mark for high-frequency heating equipment.

Thermotron. Trade mark for an early range of receiving valves.

thetatron. Linear discharge tube used at Culham Laboratories for high-temperature plasma research. The rise in temperature is attained principally by pinching the plasma filament carrying the discharge current.

thyratron.*

Tometron. U.K. trade mark for a range of electronic tubes.

Tonotron. U.S. and U.K. trade marks for storage cathode-ray tubes.

tornadotron. An electron resonance device for converting microwaves to submillimetre waves.

Touchetron. U.S. trade mark for a touch-controlled switch.

Trakatron. Trade mark for web guiding equipment.

transistron. French name for a type of transistor.

transitron.*

trigatron.*

Trignitron. Trade mark for a type of mercury-arc tube.

Triotron. U.K. trade mark for a range of radio valves.

trochotron.*

tronadotron. A type of millimetre wave generator based on combined cyclotron and betatron principles.

tropotron.*

tunneltron. A thin-film device that exhibits superconductivity, 'tunneling', and negative resistance.

Typotron. U.K. trade mark for CRTs.

Ubitron. A fast-wave tube using undulating beam interaction.

Ultron. (1) U.K. and U.S. trade marks for vinyl chloride resins. (2) U.K. trade mark for respirators.

Unitron. (1) U.K. trade mark for cathode-ray tubes. (2) U.K. trade mark for electromedical apparatus.

Vacutron. Name of a firm and its radio valves.

Vapotron.*

Vectron. Trade mark for microwave spectron analyser.

Velocitron.*

veristron.*

Veritron. Trade name for a pyrometer.

vibrotron.*

Videotron. Same as **Monotron** (1).

Viscometron. U.K. trade mark for electronic apparatus to control the viscosity of inks in printing presses.

Visitron. U.S. trade mark for a projection cathode-ray tube.

Vocatron. U.S. trade mark for an intercommunication system.

volemetron. A solid-state device for measuring human blood volume.

Voltron. (1) Trade mark for an early range of radio receiving valves. (2) U.S. trade mark for an insulating compound.

Webatron. U.K. trade mark for printing register control equipment.

windowtron. An experimental device for testing high-power microwave windows.

Zyklotron. Trade mark for a Swiss high-frequency valve.

Grateful thanks are due to Mr J. H. Jupe and to the editor of *Industrial Electronics* for permission to reproduce material from the 'Trons Dictionary' which originally appeared in *British Communications and Electronics* (March 1962).

312

ELECTROMAGNETIC DATA

Electromagnetic Spectrum

Freq. Hz	*Wavelength*	*Quantum Energy*	*Usual designation*	*Some applications*
ZERO	—	—	D.C.	
10	3×10^7 m			Motors, relays, power supplies
10^2	3×10^6 m	↑	Power transmission	Audio frequencies
10^3	3×10^5 m	Quantum effects largely negligible		Power distribution
10^4	3×10^4 m			Audio reproduction
10^5	3×10^3 m		VLF, LF	Long-range communication
10^6	3×10^2 m		MF	Public broadcasting systems
10^7	30 m			
10^8	3 m		HF, VHF — Radio frequencies	
10^9	30 cm			Television, navigation,
10^{10}	3 cm		UHF	industrial and medical
10^{11}	0·3 cm = 3000 μ		SHF, EHF — Radar	applications
10^{12}	300 μ	↓		Heating, 'dark
10^{13}	30 μ		Infrared	photography, signalling,
10^{14}	3 μ	0 3 eV		detection
10^{15}	0·3 μ = 3000 Å	3 eV	Visible light	Photography, microscopy
10^{16}	300 Å	30 eV	Ultraviolet	Medical, ultraviolet
10^{17}	30 Å	300 eV		recording
10^{18}	3 Å	3 KeV	*Soft*	
10^{19}	0·3 Å = 300 XU	30 KeV		Radiography, radio-
10^{20}	30 XU	300 KeV	X-rays — Gamma-rays	therapy, crystallo-
10^{21}	3 XU	3 MeV		graphy, radioisotope
10^{22}	0·3 XU	30 MeV	*Hard*	detection

Wavelengths above expressed in metres, centimetres, microns ($=10^{-6}$ m), ångström units ($=10^{-8}$ cm) or X-ray units ($=10^{-11}$ cm).

Quantum energies in electron-volts (eV), kilo-electron-volts (KeV), or million-electron-volts (MeV).

Frequency Band Designation in Radio Spectrum

CCIR Abbreviation	*Full Designation*	*Frequencies*	*Wavelengths*
VLF	Very low frequency	Below 30 kHz	Myriametric
LF	Low frequency	30-300 kHz	Kilometric
MF	Medium frequency	300-3000 kHz	Hectometric
HF	High frequency	3000-30,000 kHz	Decametric
VHF	Very high frequency	30,000 kHz-300 MHz	Metric
UHF	Ultra high frequency	300-3000 MHz	Decimetric
SHF	Super high frequency	3000-30,000 MHz	Centrimetric
EHF	Extreme high frequency	30,000-300,000 MHz	Millimetric

Frequencies up to 3×10^7 should be expressed in kilohertz, those above in megahertz.

The following bands of frequencies are commonly denoted by the code letter shown:

P-band	225-390 MHz	133·3-76·9 cm
L-band	390-1550 MHz	76·9-19·3 cm
S-band	1550-5200 MHz	19·3-5·77 cm
C-band*	3900-6200 MHz	7·69-4·84 cm
X-band	5200-10,900 MHz	5·77-2·75 cm
K-band	10,900-36,000 MHz	2·75-0·834 cm
Q-band	36,000-46,000 MHz	0·834-0·652 cm
V-band	46,000-56,000 MHz	0·652-0·536 cm

* C-band consists of part of S-band and part of X-band. Other bands abut without overlapping.

Frequencies above 1,000 MHz are commonly expressed in gigahertz (10^9 Hz).

For all types of electromagnetic radiation, frequency in Hz × wavelength in metres = velocity of propagation = 3×10^8 m sec^{-1}.

Allocation of HF Bands for Television and FM Radio in U.K.

(I) VHF Bands

BAND 1.	41-68 MHz	Television (405 lines)	Channels 1-5
BAND 2.	87·5-100 MHz	FM Radio	
BAND 3.	174-216 MHz	Television (405 lines)	Channels 6-13

(II) UHF Bands

BAND 4.	470-585 MHz	Television (625 lines)	Channels 21-34
BAND 5.	610-960 MHz	,, ,, ,, ,,	Channels 39-68

TV Line Standards in Various Countries

Line System		Areas
405	1	U.K. (part system), Eire (part system), Gibraltar, Hong Kong.
525	{ 2 3	All countries in N. and S. America except those listed under (6). Philippines, Thailand, Saudi Arabia, Kuwait, Iran, Cambodia, Korea, Japan, U.S. possessions and U.S. Forces stations.
625	(4 5 6 7 8	All stations in Europe except those listed under (1) and (9). Australasia. N. and S. America stations in Argentina, Venezuela, Jamaica, Trinidad, and Curaçao. China, India, Iraq, Lebanon, Singapore, Syria, Russia. All stations in Africa except Algeria.
819	9	France (part system), Belgium (part system), Monaco, Luxembourg, Algeria.

Standard Frequency Transmissions in U.K.

These take place 24 hours a day on carrier frequencies of 2·5, 5, and 10 MHz; and 1 hour each day (14.29-15.30) on 60 kHz. Each hour is divided as shown below. The 1 Hz modulation pulses consist of 5 cycles of 1 kHz tone except for the minute pulses which contain 100 cycles.

These frequencies are adjusted to ∓ 5 parts in 10^9, and calibrated against the NPL caesium clock.

The VLF telegraphy transmitter GBR has a 16 kHz carrier locked to the above signals.

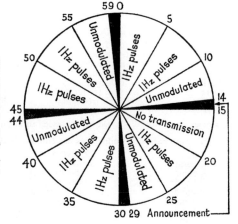

Classification of Radio Transmissions

Three sets of symbols are used to classify all standard types of transmission:

1. Indicates the type of modulation.
2. Indicates the type of signal.
3. Indicates certain important supplementary characteristics.

	Symbol	Meaning
1.	A	Amplitude-modulated CW
	F	Frequency-modulated CW
	P	Pulse-modulated CW
	B	Damped (spark) wave—now obsolete
2.	0	Absence of any modulation intended to carry information
	1	ICW telegraphy (unmodulated)
	2	Telegraphy by keying of modulated wave or of the modulation signal
	3	Telephony and broadcasting
	4	Facsimile
	5	Television video channel
	6	Four-frequency duplex telegraphy
	7	Multichannel AF telegraphy
	9	Other
3.	Blank	Double-sideband full-carrier transmissions
	A	Single-sideband reduced-carrier transmissions
	H	Single-sideband full-carrier transmissions
	J	Single-sideband suppressed-carrier transmissions
	B	Independent double-sideband transmissions
	C	Vestigial sideband transmissions
	D	Amplitude-modulated pulse transmissions
	E	Width- or time-modulated pulse transmissions
	F	Phase-modulated pulse transmissions
	G	Code-modulated pulse transmissions

Example. A band 1 or 3 television station radiates an A5C (amplitude-modulated, video, vestigial sideband) signal and an A3 (amplitude-modulated, audio broadcast) signal on the video and sound channels respectively.

Navigation Transmission

LORAN (LOng RAnge Navigation) originated during World War II at M.I.T. to provide aircraft and ships with a system for precise navigation. A loran chain consists essentially of a master and two or more slave stations. A pulse transmitted from the master is received via a ground wave by a slave and the latter in turn, at a fixed time (the 'coding delay') later, transmits its pulse. The value of the 'coding delay' time is chosen to ensure that signals in a chain will always arrive at a receiver in sequence and will not overlap. The time difference between the reception of a master pulse pair defines a hyperbolic line, and position is given by the intersection of two such lines. The transmitted pulses are of 45μ sec duration and are of bandwidth 75 kHz. The standard loran (*loran-A*) operates in the 2 MHz band. *Loran-C* has been developed to extend the loran coverage using fewer stations and employs a 100 kHz carrier frequency. The lower ground-wave attenuation allows longer station-to-station baselines to be used and clock synchronization over long distances to within ± 1 μsec. The high precision is obtained by matching both the envelope of the pulse and the phase of the cycles in the pulse.

British Waveguide Data (*Rectangular Guide*)

Guide Reference No.	6	8	10	12	14	15
External dimensions						
English (in.) {	6·660 × 3·410	4·460 × 2·310	3·000 × 1·500	2·000 × 1·000	1·500 × 0·750	1·250 × 0·625
Metric (cm.) {	16·920 × 8·661	11·330 × 5·867	7·620 × 3·810	5·080 × 2·540	3·810 × 1·905	3·175 × 1·588
Internal dimensions						
English (in.) {	6·500 × 3·250	4·300 × 2·150	2·840 × 1·340	1·872 × 0·872	1·372 × 0·622	1·122 × 0·497
Metric (cm.) {	16·510 × 8·255	10·920 × 5·416	7·214 × 3·404	4·755 × 2·215	3·485 × 1·580	2·850 × 1·262
Tolerance (in.)	∓ 0·008	∓ 0·006	∓ 0·004	∓ 0·003	∓ 0·002	∓ 0·002
Critical cutoff frequency (f_c) GHz	0·908	1·373	2·080	3·155	4·285	5·260
Nominal operating frequency (1·5 f_c) GHz	1·362	2·06	3·120	4·733	6·427	7·890
Operating frequency range { (1·25-1·9 f_c) GHz	1·14 – 1·73	1·72 – 2·61	2·60 – 3·95	3·94 – 5·99	5·38 – 8·18	6·58 – 10·00
Power rating (MW) at 1·5 f_c	13·47	5·90	2·43	1·04	0·544	0·355
Attenuation dB. per 100 ft. at 1·5 f_c						
Copper	0·154	0·286	0·555	1·047	1·700	2·338
Brass	0·239	0·443	0·860	1·62	2·635	3·625
Aluminium	0·20	0·372	0·722	1·36	2·21	3·035
Free space wavelength at cutoff frequency (cm.)	33	21·8	14·4	9·5	7·0	5·7
Free space wavelength at 1·5 f_c (cm.)	22	14·5	9·6	6·34	4·67	3·80
Guide wavelength at 1·5 f_c (cm.)	29·5	19·5	12·9	8·5	6·26	5·1
JAN. designation	RG 69/U	RG 104/U	RG 48/U	RG 49/U	RG 50/U	RG 51/U

British Waveguide Data (*Rectangular Guide*)

16	18	20	22	23	24	25	26
1·000 × 0·500 2·540 × 1·270	0·702 × 0·391 1·783 × 0·993	0·500 × 0·250 1·270 × 0·635	0·360 × 0·220 0·9144 × 0·5588	0·304 × 0·192 0·7722 × 0·4877	0·268 × 0·174 0·6807 × 0·4420	0·228 × 0·154 0·5791 × 0·3912	0·202 × 0·141 0·5131 × 0·3581
0·900 × 0·400 2·286 × 1·016 ∓0·001	0·622 × 0·311 1·580 × 0·790 ∓0·001	0·420 × 0·170 1·067 0·4318 ∓0·0008	0·280 × 0·140 0·7112 × 0·3556 ∓0·0008	0·224 × 0·112 0·5690 × 0·2845 ∓0·0008	0·188 × 0·094 0·4775 × 0·2388 ∓0·0008	0·148 × 0·074 0·3759 × 0·1880 ∓0·0008	0·122 × 0·061 0·3099 × 0·1549 ∓0·0008
6·560	9·49	14·08	21·10	26·35	31·4	39·9	48·4
9·84	14·23	21·12	31·65	39·52	47·1	59·8	72·6
8·2 − 12·5	11·9 − 18·0	17·6 − 26·7	26·4 − 40·1	33·0 − 50·1	39·3 − 59·7	49·9 − 75·8	60·5 − 92·0
0·229	0·123	0·048	0·025	0·016	0·010	0·007	0·005
3·24 5·025 4·22	5·21 8·07 6·77	10·9 16·9 14·18	17·3 26·85 22·5	24·0 37·2 31·2	31·3 48·5 40·7	44·7 69·3 58·1	59·7 92·5 77·5
4·57	3·16	2·13	1·42	1·14	0·955	0·752	0·620
3·05	2·10	1·42	0·95	0·76	0·64	0·501	0·413
4·1	2·82	1·90	1·275	1·02	0·86	0·67	0·555
RG 52/U	RG 91/U	RG 53/U	RG 96/U	RG 97/U		RG 98/U	RG 99/U

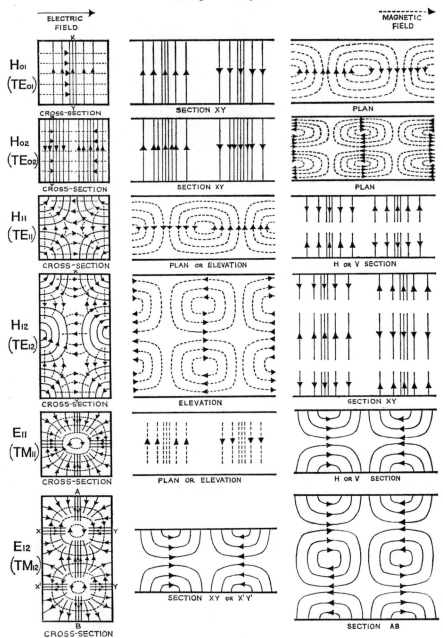

There is no field configuration corresponding to any E_{om} mode possible in a rectangular waveguide.

SMITH CHART AND ITS APPLICATIONS

(Reference: P. H. Smith, *Electronics* 1944, **17**, 130-133, 318-325)

A device used for the graphical solution of problems arising from the propagation of acoustic, electric or electromagnetic waves along some form of transmission line or guide. It is a circular chart on which a vector of length equal to the radius of the diagram and rotating in the anti-clockwise direction can be imagined to represent a wave advancing down a loss-free line. The phase of this wave at the termination is taken to be the direction when the vector lies along the *x*-axis (i.e., directed horizontally to the right). Similarly a vector rotating in the clockwise direction represents the reflected wave, its length being given by the product of the advancing wave amplitude (the radius of the diagram) and the voltage reflection coefficient for the termination.

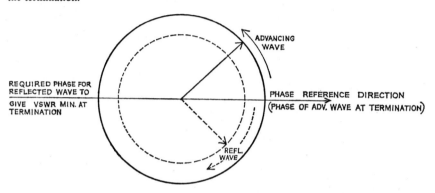

It follows that a radial cursor can be fitted to the diagram and calibrated to read either the reflection coefficient *K* or the VSWR* which is always given by $(1 + K)/(1 - K)$.

Over a line with finite losses the locus of the vector representing the wave gradually spirals in towards the centre of the diagram and the VSWR reduces as the termination is approached. This enables further radial scales, calibrated in decibels transmission loss and loss coefficient, to be used to correct a VSWR value at a point other than the termination for line attenuation.

The value of the diagram is that two sets of orthogonal circles can be superimposed on it, one set representing normalized values of the line resistance and the other normalized values of the line reactance. The terminating impedance is then given by the values of these curves at the point representing the tip of the reflected-wave vector at the termination. The impedance at any other point a distance *l* from the termination is obtained in the same way after rotating

this reflected wave vector for the termination through an angle $\dfrac{l}{\lambda/2} \times 2\pi$ (or $4\pi l/\lambda$) in a clock-

wise direction. This process is facilitated by calibrating the circumference of the diagram in wavelengths—a complete rotation corresponding to $\lambda/2$.

The chart can be used for quick graphical conversion of data from any one of the following forms to any of the others:

    (*a*) Normalized resistance and reactance of terminating impedance,

    (*b*) Coefficient of reflection and phase change at termination,

    (*c*) Position of standing-wave minimum and VSWR,

    (*d*) Line impedance at a point a specified distance from the termination.

It also enables problems on matching loads to transmission systems (i.e., reducing the VSWR to unity) to be readily solved by such methods as adjustment of the load or use of series or shunt matching stubs or other loading reactances. When it is desired to work with admittances (as, e.g., in shunt matching), use can be made of the fact that corresponding impedance and admittance values occur at points equidistant from the centre of the chart and diametrically opposite each other.

* Voltage standing wave ratio (Vol. ratio on charts)

The following brief set of rules indicate a few of the problems to which the Smith chart will give an immediate solution:—

1. From the radial parameters shown at the bottom of the diagram, reflection coefficients (voltage or power) can be converted to corresponding values of the return loss for the line or the reflection loss for the termination (in dB), and the resulting VSWR (either numerical or in dB). The value of any of these quantities defines the length of a radius vector from the centre of the main diagram whose locus will pass through all the impedance values which (ignoring losses) will be encountered along the line.

2. When the line has a known loss in dB the corresponding radial scale at the bottom enables the length of the radius vector referred to in (1) above to be corrected for loss. When standing waves are present the loss value applicable to a matched line must be multiplied by the additional factor given on the standing-wave loss coefficient scale (against the corresponding value of SWR).

3. When the impedance at any point on the line is known, that at any other point can be determined by rotating the radius vector drawn from the centre of the diagram to the point representing the specified impedance through an amount indicated by the peripheral scale, calibrated in wavelength towards generator or wavelength towards load. Multiples of $\lambda/2$ are neglected for this purpose unless the length of the vector has to be adjusted to allow for attenuation as described under (2).

4. To determine the terminating impedance from the VSWR ratio at the termination and the distance of the nearest VSWR minimum to it, use the radial VSWR scale to give the length of the required radius vector and inscribe this on the diagram in the negative direction of x (i.e., to the left along the horizontal resistance scale). Now rotate this vector counter clockwise through the fraction of a wavelength required to bring the point concerned from the VSWR minimum to the termination, using the circumferential wavelength scale. The normalized load impedance is then read off from the chart coordinates.

Example.—On a 75 ohm cable with attenuation 0·2 dB per wavelength there is a VSWR at the input end of 1·5 and a standing-wave minimum 5·1 wavelength from the load. Determine the terminating impedance and the complex reflection coefficient at the termination.

(*a*) Correct for line loss over 5 wavelengths (1 dB for $5\lambda \times$ loss coefficient 1·08). This gives a VSWR of 1·38 with the minimum 0·1λ from the termination.

(*b*) Proceeding as in (4) above, the normalized load impedance is found to be $0·87 - 0·27j$ corresponding to an actual impedance of $65·2 - 20·2j$ ohms.

(*c*) Using the voltage reflection coefficient scale and the circumferential angular scale, it is found that $Ke^{j\phi} = 0·16$ at a phase angle of $-108°$; or $0·16/-108°$.

N.B.—In practice many lines are almost correctly matched so that standing-wave ratios outside the range 1·0-1·5 are unlikely to be encountered. This means that problems of the kind indicated above have to be solved from the central portion of the Smith chart where the rulings are most crowded and the available accuracy is least. For such problems it is convenient to have this central portion of the chart enlarged to form a separate diagram (see foldout 2).

Acknowledgment.—Both of the accompanying charts [see pocket inside back cover] are reproduced by kind permission of the Kay Electric Company, Pine Brook, New Jersey.

COMPONENT CLASSIFICATION AND CODING

Colour Code for Resistors

Colour coding has been adopted for resistors to indicate their resistance value and other characteristics. The code markings on resistors are interpreted in accordance with Tables I and II below. Either of the following two methods of marking may be used:

(1) *coloured band method* (preferred). The code takes the form of circumferential colour bands on the resistor body, e.g., for 6800 ohms∓ 5%, Grade 1.

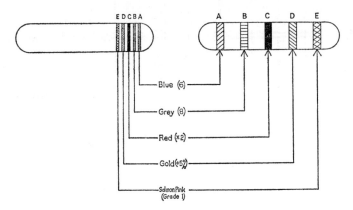

N.B.—Salmon pink may be the general body colour, not just a fifth colour band.

(2) *body, tip and central band* (or *spot*) *method*. The code takes the form of a coloured resistor body, tip and central band (or spot), e.g.,

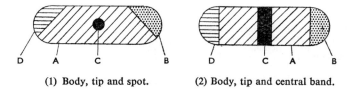

(1) Body, tip and spot.   (2) Body, tip and central band.

TABLE I. *Interpretation of Marking*

Coloured Band Method	Body, tip and Central Band (or Spot) Method		Interpretation
1st band (A)	Body	(A)	1st significant figure of resistance value.
2nd band (B)	First tip	(B)	2nd significant figure.
3rd band (C)	Central Band or Spot	(C)	Multiplier.
4th band (D)	Second tip	(D)	If present, percentage tolerance on nominal resistance value. If no colour appears in this position, tolerance ±20%.
5th band (E)			If present, grade of resistor.

L   321

TABLE II. *Colour Values*

Colour	A 1st figure	B 2nd figure	C Multiplying value	D Tolerance %	E Grade
Silver			10^{-2}	± 10	
Gold			10^{-1}	± 5	
Black		0	1		
Brown	1	1	10	± 1	
Red	2	2	10^2	± 2	
Orange	3	3	10^3		
Yellow	4	4	10^4		
Green	5	5	10^5		
Blue	6	6	10^6		
Violet	7	7	10^7		
Grey	8	8	10^8		
White	9	9	10^9		
Salmon pink					Grade 1 high- stability resistor
None				± 20	

Preferred Values of Resistances (in ohms)

Tolerance Range

The table gives preferred values between 10 and 99 for resistors and capacitors made to 5%, 10% and 20% tolerances.

Resistances are normally available in decimal multiples of these values between 10 ohms and 22 megohms although some ranges include values both above and below these limits.

Capacitors are normally supplied in these values throughout the range 10-1000 pF. Between 1000 pF and 1 μF both preferred values and strict decimal values (e.g., 0·22 μF, 0·20 μF) will be encountered.

The tolerance rating given a component is governed by its long term stability grading—any component can be given a preferred value rating in any of these three ranges within the specified accuracy limits. For example, a 42 ohm resistor would be marked 43 ohms if of 5% grade, 39 ohms if of 10% grade, and 47 ohms if of 20% grade.

5%	10%	20%
10	10	10
11		
12	12	
13		
15	15	15
16		
18	18	
20		
22	22	22
24		
27	27	
30		
33	33	33
36		
39	39	
43		
47	47	47
51		
56	56	
62		
68	68	68
75		
82	82	
91		

and decimal multiples of the above values.

PROPERTIES OF MATERIALS USED IN ELECTRONICS

Ferrite and Ferroelectric Materials

The two types of ferrite materials which are generally used for inductive components are the manganese-zinc and the nickel-zinc ferrites, the latter being chiefly employed for the higher frequency applications. Both types of material are available in various grades according to the particular application and, by kind permission of Mullard Ltd., a few typical types and their chief characteristics are given in the following tables.

Physical Properties of some Ferroelectric Materials

Material and Composition	Dielectric Constant	Curie Temp. °C	Comments
Barium-strontium titanate ($BaTiO_3$: $SrTiO_3$) 65 35	3000 and above	20°	Dielectric for capacitors
Barium-lead zirconate ($BaZrO_3$: $PbZrO_3$) 35 65	6000 and above	0°	Very high dielectric constant over ambient temperature range
Barium titanate ($BaTiO_2$)	1400	120°	Piezoelectric transducer, also non-linear dielectric
Barium-lead titanate ($BaTiO_3$: $PbTiO_3$) 96 4	1100	130°	Piezoelectric transducer with more permanent polarization than $BaTiO_3$
Lead-titanate zirconate ($PbTiO_3$: $PbZnO_3$) 45 55	600	280°	Has large electromechanical coupling factor

Physical Property		Nickel-Zinc Ferrite Grades			
		B_1	B_3	B_5	B_6
Initial Permeability		$\geqslant 500$	100 to 200	10 to 25	6 to 10
Loss Factor at Frequencies of	MHz 0·06 0·50 1·0 30·0 100·0	$\leqslant 30$ $\leqslant 80$	$\leqslant 85$	$\leqslant 800$	$\leqslant 6000$
Magnetizing Force (H) (Oersted)		10	30	80	200
Magnetic Flux (B_{sat}) Density at 25°C (Gauss)		$\geqslant 2200$	$\geqslant 2500$	$\geqslant 1500$	$\geqslant 1200$
Resistivity (ρ) (Ohm-cm)		$\geqslant 10^5$	$\geqslant 10^5$	$\geqslant 10^5$	$\geqslant 10^5$
Curie Temperature (°C)		$\geqslant 125$	$\geqslant 350$	$\geqslant 500$	$\geqslant 500$
Magnetostriction Coefficient		$(-4$ to $-6) \times 10^{-6}$	$(-13$ to $-18) \times 10^{-6}$	$(-22$ to $-32) \times 10^{-6}$	—
Comments		Used up to 1 MHz	Used in range 1 to 5 MHz	Used in range 20 to 50 MHz	Used in range 50 to 100 MHz

Piezomagnetic Materials

Material	Curie Temperature (°C)	Electrical Resistivity (Ohm cm)	Saturation Magnetostriction ($\times 10^{-6}$)	Optimum Piezomagnetic Coupling Factor (k_0)	Q-factor of completely free core	Biasing field (B) for maximum k (Amp. cm^{-1})
Nickel stack	358	7×10^{-6}	-33	0·15 to 0·31	50 to 250	8 to 12
Nickel solid core	358	7×10^{-6}	-33	small	$\geqslant 400$	8 to 12
4 Co 96 Ni	410	10×10^{-6}	-31	up to 0·51		
2 V Permendur (2V-49Co-49Fe)	980	30×10^{-6}	$+70$	0·20 to 0·37		
Ni-Cu-Co Ferrite (Ferroxcube 7A1)	530	10^2 to 10^3	-28	0·25 to 0·32	$\geqslant 2000$	12 to 20
Ni-Cu-Co Ferrite (Ferroxcube 7A2)	530	10^2 to 10^3	-28	0·20 to 0·26	$\geqslant 2500$	9 to 14
98·8-1·2 Ni-Co Ferrite (Ferroxcube 7B)	590	$> 10^4$	-28	0·19 to 0·22	$\geqslant 4000$	25 to 50

k_0 is the real part of the piezomagnetic or electromechanical coupling coefficient and is given by (stored elastic energy/stored magnetic energy) at frequencies far below mechanical resonance frequency. k_0 is a maximum when $B \simeq 0.7B$ saturation.

324

A₁	A₄	A₉	A₁₃	Physical Property
		Manganese-Zinc Ferrite Grades		
$\geqslant 500$	$\geqslant 1200$	—	$\geqslant 1840$	Initial Permeability at B_{max} < 0·5 Gauss
				Loss Factor at B_{max} < 0·5 Gauss and Frequency of kHz
			$\leqslant 1·2$	4
$\leqslant 16$	$\leqslant 35$		—	60
—	—		$\leqslant 5·0$	100
$\leqslant 27$	$\leqslant 80$			250
$\leqslant 50$	$\leqslant 120$			450
10	10	10	10	Magnetizing Force (H) (Oersted)
				Magnetic Flux Density (B_{sat}) Temperature (°C) 15 to 30
$\geqslant 3200$	$\geqslant 3600$	$\geqslant 4500$	$\geqslant 3400$	100
$\geqslant 2100$		$\geqslant 3700*$		
$\geqslant 20$	$\geqslant 20$	—	$\geqslant 100$	Resistivity (ρ) (Ohm-cm)
$\geqslant 130$	$\geqslant 140$	$\geqslant 210$	$\geqslant 170$	Curie Temperature (°C)
0·4	0·3	—	—	Coercive Force (Average) (Oersted)
Older type of material used in simple transformers	Used in communication transformers	* At 85°C standard material for television line output transformers	Very low loss and highly stable material used in pulse and wideband transformers up to 100 MHz	Comments

Piezomagnetic Materials

These correspond to the *piezoelectric materials* (see p. 326) in which the phenomena observed involve magnetic instead of electric fields. Nickel has the high Curie point of 350°C approximately and so can be used at fairly high temperatures but is restricted to employment at fairly low frequencies, i.e., ~20 kHz, because of eddy-current losses. For higher frequencies ceramic-type permanent magnet materials are used, such as nickel-copper-cobalt ferrite Ferroxcube. FXC 7A2 is a material manufactured by Philips which has a Curie temperature of 550°C, and electroacoustic efficiencies of the order of 70-90% can be obtained. This efficiency is roughly twice that of nickel, and is approximately the same as for piezoelectric ceramics compared with which Ferroxcube has the advantage of offering a lower transducer impedance. See table on opposite page.

Grateful thanks are due to Mr C. M. van der Burgt for permission to reproduce material which first appeared in *Electronic Technology* (v. 37, Sept. 1960, pp. 330-41).

Piezoelectric Materials

Piezoelectric Coefficients

The electromechanical coefficient k is defined by:

$$k^2 = \frac{\text{mechanical energy converted into electrical charge}}{\text{mechanical energy injected into the crystal}}.$$

The converse effect is also true when an electrical potential difference is applied to the crystal, whence

$$k^2 = \frac{\text{electrical energy converted into mechanical energy}}{\text{electrical energy injected into the crystal}}.$$

In defining the piezoelectric characteristics, the first figure denotes the direction of the applied electrical signal while the second figure gives the direction of the elastic stress or strain. In a polycrystalline material the electrical signal is usually applied along the axis of polarization, which is determined in piezoelectric ceramics by the initial electrostatic field applied to the material (the direction is given by the subscript 3). The electromechanical coupling factors k_{31}, k_{33} and k_p refer respectively to the transverse, the thickness and the radial modes. The radial modes are those of a disk vibrating radially but excited by electrodes on its flat surfaces. d_{31} and d_{33} give the electrical charge generated by a given applied force (or in the converse effect it is the deflection due to a given applied voltage) when the stress is applied in the transverse and in the polarization directions respectively. The same numerical coefficients apply for both effects, $d_{\text{(direct)}}$ being expressed in coulomb/metre and $d_{\text{(converse)}}$ in metre/volt. The g constant denotes the electric field produced in a piezoelectric crystal by an applied stress and is given by

$$g = \frac{\text{volt/metre}}{\text{newton/metre}^2}.$$

The constants are connected by the relation $g = d/(K\epsilon_0)$, where K is the dielectric constant and ϵ_0 is the permittivity of a vacuum, numerically equal to 9×10^{-12} farad/metre. Another relation, which involves Young's modulus E and the coupling coefficient k, is $k^2 = gdE$.

Physical Property	Quartz 0° X-cut	Lithium Sulphate 0° Y-cut	Barium Titanate (BaTiO₃)	Lead Meta-Niobate	Lead Zirconate-Titanate		Units
					LZT-4	LZT-5	
Density (ρ)	2·65	2·06	5·6	5·8	7·6	7·7	10^3 Kg/m³
Acoustic impedance (ρc)	15·2	11·2	24	16	30·0	28·0	10^6 Kg/m²s
Young's modulus (E)	8·3	9·4	11–12	2·9	8·15	6·75	10^{10} Newton/m²
Mechanical quality factor (Q)	10^6	—	400	11	500	75	—
Frequency × thickness of specimen	2870	2730	2740	1400	2000	1800	kHz mm
Dielectric constant	4·5	10·3	1700	225	1200–1300	1500–1700	—
Coupling factor k_{31}	0·1	—	0·22	—	0·36	0·32	—
k_p	0·1	—	0·35	0·07	0·58	0·58	—
k_{33}	0·1	0·35	0·50	0·42	0·64	0·675	—
Piezoelectric constants d_{31}	2·3	16	8	—	285	374	10^{-12} m/volt
d_{33}	2·3	—	150–190	85	—	—	10^{-12} m/volt
Volume resistivity (at 25°C)	$>10^{12}$	—	$>10^{11}$	10^9	$>10^{12}$	$>10^{13}$	ohm-metre
Curie temperature	575	—	115	550	320	350	°C
Safe operating temperature	550	75	85	500	270	290	°C

326

Quartz Crystal Cuts

The crystal axes are used as reference directions. The X-axis is the electrical axis and the Y-axis the mechanical axis for the associated piezoelectric properties. X-, Y- and Z-cut crystals are slices made perpendicular to these X-, Y- and Z-axes. X-cut crystals have a negative temperature coefficient of resonant frequency while Y-cut crystals have a positive coefficient. XY cuts are made at such an angle to the X- and Y-axes that the temperature variation is reduced to a minimum. Cuts denoted by AT, BT, CT, DT, ET and FT are rectangular slices made at different angles to the Z-axis ($+35° 13'$, $-49°$, $+38°$, $-52°$, $+66·5°$ and $-57°$ respectively) and with the X-axis parallel to one edge. The GT cut is made from a similar slice at $+51°$ to the Z-axis, but with the X-axis lying in the plane of the slice and orientated at an angle of 45° to one edge. Ring crystals are rings prepared from a Y-cut slice. NT cuts are orientated so that flexural vibrations result in place of the usual longitudinal vibrations. Cuts AT and BT are used for high frequencies; CT, DT, ET, and FT for low frequencies; GT cuts give the smallest temperature variation.

QUARTZ CRYSTAL SHOWING VARIOUS CUTS

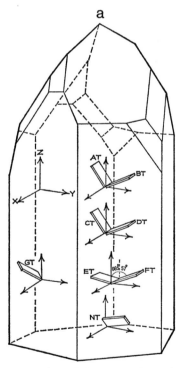

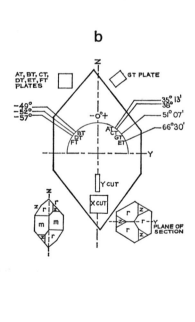

Orientation of the commonly used cuts in relation to an idealized crystal.

Various cuts as viewed along the X-axis.

Insulating Materials

Electrical and Physical Properties of Ceramic Insulators

Material	Density (g/cc)	Thermal Expansion (deg^{-1}C)	Maximum Operating Temperature (°C)	Dielectric* Strength (volt/mil)	Dielectric Constant	Power Factor at 1 MHz	Resistivity at 20°C (ohm-cm)
Alumina	2·3	$5·4 \times 10^{-6}$	1650	40-100	5-6	0·0002-0·01	10^{12}-10^{14}
Alumina porcelain	3·5	$5·4 \times 10^{-6}$	1480	250-400	8-9	0·001-0·002	10^{14}-10^{15}
Alumina silicate	2·4	$5·4 \times 10^{-6}$	1650	40-100	5-6	0·0008-0·01	10^{12}-10^{15}
Cordierite ($Mg_2Al_4Si_5O_{18}$)	2·0	$3·6 \times 10^{-6}$	1260	40-100	5	0·004-0·010	10^{12}-10^{14}
Fosterite (Mg_2SiO_4)	2·8	11×10^{-6}	1090	200-300	6	0·0003	10^{13}-10^{15}
High-voltage porcelain	2·4	7×10^{-6}	980	250-400	6-7	0·006-0·01	10^{12}-10^{14}
Low-voltage porcelain	2·3	7×10^{-6}	900	40-100	6-7	0·01-0·02	10^{12}-10^{14}
Steatite ($MgSiO_3$)	2·6	9×10^{-6}	1090	200-350	6-7	0·0008-0·0035	10^{13}-10^{15}
Zirconia porcelain	3·6	$4·5 \times 10^{-6}$	1200	250-350	8-9	0·0006-0·002	10^{13}-10^{15}

* 0·25 inch thick specimen.

Insulating Materials
Electrical Properties of Common Glasses

Type of Glass	Resistivity (ohm-cm)		*Dielectric Properties		
	20°C	290°C	Dielectric Constant (K)	Power Factor (%)	Dielectric Loss Factor (90) (δ)
Aluminosilicate	10^{17}	10^9	6·3	0·40	2·30
Borosilicate (low expansion)	10^{15}	10^7	4·6	0·50	1·30
Borosilicate (low loss)	10^{17}	10^9	4·0	0·06	0·24
Borosilicate (sealing tungsten)	10^{16}	10^7	4·9	0·30	1·30
Lead-alkali-silicate (electrical)	10^{16}	10^7	6·6	0·16	1·10
Lead-alkali-silicate (high lead)	10^{17}	10^{10}	9·5	0·09	0·85
96% Silica	10^{17}	10^9	3·8	0·05	0·10
Silica glass	—	10^{10}	3·8	0·02	—
Soda-lime	10^{18}	10^5	7·3	0·4-1·0	6·50

Dielectric strengths of glasses are sensitive to environment and vary from 80 kV/cm to 10,000 kV/cm
* Measured at 20°C and 1 MHz; $\delta = K \tan \theta$, where $\theta = $ loss angle of material.

329

L *

Dielectric Heating

This type of heating is applied to non-conductors and the heat is generated within the material. The latter should be of a polar nature for most efficient heating, which is proportional to $\dfrac{E^2 f \cdot S \phi}{t}$, where E is the applied voltage, f the frequency, S the area of the heated specimen and t its thickness. ϕ is the 'loss factor' of the material which is equal to the product of the dielectric constant and the power factor. The value of the applied voltage is limited by the breakdown strength of the dielectric. Both the dielectric constant and power factor are functions of frequency and temperature. Values for some typical materials at a frequency of one megahertz are given below.

Dielectric	Dielectric Constant	Power Factor	Loss Factor
Amber	2·8	$0·02 \times 10^{-2}$	$0·06 \times 10^{-2}$
Asbestos Board	3·0	22×10^{-2}	66×10^{-2}
Beech Wood (15% moisture)	9·4	$5·8 \times 10^{-2}$	55×10^{-2}
Ebonite	3·0	$0·9 \times 10^{-2}$	$2·7 \times 10^{-2}$
Mica	7·0	$0·02 \times 10^{-2}$	$0·14 \times 10^{-2}$
Mycalex	7·0	$0·3 \times 10^{-2}$	$2·1 \times 10^{-2}$
Paraffin Oil	2·3	$0·01 \times 10^{-2}$	$0·02 \times 10^{-2}$
Paper (dry)	3·2	$0·23 \times 10^{-2}$	$0·74 \times 10^{-2}$
Perspex	~3·1	$\sim 3·3 \times 10^{-2}$	$\sim 10 \times 10^{-2}$
Polystyrene	2·5	$0·02 \times 10^{-2}$	$0·05 \times 10^{-2}$
Polythene	2·3	$0·02 \times 10^{-2}$	$0·05 \times 10^{-2}$
PTFE	2·0	$0·02 \times 10^{-2}$	$0·04 \times 10^{-2}$
Quartz (fused)	3·8	$0·02 \times 10^{-2}$	$0·08 \times 10^{-2}$
Glass (Pyrex)	4·7	$0·3 = 10^{-2}$	$1·5 \times 10^{-2}$

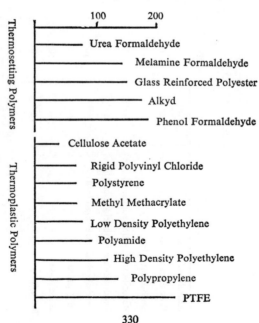

Maximum Operating Temperature (°C)

330

Electrical Properties of Gases and Semiconductors

Table of Ionization Potentials, etc., for Various Gases

Gas	Ionization Energy in Volts	Average Velocity (NTP)	Mean Free Path (NTP)
Helium	24·5	1208 m/s	28·5 $\times 10^{-8}$ m
Neon	21·5	538 m/s	19·3 $\times 10^{-8}$ m
Nitrogen	16·7	454 m/s	9·44 $\times 10^{-8}$ m
Hydrogen	15·9	1696 m/s	18·3 $\times 10^{-8}$ m
Argon	15·7	381 m/s	10·0 $\times 10^{-8}$ m
Carbon monoxide	14·2	454 m/s	9·27 $\times 10^{-8}$ m
Oxygen	13·5	425 m/s	9·95 $\times 10^{-8}$ m
Krypton	13·3	263 m/s	9·49 $\times 10^{-8}$ m
Water vapour	13·2	652 m/s	7·22 $\times 10^{-8}$ m
Xenon	11·5	210 m/s	5·61 $\times 10^{-8}$ m
Mercury vapour	10·4	170 m/s	15·6 $\times 10^{-8}$ m

N.B.—(1) Rms velocity is 1·086 times average velocity.
(2) Electron mean free paths in a gas exceed those of the gas molecules by a theoretical factor of $4\sqrt{2}\,(=5·6)$.

Electrical Properties of Germanium and Silicon

Property	Germanium		Silicon	
	at 300°K (27°C)	at 273°K (0°C)	at 300°K (27°C)	at 273°K (0°C)
Energy gap in electron-volts	0·67	0·75	1·106	1·153
Number of current carriers of either sign per cm³ in intrinsic material	$2·5 \times 10^{13}$	6×10^{12}	$1·6 \times 10^{10}$	4×10^{9}
Maximum Resistivity in ohm cm	46	200	$2·3 \times 10^{5}$	10^{6}
Mobiliy of electrons in cm² volt⁻¹ sec⁻¹	3600	4200	1200	1600
Mobility of holes in cm² volt⁻¹ sec⁻¹	1700	2100	400	520
Dielectric constant	16	—	12	—

Impurities in Germanium and Silicon

Element	Type	Ionization energy (eV)	
		In germanium	In silicon
Boron	acceptor	0·0104	0·045
Aluminium	acceptor	0·0102	0·057
Gallium	acceptor	0·0108	0·065
Indium	acceptor	0·0112	0·16
Phosphorus	donor	0·0120	0·044
Arsenic	donor	0·0127	0·049
Antimony	donor	0·0096	0·039
Lithium	donor (interstitial)	0·0093	0·033

Thermocouple Data

Type	Copper/Constantan	Iron/Constantan	Chromel/Alumel	Chromel/Constantan	Platinum/Platinum Rhodium* (10%)
Useful Working Temperature Range	−200°C to +300°C	−200°C to 1380°C	−200°C to 1200°C	0°C to 1100°C	0°C to 1450°C
Thermo-e.m.f. (reference junction at 0°C)	°C mV 100 4·24 200 9·06 300 14·42	°C mV 100 5·28 200 10·78 400 21·82 600 33·16 800 45·48 1000 58·16	°C mV 100 4·1 200 8·13 400 16·39 600 24·90 800 33·31 1000 41·31 1200 48·85 1400 55·81	°C mV 100 6·3 200 13·3 400 28·5 600 44·3	°C mV 100 0·643 200 1·436 400 3·251 600 5·222 800 7·330 1000 9·569 1200 11·924 1400 14·312 1600 16·674

* Using a platinum/platinum rhodium (13%) couple the e.m.f. per deg. C is slightly increased above that for platinum/rhodium (10%), the total e.m.f. being 18·68 mV for hot junction at 1600°C.

	Copper	Constantan (Eureka)	Iron	Chromel	Alumel	Platinum	Platinum/ Rhodium (10%)
Resistivity (10^{-6} ohm-cm)	1·75	49	10	70	29·4	10	21
Temp. Coeff. Resistance deg^{-1} C	$3·9 \times 10^{-3}$	1×10^{-5}	5×10^{-3}	$3·5 \times 10^{-4}$	$1·25 \times 10^{-4}$	3×10^{-3}	$1·8 \times 10^{-3}$
Composition		60% Copper 40% Nickel		90% Nickel 10% Chromium	94% Nickel 2% Aluminium 3% Manganese 1% Silicon		90% Platinum 10% Rhodium
Melting-point (°C)	1085	1190	1535	1400	1430	1755	1700

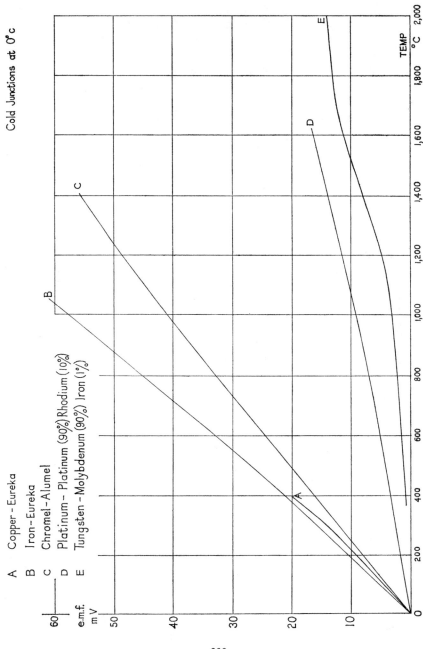

Cold Junctions at 0°c

A Copper – Eureka
B Iron – Eureka
C Chromel – Alumel
D Platinum – Platinum (90%) Rhodium (10%)
E Tungsten – Molybdenum (90%) Iron (1%)

e.m.f. mV

TEMP °C

333

Wire Data (SWG)

SWG No.	Bare Copper Wire				Covered Copper Wire (turns per inch)					Resistance Wire	
	Length/ohm in feet	Ohm/1000 yd	Maximum safe current (amps)	Diameter (mils)	Enamelled	SCC	SSC	DCC	DSC	Manganin Ohm/1000 yd	Constantan Ohm/1000 yd
14	628·1	4·776	19·0	80	—	11·4	12·1	10·6	11·8	120·0	138·1
16	403·8	7·478	13·0	64	15·0	14·1	14·9	13·2	14·6	186·5	216·0
18	226·2	13·27	7·0	48	19·8	18·5	20·0	17·2	19·4	330·1	384·0
20	127·2	23·59	4·0	36	26·1	23·8	26·3	21·7	25·3	589·8	662·0
22	76·8	39·00	2·5	28	33·3	29·4	33·3	26·3	31·8	978·3	1128
24	47·4	63·16	1·5	22	42·1	35·7	42·1	31·3	40·0	1581	1826
26	31·8	94·35	1·0	18	50·6	41·7	50·6	35·7	47·6	2359	2729
28	21·54	139·6	0·7	14·8	61·4	48·1	60·4	40·2	56·2	3493	4205
30	15·09	198·8	0·5	12·4	73·3	54·4	72·0	44·7	67·1	4983	5750
32	11·46	262·1	0·4	10·8	83·0	63·3	81·3	50·5	75·2	6564	7581

N.B.—The safe current quoted is only applicable to single copper wires; for enclosed windings, as in a transformer, the safe current would be only one third or less of the value in table.

SWG = Standard Wire Gauge

1 ft. = 30·48 cm.

1 mil = 10^{-3} in.

1 in. = 2·54 cm.

1000 yd. = 914·4 metre.

SCC Single cotton covered
SSC   ,,   silk   ,,
DCC Double cotton   ,,
DSC   ,,   silk   ,,

ELECTROACOUSTIC DATA

Acoustic Spectrum

Frequency or Wavelength	Specification	Some Applications
0-15 Hz	Infrasonic frequencies	Occur in nature, in car vibrations, etc.
15-15,000 Hz	Audio frequencies	50 Hz mains power supply, a.c. motors, audio apparatus for recording and reproduction, viz., loudspeakers, microphones, amplifiers, intercoms, etc.
15 kHz-30 kHz	Lower end of ultrasonic frequencies and of radio frequencies	Sound waves applied to ultrasonic cleaning, flow detection, sonar, etc. Electromagnetic waves for induction heating
30 kHz-5 MHz	Middle range of ultrasonic frequencies	Applied in flaw or defect testing of solids and liquids, also delay lines
>100 MHz	Ultrasonic microwave region	Used in solid-state research problems

Acoustic Absorption

The absorption coefficient is a measure of the ratio of the acoustic energy which is absorbed by a surface to that which is incident on the surface. The absorption coefficient of an open window is therefore unity, provided that the wavelength of the sound is large compared with the dimensions of the opening. In a hall or room the average absorption can vary from only 2 per cent. for a painted plaster surface to between 60 and 70 per cent. for a surface which has been acoustically treated. The absorption coefficient of a material shows a variation with the frequency of the incident sound. For example, in the case of mineral wool, it has the values of 0·15, 0·75 and 0·85 at the frequencies of 125 Hz, 500 Hz and 2000 Hz respectively. These frequencies are representative of the low, middle and high audio range.

One of the distinguishing acoustical features of a room is its reverberation time (T), which is defined as the time for sound to decay to 10^{-6} of its intensity, i.e., to fall 60 dB. The time may be calculated from the following formula

$$T = \frac{0 \cdot 049\,V}{-2 \cdot 30\,S \log_{10}(1-\alpha)},$$

where T is in seconds, V is the volume of the enclosure in c. ft., S is the total surface area of the room in sq. ft., and α is the average absorption coefficient of all the bounding surfaces of the room. The graphs below show approximately the optimum reverberation times for rooms of different volumes when used for different types of acoustic communication. In order to calculate the optimum absorption required to achieve the optimum acoustic acceptable conditions, the volume of the room or hall should first be evaluated and the optimum reverberation time found from the appropriate linear graph of the figure. The formula above, or its more approximate form $T = 0 \cdot 05\,V/(S\alpha)$, is now used to determine the optimum value of $S\alpha$. This latter value should then be compared with that obtained from summing up the product (area × absorption coefficient) for the various surfaces involved, including allowances for furniture and audience.

If the objective is to reduce the noise level in a room, $(S\alpha)_b$ and $(S\alpha)_a$ being the respective values of the total absorption before and after treatment, then the reduction of sound pressure level (SPL) is given by $10 \log_{10}[(S\alpha)_a/(S\alpha)_b]$dB. Hence if the total absorption is increased by 100 per cent, the reduction in SPL is 3 dB.

In the case of the transmission of sound from one room to another through the partition wall, the acoustic insulation is mainly a function of the mass of the wall itself. The transmission loss increases by 4 dB approximately for a doubling of the mass of the partition, but if the latter is thin and flexible it may show an "acoustic transparency" in the region of a particular frequency. This frequency may be changed by altering the mass or the stiffness of the partition, depending on whether the loss of insulation is due to the resonance or coincidence effects respectively.

335

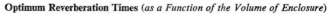

Optimum Reverberation Times (*as a Function of the Volume of Enclosure*)

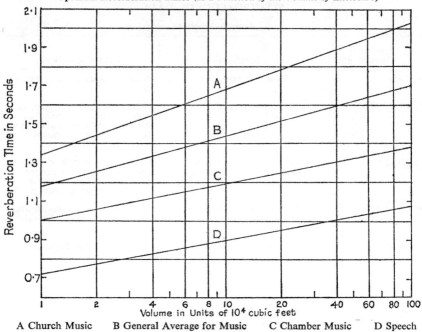

A Church Music   B General Average for Music   C Chamber Music   D Speech

Acoustic Absorption Coefficients of some Common Materials

Material	Absorption Coefficient at various frequencies (hertz)					
	125 Hz	250 Hz	500 Hz	1000 Hz	2000 Hz	4000 Hz
Brick Wall, unpainted	0·02	0·02	0·03	0·04	0·05	0·05
Brick Wall, painted (approximately the same as for unpainted concrete)	0·01	0·01	0·02	0·02	0·02	0·025
Wood Block Floor	0·05	0·03	0·06	0·09	0·10	0·22
Linoleum on Solid Floor	0·04	0·03	0·04	0·04	0·03	0·02
Wood Panelling (between ¼″ and ½″ thick with air space behind)	0·30	0·25	0·20	0·17	0·15	0·10
Carpet (with underlay)	0·20	0·25	0·35	0·40	0·50	0·75
Heavy Draped Curtains	0·07	0·31	0·49	0·75	0·70	0·60
Glass	0·04	0·04	0·03	0·03	0·02	0·02
Glass Fibre on Floor or Stuck in Wall	0·10	0·35	0·55	0·65	0·75	0·80
Acoustic Tiles: (1) Screwed to ceiling, etc.	0·10	0·20	0·40	0·50	0·45	0·50
(2) Mounted on battens	0·30	0·45	0·50	0·55	0·65	0·80

Noise Measurement

The rms sound pressure is the quantity normally measured when dealing with acoustic noise, the unit being one dyne/cm^2 ≡ one microbar (μ bar) ≡ 0·1 Newton/m^2. The smallest sound pressure detectable at 1000 Hz by the average person is 2×10^{-5} N/m^2, whereas the largest without-pain perception is of the order of 100 N/m^2. With this large dynamic range it is more convenient therefore to use a relative scale of sound pressure, the decibel (dB) scale.

Hence *sound pressure level* (SPL) $= 10 \log (p^2/p_\gamma^2) = 20 \log (p/p_\gamma)$ dB, where p is the measured sound pressure and p_γ is the reference pressure, which is usually 2×10^{-5} N/m^2. On this scale typical sound levels (dB) in various environmental conditions would be approximately as follows:—broadcasting studio 15, living room (suburb) 45, conversational speech 65, inside tube train 95, pneumatic drill 125.

As a result of numerous experiments it has been found that if the physical intensity of a sound is increased so that it appears to the listener to have doubled its loudness the actual increase is not equal to a factor of two on the dB scale. In fact the increase is approximately 10 dB over most of the audio range. The unit of loudness as perceived by the human ear is termed the *sone*, whereas the scale of loudness level is in *phons*.

According to the International Standardisation Organisation (ISO) Recommendation ISO/R 131-1959 the relationship between sones (S) and phons (P), over a range of loudness levels between 20 and 120 phons, may be expressed as $S = 2^{(P-40)/10}$ or $P = 40 + 10 \log_2 S$.

Annoyance and *noisiness* are concepts which do not seem to be predictable from physical measurements but some attempt has been made to assess the noisiness of aircraft noise by means of a scale graduated in PNdB (*perceived noise level*). This sound level is measured by the sound pressure level of a reference sound which is considered by normal observers to be equally noisy. The reference source consists of a band of random noise between one-third and one octave wide with its centre on 1000 Hz. Analogous to the relation between phon and sone is a relation between the unit of noisiness, the *noy*, and the perceived noise level in PNdB, viz., PNdB $= 40 + 10 \log_2$ (N).

In order that objective noise measurements can be interpreted in noise nuisance criteria, sound level meters are supplied with a set of frequency weighting networks which are designated A, B and C. The symbols dB(A), dB(B) and dB(C) designate sound levels relative to 2×10^{-5} N/m^2 as measured with frequency weightings A, B or C respectively, and with the stipulation that the sound level of a 1 kHz signal is equal numerically to the sound pressure level. Hence all the weighting curves coincide at the frequency of 1 kHz. The characteristic C shows only little dependence on frequency over the greater part of the audio range, whereas A has a strong frequency dependency below about 1000 Hz. In the revision of the ISO regulation for the measurement of aircaft noise it has been proposed that: (*a*) the noise be measured with a weighting network inverse to the '40 noy contour' having 0 dB insertion loss at 1 kHz; and (*b*) the reading so obtained be increased by 7 dB to approximate to the perceived noise level L_{PN} (in PNdB). Unfortunately the same 'N-weighting' has been applied to this network while the symbol 'dB(N)' has been used to designate the quantity some 7 dB higher than the measured level. In order to avoid this ambiguity a recent meeting of the International Technical Commission (I E C/TC29) suggests that the letter D be used to define the weighting curve that coincides with the A, B and C curves at the 1 kHz point. The letter N may then specifically designate a curve that is 7 dB higher. All direct measurements intended to approximate to perceived noise level should therefore be quoted to comply with the relation $L_{PN} \simeq L_D + 7$, where L_{PN} is the perceived noise level in dB, and L_D is the D-weighted pressure level in dB.

Standard Octave Bands for Noise Measurements

Range Hz	Centre Frequency Hz	Range Hz	Centre Frequency Hz
45 - 90	63	710 - 1400	1000
90 - 180	125	1400 - 2800	2000
180 - 355	250	2800 - 5600	4000
355 - 710	500	5600 - 11,200	8000

SPACE VEHICLES AND SATELLITE COMMUNICATIONS

Space Research

While it is undoubtedly true that the main motive for space research during the last twenty years has been military, there have been a very large number of important scientific applications and discoveries as well. Some of these are briefly tabulated below:

(1) the development of interplanetary flight techniques;

(2) the collection of astronomical data;

(3) upper atmospheric research leading to important discoveries relating to:
    (i) the magnetic fields of the solar system,
    (ii) the Van Allen and other radiation belts,
    (iii) the ionization of the upper atmosphere,
    (iv) the nature of primary cosmic rays, the aurora, etc.,
    (v) tidal movements of the ionosphere;

(4) the development of communications satellites;

(5) the study of how gravitational forces affect biological development;

(6) the introduction of improved navigational techniques;

(7) the vast increase in our ability to understand the factors controlling our weather and to produce more reliable short- and long-term forecasts. This may eventually lead to the development of techniques for climatic control.

This very broad research field has obviously required the development of a number of extremely large-scale independent research programmes and corresponding space vehicles. The table below lists some of the best-known of these vessels together with their country of origin and research programme.

Directory of Space Vehicles

Aeros (U.S.)	Projected synchronous orbiting meteorological satellite.
Agena (U.S.)	Component of three-stage launching vehicle based on *Atlas* and *Thor*.
Alouette (Canada-U.S.)	Joint ionospheric-sounding satellite.
Anna (U.S.)	Satellite for geomagnetic investigations.
APL (U.S.)	*A*pplied *P*hysics *L*aboratory research satellite (1963).
Apollo (U.S.)	Spacecraft programme designed for eventual manned lunar landing.
Ariel (U.K.-U.S.)	Joint scientific satellite programme; see **U.K.**
Atlas (U.S.)	First stage of two-stage launching rocket.
ATS (U.S.)	Technological (communications, meteorological, and navigational) series of satellites.
Aurora (U.S.)	Code name for manned *Mercury* spaceflight.
Beacon (U.S.)	Early experimental satellite.
Biosatellite (U.S.)	Earth-orbiting biological laboratories programme.
Blue Scout (U.S.)	Military ionospheric-sounding project.
Canary Bird	Commercial communications satellite; see **Intelsat.**
Centaur (U.S.)	Second stage launching vehicle for use with *Atlas*.
Comsat (U.S.)	Commercial satellite communication system.
Cosmos (U.S.S.R.)	Radiation-belt investigation (and possibly military) satellites.
Courier (U.S.)	Military communications satellite.
D1, D2, etc. (France)	Series of geodetic and scientific satellites.
Delta (U.S.)	Component of three-stage launching vehicle based on *Atlas*.
Discoverer (U.S.)	Series of early space capsules.
Early Bird (U.S.)	Commercial active synchronous communications satellite.
Echo (U.S.)	Passive communications satellite.
Elektron (U.S.S.R.)	Radiation-belt investigation satellite.
Esro	Series of scientific satellites (*E*uropean *S*pace *R*esearch *O*rganization).
Essa (U.S.)	Meteorological satellites developed from *Tiros*.
Explorer (U.S.)	High-altitude scientific observation satellite.
Faith (U.S.)	Code name for orbital manned *Mercury* flight.
Ferret (U.S.)	A communications monitoring offshoot of *Samos*.
Freedom (U.S.)	Code name for first manned *Mercury* flight.
Friendship (U.S.)	Code name for manned spaceflights in *Mercury*.
Gemini (U.S.)	Spacecraft programme for rendezvous between vessels in space, following *Mercury* project and preceding *Apollo*.

Geos (U.S.)	Series of geodetic satellites.
GGSE (U.S.)	Gravity gradient *stabilization experiment.*
Greb (U.S.)	Solar X-ray monitoring satellite.
Heos (ESRO)	Projected satellite for interplanetary physics and cosmic-ray study.
Injun (U.S.)	Cosmic-ray research satellite.
Intelsat	(I) International communications system using *Early Bird* satellite. (II) Commercial system (see **Nascom**) using *Canary Bird* and *Lani Bird* satellites. (III) Projected worldwide satellite system.
Isis (Canada-U.S.)	Joint ionospheric-sounding programme.
Joe (U.S.)	Suborbital test programme for *Mercury* capsule.
Jupiter (U.S.)	Biological experiment project.
Lafti (U.S.)	Satellite programme.
Lambda (Japan)	Satellite programme.
Lani Bird	Commercial communications satellite; see **Intelsat**.
Las (ESRO)	Projected astronomical satellite.
Les (U.S.)	Testing programme for military communications study.
Liberty Bell (U.S.)	Code name for suborbital manned *Mercury* flight.
Luna Orbiter (U.S.)	Programme for obtaining lunar topographical information for *Apollo* project.
Lunik (or **Luna**) (U.S.S.R.)	Lunar probe satellites.
Mariner (U.S.)	Interplanetary space probe vessels.
Mercury (U.S.)	Series of manned space capsules.
Minuteman (U.S.)	Solid-fuel rocket developed as missile.
Mol (U.S.)	Manned spacecraft programme for military investigation.
Molniya (U.S.S.R.)	Programme of space communications satellites.
Nascom	Joint programme to provide *Apollo* network communications, using *Intelsat II* satellites.
Navy Navigation Satellites (U.S.)	Operational system of navigational satellites.
NDS	See **Vela.**
Nike (U.S.)	Rocket used in investigations of upper atmosphere.
Nimbus (U.S.)	Meteorological satellites programme.
Noss (U.S.)	Orbiting space station, proposed to be eventually manned.
OAO (U.S.)	Scientific satellite programme (*O*rbiting *A*stronomical *O*bservatory).
OGO (U.S.)	Scientific satellite project (*O*rbiting *G*eophysical *O*bservatory).
Oscar	Series of small communications satellites carrying amateur radio.
OSO (U.S.)	Scientific satellites for solar flare investigation.
Pioneer (U.S.)	Series of long-range spacecraft.
Polaris (U.S.)	Submarine-launched missile.
Polyot (U.S.S.R.)	Prototype steerable spacecraft.
Prospector (U.S.)	Unmanned lunar spacecraft project.
Ranger (U.S.)	Lunar-study project.
Rebound (U.S.)	Project for launching *Echo* satellites.
Redstone (U.S.)	Launch vehicle for *Mercury* capsule.
Relay (U.S.)	Active communications satellites.
Rover (U.S.)	Projected nuclear rocket engine.
Samos (U.S.)	Photographic and electronics reconnaissance programme.
San Marco (U.S.-Italy)	Joint programme of scientific satellite series.
Saturn (U.S.)	Projected launching system for ten-ton satellites.
Score (U.S.)	Army communications satellite.
Scout (U.S.)	Launch vehicle used with *Ariel.*
Secor (U.S.)	U.S. Army geodetic programme.
Sentry	Satellite system for detecting nuclear explosions in space.
Shotput (U.S.)	Suborbital communications test project.
Sigma (U.S.)	Code name for orbital manned *Mercury* flight.
SMS (U.S.)	Projected series of *synchronous meteorological s*atellites.
Soyuz (U.S.S.R.)	Manned spaceflight programme.
Sputnik (U.S.S.R.)	Range of scientific satellites.
Starflash (U.S.)	Interagency geodetic satellite emitting coded flashes.
Surveyor (U.S.)	Unmanned spacecraft used to obtain samples of lunar surface.
Syncom (U.S.)	Projected synchronous orbiting communications satellite.
TD1, TD2 (ESRO)	Series of scientific satellites.
Telstar (U.S.)	First commercial communications satellite.
Thor (U.S.)	Launching vehicle for *Tiros* and *USAF* satellites.
Tiros (U.S.)	Series of meteorological satellites (see abbs.).

Transit (U.S.)	Navigational satellites sponsored by U.S. Navy.
U.K.	Satellites of *Ariel* programme, the third (1967) of which **(Ariel 3 or U.K. 3)** is all-British.
Union (U.S.S.R.)	Manned spacecraft programme.
USAF (U.S.)	Code name for satellites launched in *Samos* and *Discoverer* programmes.
Vanguard	Series of scientific satellites launched as part of International Geophysical Year.
Vela (U.S.)	Nuclear explosion detection satellites, now called NDS system.
Venus Probe (U.S.)	Interplanetary rocket project.
Vostok (U.S.S.R.)	Launching rocket for manned spaceflights.
Voyager (U.S.)	Manned follow-up to *Mariner*.
Zond (U.S.S.R.)	Guided space probe to Venus.

Goonhilly Earth Stations

Goonhilly 1 is a G.P.O. station for satellite communication, operating at 6 GHz as transmitter and 4 GHz as receiver. It employs an 85 ft. diameter paraboloid reflector accurate to 0·1 inch over 99 per cent. of its surface with a total weight in the rotating system of over 1,000 tons. It can transmit up to 10 kW useful power and provide useful reception for a signal of 1 picowatt, after amplification in a ruby travelling-wave laser operating at cryogenic temperatures, followed by a low-noise travelling-wave tube. Steering is carried out by a servo system controlled by punched tape which carries information on the predicted orbit of the satellite. Small steering errors or oscillations due to wind gusts are compensated for by an automatic beam swinging system controlled by the received signal and operating on the aerial feed. (A 90-ft. dish aerial is under construction.)

The performance is claimed to be comparable to that of the cornucopia horns used in the American and French stations, and to have a lower intrinsic noise temperature in rain since it does not require a radome.

A second space communication station (*Goonhilly* 2), which is linked to the recently launched Intelsat III satellite, has now become operational. It will provide telephone communication with the USA, Canada, Africa and the Middle East, with a total capacity of nearly 400 circuits as well as a colour television link. Multiple access facilities for simultaneous contact with different Earth stations have been incorporated.

The design of the Goonhilly stations is now being widely adopted for use elsewhere. The Marconi company have constructed five such stations and are currently working on a number of others for both civil and military applications.

SUMMARY OF FUNDAMENTAL CONCEPTS OF
ATOMIC AND NUCLEAR PHYSICS

The following notes have been prepared to help expand and correlate some of the more technical entries in this field. Terms first printed in heavy type below are defined in the main part of the dictionary.

Atomic Physics

All chemical **elements** have characteristic **atoms**, and differences in their chemical behaviour arise from differences in the **electron** configuration of the atom in its normal electrically neutral state.

Each atom consists of a heavy **nucleus** with a positive charge produced by a number of **protons** equal to its **atomic number**, and there are normally an equal number of electrons outside the nucleus to balance this charge. The best picture of the atom is given by **Sommerfeld's** model modified by the **wave mechanical** concept of **orbitals**. Electrons are **fermions** which must conform to the **Pauli exclusion principle**, so no two of them in the same atom can have identical energies, while possible energies are all **quantized** and so differ by discrete amounts.

Possible orbitals and their energies are classified by four **quantum numbers**. The **principal quantum number** indicates the **shell** to which the orbital belongs and varies from 1 for the K-shell (that closest to the nucleus) to 7 for the Q-shell (that most remote). In general, the closer an electron to the nucleus the greater the **coulomb** attraction between them, and therefore the consequent **binding energy** retaining the electron in the atom. Orbitals may be of varying **eccentricity** with **angular momentum** values quantized to exact multiples of the **Dirac unit** $\hbar$. These angular momentum values are indicated by the second or **orbital quantum number**. The total number of possible orbitals in any shell is theoretically equal to its principal quantum unmber although in practice only four such orbitals of varying eccentricity are occupied in any shell for **unexcited** atoms.

These have orbital quantum numbers from 0 to 3 indicating increasing orbital angular momenta (Dirac units). They are also commonly represented by the letters s, **p**, d and f. (These letters are derived from **series of spectrum lines** known as sharp, principal, diffuse and fundamental respectively.) In the **s-state** electrons can have only two energy values corresponding to two different directions of **spin**, but all other orbitals may be orientated at different angles to a preferred direction (corresponding with that of an external **magnetic field**), there being 3 such directions for a **p-state**, 5 for a **d-state** and 7 for an **f-state**. Each such direction has a slightly different energy (the difference depending upon the field value and being associated with the **Zeeman** splitting of spectrum lines) and so each such differently orientated orbital can contain its two electrons of opposite spin.

Table 1 (reproduced by kind permission of the Mullard Educational Service) shows the electron configurations of all elements; s-levels (the **energy levels** associated with s-states) contain up to 2 electrons, p-levels up to 6, d-levels up to 10 and f-levels up to 14. In the **ground state** of an atom its electrons always occupy the lowest available energy levels as indicated on this table, i.e., the 5s level in the P-shell is lower in energy (the electron is more strongly bound) than the 4d or 4f levels of the O-shell. This table also gives the chemical symbol and group of each element—these being determined purely by its atomic number and consequent electron configuration. The number of **neutrons** most commonly found in the nucleus of each atom is also given. Nuclear binding forces tend to give greatest stability when the **neutron number** and the **proton number** are approximately equal. But due to **electrostatic** repulsion between protons the heavier nuclei are most stable when less than half their **nucleons** are protons, and elements with more than 83 protons in the nucleus are unstable, eventually undergoing **radioactive disintegration**.

Those with more than 92 protons are not found naturally on earth. They can be synthesized in high-energy laboratories but have limited **half-lives**. These are the **transuranic elements**.

Most elements exist with several stable **isotopes** and the chemical **atomic weight** gives the average weight of a normal mixture of atoms of the various possible isotopes (see natural isotopic **abundance**). But where the proton-neutron ratio in a nucleus differs appreciably from the optimum the isotope will again be radioactive, converting a neutron into a proton through the emission of a negative **beta-particle**, or a proton into a neutron through either the emission of a **positron** or **electron capture**.

TABLE 1

LIST OF THE ELEMENTS

Atomic Number	Name	Symbol	Atomic Weight	K s	L s	L p	M s	M p	M d	N s	N p	N d	N f	O s	O p	O d	O f	P s	P p	P d	Q s	Group	Number of Neutrons	Atomic Number
1	HYDROGEN	H	1·008	1																		I	-	1
2	HELIUM	He	4·003	2																		I	-	2
3	LITHIUM	Li	6·939	2	1																	I	4	3
4	BERYLLIUM	Be	9·012	2	2																	II	5	4
5	BORON	B	10·81	2	2	1																III	6	5
6	CARBON	C	12·01	2	2	2																IV	6	6
7	NITROGEN	N	14·01	2	2	3																V	7	7
8	OXYGEN	O	16·00	2	2	4																VI	8	8
9	FLUORINE	F	19·00	2	2	5																VII	10	9
10	NEON	Ne	20·18	2	2	6																-	10	10
11	SODIUM	Na	22·99	2	2	6	1															I	12	11
12	MAGNESIUM	Mg	24·31	2	2	6	2															II	12	12
13	ALUMINIUM	Al	26·98	2	2	6	2	1														III	14	13
14	SILICON	Si	28·09	2	2	6	2	2														IV	14	14
15	PHOSPHORUS	P	30·97	2	2	6	2	3														V	16	15
16	SULPHUR	S	32·06	2	2	6	2	4														VI	16	16
17	CHLORINE	Cl	35·45	2	2	6	2	5														VII	18	17
18	ARGON	Ar	39·95	2	2	6	2	6														-	22	18
19	POTASSIUM	K	39·10	2	2	6	2	6	-	1												I	20	19
20	CALCIUM	Ca	40·08	2	2	6	2	6	-	2												II	20	20
21	SCANDIUM	Sc	44·96	2	2	6	2	6	1	2												IIIa	24	21
22	TITANIUM	Ti	47·90	2	2	6	2	6	2	2												IVa	26	22
23	VANADIUM	V	50·94	2	2	6	2	6	3	2												Va	28	23
24	CHROMIUM	Cr	52·00	2	2	6	2	6	5	1												VIa	28	24
25	MANGANESE	Mn	54·94	2	2	6	2	6	5	2												VIIa	30	25
26	IRON	Fe	55·85	2	2	6	2	6	6	2												VIII	30	26
27	COBALT	Co	58·93	2	2	6	2	6	7	2												VIII	32	27
28	NICKEL	Ni	58·71	2	2	6	2	6	8	2												VIII	30	28
29	COPPER	Cu	63·54	2	2	6	2	6	10	1												Ia	34	29
30	ZINC	Zn	65·37	2	2	6	2	6	10	2												IIa	34	30
31	GALLIUM	Ga	69·72	2	2	6	2	6	10	2	1											III	38	31
32	GERMANIUM	Ge	72·59	2	2	6	2	6	10	2	2											IV	42	32
33	ARSENIC	As	74·92	2	2	6	2	6	10	2	3											V	42	33
34	SELENIUM	Se	78·96	2	2	6	2	6	10	2	4											VI	46	34
35	BROMINE	Br	79·91	2	2	6	2	6	10	2	5											VII	44	35

The columns under "DISTRIBUTION OF ELECTRONS WITHIN THE SHELLS" are grouped as: K (s); L (s p); M (s p d); N (s p d f); O (s p d f); P (s p d); Q (s).

Electron configuration / periodic data table (elements 36–75). Constant inner subshells for all rows: 1s² 2s² 2p⁶ 3s² 3p⁶ 3d¹⁰ 4s² 4p⁶.

Z	N	Group	4d	4f	5s	5p	5d	6s	At. Wt.	Sym.	Element	Z
36	48	—	—	—	—	—	—	—	83·80	Kr	KRYPTON	36
37	48	I	—	—	1	—	—	—	85·47	Rb	RUBIDIUM	37
38	50	II	—	—	2	—	—	—	87·62	Sr	STRONTIUM	38
39	50	IIIa	1	—	2	—	—	—	88·91	Y	YTTRIUM	39
40	50	IVa	2	—	2	—	—	—	91·22	Zr	ZIRCONIUM	40
41	52	Va	4	—	1	—	—	—	92·91	Nb	NIOBIUM	41
42	56	VIa	5	—	1	—	—	—	95·94	Mo	MOLYBDENUM	42
43	(54)	VIIa	6	—	1	—	—	—	(97)	Tc	TECHNETIUM*	43
44	58	VIII	7	—	1	—	—	—	101·1	Ru	RUTHENIUM	44
45	58	VIII	8	—	1	—	—	—	102·9	Rh	RHODIUM	45
46	60	VIII	10	—	—	—	—	—	106·4	Pd	PALLADIUM	46
47	60	Ia	10	—	1	—	—	—	107·9	Ag	SILVER	47
48	66	IIa	10	—	2	—	—	—	112·4	Cd	CADMIUM	48
49	64	IIIa	10	—	2	1	—	—	114·8	In	INDIUM	49
50	70	IVa	10	—	2	2	—	—	118·7	Sn	TIN	50
51	70	Va	10	—	2	3	—	—	121·8	Sb	ANTIMONY	51
52	78	VIa	10	—	2	4	—	—	127·6	Te	TELLURIUM	52
53	74	VIIa	10	—	2	5	—	—	126·9	I	IODINE	53
54	78	—	10	—	2	6	—	—	131·3	Xe	XENON	54
55	78	I	10	—	2	6	—	1	132·9	Cs	CAESIUM	55
56	82	II	10	—	2	6	—	2	137·3	Ba	BARIUM	56
57	82	IIIa	10	—	2	6	1	2	138·9	La	LANTHANUM	57
58	82	IIIa	10	2	2	6	—	2	140·1	Ce	CERIUM	58
59	82	IIIa	10	3	2	6	—	2	140·9	Pr	PRASEODYMIUM	59
60	82	IIIa	10	4	2	6	—	2	144·2	Nd	NEODYMIUM	60
61	(84)	IIIa	10	5	2	6	—	2	(145)	Pm	PROMETHIUM*	61
62	90	IIIa	10	6	2	6	—	2	150·4	Sm	SAMARIUM	62
63	90	IIIa	10	7	2	6	—	2	152·0	Eu	EUROPIUM	63
64	94	IIIa	10	7	2	6	1	2	157·3	Gd	GADOLINIUM	64
65	94	IIIa	10	9	2	6	—	2	158·9	Tb	TERBIUM	65
66	98	IIIa	10	10	2	6	—	2	162·5	Dy	DYSPROSIUM	66
67	98	IIIa	10	11	2	6	—	2	164·9	Ho	HOLMIUM	67
68	98	IIIa	10	12	2	6	—	2	167·3	Er	ERBIUM	68
69	100	IIIa	10	13	2	6	—	2	168·9	Tm	THULIUM	69
70	104	IIIa	10	14	2	6	—	2	173·0	Yb	YTTERBIUM	70
71	104	IIIa	10	14	2	6	1	2	175·0	Lu	LUTECIUM	71
72	108	IVa	10	14	2	6	2	2	178·5	Hf	HAFNIUM	72
73	108	Va	10	14	2	6	3	2	180·9	Ta	TANTALUM	73
74	110	VIa	10	14	2	6	4	2	183·9	W	TUNGSTEN	74
75	110	VIIa	10	14	2	6	5	2	186·2	Re	RHENIUM	75

LIST OF THE ELEMENTS (contd.)

No.	Element	Symbol	At. Wt.	1s	2s	2p	3s	3p	3d	4s	4p	4d	4f	5s	5p	5d	5f	6s	6p	6d	7s	Group	Neutrons
76	OSMIUM	Os	190·2	2	2	6	2	6	10	2	6	10	14	2	6	6	—	2	—	—	—	VIII	116
77	IRIDIUM	Ir	192·2	2	2	6	2	6	10	2	6	10	14	2	6	7	—	2	—	—	—	VIII	116
78	PLATINUM	Pt	195·1	2	2	6	2	6	10	2	6	10	14	2	6	9	—	1	—	—	—	VIII	116
79	GOLD	Au	197·0	2	2	6	2	6	10	2	6	10	14	2	6	10	—	1	—	—	—	Ia	118
80	MERCURY	Hg	200·6	2	2	6	2	6	10	2	6	10	14	2	6	10	—	2	—	—	—	IIa	122
81	THALLIUM	Tl	204·4	2	2	6	2	6	10	2	6	10	14	2	6	10	—	2	1	—	—	III	124
82	LEAD	Pb	207·2	2	2	6	2	6	10	2	6	10	14	2	6	10	—	2	2	—	—	IV	126
83	BISMUTH	Bi	209·0	2	2	6	2	6	10	2	6	10	14	2	6	10	—	2	3	—	—	V	126
84	POLONIUM*	Po	(209)	2	2	6	2	6	10	2	6	10	14	2	6	10	—	2	4	—	—	VI	(125)
85	ASTATINE*	At	(210)	2	2	6	2	6	10	2	6	10	14	2	6	10	—	2	5	—	—	VII	(125)
86	RADON*	Rn	(222)	2	2	6	2	6	10	2	6	10	14	2	6	10	—	2	6	—	—	—	(136)
87	FRANCIUM*	Fr	(223)	2	2	6	2	6	10	2	6	10	14	2	6	10	—	2	6	—	1	I	(136)
88	RADIUM*	Ra	(226)	2	2	6	2	6	10	2	6	10	14	2	6	10	—	2	6	—	2	II	(138)
89	ACTINIUM*	Ac	(227)	2	2	6	2	6	10	2	6	10	14	2	6	10	—	2	6	1	2	IIIa	(138)
90	THORIUM*	Th	232·0	2	2	6	2	6	10	2	6	10	14	2	6	10	—	2	6	2	2	IIIa	142
91	PROTACTINIUM*	Pa	(231)	2	2	6	2	6	10	2	6	10	14	2	6	10	2	2	6	1	2	IIIa	(140)
92	URANIUM*	U	238	2	2	6	2	6	10	2	6	10	14	2	6	10	3	2	6	1	2	IIIa	146
93	NEPTUNIUM*	Np	(237)	2	2	6	2	6	10	2	6	10	14	2	6	10	4	2	6	1	2	IIIa	(144)
94	PLUTONIUM*	Pu	(244)	2	2	6	2	6	10	2	6	10	14	2	6	10	6	2	6	—	2	IIIa	(150)
95	AMERICIUM*	Am	(243)	2	2	6	2	6	10	2	6	10	14	2	6	10	7	2	6	—	2	IIIa	(148)
96	CURIUM*	Cm	(247)	2	2	6	2	6	10	2	6	10	14	2	6	10	7	2	6	1	2	IIIa	(151)
97	BERKELIUM*	Bk	(247)	2	2	6	2	6	10	2	6	10	14	2	6	10	8	2	6	1	2	IIIa	(150)
98	CALIFORNIUM*	Cf	(251)	2	2	6	2	6	10	2	6	10	14	2	6	10	10	2	6	—	2	IIIa	(153)
99	EINSTEINIUM*	Es	(254)	2	2	6	2	6	10	2	6	10	14	2	6	10	11	2	6	—	2	IIIa	(155)
100	FERMIUM*	Fm	(253)	2	2	6	2	6	10	2	6	10	14	2	6	10	12	2	6	—	2	IIIa	(153)
101	MENDELEVIUM*	Mv	(256)	2	2	6	2	6	10	2	6	10	14	2	6	10	13	2	6	—	2	IIIa	(155)
102	NOBELIUM*	No	(253)	2	2	6	2	6	10	2	6	10	14	2	6	10	14	2	6	—	2	IIIa	(151)
103	LAWRENCIUM*	Lw	(257)	2	2	6	2	6	10	2	6	10	14	2	6	10	14	2	6	1	2	IIIa	(154)
104				2	2	6	2	6	10	2	6	10	14	2	6	10	14	2	6	2	2		
105				2	2	6	2	6	10	2	6	10	14	2	6	10	14	2	6	3	2		
106																							
107																							
108																							
109																							
110																							

NOTES: 1. Elements marked * are unstable.
2. In all cases, the atomic weight is shown to four significant figures where these are available.
3. Room has been left at the bottom of the table for newly discovered elements to be inserted.
4. The number of neutrons in the nucleus is calculated for the most commonly found isotope.

Table 2 (A) (a **Segrè chart**) shows the various stable isotopes of each element which exist, also many of the principle unstable ones which appear as **fission products** or can be created, e.g., by neutron **irradiation** in a nuclear **reactor**. The chemical behaviour of all isotopes of a given element is identical regardless of whether or not they are radioactive. The part reproduced in more detail in Table 2 (B) covers elements with atomic numbers between 1 and 10.

Nuclear energy arises from the **annihilation** of matter (see **mass-energy equivalence**). The mass of a nucleus is less than the sum of the masses of its **nucleons** in free space—the difference for a given nucleus being its **mass defect**. An amount of energy sufficient to create this extra mass would have to be supplied before a nucleus could be separated into its component nucleons. The mass defect is therefore a measure of the stability of the nucleus against total disruption. Its value is greatest per nucleon for elements of medium atomic weight. This is shown in Table 3. Hence if two light atoms can be fused together to form one of medium atomic weight the sum of the masses of the interacting atoms will be greater than that of the composite atom and the difference will be released as **fusion energy**. Similarly, if a very heavy atom divides into two lighter ones some mass is released as **fission energy**. The former process is most important in the case of helium formed from the fusion of hydrogen isotopes—the source of stellar energy (see **carbon cycle**) and also the principle of the H-bomb. The process occurs automatically at temperatures above 20 million degrees centigrade (see **thermonuclear reaction**). The latter process occurs with a number of heavy isotopes referred to as **fissile**, ^{238}U, ^{235}U, ^{232}Th and ^{239}Pu being the most widely used. Absorption of a neutron by these isotopes leads to a very unstable nucleus which quickly undergoes fission. This is associated with the release of several isolated neutrons which may initiate further fission leading to a **chain reaction** in a quantity of fissile material of more than a certain **critical mass**. The fission products produced are usually two atoms of rather unequal weight. Table 4 shows **fission yields** plotted against the atomic numbers of the products.

Spontaneous fission of relatively stable heavy nuclei (^{232}Th and above) also occurs very occasionally.

TABLE 2 (A)

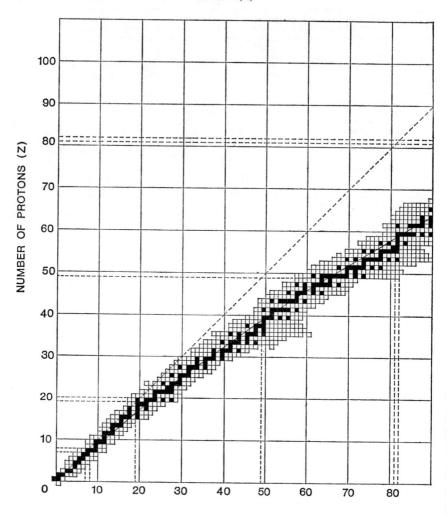

Table 2 (A) shows a complete Segrè chart reduced in scale so that no data about individual nuclides can be represented. On it stable nuclides are shown as solid black squares and principal radioactive nuclides as open squares. The line $Z = N$ at 45° to the origin would indicate the locus of nuclides with equal proton and neutron numbers. The extent to which the stability

TABLE 2 (A)

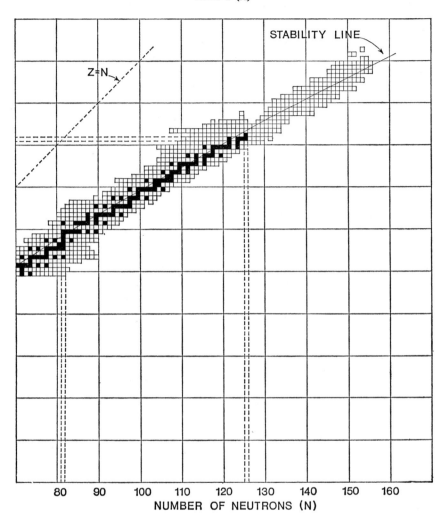

line curves away from this indicates the effect of coulomb repulsion in heavy nuclei. The positions of nuclei for which the proton or neutron number corresponds to a magic number are indicated by dotted lines. Exceptionally large numbers of stable isotopes tend to occur at these values. (Complete Segrè charts are available from Gersbach und Sohn Verlag of Munich.)

TABLE 2 (B)

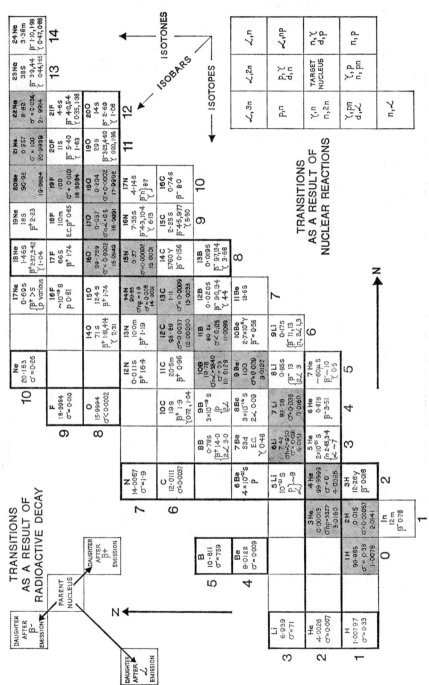

See *Notes* on page 350

TABLE 3

Mean Binding Energy per Nucleon for Stable Isotopes

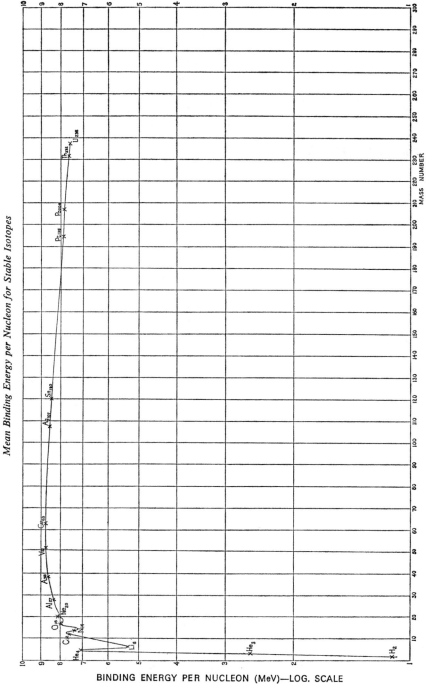

See *Notes* on page 350

349

Table 2 (B) shows the portion of the Segrè chart for atomic numbers less than 10, enlarged to normal size. The extreme left-hand square in each row gives the chemical symbol for the element concerned, its normal atomic weight and its thermal neutron absorption cross-section in barns. To the right of this each known isotope occupies one square.

Stable isotopes are shown lightly shaded. The data gives their mass number and normal percentage abundance, the activation cross-section for thermal neutrons in barns (for an n, γ reaction unless some other reaction is specified) and the atomic weight of the isotope.

For radioactive isotopes the mass number, the half-life and the mode and energy of decay are normally given unless excessively complex or not known.

Rows and columns for which the proton or neutron number corresponds to a magic number are enclosed by heavy 'box' lines.

The two supplementary diagrams show the transitions which arise either as a result of radioactive decay or of a nuclear reaction.

This energy is the resultant of five contributing factors usually considered separately. They then form the individual terms of the traditional semi-empirical mass formula.

(1) *Volume energy*, due to short-range attractive forces. This is calculated from the total nuclear volume, assuming each nucleon is surrounded by other nucleons.

(2) *Surface energy*, a correcting term applied to (1) to allow for nucleons on the surface not being surrounded by others.

(3) *Coulomb energy*, allows for the disruptive effect of coulomb electrostatic repulsion.

(4) *Symmetry energy*, allows for the reduction in stability arising from the neutron excess in most nuclei.

(5) *Pairing energy*, allows for a reduction in stability when there are unpaired protons or neutrons.

Term (1) corresponds to a positive binding energy of about 14 MeV per nucleon. All other terms are disruptive and reduce the resultant binding energy to a typical value of about 8 MeV per nucleon which varies with the total mass in the manner shown on the graph. The probable grouping of nucleons into shells (see **magic numbers**) would introduce further small correcting terms beyond those discussed above.

TABLE 4

DIRECT FISSION YIELDS

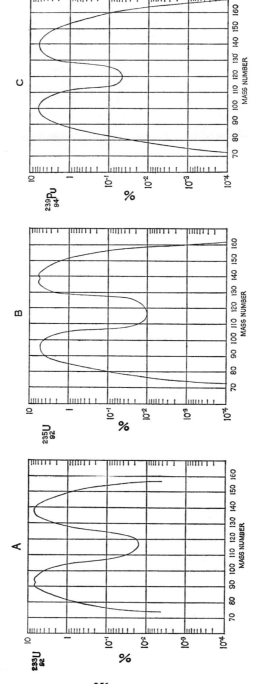

Less is known about nuclear structure than about atomic structure. During the past 30 years an increasing number of apparently **fundamental particles**, mostly unstable, have been detected in high-energy physics laboratories. Many attempts have been made to explain and classify these but usually further discoveries have quickly rendered such attempts obsolescent.

It is only within the last few years that the application of **Lie algebra** (or **group theory**) to the quantum mechanical relationships which appear to be observed experimentally, has led to an apparently coherent system of classification and to predictions about the existence and properties of other undetected particles that have subsequently been verified experimentally.

The following notes and Table 6 listing such particles must be regarded as a very tentative attempt to summarize the present position.

Four types of **field of force** appear to be associated with interactions between subatomic particles. Between nucleons the strongest are very short-range forces known as the **strong interactions**. They are believed to be **exchange forces** (Table 5) associated with the exchange of particles termed **mesons** between the nucleons of a nucleus (in the same way that the exchange of electrons between atoms produces chemical bonds). Mesons are **bosons** not subject to the Pauli exclusion principle and unlike fermions they can be freely created or destroyed in nuclear reactions. Those reactions taking place through the agency of the strong interactions are completed in times of the order of 10^{-23} sec. Apparently new particles often appear only while such interactions are taking place; these are called **resonances**. In other cases the new particles cannot disappear by means of any strong interaction because all those which are possible from the point of view of conservation of energy contravene some not yet fully understood quantum **selection rule**. They are then said to be **metastable** particles and decay through a **weak interaction** force about 10^{13} times weaker than the strong force. Such decay takes a proportionately longer time, the half-life of these metastable particle configurations being of the order of 10^{-10} sec.

Electromagnetic fields form the third type of force and the coulomb repulsion of protons in a nucleus is about 1 per cent. of the short-range attractive force. Since it acts between all the protons, while the short-range forces act only between adjacent nucleons, it is this repulsion which makes atoms of atomic number greater than 82 unstable. The fourth type of force is that associated with the gravitational field—significant only with matter on a macroscopic scale, although attempts to explain it on a quantum mechanical basis similar to other fields have been made (see **graviton**). No explanation of the origin of the weak interaction is yet known, but again it is thought it may be an exchange force associated with a so far undetected particle—the **intermediate charged vector boson**. The **photon**, the agent of the electromagnetic field, is a massless boson.

TABLE 5

Distinction between Four Classes of Exchange Force Which Act between Nucleons

Force	Property Exchanged	Diagrammatic Representation
Wigner	No exchange (or spin, charge and position)	
Heisenberg	Charge (or spin and position)	
Majorana	Position (or charge and spin)	
Bartlett	Spin (or charge and position)	

TABLE 6

LIST OF PARTICLES SO FAR DETECTED — LOWEST MASS STATES ONLY SHOWN

		Y	I	J	Rest Mass (MeV)	S, M or R	+2	+1	0	−1	CHARGE (0)	+1	0	−1	−2
							← — — — PARTICLES — — — →					← — — ANTI-PARTICLES — — →			
BARYONS (STRONGLY INTERACTING FERMIONS)	OMEGA	−2	0	3/2	1,675	M				Ω^- (0)		$\tilde{\Omega}^+$ (0)			
	Xi	−1	1/2	1/2	1,318	M			$\Xi^0\,(+\tfrac{1}{2})$	$\Xi^-\,(-\tfrac{1}{2})$		$\tilde{\Xi}^+\,(+\tfrac{1}{2})$	$\tilde{\Xi}^0\,(-\tfrac{1}{2})$		
	SIGMA	0	1	1/2	1,193	M		$\Sigma^+\,(+1)$	$\Sigma^0\,(0)$	$\Sigma^-\,(-1)$		$\tilde{\Sigma}^+\,(+1)$	$\tilde{\Sigma}^0\,(0)$	$\tilde{\Sigma}^-\,(-1)$	
	LAMBDA	0	0	1/2	1,115	M			$\Lambda^0\,(0)$				$\tilde{\Lambda}^0\,(0)$		
	DELTA	1	3/2	3/2	1,236	R	$\Delta^{++}\,(+\tfrac{3}{2})$	$\Delta^+\,(+\tfrac{1}{2})$	$\Delta^0\,(-\tfrac{1}{2})$	$\Delta^-\,(-\tfrac{3}{2})$		$\tilde{\Delta}^+\,(+\tfrac{3}{2})$	$\tilde{\Delta}^0\,(+\tfrac{1}{2})$	$\tilde{\Delta}^-\,(-\tfrac{1}{2})$	$\tilde{\Delta}^{--}\,(-\tfrac{3}{2})$
	NUCLEON	1	1/2	1/2	939	n M / p S		$N^+\,(+\tfrac{1}{2})$	$N^0\,(-\tfrac{1}{2})$				$\tilde{N}^0\,(+\tfrac{1}{2})$	$\tilde{N}^-\,(-\tfrac{1}{2})$	
HEAVY INTERMEDIATE BOSON	HYPOTHETICAL				~1,000	R									
MESONS (STRONGLY INTERACTING BOSONS)	η MESON	0	0	0	548	M					η_0				
	K Meson (Kaon)	1	1/2	0	496	M		$K^+\,(+\tfrac{1}{2})$	$K^0\,(-\tfrac{1}{2})$				$\tilde{K}^0\,(+\tfrac{1}{2})$	$\tilde{K}^-\,(-\tfrac{1}{2})$	
	π Meson (Pion)	0	1	0	137	M		$\pi^+\,(+1)$			$\pi_0\,(0)$			$\pi^-\,(-1)$	
LEPTONS (WEAKLY INTERACTING FERMIONS)	MUON			1/2	106	M			μ^-			μ^+			
	ELECTRON			1/2	0·5	S			e^-			e^+			
	NEUTRINO (MUON)			1/2	0	S			ν_μ				$\tilde{\nu}_\mu$		
	" (ELECTRON)			1/2	0	S			ν_e				$\tilde{\nu}_e$		
MASSLESS BOSONS	PHOTON			1	0	S					γ				
	GRAVITON	HYPOTHETICAL		2	0	S									

| | 1 | 2 | 3 | 4 | 5 | 6 | 7 | 8 | 9 | 10 | 11 | 12 | 13 | 14 |

N.B.—Column 1.   Hypercharge, Y. This is twice the displacement of the charge centre of a multiplet from the zero line.

Column 2.   Isotopic Spin, I. The isotopic spin quantum number of each individual member of a multiplet is given in brackets below it.

Column 3.   Spin Angular momentum, J. This will not be the same for all mass recurrences of particles with corresponding Y and I.

Column 4.   Rest Mass in MeV. The rest mass of an electron is 0·511 MeV.

Column 5.   Stable (S), Metastable (M), or Resonance (R).

Columns 6-9.   Particles grouped in multiplets.

Column 10. η_0 and π_0 mesons and photons have zero charge and zero hypercharge. They act as their own anti-particles.

Columns 11-14.   Anti-particles grouped in multiplets. Values of charge and hypercharge are reversed in sign for anti-particles.

Indicates position of charge centre of multiplets along charge scale when this is not zero ($Y \neq 0$).

Table 6 groups the particles so far detected or thought to exist in terms of what appear to be their most important quantum numbers—**isotopic spin** and **hypercharge**. The fermions are in two groups, **baryons** and **leptons**, and are subject to a numerical conservation law, being created or destroyed only in conjunction with an **anti-particle** from the same group. The lightest baryon is the proton—the only one stable in isolation (neutrons appear stable in a nucleus but decay into a proton, electron and **anti-neutrino** in free space—an example of a weak interaction). Other unstable baryons are the various groups of **hyperons** existing as charge **multiplets** with closely corresponding masses. These can be created only by very high energy **cosmic radiation**, or by powerful **particle accelerators**, considerable energy being required to create the extra rest mass of the heavier particles. (Nuclear reactions are all reversible so these particles are created from lighter ones by supplying sufficient energy to reverse their normal decay process.) Many of the heavier particles are high-energy **recurrences** of the lighter ones—the greater energy also being associated with a greater angular momentum. A specific nuclear reaction which has a high **cross-section** for incident particles of one energy will show a large cross-section again for a higher energy—the resulting reaction leading to the emission of the higher mass particle. The concept of the **Regge trajectory** has enabled some of these recurrences to be simply explained and the existence of others to be predicted. As with the mesons, many of these heavier baryons are resonances decaying again almost instantaneously through a strong interaction. It thus appears that as greater energies become available from bigger accelerating machines the production of more and more particles may be possible, many so short-lived that they may be almost indistinguishable experimentally from the components with which they are associated during production or decay. The description **elementary** (or **fundamental**) **particle** is consequently deprecated today and attempts are being made to account for all the properties of strongly interacting particles in terms of still more fundamental components which have been termed **quarks**.

Table 7 shows recurrences of mesons and baryons which are excluded from Table 5, but which have been detected. (*N.B.*—The term recurrence is applied here to all higher mass multiplets of groups with different masses but corresponding hypercharge and isotopic spin—not just to Regge recurrences for which the difference in spin angular momentum must be 2 units.)

The most important recent development has been the grouping of these strongly interacting particles into **supermultiplets** with their differences in charge producing mass differences of the order of 0·1 per cent. and their differences in hypercharge producing mass differences of about 10 per cent. Lie group S.U.(3) represents symmetry relationships between eight independent components (the **eightfold way**). There appear to be at least two octets of particles interrelated in a similar way among both the mesons and the baryons. Lie algebra also accounts for the existence of singlet particles and of decuplets. Members of a baryon decuplet of this type were all known except for one when this classification was introduced. The missing particle, the **omega minus**, has rather distinctive properties and was quickly isolated in 1964 once these had been predicted. The mass relationships required by this grouping correlated uniformly accurately with the experimental values both for the decuplet and the various octets. Two of the octets and the decuplet are shown in Table 8.

Key to Table 7

— Mass level

Spin angular momentum given on
    left

Numerical mass value given on
    right

M = Metastable particle

R = Resonance

Multiplet structure given at top
    of column

TABLE 7

Mass Recurrences of Particle Multiplets with the Same Isotopic Spin and Hypercharge

	BARYONS						MESONS		
PARTICLE	Ω OMEGA	Ξ XI	Σ SIGMA	Λ LAMBDA	Δ DELTA	N NUCLEON	π PI	η ETA	K K̄ (KAPPA ANTI KAPPA)
HYPERCHARGE Y	-2	-1	0	0	+1	+1	0	0	∓1
ISOTOPIC SPIN I	0	½	1	0	3/2	½	1	0	½
MULTIPLET STRUCTURE (2I+1)	SINGLET	DOUBLET	TRIPLET	SINGLET	QUADRUPLET	DOUBLET	TRIPLET	SINGLET	DOUBLET
2,800					? —R 2825				
2,600						? —2645			
2,400									
2,360					11/2 R 2360				
2,190						9/2 2190			
2,000									
1,924					7/2 R 1924				
1,820		3/2 R 1820							
1,815				5/2 R 1815					
1,800									
1,765			5/2 R 1765						
1,688						5/2 1688			
1,675	3/2 M 1675								
1,660			? R 1660						
1,600									
1,530		3/2 R 1530							
1,520				3/2 R 1520					
1,518						3/2 R 1518			
1,420								? R 1420	
1,410									2 —1410 R
1,405				½ 1405 R					
1,400									
1,382			3/2 R 1382						
1,324							2 R 1324		
1,318		½ M 1318							
1,253								2 R 1253	
1,236					3/2 R 1236				
1,220							? R 1220		
1,215									? —1215 R
1,200									
1,193			½ M 1193						
1,115				½ M 1115					
1,072							1 R 1072		
1,019								1 R 1019	
1,000									
959								0 R 959	
939						½ M 939			
891									1 R 891
800									
782								1 R 782	
769							1 R 769		
725									0 —725 R
600									
548								0 M 548	
496									0 M 496
400									
200									
137							0 M 137		
0									

REST MASS (MeV)

TABLE 8

Grouping of Supermultiplets

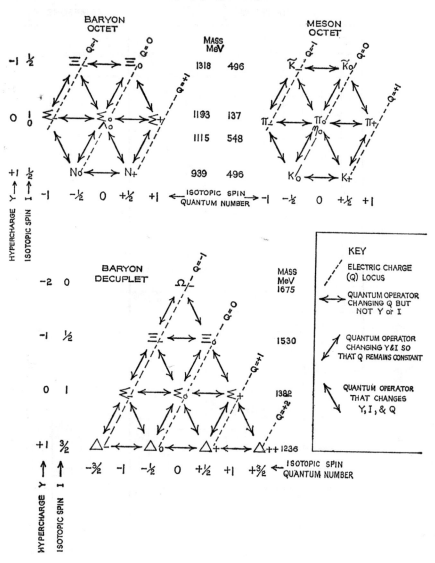

IONIZING RADIATIONS

Radiation Hazards

No immediate ill effect is felt by a person experiencing a dangerous dose of ionizing radiation, so extreme care must be taken to avoid this occurring unwittingly. The ionization produced in the tissue cells subsequently leads to the formation of damaging chemical radicals which may poison the cell or otherwise affect the organism, leading in extreme cases to the condition known as radiation sickness. Other harmful permanent or semi-permanent effects may be brought on by repeated doses below the level which would bring on noticeable radiation sickness.

Radiation hazards are usually classed as internal or external. Internal hazard arises from the ingestion of radioactive gases, radioactive particles in the air, etc. When the isotope concerned has a long half-life and long biological half-life it leads to continuous internal dosage and can be very serious. Consequently eating and drinking are strictly forbidden in laboratories where radioactive materials are handled. External hazard arises from radiation emitted by sources outside the body and will normally be very slight for the quantities of radioactive materials handled in chemical operations. It is most serious in the case of direct exposure to the beam from an X-ray tube but may also be appreciable with high intensity radioactive sources such as those used in radiography, or at close distances with relatively weak beta-ray sources.

Reference: International Commission on Radiological Protection (1959 Report).

Recommendations

EXTERNAL RADIATION

Definitions

The permissible dose for an individual is that dose, accumulated over a long period of time or resulting from a single exposure, which, in the light of present knowledge, carries a negligible probability of severe somatic or genetic injuries; furthermore it is such a dose that any effects that ensue more frequently are limited to those of a minor nature that would not be considered unacceptable by the exposed individual and by competent medical authorities.

The permissible dose to the gonads for the whole population is limited primarily by considerations with respect to genetic effects.

Exposure of an individual who normally works in a controlled area constitutes *occupational exposure*. A controlled area shall be established where persons occupationally exposed could receive doses in excess of 1·5 rems per year. A *controlled area* is an area in which the exposure of personnel to radiation or radioactive material is under the supervision of a *radiation protection officer*.

Maximum Permissible Levels for Occupational Exposure

Exposure of the gonads, blood-forming organs and the lenses of the eyes.
$D = 5(N - 18)$, where D is tissue dose in rems.
N is age in years.
Rate must not exceed 3 rems during any 13 consecutive weeks.
Exposure of the skin and thyroid.
8 rems in any 13 consecutive weeks.
Derived from 0·6 rem per week, therefore 30 rems per annum (50 weeks).
Hands and forearms, feet and ankles.
20 rems in 13 weeks.
Derived from 1·5 rems per week, therefore 75 rems per annum.
Internal organs other than thyroid, gonads and blood-forming organs.
4 rems in 13 weeks.
Derived from 0·3 rem per week, therefore 15 rems per annum.

Maximum Permissible Levels for Other Special Groups

Adults who work in the vicinity of, or who occasionally enter, controlled areas may not receive more than 1·5 rem per annum to the gonads, the blood-forming organs and the lenses of the eye. 3 rem are permitted for skin and thyroid.

Members of the public living in the neighbourhood of controlled areas are restricted to one-third of the above doses because of the presence of children in this group.

The latest report of the International Commission on Radiological Protection makes a number of minor changes in the recommendations based on the 1959 Report given above, and some rather greater changes in the approach it proposes for assessing exposure risk. It is important to remember that legislation on radiation hazards in most countries is based on the 1959 Report and not this more recent one. The changes suggested can be summarized as follows:—

Only two classes of persons are considered—those occupationally exposed and the general public. Occupationally-exposed workers most unlikely to receive radiation doses in excess of three-tenths of the permissible annual dose need not be subject to personnel monitoring and special health supervision.

Further restrictions are introduced in the cases of neutron irradiation of the eye and of women of child-bearing age, and certain relaxations are permitted whereby classified radiation workers may be subject to planned special exposures in exceptional circumstances.

Quality Factors of Commonly-encountered Radiations

Radiation	Generally-accepted quality factor	Typical linear energy transfer KeV per micron soft tissue
Medium-energy X-, γ-, β-rays	1	0·2-1
Low-energy β-rays (< 1·7 KeV)	2	1-50
Medium-energy protons	5	5
Slow neutrons	5	3
Fast neutrons	10	7-10
Low-energy α-rays	20	20

The quality factor used in radiological protection is a " rounded-off " and averaged value of the *relative biological effectiveness* for the type of radiation concerned. Exact values of *rbe* depend not only on the energy of the radiation but also on such factors as the site irradiated, the type of biological consequence being studied, the dose rate and degree of fractionation of the radiation, and the age and physical condition (oxygenation, etc.) of the animal irradiated.

CHART OF RADIATION EFFECTS

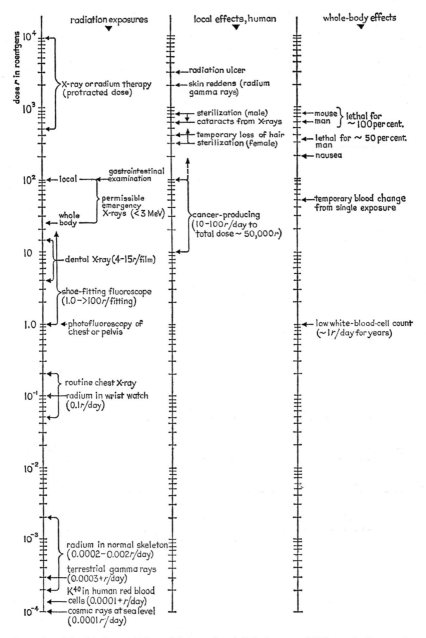

Biological and Chemical Effects of Radiation on Materials

These materials vary widely in their susceptibility to radiation, and their relative sensitivity is shown in the diagram below. By comparison it should be noted that while a dose of 10^3 rads is lethal to man, the majority of ceramics and metals are usable after doses of up to 10^{12} rads or greater.

There are two main effects on organic materials arising from irradiation, viz. cross-linking (or polymerization) and scission (or cleavage). In the process hydrogen is evolved and, depending on the nature of the molecular structure, side reactions may result from the reactive sites created in the residual organic molecule. All properties of the material are not changed to the same degree so that the useful life of a substance should be quoted in terms of that of its most 'critical' property. In general, longest lives are usually obtained with aromatic-rich compounds and when the radiation exposure takes place at ordinary temperatures and in an inert gas atmosphere. In this absence of oxidation and other time-dependent factors, it appears that the overall effects of radiation are independent of dose rate, being dependent only on the total dose. The statement 'equal-energy-equal-damage' appeared to be true, but recent work has indicated that neutrons are more damaging than electrons.

In liquids, the existence of cross-linking is shown by an increase in viscosity while its effect in solids is to increase hardness and brittleness. On the other hand, cleavage leads to the formation of lower molecular weight materials and results in a lowered viscosity of liquids and a softening of solids. Cross-linking only occurs in amorphous or non-crystalline regions of a polymer so that the response to radiation is strongly dependent on the date of crystallization.

The plastic which has received most attention commercially is polyethylene. When irradiation is carried out in the amorphous state the product has properties superior to those obtained with similar but chemically-produced substances. On irradiating a partially crystalline sample the effect of increasing dosage is to change the material into a rubbery state and finally back to a solid and even a very brittle material. This is accompanied by a change of colour to dark brown or black.

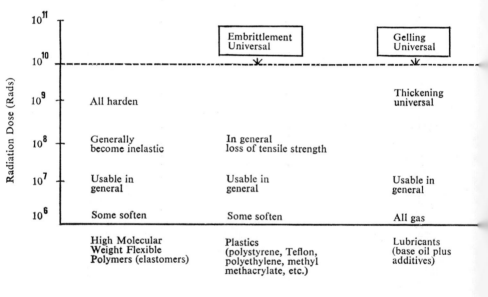

Cross-sections of Nuclear Fuels

(Grateful acknowledgement is made to the Westinghouse Electrical Institute of New York for permission to reproduce the following text and graphs, as well as Table 4 on Fission Yields (p. 351) and the graphs on gamma-ray Shielding (p. 364). The graphs were originally produced for the Institute under the guidance of Dr W. E. Shoupp.)

The cross-section curves are given for total cross-section whenever possible, otherwise for fission as indicated. Solid lines give the cross-section curve from 0·01 to 1000 eV, and broken lines from 0·001 to 100 MeV.

360

CROSS - SECTIONS

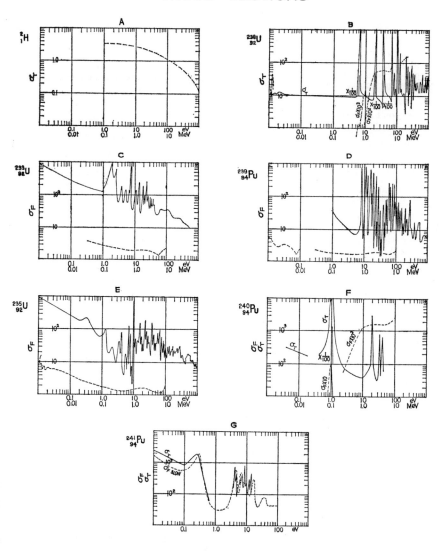

Calculations of Radiation Dose Rates and Shielding in Simple Cases

The following approximate formulae are sufficiently accurate for estimates of received dose rate during experimental work:

(a) *Alpha Emitters*

Radiation-sensitive tissues in the human skin extend from a depth of 0·07 mm downwards. Alpha-particles of energy less than 6·5 MeV cannot penetrate to this depth and further shielding is therefore unnecessary.

(b) *Beta Emitters*

For point sources of strength C curies the dose rate is 300 C r/h at 1 foot and 3000 C r/h at 10 cm, neglecting air absorption.

For extended sources the dose rate at or near the surface is given by the product of the specific activity in microcuries per gram and the average energy released per disintegration.

The inverse square law cannot be applied to beta radiation, and particles of energy up to 2 MeV will be completely absorbed by a sheet of material weighing 1 gm per sq cm. However, with strong sources the resulting bremsstrahlung must be considered. Where the beta radiation is absorbed in material of fairly low atomic number, approximately 1 % of the energy appears in this form. Thus a bottle containing 1 Ci of a source emitting beta-particles of average energy 1 MeV will emit radiation similar to 10 mCi of a 1 MeV gamma emitter.

(c) *Gamma Emitters*

Gamma-rays obey the inverse-square law but the dose rate depends upon the average total energy released per disintegration which is not always easily found. Calculation are therefore based on a constant known as the K-factor for the isotope. This is the dose rate (in r/h) which would be experienced 1 cm from a source of 1 mCi. It corresponds approximately to the dose in mr/h 1 foot from the same source.

The dose for other sources and distances is calculated by multiplying the K-factor by C/d^2, where C is the source strength in mCi and d the distance in cm.

An alternative expression for calculating dose rates from gamma emitters is:

$$D.R \ (r/h) = \frac{SE}{4\pi x^2} \times 2 \times 10^{-6},$$

where S is the source strength in disintegrations per second, E is the total photon energy released per disintegration in MeV, and x is the distance in cm.

(d) *Gamma Dose Rate from Chemical Neutron Sources*

Most (αn) and (γn) chemical neutron sources emit intense gamma radiation. The dose rates 1 metre from unshielded sources with a total emission of 10^6 neutrons per second are as follows:

Type of Source	Gamma dose rate at 1m. (mr/hr.)
Actinium—Beryllium	8
Americium—Beryllium	0·1
Lead-210—Beryllium	9
Plutonium—Beryllium	0·1
Polonium—Beryllium	0·1
Radium—Beryllium	60
Thorium-228—Beryllium	30
Antimony—Beryllium	950

(e) *Gamma-ray Shielding Calculations*

A wall of standard 2″ lead bricks reduces the gamma dose rate for medium energy emitters (e.g., ^{60}Co) by a factor of about 10. Text following reproduced from Westinghouse Institute material.

In the curves giving γ-ray attenuation for various materials it is important to note that the average 10-fold attenuation thickness was used. Due to the fact that the build-up factor does not vary linearly with energy, the dose rate versus thickness is not a pure exponential. It may be approximated as the sum of two exponentials. To obtain data for these curves, the thickness required to reduce the intensity by the first factor of 10 and the thickness for the last factor of 10 were obtained, and a linear average was taken of the two values. Thus, for small thicknesses of material, the 10-fold attenuation thickness given is slightly less than the thickness which should be used, while for large thicknesses the values are conservative. It should also be remembered that the calculation is for an infinite-plane monodirectional source, whereas in most experimental arrangements, the actual source will be something between a monodirectional and a point isotropic source, requiring less shielding than is indicated by the curve.

The uncertainty in using the average 10-fold attenuation thickness is a maximum of 25 % at low energies, decreasing with higher energies and more dense materials. For very thin shields, it is safe to add 20 to 25 % to the indicated thickness; while for very thick shields, it would be permissible to subtract 15 to 20 %.

INTERNATIONAL RADIATION HAZARD SIGN

The B.S. (Standard 3510, 1962) recommend that this symbol shall be coloured black on a yellow ground, the colour approximating to No. 309 of Standard 381C (*Colours for ready-mixed paints*). No preferred orientation is specified.

SHIELDING INFORMATION

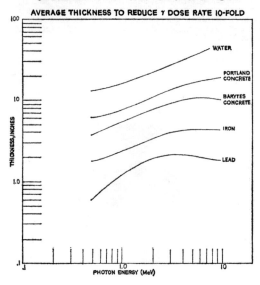

AVERAGE THICKNESS TO REDUCE γ DOSE RATE 10-FOLD

WATER

PORTLAND
CONCRETE

BARYTES
CONCRETE

IRON

LEAD

THICKNESS, INCHES

PHOTON ENERGY (MeV)

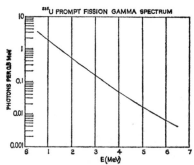

^{235}U PROMPT FISSION GAMMA SPECTRUM

PHOTONS PER 0.5 MeV

E (MeV)

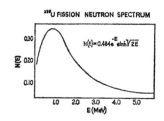

^{235}U FISSION NEUTRON SPECTRUM

$$N(E) = 0.484e^{-E} \sinh\sqrt{2E}$$

N(E)

E (MeV)

Data on Commonly-encountered Artificial Radioisotopes

Notes

Column 2—Mode of production, gives reaction usually employed and indicates method by R (neutron activation in reactor), C (cyclotron bombardment by charged particles, and F (fission product).

Column 3—Half-life; s = seconds, m = minutes, h = hours, d = days, y = years, (m) = metastable decay.

Column 4—Mode of decay, α (alpha-particle emission), β (beta-particle emission), β^+ (positron emission) or EC (electron capture by nucleus from K- or L-shell). The limiting energy or energies of the particles in MeV are also given with the percentage, having each energy indicated in brackets when not 100%. This available energy is shared in an arbitrary proportion between the beta-particle and the neutrino.

Column 5—Gamma-ray energies in MeV with the percentage of disintegrations in which these photon energies appear indicated in the same way. Two 0·51 MeV quanta are always associated with the annihilation of emitted positrons. These are indicated by A. X-rays emitted following electron capture are also listed.

Column 6—Specific gamma-ray emission for gamma-emitting isotopes. (The gamma dose rate in r/h 1 cm from an unshielded point source of 1 mCi.)

Data from AERE Publication R. 2938.

1	2	3	4	5	6
Aluminium-28	^{27}Al(nγ) ^{28}Al	R	2·27 m	β 2·87	1·78	8·5
Antimony-122	^{121}Sb(nγ) ^{122}Sb	R	2·74 d	β 1·42(63%) 1·99(30%) 0·73(4%) EC(3%)	0·57(66%) 0·69(3%) 1·26(1%) 1·14(1%)	2·4
Antimony-124	^{123}Sb(nγ) ^{124}Sb	R	60 d	β 0·61(51%) 2·31(23%) 0·22(11%) 1·60(7%) 0·95(6%) 1·66(2%)	0·60(99%) 1·70(50%) 0·72(14%) 0·65(8%) 2·09(6%) 1·37(5%) 0·97(3%) 1·05(2%) 1·45(2%) 1·33(1%)	9·8
Antimony-125	^{124}Sn(nγ) ^{125}Sn β ^{125}Sb	R	2·0 y	β 0·30(45%) 0·12(29%) 0·61(14%) 0·44(12%)	various in range 0·035-0·637	
Argon-37	^{40}Ca(nα) ^{37}Al	R	34·5 d	EC	0·0026	
Argon-41	^{40}A(nγ) ^{41}A	R	110 m	β 1·20(99·1%) 2·48(0·9%)	1·29(99·1%)	6·6
Arsenic-72	^{69}Ga(αn) ^{72}As	C	26 h	β$^+$ 2·50(56%) 3·34(17%) 1·84(3%) other 2% EC(22%)	0·835(78%) 0·63(8%) 0·51 A 0·01 following EC	10·1
Arsenic-74	^{74}Ge(d2n) ^{74}As	C	18 d	β$^+$ 0·91(26%) 1·51(4%) β$^-$1·36(18%) 0·72(15%) EC(37%)	0·596(62%) (with 0·91 β$^+$ or EC) 0·51 A 0·635(15%) (with 0·72 β$^-$ 0·01 following EC)	4·4
Arsenic-76	^{75}As(nγ) ^{76}As	R	26·5 h	β 2·97(56%) 2·41(31%) 1·20(6·5%) 1·75(3·5%) 0·35(3%)	0·56(45%) 0·66(6·3%) 1·21(5·3%) 2·08(1·0%) 1·44(0·8%) 1·79(0·3%)	2·4

365

1	2		3	4	5	6
Arsenic-77	${}_{76}$Ge$(n\gamma)$ ${}_{77}$Ge β ${}_{77}$As	R	38·7 h	β 0·68(94%) 0·44(3%) 0·16(3%)	0·245(2·5%) 0·525(0·8%) with others very weak	
Barium-131	${}_{130}$Ba$(n\gamma)$ ${}_{131}$Ba	R	11·5 d	EC	various up to 0·50	
Barium-133	${}_{132}$Ba$(n\gamma)$ ${}_{133}$Ba	R	7·5 y	EC	0·358(70%) 0·079(29%) 0·302(26%) 0·381(10%) 0·056(2%) 0·274(1%) 0·031 following EC	
Barium-140	U(nf) ${}_{140}$Ba	F	12·8 d	β 1·02(60%) 0·48(25%) 0·59(10%) 0·83(5%)	0·54(26%) 0·16(10%) 0·44(5%) 0·30(5%) 0·13(1·4%). See also ${}_{140}$La (daughter)	12·4 with ${}_{140}$La
Beryllium-7	${}_{12}$C(p, 3p 3n) ${}_{7}$Be	C	53 d	EC	0·48(12%) 0·052 following EC	
Bismuth-206	${}_{206}$Pb(d2n) ${}_{206}$Bi	C	6·3 d	EC	0·803(100%) and various others up to 1·72 (30%)	
Bismuth-207	${}_{207}$Pb(pn) ${}_{207}$Bi	C	28 y	EC	0·57(95%) 1·06(87%) 1·77(8%)	
Bismuth-210 (Radium E)	${}_{209}$Bi$(n\gamma)$ ${}_{210}$Bi	R	5 d	β 1·17		
Bromine-82	${}_{81}$Br$(n\gamma)$ ${}_{82}$Br	R	36 h	β 0·44	0·78(83%) 0·55(75%) 0·62(42%) 1·04(29%) 1·32(28%) 0·70(28%) 0·83(25%) 1·48(17%)	14·6
Cadmium-109	${}_{109}$Ag(d2n) ${}_{109}$Cd	C	470 d	EC	0·088(4%) 0·022 0·003 following EC	
Cadmium-115	${}_{114}$Cd$(n\gamma)$ ${}_{115}$Cd	R	43 d(m) 2·3 d	β 1·61(97%) 0·68(2%) 0·31(1%) β 1·11(61·5%) 0·59(25%) 0·63(12%) 0·84(5·5%) 0·86(1·5%)	0·94(2·3%) 1·30(1%) 0·52(25%) 0·49(12%) 0·34(2·7%)	
Caesium-131	${}_{130}$Ba$(n\gamma)$ ${}_{131}$Ba EC ${}_{131}$Cs	R	10 d	EC	0·030 0·004 following EC	

1	2		3	4	5	6
Caesium-134	$_{133}Cs(n\gamma)$ $_{134}Cs$	R	2·19 y	β 0·65(75%) 0·09(20%) 0·28(3%) 0·89(2%)	0·605(98%) 0·80(90%) 0·57(24%) and others (3·3% or less) to 1·37	8·7
Caesium-137	U(nf) $_{137}Cs$	F	30 y	β 0·51(92%) 1·17(8%)	0·662(82%)	3·1
Calcium-45	$_{44}Ca(n\gamma)$ $_{45}Ca$	R	165 d	β 0·25		
Calcium-47	$_{46}Ca(n\gamma)$ $_{47}Ca$	R	4·7 d	β 0·66(83%) 1·94(17%)	1·31(77%) 0·83(6%) 0·48(6%)	5·7
Carbon-11	$_{10}B(dn)$ $_{11}C$	C	20 m	β+ 0·97	0·51 A	
Carbon-14	$_{14}N(np)$ $_{14}C$	R	5730 y	β 0·155		
Cerium-141	$_{140}Ce(n\gamma)$ $_{141}Ce$	R	32·5 d	β 0·44(70%) 0·58(30%)	0·145(49%)	0·35
Cerium-143	$_{142}Ce(n\gamma)$ $_{143}Ce$	R	33 h	β 1·11(40%) 1·40(37%) 0·54(12%) 0·30(6%) 0·73(5%)	various to 1·1 ·See also $_{143}Pr$ (daughter)	
Cerium-144	U(nf) $_{144}Ce$	F	285 d	β 0·32(76%) 0·19(19·5%) 0·24(4·5%). See also $_{144}Pr$ (daughter)	various of low energy (most abundant and highest energy 0·133(11%))	
Chlorine-36	$_{35}Cl(n\gamma)$ $_{36}Cl$	R	3×10⁵ y	β 0·714(98·3%) EC(1·7%)		
Chromium-51	$_{50}C(n\gamma)$ $_{51}C$	R	27·8 d	EC	0·323(8%) 0·005 following EC	0·15
Cobalt-56	$_{56}Fe(d2n)$ $_{56}Co$	C	77 d	β+ 1·50(18%) EC (82%)	0·845(100%) 1·24(70%) 1·75(18%) 1·03(16%) 2·60(16%) 3·25(12%) 2·02(11%) 1·36(5%) 2·99(1%) 3·47(1%) 0·51 A	17·6
Cobalt-57	{ $_{56}Fe(dn)$ $_{57}Co$ / $_{60}Ni(p\alpha)$ $_{57}Co$ }	C C	270 d	EC	0·122(88%) 0·136(10%) 0·014(6%)	
Cobalt-58	$_{58}Ni(np)$ $_{58}Co$	R	71 d	EC(85%) β+ 0·47(15%)	0·81, 0·51 A 0·006 following EC	5·5
Cobalt-60	$_{59}Co(n\gamma)$ $_{60}Co$	R	5·27 y	β 0·31	1·17, 1·33	13·2

367

1	2	3		4	5	6
Copper-64	$_{63}$Cu(nγ) $_{64}$Cu	R	12·8 h	β 0·57(38%) β^+ 0·66(19%) EC (43%)	0·51 A 1·34 (weak) 0·007 following EC	1·2
Erbium-169	$_{168}$Er(nγ) $_{169}$Er	R	9·4 d	β 0·34		
Europium-152	$_{151}$Eu(nγ) $_{152}$Eu	R	13 y	β 0·71(12%) 1·47(7%) 0·36(3%) 0·22(2%) 1·04(2%) EC (74%)	0·34(26%) 0·78(12%) 1·10(3%) 0·122(33%) 1·41(25%) 0·96(15%) 1·11(13%) 1·09(12%) 0·44(5%) 0·245(4%) 0·87(4%) 0·040 following EC	5·8
Europium-154	$_{153}$Eu(nγ) $_{154}$Eu	R	16 y	β 0·55(30%) 0·25(28%) 0·83(20%) 0·15(12%) 1·84(7%) 1·60(3%)	various to 1·28	
Europium-155	$_{154}$Sm(nγ) $_{155}$Sm β $_{155}$Eu	R	1·7 y	β 0·16(40%) 0·15(30%) 0·25(20%) 0·19(10%)	various—all weak	
Gadolinium-153	$_{152}$Gd(nγ) $_{153}$Gd	R	236 d	EC	0·103(25%) 0·097(21%) 0·070(2%) 0·041 and 0·006 following EC	
Gallium-67	$_{66}$Zn(dn) $_{67}$Ga	C	78 h	EC	0·092(69%) 0·182(24%) 0·30(22%) and others all very weak 0·008 following EC	
Gallium-68	$_{69}$Ga(p,2n) $_{68}$Ge EC $_{68}$Ga	C	68 m	β^+ 1·89(86%) 0·82(1%) EC (13%)	1·08(4%) 0·51 A 0·008 following EC	
Germanium-68	$_{69}$Ga(p,2n) $_{68}$Ge	C	280 d	EC	0·009 following EC	
Germanium-71	$_{70}$Ge(nγ) $_{71}$Ge	R	11 d	EC	0·009 following EC	
Gold-195	Pt(d,xn) $_{195}$Au	C	185 d	EC	weak gammas and X-rays (0·06, 0·009) following EC only	
Gold-198	$_{197}$Au(nγ) $_{198}$Au	R	2·7 d	β 0·96(99%) 0·29(1%)	0·412(95·6%) 0·68(1·1%)	2·3
Gold-199	$_{198}$Pt(nγ) $_{199}$Pt β $_{199}$Au	R	3·15 d	β 0·30(70%) 0·25(24%) 0·46(6%)	0·159(42%) 0·209(9%)	

1	2	3		4	5	6
Hafnium-175	$_{174}$Hf(nγ) $_{175}$Hf	R	70 d	EC	0·343(86%) 0·089(4%) 0·433(1·4%) 0·054 and 0·008 following EC	
Hafnium-181	$_{180}$Hf(nγ) $_{181}$Hf	R	42·5 d	β 0·41(92%) 0·40(4%) and others (4%)	0·482(81%) 0·133(40%) 0·346(13%) 0·136(6%) 0·137(2%) 0·476(2%)	
Holmium-166	$_{165}$Ho(nγ) $_{166}$Ho	R	27 h	β 1·84(47%) 1·76(37%) 0·87(9%) 0·41(5%) 0·23(2%)	0·08(6%) 1·38(1%) and others (very weak)	
Hydrogen-3 (Tritium)	$_{6}$Li(nα) $_{3}$H	R	12·26 y	β 0·018		
Indium-114	$_{113}$In(nγ) $_{114}$In(m) $_{114}$In	R	50 d(m) 72 s	EC (3·5%) transition to $_{114}$In(96·5%) β 1·98(99%) EC(0·7%)	0·56(3·5%) 0·72(3·5%) 0·19(18·5%) 0·024(In X-rays)	
Iodine-124	$_{121}$Sb(αn) $_{124}$I	C	4 d	β^{+} 1·6(14%) 2·2(11%) EC(75%)	0·60(66%) 0·73(14%) 1·70(14%) 0·64(12%) 1·51(4%) 1·37(3%) 2·09(2%) 2·26(2%) and others very weak. 0·51 A	7·2
Iodine-125	$_{123}$Sb(α2n) $_{125}$I	C	60 d	EC	0·035(7%) 0·027 following EC	
Iodine-129	U(nf) $_{129}$I	F	1·6 × 10⁷ y	β 0·15	0·04 (10%)	
Iodine-130	$_{129}$I(nγ) $_{130}$I	R	12·5 h	β 0·60(54%) 1·02(46%)	0·66, 0·53, 0·74(70%) 1·15(30%) 0·41(24%)	2·2
Iodine-131	U(nf) $_{131}$Te β $_{131}$I	F	8 d	β 0·61(87%) 0·33(9%) 0·25(3%) 0·81(1%)	0·36(80%) 0·64(9%) 0·28(5%) 0·72(3%) 0·08(2%)	
Iodine-132	U(nf) $_{132}$Te β $_{132}$I	F	2·26 h	β 1·53(24%) 1·16(23%) 0·90(20%) 2·12(18%) 0·73(15%)	0·67, 0·78(85%) 0·53(27%) 0·96(21%) 1·40(12%) 1·16(9%) 0·62(7%) 1·96(5%) 2·20(2%)	11·8
Iridium-192	$_{191}$Ir(nγ) $_{192}$Ir	R	74·4 d	β 0·67(50%) 0·54(40%) 0·24(6%) EC(4%)	0·316(83%) 0·468(53%) 0·296(30%) 0·308(29%) 0·605(12%) 0·613(7%) 0·588(6%)	4·8

369

1	2	3	4	5	6
Iron-52	$_{52}$Cr(α 4n) $_{52}$Fe	C 8 h	β^+ 0·80(57%) EC (43%)	0·165, 0·51 A 0·006 following EC	
Iron-55	$_{54}$Fe(nγ) $_{55}$Fe	R 2·7 y	EC	0·006 following EC	
Iron-59	$_{58}$Fe(nγ) $_{59}$Fe	R 45 d	β 0·46(53%) 0·27(46%) 0·13(1%)	1·10(56%) 1·29(44%) 0·19(2·4%)	6·4
Krypton-85	U(nf) $_{85}$Kr	F 10·6 y	β 0·67(99·6%) 0·15(0·4%)	0·51(0·4%)	
Lanthanum-140	$_{139}$La(nγ) $_{140}$La	R 40·2 h	β 1·38(45%) 1·10(26%) 0·83(12%) 1·71(10%) 2·20(7%)	1·60(95%) 0·49(41%) 0·82(27%) 0·33(19%) 0·92(11%) 2·54(4%) others very weak	11·3
Lutecium-177	$_{176}$Lu(nγ) $_{177}$Lu	R 6·75 d	β 0·50(90%) 0·17(7%) 0·38(3%)	0·21(6·7%) 0·11(3·2%)	0·09
Magnesium-28	$_{37}$Cl(p, 6p4n) $_{28}$Mg	C 21·4 h	β 0·42. See also $_{28}$Al (daughter)	0·032(96%) 1·35(70%) 0·40(30%) 0·95(30%)	15·7 with $_{28}$Al
Manganese-52	$_{56}$Fe(p, αn) $_{52}$Mn	C 5·7 d	β^+ 0·58(33%) EC (67%)	0·73, 0·94, 1·45 0·51 A 0·005 following EC	18·6
Manganese-54	$_{56}$Fe(dα) $_{54}$Mn	C 291 d	EC	0·84, 0·005 following EC 0·133 (31%) 0·164 (4·5%)	4·7
Mercury-197	$_{196}$Hg(nγ) $_{197}$Hg(m) $_{197}$Hg	R 24 h 65 h	EC	0·077(20%) 0·07 and 0·01 following EC	1·3
Mercury-203	$_{202}$Hg(nγ) $_{203}$Hg	R 47 d	β 0·21	0·279(83%)	
Molybdenum-99	$_{98}$Mo(nγ) $_{99}$Mo	R 67 h	β 1·23(85%) 0·45(14%) 0·87(1%)	various to 0·78	
Neodymium-147	$_{146}$Nd(nγ) $_{147}$Nd	R 11·1 d	β 0·81(77%) 0·37(20%) 0·21(3%)	0·091(30%) 0·53(15%) 0·32(2·9%) 0·44(1·9%) 0·40(1·6%) 0·28(1·5%) 0·12(1%) 0·69(1%)	0·8
Nickel-63	$_{62}$Ni(nγ) $_{63}$Ni	R 125 y	β 0·067		
Niobium-95	$_{94}$Zr(nγ) $_{95}$Zr β $_{95}$Nb	R 35 d	β 0·16(99%) 0·93(1%)	0·76(99%)	4·2
Osmium-191	$_{190}$Os(nγ) $_{191}$Os	R 15 d	β 0·143	0·129(20%) 0·065, 0·01 (Ir X-rays from $_{191}$Ir(m) (daughter)	

370

1	2		3	4	5	6
Osmium-193	$_{192}$Os($n\gamma$) $_{193}$Os	R	31 h	β 1·11(67%) 0·65(13%) 1·03(6%) 0·55(5%) 0·97(4%) 0·72(3%) 0·75(2%)	0·46(4%) 0·139(3%) 0·28(1·3%) others very weak	
Palladium-103	$_{102}$Pd($n\gamma$) $_{103}$Pd	R	17 d	EC	0·020 following EC	
Phosphorus-32	$_{31}$P($n\gamma$) $_{32}$P	R	14·2 d	β 1·71		
Potassium-42	$_{41}$K($n\gamma$) $_{42}$K	R	12·45 h	β 3·6(82%) 2·0(18%)	1·52(18%)	1·4
Potassium-43	$_{40}$A(αp) $_{43}$K	C	22 h	β 0·83(87%) 0·47(8%) 1·24(3·5%) 1·81(1·5%)	0·37(85%) 0·61(81%) 0·39(18%) 0·59(13%) 0·22(3%) 1·01(2%)	5·6
Praseodymium-142	$_{141}$Pr($n\gamma$) $_{142}$Pr	R	19·2 h	β 2·15(96%) 0·58(4%)	1·57(4%)	
Praseodymium-143	$_{142}$Ce($n\gamma$) $_{143}$Ce β $_{143}$Pr	R	13·8 d	β 0·93		
Praseodymium-144	U(nf) $_{144}$Ce β $_{144}$Pr	F	17·5 m	β 2·98(97·7%) 2·29(1·3%) 0·80(1%)	0·69(1·6%)	
Promethium-147	$_{146}$Nd($n\gamma$) $_{147}$Nd β $_{147}$Pm	R	2·6 y	β 0·22		
Rhenium-186	$_{185}$Re($n\gamma$) $_{186}$Re	R	3·7 d	β 1·07(73%) 0·93(23%) EC(4%)	0·137(11%) 0·122(2%)	1·7
Rhodium-106	U(nf) $_{106}$Ru β $_{106}$Rh	F	30 s	β 3·6(70%) 2·4(12%) 3·1(12%) 2·0(3%) other 3%	0·51(30%) 0·62(15%) others up to 2·9	
Rubidium-86	$_{85}$Rb($n\gamma$) $_{86}$Rb	R	18·7 d	β 1·77(91·5%) 0·68(8·5%)	1·08(8·5%)	0·5
Ruthenium-103	$_{102}$Ru($n\gamma$) $_{103}$Ru	R	40 d	β 0·21(90%) 0·12(7%) 0·70(3%)	0·50(90%) 0·61(7%) others weak	
Ruthenium-106	U(nf) $_{106}$Ru	F	1·0 y	β 0·039. See also $_{106}$Rh		
Samarium-153	$_{152}$Sm($n\gamma$) $_{153}$Sm	R	46·2 h	β 0·70(53%) 0·63(26%) 0·80(20%)	0·10(37%) 0·07(4%) others very weak	
Scandium-46	$_{45}$Sc($n\gamma$) $_{46}$Sc	R	84 d	β 0·36	0·89 1·12	10·9

1	2		3	4	5	6
Scandium-47	$_{46}$Ca(nγ) $_{47}$Ca β $_{47}$Sc	R	3·4 d	β 0·45(74%) 0·61(26%)	0·16(74%)	0·59
Selenium-75	$_{74}$Se(nγ) $_{75}$Se	R	121 d	EC	0·27(56%) 0·14(54%) 0·28(23%) 0·12(15%) 0·40(12·5%) 0·096(3%) 0·20(1·5%) 0·31(1·4%) 0·066(1%) 0·01 following EC	2·0
Silver-110	$_{109}$Ag(nγ) $_{110}$Ag	R	253 d(m)	β 0·087(55%) 0·53(43%) Transition (2%) to $_{110}$Ag 24 s (β 2·87)	0·66(94%) 0·88(69%) 0·94(29%) 1·38(26%) 0·76(21%) 0·71(17%) 1·51(14%) 0·68(12%) 0·81(8%) 0·74(5%) 1·48(5%) and 0·116 from transition 0·66(5%)	14·3
			24 s	β 2·87(95%) 2·24(5%)		
Silver-111	$_{110}$Pd(nγ) $_{111}$Pd β $_{111}$Ag	R	7·5 d	β 1·04(93%) 0·69(6%) 0·79(1%)	0·34(6%) 0·247(1%)	
Sodium-22	$_{24}$Mg(dα) $_{22}$Na	C	2·6 y	β^+ 0·54(89%) EC (11%)	1·28 0·51 A	12·0
Sodium-24	$_{23}$Na(nγ) $_{24}$Na	R	15 h	β 1·39	1·37 2·75	18·4
Strontium-85	$_{84}$Sr(nγ) $_{85}$Sr	R	65 d	EC	0·513 0·013 following EC	3·0
Strontium-89	$_{88}$Sr(nγ) $_{89}$Sr	R	51 d	β 1·46	0·91 very weak	
Strontium-90	U(nf) $_{90}$Sr	F	28 y	β 0·54. See also $_{90}$Y (daughter).		
Sulphur-35	$_{35}$Cl(np) $_{35}$S	R	87·2 d	β 0·167		
Tantalum-182	$_{181}$Ta(nγ) $_{182}$Ta	R	115 d	β 0·18(38%) 0·44(23%) 0·36(20%) 0·51(8%) 0·25(5%) 0·48(4%) 0·33(2%)	1·12(33%) 0·068(31%) 1·22(28%) 1·19(15%) 0·222(12%) 1·23(11%) and twelve other energies including 0·059 W X-rays	6·8
Technetium-99	$_{98}$Mo(nγ) $_{99}$Mo β $_{99}$Tc	R	2·12 × 10^5 y	β 0·29		
Tellurium-129	U(nf) $_{129}$Te(m) → $_{129}$Te	F	41 d(m) 74 m	Transition to $_{129}$Te β 1·45(71%) 0·99(15%) 0·29(10%) 0·69(4%)	various to 1·12 all weak	

1	2	F	3	4	5	6
Tellurium-132	U(nf) $_{132}$Te	F	78 h	β 0·22. See also $_{132}$I	0·23(95 %) 0·028 from I X-rays	2·2 including $_{132}$I (daughter)
Terbium-160	$_{159}$Tb(nγ) $_{160}$Tb	R	73 d	β 0·56(38 %) 0·86(20 %) 0·46(19 %) 0·3(12 %) 0·76(11 %)	various up to 1·27	
Thallium-204	$_{203}$Tl(nγ) $_{204}$Tl	R	3·9 y	β 0·77(98 %) EC (2 %)		
Thulium-170	$_{169}$Tm(nγ) $_{170}$Tm	R	127 d	β 0·97(78 %) 0·88(22 %) EC (0·15 %)	0·084(3 %) and 0·052(4 %) 0·007(3 %) from Yb X-rays	0·025
Tin-113	$_{112}$Sn(nγ) $_{113}$Sn	R	119 d	EC	0·26(2 %) 0·024, 0·003 following EC 0·39(64 %) from $_{113}$In(m) (daughter)	
Tungsten-185	$_{184}$W(nγ) $_{185}$W	R	73 d	β 0·43		
Tungsten-187	$_{186}$W(nγ) $_{187}$W	R	24 h	β 0·62(64 %) 1·31(20 %) 0·69(7 %) 0·54(4 %) 0·63(4 %) others (1 %)	0·686(32 %) 0·479(28 %) 0·072(13 %) 0·134(10 %) 0·618(8 %) 0·552(6 %) 0·773(5 %)	3·0
Vanadium-48	Ti(pxn) $_{48}$V	C	16·2 d	β+ 0·70(56 %) EC(44 %)	0·99, 1·31(97·5 %) 2·25(2·5 %) 0·51 A	15·6
Xenon-133	$_{132}$Xe(nγ) $_{133}$Xe	R	5·27 d	β 0·34	0·081(32 %) 0·23 from $_{133}$Xe(m)	
Ytterbium-169	$_{168}$Yb(nγ) $_{169}$Yb	R	31 d	EC	various to 0·31, 0·051, 0·007 following EC	
Yttrium-88	$_{88}$Sr(d2n) $_{88}$Y	C	105 d	β+ 0·6(0·2 %) EC (99·8 %)	1·85(99 %) 0·91(90 %) 0·51 A	
Yttrium-90	$_{89}$Y(nγ) $_{90}$Y	R	64·2 h	β 2·25		
Yttrium-91	U(nf) $_{91}$Y	F	58 d	β 1·53	1·21(0·3 %)	
Zinc-65	$_{64}$Zn(nγ) $_{65}$Zn	R	245 d	β+ 0·325(1·5 %) EC (98·5 %)	1·11(45 %) 0·51 A 0·008 following EC	2·7
Zirconium-95	$_{94}$Zr(nγ) $_{95}$Zr	R	65 d	β 0·40(55 %) 0·36(43 %) 0·88(2 %)	0·73(55 %) 0·76(43 %)	4·1

DECIBEL TABLES

Voltage and Current Ratios Loss

dB	0	1	2	3	4	5	6	7	8	9	mean differences								
											1	2	3	4	5	6	7	8	9
0·0	1·00000	99886	99770	99655	99541	99426	99312	99197	99083	98969	12	23	35	46	57	68	80	91	103
0·1	0·98855	98742	98628	98515	98401	98289	98175	98063	97949	97838	12	23	35	46	57	68	80	91	103
0·2	0·97724	97610	97499	97386	97275	97162	97051	96939	96828	96716	11	22	33	44	56	67	78	89	100
0·3	0·96605	96494	96383	96272	96161	96051	95940	95830	95719	95610	11	22	33	44	56	67	78	89	100
0·4	0·95499	95399	95280	95169	95060	94933	94824	94733	94624	94515	11	22	33	43	54	65	76	87	98
0·5	0·94406	94298	94189	94081	93972	93865	93756	93650	93541	93434	11	22	33	43	54	65	76	87	98
0·6	0·93325	93217	93111	93003	92897	92789	92683	92576	92470	92363	11	21	32	42	53	64	75	85	96
0·7	0·92257	92151	92045	91939	91833	91728	91621	91517	91411	91307	11	21	32	42	53	64	75	85	96
0·8	0·91201	91095	90991	90886	90782	90677	90573	90469	90365	90261	11	21	32	42	52	62	73	83	94
0·9	0·90157	90054	89950	89807	89743	89640	89536	89435	89331	89229	11	21	32	42	52	62	73	83	94
1·0	0·89125	89022	88920	88818	88716	88614	88512	88410	88308	88207	10	20	31	41	51	61	71	81	92
1·1	0·88105	88004	87902	87802	87700	87600	87498	87399	87297	87198	10	20	31	41	51	61	71	81	92
1·2	0·87096	86995	86896	86795	86696	86596	86497	86397	86298	86198	10	20	30	40	50	60	70	79	89
1·3	0·86099	86000	85901	85803	85704	85606	85507	85409	85310	85213	10	20	30	40	50	60	70	79	89
1·4	0·85114	84935	84918	84820	84723	84625	84528	84430	84333	84237	10	19	29	39	49	58	68	78	88
1·5	0·84140	84043	83946	83850	83753	83657	83560	83465	83368	83273	10	19	29	39	49	58	68	78	88
1·6	0·83176	83080	82985	82889	82794	82699	82604	82509	82414	82319	10	19	29	38	48	57	67	76	86
1·7	0·82224	82130	82035	81941	81846	81753	81658	81565	81470	81378	10	19	29	38	48	57	67	76	86
1·8	0·81283	81189	81096	81003	80910	80817	80724	80631	80538	80446	10	19	28	37	47	56	65	74	84
1·9	0·80353	80261	80168	80076	79983	79892	79799	79709	79616	79526	10	19	28	37	47	56	65	74	84
2·0	0·79433	79341	79250	79159	79068	78977	78886	78796	78705	78615	9	18	27	36	45	54	63	72	82
2·1	0·78524	78434	78343	78254	78163	78074	77983	77895	77804	77716	9	18	27	36	45	54	63	72	82
2·2	0·77625	77535	77446	77357	77268	77179	77090	77002	76913	76825	9	18	27	35	44	53	62	71	80
2·3	0·76736	76649	76560	76473	76384	76297	76208	76122	76033	75947	9	18	27	35	44	53	62	71	80
2·4	0·75858	75770	75683	75596	75509	75423	75336	75249	75162	75076	9	17	26	35	44	52	61	69	78
2·5	0·74989	74904	74817	74732	74645	74560	74473	74389	74302	74218	9	17	26	35	43	52	61	69	78
2·6	0·74131	74046	73961	73875	73790	73706	73621	73536	73451	73367	9	17	26	34	43	51	60	68	77
2·7	0·73282	73199	73114	73031	72946	72863	72778	72696	72611	72529	9	17	26	34	43	51	60	68	77
2·8	0·72444	72360	72277	72194	72111	72028	71945	71862	71779	71697	9	17	25	34	42	51	60	66	75
2·9	0·71614	71533	71450	71368	71285	71204	71121	71041	70958	70878	9	17	25	33	42	50	58	66	75

Numbers in mean difference columns to be *subtracted*.

Voltage and Current Ratios (*continued*)

Loss

dB	0	1	2	3	4	5	6	7	8	9		mean differences							
											1	2	3	4	5	6	7	8	9
3·0	0·70795	70713	70632	70550	70469	70388	70307	70227	70146	70065	8	16	24	32	40	48	56	64	73
3·1	0·69984	69904	69823	69744	69663	69584	69503	69424	69343	69264	8	16	24	32	40	48	56	64	73
3·2	0·69183	69103	69024	68944	68865	68786	68707	68628	68549	68470	8	16	24	32	40	47	55	63	71
3·3	0·68391	68313	68234	68156	68077	67999	67920	67843	67764	67687	8	16	24	32	40	47	55	63	71
3·4	0·67608	67530	67453	67375	67298	67220	67143	67065	66988	66911	8	15	23	31	39	46	54	62	70
3·5	0·66834	66758	66681	66604	66527	66451	66374	66299	66222	66146	8	15	23	31	39	46	54	62	70
3·6	0·66069	65992	65917	65841	65766	65690	65615	65539	65464	65388	8	15	23	30	38	45	53	60	68
3·7	0·65313	65238	65163	65088	65013	64938	64863	64789	64714	64640	8	15	23	30	38	45	53	60	68
3·8	0·64565	64491	64417	64343	64269	64195	64121	64047	63973	63900	8	15	22	29	37	44	52	59	67
3·9	0·63826	63754	63680	63607	63533	63461	63387	63315	63241	63170	8	15	22	29	37	44	52	59	67
4·0	0·63096	63023	62951	62888	62806	62733	62661	62589	62517	62445	7	14	22	29	36	43	51	58	65
4·1	0·62373	62302	62230	62159	62087	62016	61944	61874	61802	61731	7	14	22	29	36	43	51	58	65
4·2	0·61659	61588	61518	61446	61376	61335	61235	61164	61094	61024	7	14	21	28	35	42	49	56	63
4·3	0·60954	60884	60814	60744	60674	60604	60534	60465	60395	60326	7	14	21	28	35	42	49	56	63
4·4	0·60256	60186	60117	60048	59979	59910	59841	59773	59704	59635	7	14	21	27	34	41	48	55	62
4·5	0·59566	59498	59429	59362	59293	59225	59156	59089	59020	58953	7	14	21	27	34	41	48	55	62
4·6	0·58884	58816	58749	58681	58614	58546	58479	58412	58345	58277	7	13	20	27	33	40	47	54	61
4·7	0·58210	58143	58075	58017	57943	57877	57810	57744	57677	57614	7	13	20	27	33	40	47	54	61
4·8	0·57544	57478	57412	57364	57280	57214	57148	57082	57016	56951	7	13	20	26	33	39	46	52	57
4·9	0·56885	56820	56754	56690	56624	56560	56494	56430	56364	56300	7	13	20	26	33	39	46	52	57
5·0	0·56234	56169	56105	56040	55976	55911	55847	55783	55719	55654	7	13	20	26	32	38	45	51	58
5·1	0·55590	55527	55463	55400	55336	55272	55208	55145	55081	55018	7	13	20	26	32	38	45	51	58
5·2	0·54954	54891	54828	54765	54702	54639	54576	54513	54450	54388	7	13	19	25	31	38	44	50	57
5·3	0·54325	54263	54200	54138	54075	54014	53951	53890	53827	53766	7	13	19	25	31	37	44	50	57
5·4	0·53703	53641	53580	53517	53456	53394	53333	53272	53211	53149	6	12	18	24	30	37	43	49	55
5·5	0·53088	53027	52966	52906	52845	52784	52723	52663	52602	52542	6	12	18	24	30	37	43	49	55
5·6	0·52481	52420	52360	52300	52240	52179	52119	52060	52000	51940	6	12	18	24	30	36	42	48	54
5·7	0·51880	51820	51761	51702	51642	51583	51523	51464	51404	51346	6	12	18	24	29	36	42	48	54
5·8	0·51286	51226	51168	51118	51050	50991	50933	50874	50816	50757	6	12	18	23	29	35	41	47	53
5·9	0·50699	50640	50582	50524	50466	50408	50350	50292	50234	50177	6	12	18	23	29	35	41	47	53
6·0	0·50119	50060	50003	49945	49888	49831	49774	49716	49659	49602	6	11	17	23	28	34	40	46	52
6·1	0·49545	49488	49431	49374	49317	49261	49204	49148	49091	49035	6	11	17	23	28	34	40	46	52
6·2	0·48978	48921	48865	48809	48753	48697	48641	48585	48529	48473	6	11	17	22	28	33	39	45	51
6·3	0·48417	48362	48306	48251	48195	48140	48084	48029	47973	47919	6	11	17	22	28	33	39	45	51
6·4	0·47863	47807	47753	47697	47641	47588	47534	47478	47424	47369	6	11	17	22	28	33	39	44	49

Numbers in mean difference columns to be *subtracted*.

Voltage and Current Ratios (continued)

Loss

dB	0	1	2	3	4	5	6	7	8	9	mean differences								
											1	2	3	4	5	6	7	8	9
6·5	0·47315	47260	47206	47152	47098	47043	46989	46935	46881	46828	6	11	17	22	28	33	39	44	49
6·6	0·46774	46719	46666	46612	46559	46507	46452	46398	46345	46291	6	11	16	21	27	32	38	43	48
6·7	0·46238	46185	46132	46079	46026	45973	45920	45867	45814	45762	5	11	16	21	27	32	38	43	48
6·8	0·45709	45656	45604	45551	45499	45446	45394	45342	45290	45238	5	10	16	21	26	31	37	42	47
6·9	0·45186	45134	45082	45030	44978	44927	44875	44823	44771	44720	5	10	16	21	26	31	37	42	47
7·0	0·44668	44617	44566	44514	44463	44412	44361	44310	44259	44208	5	10	15	20	25	30	36	41	46
7·1	0·44157	44106	44055	44005	43954	43904	43853	43803	43752	43703	5	10	15	20	25	30	36	41	46
7·2	0·43652	43601	43551	43501	43451	43401	43351	43301	43251	43202	5	10	15	20	25	30	35	40	45
7·3	0·43152	43103	43053	43004	42954	42905	42855	42806	42756	42708	5	10	15	20	25	30	35	40	45
7·4	0·42658	42609	42560	42511	42462	42413	42364	42316	42267	42219	5	10	15	19	24	29	34	39	44
7·5	0·42170	42122	42073	42025	41976	41928	41879	41832	41783	41736	5	10	15	19	24	29	34	39	44
7·6	0·41687	41638	41591	41542	41495	41447	41400	41352	41305	41257	5	9	14	19	24	28	33	38	43
7·7	0·41210	41162	41115	41067	41021	40973	40926	40879	40832	40785	5	9	14	19	24	28	33	38	43
7·8	0·40738	40690	40644	40597	40551	40504	40458	40411	40365	40318	5	9	14	18	23	28	32	37	42
7·9	0·40272	40225	40179	40133	40087	40040	39994	39948	39902	39857	5	9	14	18	23	28	32	37	42
8·0	0·39811	39764	39719	39673	39628	39582	39537	39491	39446	39400	5	9	14	18	22	27	31	36	41
8·1	0·39355	39309	39264	39219	39174	39129	39084	39039	38994	38950	5	9	14	18	22	27	31	36	41
8·2	0·38905	38859	38815	38770	38726	38681	38637	38592	38548	38503	5	9	13	18	22	27	31	35	40
8·3	0·38459	38415	38371	38326	38282	38238	38194	38151	38107	38063	5	9	13	18	22	27	31	35	40
8·4	0·38019	37974	37931	37887	37844	37800	37757	37713	37670	37627	5	9	13	17	22	26	30	35	39
8·5	0·37584	37540	37497	37454	37411	37368	37325	37282	37239	37197	5	9	13	17	21	26	30	35	39
8·6	0·37154	37110	37068	37025	36983	36940	36898	36855	36813	36770	4	9	12	17	20	25	29	34	38
8·7	0·36728	36686	36644	36601	36559	36517	36475	36434	36392	36350	4	8	12	17	20	25	29	34	38
8·8	0·36308	36265	36224	36182	36141	36099	36058	36016	35975	35933	4	8	12	16	20	25	29	33	38
8·9	0·35892	35851	35810	35768	35727	35686	35645	35604	35563	35522	4	8	12	16	20	25	29	33	38
9·0	0·35481	35440	35400	35358	35318	35277	35237	35196	35156	35115	4	8	12	16	20	24	28	32	36
9·1	0·35075	35035	34995	34954	34914	34874	34834	34794	34754	34714	4	8	12	16	20	24	28	32	36
9·2	0·34674	34634	34594	34554	34514	34475	34435	34396	34356	34317	4	8	12	16	20	24	28	32	36
9·3	0·34277	34238	34198	34159	34119	34081	34041	34003	33963	33924	4	8	12	16	20	24	28	32	36
9·4	0·33884	33845	33806	33767	33729	33690	33651	33613	33574	33536	4	8	12	15	19	23	27	31	35
9·5	0·33497	33459	33420	33382	33343	33305	33266	33228	33189	33152	4	8	12	15	19	23	27	31	35
9·6	0·33113	33075	33037	32999	32961	32923	32885	32847	32809	32773	4	8	12	15	19	23	27	30	34
9·7	0·32735	32697	32659	32622	32584	32547	32509	32472	32434	32397	4	8	11	15	19	23	27	30	34
9·8	0·32359	32322	32285	32248	32211	32174	32137	32100	32063	32026	4	7	11	15	19	22	26	29	33
9·9	0·31989	31953	31916	31879	31842	31806	31769	31733	31696	31660	4	7	11	15	19	22	26	29	33

Numbers in mean difference columns to be *subtracted*.

Voltage and Current Ratios (*continued*)

Loss

dB	0	1	2	3	4	5	6	7	8	9	mean differences								
											1	2	3	4	5	6	7	8	9
10·0	0·31623	31586	31550	31513	31477	31441	31405	31369	31333	31297	4	7	11	14	18	22	25	29	32
10·1	0·31261	31225	31189	31153	31117	31082	31046	31010	30974	30939	4	7	11	14	18	22	25	29	32
10·2	0·30903	30867	30832	30796	30761	30725	30690	30655	30620	30584	4	7	11	14	18	21	24	28	32
10·3	0·30549	30514	30479	30444	30409	30374	30339	30304	30269	30235	4	7	11	14	18	21	24	28	32
10·4	0·30200	30164	30130	30095	30061	30026	29992	29957	29923	29888	4	7	11	14	18	21	24	28	31
10·5	0·29854	29819	29785	29751	29717	29682	29648	29614	29580	29546	4	7	11	14	18	21	24	28	31
10·6	0·29512	29478	29444	29410	29376	29343	29309	29276	29242	29208	4	7	10	13	17	20	23	27	30
10·7	0·29174	29141	29107	29074	29040	29007	28973	28941	28907	28874	4	7	10	13	17	20	23	27	30
10·8	0·28840	28807	28774	28741	28708	28675	28642	28609	28576	28543	4	7	10	13	17	20	23	26	30
10·9	0·28510	28478	28445	28412	28379	28347	28314	28282	28249	28217	4	7	10	13	17	20	23	26	30
11·0	0·28184	28151	28118	28086	28053	28022	27989	27957	27924	27893	3	6	10	13	16	19	23	26	29
11·1	0·27861	27829	27797	27765	27733	27601	27669	27638	27606	27574	3	6	10	13	16	19	23	26	29
11·2	0·27542	27510	27479	27447	27416	27384	27353	27321	27290	27258	3	6	10	13	16	19	22	25	28
11·3	0·27227	27195	27164	27133	27102	27071	27040	27008	26977	26946	3	6	10	13	16	19	22	25	28
11·4	0·26915	26884	26853	26823	26792	26761	26730	26700	26669	26638	3	6	9	12	16	18	21	24	28
11·5	0·26607	26577	26546	26516	26485	26455	26424	26394	26363	26334	3	6	9	12	16	18	21	24	28
11·6	0·26303	26272	26242	26212	26182	26152	26122	26092	26062	26032	3	6	9	12	16	18	21	24	27
11·7	0·26002	25972	25942	25912	25882	25853	25823	25793	25763	25734	3	6	9	12	16	18	21	24	27
11·8	0·25704	25674	25645	25615	25586	25556	25527	25497	25468	25439	3	6	9	12	16	18	21	23	26
11·9	0·25410	25380	25351	25322	25293	25265	25236	25206	25177	25148	3	5	9	12	16	18	21	23	26
12·0	0·25119	25090	25061	25032	25003	24975	24946	24918	24889	24860	3	6	9	11	14	17	20	23	26
12·1	0·24831	24803	24774	24746	24717	24689	24660	24633	24604	24576	3	6	9	11	14	17	20	23	26
12·2	0·24547	24519	24490	24462	24433	24406	24377	24350	24321	24294	3	6	9	11	14	17	20	22	25
12·3	0·24266	24238	24210	24183	24155	24127	24099	24072	24044	24016	3	6	9	11	14	17	20	22	25
12·4	0·23988	23960	23933	23905	23878	23850	23823	23795	23768	23741	3	5	8	11	14	16	19	22	24
12·5	0·23714	23686	23659	23632	23605	23577	23550	23523	23496	23469	3	5	8	11	14	16	19	22	24
12·6	0·23442	23415	23388	23363	23336	23308	23281	23254	23227	23201	3	5	8	11	14	16	19	21	24
12·7	0·23174	23148	23121	23094	23067	23041	23014	22988	22961	22936	3	5	8	11	13	16	19	21	24
12·8	0·22909	22882	22856	22829	22803	22777	22751	22725	22699	22672	3	5	8	10	13	16	19	21	24
12·9	0·22646	22620	22594	22568	22542	22517	22491	22465	22439	22413	3	5	8	10	13	16	19	21	24
13·0	0·22387	22361	22336	22309	22284	22258	22233	22207	22182	22156	3	5	8	10	13	15	18	20	23
13·1	0·22131	22105	22080	22054	22029	22004	21979	21953	21928	21903	3	5	8	10	13	15	18	20	23
13·2	0·21878	21852	21827	21802	21777	21752	21727	21702	21677	21652	3	5	8	10	13	15	18	20	23
13·3	0·21627	21602	21577	21553	21528	21503	21478	21454	21429	21405	3	5	8	10	13	15	18	20	23
13·4	0·21380	21354	21330	21305	21281	21256	21232	21208	21184	21159	3	5	8	10	13	15	17	19	22

Numbers in mean difference columns to be *subtracted*.

Voltage and Current Ratios (*continued*)

Loss

dB	0	1	2	3	4	5	6	7	8	9	mean differences								
											1	2	3	4	5	6	7	8	9
13·5	0·21135	21110	21086	21062	21038	21013	20989	20965	20941	20917	3	5	8	10	13	15	17	19	22
13·6	0·20893	20869	20845	20821	20797	20773	20749	20725	20701	20678	3	5	8	10	12	14	16	19	22
13·7	0·20654	20630	20606	20583	20559	20536	20512	20488	20464	20441	3	5	8	10	12	14	16	19	22
13·8	0·20417	20393	20370	20347	20324	20300	20277	20253	20230	20207	3	5	7	9	11	14	16	19	21
13·9	0·20184	20160	20137	20114	20091	20068	20045	20022	19999	19976	3	5	7	9	11	14	16	19	21
14·0	0·19953	19930	19907	19884	19861	19838	19815	19793	19770	19747	3	5	7	9	11	14	16	18	21
14·1	0·19724	19702	19679	19667	19634	19611	19588	19566	19543	19521	3	5	7	9	11	14	16	18	21
14·2	0·19498	19476	19454	19431	19409	19386	19364	19342	19320	19297	2	4	7	9	11	13	16	18	20
14·3	0·19275	19253	19231	19209	19189	19165	19143	19121	19099	19077	2	4	7	9	11	13	16	18	20
14·4	0·19055	19033	19011	18989	18967	18945	18923	18902	18880	18858	2	4	7	9	11	13	15	17	20
14·5	0·18836	18815	18793	18772	18750	18729	18707	18686	18664	18643	2	4	7	9	11	13	15	17	20
14·6	0·18621	18599	18578	18556	18535	18514	18493	18471	18450	18429	2	4	6	8	10	13	15	17	19
14·7	0·18408	18386	18365	18344	18323	18302	18281	18260	18239	18218	2	4	6	8	10	13	15	17	19
14·8	0·18197	18176	18155	18134	18113	18093	18072	18051	18030	18010	2	4	6	8	10	12	15	17	19
14·9	0·17989	17968	17947	17927	17906	17886	17865	17845	17824	17804	2	4	6	8	10	12	14	17	19
15·0	0·17783	17762	17742	17721	17701	17680	17659	17640	17619	17599	2	4	6	8	10	12	14	16	18
15·1	0·17579	17559	17539	17518	17498	17478	17458	17438	17418	17398	2	4	6	8	10	12	14	16	18
15·2	0·17378	17358	17338	17318	17298	17278	17258	17239	17219	17199	2	4	6	8	10	12	14	16	18
15·3	0·17179	17160	17140	17120	17100	17081	17061	17042	17022	17002	2	4	6	8	10	12	14	16	18
15·4	0·16982	16962	16943	16923	16904	16885	16866	16846	16827	16807	2	4	6	8	10	12	14	15	17
15·5	0·16788	16768	16749	16730	16711	16691	16672	16653	16634	16615	2	4	6	8	10	12	14	15	17
15·6	0·16596	16577	16558	16539	16520	16501	16482	16463	16444	16425	2	4	6	8	10	11	13	15	17
15·7	0·16406	16387	16368	16350	16331	16312	16293	16274	16255	16237	2	4	6	7	9	11	13	15	17
15·8	0·16218	16199	16181	16162	16144	16124	16106	16087	16069	16050	2	4	6	7	9	11	13	15	17
15·9	0·16032	16014	15996	15977	15959	15940	15922	15903	15885	15867	2	4	6	7	9	11	13	15	17
16·0	0·15849	15830	15812	15794	15776	15758	15740	15722	15704	15686	2	4	6	7	9	11	13	14	16
16·1	0·15668	15649	15631	15614	15596	15578	15560	15542	15524	15506	2	4	6	7	9	11	13	14	16
16·2	0·15488	15471	15453	15435	15417	15400	15382	15364	15346	15329	2	4	6	7	9	11	13	14	16
16·3	0·15311	15294	15276	15259	15241	15223	15205	15189	15171	15154	2	4	5	7	9	11	13	14	16
16·4	0·15136	15118	15101	15083	15066	15048	15031	15014	14997	14979	2	3	5	7	9	10	12	14	16
16·5	0·14962	14945	14928	14911	14894	14876	14859	14842	14825	14808	2	3	5	7	9	10	12	14	16
16·6	0·14791	14774	14757	14740	14723	14706	14689	14672	14655	14639	2	3	5	7	9	10	12	13	15
16·7	0·14622	14605	14588	14572	14555	14538	14521	14505	14488	14471	2	3	5	7	9	10	12	13	15
16·8	0·14454	14437	14421	14404	14388	14371	14355	14338	14322	14305	2	3	5	7	9	10	12	13	15
16·9	0·14289	14272	14256	14239	14223	14207	14191	14174	14158	14141	2	3	5	7	9	10	12	13	15

Numbers in mean difference columns to be *subtracted*.

Voltage and Current Ratios (*continued*)

Loss

dB	0	1	2	3	4	5	6	7	8	9	1	2	3	4	5	6	7	8	9
														mean differences					
17·0	0·14125	14109	14093	14076	14060	14044	14028	14012	13996	13980	2	3	4	6	8	10	12	13	14
17·1	0·13964	13948	13932	13916	13900	13884	13868	13852	13836	13820	2	3	4	6	8	10	12	13	14
17·2	0·13804	13788	13772	13756	13740	13725	13709	13693	13677	13662	2	3	4	6	8	9	11	13	14
17·3	0·13646	13630	13614	13599	13583	13568	13552	13537	13521	13506	2	3	4	6	8	9	11	13	14
17·4	0·13490	13474	13459	13443	13428	13412	13397	13382	13366	13350	2	3	4	6	8	9	11	12	14
17·5	0·13335	13320	13305	13289	13274	13258	13243	13228	13213	13198	2	3	4	6	7	9	11	12	14
17·6	0·13183	13167	13152	13137	13122	13107	13092	13077	13062	13047	2	3	4	6	7	9	11	12	14
17·7	0·13032	13017	13002	12987	12972	12957	12942	12927	12912	12897	2	3	4	6	7	9	11	12	14
17·8	0·12882	12868	12853	12838	12823	12809	12794	12779	12764	12749	2	3	4	6	7	9	11	12	14
17·9	0·12735	12721	12706	12691	12677	12662	12647	12633	12618	12604	2	3	4	6	7	9	11	12	14
18·0	0·12589	12574	12560	12545	12531	12517	12503	12488	12474	12459	2	3	4	6	7	9	10	11	13
18·1	0·12445	12431	12417	12402	12388	12373	12359	12345	12331	12317	2	3	4	6	7	9	10	11	13
18·2	0·12303	12288	12274	12260	12246	12232	12218	12204	12190	12176	2	3	4	6	7	8	10	11	13
18·3	0·12162	12148	12134	12120	12106	12092	12078	12064	12050	12037	2	3	4	6	7	8	10	11	13
18·4	0·12023	12009	11995	11981	11967	11954	11940	11926	11912	11899	2	3	4	5	6	8	9	11	13
18·5	0·11885	11872	11858	11844	11830	11817	11803	11790	11776	11763	2	3	4	5	6	8	9	11	13
18·6	0·11749	11735	11722	11708	11695	11681	11668	11654	11641	11627	2	3	4	5	6	8	9	11	12
18·7	0·11615	11601	11588	11574	11561	11548	11535	11521	11508	11495	2	3	4	5	6	8	9	11	12
18·8	0·11482	11468	11455	11442	11429	11415	11402	11389	11376	11363	2	3	4	5	6	8	9	11	12
18·9	0·11350	11337	11324	11311	11298	11285	11272	11259	11246	11233	2	3	4	5	6	8	9	11	12
19·0	0·11220	11207	11194	11182	11169	11156	11143	11130	11117	11105	2	3	4	5	6	8	9	10	11
19·1	0·11092	11079	11066	11054	11041	11028	11015	11003	10990	10978	2	3	4	5	6	8	9	10	11
19·2	0·10965	10953	10940	10927	10914	10902	10889	10877	10864	10852	2	3	4	5	6	8	9	10	11
19·3	0·10839	10827	10814	10802	10789	10778	10765	10753	10740	10728	2	3	4	5	6	8	9	10	11
19·4	0·10715	10703	10691	10678	10666	10653	10641	10629	10617	10605	1	2	3	5	6	7	8	10	11
19·5	0·10593	10580	10568	10556	10544	10532	10520	10507	10495	10483	1	2	3	5	6	7	8	10	11
19·6	0·10471	10459	10447	10435	10423	10411	10399	10387	10375	10363	1	2	3	5	6	7	8	10	11
19·7	0·10351	10340	10328	10316	10304	10292	10280	10269	10257	10245	1	2	3	5	6	7	8	10	11
19·8	0·10233	10221	10209	10198	10186	10174	10162	10151	10139	10128	1	2	3	5	6	7	8	9	10
19·9	0·10117	10105	10093	10081	10069	10058	10046	10035	10023	10012	1	2	3	5	6	7	8	9	10

Numbers in mean difference columns to be *subtracted*.

Power Ratios

Loss

dB	0	1	2	3	4	5	6	7	8	9		mean differences							
											1	2	3	4	5	6	7	8	9
0·0	1·00000	99770	99541	99312	99083	98855	98628	98401	98175	97949	23	46	68	91	114	137	160	182	205
0·1	0·97724	97499	97275	97051	96828	96605	96383	96161	95940	95719	22	44	67	89	111	133	155	178	200
0·2	0·95499	95280	95060	94842	94624	94406	94189	93972	93756	93541	22	43	65	87	109	130	152	174	195
0·3	0·93325	93111	92897	92683	92470	92257	92045	91833	91622	91411	21	42	64	85	106	127	149	170	191
0·4	0·91201	90991	90782	90573	90365	90157	89950	89743	89536	89331	21	42	62	83	104	125	146	166	187
0·5	0·89125	88920	88716	88512	88308	88105	87902	87700	87498	87297	20	41	61	81	102	122	142	162	183
0·6	0·87096	86896	86696	86497	86298	86099	85901	85704	85507	85310	20	40	60	79	99	119	139	158	178
0·7	0·85114	84918	84723	84528	84333	84140	83946	83753	83560	83368	19	39	58	78	97	116	136	155	175
0·8	0·83176	82985	82794	82604	82414	82224	82035	81846	81658	81470	19	38	57	76	95	113	132	151	170
0·9	0·81283	81096	80910	80724	80538	80353	80168	79983	79799	79616	19	37	56	74	93	111	130	148	167
1·0	0·79433	79250	79068	78886	78705	78524	78343	78163	77983	77804	18	36	54	72	91	109	127	145	163
1·1	0·77625	77446	77268	77090	76913	76736	76560	76384	76208	76033	18	35	53	71	89	107	125	142	159
1·2	0·75858	75683	75509	75336	75162	74989	74817	74645	74473	74302	17	35	52	69	87	104	121	138	156
1·3	0·74131	73961	73790	73621	73451	73282	73114	72946	72778	72611	17	34	51	68	85	101	118	135	152
1·4	0·72444	72277	72111	71945	71779	71614	71450	71285	71121	70958	17	33	50	66	83	99	116	132	149
1·5	0·70795	70632	70469	70307	70146	69984	69823	69663	69503	69343	16	32	48	64	81	97	113	129	145
1·6	0·69183	69024	68865	68707	68549	68391	68234	68077	67920	67764	16	31	47	63	79	95	110	126	142
1·7	0·67608	67453	67298	67143	66988	66843	66681	66527	66374	66222	15	31	46	62	77	93	108	123	139
1·8	0·66069	65917	65766	65615	65464	65313	65163	65013	64863	64714	15	30	45	60	75	90	105	120	135
1·9	0·64565	64417	64269	64121	63973	63826	63680	63533	63387	63241	15	29	44	59	74	88	103	118	132
2·0	0·63069	62951	62806	62661	62517	62373	62230	62087	61944	61802	14	29	43	58	72	86	101	115	130
2·1	0·61659	61518	61376	61235	61094	60954	60814	60674	60534	60395	14	28	42	56	70	84	98	113	126
2·2	0·60256	60117	59979	59841	59704	59566	59429	59293	59156	59020	14	27	41	55	69	82	96	110	123
2·3	0·58884	58749	58614	58479	58345	58210	58076	57943	57810	57677	13	27	40	54	67	80	94	107	121
2·4	0·57544	57412	57280	57148	57016	56885	56754	56624	56494	56364	13	26	39	52	66	79	92	105	118
2·5	0·56234	56105	55976	55847	55719	55591	55463	55336	55208	55081	13	26	38	51	64	77	90	102	115
2·6	0·54954	54828	54702	54576	54450	54325	54200	54075	53951	53827	13	25	38	50	63	75	88	100	113
2·7	0·53703	53580	53456	53333	53211	53088	52966	52845	52723	52602	12	24	37	49	61	73	85	98	110
2·8	0·52481	52360	52240	52119	52000	51880	51761	51642	51523	51404	12	24	36	48	60	72	84	96	108
2·9	0·51286	51168	51050	50933	50816	50699	50582	50466	50350	50234	12	23	35	47	58	70	82	93	105
3·0	0·50119	50003	49888	49774	49659	49545	49431	49317	49204	49091	11	23	34	46	57	68	80	91	103
3·1	0·48865	48865	48753	48641	48529	48417	48306	48195	48084	47973	11	22	33	45	56	67	78	89	100
3·2	0·47863	47753	47643	47534	47424	47315	47206	47098	46989	46881	11	22	33	44	54	65	76	87	98
3·3	0·46774	46666	46559	46452	46345	46238	46132	46026	45920	45814	11	21	32	43	53	64	75	85	96
3·4	0·45709	45604	45499	45394	45290	45186	45082	44978	44875	44771	10	21	31	42	52	62	73	83	94

Numbers in mean difference columns to be *subtracted*.

380

Loss

dB	0	1	2	3	4	5	6	7	8	9		mean differences								
												1	2	3	4	5	6	7	8	9
3·5	0·44668	44566	44463	44361	44259	44157	44055	43954	43853	43752		10	20	30	41	51	61	71	81	91
3·6	0·43652	43551	43451	43351	43251	43152	43053	42954	42855	42756		10	20	30	40	50	60	70	80	89
3·7	0·42658	42560	42462	42364	42267	42170	42073	41976	41879	41783		10	19	29	39	49	58	68	78	87
3·8	0·41687	41591	41495	41400	41305	41210	41115	41021	40926	40832		9	19	28	38	47	57	66	76	85
3·9	0·40738	40644	40551	40458	40365	40272	40179	40087	39994	39902		9	19	28	37	46	56	65	74	83
4·0	0·39811	39719	39628	39537	39446	39355	39264	39174	39084	38994		9	18	27	36	45	54	63	72	82
4·1	0·38905	38815	38726	38637	38548	38459	38371	38282	38194	38107		9	18	27	35	44	53	62	71	80
4·2	0·38019	37931	37844	37757	37670	37584	37497	37411	37325	37239		9	18	26	35	43	52	61	69	78
4·3	0·37154	37068	36983	36898	36813	36728	36644	36559	36475	36392		8	17	25	34	42	51	59	68	76
4·4	0·36308	36224	36141	36058	35975	35892	35810	35727	35645	35563		8	16	25	33	41	50	58	66	74
4·5	0·35481	35400	35318	35237	35156	35075	34995	34914	34834	34754		8	16	24	32	40	48	56	65	73
4·6	0·34674	34594	34514	34435	34356	34277	34198	34119	34041	33963		8	16	24	32	40	47	55	63	71
4·7	0·33884	33806	33729	33651	33574	33497	33420	33343	33266	33189		8	16	23	31	39	46	54	62	69
4·8	0·33113	33037	32961	32885	32809	32735	32659	32584	32509	32434		8	15	23	30	38	45	53	60	68
4·9	0·32359	32285	32211	32137	32063	31989	31916	31842	31769	31696		7	15	22	29	37	44	52	59	66
5·0	0·31623	31550	31477	31405	31333	31261	31189	31117	31046	30974		7	14	22	29	36	43	50	58	65
5·1	0·30903	30832	30761	30690	30620	30549	30479	30409	30339	30269		7	14	21	28	35	42	49	56	63
5·2	0·30200	30130	30061	29992	29923	29854	29785	29717	29648	29580		7	14	21	28	35	41	48	55	62
5·3	0·29512	29444	29376	29309	29242	29174	29107	29040	28973	28907		7	13	20	27	34	41	47	54	60
5·4	0·28840	28774	28708	28642	28576	28510	28445	28379	28314	28249		7	13	20	26	33	39	46	52	59
5·5	0·28184	28119	28054	27990	27925	27861	27797	27733	27669	27606		6	13	19	26	32	39	45	51	58
5·6	0·27542	27479	27416	27353	27290	27227	27164	27102	27040	26977		6	13	19	25	31	38	44	50	56
5·7	0·26915	26853	26792	26730	26669	26607	26546	26485	26424	26363		6	13	18	25	31	37	43	49	55
5·8	0·26303	26242	26182	26122	26062	26002	25942	25882	25823	25763		6	12	18	24	30	36	42	48	54
5·9	0·25704	25645	25586	25527	25468	25410	25351	25293	25236	25177		6	12	18	23	29	35	41	47	53
6·0	0·25119	25061	25003	24946	24889	24831	24774	24717	24660	24604		6	11	17	23	29	34	40	46	51
6·1	0·24547	24491	24434	24378	24322	24266	24210	24155	24099	24044		6	11	17	22	28	34	39	45	50
6·2	0·23988	23933	23878	23823	23768	23714	23659	23605	23550	23496		5	11	16	22	27	33	38	44	49
6·3	0·23442	23388	23336	23281	23227	23174	23121	23067	23014	22961		5	11	16	21	27	32	37	43	48
6·4	0·22909	22856	22803	22751	22699	22646	22594	22542	22491	22439		5	10	16	21	26	31	37	43	47
6·5	0·22387	22336	22284	22233	22182	22131	22080	22029	21979	21928		5	10	15	20	25	31	36	41	46
6·6	0·21878	21827	21777	21727	21677	21627	21577	21528	21478	21429		5	10	15	20	25	30	35	40	45
6·7	0·21380	21330	21281	21232	21184	21135	21086	21038	20989	20941		5	10	15	19	24	29	34	39	44
6·8	0·20893	20845	20797	20749	20701	20654	20606	20559	20512	20464		5	10	14	19	24	29	33	38	43
6·9	0·20417	20370	20324	20277	20230	20184	20137	20091	20045	19999		5	9	14	19	23	28	32	37	42

Numbers in mean difference columns to be *subtracted*.

Loss

dB	0	1	2	3	4	5	6	7	8	9	mean differences								
											1	2	3	4	5	6	7	8	9
7·0	0·19953	19907	19861	19815	19770	19724	19679	19634	19588	19543	5	9	14	18	23	27	32	36	41
7·1	0·19498	19454	19409	19364	19320	19275	19231	19187	19143	19099	4	9	13	18	22	26	31	35	40
7·2	0·19055	19011	18967	18923	18880	18836	18793	18750	18707	18664	4	9	13	17	22	26	30	35	39
7·3	0·18621	18578	18535	18493	18450	18408	18365	18323	18281	18239	4	8	13	17	21	25	30	34	38
7·4	0·18197	18155	18113	18072	18030	17989	17947	17906	17865	17824	4	8	12	17	21	25	29	33	37
7·5	0·17783	17742	17701	17660	17620	17579	17539	17498	17458	17418	4	8	12	16	20	24	28	32	36
7·6	0·17378	17338	17298	17258	17219	17179	17140	17100	17061	17022	4	8	12	16	20	24	28	32	36
7·7	0·16982	16943	16904	16866	16827	16788	16749	16711	16672	16634	4	8	12	15	19	23	27	31	35
7·8	0·16596	16558	16520	16482	16444	16406	16368	16331	16293	16255	4	8	11	15	19	23	26	30	34
7·9	0·16218	16181	16144	16106	16069	16032	15996	15959	15922	15885	4	7	11	15	18	22	26	30	33
8·0	0·15849	15812	15776	15740	15704	15668	15631	15596	15560	15524	4	7	11	14	18	22	25	29	32
8·1	0·15488	15453	15417	15382	15346	15311	15276	15241	15205	15171	4	7	11	14	18	21	25	28	32
8·2	0·15136	15101	15066	15031	14997	14962	14928	14894	14859	14825	3	7	10	14	17	21	24	28	31
8·3	0·14791	14757	14723	14689	14655	14622	14588	14555	14521	14488	3	7	10	13	17	20	24	27	30
8·4	0·14454	14421	14388	14355	14322	14289	14256	14223	14191	14158	3	7	10	13	16	20	23	26	30
8·5	0·14125	14093	14060	14028	13996	13964	13932	13900	13868	13836	3	6	10	13	16	19	22	26	29
8·6	0·13804	13772	13740	13709	13677	13646	13614	13583	13552	13521	3	6	9	13	16	19	22	25	28
8·7	0·13490	13459	13428	13397	13366	13335	13305	13274	13243	13213	3	6	9	12	15	18	21	25	28
8·8	0·13183	13152	13122	13092	13062	13032	13002	12972	12942	12912	3	6	9	12	15	18	21	24	27
8·9	0·12882	12853	12823	12794	12764	12735	12706	12677	12647	12618	3	6	9	12	15	18	21	24	26
9·0	0·12589	12560	12531	12503	12474	12445	12417	12388	12359	12331	3	6	9	11	14	17	20	23	26
9·1	0·12303	12274	12246	12218	12190	12162	12134	12106	12078	12050	3	6	8	11	14	17	20	22	25
9·2	0·12023	11995	11976	11940	11912	11885	11858	11830	11803	11776	3	5	8	11	13	16	19	22	25
9·3	0·11749	11722	11695	11668	11641	11614	11588	11561	11535	11508	3	5	8	11	13	16	19	21	24
9·4	0·11482	11455	11429	11402	11376	11350	11324	11298	11272	11246	3	5	8	11	13	16	18	21	24
9·5	0·11220	11194	11169	11143	11117	11092	11066	11041	11015	10990	3	5	8	10	13	15	18	20	23
9·6	0·10965	10940	10914	10889	10864	10839	10814	10789	10765	10740	3	5	8	10	13	15	18	20	23
9·7	0·10715	10691	10666	10641	10617	10593	10568	10544	10520	10495	3	5	7	10	12	15	17	20	22
9·8	0·10471	10447	10423	10399	10375	10351	10328	10304	10280	10257	2	5	7	10	12	14	17	19	21
9·9	0·10233	10209	10186	10162	10139	10116	10093	10069	10046	10023	2	5	7	9	12	14	16	19	21

Numbers in mean difference columns to be *subtracted*.

Voltage and Current Ratios Gain

dB	0	1	2	3	4	5	6	7	8	9	mean differences								
											1	2	3	4	5	6	7	8	9
0·0	1·0000	1·0012	1·0023	1·0035	1·0046	1·0058	1·0069	1·0081	1·0093	1·0105	1	2	3	5	6	7	8	9	10
0·1	1·0116	1·0128	1·0139	1·0151	1·0162	1·0174	1·0186	1·0198	1·0209	1·0221	1	2	3	5	6	7	8	9	10
0·2	1·0233	1·0245	1·0257	1·0269	1·0280	1·0292	1·0304	1·0316	1·0328	1·0340	1	2	3	5	6	7	8	9	10
0·3	1·0351	1·0363	1·0375	1·0389	1·0399	1·0411	1·0423	1·0435	1·0447	1·0459	1	2	3	5	6	7	8	9	10
0·4	1·0471	1·0483	1·0495	1·0507	1·0520	1·0532	1·0544	1·0556	1·0568	1·0580	1	2	3	5	6	7	8	10	11
0·5	1·0593	1·0605	1·0617	1·0629	1·0641	1·0653	1·0666	1·0678	1·0691	1·0703	1	2	3	5	6	7	8	10	11
0·6	1·0715	1·0723	1·0740	1·0753	1·0765	1·0778	1·0789	1·0802	1·0814	1·0827	2	3	4	5	6	8	9	10	11
0·7	1·0839	1·0852	1·0864	1·0877	1·0889	1·0901	1·0914	1·0927	1·0940	1·0953	2	3	4	5	6	8	9	10	11
0·8	1·0965	1·0978	1·0990	1·1003	1·1015	1·1028	1·1041	1·1054	1·1066	1·1079	2	3	4	5	6	8	9	10	11
0·9	1·1092	1·1105	1·1117	1·1130	1·1143	1·1156	1·1169	1·1182	1·1194	1·1207	2	3	4	5	6	8	9	10	11
1·0	1·1220	1·1233	1·1246	1·1259	1·1272	1·1285	1·1298	1·1311	1·1324	1·1337	2	3	4	5	6	8	9	11	12
1·1	1·1350	1·1363	1·1376	1·1389	1·1402	1·1415	1·1429	1·1442	1·1455	1·1468	2	3	4	5	6	8	9	11	12
1·2	1·1482	1·1495	1·1508	1·1521	1·1535	1·1548	1·1561	1·1574	1·1588	1·1601	2	3	4	5	6	8	9	11	12
1·3	1·1614	1·1627	1·1641	1·1654	1·1668	1·1681	1·1695	1·1708	1·1722	1·1735	2	3	4	5	6	8	9	11	12
1·4	1·1749	1·1763	1·1776	1·1790	1·1803	1·1817	1·1830	1·1844	1·1858	1·1872	2	3	4	5	6	8	9	11	12
1·5	1·1885	1·1899	1·1912	1·1926	1·1940	1·1954	1·1967	1·1981	1·1995	1·2009	2	3	4	6	7	8	9	11	12
1·6	1·2023	1·2037	1·2050	1·2064	1·2078	1·2092	1·2106	1·2120	1·2134	1·2148	2	3	4	6	7	8	9	11	12
1·7	1·2162	1·2176	1·2190	1·2204	1·2218	1·2232	1·2246	1·2260	1·2274	1·2288	2	3	4	6	7	8	9	11	13
1·8	1·2303	1·2317	1·2331	1·2344	1·2359	1·2373	1·2388	1·2402	1·2417	1·2431	2	3	4	6	7	8	10	11	13
1·9	1·2445	1·2459	1·2474	1·2488	1·2503	1·2517	1·2531	1·2544	1·2560	1·2574	2	3	4	6	7	9	10	11	13
2·0	1·2589	1·2603	1·2618	1·2633	1·2647	1·2662	1·2677	1·2692	1·2706	1·2721	2	3	4	6	7	9	10	12	13
2·1	1·2735	1·2750	1·2764	1·2779	1·2794	1·2809	1·2823	1·2838	1·2853	1·2868	2	3	4	6	7	9	10	12	13
2·2	1·2882	1·2897	1·2912	1·2927	1·2942	1·2957	1·2972	1·2987	1·3002	1·3017	2	3	4	6	7	9	10	12	13
2·3	1·3032	1·3047	1·3063	1·3077	1·3092	1·3107	1·3122	1·3137	1·3152	1·3167	2	3	4	6	7	9	10	12	13
2·4	1·3183	1·3198	1·3213	1·3228	1·3243	1·3258	1·3274	1·3289	1·3305	1·3320	2	3	4	6	7	9	10	12	13
2·5	1·3335	1·3350	1·3366	1·3381	1·3397	1·3412	1·3428	1·3443	1·3459	1·3474	2	3	4	6	7	9	10	12	13
2·6	1·3490	1·3506	1·3521	1·3537	1·3552	1·3568	1·3583	1·3599	1·3614	1·3630	2	3	4	6	7	9	11	13	13
2·7	1·3646	1·3662	1·3677	1·3693	1·3709	1·3725	1·3740	1·3756	1·3772	1·3788	2	3	4	6	7	9	11	13	14
2·8	1·3804	1·3820	1·3836	1·3852	1·3868	1·3884	1·3900	1·3916	1·3932	1·3948	2	3	4	6	7	9	11	13	14
2·9	1·3964	1·3980	1·3996	1·4012	1·4028	1·4044	1·4060	1·4076	1·4093	1·4109	2	3	4	6	7	9	11	13	14
3·0	1·4125	1·4141	1·4158	1·4174	1·4191	1·4207	1·4223	1·4239	1·4256	1·4272	2	3	5	7	8	10	11	13	14
3·1	1·4289	1·4305	1·4322	1·4338	1·4355	1·4371	1·4388	1·4404	1·4421	1·4437	2	3	5	7	8	10	11	13	14
3·2	1·4454	1·4471	1·4488	1·4505	1·4521	1·4538	1·4555	1·4572	1·4588	1·4605	2	3	5	7	8	10	11	13	15
3·3	1·4622	1·4639	1·4655	1·4672	1·4689	1·4706	1·4723	1·4740	1·4757	1·4774	2	3	5	7	8	10	11	13	15
3·4	1·4791	1·4808	1·4825	1·4842	1·4859	1·4876	1·4894	1·4911	1·4928	1·4945	2	3	5	7	8	10	12	14	15

Voltage and Current Ratios *(continued)*

Gain

dB	0	1	2	3	4	5	6	7	8	9	mean differences								
											1	2	3	4	5	6	7	8	9
3·5	1·4962	1·4979	1·4997	1·5014	1·5031	1·5048	1·5066	1·5083	1·5101	1·5118	2	3	5	7	8	10	12	14	15
3·6	1·5136	1·5154	1·5171	1·5189	1·5205	1·5223	1·5241	1·5259	1·5276	1·5294	2	4	6	7	9	11	12	14	16
3·7	1·5311	1·5329	1·5346	1·5364	1·5382	1·5400	1·5417	1·5435	1·5453	1·5471	2	4	6	7	9	11	12	14	16
3·8	1·5488	1·5506	1·5524	1·5542	1·5560	1·5578	1·5596	1·5614	1·5631	1·5649	2	4	6	7	9	11	12	14	16
3·9	1·5668	1·5686	1·5704	1·5722	1·5740	1·5758	1·5776	1·5794	1·5812	1·5830	2	4	6	7	9	11	12	14	16
4·0	1·5849	1·5867	1·5885	1·5903	1·5922	1·5940	1·5959	1·5977	1·5996	1·6014	2	4	6	7	9	11	13	15	16
4·1	1·6032	1·6050	1·6069	1·6087	1·6106	1·6124	1·6144	1·6162	1·6181	1·6199	2	4	6	8	9	11	13	15	16
4·2	1·6218	1·6237	1·6255	1·6274	1·6293	1·6312	1·6331	1·6350	1·6368	1·6387	2	4	6	8	9	11	13	15	17
4·3	1·6406	1·6425	1·6444	1·6463	1·6482	1·6501	1·6520	1·6539	1·6558	1·6577	2	4	6	8	9	11	13	15	17
4·4	1·6596	1·6615	1·6634	1·6653	1·6672	1·6691	1·6711	1·6730	1·6749	1·6768	2	4	6	8	10	12	13	15	17
4·5	1·6788	1·6807	1·6827	1·6846	1·6866	1·6885	1·6904	1·6923	1·6943	1·6962	2	4	6	8	10	12	13	15	17
4·6	1·6982	1·7002	1·7022	1·7042	1·7061	1·7081	1·7100	1·7120	1·7140	1·7160	2	4	6	8	10	12	14	16	18
4·7	1·7179	1·7199	1·7219	1·7239	1·7258	1·7278	1·7298	1·7318	1·7338	1·7358	2	4	6	8	10	12	14	16	18
4·8	1·7378	1·7398	1·7418	1·7438	1·7458	1·7478	1·7498	1·7518	1·7539	1·7559	2	4	6	8	10	12	14	16	18
4·9	1·7579	1·7599	1·7620	1·7640	1·7660	1·7680	1·7701	1·7721	1·7742	1·7762	2	4	6	8	10	12	14	16	18
5·0	1·7783	1·7804	1·7824	1·7845	1·7865	1·7886	1·7906	1·7927	1·7947	1·7968	2	4	6	8	10	12	14	17	19
5·1	1·7989	1·8010	1·8030	1·8051	1·8072	1·8093	1·8113	1·8134	1·8155	1·8176	2	4	6	8	10	12	15	17	19
5·2	1·8197	1·8218	1·8239	1·8260	1·8281	1·8302	1·8323	1·8344	1·8365	1·8386	2	4	6	8	11	13	15	17	19
5·3	1·8408	1·8429	1·8450	1·8471	1·8493	1·8514	1·8535	1·8556	1·8578	1·8599	2	4	6	8	11	13	15	17	19
5·4	1·8621	1·8643	1·8664	1·8686	1·8707	1·8729	1·8750	1·8772	1·8793	1·8815	2	4	7	9	11	13	15	17	20
5·5	1·8836	1·8858	1·8880	1·8902	1·8923	1·8945	1·8967	1·8989	1·9011	1·9033	2	4	7	9	11	13	15	17	20
5·6	1·9055	1·9077	1·9099	1·9121	1·9143	1·9165	1·9187	1·9209	1·9231	1·9253	2	4	7	9	11	13	16	18	20
5·7	1·9275	1·9297	1·9320	1·9342	1·9364	1·9386	1·9409	1·9431	1·9454	1·9476	2	5	7	9	11	14	16	18	21
5·8	1·9498	1·9521	1·9543	1·9566	1·9588	1·9611	1·9634	1·9657	1·9679	1·9702	2	5	7	9	11	14	16	18	21
5·9	1·9724	1·9747	1·9770	1·9793	1·9815	1·9838	1·9861	1·9884	1·9907	1·9930	2	5	7	9	11	14	16	18	21
6·0	1·9953	1·9976	1·9999	2·0022	2·0045	2·0068	2·0091	2·0114	2·0137	2·0160	2	5	7	9	12	14	16	19	21
6·1	2·0184	2·0207	2·0230	2·0253	2·0277	2·0300	2·0324	2·0347	2·0370	2·0393	3	5	7	10	12	14	16	19	21
6·2	2·0417	2·0441	2·0464	2·0488	2·0512	2·0535	2·0559	2·0583	2·0606	2·0630	3	5	7	10	12	14	16	19	21
6·3	2·0654	2·0678	2·0701	2·0725	2·0749	2·0773	2·0797	2·0821	2·0845	2·0869	3	5	8	10	13	15	16	19	22
6·4	2·0893	2·0917	2·0941	2·0965	2·0989	2·1013	2·1038	2·1062	2·1086	2·1110	3	5	8	10	13	15	16	19	22
6·5	2·1135	2·1159	2·1184	2·1208	2·1232	2·1256	2·1281	2·1305	2·1330	2·1354	3	5	8	10	13	15	17	19	22
6·6	2·1380	2·1405	2·1429	2·1454	2·1478	2·1503	2·1528	2·1553	2·1577	2·1602	3	5	8	10	13	15	17	20	22
6·7	2·1627	2·1652	2·1677	2·1702	2·1727	2·1752	2·1777	2·1802	2·1827	2·1852	3	5	8	10	13	15	17	20	22
6·8	2·1878	2·1903	2·1928	2·1953	2·1979	2·2004	2·2029	2·2054	2·2080	2·2105	3	5	8	10	13	15	17	20	22
6·9	2·2131	2·2156	2·2182	2·2207	2·2233	2·2258	2·2284	2·2309	2·2336	2·2361	3	5	8	10	13	15	17	20	22

Voltage and Current Ratios *(continued)* — Gain

dB	0	1	2	3	4	5	6	7	8	9	mean differences 1	2	3	4	5	6	7	8	9
7·0	2·2387	2·2413	2·2439	2·2465	2·2491	2·2515	2·2542	2·2568	2·2594	2·2620	3	5	8	10	13	16	18	21	23
7·1	2·2646	2·2672	2·2699	2·2725	2·2751	2·2777	2·2803	2·2829	2·2856	2·2882	3	5	8	11	13	16	18	21	23
7·2	2·2909	2·2936	2·2961	2·2988	2·3014	2·3041	2·3067	2·3094	2·3121	2·3148	3	5	8	11	14	16	18	21	23
7·3	2·3174	2·3201	2·3227	2·3254	2·3281	2·3308	2·3336	2·3363	2·3388	2·3415	3	5	8	11	14	16	18	21	24
7·4	2·3442	2·3469	2·3496	2·3523	2·3550	2·3577	2·3605	2·3632	2·3659	2·3686	3	5	8	11	14	16	19	22	24
7·5	2·3714	2·3741	2·3768	2·3795	2·3823	2·3850	2·3878	2·3905	2·3933	2·3960	3	5	8	11	14	16	19	22	24
7·6	2·3988	2·4016	2·4044	2·4072	2·4099	2·4127	2·4155	2·4183	2·4210	2·4238	3	6	8	11	14	17	19	22	25
7·7	2·4266	2·4294	2·4322	2·4350	2·4378	2·4406	2·4434	2·4462	2·4491	2·4519	3	6	8	11	14	17	20	22	25
7·8	2·4547	2·4576	2·4604	2·4633	2·4660	2·4689	2·4717	2·4746	2·4774	2·4803	3	6	8	11	14	17	20	23	26
7·9	2·4831	2·4860	2·4889	2·4918	2·4946	2·4975	2·5003	2·5032	2·5061	2·5090	3	6	9	11	14	17	20	23	26
8·0	2·5119	2·5148	2·5177	2·5206	2·5236	2·5265	2·5293	2·5322	2·5351	2·5380	3	6	9	12	15	18	20	23	26
8·1	2·5410	2·5439	2·5468	2·5497	2·5527	2·5556	2·5586	2·5615	2·5645	2·5674	3	6	9	12	15	18	21	23	26
8·2	2·5704	2·5734	2·5763	2·5793	2·5823	2·5853	2·5882	2·5912	2·5942	2·5972	3	6	9	12	15	18	21	24	27
8·3	2·6002	2·6032	2·6062	2·6092	2·6122	2·6152	2·6182	2·6212	2·6242	2·6272	3	6	9	12	15	18	21	24	27
8·4	2·6303	2·6334	2·6363	2·6394	2·6424	2·6455	2·6485	2·6516	2·6546	2·6577	3	6	9	12	15	18	21	24	27
8·5	2·6607	2·6638	2·6669	2·6700	2·6730	2·6761	2·6792	2·6823	2·6853	2·6884	3	6	9	12	15	18	22	25	28
8·6	2·6915	2·6946	2·6977	2·7008	2·7040	2·7071	2·7102	2·7133	2·7164	2·7195	3	6	9	12	16	19	22	25	28
8·7	2·7227	2·7258	2·7290	2·7321	2·7353	2·7384	2·7416	2·7447	2·7479	2·7510	3	6	9	13	16	19	22	25	28
8·8	2·7542	2·7574	2·7606	2·7638	2·7669	2·7701	2·7733	2·7765	2·7797	2·7829	3	6	10	13	16	19	22	26	29
8·9	2·7861	2·7893	2·7925	2·7957	2·7990	2·8022	2·8054	2·8086	2·8119	2·8151	3	6	10	13	16	19	23	26	29
9·0	2·8184	2·8217	2·8249	2·8282	2·8314	2·8347	2·8379	2·8412	2·8445	2·8478	3	7	10	13	16	20	23	26	29
9·1	2·8510	2·8543	2·8576	2·8609	2·8642	2·8675	2·8708	2·8741	2·8774	2·8807	3	7	10	13	17	20	23	26	30
9·2	2·8840	2·8874	2·8907	2·8941	2·8973	2·9007	2·9040	2·9074	2·9107	2·9141	3	7	10	13	17	20	23	27	30
9·3	2·9174	2·9207	2·9242	2·9276	2·9309	2·9343	2·9376	2·9410	2·9444	2·9478	3	7	10	14	17	20	24	27	30
9·4	2·9512	2·9546	2·9580	2·9614	2·9648	2·9682	2·9717	2·9751	2·9785	2·9819	3	7	10	14	17	20	24	27	31
9·5	2·9854	2·9888	2·9923	2·9957	2·9992	3·0026	3·0061	3·0095	3·0130	3·0164	3	7	10	14	17	21	24	28	31
9·6	3·0200	3·0235	3·0269	3·0304	3·0339	3·0374	3·0409	3·0444	3·0479	3·0514	3	7	10	14	17	21	25	28	31
9·7	3·0549	3·0584	3·0620	3·0655	3·0690	3·0725	3·0761	3·0796	3·0832	3·0867	4	7	11	14	18	21	25	28	32
9·8	3·0903	3·0939	3·0974	3·1010	3·1046	3·1079	3·1117	3·1153	3·1189	3·1225	4	7	11	14	18	22	26	29	33
9·9	3·1261	3·1297	3·1333	3·1369	3·1405	3·1441	3·1477	3·1513	3·1550	3·1586	4	7	11	14	18	22	26	29	33
10·0	3·1623	3·1660	3·1696	3·1733	3·1769	3·1806	3·1842	3·1879	3·1916	3·1953	4	7	11	15	18	22	26	29	33
10·1	3·1989	3·2026	3·2063	3·2100	3·2137	3·2174	3·2211	3·2248	3·2285	3·2322	4	7	11	15	19	22	26	30	33
10·2	3·2359	3·2397	3·2434	3·2472	3·2509	3·2547	3·2584	3·2622	3·2659	3·2697	4	8	11	15	19	23	27	30	34
10·3	3·2735	3·2773	3·2809	3·2847	3·2885	3·2923	3·2961	3·2999	3·3037	3·3085	4	8	12	15	19	23	27	31	34
10·4	3·3113	3·3152	3·3189	3·3228	3·3266	3·3305	3·3343	3·3382	3·3420	3·3459	4	8	12	15	19	23	27	31	35

Voltage and Current Ratios (*continued*)

Gain

dB	0	1	2	3	4	5	6	7	8	9	mean differences 1	2	3	4	5	6	7	8	9
10·5	3·3497	3·3536	3·3574	3·3613	3·3651	3·3690	3·3729	3·3768	3·3806	3·3845	4	8	12	15	19	23	27	31	35
10·6	3·3884	3·3924	3·3963	3·4003	3·4041	3·4081	3·4119	3·4159	3·4198	3·4238	4	8	12	16	20	24	28	32	36
10·7	3·4277	3·4317	3·4356	3·4396	3·4435	3·4475	3·4514	3·4554	3·4594	3·4634	4	8	12	16	20	24	28	32	36
10·8	3·4674	3·4714	3·4754	3·4794	3·4834	3·4874	3·4914	3·4954	3·4995	3·5035	4	8	12	16	20	24	28	32	36
10·9	3·5075	3·5115	3·5156	3·5196	3·5237	3·5277	3·5318	3·5358	3·5400	3·5440	4	8	12	16	20	24	28	32	36
11·0	3·5481	3·5522	3·5563	3·5604	3·5645	3·5686	3·5727	3·5768	3·5810	3·5851	4	8	12	16	21	25	29	33	37
11·1	3·5892	3·5933	3·5975	3·6016	3·6058	3·6099	3·6141	3·6182	3·6224	3·6265	4	8	12	17	21	25	29	33	37
11·2	3·6308	3·6350	3·6392	3·6434	3·6475	3·6517	3·6559	3·6601	3·6644	3·6686	4	8	13	17	21	25	29	34	38
11·3	3·6728	3·6770	3·6813	3·6855	3·6898	3·6940	3·6983	3·7025	3·7068	3·7110	4	8	13	17	21	25	30	34	38
11·4	3·7154	3·7197	3·7239	3·7282	3·7325	3·7368	3·7411	3·7454	3·7497	3·7540	4	9	13	17	22	26	30	35	39
11·5	3·7584	3·7627	3·7670	3·7713	3·7757	3·7800	3·7844	3·7887	3·7931	3·7974	5	9	13	17	22	26	30	35	39
11·6	3·8019	3·8063	3·8107	3·8151	3·8194	3·8238	3·8282	3·8326	3·8371	3·8415	5	9	13	18	22	27	31	35	40
11·7	3·8459	3·8503	3·8548	3·8592	3·8637	3·8681	3·8726	3·8770	3·8815	3·8859	5	9	13	18	22	27	31	35	40
11·8	3·8905	3·8950	3·8994	3·9039	3·9084	3·9129	3·9174	3·9219	3·9264	3·9309	5	9	14	18	23	27	32	36	41
11·9	3·9355	3·9400	3·9446	3·9491	3·9537	3·9582	3·9628	3·9673	3·9719	3·9764	5	9	14	18	23	27	32	36	41
12·0	3·9811	3·9857	3·9902	3·9948	3·9994	4·0040	4·0087	4·0133	4·0179	4·0225	5	9	14	19	24	28	33	37	42
12·1	4·0272	4·0318	4·0365	4·0411	4·0458	4·0504	4·0551	4·0597	4·0644	4·0690	5	10	14	19	24	28	33	37	42
12·2	4·0738	4·0785	4·0832	4·0879	4·0926	4·0973	4·1020	4·1067	4·1115	4·1162	5	9	14	19	24	28	33	38	43
12·3	4·1210	4·1257	4·1305	4·1352	4·1400	4·1447	4·1495	4·1542	4·1591	4·1638	5	9	14	19	24	28	33	38	43
12·4	4·1687	4·1736	4·1783	4·1832	4·1879	4·1928	4·1976	4·2025	4·2073	4·2122	5	10	15	19	24	29	34	39	44
12·5	4·2170	4·2219	4·2267	4·2316	4·2364	4·2413	4·2462	4·2511	4·2560	4·2609	5	10	15	19	24	29	34	39	44
12·6	4·2658	4·2708	4·2756	4·2806	4·2855	4·2905	4·2954	4·3004	4·3053	4·3103	5	10	15	20	25	30	35	40	45
12·7	4·3152	4·3202	4·3251	4·3301	4·3351	4·3401	4·3451	4·3501	4·3551	4·3601	5	10	15	20	25	30	35	40	45
12·8	4·3652	4·3703	4·3752	4·3803	4·3853	4·3904	4·3954	4·4005	4·4055	4·4106	6	11	16	20	25	30	36	41	46
12·9	4·4157	4·4208	4·4259	4·4310	4·4361	4·4412	4·4463	4·4514	4·4566	4·4617	6	11	16	20	25	30	36	41	46
13·0	4·4668	4·4720	4·4771	4·4823	4·4875	4·4927	4·4978	4·5030	4·5082	4·5134	5	10	16	21	26	31	37	42	47
13·1	4·5186	4·5238	4·5290	4·5342	4·5394	4·5446	4·5499	4·5551	4·5604	4·5656	5	10	16	21	26	31	37	42	47
13·2	4·5709	4·5762	4·5814	4·5867	4·5920	4·5973	4·6026	4·6079	4·6132	4·6185	6	11	16	21	27	32	38	43	48
13·3	4·6238	4·6291	4·6345	4·6398	4·6452	4·6505	4·6559	4·6612	4·6666	4·6719	6	11	16	21	27	32	38	43	48
13·4	4·6774	4·6828	4·6881	4·6935	4·6989	4·7043	4·7098	4·7152	4·7206	4·7260	6	11	17	22	28	33	39	44	49
13·5	4·7315	4·7369	4·7424	4·7478	4·7534	4·7588	4·7643	4·7697	4·7753	4·7807	6	11	17	22	28	33	39	44	49
13·6	4·7863	4·7919	4·7973	4·8029	4·8084	4·8140	4·8195	4·8251	4·8306	4·8362	6	11	17	22	28	33	39	45	51
13·7	4·8417	4·8473	4·8529	4·8585	4·8641	4·8697	4·8753	4·8809	4·8865	4·8921	6	11	17	22	28	33	39	45	51
13·8	4·8978	4·9035	4·9091	4·9148	4·9204	4·9261	4·9317	4·9374	4·9431	4·9488	6	11	17	23	29	34	40	46	52
13·9	4·9545	4·9602	4·9659	4·9716	4·9774	4·9831	4·9888	4·9945	5·0003	5·0060	6	11	17	23	29	34	40	46	52

Voltage and Current Ratios (continued)

Gain

dB	0	1	2	3	4	5	6	7	8	9	md 1	2	3	4	5	6	7	8	9
14·0	5·0119	5·0177	5·0234	5·0292	5·0350	5·0408	5·0466	5·0524	5·0582	5·0640	6	12	18	23	29	35	41	47	53
14·1	5·0699	5·0757	5·0816	5·0874	5·0933	5·0991	5·1050	5·1108	5·1168	5·1226	6	12	18	23	29	35	41	47	53
14·2	5·1286	5·1346	5·1404	5·1464	5·1523	5·1583	5·1642	5·1702	5·1761	5·1821	6	12	18	24	30	36	42	48	54
14·3	5·1880	5·1940	5·2000	5·2060	5·2119	5·2179	5·2240	5·2300	5·2360	5·2420	6	12	18	24	30	36	42	48	54
14·4	5·2481	5·2542	5·2602	5·2663	5·2723	5·2784	5·2845	5·2906	5·2966	5·3027	6	12	18	24	30	37	43	49	55
14·5	5·3088	5·3149	5·3211	5·3272	5·3333	5·3394	5·3456	5·3517	5·3580	5·3641	6	12	18	24	30	37	43	49	55
14·6	5·3703	5·3766	5·3827	5·3890	5·3951	5·4014	5·4075	5·4138	5·4200	5·4263	7	13	19	25	31	38	44	50	56
14·7	5·4325	5·4388	5·4450	5·4513	5·4576	5·4639	5·4702	5·4765	5·4828	5·4891	7	13	19	25	31	38	44	50	56
14·8	5·4954	5·5020	5·5081	5·5145	5·5208	5·5272	5·5336	5·5400	5·5463	5·5527	7	13	20	26	32	38	45	51	58
14·9	5·5590	5·5654	5·5719	5·5783	5·5847	5·5911	5·5976	5·6040	5·6105	5·6169	7	13	20	26	32	38	45	51	58
15·0	5·6234	5·6300	5·6364	5·6430	5·6494	5·6560	5·6624	5·6690	5·6754	5·6820	7	13	20	26	33	39	46	52	59
15·1	5·6885	5·6951	5·7016	5·7082	5·7148	5·7214	5·7280	5·7346	5·7412	5·7478	7	13	20	26	33	39	46	52	59
15·2	5·7544	5·7611	5·7677	5·7744	5·7810	5·7877	5·7943	5·8010	5·8076	5·8143	7	13	20	27	34	40	47	54	61
15·3	5·8210	5·8277	5·8345	5·8412	5·8479	5·8546	5·8614	5·8681	5·8749	5·8816	7	13	20	27	34	40	47	54	61
15·4	5·8884	5·8953	5·9020	5·9089	5·9156	5·9225	5·9293	5·9362	5·9429	5·9498	7	14	21	27	34	41	48	55	62
15·5	5·9566	5·9635	5·9704	5·9773	5·9841	5·9910	5·9979	6·0048	6·0117	6·0186	7	14	21	27	34	41	48	55	62
15·6	6·0256	6·0326	6·0395	6·0465	6·0534	6·0604	6·0674	6·0744	6·0814	6·0884	7	14	21	28	35	42	49	56	63
15·7	6·0954	6·1024	6·1094	6·1164	6·1235	6·1305	6·1376	6·1446	6·1518	6·1588	7	14	21	28	35	42	49	56	63
15·8	6·1659	6·1731	6·1802	6·1874	6·1944	6·2016	6·2087	6·2159	6·2230	6·2302	7	14	22	29	36	43	50	58	65
15·9	6·2373	6·2445	6·2517	6·2589	6·2661	6·2733	6·2806	6·2878	6·2951	6·3023	7	14	22	29	36	43	50	58	65
16·0	6·3096	6·3170	6·3241	6·3315	6·3387	6·3461	6·3533	6·3607	6·3680	6·3754	8	15	22	29	37	44	51	59	66
16·1	6·3826	6·3900	6·3973	6·4047	6·4121	6·4195	6·4269	6·4343	6·4417	6·4491	8	15	22	29	37	44	51	59	66
16·2	6·4565	6·4640	6·4714	6·4789	6·4863	6·4938	6·5013	6·5088	6·5163	6·5238	8	15	23	30	38	45	53	60	68
16·3	6·5313	6·5388	6·5464	6·5539	6·5615	6·5690	6·5766	6·5841	6·5917	6·5992	8	15	23	30	38	45	53	60	68
16·4	6·6069	6·6146	6·6222	6·6299	6·6374	6·6451	6·6527	6·6604	6·6681	6·6758	8	15	23	31	39	46	54	62	69
16·5	6·6834	6·6911	6·6988	6·7065	6·7143	6·7220	6·7298	6·7375	6·7453	6·7530	8	15	23	31	39	46	54	62	69
16·6	6·7608	6·7687	6·7764	6·7843	6·7920	6·7999	6·8077	6·8156	6·8234	6·8313	8	16	24	32	40	47	55	63	71
16·7	6·8391	6·8470	6·8549	6·8628	6·8707	6·8786	6·8865	6·8944	6·9024	6·9103	8	16	24	32	40	47	55	63	71
16·8	6·9183	6·9264	6·9343	6·9424	6·9503	6·9584	6·9663	6·9744	6·9823	6·9904	8	16	24	32	40	48	56	64	72
16·9	6·9984	7·0065	7·0146	7·0227	7·0307	7·0388	7·0469	7·0550	7·0632	7·0713	8	16	24	32	40	48	56	64	72
17·0	7·0795	7·0878	7·0958	7·1041	7·1121	7·1204	7·1285	7·1368	7·1450	7·1533	9	17	25	33	41	50	58	66	74
17·1	7·1614	7·1697	7·1779	7·1862	7·1945	7·2028	7·2111	7·2194	7·2277	7·2360	9	17	25	33	41	50	58	66	74
17·2	7·2444	7·2529	7·2611	7·2696	7·2778	7·2863	7·2946	7·3031	7·3114	7·3199	9	17	26	34	42	51	59	68	76
17·3	7·3282	7·3367	7·3451	7·3536	7·3621	7·3706	7·3790	7·3875	7·3961	7·4046	9	17	26	34	42	51	60	68	76
17·4	7·4131	7·4218	7·4302	7·4389	7·4473	7·4560	7·4645	7·4732	7·4817	7·4904	9	17	26	35	43	52	61	69	78

Voltage and Current Ratios (continued)

Gain

dB	0	1	2	3	4	5	6	7	8	9		mean differences								
												1	2	3	4	5	6	7	8	9
17·5	7·4989	7·5076	7·5162	7·5249	7·5336	7·5423	7·5509	7·5596	7·5683	7·5770		9	17	26	35	43	52	61	69	78
17·6	7·5858	7·5947	7·6033	7·6122	7·6208	7·6297	7·6384	7·6473	7·6560	7·6649		9	18	27	35	44	53	62	71	80
17·7	7·6736	7·6825	7·6913	7·7002	7·7090	7·7179	7·7268	7·7357	7·7446	7·7535		9	18	27	35	44	53	62	71	80
17·8	7·7625	7·7716	7·7804	7·7895	7·7983	7·8074	7·8163	7·8254	7·8343	7·8434		9	18	27	36	45	54	63	72	81
17·9	7·8524	7·8615	7·8705	7·8796	7·8886	7·8977	7·9068	7·9159	7·9250	7·9341		9	18	27	36	45	54	63	72	81
18·0	7·9433	7·9526	7·9616	7·9709	7·9799	7·9892	7·9983	8·0076	8·0168	8·0261		10	19	28	37	46	56	65	74	83
18·1	8·0353	8·0446	8·0538	8·0631	8·0724	8·0817	8·0910	8·1003	8·1096	8·1189		10	19	28	37	46	56	65	74	83
18·2	8·1283	8·1378	8·1470	8·1565	8·1658	8·1753	8·1846	8·1941	8·2035	8·2130		10	19	29	38	47	57	67	76	85
18·3	8·2224	8·2319	8·2414	8·2509	8·2604	8·2699	8·2794	8·2889	8·2985	8·3080		10	19	29	38	47	57	67	76	85
18·4	8·3176	8·3273	8·3368	8·3465	8·3560	8·3657	8·3753	8·3850	8·3946	8·4043		10	19	29	39	48	58	68	78	87
18·5	8·4140	8·4237	8·4333	8·4430	8·4528	8·4625	8·4723	8·4820	8·4918	8·5015		10	19	29	39	48	58	68	78	87
18·6	8·5114	8·5213	8·5310	8·5409	8·5507	8·5606	8·5704	8·5803	8·5901	8·6000		10	20	30	40	50	60	70	79	89
18·7	8·6099	8·6198	8·6298	8·6397	8·6497	8·6596	8·6696	8·6795	8·6896	8·6995		10	20	30	40	50	60	70	79	89
18·8	8·7096	8·7198	8·7297	8·7399	8·7498	8·7600	8·7700	8·7802	8·7902	8·8004		10	20	30	41	51	61	71	81	92
18·9	8·8105	8·8207	8·8308	8·8410	8·8512	8·8614	8·8716	8·8818	8·8920	8·9022		10	20	30	41	51	61	71	81	92
19·0	8·9125	8·9229	8·9331	8·9435	8·9536	8·9640	8·9743	8·9847	8·9950	9·0054		11	21	31	42	52	62	72	83	94
19·1	9·0157	9·0261	9·0365	9·0469	9·0573	9·0677	9·0782	9·0886	9·0991	9·1095		11	21	31	42	52	62	72	83	94
19·2	9·1201	9·1307	9·1411	9·1517	9·1622	9·1728	9·1833	9·1939	9·2045	9·2151		11	21	31	42	53	64	74	85	97
19·3	9·2257	9·2363	9·2470	9·2576	9·2683	9·2789	9·2897	9·3003	9·3111	9·3217		11	21	31	42	53	64	74	85	97
19·4	9·3325	9·3434	9·3541	9·3650	9·3756	9·3865	9·3972	9·4061	9·4189	9·4298		11	22	32	43	54	65	76	87	98
19·5	9·4406	9·4515	9·4624	9·4733	9·4842	9·4951	9·5060	9·5169	9·5280	9·5389		11	22	32	43	54	65	76	87	98
19·6	9·5499	9·5610	9·5719	9·5830	9·5940	9·6051	9·6161	9·6272	9·6383	9·6494		11	22	33	44	56	67	77	89	101
19·7	9·6605	9·6716	9·6828	9·6939	9·7051	9·7162	9·7275	9·7386	9·7499	9·7610		11	22	33	44	56	67	77	89	101
19·8	9·7724	9·7838	9·7949	9·8063	9·8175	9·8289	9·8401	9·8515	9·8628	9·8742		12	23	35	46	57	68	79	91	103
19·9	9·8855	9·8969	9·9083	9·9197	9·9312	9·9426	9·9541	9·9655	9·9770	9·9884		12	23	35	46	57	68	79	91	103

Gain

Power Ratios

The last nine columns are *mean differences*.

dB	0	1	2	3	4	5	6	7	8	9	1	2	3	4	5	6	7	8	9
0·0	1·0000	1·0023	1·0046	1·0069	1·0093	1·0116	1·0139	1·0162	1·0186	1·0209	2	5	7	9	12	14	16	19	21
0·1	1·0233	1·0257	1·0280	1·0304	1·0328	1·0351	1·0375	1·0399	1·0423	1·0447	2	5	7	10	12	14	17	19	21
0·2	1·0471	1·0495	1·0520	1·0544	1·0568	1·0593	1·0617	1·0641	1·0666	1·0690	2	5	7	10	12	15	17	20	22
0·3	1·0715	1·0740	1·0765	1·0789	1·0814	1·0839	1·0864	1·0889	1·0914	1·0940	3	5	8	10	13	15	18	20	23
0·4	1·0965	1·0990	1·1015	1·1041	1·1066	1·1092	1·1117	1·1143	1·1169	1·1194	3	5	8	10	13	15	18	20	23
0·5	1·1220	1·1246	1·1272	1·1298	1·1324	1·1350	1·1376	1·1402	1·1429	1·1455	3	5	8	11	13	16	18	21	24
0·6	1·1482	1·1508	1·1535	1·1561	1·1588	1·1614	1·1641	1·1668	1·1695	1·1722	3	5	8	11	13	16	19	21	24
0·7	1·1749	1·1776	1·1803	1·1830	1·1858	1·1885	1·1912	1·1940	1·1967	1·1995	3	5	8	11	14	16	19	22	25
0·8	1·2023	1·2050	1·2078	1·2106	1·2134	1·2162	1·2190	1·2218	1·2246	1·2274	3	5	8	11	14	17	19	22	25
0·9	1·2303	1·2331	1·2359	1·2388	1·2417	1·2445	1·2474	1·2503	1·2531	1·2560	3	6	9	11	14	17	20	23	26
1·0	1·2589	1·2618	1·2647	1·2677	1·2706	1·2735	1·2764	1·2794	1·2823	1·2853	3	6	9	12	15	18	21	24	26
1·1	1·2882	1·2912	1·2942	1·2972	1·3002	1·3032	1·3062	1·3092	1·3122	1·3152	3	6	9	12	15	18	21	24	27
1·2	1·3183	1·3213	1·3243	1·3274	1·3305	1·3335	1·3366	1·3397	1·3428	1·3459	3	6	9	12	15	18	22	25	28
1·3	1·3490	1·3521	1·3552	1·3583	1·3614	1·3646	1·3677	1·3709	1·3740	1·3772	3	6	9	13	16	19	22	25	28
1·4	1·3804	1·3836	1·3868	1·3900	1·3932	1·3964	1·3996	1·4028	1·4060	1·4093	3	6	10	13	16	19	22	26	29
1·5	1·4125	1·4158	1·4191	1·4223	1·4256	1·4289	1·4322	1·4355	1·4388	1·4421	3	7	10	13	16	20	23	26	30
1·6	1·4454	1·4488	1·4521	1·4555	1·4588	1·4622	1·4655	1·4689	1·4723	1·4757	3	7	10	13	17	20	24	27	30
1·7	1·4791	1·4825	1·4859	1·4894	1·4928	1·4962	1·4997	1·5031	1·5066	1·5101	3	7	10	14	17	21	24	28	31
1·8	1·5136	1·5171	1·5205	1·5241	1·5276	1·5311	1·5346	1·5382	1·5417	1·5453	4	7	11	14	18	21	25	28	32
1·9	1·5488	1·5524	1·5560	1·5596	1·5631	1·5668	1·5704	1·5740	1·5776	1·5812	4	7	11	14	18	22	25	29	32
2·0	1·5849	1·5885	1·5922	1·5959	1·5996	1·6032	1·6069	1·6106	1·6144	1·6181	4	7	11	15	18	22	26	30	33
2·1	1·6218	1·6255	1·6293	1·6331	1·6368	1·6406	1·6444	1·6482	1·6520	1·6558	4	8	11	15	19	23	26	30	34
2·2	1·6596	1·6634	1·6672	1·6711	1·6750	1·6788	1·6827	1·6866	1·6904	1·6943	4	8	12	15	19	23	27	31	35
2·3	1·6982	1·7022	1·7061	1·7100	1·7140	1·7179	1·7219	1·7258	1·7298	1·7338	4	8	12	16	20	24	28	32	36
2·4	1·7378	1·7418	1·7458	1·7498	1·7539	1·7579	1·7620	1·7660	1·7701	1·7742	4	8	12	16	20	24	28	32	36
2·5	1·7783	1·7824	1·7865	1·7906	1·7947	1·7989	1·8030	1·8072	1·8113	1·8155	4	8	12	17	21	25	29	33	36
2·6	1·8197	1·8239	1·8281	1·8323	1·8365	1·8408	1·8450	1·8493	1·8535	1·8578	4	8	12	17	21	25	29	34	38
2·7	1·8621	1·8664	1·8707	1·8750	1·8793	1·8836	1·8880	1·8923	1·8967	1·9011	4	9	13	17	22	26	30	35	39
2·8	1·9055	1·9099	1·9143	1·9187	1·9231	1·9275	1·9320	1·9364	1·9409	1·9454	4	9	13	18	22	26	31	35	40
2·9	1·9498	1·9543	1·9588	1·9634	1·9679	1·9724	1·9770	1·9815	1·9861	1·9907	5	9	14	18	23	27	32	36	41
3·0	1·9953	1·9999	2·0045	2·0091	2·0137	2·0184	2·0230	2·0277	2·0324	2·0370	5	9	14	19	23	28	32	37	42
3·1	2·0417	2·0464	2·0512	2·0559	2·0606	2·0654	2·0701	2·0749	2·0797	2·0845	5	10	14	19	24	29	33	38	43
3·2	2·0893	2·0941	2·0989	2·1038	2·1086	2·1135	2·1184	2·1233	2·1281	2·1330	5	10	15	19	24	29	34	39	44
3·3	2·1380	2·1429	2·1478	2·1528	2·1577	2·1627	2·1677	2·1727	2·1777	2·1827	5	10	15	20	25	30	35	40	45
3·4	2·1878	2·1928	2·1979	2·2029	2·2080	2·2131	2·2182	2·2233	2·2284	2·2336	5	10	15	20	25	31	36	41	46

Power Ratios (continued)

dB	0	1	2	3	4	5	6	7	8	9	mean differences 1	2	3	4	5	6	7	8	9
3·5	2·2387	2·2439	2·2491	2·2542	2·2594	2·2646	2·2699	2·2751	2·2803	2·2856	5	10	16	21	26	31	37	42	47
3·6	2·2909	2·2961	2·3014	2·3067	2·3121	2·3174	2·3227	2·3281	2·3336	2·3388	5	11	16	21	27	32	37	43	48
3·7	2·3442	2·3496	2·3550	2·3605	2·3659	2·3714	2·3768	2·3823	2·3878	2·3933	5	11	16	22	27	33	38	44	49
3·8	2·3988	2·4043	2·4099	2·4155	2·4210	2·4266	2·4322	2·4378	2·4434	2·4491	6	11	17	22	28	33	39	45	50
3·9	2·4547	2·4604	2·4660	2·4717	2·4774	2·4831	2·4889	2·4946	2·5003	2·5061	6	11	17	23	29	34	40	46	51
4·0	2·5119	2·5177	2·5236	2·5293	2·5351	2·5410	2·5468	2·5527	2·5586	2·5645	6	12	18	23	29	35	41	47	53
4·1	2·5704	2·5763	2·5823	2·5882	2·5942	2·6002	2·6062	2·6122	2·6182	2·6242	6	12	18	24	30	36	42	48	54
4·2	2·6303	2·6363	2·6424	2·6485	2·6546	2·6607	2·6669	2·6730	2·6792	2·6853	6	12	18	24	31	37	43	49	55
4·3	2·6915	2·6977	2·7040	2·7102	2·7164	2·7227	2·7290	2·7353	2·7416	2·7479	6	13	19	25	31	38	44	50	56
4·4	2·7542	2·7606	2·7669	2·7733	2·7797	2·7861	2·7925	2·7990	2·8054	2·8119	6	13	19	26	32	39	45	51	58
4·5	2·8184	2·8249	2·8314	2·8379	2·8445	2·8510	2·8576	2·8642	2·8708	2·8774	7	13	20	26	33	39	46	52	59
4·6	2·8840	2·8907	2·8973	2·9040	2·9107	2·9174	2·9242	2·9309	2·9376	2·9444	7	13	20	27	34	40	47	54	60
4·7	2·9512	2·9580	2·9648	2·9717	2·9785	2·9854	2·9923	2·9992	3·0061	3·0130	7	14	21	28	34	41	48	55	62
4·8	3·0200	3·0269	3·0339	3·0409	3·0479	3·0549	3·0620	3·0690	3·0761	3·0832	7	14	21	28	35	42	49	56	63
4·9	3·0903	3·0974	3·1046	3·1117	3·1189	3·1261	3·1333	3·1405	3·1477	3·1550	7	14	22	29	36	43	50	58	65
5·0	3·1623	3·1696	3·1769	3·1842	3·1916	3·1989	3·2063	3·2137	3·2211	3·2285	7	15	22	29	37	44	52	59	66
5·1	3·2359	3·2434	3·2509	3·2584	3·2659	3·2735	3·2809	3·2885	3·2961	3·3037	8	15	23	30	38	45	53	60	68
5·2	3·3113	3·3189	3·3266	3·3343	3·3420	3·3497	3·3574	3·3651	3·3729	3·3806	8	15	23	31	39	46	54	62	69
5·3	3·3884	3·3963	3·4041	3·4119	3·4198	3·4277	3·4356	3·4435	3·4514	3·4594	8	16	24	32	40	47	55	63	71
5·4	3·4674	3·4754	3·4834	3·4914	3·4995	3·5075	3·5156	3·5237	3·5318	3·5400	8	16	24	32	40	48	56	65	73
5·5	3·5481	3·5563	3·5645	3·5727	3·5810	3·5892	3·5975	3·6058	3·6141	3·6224	8	16	25	33	41	50	58	66	74
5·6	3·6308	3·6392	3·6475	3·6559	3·6644	3·6728	3·6813	3·6898	3·6983	3·7068	8	17	25	34	42	51	59	68	76
5·7	3·7154	3·7239	3·7325	3·7411	3·7497	3·7584	3·7670	3·7757	3·7844	3·7931	9	17	26	35	43	52	61	69	78
5·8	3·8019	3·8107	3·8194	3·8282	3·8371	3·8459	3·8548	3·8637	3·8726	3·8815	9	18	27	35	44	53	62	71	80
5·9	3·8905	3·8994	3·9084	3·9174	3·9264	3·9355	3·9446	3·9537	3·9628	3·9719	9	18	27	36	45	54	63	72	82
6·0	3·9811	3·9902	3·9994	4·0087	4·0179	4·0272	4·0365	4·0458	4·0551	4·0644	9	19	28	37	46	56	65	74	83
6·1	4·0738	4·0832	4·0926	4·1020	4·1115	4·1210	4·1305	4·1400	4·1495	4·1591	9	19	28	38	47	57	66	76	85
6·2	4·1687	4·1783	4·1879	4·1976	4·2073	4·2170	4·2267	4·2364	4·2462	4·2560	10	19	29	39	49	58	68	78	87
6·3	4·2658	4·2756	4·2855	4·2954	4·3053	4·3152	4·3251	4·3351	4·3451	4·3551	10	20	30	40	50	60	70	80	89
6·4	4·3652	4·3752	4·3853	4·3954	4·4055	4·4157	4·4259	4·4361	4·4463	4·4566	10	20	30	41	51	61	72	81	91
6·5	4·4668	4·4771	4·4875	4·4978	4·5082	4·5186	4·5290	4·5394	4·5499	4·5604	10	21	31	42	52	62	73	83	94
6·6	4·5709	4·5814	4·5920	4·6026	4·6132	4·6238	4·6345	4·6452	4·6559	4·6666	11	21	32	43	53	64	75	85	96
6·7	4·6774	4·6881	4·6989	4·7098	4·7206	4·7315	4·7424	4·7534	4·7642	4·7753	11	22	33	44	54	65	76	87	98
6·8	4·7863	4·7973	4·8084	4·8195	4·8306	4·8417	4·8529	4·8641	4·8753	4·8865	11	22	33	45	56	67	78	89	100
6·9	4·8978	4·9091	4·9204	4·9317	4·9431	4·9545	4·9659	4·9774	4·9888	5·0003	11	23	34	46	57	68	80	91	103

Power Ratios (continued)

Gain

dB	0	1	2	3	4	5	6	7	8	9	mean differences								
											1	2	3	4	5	6	7	8	9
7·0	5·0119	5·0234	5·0350	5·0466	5·0582	5·0699	5·0816	5·0933	5·1050	5·1168	12	23	35	47	58	70	82	93	105
7·1	5·1286	5·1404	5·1523	5·1642	5·1761	5·1880	5·2000	5·2119	5·2240	5·2360	12	24	36	48	60	72	84	96	108
7·2	5·2481	5·2602	5·2723	5·2845	5·2966	5·3088	5·3211	5·3333	5·3456	5·3580	12	25	37	49	61	73	85	98	110
7·3	5·3703	5·3827	5·3951	5·4075	5·4200	5·4325	5·4450	5·4576	5·4702	5·4828	13	25	38	50	63	75	88	100	113
7·4	5·4954	5·5081	5·5208	5·5336	5·5463	5·5590	5·5719	5·5847	5·5976	5·6105	13	26	38	51	64	77	90	102	115
7·5	5·6234	5·6364	5·6494	5·6624	5·6754	5·6885	5·7016	5·7148	5·7280	5·7412	13	26	39	52	66	79	92	105	118
7·6	5·7544	5·7677	5·7810	5·7943	5·8076	5·8210	5·8345	5·8479	5·8614	5·8749	13	27	40	54	67	80	94	107	121
7·7	5·8884	5·9020	5·9156	5·9293	5·9429	5·9566	5·9704	5·9841	5·9979	6·0117	14	27	41	55	69	82	96	110	123
7·8	6·0256	6·0395	6·0534	6·0674	6·0814	6·0954	6·1094	6·1235	6·1376	6·1518	14	28	42	56	70	84	98	112	126
7·9	6·1659	6·1802	6·1944	6·2087	6·2230	6·2373	6·2517	6·2661	6·2806	6·2951	14	29	43	58	72	86	101	115	130
8·0	6·3096	6·3241	6·3387	6·3533	6·3680	6·3826	6·3973	6·4121	6·4269	6·4417	15	29	44	59	74	88	103	118	132
8·1	6·4565	6·4714	6·4863	6·5013	6·5163	6·5313	6·5464	6·5615	6·5766	6·5917	15	30	45	60	75	90	105	120	135
8·2	6·6069	6·6222	6·6374	6·6527	6·6681	6·6834	6·6988	6·7143	6·7298	6·7453	15	31	46	62	77	92	108	123	139
8·3	6·7608	6·7764	6·7920	6·8077	6·8234	6·8391	6·8549	6·8707	6·8865	6·9024	16	32	47	63	79	95	110	126	142
8·4	6·9183	6·9343	6·9505	6·9663	6·9823	6·9984	7·0146	7·0307	7·0469	7·0623	16	32	48	64	81	97	113	129	145
8·5	7·0795	7·0958	7·1121	7·1285	7·1450	7·1614	7·1779	7·1945	7·2111	7·2277	16	33	50	66	83	99	116	132	149
8·6	7·2444	7·2611	7·2778	7·2946	7·3114	7·3282	7·3451	7·3621	7·3790	7·3961	17	34	51	68	85	101	118	135	152
8·7	7·4131	7·4302	7·4473	7·4645	7·4817	7·4989	7·5162	7·5336	7·5509	7·5683	17	35	52	69	87	104	121	138	156
8·8	7·5858	7·6033	7·6208	7·6384	7·6560	7·6736	7·6913	7·7090	7·7268	7·7446	18	35	53	71	89	107	125	142	159
8·9	7·7625	7·7804	7·7983	7·8163	7·8343	7·8524	7·8705	7·8886	7·9068	7·9250	18	36	54	72	91	109	127	145	163
9·0	7·9433	7·9616	7·9799	7·9983	8·0168	8·0353	8·0538	8·0724	8·0910	8·1096	19	37	56	74	93	111	130	148	167
9·1	8·1283	8·1470	8·1658	8·1846	8·2035	8·2224	8·2414	8·2604	8·2794	8·2985	19	38	57	76	95	113	132	151	170
9·2	8·3176	8·3368	8·3560	8·3753	8·3946	8·4140	8·4333	8·4528	8·4723	8·4918	19	39	58	78	97	116	136	155	175
9·3	8·5114	8·5310	8·5507	8·5704	8·5901	8·6099	8·6298	8·6497	8·6696	8·6896	20	40	60	79	99	119	139	158	178
9·4	8·7096	8·7297	8·7498	8·7700	8·7902	8·8105	8·8308	8·8512	8·8716	8·8920	20	41	61	81	102	122	142	162	183
9·5	8·9125	8·9331	8·9536	8·9743	8·9950	9·0157	9·0365	9·0573	9·0782	9·0991	21	42	62	83	104	125	146	166	187
9·6	9·1201	9·1411	9·1622	9·1833	9·2045	9·2257	9·2470	9·2683	9·2896	9·3111	21	42	64	85	106	127	149	170	191
9·7	9·3325	9·3541	9·3756	9·3972	9·4189	9·4406	9·4624	9·4842	9·5060	9·5280	22	43	65	87	109	130	152	174	195
9·8	9·5499	9·5719	9·5940	9·6161	9·6383	9·6605	9·6828	9·7051	9·7275	9·7499	22	44	67	89	111	133	155	178	200
9·9	9·7724	9·7949	9·8175	9·8401	9·8628	9·8855	9·9083	9·9312	9·9541	9·9770	23	46	68	91	114	137	160	182	205

Decibels (r = voltage or current ratio) **Gain**

r	0	1	2	3	4	5	6	7	8	9	1	2	3	4	5	6	7	8	9
														mean differences					
1·0	0·0000	0·0864	0·1720	0·2568	0·3406	0·4238	0·5062	0·5876	0·6684	0·7486	84	170	254	340	424	508	594	678	762
											80	162	242	324	404	484	566	646	728
1·1	0·8278	0·9064	0·9944	1·0616	1·1380	1·2140	1·2892	1·3638	1·4376	1·5110	74	154	232	308	386	464	540	618	696
											74	148	222	296	370	444	518	592	666
1·2	1·5836	1·6558	1·7272	1·7982	1·8684	1·9382	2·0074	2·0760	2·1442	2·2118	72	142	212	284	354	426	496	568	638
											68	136	204	272	340	408	476	544	614
1·3	2·2788	2·3454	2·4114	2·4770	2·5420	2·6066	2·6708	2·7344	2·7976	2·8602	66	132	196	262	328	394	458	524	590
											64	126	190	252	316	380	442	506	568
1·4	2·9226	2·9844	3·0458	3·1066	3·1672	3·2274	3·2870	3·3464	3·4052	3·4638	60	122	182	244	304	366	426	488	548
											58	118	176	236	294	354	412	472	530
1·5	3·5218	3·5796	3·6368	3·6938	3·7504	3·8066	3·8624	3·9180	3·9732	4·0280	56	114	170	228	284	342	398	456	512
											56	110	166	220	276	330	386	442	496
1·6	4·0824	4·1366	4·1802	4·2438	4·2968	4·3496	4·4022	4·4544	4·5062	4·5578	54	106	160	214	268	320	374	428	480
											52	104	156	208	260	312	364	416	466
1·7	4·6090	4·6600	4·7106	4·7610	4·8110	4·8608	4·9112	4·9594	5·0084	5·0570	52	100	152	202	252	302	352	402	454
											50	98	146	196	244	294	342	392	440
1·8	5·1054	5·1536	5·2014	5·2490	5·2964	5·3434	5·3902	5·4368	5·4832	5·5292	48	96	142	190	238	286	334	380	428
											46	92	138	186	232	278	324	370	416
1·9	5·5750	5·6206	5·6660	5·7112	5·7560	5·8006	5·8452	5·8894	5·9334	5·9770	46	90	136	180	226	270	316	360	406
											44	88	132	176	220	264	308	352	396
2·0	6·0206	6·0640	6·1070	6·1500	6·1926	6·2350	6·2774	6·3194	6·3612	6·4030	42	86	128	170	212	254	296	340	380
2·1	6·4444	6·4856	6·5268	6·5676	6·6082	6·6488	6·6890	6·7292	6·7692	6·8088	40	82	122	162	202	242	282	324	364
2·2	6·8484	6·8878	6·9270	6·9660	7·0050	7·0436	7·0822	7·1206	7·1586	7·1968	40	78	116	154	194	232	270	308	348
2·3	7·2346	7·2722	7·3098	7·3472	7·3844	7·4214	7·4582	7·4950	7·5316	7·5680	38	74	112	148	186	222	260	296	334
2·4	7·6042	7·6404	7·6764	7·7122	7·7478	7·7834	7·8188	7·8540	7·8890	7·9240	36	70	106	142	178	212	248	284	318

Gain

(r = voltage or current ratio)

The last nine columns are *mean differences*.

r	0	1	2	3	4	5	6	7	8	9	1	2	3	4	5	6	7	8	9
2·5	7·9588	7·9934	8·0280	8·0624	8·0966	8·1308	8·1648	8·1986	8·2324	8·2660	34	68	102	136	170	204	238	272	306
2·6	8·2994	8·3328	8·3660	8·3992	8·4320	8·4650	8·4976	8·5302	8·5626	8·5950	32	66	98	132	164	196	230	262	296
2·7	8·6272	8·6594	8·6914	8·7232	8·7550	8·7866	8·8182	8·8496	8·8808	8·9120	32	64	94	126	158	190	222	252	284
2·8	8·9432	8·9742	9·0050	9·0358	9·0664	9·0968	9·1274	9·1576	9·1878	9·2180	30	60	92	122	152	182	214	244	274
2·9	9·2480	9·2778	9·3076	9·3374	9·3670	9·3964	9·4258	9·4552	9·4844	9·5134	30	58	88	118	148	176	206	236	264
3·0	9·5424	9·5714	9·6002	9·6288	9·6574	9·6860	9·7144	9·7428	9·7710	9·7992	28	58	86	114	144	172	200	228	258
3·1	9·8272	9·8552	9·8831	9·9108	9·9386	9·9662	9·9938	10·0212	10·0486	10·0758	28	56	82	110	138	166	194	220	248
3·2	10·1030	10·1300	10·1572	10·1840	10·2108	10·2376	10·2644	10·2910	10·3174	10·3440	26	54	80	108	134	160	188	214	242
3·3	10·3702	10·3966	10·4228	10·4488	10·4750	10·5008	10·5268	10·5526	10·5784	10·6040	26	52	78	104	130	156	182	208	234
3·4	10·6296	10·6550	10·6806	10·7059	10·7312	10·7564	10·7816	10·8066	10·8316	10·8566	26	50	76	100	126	152	176	202	226
3·5	10·8814	10·9062	10·9308	10·9554	10·9800	11·0046	11·0290	11·0534	11·0776	11·1018	24	48	74	98	122	146	170	196	220
3·6	11·1260	11·1502	11·1742	11·1982	11·2220	11·2458	11·2696	11·2934	11·3170	11·3406	24	48	72	96	120	142	166	190	214
3·7	11·3640	11·3874	11·4108	11·4342	11·4574	11·4806	11·5038	11·5268	11·5498	11·5728	24	46	70	92	116	140	162	186	208
3·8	11·5956	11·6184	11·6412	11·6640	11·6866	11·7092	11·7318	11·7542	11·7766	11·7990	22	46	68	90	114	136	158	180	204
3·9	11·8212	11·8436	11·8658	11·8878	11·9100	11·9320	11·9540	11·9758	11·9976	12·0192	22	44	66	88	110	132	154	176	198
4·0	12·0412	12·0628	12·0846	12·1062	12·1276	12·1491	12·1706	12·1918	12·2132	12·2344	22	42	64	86	108	128	150	172	194
4·1	12·2556	12·2768	12·2980	12·3190	12·3400	12·3610	12·3818	12·4028	12·4236	12·4442	20	42	62	84	106	126	148	168	190
4·2	12·4650	12·4856	12·5062	12·5268	12·5474	12·5678	12·5882	12·6086	12·6288	12·6492	20	40	62	82	102	122	142	164	184
4·3	12·6694	12·6886	12·7096	12·7298	12·7498	12·7698	12·7898	12·8096	12·8294	12·8492	20	40	60	80	100	120	140	160	180
4·4	12·8690	12·8888	12·9084	12·9280	12·9476	12·9672	12·9866	13·0062	13·0256	13·0450	20	40	58	78	98	118	136	156	176
4·5	13·0642	13·0836	13·1028	13·1220	13·1412	13·1602	13·1792	13·1984	13·2174	13·2362	20	38	58	76	96	114	134	152	172
4·6	13·2552	13·2740	13·2928	13·3116	13·3304	13·3490	13·3678	13·3864	13·4050	13·4234	18	38	56	74	94	112	130	148	168
4·7	13·4420	13·4604	13·4788	13·4972	13·5156	13·5338	13·5522	13·5704	13·5886	13·6068	18	36	54	72	92	110	128	144	164
4·8	13·6248	13·6430	13·6610	13·6790	13·6970	13·7148	13·7328	13·7506	13·7684	13·7862	18	36	54	72	90	106	126	144	162
4·9	13·8040	13·8216	13·8394	13·8570	13·8746	13·8922	13·9096	13·9272	13·9446	13·9620	18	36	52	70	88	106	124	140	158
5·0	13·9794	13·9968	14·0140	14·0314	14·0486	14·0658	14·0830	14·1002	14·1172	14·1344	18	34	52	68	86	104	120	138	154
5·1	14·1514	14·1682	14·1854	14·2042	14·2192	14·2361	14·2530	14·2698	14·2866	14·3034	16	34	50	68	84	100	118	134	152
5·2	14·3200	14·3368	14·3534	14·3700	14·3866	14·4032	14·4198	14·4362	14·4526	14·4692	16	34	50	66	84	100	116	132	150
5·3	14·4855	14·5018	14·5182	14·5346	14·5508	14·5670	14·5832	14·5994	14·6156	14·6318	16	32	48	64	82	98	114	130	146
5·4	14·6478	14·6640	14·6800	14·6960	14·7120	14·7280	14·7438	14·7598	14·7756	14·7914	16	32	48	64	80	96	112	128	144

N*

Decibels (*continued*) (r = voltage or current ratio) Gain

r	0	1	2	3	4	5	6	7	8	9	md 1	2	3	4	5	6	7	8	9
5·5	14·8072	14·8230	14·8388	14·8546	14·8702	14·8858	14·9014	14·9162	14·9326	14·9482	16	32	46	62	78	94	110	126	140
5·6	14·9638	14·9792	14·9948	15·0102	15·0256	15·0410	15·0564	15·0716	15·0870	15·1022	16	30	46	62	78	92	108	124	138
5·7	15·1174	15·1328	15·1480	15·1630	15·1782	15·1934	15·2084	15·2236	15·2386	15·2536	16	30	46	60	76	90	106	120	136
5·8	15·2686	15·2836	15·2984	15·3134	15·3282	15·3432	15·3580	15·3728	15·3876	15·4024	14	30	44	60	74	88	104	118	134
5·9	15·4160	15·4318	15·4464	15·4610	15·4758	15·4904	15·5050	15·5194	15·5340	15·5486	14	30	44	58	74	88	102	116	132
6·0	15·5630	15·5774	15·5920	15·6064	15·6208	15·6352	15·6494	15·6638	15·6780	15·6924	14	28	44	58	72	86	100	116	130
6·1	15·7066	15·7208	15·7350	15·7492	15·7634	15·7776	15·7916	15·8058	15·8198	15·8338	14	28	42	56	72	86	100	114	128
6·2	15·8478	15·8618	15·8758	15·8898	15·9036	15·9176	15·9314	15·9454	15·9592	15·9730	14	28	42	56	70	82	96	110	124
6·3	15·9868	16·0006	16·0144	16·0280	16·0418	16·0554	16·0692	16·0828	16·0964	16·1100	14	28	40	54	68	82	96	108	122
6·4	16·1236	16·1372	16·1508	16·1642	16·1778	16·1912	16·2046	16·2180	16·2316	16·2448	14	26	40	54	68	80	96	108	120
6·5	16·2582	16·2716	16·2850	16·2982	16·3116	16·3248	16·3380	16·3514	16·3646	16·3778	14	26	40	52	66	80	92	106	118
6·6	16·3808	16·4040	16·4172	16·4302	16·4434	16·4564	16·4694	16·4826	16·4956	16·5086	14	26	40	52	66	80	92	104	116
6·7	16·5214	16·5344	16·5464	16·5604	16·5732	16·5860	16·5990	16·6118	16·6246	16·6374	12	26	38	52	64	76	90	102	116
6·8	16·6502	16·6630	16·6756	16·6884	16·7012	16·7138	16·7264	16·7392	16·7518	16·7644	12	26	38	50	64	76	88	100	114
6·9	16·7770	16·7896	16·8022	16·8146	16·8272	16·8396	16·8522	16·8646	16·8762	16·8896	12	24	38	50	62	74	86	100	112
7·0	16·9020	16·9144	16·9268	16·9392	16·9514	16·9638	16·9760	16·9884	17·0006	17·0130	12	24	38	50	62	74	86	100	112
7·1	17·0252	17·0374	17·0496	17·0618	17·0740	17·0862	17·0982	17·1104	17·1224	17·1346	12	24	36	48	62	74	86	98	110
7·2	17·1466	17·1588	17·1708	17·1828	17·1948	17·2068	17·2188	17·2306	17·2426	17·2546	12	24	36	48	60	72	84	96	108
7·3	17·2664	17·2784	17·2902	17·3020	17·3140	17·3258	17·3376	17·3494	17·3612	17·3728	12	24	36	48	60	72	82	94	106
7·4	17·3846	17·3964	17·4080	17·4198	17·4314	17·4432	17·4548	17·4664	17·4780	17·4896	12	24	36	46	58	70	82	92	104
7·5	17·5012	17·5128	17·5244	17·5358	17·5474	17·5590	17·5704	17·5820	17·5934	17·6048	12	24	34	46	58	70	82	92	104
7·6	17·6162	17·6276	17·6390	17·6504	17·6618	17·6732	17·6846	17·6960	17·7072	17·7186	12	22	34	46	58	68	80	92	102
7·7	17·7298	17·7410	17·7524	17·7636	17·7748	17·7860	17·7972	17·8084	17·8196	17·8308	12	22	34	44	56	68	78	90	102
7·8	17·8418	17·8530	17·8642	17·8752	17·8864	17·8974	17·9084	17·9194	17·9306	17·9416	12	22	34	44	56	66	78	88	100
7·9	17·9526	17·9636	17·9746	17·9854	17·9964	18·0074	18·0182	18·0292	18·0400	18·0510	12	22	34	44	56	66	78	88	100
8·0	18·0618	18·0726	18·0834	18·0944	18·1052	18·1160	18·1268	18·1374	18·1482	18·1590	10	22	32	44	54	64	76	86	98
8·1	18·1696	18·1804	18·1912	18·2018	18·2124	18·2232	18·2338	18·2444	18·2550	18·2656	10	22	32	42	54	64	74	84	96
8·2	18·2762	18·2868	18·2974	18·3080	18·3186	18·3290	18·3396	18·3502	18·3606	18·3710	10	22	32	42	54	64	74	84	96
8·3	18·3816	18·3920	18·4024	18·4128	18·4232	18·4336	18·4442	18·4546	18·4648	18·4752	10	20	32	42	52	62	72	82	94
8·4	18·4856	18·4960	18·5062	18·5166	18·5268	18·5372	18·5474	18·5576	18·5680	18·5782	10	20	30	40	52	62	72	82	92

(columns 1–9 following the main value columns are **mean differences**)

Decibels (*continued*)   (r = voltage or current ratios)   **Gain**

r	0	1	2	3	4	5	6	7	8	9	1	2	3	4	5	6	7	8	9
														mean differences					
8·5	18·5884	18·5986	18·6088	18·6190	18·6292	18·6394	18·6494	18·6596	18·6698	18·6798	10	20	30	40	52	62	72	82	92
8·6	18·6900	18·7000	18·7102	18·7202	18·7302	18·7404	18·7504	18·7604	18·7704	18·7804	10	20	30	40	50	60	70	80	90
8·7	18·7904	18·8004	18·8104	18·8202	18·8302	18·8402	18·8500	18·8600	18·8698	18·8798	10	20	30	40	50	60	70	80	90
8·8	18·8896	18·8996	18·9094	18·9192	18·9290	18·9388	18·9486	18·9584	18·9682	18·9780	10	20	30	40	50	58	68	78	88
8·9	18·9878	18·9976	19·0072	19·0170	19·0268	19·0364	19·0462	19·0558	19·0656	19·0752	10	20	30	38	48	58	68	78	88
9·0	19·0848	19·0944	19·1042	19·1138	19·1234	19·1330	19·1426	19·1522	19·1618	19·1712	10	20	28	38	48	58	68	76	86
9·1	19·1808	19·1904	19·1998	19·2094	19·2190	19·2284	19·2380	19·2474	19·2568	19·2664	10	18	28	38	48	56	66	76	84
9·2	19·2758	19·2852	19·2946	19·3040	19·3134	19·3228	19·3322	19·3416	19·3510	19·3604	10	18	28	38	48	56	66	76	84
9·3	19·3696	19·3790	19·3884	19·3976	19·4070	19·4162	19·4256	19·4348	19·4440	19·4536	10	18	28	36	46	56	64	76	84
9·4	19·4626	19·4718	19·4810	19·4902	19·4994	19·5086	19·5178	19·5270	19·5362	19·5454	10	18	28	36	46	56	64	74	84
9·5	19·5544	19·5636	19·5728	19·5818	19·5910	19·6000	19·6092	19·6182	19·6274	19·6364	10	18	28	36	46	54	64	72	82
9·6	19·6454	19·6544	19·6636	19·6726	19·6816	19·6906	19·6996	19·7086	19·7176	19·7264	10	18	28	36	46	54	64	72	82
9·7	19·7354	19·7444	19·7534	19·7622	19·7712	19·7800	19·7890	19·7978	19·8068	19·8156	8	18	26	36	44	54	62	72	80
9·8	19·8246	19·8334	19·8422	19·8510	19·8600	19·8688	19·8776	19·8864	19·8952	19·9040	8	18	26	36	44	52	62	72	80
9·9	19·9128	19·9214	19·9302	19·9390	19·9478	19·9564	19·9652	19·9740	19·9826	19·9914	8	18	26	34	44	52	62	70	78

Decibels

Gain

(r = power ratio)

r	0	1	2	3	4	5	6	7	8	9
1·0	0·0000	0·0432	0·0860	0·1284	0·1703	0·2119	0·2531	0·2938	0·3342	0·3773
1·1	0·4139	0·4532	0·4922	0·5308	0·5690	0·6070	0·6446	0·6819	0·7188	0·7555
1·2	0·7918	0·8279	0·8636	0·8991	0·9342	0·9691	1·0037	1·0380	1·0721	1·1059
1·3	1·1394	1·1727	1·2057	1·2385	1·2710	1·3033	1·3354	1·3672	1·3988	1·4301
1·4	1·4613	1·4922	1·5229	1·5534	1·5836	1·6137	1·6435	1·6732	1·7026	1·7319
1·5	1·7609	1·7898	1·8184	1·8469	1·8752	1·9033	1·9312	1·9590	1·9866	2·0140
1·6	2·0412	2·0683	2·0951	2·1219	2·1484	2·1748	2·2011	2·2272	2·2531	2·2786
1·7	2·3045	2·3300	2·3553	2·3805	2·4055	2·4304	2·4551	2·4797	2·5042	2·5285
1·8	2·5527	2·5768	2·6007	2·6245	2·6482	2·6717	2·6951	2·7184	2·7416	2·7646
1·9	2·7875	2·8103	2·8330	2·8556	2·8780	2·9003	2·9226	2·9447	2·9667	2·9885
2·0	3·0103	3·0320	3·0535	3·0750	3·0963	3·1175	3·1387	3·1597	3·1806	3·2015
2·1	3·2222	3·2428	3·2634	3·2838	3·3041	3·3244	3·3445	3·3646	3·3846	3·4044
2·2	3·4242	3·4439	3·4635	3·4830	3·5025	3·5218	3·5411	3·5603	3·5793	3·5984
2·3	3·6173	3·6361	3·6549	3·6736	3·6922	3·7107	3·7291	3·7475	3·7658	3·7840
2·4	3·8021	3·8202	3·8382	3·8561	3·8739	3·8917	3·9094	3·9270	3·9445	3·9620

mean differences

r	1	2	3	4	5	6	7	8	9
1·0	42 / 40	85 / 81	127 / 121	170 / 162	212 / 202	254 / 242	297 / 283	339 / 323	381 / 364
1·1	37 / 37	77 / 74	116 / 111	154 / 148	193 / 185	232 / 222	270 / 259	309 / 296	348 / 333
1·2	36 / 34	71 / 68	106 / 102	142 / 136	177 / 170	213 / 204	248 / 238	284 / 272	319 / 307
1·3	33 / 32	66 / 63	98 / 95	131 / 126	164 / 158	197 / 190	229 / 221	262 / 253	295 / 284
1·4	30 / 29	61 / 59	91 / 88	122 / 118	152 / 147	183 / 177	213 / 206	244 / 236	274 / 265
1·5	28 / 28	57 / 55	85 / 83	114 / 110	142 / 138	171 / 165	199 / 193	228 / 221	256 / 248
1·6	27 / 26	53 / 52	80 / 78	107 / 104	134 / 130	160 / 156	187 / 182	214 / 208	240 / 233
1·7	26 / 25	50 / 49	76 / 73	101 / 98	126 / 122	151 / 147	176 / 171	201 / 196	227 / 220
1·8	24 / 23	48 / 46	71 / 69	95 / 93	119 / 116	143 / 139	167 / 162	190 / 185	214 / 208
1·9	23 / 22	45 / 44	68 / 66	90 / 88	113 / 110	135 / 132	158 / 154	180 / 176	203 / 198
2·0	21	43	64	85	106	127	148	170	190
2·1	20	41	61	81	101	121	141	162	182
2·2	20	39	58	77	97	116	135	154	174
2·3	19	37	56	74	93	111	130	148	167
2·4	18	35	53	71	89	106	124	142	159

Decibels (continued) **Gain** (r = power ratio)

r	0	1	2	3	4	5	6	7	8	9	mean differences 1	2	3	4	5	6	7	8	9
2·5	3·9794	3·9967	4·0140	4·0312	4·0483	4·0654	4·0824	4·0993	4·1162	4·1330	17	34	51	68	85	102	119	136	153
2·6	4·1497	4·1664	4·1830	4·1996	4·2160	4·2325	4·2488	4·2651	4·2813	4·2975	16	33	49	66	82	98	115	131	148
2·7	4·3136	4·3297	4·3457	4·3616	4·3775	4·3933	4·4091	4·4248	4·4404	4·4560	16	32	47	63	79	95	111	126	142
2·8	4·4716	4·4871	4·5025	4·5179	4·5332	4·5484	4·5637	4·5788	4·5939	4·6090	15	30	46	61	76	91	107	122	137
2·9	4·6240	4·6389	4·6538	4·6687	4·6835	4·6982	4·7129	4·7276	4·7422	4·7567	15	29	44	59	74	88	103	118	132
3·0	4·7712	4·7857	4·8001	4·8144	4·8287	4·8430	4·8572	4·8714	4·8855	4·8996	14	29	43	57	72	86	100	114	129
3·1	4·9136	4·9276	4·9415	4·9554	4·9693	4·9831	4·9969	5·0106	5·0243	5·0379	14	28	41	55	69	83	97	110	124
3·2	5·0515	5·0650	5·0786	5·0920	5·1054	5·1188	5·1322	5·1455	5·1587	5·1720	13	27	40	54	67	80	94	107	121
3·3	5·1851	5·1983	5·2114	5·2244	5·2375	5·2504	5·2634	5·2763	5·2892	5·3020	13	26	39	52	65	78	91	104	117
3·4	5·3148	5·3275	5·3403	5·3529	5·3656	5·3782	5·3908	5·4033	5·4158	5·4283	13	25	38	50	63	76	88	101	113
3·5	5·4407	5·4531	5·4654	5·4777	5·4900	5·5023	5·5145	5·5267	5·5388	5·5509	12	24	37	49	61	73	85	98	110
3·6	5·5630	5·5751	5·5871	5·5991	5·6110	5·6229	5·6348	5·6467	5·6585	5·6703	12	24	36	48	60	71	83	95	107
3·7	5·6820	5·6937	5·7054	5·7171	5·7287	5·7403	5·7519	5·7634	5·7749	5·7864	12	23	35	46	58	70	81	93	104
3·8	5·7978	5·8092	5·8206	5·8320	5·8433	5·8546	5·8659	5·8771	5·8883	5·8995	11	23	34	45	57	68	80	90	102
3·9	5·9106	5·9218	5·9329	5·9439	5·9550	5·9660	5·9770	5·9879	5·9988	6·0097	11	22	33	44	55	66	77	88	99
4·0	6·0206	6·0314	6·0423	6·0531	6·0638	6·0746	6·0853	6·0959	6·1066	6·1172	11	21	32	43	54	64	75	86	97
4·1	6·1278	6·1384	6·1490	6·1595	6·1700	6·1805	6·1909	6·2014	6·2118	6·2221	10	21	31	42	53	63	74	84	95
4·2	6·2325	6·2428	6·2531	6·2634	6·2737	6·2839	6·2941	6·3043	6·3144	6·3246	10	20	31	41	51	61	71	82	92
4·3	6·3347	6·3448	6·3548	6·3649	6·3749	6·3849	6·3949	6·4048	6·4147	6·4246	10	20	30	40	50	60	70	80	90
4·4	6·4345	6·4444	6·4542	6·4640	6·4738	6·4836	6·4933	6·5031	6·5128	6·5225	10	20	29	39	49	59	68	78	88
4·5	6·5321	6·5418	6·5514	6·5610	6·5706	6·5801	6·5896	6·5992	6·6087	6·6181	10	19	29	38	48	57	67	76	86
4·6	6·6276	6·6370	6·6464	6·6558	6·6652	6·6745	6·6839	6·6932	6·7025	6·7117	9	19	28	37	47	56	65	74	84
4·7	6·7210	6·7302	6·7394	6·7486	6·7578	6·7669	6·7761	6·7852	6·7943	6·8034	9	18	27	36	46	55	64	73	82
4·8	6·8124	6·8215	6·8305	6·8395	6·8485	6·8574	6·8664	6·8753	6·8842	6·8931	9	18	27	36	45	53	63	72	81
4·9	6·9020	6·9108	6·9197	6·9285	6·9373	6·9461	6·9548	6·9636	6·9723	6·9810	9	18	26	35	44	53	62	70	79
5·0	6·9897	6·9984	7·0070	7·0157	7·0243	7·0329	7·0415	7·0501	7·0586	7·0672	9	17	26	34	43	52	60	69	77
5·1	7·0757	7·0842	7·0927	7·1012	7·1096	7·1181	7·1265	7·1349	7·1433	7·1517	8	17	25	34	42	50	59	67	76
5·2	7·1600	7·1684	7·1767	7·1850	7·1933	7·2016	7·2099	7·2181	7·2263	7·2346	8	17	25	33	42	50	58	66	75
5·3	7·2428	7·2509	7·2591	7·2673	7·2754	7·2835	7·2916	7·2997	7·3078	7·3159	8	16	24	32	41	49	57	65	73
5·4	7·3239	7·3320	7·3400	7·3480	7·3560	7·3640	7·3719	7·3799	7·3878	7·3957	8	16	24	32	40	48	56	64	72

Decibels (*continued*)

(r = power ratio)

r	0	1	2	3	4	5	6	7	8	9	mean differences 1	2	3	4	5	6	7	8	9
5·5	7·4036	7·4115	7·4194	7·4273	7·4351	7·4429	7·4507	7·4586	7·4663	7·4741	8	16	23	31	39	47	55	63	70
5·6	7·4819	7·4896	7·4974	7·5051	7·5128	7·5205	7·5282	7·5358	7·5435	7·5511	8	15	23	31	39	46	54	62	69
5·7	7·5587	7·5664	7·5740	7·5815	7·5891	7·5967	7·6042	7·6118	7·6193	7·6268	8	15	23	30	38	45	53	60	68
5·8	7·6343	7·6418	7·6492	7·6567	7·6641	7·6716	7·6790	7·6864	7·6938	7·7012	7	15	22	30	37	45	53	59	67
5·9	7·7085	7·7159	7·7232	7·7305	7·7379	7·7452	7·7525	7·7597	7·7670	7·7743	7	15	22	29	37	44	51	58	66
6·0	7·7815	7·7887	7·7960	7·8032	7·8104	7·8176	7·8247	7·8319	7·8390	7·8462	7	14	22	28	36	43	50	58	65
6·1	7·8533	7·8604	7·8675	7·8746	7·8817	7·8888	7·8958	7·9029	7·9099	7·9169	7	14	21	28	36	43	50	57	64
6·2	7·9239	7·9309	7·9379	7·9449	7·9518	7·9588	7·9657	7·9727	7·9796	7·9865	7	14	21	28	35	41	48	55	62
6·3	7·9934	8·0003	8·0072	8·0140	8·0209	8·0277	8·0346	8·0414	8·0482	8·0550	7	14	20	27	34	41	48	54	61
6·4	8·0618	8·0686	8·0754	8·0821	8·0889	8·0956	8·1023	8·1090	8·1158	8·1224	7	13	20	27	34	40	47	54	60
6·5	8·1291	8·1358	8·1425	8·1491	8·1558	8·1624	8·1690	8·1757	8·1823	8·1889	7	13	20	26	33	40	46	53	59
6·6	8·1954	8·2020	8·2086	8·2151	8·2217	8·2282	8·2347	8·2413	8·2478	8·2543	7	13	20	26	33	39	46	52	59
6·7	8·2607	8·2672	8·2737	8·2802	8·2866	8·2930	8·2995	8·3059	8·3123	8·3187	6	13	19	25	32	38	45	51	58
6·8	8·3251	8·3315	8·3378	8·3442	8·3506	8·3569	8·3632	8·3696	8·3759	8·3822	6	13	19	25	32	38	44	51	57
6·9	8·3885	8·3948	8·4011	8·4073	8·4136	8·4198	8·4261	8·4323	8·4386	8·4448	6	12	19	25	31	37	43	50	56
7·0	8·4510	8·4572	8·4634	8·4696	8·4757	8·4819	8·4880	8·4942	8·5003	8·5065	6	12	19	25	31	37	43	50	56
7·1	8·5126	8·5187	8·5248	8·5309	8·5370	8·5431	8·5491	8·5552	8·5612	8·5673	6	12	18	24	31	37	43	49	55
7·2	8·5733	8·5794	8·5854	8·5914	8·5974	8·6034	8·6094	8·6153	8·6213	8·6273	6	12	18	24	30	36	42	48	54
7·3	8·6332	8·6392	8·6451	8·6510	8·6570	8·6629	8·6688	8·6747	8·6806	8·6864	6	12	18	24	30	35	41	47	53
7·4	8·6923	8·6982	8·7040	8·7099	8·7157	8·7216	8·7274	8·7332	8·7390	8·7448	6	12	17	23	29	35	41	46	52
7·5	8·7506	8·7564	8·7622	8·7679	8·7737	8·7795	8·7852	8·7910	8·7967	8·8024	6	12	17	23	29	35	41	46	52
7·6	8·8081	8·8138	8·8195	8·8252	8·8309	8·8366	8·8423	8·8480	8·8536	8·8593	6	11	17	23	29	34	40	46	51
7·7	8·8649	8·8705	8·8762	8·8818	8·8874	8·8930	8·8986	8·9042	8·9098	8·9152	6	11	17	22	28	34	39	45	50
7·8	8·9209	8·9265	8·9321	8·9376	8·9432	8·9487	8·9542	8·9597	8·9653	8·9708	6	11	17	22	28	33	39	44	50
7·9	8·9763	8·9818	8·9873	8·9927	8·9982	9·0037	9·0091	9·0146	9·0200	9·0255	6	11	17	22	28	33	39	44	50
8·0	9·0309	9·0363	9·0417	9·0472	9·0526	9·0580	9·0634	9·0687	9·0741	9·0795	5	11	16	22	27	32	38	43	49
8·1	9·0848	9·0902	9·0956	9·1009	9·1062	9·1116	9·1169	9·1222	9·1275	9·1328	5	11	16	21	27	32	37	42	48
8·2	9·1381	9·1434	9·1487	9·1540	9·1593	9·1645	9·1698	9·1751	9·1803	9·1855	5	11	16	21	27	32	37	42	48
8·3	9·1908	9·1960	9·2012	9·2064	9·2117	9·2169	9·2221	9·2273	9·2324	9·2376	5	10	16	21	26	31	36	42	47
8·4	9·2428	9·2480	9·2531	9·2583	9·2634	9·2686	9·2737	9·2788	9·2840	9·2891	5	10	15	20	26	31	36	41	46

Decibels (continued)

(r = power ratio)

Gain

r	0	1	2	3	4	5	6	7	8	9	mean differences								
											1	2	3	4	5	6	7	8	9
8·5	9·2942	9·2993	9·3044	9·3095	9·3146	9·3197	9·3247	9·3298	9·3349	9·3399	5	10	15	20	26	31	36	41	46
8·6	9·3450	9·3500	9·3551	9·3601	9·3651	9·3702	9·3752	9·3802	9·3852	9·3902	5	10	15	20	25	30	35	40	45
8·7	9·3952	9·4002	9·4052	9·4101	9·4151	9·4201	9·4250	9·4300	9·4349	9·4399	5	10	15	20	25	30	35	40	45
8·8	9·4448	9·4498	9·4547	9·4596	9·4645	9·4694	9·4743	9·4792	9·4841	9·4890	5	10	15	20	25	29	34	39	44
8·9	9·4939	9·4988	9·5036	9·5085	9·5134	9·5182	9·5231	9·5279	9·5328	9·5376	5	10	15	19	24	29	34	39	44
9·0	9·5424	9·5472	9·5521	9·5569	9·5617	9·5665	9·5713	9·5761	9·5809	9·5856	5	10	14	19	24	29	34	38	43
9·1	9·5904	9·5952	9·5999	9·6047	9·6095	9·6142	9·6190	9·6237	9·6284	9·6332	5	9	14	19	24	28	33	38	42
9·2	9·6379	9·6426	9·6473	9·6520	9·6567	9·6614	9·6661	9·6708	9·6755	9·6802	5	9	14	19	24	28	33	38	42
9·3	9·6848	9·6895	9·6942	9·6988	9·7035	9·7081	9·7128	9·7174	9·7220	9·7267	5	9	14	18	23	28	32	38	42
9·4	9·7313	9·7359	9·7405	9·7451	9·7497	9·7543	9·7589	9·7635	9·7681	9·7727	5	9	14	18	23	28	32	37	42
9·5	9·7772	9·7818	9·7864	9·7909	9·7955	9·8000	9·8046	9·8091	9·8137	9·8182	5	9	14	18	23	27	32	36	41
9·6	9·8227	9·8272	9·8318	9·8363	9·8408	9·8453	9·8498	9·8543	9·8588	9·8632	5	9	14	18	23	27	32	36	41
9·7	9·8677	9·8722	9·8767	9·8811	9·8856	9·8900	9·8945	9·8989	9·9034	9·9078	4	9	13	18	22	27	31	36	40
9·8	9·9123	9·9167	9·9211	9·9255	9·9300	9·9344	9·9388	9·9432	9·9476	9·9520	4	9	13	18	22	26	31	35	40
9·9	9·9564	9·9607	9·9651	9·9695	9·9739	9·9782	9·9826	9·9870	9·9913	9·9957	4	9	13	17	22	26	31	35	39

These decibel tables were computed by Dr J. de Klerk and grateful acknowledgment is made for permission to reproduce them.

SYMBOLS, UNITS AND NOMENCLATURE IN PHYSICS
(Summarized from the 1965 Report of the International Union of
Pure and Applied Physics.)
Physical Quantities

The symbol for a **physical quantity** (sometimes 'physical magnitude' in U.S.) is equivalent to the product of the *numerical value* (or the measure), a pure number, and a *unit*, i.e.,

$$\text{physical quantity} = \text{numerical value} \times \text{unit}.$$

For some dimensionless physical quantities the unit has no name or symbol and is not explicitly indicated.

Examples:

$$E = 200 \text{ erg} \qquad n = 1 \cdot 55 \text{ (for quartz)}$$
$$F = 27\text{N} \qquad v = 3 \times 10^8 \text{ s}^{-1}.$$

SYMBOLS FOR PHYSICAL QUANTITIES—GENERAL RULES

Symbols for physical quantities should be *single letters* of the Latin or Greek alphabet with or without modifying signs, subscripts, superscripts, dashes, etc.

Remarks:

(*a*) An exception to this rule consists of the two-letter symbols, which are sometimes used to represent dimensionless combinations of physical quantities. If such a symbol, composed of two letters, appears as a factor in a product, it is recommended to separate this symbol from the other symbols by a dot or by brackets or by a space.

(*b*) Abbreviations, i.e., shortened forms of names or expressions, such as p.f. for partition function, should not be used in physical equations. These abbreviations in the text should be written in ordinary roman type.

Symbols for physical quantities should be printed in *italic type*.

Remark:

It is recommended to consider as a guiding principle for the printing of indices the criterion: only indices which are symbols for physical quantities should be printed in italic type.

Examples:

	Upright indices	Sloping indices
	C_g (g = gas)	p in C_p
	g_n (n = normal)	n in $\Sigma_n a_n \phi_n$
	μ_r (r = relative)	x in $\Sigma_x a_x b_x$
	E_k (k = kinetic)	i, k in g_{ik}
	χ_e (e = electric)	x in p_x

Symbols for vectors and tensors. To avoid the usage of subscripts it is often convenient to indicate vectors and tensors of the second rank by letters of a special type. The following choice is recommended:

(*a*) Vectors should be printed in bold type, by preference bold italic type, e.g., *A, a*.

(*b*) Tensors of the second rank should be printed in bold face sans serif type, e.g., **S, T**.

Units

SYMBOLS FOR UNITS—GENERAL RULES

Symbols for units of physical quantities should be printed in *roman type*.

Symbols for units should not contain a final full stop and should remain unaltered in the plural, e.g. 7 cm and *not* 7 cms.

Symbols for units should be printed in *lower case* roman type. However, the symbol for a unit derived from a proper name should start with a capital roman letter, e.g., m (metre); A (ampere); Wb (weber); Hz (hertz).

The following *prefixes* should be used to indicate decimal fractions or multiples of a unit:

deci	$(= 10^{-1})$	d			
centi	$(= 10^{-2})$	c			
milli	$(= 10^{-3})$	m	kilo	$(= 10^{3})$	k
micro	$(= 10^{-6})$	μ	mega	$(= 10^{6})$	M
nano	$(= 10^{-9})$	n	giga	$(= 10^{9})$	G
pico	$(= 10^{-12})$	p	tera	$(= 10^{12})$	T
femto	$(= 10^{-15})$	f			
atto	$(= 10^{-18})$	a			

The use of *double prefixes* should be avoided when single prefixes are available:

Not mμs   *but* ns   (nanosecond)
Not kMW   *but* GW   (gigawatt)
Not $\mu\mu$F   *but* pF   (picofarad)

When a prefix is placed before the symbol of a unit, the *combination of prefix and symbol* should be considered as *one new symbol*, which can be raised to a positive or negative power without using brackets.

Examples: cm³,   mA², μs^{-1}.

Remark:

cm³ *always means* $(0.01 \text{ m})^3$ *not* 0.01 m^3
μs^{-1} *always means* $(10^{-6} \text{ s})^{-1}$ *not* 10^{-6} s^{-1}.

MATHEMATICAL OPERATIONS

Multiplication of two units may be indicated in *one* of the following ways:

Nm   N m   N · m   N . m

Division of one unit by another unit may be indicated in *one* of the following ways:

$$\frac{m}{s} \qquad m/s \qquad m\,s^{-1}$$

or by any other way of writing the product of m and s^{-1}. Not more than one solidus should be used.

Examples: *Not* cm/s/s   *but* cm/s² = cm s^{-2}
          Not 1 poise = 1 g/s/cm   *but* 1 poise = 1 g/s cm
                                          = 1 g s^{-1} cm^{-1}
          Not J/K/mol   *but* J/K mol = J/K^{-1} mol^{-1}.

Numbers

Numbers should be printed in *upright type*.
The *decimal* sign between digits in a number should be a point (·).
The *multiplication* sign between numbers should be a cross ($\times$), e.g., 2.3×3.4.
Division of one number by another number may be indicated in the following ways:

$$\frac{136}{273.15} \qquad 136/273.15$$

or by writing it as the product of numerator and the inverse first power of the denominator. In such cases the number under the inverse power should always be placed between brackets.

Remark:
When the solidus is used and when there is any doubt where the numerator starts or the denominator ends, brackets should be used.

To facilitate the reading of *long numbers*, the digits may be grouped in *groups of three*, but *no* comma or point should be used except for the decimal sign.

Example: 2 573·421 736.

Symbols for Chemical Elements, Nuclides and Particles

Symbols for chemical elements should be written in *roman type*. The symbol is not followed by a full stop.

Examples: Ca   C   H   He

The attached numerals specifying a *nuclide* are:

mass number $^{14}N_2$ atoms/molecule

Remark:

The atomic number may be placed as a left subscript, if desired. The right superscript position should be used, if required, for indicating a state of ionization (e.g., Ca^{2+}, PO_4^{3-}) or an excited state (e.g., $^{110}Ag^m$, He*).

Symbols for particles and quanta

It is recommended that the following notation be used:

proton	p	α-particle	α
neutron	n	pion	π
λ-particle	Λ	K-meson	K
Σ-particle	Σ	electron	e
Ξ-particle	Ξ	muon	μ
deuteron	d	neutrino	ν
triton	t	photon	γ

A nucleon (proton or neutron) is indicated by N.

It is recommended that the charge of particles be indicated by adding the superscript $+$, $-$ or 0.

Examples: π^+ π^- π^0, p^- p^+, e^+ e^-.

If in connection with the symbols p and e no charge is indicated, these symbols should refer to the positive proton and the negative electron respectively.

The tilde $\sim$ above the symbol of a particle is used to indicate the corresponding anti-particle.

Examples: $\tilde{n}$, $\tilde{\nu}$, $\tilde{p}$, $\tilde{N}$.

Quantum States

GENERAL RULES

A letter symbol indicating the quantum state of *a system* should be printed in capital roman type.

A letter symbol indicating the quantum state of *a single particle* should be printed in lower case roman type.

ATOMIC SPECTROSCOPY

The letter symbols indicating atomic quantum states are:

$L, l = 0$: S, s	$L, l = 4$: G, g	$L, l = 8$: L, l
$= 1$: P, p	$= 5$: H, h	$= 9$: M, m
$= 2$: D, d	$= 6$: I, i	$= 10$: N, n
$= 3$: F, f	$= 7$: K, k	$= 11$: O, o

A right-hand subscript indicates the total angular momentum quantum number J or j. A left-hand superscript indicates the spin multiplicity $2S + 1$.

Example: $^2P_{3/2}$—state   ($J = \frac{3}{2}$, multiplicity 2)
$p_{3/2}$—electron ($j = \frac{3}{2}$).

An atomic electron configuration is indicated symbolically by

$$(nl)^\kappa \quad (n'l')^{\kappa'} \ldots.$$

Instead of $l = 0, 1, 2, 3, \ldots$ one uses the quantum state symbol s, p, d, f, $\ldots$

Example: the atomic configuration: $(1s)^2 (2s)^2 (2p)^3$.

MOLECULAR SPECTROSCOPY

The letter symbols, indicating molecular electronic quantum states, are in the case of *linear molecules*

$$\Lambda, \lambda = 0:\ \Sigma, \sigma$$
$$= 1:\ \Pi, \pi$$
$$= 2:\ \Delta, \delta$$

and for *non-linear molecules*

$$A, a;\quad B, b;\quad E, e;\quad \text{etc.}$$

Remarks:

A left-hand superscript indicates the spin multiplicity. For molecules having a symmetry centre the parity symbol g or u, indicating respectively symmetric or anti-symmetric behaviour on inversion, is attached as a right-hand subscript. A $+$ or $-$ sign attached as a right-hand superscript indicates the symmetry as regards reflection in any plane through the symmetry axis of the molecules.

Examples: $\qquad \Sigma_g^+, \qquad \Pi_u, \qquad {}^2\Sigma, \qquad {}^3\Pi, \qquad$ etc.

The letter symbols indicating the vibrational angular momentum states in the case of *linear molecules* are

$$l = 0:\ \Sigma$$
$$= 1:\ \Pi$$
$$= 2:\ \Delta$$

NUCLEAR SPECTROSCOPY

The spin and parity assignment of a nuclear state is

$$J^\pi$$

where the parity symbol π is $+$ for even and $-$ for odd parity.

Examples: $\qquad 3^+, \quad 2^-, \quad$ etc.

A shell model configuration is indicated symbolically by

$$(nlj)^\kappa\ (n'l'j')^{\kappa'}$$

where the first bracket refers to the proton shell and the second to the neutron shell. Negative values of κ or κ' indicate holes in a completed shell. Instead of $l = 0, 1, 2, 3, \ldots$ one uses the quantum state symbol s, p, d, f, ...

Example: the nuclear configuration: $(1\,d\,\tfrac{3}{2})^3\,(1\,f\,\tfrac{7}{2})^2$.

SPECTROSCOPIC TRANSITIONS

The upper level and the lower level are indicated by ′ and ″ respectively.

Examples: $\qquad h\nu = E' - E'' \qquad \sigma = T' - T''$.

A spectroscopic transition should be indicated by writing the upper state first and the lower state second, connected by a dash in between.

Examples:
$\qquad {}^2P_{1/2} - {}^2S_{1/2} \qquad$ for an electronic transition
$\qquad (J', K') - (J'', K'') \qquad$ for a rotational transition
$\qquad v' - v'' \qquad$ for a vibrational transition.

Absorption transition and emission transition may be indicated by arrows $\leftarrow$ and $\rightarrow$ respectively.

Examples:
$\qquad {}^2P_{1/2} \rightarrow {}^2S_{1/2}$ emission from ${}^2P_{1/2}$ to ${}^2S_{1/2}$
$\qquad (J', K') \leftarrow (J'', K'')$ absorption from (J'', K'') to (J', K').

The difference Δ between two quantum numbers should be that of the upper state minus that of the lower state.

403

Example: $$\Delta J = J' - J''.$$

The indications of the branches of the rotation band should be as follows:

$$\Delta J = J' - J'' = -2: \quad \text{O branch}$$
$$= -1: \quad \text{P branch}$$
$$= 0: \quad \text{Q branch}$$
$$= +1: \quad \text{R branch}$$
$$= +2: \quad \text{S branch}$$

Nomenclature

USE OF THE WORD SPECIFIC

The word 'specific' in English names for physical quantities should be restricted to the meaning 'divided by mass'.

Examples:

specific volume	volume/mass
specific energy	energy/mass
specific heat capacity	heat capacity/mass.

ABBREVIATED NOTATION FOR NUCLEAR REACTION

The meaning of the symbolic expression indicating a nuclear reaction should be the following:

$$\text{initial nuclide} \left(\begin{matrix} \text{incoming particle(s)} \\ \text{or quanta} \end{matrix} , \begin{matrix} \text{outgoing particle(s)} \\ \text{or quanta} \end{matrix} \right) \text{final nuclide}$$

Examples:

$$^{14}\text{N}(\alpha, p)^{17}\text{O} \qquad ^{59}\text{Co}(n, \gamma)^{60}\text{Co}$$
$$^{23}\text{Na}(\gamma, 3n)^{20}\text{Na} \qquad ^{31}\text{P}(\gamma, pn)^{29}\text{Si}.$$

NUCLIDE

A species of atoms, identical as regards atomic number (proton number) and mass number (nucleon number) should be indicated by the word *nuclide*, not by the word isotope.

Different nuclides having the same atomic number should be indicated as *isotopes* or *isotopic nuclides*.

Different nuclides having the same mass number should be indicated as *isobars* or *isobaric nuclides*.

International Symbols for Units

INTERNATIONAL SYSTEMS OF UNITS

The MKSA SYSTEM is a coherent system of units for mechanics, electricity and magnetism, based on *four basic units* for the four basic quantities length, mass, time and electric current intensity:

metre	m
kilogramme	kg
second	s
ampere	A

The following units of the MKSA system have special names and symbols, which have been approved by the Conférence Générale des Poids et Mesures:

Length	l, b, h	metre	m
Time	t	second	s
Mass	m	kilogramme	kg
Frequency	v, f	hertz (= c/s)	Hz
Force	F	newton ($= \text{kg m s}^{-2}$)	N
Energy	E	joule ($= \text{kg m}^2 \text{ s}^{-2}$)	J
Power	P	watt ($= \text{J s}^{-1}$)	W
Current	I	ampere	A
Electric Charge	Q	coulomb ($= \text{A s}$)	C
Electric Potential Difference	V	volt ($= \text{W A}^{-1}$)	V

Electric Capacitance	C	farad ($= C V^{-1}$)	F
Electric Resistance	R	ohm ($= V A^{-1}$)	Ω
Inductance	L	henry ($= V s A^{-1}$)	H
Magnetic Flux	Φ	weber ($= V s$)	Wb
Magnetic Flux Density	B	tesla ($= Wb\ m^{-2}$)	T

The KELVIN. In the field of *thermodynamics* one introduces an additional basic unit, corresponding to the basic quantity:
thermodynamic temperature, the unit being the kelvin, symbol: K.

When the *customary temperature* is used, defined by $t = T - T_0$, where $T_0 = 273 \cdot 15$ K, this is usually expressed in degree Celsius, symbol: °C. For *temperature interval* the name degree, symbol deg, is often used, the indications 'Kelvin' or 'Celsius', indicating the zero point of the temperature scale used, being irrelevant in this case.

The CANDELA. In the field of *photometry* one introduces an additional basic unit, corresponding to the basic quantity *luminous intensity*, this unit being the candela, symbol cd.

Special names and symbols for units in this field are:

I	candela	cd
Φ	lumen	lm
E	lux ($= lm\ m^{-2}$)	lx

The INTERNATIONAL SYSTEM OF UNITS (SI). For the coherent system based on the six basic units:

metre	m	ampere	A
kilogramme	kg	kelvin	K
second	s	candela	cd

the name *International System of Units* has been recommended by the Conférence Générale des Poids et Mesures in 1960. The units of this system are called *SI units*.

The MOLE. In the field of *chemical and molecular physics*, in addition to the basic quantities defined above having units defined by the Conférence Générale des Poids et Mesures, *amount of substance* is also treated as a basic quantity. The recommended basic unit is the mole, symbol mol. The mole is defined as the amount of substance of a system, which contains the same number of molecules (or ions, or atoms, or electrons, as the case may be), as there are atoms in exactly 12 gramme of the pure carbon nuclide ^{12}C.

OTHER UNITS

	ångström (10^{-10} m)	Å
σ	barn ($= 10^{-28}\ m^2$)	b
V	litre ($10^{-3}\ m^3$)	l
$t, \tau, T_{1/2}$	minute (60 s)	min
$t, \tau, T_{1/2}$	hour (3600 s)	h
$t, \tau, T_{1/2}$	day	d
$t, \tau, T_{1/2}$	year	a
p	atmosphere	atm
E	kilowatt-hour ($3 \cdot 6 \times 10^6$ J)	kWh
Q	calorie	cal
E, Q	electron-volt	eV
m	tonne ($= 1000$ kg)	t
m_a	(unified) atomic mass unit	u
p	bar ($= 10^5$ N m^{-2})	bar
A	curie	Ci

Remark:

The (unified) atomic mass unit is defined as $^1/_{12}$th of the mass of an atom of the ^{12}C nuclide.

(End of IUPAP Report)

Physical Constants: Values in SI Units

Quantity		International System
speed of light	c	$2.997\ 925\ \times 10^8$ m s^{-1}
Boltzmann constant	k	$1.380\ 54\ \times 10^{-23}$ J K^{-1}
mass of hydrogen atom	m_H	$1.673\ 43\ \times 10^{-27}$ kg
proton mass	m_p	$1.672\ 52\ \times 10^{-27}$ kg
neutron mass	m_n	$1.674\ 82\ \times 10^{-27}$ kg
electron mass	m_e	$9.109\ 1\ \times 10^{-31}$ kg
	m_p/m_e	$1.836\ 10\ \times 10^3$
charge of positron	e	$1.602\ 10\ \times 10^{-19}$ C
charge to mass ratio	e/m	$1.758\ 796\ \times 10^{11}$ C kg^{-1}
Zeeman splitting constant	$e/4\pi mc$	$4.668\ 58\ \times 10$ m^{-1} T^{-1}
Planck constant	h	$6.625\ 6\ \times 10^{-34}$ J s
	$h/2\pi = \hbar$	$1.054\ 50\ \times 10^{-34}$ J s
	h/e	$4.135\ 56\ \times 10^{-15}$ J s C^{-1}
	h/k	$4.799\ 3\ \times 10^{-11}$ s K
1st radiation constant	$c_1 = 2\pi hc^2$	$3.741\ 5\ \times 10^{-16}$ W m^2
2nd radiation constant	$c_2 = hc/k$	$1.438\ 79\ \times 10^{-2}$ m K
Wien's radiation law		
	$\lambda_{max}T = c_2/4.965\ 114$	$2.897\ 8\ \times 10^{-3}$ m K
Stefan-Boltzmann constant	σ	$5.669\ 7\ \times 10^{-8}$ Wm^{-2} K^{-4}
fine structure constant	α	$7.297\ 20\ \times 10^{-3}$
	α^{-1}	$1.370\ 388\ \times 10^2$
	α^2	$5.324\ 92\ \times 10^{-5}$
Bohr radius	a_0	$5.291\ 67\ \times 10^{-11}$ m
Rydberg constant	R_∞	$1.097\ 373\ 1 \times 10^7$ m^{-1}
	R_H	$1.096\ 775\ 8 \times 10^7$ m^{-1}
	$R_\infty c$	$3.289\ 842\ \times 10^{15}$ s^{-1}
	$R_\infty hc$	$2.179\ 72\ \times 10^{-18}$ J
Bohr magneton	μ_B	$9.273\ 2\ \times 10^{-24}$ J T^{-1}
magnetic moment of electron	μ_e	$9.284\ 0\ \times 10^{-24}$ J T^{-1}
nuclear magneton	μ_N	$5.050\ 5\ \times 10^{-27}$ J T^{-1}
magnetic moment of proton	μ_p	$1.410\ 49\ \times 10^{-26}$ J T^{-1}
gyromagnetic ratio of proton	γ_p	$2.675\ 19\ \times 10^8$ s^{-1} T^{-1}
Compton wavelengths:		
of electron	$\lambda_C = h/m_e$	$2.426\ 21\ \times 10^{-12}$ m
	$\lambda_C/2\pi$	$3.861\ 44\ \times 10^{-13}$ m
of proton	$\lambda_{Cp} = h/m_p c$	$1.321\ 40\ \times 10^{-15}$ m
	$\lambda_{Cp}/2\pi$	$2.103\ 07\ \times 10^{-16}$ m
of neutron	$\lambda_{Cn} = h/m_n c$	$1.319\ 58\ \times 10^{-15}$ m
	$\lambda_{Cn}/2\pi$	$2.100\ 18\ \times 10^{-16}$ m
Avogadro constant	N_A	$6.022\ 52\ \times 10^{23}$ mol^{-1}
molar volume of ideal gas at s.t.p.		
(1 atm and 0°C)	V_m	$2.241\ 36\ \times 10^{-2}$ m^3 mol^{-1}
molar gas constant	R	$8.314\ 3\ \times$ J mol^{-1} K^{-1}
Faraday constant	$F = N_A e$	$9.648\ 70\ \times 10^4$ C mol^{-1}

N.B.—In the *unified scale* (see entry in main text) u is the suggested symbol for the unit which replaces the atomic mass unit (amu) of the physical scale. The conversion factor is given by $1u = 1.0003179$ amu. To derive the proton mass (M) in g it is to be noted that $N_A u = 1$g, where N_A is expressed as above on the unified scale.

Physical Concepts in Rationalized MKS Units

Concept	Symbol	Name of Unit	Abbreviation of Unit Name	Definition or Defining Equation	Explanations; Equivalent Units; Alternative Definitions; etc.
Length	l	metre	m	1 m = 1 650 763·73 wavelengths of radiation $(2p_{10} -5d_5)$ of Kr 86.	
Mass	m	kilogramme	kg	International Prototype Kilogramme. 1 kg pure water at 4°C and 760 mm pressure occupies 1 litre.	
Time	t	second	s	Mean solar second.	
The above are internationally agreed basic units.					
Area	A, a	square metre	m^2	$a = l^2$	(All concepts except the electrical concepts are derived from the three above. All concepts are capable of being written in terms of the first three; but electrical quantities need an additional fundamental concept.)
Volume	V, v	cubic metre	m^3	$V = l^3$	
Velocity	v, u	metre/second	$m\ s^{-1}$	$v = dl/dt$	
Acceleration	a	metre/second²	$m\ s^{-2}$	$a = d^2l/dt^2$	
Density	ρ	kilogramme/metre³	$kg\ m^{-3}$	$\rho = m/V$	
Mass rate of flow		kilogramme/sec	$kg\ s^{-1}$	dm/dt	
Volume rate of flow		cubic metre/sec	$m^3\ s^{-1}$	dV/db	
Moment of inertia	I	kilogramme metre²	$kg\ m^2$	$I = Mk^2$	
Momentum	p	kilogramme metre/sec	$kg\ m\ s^{-1}$	$p = mv$	
Angular momentum	I, ω	kilogramme metre²/sec	$kg\ m^2\ s^{-1}$	dI/dt	
Kinetic energy	$T, (W)$	kilogramme metre²/sec²	$kg\ m^2\ s^{-2}$	$T = \tfrac{1}{2}mv^2$	Newton metre (N m)
Force	F	Newton	N	$F = ma$	$kg\ m\ s^{-2}$
Torque (Moment of force)	$T, (M)$	Newton metre	N m	$T = Fl$	$T = P/2\pi n.$ n in rev/s
Potential energy	$V, (w)$	Newton metre	N m	$V = \int Fdl$	$kg\ m^2\ s^{-2}$
Work (Energy, Heat)	$W, (U)$	Joule	J	$W = \int Fdl$	N m (1 J = 1 N m by definition)
Heat (Enthalpy)	$Q, (H)$	Joule	J	$H = U + pv$	Definition for a fluid. U = Internal energy
Power	P	Watt	W	$P = dW/dt$	1 W = 1 J s⁻¹ by definition
Pressure (Stress)	$p\ (\sigma, f)$	Newton/metre²	$N\ m^{-2}$	$p = F/A$	Usually pressure in fluid, stress in solids.
Surface tension	$\gamma\ (\sigma)$	Newton/metre²	$N\ m^{-2}$	$\gamma = F/l$	Free surface energy.
Viscosity, dynamic	η, μ	Poise	P	$\dfrac{P}{A} = \eta dv/dl$	$10^{-1}\ N\ s\ m^{-2}$. Defined in CGS units
Viscosity, kinematic	ν	Stokes	S	$\nu = \eta/\rho$	$10^{-1}\ N\ s\ m\ kg^{-1}$. Defined in CGS units
Action			J s	$\int Wdt$	
Temperature	θ, T	degree C, degree K	°C, °K	$T°K = (\theta + 273\cdot16)°C$	International Temperature Scale
Velocity of light	c	metre/second	$m\ s^{-1}$	Fundamental, measured, constant	

Physical Concepts in Rationalized MKS Units

Concept	Symbol	Name of Unit	Abbreviation of Unit Name	Definition or Defining Equation	Explanations; Equivalent Units; Alternative Definitions; etc.
Permeability of vacuum	μ_0	Henry/metre	H m^{-1}	$\mu_0 = 4\pi \times 10^{-7}$ H/m	Defined value to give coherent rationalized electrical units.
Permittivity of vacuum	ϵ_0	Farad/metre	F m^{-1}	$\epsilon_0 = 1/\mu_0 c^2$	Derived in Maxwell's theory of EM radiation.
Electric charge	Q	Coulomb	C	$F = (Q_1 Q_2)/(4\pi\epsilon_0 r^2)$	A s Coulomb's Law. Also $Q = \int I\,dt$
Electric current	I	Ampere	A	$F = (2\mu_0 I_1 I_2)/(4\pi r^2)$	Also $I = dQ/dt$
Electric potential (Potential difference)	V	Volt	V	$V_r = \int_a^\infty F\,dl \quad (V_{ab} = \int_a F\,dl)$	N m C^{-1} or J C^{-1}
Electric field-strength (Electric force)	E	Volt/metre	V m^{-1}	$E = dV/dl$	N C^{-1} E = Force on unit point charge J/A
Electric resistance	R	Ohm	Ω	$R = V/I$	
Electric conductance	S	Siemens	Ω^{-1}	$\dfrac{1}{R} = \dfrac{I}{V}$	
Electric flux	Ψ	Coulomb	$\Psi = Q$		
Electric flux density (Displacement)	D	Coulomb/metre^2	C m^{-2}	$D = d\Psi/dA$	
Permittivity	ϵ	Farad/metre	F m^{-1}	$\epsilon = D/E$	
Relative permittivity	ϵ_r			$\epsilon_r = \epsilon/\epsilon_0$	A numeric.
Magnetic field-strength (Magnetic force)	H	Amp. turn/metre	AT m^{-1}	$dH = I\,dl \sin\theta/4\pi r^2$	The turn is a numeric not a unit.
Magnetic flux	Φ	Weber	Wb	$\Phi = -\int e\,dt$	V s Faraday-Lenz Law.
Magnetic flux density	B	Weber/metre^2, Tesla	Wb/m^2, T	$B = d\Phi/dA$	V s m^{-1}
Permeability	μ	Henry/metre	H/m	$\mu = B/H$	
Relative permeability	μ_r			$\mu_r = \mu/\mu_0$	A numeric
Coefficient of mutual induction	M	Henry	H	$e_2 = -M\,dI_1/dt$	Wb/A
Coefficient of self-induction	L	Henry	H	$e = -L\,dI/dt$	Wb/A
Capacitance	C	Farad	F	$C = Q/V$	CV^{-1}
Reactance	X	Equivalent ohm	Ω	$X = \omega L$ or $\dfrac{1}{\omega C}$	Sinusoidal a.c. Also $\omega = 2\pi \times$ frequency
Impedance	Z	Equivalent ohm	Ω	$Z = \sqrt{R^2 + X^2}$	

Names in parentheses with upper case initial, e.g., (Energy), are alternatives. Words in parentheses with l.c. initial, e.g., (dynamic), are adjectival. Symbols and names are in accordance with BS 1991, BS 350, and BS 1637.
(Reproduced from Chambers's *Six-Figure Mathematical Tables*.)

408

Standard Values and Equivalents

Gravitational constant G: $= (6 \cdot 668 \pm 0 \cdot 005) \times 10^{-11}$ N m²kg^{-2}

Standard acceleration of gravity: $g = 32 \cdot 174$ 0 ft s^{-2} $= 9 \cdot 806$ 65 m s^{-2}

Acceleration of gravity at Greenwich: $g = 9 \cdot 818$ 83 m s^{-2}

Standard atmosphere ($\equiv 760$ mm Hg to 1 in 7×10^{6}): 1 atm $= 101 \cdot 325$ k N m^{-2}

Solar year: 365 d 5 h 48 m 45·5 s

Sidereal year contains: 365·256 360 42 mean solar days

Mean solar second: 1/86400 mean solar day

International Temperature Scale of 1948
(all at a pressure of one standard atmosphere):

b.p. Oxygen	$-182 \cdot 970°$C	b.p. Sulphur	$444 \cdot 600°$C
m.p. Ice	$0°$C	f.p. Silver	$960 \cdot 8°$C
b.p. Water	$100°$C	f.p. Gold	$1063 \cdot 0°$C

m.p. Ice on Kelvin Scale: $0°$C $= 273 \cdot 15°$K

Triple point of water: $0 \cdot 0100°$C

International Table calorie: 1 cal$_{IT}$ $= 4 \cdot 186$ 8 J

15°C calorie: 1 cal$_{15}$ ' =. $4 \cdot 185$ 5 J

British thermal unit: 1 B.t.u. $= 778 \cdot 169$ ft lbf $= 1055 \cdot 06$ J

Therm: 1 therm $= 10^{5}$ B.t.u.

Horsepower: 1 hp $= 550$ ft lbf s^{-1} $= 745 \cdot 700$ W

CGS units: Force 1 dyn $= 10^{-5}$ N; Work 1 erg $= 10^{-7}$ J

Electron-volt: 1 eV $= 1 \cdot 6021 \times 10^{-19}$ J

e.m.f. of Normal Weston cadmium cell at 20°C: $E = (1 \cdot 018$ 74 $\pm 0 \cdot 000$ 003$)$ V

Velocity of sound at sea level at 0°C: $V = 1088$ ft s^{-1}
$$= 331 \cdot 7 \text{ m s}^{-1}$$

Standard concert pitch: A $= 440$ Hz.

Scientific pitch: C $= 256$ Hz.

CONVERSION TABLES
Conversion Factors for British and Metric Mechanical Units

Metric to British
British to Metric

Length

Metric to British	British to Metric
1 cm = 0·393 701 in	1 in = **2·54** cm
1 m = 1·093 61 yd	1 yd = **0·914 4** m
1 km = 0·621 371 mile	1 mile = 1·609 34 km

Area

Metric to British	British to Metric
1 cm^2 = 0·155 000 in^2	1 in^2 = **6·451 6** cm^2
1 m^2 = 1·195 99 yd^2	1 yd^2 = 0·836 127 m^2
1 km^2 = 0·386 103 mile^2	1 mile^2 = 2·589 99 km^2
1 hectare = 2·471 05 acre	1 acre = 4046·86 m^2

Volume

Metric to British	British to Metric
1 cm^3 = 0·061 023 7 in^3	1 in^3 = 16·387 1 cm^3
1 m^3 = 1·307 95 yd^3	1 yd^3 = 0·764 555 m^3
1 litre* = 1·759 80 U.K. pt	1 U.K.pt = 0·568 litre

Mass

Metric to British	British to Metric
1 kg = 2·204 62 lb	1 lb = **0·453 592 37** kg
1 kg = 0·984 205 × 10^{-3} ton (U.K.)	1 ton = 1016·05 kg
1 g = 0·035 247 0 oz	1 oz = 28·349 5 g

Force

Metric to British	British to Metric
1 kgf = 2·204 62 lbf	1 lbf = 0·453 592 kgf
= 70·931 6 pdl	= 4·448 22 N
1 kgf = **9·806 65** N	1 pdl = 0·014 098 1 kgf
1 N = 0·224 809 lbf	= 0·138 255 N
= 7·233 01 pdl	1 lbf = 32·174 4 pdl

Work

Metric to British	British to Metric
1 J = 0·737 562 ft lbf	1 ft lbf = 1·355 82 N/m
= 23·730 4 ft pdl	1 ft pdl = 0·042 140 1 N/m

Technical Units of Mass

1 Metric Technical Unit of Mass = **9·806 65** kg
1 British Technical Unit of Mass (slug) = 32·174 0 lb

1 micron (μ)	= 10^{-6} m	1 parsec	= 3·263 light-y
1 Ångström (Å)	= 10^{-10} m	1 barn	= 10^{-28} m^2
1 micro-inch (μ in)	= 10^{-6} in	1 torr	= 1/760 standard atmosphere
1 milli-inch = 1 mil	= 10^{-3} in	1 bar (b)	= 10^6 dyn cm^{-2}
1 light-year	= 5·880 × 10^{12} miles	1 millibar (mb)	= 0·7501 torr

Exact values in **bold** type

* 1 litre = 1·000 028 dm^3, but the word *litre* and symbol l is conveniently taken as a special name for the cubic decimetre, though not for precise measurements.

Illumination and Luminance Conversion Table

There is some multiplicity of photometric units—partly because of the still widespread use of obsolescent units. The preferred units, defined in BSI 233 1953, are the *candela, lux, lumen* and *nit*. Other metric units are the *stilb* and the *phot* (CGS units).

Use of non-metric units should be deprecated. The conversion factors between the various units for *illumination* and *luminance* are given below:

ILLUMINATION

		lux	phot	foot-candle
1 lux (1m/m²)	=	1	10^{-4}	$9{\cdot}29 \times 10^{-2}$
1 phot (1m/cm²)	=	10^4	1	929
1 foot-candle (1m/ft²)	=	$10{\cdot}76$	$10{\cdot}76 \times 10^{-4}$	1

LUMINANCE

		nit	stilb	cd/ft²	apostilb	lambert	foot-lambert
1 nit (cd/m²)	=	1	10^{-4}	$9{\cdot}29 \times 10^{-2}$	π	$\pi \times 10^{-4}$	$0{\cdot}292$
1 stilb (cd/cm²)	=	10^4	1	929	$\pi \times 10^{-4}$	π	2920
1 cd/ft²	=	$10{\cdot}76$	$1{\cdot}076 \times 10^{-3}$	1	$33{\cdot}8$	$3{\cdot}38 \times 10^{-3}$	π
1 apostilb (1m/m²)	=	$1/\pi$	$1/(\pi \times 10^{-4})$	$2{\cdot}96 \times 10^{-2}$	1	10^{-4}	$9{\cdot}29 \times 10^{-2}$
1 lambert (1m/cm²)	=	$1/(\pi \times 10^{-4})$	$1/\pi$	296	10^4	1	929
1 foot-lambert or 'equivalent foot-candle' (1m/ft²)	=	$3{\cdot}43$	$3{\cdot}43 \times 10^{-4}$	$1/\pi$	$10{\cdot}76$	$1{\cdot}076 \times 10^{-3}$	1

Conversion Factors required in Nuclear Reactor Physics

Multiply	*By*	*To obtain*
Atomic mass units	$1{\cdot}659\ 9 \times 10^{-24}$	grams
Atomic mass units	$931{\cdot}2$	MeV
Atomic weight (Chemical scale $_{Nat}O = 16$)	$1{\cdot}000\ 272$	atomic weight (Physical scale $_{16}O = 16$)
Atomic weight (Unified scale $_{12}C = 12$)	$1{\cdot}000\ 32$	atomic weight (Physical scale $_{16}O = 16$)
Barns	1×10^{-24}	square centimetre
B.t.u.	$1{\cdot}28 \times 10^{-8}$	gm ^{235}U fissioned*
B.t.u.	$2{\cdot}930 \times 10^{-4}$	kilowatt-hours
B.t.u.	$3{\cdot}29 \times 10^{13}$	fissions
B.t.u./h	$0{\cdot}292\ 93$	Watts
Curies	$3{\cdot}70 \times 10^{10}$	disintegrations/sec
Electron mass	$0{\cdot}510\ 98$	MeV
Energy equivalence of electron mass	$0{\cdot}510\ 98$	MeV
Electron-volts	$1{\cdot}782\ 5 \times 10^{-33}$	grams
Electron-volts	$1{\cdot}602\ 1 \times 10^{-12}$	ergs
Fission of 1gm ^{235}U	1	$\sim$ megawatt-days
No. of fissions*	$8{\cdot}905\ 8 \times 10^{-18}$	kilowatt-hours
No. of fissions*	$3{\cdot}204\ 1 \times 10^{-4}$	ergs
Grams	$8{\cdot}987\ 6 \times 10^{20}$	ergs
Half-life	$1{\cdot}443$	mean life [= (decay constant) $^{-1}$]
Kilowatt-hour	$2{\cdot}786\ 5 \times 10^{17}$	^{235}U fission neutrons*
Kilowatt-hour	$2{\cdot}247 \times 10^{19}$	MeV
Kilowatts per kilogram ^{235}U	$2{\cdot}43 \times 10^{10}$	average thermal neutron flux in fuel*†
Megawatt-days per ton U	$1{\cdot}174 \times 10^{-4}$	% U atoms fissioned‡
Megawatts per ton U	$2{\cdot}68 \times 10^{10}/E$	average thermal neutron flux in fuel*†
MeV	$3{\cdot}82 \times 10^{-14}$	gm-calories
MeV	$1{\cdot}517 \times 10^{-16}$	B.t.u.
Röntgens	$2{\cdot}58 \times 10^{-4}$	Coulombs of ion charge/kg of standard air
Röntgens	$5{\cdot}24 \times 10^7$	MeV absorbed/gm air
REP (*röntgen equivalent physical*)	$93{\cdot}1$	ergs absorbed/gm soft tissue
Rad	100	ergs absorbed/gm (any medium)
Watts	$3{\cdot}121 \times 10^{10}$	fissions/s

* denotes at 200 MeV/fission.
† $\bar{\sigma}$ (fission = 500 barn).
‡ at 200 MeV/fission in ^{235}U-^{238}U mixture having low ^{235}U content.
E enrichment in ^{235}U/total mass, both in grams, and assuming absence of any other fissionable isotope.

Unit Conversion Table

| Quantity | Practical Units | Absolute Systems of Units | | | | | |
| | Unit | Rationalized MKSA Units | | CGS Electromagnetic Units | | CGS Electrostatic Units | |
		Unit	Multiple	Unit	Multiple	Unit	Multiple
Energy	Joule	Joule	$\times 1$	Erg	(G) $\times 10^{-7}$	Erg	(G) $\times 10^{-7}$
Power	Watt	Watt	$\times 1$	Erg/s	(G) $\times 10^{-7}$	Erg/s	(G) $\times 10^{-7}$
Electric Charge	Coulomb	Coulomb	$\times 1$	Abcoulomb	$\times 10$	Statcoulomb	$\div 3 \times 10^{9}$
Polarization	Coulomb/cm²	Coulomb/m²	$\times 10^{-4}$	Abcoulomb/cm²	$\times 10$	Statcoulomb/cm²	$\div 3 \times 10^{9}$
Electric Potential	Volt	Volt	$\times 1$	Abvolt	$\times 10^{-8}$	Statvolt	$\times 300$
Electric Field	Volt/cm	Volt/m	$\times 10^{-2}$	Abvolt/cm	$\times 10^{-8}$	Statvolt/cm	$\times 300$
Permittivity		Farad/m	$\times 4\pi \times 10^{-9}$		$\times 10^{2}$		$\times 9 \times 10^{18}$
Capacitance	Farad	Farad	$\times 1$	Abfarad	$\times 10^{9}$	Cm or Statfarad	$\div 9 \times 10^{11}$
Displacement		Coulomb/m²	$\times 4\pi \times 10^{-4}$		$\times 10$		$\div 3 \times 10^{9}$
Electric Flux		Coulomb	$\times 4\pi$		$\times 10$		$\div 3 \times 10^{9}$
Current	Ampere	Ampere	$\times 1$	Abampere	$\times 10$	Statampere	$\times 3 \times 10^{9}$
Resistance	Ohm	Ohm	$\times 1$	Abohm	$\times 10^{-9}$	Statohm	$\times 9 \times 10^{11}$
Resistivity	Ohm/cm	Ohm/m	$\times 10^{2}$	Abohm/cm	$\times 10^{-9}$	Statohm/cm	$\times 9 \times 10^{11}$
Inductance	Henry	Henry	$\times 1$	Abhenry	$\times 10^{-9}$	Stathenry	$\times 9 \times 10^{11}$
Magnetic Pole		Weber	$\div 4\pi \times 10^{-8}$		(G) $\times 1$		
Magnetization		Weber/m²	$\div 4\pi \times 10^{-4}$		(G) $\times 1$		
Magnetic Field	Oersted	Ampere turn/m	$\times 4\pi \times 10^{-3}$	Oersted	(G) $\times 1$		
Permeability	Gauss/Oersted	Henry/m	$\div 4\pi \times 10^{-7}$	Gauss/Oersted	(G) $\times 1$		
Magnetic Induction	Gauss	Weber/m² (or Tesla)	$\times 10^{4}$	Gauss	(G) $\times 1$		
Magnetic Flux	Maxwell	Weber	$\times 10^{8}$	Maxwell	(G) $\times 1$		
Magnetic Potential	Gilbert	Ampere turn	$\times 4\pi \times 10^{-1}$	Gilbert	(G) $\times 1$		
Reluctance	Gilbert/Maxwell	Ampere turn/Weber	$\times 4\pi \times 10^{-9}$	Gilbert/Maxwell	(G) $\times 1$		
Electric Susceptibility	}Dimensionless		$\div 4\pi$		$\times 1$		
Magnetic Susceptibility	}Dimensionless		$\div 4\pi$		$\times 1$		

412

Note. The multiple column gives the ratio of the size of the absolute unit to that of the corresponding practical unit. To convert the numerical value of a quantity expressed in practical units to absolute units it must be multiplied by the reciprocal of this factor.

Where the practical unit is not the same as the corresponding MKSA absolute unit the ratio of the size of a CGS absolute unit to the MKSA absolute unit can be obtained by *dividing* the multiple given in the CGS column by that given in the MKSA column.

The conversion factors between the two CGS systems of units have been calculated by taking the velocity of electromagnetic waves in free space as 3×10^{10} cm/sec.

The symbol G indicates a Gaussian unit.

No generally accepted names exist for some of the units in this table.

SUPPLEMENT

absolute code. Computer code using absolute addresses and written in machine language.

absolute-value computer. One in which all variables are processed in terms of their absolute value; cf. *incremental computer.*

acceleration space. That in which an accelerating field acts on a beam of charged particles.

accumulation factor. That by which the concentration of a radioactive nuclide in air or water may be increased as a result of selective absorption by a living organism. AFs of up to 10^6 occur with some forms of marine life.

acoustic amplification. That occurring in a CdS piezoelectric crystal when the drift velocity of the electrons under the applied electric field is greater than the speed of sound in the crystal.

acoustic amplifier. One amplifying mechanical vibrations.

acoustic lens. A system of slats or disks to spread or converge sound waves. In liquids, optical forms of lenses in polystyrene are used, but due to different velocities, they are concave for sound-converging systems and vice versa.

acoustic resistance unit. A device to control the Q-factor of a loudspeaker enclosure system. It consists of sound-absorbing material placed over the port in a reflex loudspeaker enclosure.

acoustic suspension. Sealed-cabinet system of loudspeakers in which the main restoring force of the diaphragm is provided by the acoustic stiffness of the enclosed air.

actino-uranium. Term sometimes used for uranium-235 since the latter is parent of the radioactive family of actinium.

action potential. That produced in a nerve by a stimulus. It is a voltage pulse arising from sodium ions entering the axon and changing its potential from -70 mV to $+40$ mV. With a continuing stimulus the pulses are repeated at up to several hundred times a second, leading in motor nerves to continuous muscular response (tetanus).

addend. Computing term for a number which is to be added to a second number, the *augend,* to produce a sum.

adiabatic light guide. One in which total internal reflection is applied to the coupling of a scintillator to a photomultiplier so that no energy losses occur.

Adler tube. A cyclotron-wave electron-beam parametric amplifier used for very low noise microwave amplification. Unlike other parametric amplifiers, the Adler tube can be operated over a range of signal frequencies with a fixed pump frequency.

air position indicator. Navigational system

indicating aircraft positions with reference to the air mass in which it is moving. Cf. *ground position indicator.*

alkali metals. Those belonging to Group I of the periodic table, viz., lithium, sodium, potassium, rubidium and caesium. The oxides of these metals are used as sources of thermionic emission. See *oxide-coated cathode* (in main text).

alkaline earth metals. Those belonging to Group II of the periodic table, viz., beryllium, magnesium, calcium, strontium, barium, and radium. The oxides of these metals are used as sources of thermionic emission. See *oxide-coated cathode* (in main text).

altimeter. Instrument recording the altitude of an aircraft. Traditionally these are pressure-operated, but electronic (earth-capacitance) altimeters have also been developed.

aluminizing. Vaporizing of aluminium for its deposition on glass to give a mirror which is highly reflective, particularly in the ultra-violet region. Also used as coating on phosphor screens of picture tubes.

ammonia clock. An accurate clock which depends for its action on the frequency of vibration of the nitrogen atom in ammonia.

anechoic tank. Low acoustic reverberation water tank with lining of suitable impedance.

anode breakdown voltage. That required to trigger a discharge in a cold-cathode glow tube when the starter gap (if any) is not conducting. It is measured with any grids or other electrodes earthed to cathode.

anode shield. Electrode used in high-power gas tubes to shield the anode from damage by ion bombardment.

Applegate diagram. Presentation of the bunching and debunching of an electron beam in a velocity-modulation tube, e.g., a klystron.

arc discharge. That through a heavily-ionized gas (plasma), characterized by high-current density and low residual field.

arc drop. The residual voltage drop after an arc discharge has struck.

arch. A system of interlinked computers in which each computer is controlled by the one preceding it. (*Ar*ticulated *c*omputing *h*ierarchy.)

argon laser. One using singly-ionized argon. It gives strong emission at 4880Å, 5145Å and 4965Å.

askania. Combination of telescope, theodolite and ciné-camera for recording track of missiles, together with data, on film; timing marks are recorded by a small discharge lamp.

assumed decimal point. In fixed-point notation, the decimal or radix point assumed to be implicitly located at some predetermined position and not allocated a storage location; cf. *actual decimal point* (in main text).

astigmatism. Linear defocus of image or trace because of asymmetry in the focusing elements.

audio-frequency peak limiter. Device for cutting-off, above a certain maximum, the amplitudes in an audio-frequency system.

augend. See **addend.**

Autocomm. Business computer system which is designed for commercial data-processing problems.

autocontrol. Self-control of a mechanism involving its own radio transmission.

autoflare. An aircraft automatic landing system which operates on the critical 'flare-out' stage of the landing, i.e., from an altitude of 50 feet to touchdown.

autolector. A machine which reads hand-written and computer-printed forms into a computer store at high speeds.

automatic computer. One operated under the control of a monitor.

automatic dictionary. The stored word for word, and phrase for phrase, equivalents used in language translation by computer.

automatic focusing. Electrostatic focusing in a TV tube in which the anode is internally connected via a resistor to the cathode.

automatic interruption and time-sharing. A large computer controls many peripheral units operating at their own speed with maximum efficiency. These all interrupt the main sequence of processing in some predetermined way when one task is finished and are reactivated by the central computer when a new one is to be taken up.

auxiliary routine. One prepared to check the operation of a computing system or to help in debugging other routines.

background processing. Routine work carried out by a computer only during periods when higher priority requirements do not arise.

backward-wave amplification. See **slow-wave structure.**

balance coil. Centre-tapped coil used to make connection to a balanced line.

band expansion factor. Ratio of bandwidth to highest modulation frequency for an FM transmission. Equal to twice the deviation ratio.

bank winding. Method of coil winding developed to reduce self-capacitance.

Bankpac. Specialized programme for banking industry.

bantam. Midget vacuum valve.

bar. (1) Unit of pressure equal to 10^6 dyne/cm^2. (2) Sometimes used (deprecated) as abb. for *barye* (in main text).

bark. See under **critical band.**

barrage reception. One in which interference of radio signals from any particular direction is minimized by selecting the appropriate directional aerial to give maximum signal/interference ratio.

barrier-layer rectifier. One using the non-symmetrical properties of a semiconductor barrier layer.

baseband. The frequency band occupied by the signal in modulation.

batch processing. Data reduction which incorporates coding and collection into groups, prior to other processing.

bead. An electrical insulator in the form of a bead, used, e.g., to insulate the inner wire of a coaxial line or the leads of a thermocouple. See **ferrite-.**

bead thermistor. One in which two wire leads are inserted into a small beadlike piece of semiconducting material.

beam alignment. That of the electron beam perpendicular to the mosaic in a TV camera tube.

beam-coupling coefficient. The ratio of the a.c. signal current produced to the d.c. beam current in beam coupling.

bidirectional waveform. One which shows reversal of polarity, e.g., a bidirectional pulse generator produces both positive and negative pulses.

bifurcated contact. One which is forked or pronged.

Binac. The first U.S. high-speed digital computer which operated serially on binary numbers. (*Bin*ary *a*utomatic *c*omputer.)

binary point. The radix point in calculations carried out on the *binary scale.*

bionics. Various phenomena and functions which characterize biological systems, with particular reference to electronic systems.

biquinary. A form of binary-coded decimal notation in which the first digit indicates 0 or 5 and the next three indicate 0 to 4 (represented by conventional binary notation). These must be added to cover the range 0 to 9.

birefringent filter. One based on the polarization of light which enables a narrow spectral band of 1Å or less to be isolated, i.e., effectively a *monochromator.* Used for photographing solar flares, etc.

bistatic radar. Radar system in which the transmitter and receiver are situated at a distance apart, each using its own aerial. Cf. *monostatic radar.*

black after white. Defect produced by ringing in the signals circuit of a TV receiver leading to white vertical edges being bounded by a dark line. In moderation this improves the apparent horizontal resolution of the receiver.

black compression. Electronic reduction of contrast at low intensity levels in TV picture reproduction. See also **white compression.**

black screen. A TV picture tube in which the screen is covered with a light-absorbing neutral filter, or the phosphor is a dark grey colour. This increases contrast by reduction of ambient illumination reflected from the face of the tube.

blacker than black. The range of TV or facsimile signal levels in which synchronizing and control signals are transmitted. These cut off the electron beam and are therefore not visible.

block diagram. A simplified diagram showing the basic units which form a complex circuit or control system as a series of boxes. The lines linking these indicate the mode of operation of the system more clearly than a full circuit.

Bohm diffusion formula. Empirical relationship for diffusion coefficient of plasma in ion source.

Bragg-Gray cavity. Gas-filled enclosure in a medium which can be used to study the radiation dose associated with the medium as long as certain requirements laid down by Bragg and Gray are satisfied.

brain wave. The waveform of the *brain voltage* which characterizes an individual and his health condition.

break point. One in a computer programme where a conditional interruption may be made for a visual check or print-out.

Brewster windows. Windows attached in certain designs of gas laser to reduce the reflection losses which would arise from the use of external mirrors. Their operation depends on the setting of the windows at the Brewster angle to the incident light.

bridging loss. That resulting from bridging a system with an impedance, such as the input impedance of a measuring or monitoring device.

Brillouin scattering. The interaction of sound waves at microwave frequencies and electromagnetic waves at optical frequencies. The advent of the laser as a coherent light source has permitted a wider field of application of this interaction. Brillouin scattering from the thermal (Debye) waves in a transparent liquid enables the velocity and attenuation of sound in the fluid to be measured at microwave frequencies.

broadband. Said of a valve or circuit with no resonant elements and therefore capable of operating with similar efficiency over a wide range of frequencies.

roca galvanometer. An early type of moving-magnet galvanometer.

buffer solution. One in which the hydrogen concentration, and hence its acidity or alkalinity, is almost unchanged by dilution. It also resists a change of pH when acid or alkali is added.

bunching parameter. One indicating the efficiency of the bunching action in a klystron.

CP isotope. See **cosmogenic.**

cable code. An adaptation of the Morse code, used in submarine cable signalling and in some radio links. In cable signalling, the dots and dashes are given respectively by positive- and negative-going currents. For radio, two frequencies are displaced slightly below and above a middle frequency.

cadmium-sulphide detector. Radiation detector equivalent to a solid-state ionization chamber but with amplified current level (due to hole trapping). Its main drawback is a slow response time.

calandria. Sealed vessel used in core of certain types of reactor.

calculating machine. Mechanical device for performing arithmetical operations.

calling sequence. The set of instructions which define the values of parameters to be used in a closed subroutine and transfer control to it, or return control to the main routine.

capacitance integrator. Resistance-capacitance circuit with output proportional to the capacitor current.

capacitance potentiometer. A series combination of capacitances across an a.c. supply which allows the desired p.d. to be tapped-off.

carbon-dioxide (CO_2) laser. One in which the active gaseous medium is a mixture of carbon dioxide and other gases. It is excited by a glow discharge and operates at 10·6 microns wavelength.

carbon fibres. Very fine filaments of the order of 8 microns diameter which are used in composite materials, being bound together with epoxy or polyester resins. These fibre-reinforced materials have in general a much higher strength/weight ratio than metals. Boron and glass are alternative fibre materials.

carbon rheostat. High-current variable resistor controlled by altering the pressure on a pile of carbon plates.

carbonized filament. Thoriated tungsten coated with tungsten carbide to reduce the loss of thorium from the surface.

card deck. A set of punched cards comprising (part of) a programme.

card feed. The mechanism which moves cards sequentially into read or print units.

card stacker or **hopper.** The unit in which cards are stacked before or after feeding.

cardiac pacemaker. See **pacemaker.**

carrier beat. Interference produced by heterodyne note generated between two carriers or between a carrier and a reference oscillator.

carrier modulation. The process of impressing information on to a carrier wave for transmission. See *amplitude*, *frequency*, *phase* (in main text).

carrier telephony. The use of a modulated carrier wave for cable transmission in telephony.

catadioptric. Said of an optical system in projection TV receivers which uses reflecting mirrors as well as lenses. See *Schmidt optical system* (in main text).

cathode poisoning. Reduction of thermionic emission from a cathode as a result of minute traces of adsorbed impurities.

cathode sputtering. Production of thin films of metal or semiconductor by electric discharge under vacuum. The cathode is formed from the material to be deposited and the discharge produces ions which are carried across on to both the anode and nearby insulating surfaces.

cathodoelectroluminescence. Normally the

analogous effect to *photoelectroluminescence* (q.v.) for a fluorescent screen excited by cathode rays and not photons. It is also used interchangeably with *electro-cathodoluminescence* (where the excitation precedes the read-out obtained by subsequently applying the field).

central processing unit. The 'heart' of a computer, comprising the arithmetic unit, associated special registers and main fast storage together with circuits for controlling the execution of instructions.

central telegraphic office. Control office for long-distance radio telegraphy, messages being relayed by landlines or high-frequency radio links to or from outlying receiving or transmitting stations.

centre tap. Connection to the electric midpoint of a component such as a resistor or inductor.

ceramic photocell. One using a light-sensitive resistor, comprising titanium oxide and various titanates and metal oxides.

chain printer. High-speed printer in which type is attached to the links of a closed chain.

channel multiplier. A form of electron multiplier in which separate dynodes are replaced by a tube of high-resistivity material along which there is a continuous potential gradient.

channeling. Penetration of impurity atoms relatively deep into a monocrystalline sample because they are injected parallel to one of the crystal axes.

channelized transmitter. One providing more than one communication channel simultaneously, each at a different frequency and separately modulated but operating on a common aerial and power supply.

Chapman region. Hypothetical region of the ionosphere where the variation of electron density with height obeys an approximately parabolic law.

character density. The number of characters which can be stored per unit length along a magnetic tape, drum or disk track.

chart comparison unit. Type of radar display superimposed upon navigational chart.

check digit. Redundant bit in stored word, used in self-checking procedures such as a parity check. If there is more than one check digit, a **check number** is formed.

check indicator. One which displays an error warning when a computer fails a check routine.

checkout. A set of routines used to evaluate a computer programme under operating conditions.

checkpoint. One in a computer programme at which a record is made of all parameter values so far obtained and/or all machine variables such as the location of storage.

Chinese binary. A form of coding for punched cards in which the card is read in columns rather than rows.

chock automatic system. An electronic transistorized device for automatically advancing the roof supports in automatic coal-mining after each run of the coal face shearer.

choke input filter. In a rectifying circuit, a filter which has a series choke for its first component.

chromoscope. A type of *colorimeter.*

chronistor. Electrochemical elapsed-time indicator.

chronograph. Recording chronometer.

circular shift. See **end-around shift.**

Clasp. An experiment on the field configurations inside a stellerator discharge tube, carried out at Culham Laboratories. It confirmed that stellerator geometry should provide a practical basis for the construction of a fusion reactor.

clear band. Term used in optical character recognition for one which is free of all print.

Climax. A plasma containment experiment carried out at Culham Laboratories using a toroidal quadrupole magnetic trap.

clock. Unit for regular pulses in a synchronous computer, for timing all operations, gating, recording, printing.

clockspring core. One in which thin magnetic tape is wound to form a tight coil.

closed magnetic circuit. The magnetic core of an inductor or transformer without any air gap.

clutch. Unit used to adjust the operating speed of peripheral computing equipment when this must be operated synchronously with other equipment.

coastal reflection. Reflection of signal by a land mass so that the resultant received signal consists of direct and reflected waves. This effect causes errors in direction-finding.

coated lens. A controlled evaporated coating on a lens of magnesium-(or calcium-) fluoride which minimizes loss by reflection of light. Its refractive index should be approximately equal to the square root of that of the glass.

coaxial terminal. One to which a coaxial line may be connected.

coherent radiation. Electromagnetic radiation for which there is a definite phase relationship between different points in the beam. This is essential before any interference effects can be observed. Traditionally in optics a beam from a single source must be split and then recombined to show these effects. Lasers now provide an intrinsically coherent source.

coincidence tuning. Tuning of all stages to the mid-band frequency in contradistinction to *staggered tuning* (in main text).

collimator. Device used for producing a parallel or near parallel beam of radiation.

collinear array. A directional system consisting of two or more half-wavelength aerial radiators, excited in phase, and placed end to end horizontally. The radiation is a maximum in the perpendicular direction to the line of the array.

collision frequency. An important constant for

ions in the ionosphere. The term is also used in other branches of physics and electronics, e.g., kinetic theory.

collisional excitation. The transfer of energy when an atom is raised to an excited state by collision with another particle.

colour saturation. In colour TV, the complementary and primary hues are termed *saturated hues* or *colours* but in nature very few of these are fully saturated in the same way. Since, therefore, mostly desaturated colours should be displayed on the colour tube, some white colour must be added to the saturated hue.

Columbus. Type of linear-plasma machine, characterized by the attainment of very high electric fields, for studying pinch effect.

column binary. See **Chinese binary.**

commag. International code name for a picture film combined with a magnetic sound-track.

command guidance. The guidance of guided missiles or aircraft by electronic means through signals from an external source.

common block. A block of storage locations in a digital computer which is associated with information or data required both in the main programme and in a specific subprogramme.

common-channel interference. Interference between two broadcast stations using identical or nearly similar carrier frequencies or whose sideband frequencies encroach on one another.

common language. A machine code which is common to different computers or data-processing devices, e.g., tape-operated print-out units.

communications satellite. Artificial satellite which is projected into a predetermined orbit around the earth to function as a relay station. See **synchronous-**.

commutation switch. That controlling the sequential switching operations required for multichannel pulse communication systems.

comopt. International code name for a picture film combined with an optical sound-track.

complete operation. In computing, the processing of one complete instruction, from assembly of all operands from storage to return there of the processed results.

complex wave. One with a non-sinusoidal waveform which can be resolved into a fundamental with superimposed harmonics. See *Fourier principle* (in main text).

compound connection. A high-gain amplifier stage which uses two transistors in which the emitter of the first feeds directly the base of the second. The output may be taken from a load resistor which is common to the two collectors.

Compton absorption. That part of the absorption of a beam of X-rays or γ-rays associated with Compton scattering processes. In general, it is greatest for medium-energy quanta and in absorbers of low atomic

weight. At lower energies *photoelectric absorption* is more important and at high energies *pair production* predominates.

computer control. Automatic control of, e.g., a tool, a factory, a processing plant, etc., by a computer operating to a predetermined programme.

concatenate. To connect separate units in series, forming a complete system.

concertina amplifier. Valve amplifier which has load resistors in both the cathode and anode circuits. In this way outputs of opposite phase are provided for driving a push-pull stage.

concurrent processing. The simultaneous solution of more than one programme by a computer using time-sharing techniques.

cone break-up. The effect when a loudspeaker cone ceases to move as a whole. Usually applies to the higher frequencies within the operating band of a drive unit.

cone surround. The strip of compliant material fitted to the periphery of a loudspeaker cone to seal it to the frame and yet allow an axial movement.

constant amplitude/velocity recording. Methods of recording in which, for a fixed amplitude signal, the recorded amplitude is respectively independent of frequency, or inversely proportional to frequency.

constant-current generator. A high internal impedance electric source which will deliver a virtually constant output current into a wide range of load impedances.

continuous loading. Loading inductances distributed along conductor.

contourgraph. Device which uses the three-dimensional aspects of a CRT, viz., vertical signal deflection, intensity modulation and flexibility of time base. It is essentially a compact display of non-identical, quasi-periodic data and usually comprises a time-exposure photograph of a number of oscillograph traces. In order to emphasize and make for ease of study the similarities or differences between cycles, the data are arranged about an event common to each cycle and collected over extended time periods.

Contran. A computer language combining the most desirable features of *Fortran* IV and *Algol* 60. (*Control trans*lator.)

control instructions. The part of a computer programme which governs the manipulation of data and interpretation of instructions but does not contribute to the actual processing.

control signal. The actuating signal in automatic control systems.

control word. One which transmits instructions to a digital computer, e.g., XEQ for execute.

convergence surface. Ideally, that generated by the point on which the electron beams in a multibeam CRT converge. In practice this can only be regarded as a point to a first approximation.

conversion. Component, auxiliary to a computer system, which accepts decimal digits

O

and makes a unique conversion to binary digits in the output, or vice versa.

coordinated transposition. The reduction of mutual inductive effects in multiline transmission systems (telephony or power) by periodically interchanging positions.

coronium. Originally thought to be a new element causing emission lines in corona spectrum of sun, but found to arise from highly-ionized iron, calcium and nickel.

Corosil. RTM for a grain-orientated silicon-iron alloy for cores.

cosmogenic. Said of an isotope capable of being produced by the interaction of cosmic radiation with the atmosphere or the surface of the earth. It is sometimes called a CP (*cosmic-ray produced*) *isotope.*

Cottrell effect. Gradual loss of elasticity of uranium under conditions of intense radiation.

count-down. General term for counting backwards, in seconds, to zero for the initiation of an event.

counter telescope. A column of radiation detectors connected in coincidence so that a count is only recorded if an ionizing track passes through them all. Used especially to study the directional properties of cosmic rays.

counterweight. In a gramophone, the mass at back of the pickup arm, situated behind the pivot, to 'counter' or 'balance' the mass of the arm and head assembly. It allows the adjustment of the stylus force to a suitable value.

cramping. The contraction of either side or of the central section of a TV picture which may arise, e.g., from poor emission of certain thermionic valves.

critical band. In a complex sound it is not only the loudness and pitch which determine the human perception but also its harmonic content and transient behaviour. In order to take this effect into account, investigations have shown that certain critical bands of frequency have a definite relationship with vibration maxima on the basilar membrane. Based on these results the audio range from 50 Hz to 13,000 Hz has been divided into 24 critical bands. One critical band corresponds to a distance of 0·13 cm along the basilar membrane and defines the unit of one *bark*. The loudness of the sound is mainly proportional to the rms of the sound pressure within a critical band but the loudness of the various bands are summed somewhat differently.

critical dimensions. Those at which a mass of fissile material may start to undergo a chain reaction.

Crolite. RTM for a ferrite.

Crolyn. TN for a recording tape coated with chromium dioxide.

cross-field bias. A system of tape recording in which a separate head is used for the application of the high-frequency bias. By suitable positioning with reference to the recording head, the demagnetizing effect of

the AF magnetic field at high audio frequencies, as compared with the conventional arrangement, is reduced. By this arrangement, the upper end of the audio response is extended.

cross-field reactor. A magnetically-operated frequency doubler, consisting of a compact toroidal coil carrying the primary current and a secondary coil wound round the first so that the axes of the coils are perpendicular at every point. When a d.c. bias current is passed through the secondary this is modulated at twice the frequency of the primary supply.

cross-linked molecule. One in which the adjacent polymer chains are joined together by short molecular chains. This network structure increases the mechanical strength and heat resistance of a substance, e.g., Bakelite.

cross-over distortion. That arising in push-pull stages when the valves (or transistors) are operating out of correct phase with one another.

crossed-pair. Arrangement of two directional microphones, either close together or one above the other, with their axes diverging at approx. 90°, for use in stereo recording.

cryosorption pump. Vacuum pump used on cryogenic equipment and operated automatically by the available temperature difference.

Cuccia coupler. One used for launching cyclotron waves on to an electron beam in a microwave tube by the application of a high-frequency transverse electric field and a longitudinal magnetic field.

current balance. A form of balance in which the force required to prevent the movement of one current-carrying coil in the magnetic field of a second coil carrying the same current is measured by means of a balancing mass. Cf. *magnetic balance.*

cycle counter. See index register.

cyclotron wave. One imposed on an electric beam in a microwave tube such that the electrons follow helical paths rotating at the cyclotron frequency.

DUF. Abb. for *diffusion under the epitaxial film* (q.v.).

Dalitz pair. Electron-positron pair produced by the decay of a free neutral pion (instead of one of the two gamma quanta normally produced).

Dalitz plot. A type of graphical diagram used in the study of decay processes leading to particle triplets. Energies of two of the particles are measured for a large number of disintegrations, using any suitable track chamber, and are plotted along different axes of the diagram.

damper winding. A permanently short-circuited winding incorporated into some electrical machines, such as synchros, in order to damp their response.

Dataphone adaptor. One for operation of a

central computer from a remote access point using a telephone line (U.S.).

Datel. A G.P.O. system for digital data transmission over telephone lines.

decay time. That in which the amplitude of an exponentially decaying quantity reduces to e^{-1} of its original value.

decision. Use of processed data by computer in selecting which of two possible operations to perform next.

decoder. A circuit used to derive stereo information from a multiplex signal in a radio tuner.

deflection defocusing. Loss of focus of a CRT spot as deflection from the centre of the screen increases.

deflection yoke. Assembly of mounted deflector coils.

Delbruck scattering. Elastic coherent scattering of gamma-rays in the coulomb field of a nucleus. The effect is small and so far has not been conclusively detected.

delete. To wipe out information by clearing specific storage locations.

delimit. In computing, to fix the boundaries of a sequence of characters or data. *Delimiters* are special characters used to do this.

demand meter. One reading or recording the total loading on an electrical system.

Destriau effect. A form of electroluminescence arising from localized regions of very intense electric field associated with impurity centres in the phosphor.

destructive read. Clearing of a storage location simultaneously with reading its contents.

diagnostic routine. Computer programme designed to aid debugging or to locate cause of malfunction.

diathermanous. Relatively transparent to radiant heat.

diathermic surgery. The use of an electric arc in surgery in preference to a knife. This has the advantage of sealing cuts and reducing bleeding.

dielectric loss angle. Complement of *dielectric phase angle.*

dielectric phase angle. That between an applied electric field and the corresponding conduction-current vector. The cosine of this angle is the power factor of the dielectric.

difference amplifier. One with three input terminals, giving an output proportional to the difference between two input signals. Sometimes called **differential amplifier.**

difference detector. One for which the output is proportional to the difference in amplitude of two input signals.

differential amplifier. See **difference amplifier.**

differentiating time. The time constant of a differentiating circuit.

diffraction grating. A series of ruled lines on a transparent or reflecting plate. If a beam of electromagnetic or ultrasonic waves falls on to such a grating, the different spectral components can be separated when the line spacings are somewhat greater than the wavelengths present. Crystal lattice struc-

tures can form three-dimensional gratings for X-rays.

diffraction pattern. That formed by equal intensity contours as a result of diffraction effects, e.g., in optics or radio transmission.

diffusion under the epitaxial film (DUF). A method of introducing by diffusion a low resistance path between the active region of an integrated circuit transistor and the contact electrode at the surface.

digital differential analyser. A special electronic computer for solution of differential equations by incremental means.

digital incremental plotter. A plotter capable of direct control by processed data from a digital computer.

digital thermometer. One using the beat note produced by temperature-sensitive and temperature-insensitive quartz-controlled oscillators, to give digital representation of temperature changes.

dineutron. Assumed transient existence of a set of two neutrons in order to explain certain nuclear reactions.

diode isolation. Isolation of the circuit elements in a microelectronic circuit by using the very high resistance of a reverse-biased *p-n* junction.

direct effect. Effect of primary ionization of vital molecules (such as nucleic acids) in a living cell as a result of exposure to radiation; cf. *indirect effect.*

direct-reading system. Instrument or measuring system set up so that the desired quantity can be obtained direct from a read-out or scale without calculation or calibration.

direct transition. Relating to the recombination processes in the energy-band structure of a semiconductor; if the transition of a charge carrier takes place between the lowest conduction-band minimum and the highest valence-band maximum without a change in the direction of motion of the electron and hole, it is called a *direct transition,* e.g., Ga As. Cf. *indirect transition.*

disk armature. One for a motor or generator wound to a large diameter on a short axial length.

distribution factor. A *modifying factor* (q.v.) used in calculating biological radiation doses which allows for the non-uniform distribution of an internally-absorbed radioisotope.

dolly. Mobile mounting for a soundfilm or television camera and its operator, with or without means of vertical or side motion (*crabbing*).

Doploc. Audio filter for tracking circuits.

dose equivalent. A radiation dose corrected for factors which would vary its biological effect (see **modifying factor**) and expressed in *rem.*

dose reduction factor. A factor giving the reduction in radiation sensitivity for a cell or organism which results from some chemical protective agent.

double-pole. Said of switch or relay contacts which operate simultaneously in two different conductors.

double-six array. A multiple aerial system responsive to a number of channels. It consists of a double array having a folded dipole, each array having in addition one reflector and four directors.

double-stream amplifier. A type of travelling-wave tube in which the operation depends upon the interaction of two electron beams of differing velocities.

double superheterodyne receiver. Radio receiver in which amplification takes place at both high and low intermediate frequencies, thus requiring two frequency-changing stages.

doubling dose. Estimated radiation dose which would double the natural mutation rate.

Drysdale potentiometer. An a.c. potentiometer of the polar type, comprising a phase-shifting transformer and resistive voltage divider. It is calibrated against a standard cell with a d.c. current and the a.c. current is then set to the same value using an electrodynamic indicator.

dynamic convergence. The edge corrections necessary after the conditions of static convergence have been applied. It is carried out by passing currents derived from the line and field time bases through the coil windings in the assembly.

dynamic performance. That during actual operation rather than during static tests.

effective address. One which has been modified during the process of a computation.

effective mass. For electrons and/or holes in a semiconductor the effective mass is a parameter which may differ appreciably from the mass of a free electron, and which depends to some extent on the position of the particle in its energy band. This modifies the mobility and hence the resulting current.

eikonal equation. The fundamental equation which determines the path of a ray.

Einstein mobility equation. In a semiconductor (or ionic solution) the mobility (μ) of the charges is given by $\mu = eD/kT$, where e is the magnitude of the charge, D is the diffusion coefficient, k is Boltzmann's constant, and T is the absolute temperature.

electrocathodoluminescence. See **cathodoelectroluminescence.**

electroluminescence. The conversion of electrical into luminous energy (see main text). It may arise, as in *cathodoluminescence* (also has another meaning as in main text) or *electrofluorescence* due to accelerated electrons striking the phosphor of a television tube, or as in the application of a.c. to certain microcrystalline insulating solids embedded in chosen dielectric media. A further form results from the application of a direct current at a low voltage (1·5 to 2·5 volts according to material) to a suitably doped crystal containing a *p-n* junction. Suitable semiconductor materials are gallium phosphide (Ga P), silicon carbide (Si C) and the alloy gallium arsenide phosphide (Ga As P). Ga S is a direct semiconductor (see *direct transition*) while Ga P

is an indirect semiconductor (see *indirect transition*), but for compositions of about 45% Ga P and less, the alloy Ga As P is a direct band semiconductor and may be used as an efficient light-emitting device. For 48% Ga P material the energy gap (E) is about 1·975 eV and hence by Einstein's equation the emitted peak wavelength $\lambda = hc/E = (12400/1\cdot975) = 6300\text{Å}$. [See table below for relevant properties of some semiconductors.] These light-emitting diodes can be made small enough to be regarded as point sources of light which itself is also fairly monochromatic, and if coherent (laser) light is not required the operating current is only of the order of ten mA. Since these diodes can be switched very rapidly they have a promising application in data processing.

Characteristics of some common semiconducting materials (at 27°C)

Material	Band-gap Energy (eV)	Wave-length (Å)	Transition $\left(\begin{array}{l} i = \text{indirect} \\ d = \text{direct} \end{array}\right)$	Doping
Ge	0·66	18,800	i	n,p
Si	1·09	11,400	i	n,p
In Sb	0·18	69,000	d	n,p
In As	0·36	34,500	d	n,p
Ga Sb	0·7	17,700	d	n,p
In P	1·26	9850	d	n,p
Ga As	1·43	8680	d	n,p
Al Sb	1·6	7750	i	n,p
Ga P	2·24	5540	i	n,p
Cd Te	1·44	8620	d	n,p
Si C	2·2–3·0	5630–4130	i	n,p

electron current. That due to electron flow as distinct from the flow of ions or 'holes'.

electrophotoluminescence. The emission of visible light from a phosphor, that has been previously or simultaneously excited by ultraviolet- or X-radiation, when an electric field is applied. See **Gudden-Pohl effect.**

electroradiescence. Emission of ultraviolet or infrared radiation from dielectric phosphors upon the application of an electric field.

electrostatic charge. Electric charge at rest on the surface of an insulator or insulated body and consequently leading to the establishment of an adjacent electrostatic field system.

electrothermoluminescence. Changes in electroluminescent radiation resulting from changes of dielectric temperature. (Some dielectrics show a series of maxima and minima when heated.) The complementary arrangement of observing changes in thermoluminescent radiation when an electric field is applied is termed **thermoelectroluminescence.**

electrum. Alloy of copper, zinc and nickel with appearance of silver. Also **German silver.**

Elektron metal. An alloy based on aluminium, used for the walls of ionization chambers and designed to be approximately air-equivalent in its properties.

Elf. TN for electroluminescent display screen,

having high-speed scanning, good half-tone, and brightness characteristics.

encapsulating. The enclosure of, e.g., an electronic component in a resin to protect it against the environment.

end-around shift. One in which digits drop off one end of a word and return at the other. Also **circular shift, logical shift, ring shift.**

end-of-file marks. Those used to indicate the termination of an input tape or quantity of data.

endoradiosondes. Electronic probes and transducers for insertion into the human body without interfering appreciably with its function, and which transmit data to receivers external to the body.

ephemeris time. See time (1).

erbium laser. Laser using erbium in YAG (*yttrium - aluminium - garnet*) glass. It has the advantage of operating between 1·53 and 1·64 micron, a range in which there is a high attenuation in water. This feature is of particular importance in laser applications to eye investigations, since a great deal of energy absorption will now occur in the cornea and aqueous humour before reaching the delicate retina.

ergonomics. The scientific study of the inter-relationship between employees and their working environment, together with its technological implications.

eriometer. A device for the measurement of the size of small particles, using optical diffraction.

Ernie. A special-purpose computer used by the G.P.O. to select winning numbers in the Premium Bond lottery. Similar random generators are used in the *Monte Carlo method*. (*Electronic random number indicating equipment.*)

error diagnosis. Compilers preparing a machine programme for a computer frequently print out the programme language statements with any erroneous instructions clearly indicated.

error routine. Diagnostic routine carried out automatically by computer if a programme check establishes an error. Only if the error persists is programming interrupted.

exciting current. That drawn by a trans-former, magnetic amplifier, or other electric machine under no-load conditions.

exponent. The power of ten by which a normalized number must be multiplied to produce the actual number. Also **index.**

f-factor. The ratio of absorbed dose (*rads*) to exposure dose (*röntgens*) for a given material and X-ray energy. The term is applied particularly to different tissues, e.g., fat, bone, muscle, etc.

false curvature. That of particle tracks (e.g., in cloud chambers, bubble chambers, spark chambers, or photographic emulsions) which results from undetected interactions and not from an applied magnetic field.

false line lock. In a television receiver a

critical setting of the line-hold control, which results in a line lock with a part of the picture to the left and a part to the right of a dark and wide vertical bar.

false retrievals. In literary searches by com-puter, references that are not required but are sufficiently closely related to the subject to be listed by the automatic search procedure.

Fano factor. One by which the intrinsic noise limit to the resolution of a gas or solid-state ionization chamber is reduced because the individual ionizing events in the energy loss process are not completely independent.

faradism. The use of an induction coil to generate high voltages for therapeutic use. See *faradic currents* (in main text).

faradmeter. Generic name for direct-reading capacitance meter. Typically it uses the mains voltage in series with an a.c. milliammeter.

fast fission factor. Ratio of total number of neutrons produced by fission in an infinite thermal reactor system to number produced by thermal fission. It represents the fractional increase in neutron population in any generation which results from fast fission, and is usually about 1·028 for natural uranium graphite power reactors.

fast-time constant circuits. Those for which the circuit parameters (particularly resis-tance and capacitance) permit a very rapid response to a step signal. Such circuits should be used exclusively in the case of pulse signals.

fast wave. See space-charge waves.

feasibility study. The first part of a system analysis directed toward determining the practicability of using computing methods.

Felici balance. An a.c. electrical measuring bridge for determining mutual inductance between windings.

Felici generator. A modern form of electro-static high-voltage generator developed in France and comparable with a Van de Graaff generator.

ferrite bead. Small element of ferrite material used particularly for threading on to wire of transmission line to increase the series inductance. See *ferrite-bead memory* (in main text).

ferrodynamometer. Any dynamometer incor-porating ferromagnetic material to enhance the torque.

ferromagnetography. A printing technique involving the application of magnetic iron particles to a latent image magnetized on to a metal drum (or sheet) and afterwards transferred to paper and fixed by pressure or heat. If the image is formed by a row of tiny electromagnets which are under com-puter control, then a high-speed print-out of characters can be obtained.

Feynman diagram. Graph relating time and displacement for the particles involved in a nuclear reaction. Photons are represented by a wavy line, fermions by a continuous line, and bosons by a broken line.

fibre optics. That concerned with the use of

suitably-coated bundles of long, thin, flexible, transparent fibres for the maximum transmission of light, i.e., as light guides.

Fibreglass. TN for glass-fibre materials used for acoustic and thermal insulation.

field coil. That placed round a magnetic core in order to carry the energizing current.

field-effect transistor. One consisting of a conducting channel formed from a strip of n- (or p-) type semiconductor with a gate, typically formed by a collar of p- (or n-) type material, the voltage applied to which controls the resistivity of the whole strip. The input impedance to the gate is very high and signals applied to this are amplified in a similar way to those applied to the grid of a triode.

field magnet. One which establishes the magnetic field required by an instrument, machine or system.

field-strength meter. A stable radio receiver calibrated for the measurement of field strength.

filament transformer. One designed specifically to supply the filaments of electron tubes.

filament windings. Those on a power transformer designed to supply the filaments of electron tubes.

file. General term for a set of data stored sequentially, e.g., along one track of a magnetic memory.

filled band. An energy-level band in which there are no vacancies and hence for which the electrons are unable to participate in conduction processes. See *hole* (in main text).

film circuits. Microelectronic circuits in which the passive components and their metallic interconnections are formed directly on an insulating substrate and the active semiconductor devices (usually in wafer form) are subsequently added.

filter discrimination constant. The difference, in dB, between the minimum attenuation in the specified rejection band and the maximum attenuation in the specified pass band.

Fingal. A UKAEA development project for processing high-activity radioactive wastes into leach-resistant glass blocks which can be disposed of with far less hazard than the original liquid. The pilot plant (*Fingal's Cave*) deals with 1000 curie per run of typical waste acids from reactor fuel rod processing. (*F*ixation *in* g*l*ass of *a*ctive *l*iquids.)

Fireflash. An air-to-air RAF missile.

fissiogenic. Said of an isotope capable of being produced as a result of nuclear fission.

fission parameter. The square root of the atomic number of a fissile element divided by its atomic weight.

Fitzgerald-Lorentz contraction. See Lorentz contraction.

fixed capacitor, inductor, resistor. Component of approximately constant and usually specified value for use in any electric circuit.

fixed-cycle operation. In computers with *fixed word length*, one operation on one word requires a fixed-cycle time.

fixed word length. Computing procedure where all word lengths contain the same number of bits regardless of their information content.

flare cutoff. The frequency, determined at the rate of flare, at which the acoustic impedance at the throat of a horn falls to zero.

flashback voltage. The inverse peak voltage in a gas tube at which ionization occurs.

flicker photometer. One designed for use with light sources of different colour which are viewed successively in rapid alternation. The colours merge but when the intensities are different a flicker is seen.

floating-point arithmetic. That of a computer which handles numbers in the form of a mantissa and integral exponent.

flowcharts (See also main text). These charts develop the logical details of the runs which are depicted by process charts from the comprehensive *block diagrams* of computer systems. A standard set of flowchart symbols has been published by the American Standards Association (ASA) Committee on Computers and Information Processing (1963). In a system flowchart which describes the flow of data through all parts of a system, the flow may be described in terms of fourteen flowchart symbols. The programme flowchart describes what occurs in a stored programme and utilizes six flowchart symbols, the use of which follows certain conventions, e.g., the normal direction of flow is taken from left to right and from top to bottom. A list of less evident flowchart abbreviations follows:

ACC	accumulator	INDR	indicator
ACCT	account	INTRPT	interrupt
ACT	actual	INVAL	invalid
ADJ	adjust	LBL	label
ALG	algebraic	MPXR	multiplexor
ALPHA	alphabetic	NUM	numeric
ASDNG	ascending	OPND	operand
ASGN	assign	OVFLO	overflow
BFR	buffer	OVLY	overlay
BLK CNT	block count	PGLIN	page and line
BM	buffer mark	PNCH	punch
CC	card column	PRT	printer
CHG	change	PT	point
CHKPT	check point	R+S	reset and
CLR	clear		start
CLS	close	RM	record mark
CMPL	complement	RSTRT	restart
CNT	count	RTN	routine
CON	constant	R/W	read/write
CPLD	coupled	SM	storage mark
CURR	current	SUMM	summarize
DCMT	document	TBL	table
DECR	decrement	TM	tapemark
DESCG	descending	TMT	transmit
DIM	dimension	TP	tape
EOF	end of file	TRK	track
EOJ	end of job	TST	test
EOR	end of reel	TU	tape unit
ERR	error	TW	typewriter
FLDL	field length	UNLD	unload
FLT	floating	WM	wordmark
HDR	header	XPL	explain
HI	high	Z	zero
HSK	housekeeping	ZN	zone.
IC	instruction counter		

flowmeter. Device for measuring, or giving an output signal proportional to, the rate of flow of a fluid in a pipe.

fluidics. The science of liquid flow in tubes, etc., which strongly simulates electron flow in conductors and conducting plasma.

fluorescent lamp. One in which most of the light output results from the excitation of a phosphor coating into fluorescence.

fluorimeter. One used for measuring the intensity of fluorescent radiation.

fluorod. Small silver activated phosphate glass cylinder used for personal radiation dosimetry.

flywheel time base. One in which synchronizing pulses control a pulse generator which is connected to the actual CRT. If reception is poor much better synchronization is achieved because of the *flywheel effect.*

focal point. The focal spot formed on the axis of a lens or curved mirror by a parallel beam of incident radiation. In its general form this definition includes acoustic lenses, and lenses or mirrors designed for use with radio waves and infrared or ultraviolet radiation.

focal spot. A spot on to which a beam of light or charged particles converges. See also *X-ray focal spot* (in main text).

focusing electrode. One used to control the focusing of an electron beam by variation of potential.

fold-over. Distortion in a TV picture showing itself as a white line in one of the edges as a result of non-linear operation in the deflection circuits of the receiver.

format. The layout of written data processed by a computer.

Fortran. A problem-orientated computer language widely used in scientific work. The machine instructions for a specific computer are prepared from a Fortran programme in an automatic compiler forming part of the computer system. (*For*mula *tran*slation.)

forward-wave amplification. See **slow-wave structure.**

four-address instruction. Computer instruction which includes the addresses of the operator, the operand and the storage location for the result, together with the address of the next instruction. Also **3 + 1 address instruction.**

four-factor formula. That giving the multiplication constant of an infinite thermal reactor as the product of the fast fission factor, the resonance escape probability, the thermal utilization factor, and the neutrons emitted per thermal neutron absorbed for the fuel material.

four-wire system. A communication system in which transmissions in opposite directions always take place along independent lines.

fox message. One which includes all the alphanumerics on a teletypewriter in addition to most of the function characters and is used for testing machines and teletype circuits. The standard form of the message is 'The quick brown fox jumped over the lazy dog's back 1234567890 – – – sending'. The blanks are for insertion of identifying symbols of sending station.

frame-grid. Said of a rugged high-performance thermionic valve in which the grid is held very rigid and close to the cathode surface by winding it on stiff rods.

frame-synchronizing pulse. That used to synchronize the vertical time base of a television or facsimile system. It must be separable from the line-synchronizing pulse, e.g., by a different duration or amplitude.

free-air ionization chamber. One of a type developed for absolute standardization purposes. The sensitive volume is defined by the window through which radiation enters the chamber and the associated electric field configuration, in such a way that it is limited to the central region of a much larger volume of air.

Freon. TN for halogenated hydrocarbon gases with exceptionally good insulating properties; also used in refrigeration and for wind tunnels. *Freon*-12 is dichlorodifluoromethane.

frequency range. That over which an instrument or device normally operates, or over which it is specified to operate.

frequency stability. The extent to which an instrument or system is free of frequency drift.

Frigistor. Thermoelectric device for cooling electronic components.

functional diagram. Block diagram showing functional relationships between units of a system.

fuse rating. The maximum current a fuse will carry continuously and/or (less frequently) the minimum current at which it can be relied upon to 'blow'.

G-line. Coated wire used to transmit microwave energy.

G-value. A constant in radiation chemistry denoting the number of molecules reacting as a result of the absorption of 100 eV radiation energy.

galvanism. Electrotherapy carried out with d.c. supplies.

gamma camera. *Scintillation camera* used to study distribution of gamma-emitting isotopes in some organ of the body.

gap breakdown. Cumulative ionization of the gas between electrodes, leading to a breakdown of insulation and a Townsend avalanche.

gap coding. Imposing of telegraphic message on navigational signal by means of interrupter.

gap factor. Ratio of energy gain in electronvolts for electrons traversing a gap across which an accelerating field acts, to the actual voltage across the gap.

gap length. The distance between adjacent surfaces of the poles in a longitudinal magnetic recording system.

gas maser. One in which the interaction takes

o**

place between molecules of gas and the microwave signal.

gas ratio. That of ion current to electron current in a gas discharge. See also *gas amplification* (1) (in main text).

gassing. Evolution of gas from electrodes of voltameter or hard vacuum tube.

gate detector. One whose operation is controlled by an external gating signal.

gate impedance. That of the gate winding of a magnetic amplifier.

gate voltage. (1) That required to operate an electronic gate. (2) That across the gate winding of a magnetic amplifier.

Gauss' laws of electrostatics and **magneto-statics.** The surface integral of the normal component of electric displacement (or magnetic flux) over any closed surface in a dielectric is equal to the total electric charge enclosed (or to zero in the magnetic case). Differential forms of these laws comprise two of *Maxwell's field equations*; see also *Poisson's equation* (in main text).

generating routine. A computer programme that can be used to write further programmes covering problems that may not be well-defined.

German silver. See *electrum*.

gimbal mount. Mount giving rotational freedom about two perpendicular axes—as used for gyroscope and nautical compass.

gimmick. Colloquialism for small capacitor formed by twisting insulated wires.

Globar. TN for silicon carbide rod which, when heated by a current, acts as a radiating black body.

gloss factor. The ratio of specular reflection to diffuse reflection for a surface (optics).

glossmeter. Electronic instrument used to measure the degree of polish of a surface.

glow-discharge microphone. One in which the speech signals modulate the current through a glow discharge directly.

glow lamp. One in which most of the luminous output arises from excitation of the gas molecules.

Goto pair. A pair of tunnel diodes connected in opposition (so that one is in the forward conduction mode when the other conducts by tunnelling) and used in high-speed gate circuits.

Grace. Automatic telephone exchange unit used in subscriber trunk-dialling system. (Group routing *and* charging *e*quipment.)

ground position indicator. Navigational system indicating aircraft position with reference to an origin on the ground. Cf. *air position indicator*.

group mark. Character forming part of a computer write instruction and indicating the end of a specific record.

Gudden-Pohl effect. A form of *electro-photoluminescence* which follows metastable excitation of a phosphor by ultraviolet light.

guided rocket. A *guided missile* having rocket propulsion.

Halden project. The world's first boiling heavy-water reactor, erected at Halden,

Norway; part of a joint European research programme.

hard copy. A written record of output data to be stored, e.g., on magnetic or punched tape.

hardware. General term for all mechanical and electrical component parts of a computer or data-processing system. Cf. *software*.

Harris instability. Type of plasma instability due to anisotropic velocity distribution of particles.

head-to-tape velocity. In transverse scan recording, in order to record and play back up to 2·5 MHz, a linear tape velocity of about 25 m sec^{-1} is required. This high velocity is attained by the use of four recording heads equally spaced around a drum which can rotate at 1500 rmp.

heading marker. Radial trace on PPI-type radar display, used to indicate the heading of the vessel.

heat coil. That operating the cut-out on some protective devices using low melting-point solders to maintain the contact.

heating element. The load resistor in any form of electric heating system or unit.

heating pattern. The spacial temperature distribution in the load during RF heating.

helium-neon (He-Ne) laser. One using a mixture of helium and neon which is energized electrically. It can give continuous operation in powers of the order of milliwatts.

heterokinetic. Of an assemblage of particles with different velocities.

heuristic routine. Intuitive rather than algorithmic routine for tackling a specific problem with a computer (i.e., one based essentially on trial and error).

high-frequency therapy. The treatment of diseases using low-current but high-frequency isolated trains of heavily damped high-voltage oscillations. Can also refer to application of ultrasonic vibrations.

high-level RF signal. An RF signal having sufficient power to fire a switching tube.

high-resistance voltmeter. One drawing negligible current and typically having a resistance in excess of 1000 ohm/volt.

Hilbert transformer. A device for obtaining a phase shift of 90°. It consists of a delay line, fed from a travelling-wave source and terminated by a negligibly small resistor, the p.d. across this forming the 0° output signal. The 90° signal is obtained by integrating the voltage along the line with a weighting function inversely proportional to the distance from the termination.

histogram. Graphical representation of statistics in which rectangles are used to represent frequency distribution.

Hollerith code. Very widely used alphanumeric punched card code.

hologram. Means of optical 'imaging' without the use of lenses, now a practical reality with the advent of the laser. The laser beam is split into two portions, one part directly illuminating a photographic film (or plate) while the other first illuminates the scene. The two portions produce an optical interference pattern on the film which, when illuminated by a laser beam, will produce two images of the original scene. One of these is virtual but the other is real and may be viewed without a lens.

homokinetic. An assemblage of particles in which all have approximately the same velocity. Also **monokinetic.**

homologous pair. In spectrographic analysis, a spectral line (used in determining the concentration of an element) and an internal standard line, such that the intensity ratio of the two lines is unaffected by changes in excitation conditions.

horizontal centring control. In a CRT, that which shifts on the tube face the starting point of the horizontal sweep (along the horizontal trace.)

horizontal resolution. The number of picture elements resolved along the scanning line in any form of facsimile reproduction.

hot electrons. Thermal electrons travelling in the solid lattice. As a result of the inability of relaxation processes to carry away the energy received from the electric field, the electrons become heated.

hot laboratory. One in which strong radioactive sources are used.

hot line. A permanent line in a submarine telephone cable system which is leased to governments and military users.

Huyghens' construction. The continuous development of wavefronts, each point on a given wavefront being regarded as a new source of secondary wavelets.

hybrid computer. One using both analogue and digital elements and techniques.

hybrid integrated circuit. A complete circuit formed by combining different types of integrated electronic sections.

hydrated electron. Very reactive free electron released in aqueous solutions by the action of ionizing radiations.

hydrodynamics. The mathematical study of the physical characteristics of liquids in motion.

hydrogen maser. A frequency-determining device which offers a stability comparable with the caesium clock and may eventually replace or supplement it as the final standard of time.

hydrogenoid. Atomic system, e.g., hydrogen, formed by a central positive nucleus with a single negatively-charged electron gravitating around it.

hydrostatics. The mathematical study of the physical characteristics of liquids at rest.

ideal transducer. One which is loss-free and matches the impedances of the source and load correctly.

identify. In computing, to attach a unique reference number or symbol to a unit of information.

identity. A statement of equality between

known or unknown quantities which is true for all values of the unknown quantities.

ignitor drop. The voltage drop between cathode and anode of the ignitor discharge in a switching tube.

illegal. Said of any character or instruction that a specific computer does not interpret as a unique valid item of data or operation.

image. In computing, a duplicated set of logical data stored in a different medium.

image interference. That produced by any signal received with a frequency at or near the image frequency.

Image Isocon. TN for very sensitive TV camera tube designed for use at low illumination levels. It has been developed from the *image orthicon* but derives the signal from electrons scattered at the target instead of from those reflected. This gives improved signal/noise ratio and much greater dynamic gain.

image-retaining panel. An electroluminescent screen which retains a fluorescent image produced by X-rays for periods of up to half an hour, if the energizing voltage on the screen is maintained. This enables the image to be studied or photographed after the X-ray unit has been turned off and the object removed. The record on the screen is wiped off as soon as the applied voltage is interrupted.

imperfect dielectric. One in which there is a loss element resulting in part of electric energy of applied field being used in heating the medium.

implicit function. A variable quantity f is said to be an implicit function of t if f and t are connected by a relation which is not explicit.

impregnated cathode. One in which the low work function compound has been impregnated into a porous tungsten cathode rather than merely coating the surface. This gives an improved cathode life.

incoherent radiation. Said especially of light beams in which the radiation is not coherent. Interference between such beams cannot be observed. See **coherent radiation.**

incremental computer. A special purpose computer (e.g., a digital differential analyser) which is specifically designed to process changes in the values of the variables.

index. (1) General term for intermittent motion of tape or mechanism under control. See *Geneva movement* (in main text). (2) See **exponent.**

index register. One containing an integer number which changes regularly during the processing of a computer programme—usually by unity each time a specific operation is completed. Also **cycle counter.**

indicating instrument. One in which the immediate value only of the measured quantity is visually indicated.

indicator gate. A step signal applied to an indicator tube to control its sensitivity.

indirect effect. Effect of ionizing radiation on a living cell as a result of chemical damage to vital molecules, produced by free radicals released through decomposition of water molecules by the radiation.

indirect transition. In semiconductors the conduction band can be quite complicated and may have minima in various directions of motion. If the highest valence-band maximum and the lowest conduction-band minimum do not occur for the same direction of motion the resulting recombination is termed an *indirect transition*, e.g., Ga P. These recombination rates are much slower than for *direct transition* (q.v.).

inductance pickup. One which depends for its operation on variation of the inductance of a tuned circuit oscillator and which therefore generates a variable frequency output signal.

induction generator. Machine similar to induction motor, but driven above its synchronous speed from a external source of power and used, e.g., as an element in a servo control system.

induction motor. A very widely used class of a.c. electric motor. Under light load conditions electromagnetic induction drives the short-circuited motor round at very nearly the frequency of the a.c. supply.

inertial control. Control of guided missiles or space capsules by means of signals obtained from the inertial forces acting on the vessel when it deviates from an existing flight path.

information capacity (of a channel). Given in bits/second by Shannon as $C = B \log_2 (1 + S/N)$, where S and N are respectively the signal and noise power, and B is the frequency bandwidth (Hz).

information retrieval. Computer location of required information previously classified and stored.

infrared homing. A guidance system based on heat radiation emitted by the target.

infrared image converter. Electron tube producing visible image on fluorescent screen from an object field emitting or exposed to infrared radiation.

input register. One which accepts a word into storage from the input at one speed and then transcribes it elsewhere into the memory at a much higher speed.

inside spider. A flexible device inserted within a voice coil so as to centre the coil accurately between the pole pieces of a dynamic loudspeaker.

instruction number. One used to label a specific instruction in a computer programme so that the machine can refer back to it subsequently. Such numbers do not indicate the order or sequence in which instructions must be carried out, or the number of the address at which the instruction will be stored. Also **statement number.**

instrument approach. Aircraft approach made using *instrument landing system*.

integrated circuit. A circuit which is fabricated as an assembly of electronic elements in a single structure, which cannot be subdivided without destroying its intended electronic function.

interaction gap. The space in a microwave tube in which the electron beam interacts with the wave system.

interchange instability. Instability in the plasma of a thermonuclear device, so called because the plasma and magnetic field exchange places.

interdigital filter. A high-frequency distributed filter formed of conducting sheet or strips. The inductors are interleaved lengths ($< \frac{\lambda}{4}$) of short-circuited line attached to a base electrode. They simultaneously form the plates of the tuning capacitors.

interdigital line. A combined vacuum tube and folded waveguide such that the phase velocity of the electromagnetic wave relative to the axis of the system is reduced to a value corresponding to the velocity of the electron beam. This principle is used in the linear accelerator and as an alternative to the helix in O-type travelling-wave tubes.

internal radiation hazard. Hazard due to ingested or adsorbed radioactive material.

internal standard line. One within the line spectrum of material being spectrographically analysed. It arises from the introduction into the material of a known amount of a particular element.

internal stored programme. One normally stored within the computer internal memory, to be used in conjunction with the external programme during processing.

internal symmetries. Invariance properties of strong interactions under such operations as charge reversal or particle/anti-particle conjugation, both of which are internal operations on the system and do not move around it in space-time.

interruption. Signal-controlled temporary halt to a specific processing routine, used, e.g., in time-sharing or if higher priority material is introduced.

intrinsic energy. In a chemical reaction, the energy inherent in a substance which may be partly released as heat. In thermodynamics, the intrinsic energy of a material system is taken as its total store of all kinds of energy.

Invar. TN for iron-nickel alloy (63·8 % Fe, 36 % Ni, 0·2 % C) with very small coefficient of expansion—widely used in strain-gauge elements.

inverted triode. One in which the anode or cathode is in the centre.

ion getter pump. Form of high-vacuum pump capable of achieving pressures lower than 10^{-8} torr.

ion implantation. A technique for modifying the electrical properties of monocrystalline materials by bombarding with charged ions in a particle accelerator.

ionium age. One calculated for geological samples (especially sea-bed deposits) from measurements of the ratio between ionium and its daughter products formed by radioactive decay.

iron-dust core. One used in a high-frequency

transformer or inductance to minimize eddy-current losses. It consists of minute magnetic particles bonded in an insulating matrix.

Isocon. See **Image Isocon.**

isodynamic line. One passing through points on the earth's surface having same value of the horizontal component of the earth's magnetic field.

isomeric polymers. Molecules belonging to families that contain exactly the same type and number of atoms but with these arranged differently in each case, e.g., *iso*butane and butane have same formula (C_4H_{10}) but their atoms are placed differently. One type of paraffin molecule ($C_{40}H_{82}$) has over 50 million million possible isomers.

isosteric. Said of substances whose molecules have the same number of atoms and the same total number of electrons, and hence have similarity in their physical properties, e.g., nitrous oxide, carbon dioxide.

isotonic. Said of two solutions which have the same osmotic pressure, i.e., the same molecular concentration.

iteration. One repetition of a group of instructions forming a subroutine which is to be carried out repeatedly in a particular computer programme.

iterative process. One involving *iteration*, implemented by a loop instruction in a programme.

jet engine. One in which propulsion is obtained from the reaction force which results from the acceleration of a fluid stream, as in jet aircraft.

jet noise. The aerodynamic generation of noise in the jet stream from a jet engine.

job. Normal term for the compiling (or assembly) and/or processing of a particular computer programme. *Job control* is used if these are not to be performed sequentially.

Jodrell Bank. The British radioastronomy observatory operated by Manchester University and provided with a 250 ft-diameter steerable paraboloid antenna.

K aerial. A K-shaped TV reception aerial, the 2 short arms forming a dipole and the vertical element acting as a reflector.

Karnaugh map. A graphical technique used for minimizing logical functions in circuit design.

Kew magnetometer. A precision instrument for the measurement of the horizontal component of the earth's magnetic field.

keyboard. An array of control keys required for the operation of a piece of equipment. On a computer the keyboard normally includes a typewriter which can be used for the direct encoding of data, or which can type out instructions to the operator at specific points in a programme.

keypunch. The device used to punch data cards or tape by operation of a keyboard.

keywords. The most informative of significant words in the title of a document. These are the elements stored in the memory in most information retrieval systems.

kilometric waves. Those with wavelengths between 1,000 and 10,000 m.

kink. The abrupt reversal of slope which occurs, e.g., in the anode voltage/anode current characteristic for a tetrode.

kludge. Colloquialism for *computer.*

klydonograph. A surge-voltage recorder.

klystron frequency multiplier. A two-cavity klystron whose output cavity is tuned to an integral multiple of fundamental frequency.

Knudsen flow. Molecular gas flow along a tube when the mean free path of the molecules is much larger than the radius of the tube.

Kopp's law. That the molar heat of a solid compound is approximately equal to the sum of the constituent atomic heats.

Kossel-Sommerfeld displacement law. That relating the arc spectrum of an element to spark spectra of elements placed higher in the periodic table.

Kostinsky effect. An effect associated with the photographic images of high density whereby their recorded separation is greater than is actually the case.

Kramers-Kronig dispersion formula. Formula relating the real and imaginary parts of the refractive index of electromagnetic waves.

Kronecker delta. A mixed tensor of second rank.

Kundt constant. Defined as the ratio of Verdet's constant to magnetic susceptibility.

Kundt rule. Rule governing the displacement of optical absorption bands towards the red, arising from changes in the refractive index due to alterations in composition, etc.

kurchatovium. The latest transuranic element reported to have been isolated. At. no. 104.

Kwic. A classification system used in data retrieval systems. (*Key*word *in* context.)

LASA (*large aperture seismic array*). A giant array of 525 seismometers installed across the 'steppes' of eastern Montana, U.S.A. The diameter of the array is 125 miles, covering an area of 10,000 square miles, the seismometers being arranged in 21 sub-arrays. Each instrument was buried in a 200ft. deep hole to reduce the effects of local noise. The individual outputs are transmitted to a data centre for recording on magnetic tape. The important feature of LASA is its directivity as the array can be steered electronically to look in any direction for earth disturbances of interest.

LSA mode. The fundamental oscillation mode of the *Gunn effect* (in main text). The term originally signified *large signal amplification* but more recently *limited space-charge accumulation*. 150 GHz is the highest frequency so far obtained with this mode.

L-band. Frequency band between 390 and 1,550 MHz.

L-ring. A method of tuning a magnetron.

labile. Said of an oscillator which can be synchronized from a remote source.

lacing. Extensive multiple punching of a data

card, usually to signify the end of a complete run.

lag filter. A radio-frequency filter in close electrical connection with the contacts of a Morse key, to minimize sparking and transients during the keying of the transmitter.

Lambert's law of illumination. That the illumination of a surface is proportional to cos θ where θ is the angle the light rays make with the normal to surface at point of incidence.

laminar flow. Fluid flow without turbulence in which the shape of a contained obstacle is closely followed by streamlines.

Langevin ion. Heavy gaseous ion formed by the clustering of molecules.

language converter. Device for producing an *image* (q.v.) of specific data, using a different medium.

lap joint. The joining of magnetic tape by splicing and then cementing the ends together.

lapel microphone. Small microphone which does not impede vision of speaker.

large aperture seismic array. See LASA.

laser gyro. An integrating-rate gyroscope which combines the properties of the optical oscillator, the laser and general relativity. It differs from conventional gyros in the absence of a spinning mass and so its performance is not affected by accelerations. It has a low power consumption and is not subject to prolonged starting time and hence is an ideal device for space-age aeronautics. Modern versions can measure rates as low as 0·1 degree per hour.

laser intrusion alarm. An obvious application of the laser because of its low-energy consumption and the simplicity of the required optical system, since the source is so directional. In turn the good directionality of the beam permits its use over a long path length, so by use of multiple reflectors a tortuous path or the area of a door may be scanned against intrusion.

laser range-finder. Device in which the range is determined by the time-of-flight measurement of a light pulse and its reflection. The data is displayed in digital form.

laser threshold. The minimum pumping power (or energy) required to operate a laser.

latent period. That between exposure to radiation and its effect.

lattice energy. Energy required to separate the ions of a crystal from each other to an infinite distance.

Lavalier cord. Microphone neck cord used by commentators when their hands must be left free.

leader. Length of blank ciné film, or magnetic or paper tape, at the beginning of a reel.

leak. (1) A path between a conductor and earth (or another conductor) which has a lowered insulation resistance. (2) A high resistance, such as the grid leak, used for the timed discharge of a capacitance, which is in series with a valve control grid.

least significant digit or **character.** That occupying the extreme right-hand position in a number or word.

Le Chatelier principle. When a system in equilibrium is subject to stress, it reacts in such a way as to oppose the application of the stress.

Lector. The code name for an auto-punching system used with Fortran.

Leddicon. TN for a class of lead-oxide photoconductive TV camera tubes.

left/right justified. The entry of data into a field in such a way that the left-hand/right-hand digit or character occupies the extreme left-hand/right-hand space.

length. The number of bits in a computer word, or the number of columns or spaces in a field.

Lenin. Russian nuclear-powered icebreaker.

light-emitting diode. See electroluminescence.

lightning surge. The current surge in a communication system resulting from a lightning discharge.

line source. In a loudspeaker enclosure, a number of forward-facing drive units are set in a vertical line to confine the vertical field but to broaden the lateral polar response.

linearity sleeve. A loop of metal foil suitably mounted in a TV receiver tube to provide horizontal linearity adjustment. It operates by modifying the current in the scanning coils through the current which is induced in the foil by the magnetic field.

link resonance. That resulting in a repeated network when the sections are coupled together.

Lipowitz alloy. Fusible alloy, 50 % Bi, 27 % Pb, 13 % Sn, 10 % Cd, m.p. 65°–70°C.

liquid-state electronics. The biological use of ions in solution to achieve similar results to man-made devices using electrons in solids.

list. Information printed out by a compiler or assembler, comprising a list of the actual machine language programme (*object deck*) storage allocations, etc.

live chassis. A broadcast receiver in which there is a direct connection between one side of the mains supply and the receiver chassis.

load and go. Said of computer operation when a symbolic language is assembled or compiled into machine language and immediately processed without a physical machine language programme (*object deck*) being produced.

load curve. One showing the variation of power consumed with time.

load efficiency. Ratio of the useful power delivered by the output stage to a specified load and the d.c. input power to the stage.

local-carrier reception. Method of receiving single-sideband transmission in which the carrier is partially suppressed and replaced at the receiver.

localizing faculty. The psychophysiological mechanism which enables the hearing system to determine the direction of a sound.

locked-oscillator detector. Form of FM detector using a pentode valve which incorporates a self-oscillating tuned circuit connected to the suppressor grid. The FM input signal is applied to a resonant tuned circuit which is connected to the control grid. The tuned circuit of the suppressor grid locks to the frequency of the incoming signal and leads to an AF anode current proportional to the FM input.

locking relay. One in which contacts made or broken when the relay is energized are unaffected by de-energization. Cf. *non-locking relay* (in main text).

logical instruction. One which leads to a *logical operation* being carried out by a computer.

logical shift. See end-around shift.

long-throw loudspeaker. A moving-coil loudspeaker in which the coil cylinder is longer than the magnetic gap, thus permitting large cone movements.

loop test. Insulation test on transmission line or cable made by connecting the conductors to form a closed loop. See *Murray-*, *Varley-* (in main text).

Lorentz contraction. The change in dimensions (or time scale) for a body moving with a velocity approaching that of light, relative to the frame of reference (*Lorentz frame*) from which measurements are made. Also **Fitzgerald-Lorentz contraction.**

lowest useful frequency. The lowest frequency (in 3–30 MHz band) which can be used for communication by means of the indirect wave at a particular time of year.

luminance flicker. That of a colour TV picture arising from variations in luminance only. Cf. *chromaticity flicker*, *colour flicker* (in main text).

lumister. Electroluminescent device for amplifiers, controls, relays, etc.

Mac. Acronym, used by both Manchester University and Massachusetts Institute of Technology, for programme-orientated computer languages developed for their own use, but also adopted by other organizations in the U.K. and the U.S. respectively.

machine code. Programme instructions coded in the form that is directly obeyed by a specific computer (i.e., without assembly or compiling).

machine error. The error in data which results from an equipment failure.

machine instruction. One which a particular machine is able to recognize and hence carry out.

macrocode. Any programme code in which a single instruction may initiate a series of individual computer operations. Cf. *microcode.*

Mad. A dialect of *Algol*, used in a number of U.S. computing centres, for which some computers can be supplied with commercial compilers. (*M*ichigan *A*lgol *d*ecoder.)

Maddida. One of the first digital computers. It was restricted to specific purposes, and

provided high accuracy. (*M*agnetic *d*rum *d*igital *d*ifferential *a*nalyser.)

magnetic balance. A form of fluxmeter in which the force required to prevent the movement of a current-carrying coil in a magnetic field is measured. Cf. *current balance.*

magnetic bearing. That indicated by a magnetic compass, estimated in degrees of arc from magnetic north clockwise.

magnetic card. One with a suitable surface which can be magnetized in selective areas so that data can be stored.

magnetic coupling. The magnetic flux linkage between one circuit and another.

magnetic ink. Ink containing particles of ferromagnetic material, used for printing data so that magnetic character recognition is possible.

magnetic screen. An enclosure made from a high magnetic permeability material such as Mumetal; used for protecting electronic components from external magnetic fields.

magnetic-tape plotting system. System producing an XY plot from data recorded on magnetic tape by a digital computer.

magnetic thermometer. One based on the use of a paramagnetic salt and the application of Curie's law that the magnetic susceptibility is inversely proportional to the absolute temperature. This provides a very sensitive thermometer at very low temperatures.

magnetoelectric. Of certain materials, e.g., chromium oxide, the property of becoming magnetized when placed in an electric field. Conversely, they are electrically-polarized when placed in a magnetic field. Such materials may be used for measuring pulse electric or magnetic fields.

magnetron arcing. Arcing between anode and cathode as a result of desorption of gas during operation. Intermittent arcing is not uncommon with high-power pulse magnetrons.

magnetron critical field. That which would just prevent an electron emitted from the cathode with zero energy from reaching the anode at a given value of anode voltage.

magnetron critical voltage. That which would just enable an electron emitted from the cathode with zero velocity to reach the anode at a given value of magnetic flux density.

magnettor. A second-harmonic type of magnetic modulator which uses a saturable reactor to amplify direct-current, or low alternating frequency, signals.

magnickel. An alloy of magnesium and nickel used for delay fuses, which have a much longer delay time than the normal type and will carry up to *ca.* twenty times the normal rated current for 10^{-2} sec.

Magnistor. TN for type of toroidal reactor, which is used for specific circuit applications, e.g., in gating, switching and counting at high frequencies.

magnitude. Size of a quantity as distinct from its sign, e.g., -5 and $+5$ have the same magnitude.

Magslip. TN for a synchro system for remote control or indication.

main anode. That carrying the load current in mercury-arc rectifiers which have an independent *excitation anode.*

main bang. Colloquialism for the transmitted pulse when picked up and displayed on a radar screen.

main path. The principal course pursued by a computer following a routine which is directed by the nature of the data and the logic of the programme.

mains aerial. A mains conductor used as a broadcast-receiving aerial. Because of its susceptibility to mains-borne electrical interference, it has largely been replaced by the ferrite-rod aerial.

mains hold. Method of synchronizing the frame frequency in TV reception with a stabilized mains frequency, e.g., 50 Hz in U.K.

mains unit. The power-supply equipment for amplifiers, etc., usually comprising mains transformer, rectifiers and smoothing filters.

major bend. The bend in a rectangular waveguide where the longitudinal axis remains in a plane parallel to the broadest side of the guide along the whole length.

major cycle. The access time to a consecutive access memory, such as a magnetic disk or tape loop.

major lobe. That containing the direction of highest detection sensitivity or maximum radiation for any form of *polar diagram.*

majority emitter. Electrode releasing majority carriers into a region of semiconductor.

malfunction. See mistake.

man-made noise. Locally-generated noise as distinct from *cosmic noise* and *circuit noise.*

many-one function switch. The converse of a *one-many function switch* in which a number of inputs must be excited before the one output is energized.

mapping. Transforming of information from one form to another or establishing a correspondence between two sets of elements.

margin-punched card. One in which the centre is left free for written (or printed) information, the edges containing the punched holes for data.

mark-sensing. Detection and automatic translation of special pencil marks at specific places on a card into punched holes.

marker pulses. (1) Pulses superimposed on a CRT display, e.g., in order to facilitate time measurements. (2) Pulses used to synchronize transmitter and receiver, e.g., in time-division multiplex.

maser relaxation. The process by which excited molecules in the higher energy state revert spontaneously or under external stimulation to their equilibrium or ground state. Maser action arises when energy is released in this process to a stimulating microwave field which is thereby reinforced.

mask microphone. Microphone designed for use inside a respirator.

mast aerial. A particular design of steel mast used for broadcasting medium waves, the radiation being directed at low angles to reduce interference from the sky wave.

master gain control. (1) In a broadcasting studio the attenuator connected between the programme input and main amplifiers in order to regulate the gain as desired. (2) In a stereo amplifier the control which adjusts simultaneously the gains of both channels to be equal.

master instruction tape. One on which all the programmes for a given system of runs are recorded.

master-slave manipulators. Remote-handling tools which duplicate inside a shielded cavity the movements carried out by the operator outside.

master trigger. The basic oscillator (usually a multivibrator) which determines the pulse repetition frequency of a radar system.

mathematical check. One introduced into a computer programme, which depends upon known mathematical relationships between the input data.

mathematical logic. The mathematical concepts, techniques and languages as used in symbolic logic regardless of their specific applications.

matrixing. The processes of combining and de-combining the three primary signals in colour TV. They represent one luminance and two chrominance components.

maximum power output. The greatest signal output which can be delivered by a transducer without exceeding a specified distortion level.

Maxwell turns. The total flux linkage in a coil expressed in CGS units.

Maxwell's circuital theorems. Generalized forms of Faraday's law of induction and Ampère's law (modified to incorporate the concept of displacement current). Two of *Maxwell's field equations* are direct developments of the circuital theorems.

mean-level AGC. An AGC system used in TV receivers, in which the AGC potential is obtained from the mean level of the signal at the grid of the synchronizing separator stage.

mean residence time. Mean period during which radioactive débris from nuclear weapon tests remains in stratosphere.

mean solar time. See time (1).

mechanical filter. One consisting of a series of mechanical resonators in the form of specially-shaped metal rods which are coupled to the input and output by means of piezoelectric crystals. These filters are often utilized in RF amplifier stages of high-grade superhet communication receivers.

median. The central one of a sequence of numbers or values.

megaparsec. Unit used to define distance of extragalactic objects. 1 Mpc $= 3 \times 10^6$ light-years.

Megger. TN for high-tension (bridge) megohmmeters, etc.

melamine resins. Resins similar to *urea resins,*

but with better stability and high-temperature resistance and therefore yielding better-quality electrical fittings.

memory fill. The placing of patterns of characters in vacant memory registers to prevent the possibility of a computer, through error, seeking instructions from forbidden registers.

memory guard. Means of preventing access to certain sections of the storage devices of a computer.

memory map. A list of all constants, variables and statement identifiers contained in a Fortran programme, with the storage locations assigned to each.

mercury delay line. One in which mercury is used as the medium for sound transmission, conversion from and to electrical energy being through suitable transducers at the ends of the mercury column.

metal-oxide semiconductor transistor. An active semiconductor device in which a conducting channel is induced in the region between two electrodes, by applying a voltage to an insulated electrode placed on the surface in this region. It is self-isolating by virtue of its construction, and so can be fabricated in a smaller area than a *bipolar transistor*.

meter-protection circuit. One designed to avoid transient overloads damaging a meter. It may employ a gas-discharge tube which breaks down at a dangerous voltage or a Zener diode or similar device.

microcircuit isolation. The electrical insulation of circuit elements from the electrically conducting silicon wafer. The two main techniques are *oxide isolation* and *diode isolation* (qq.v.).

microcode. Coding in the individual steps obeyed by a specific computer involving, e.g., the breaking-down of a multiplication routine into successive processes of addition. Cf. *macrocode*.

microfocus X-ray tube. A low power tube in which the cathode rays incident on the target are focused on to a very small spot, to provide a near point source of X-rays. This is important, e.g., in the X-ray microscope.

microstrip. A microwave transmission line consisting basically of a dielectric sheet carrying a conducting strip on one side and an earthed conducting plane on the other. The strip with its image in the plane (see *image charge* in main text) forms a parallel strip transmission line.

mil. A unit of length equal to 10^{-3} in., used in measurements of small thicknesses, i.e., thin sheets.

mill. The arithmetical unit of a computer.

Miller-Pierce oscillator. A crystal-stabilized oscillator in which the quartz crystal is connected between the grid and cathode of a triode valve, a parallel resonant circuit being in series with the anode.

mine detector. An electronic device for the detection of buried explosive mines (or buried metal) depending on the change in the electromagnetic coupling between coils in the search head.

miniaturization. Reduction of size of electrical components to contain circuits in a small space.

minor bend. The bend in a rectangular waveguide in which the longitudinal axis remains in a plane parallel to the shorter side of the guide along the whole length.

minority emitter. Electrode releasing minority carriers into a region of semiconductor.

minuend. The quantity from which another quantity is to be subtracted.

mirror galvanometer. Sensitive moving-coil instrument in which a beam of light reflected from a mirror attached above the coil former replaces the mechanical pointer, so reducing the inertia of the moving system.

misch metal. Alloy comprising cerium, lanthium, neodymium and praesodymium, used as a coating for the cathodes of voltage regulator tubes in order to reduce cathode drop.

mistake. In computing, error resulting from a programme or operator fault. When one arises from a machine fault it is a **malfunction**.

mixed base notation. A numerical coding in which not all characters have the same radix, e.g., *biquinary* (q.v.).

mixed high technique. In colour TV, high-modulation frequencies (representing fine detail in the picture) transmitted as a mixed or achromatic signal in order to conserve bandwidth.

mode. A well-defined distribution of the radiation amplitude in a cavity which results in the corresponding distribution pattern in the laser output beam. In a multimodal system the beam will tend to diverge. See also main text.

mode number(s). These indicating the mode in which devices capable of operating with more than one field configuration are actually being used, e.g., in a cavity resonator the mode numbers indicate the number of half-wavelengths in the field pattern parallel to the three axes; in a magnetron the mode number gives the number of cycles through which the phase shifts in one circuit of the anode; and in a klystron it gives the number of cycles of the field which occur while an electron is in the field-free drift space.

mode separation. The frequency difference between operation of a microwave tube in adjacent modes. See *mode jump* (in main text).

mode skip. Failure of a magnetron to fire.

mode transformer. A device which converts from one mode to another (e.g., H- to E-mode in a waveguide.)

modifying factor. General term applied to any factor used for calculating a biological radiation dose (or *dose equivalent*) in rem from the corresponding physical dose in rad. See **distribution factor**, *quality factor* (in main text). A *special modifying factor* is one which is used only under particular specified circumstances, e.g., for irradiation of a particular organ by particular types of radiation (there is a SMF of 3 for irradiation of the eye by high LET radiation) or under particular conditions (such as pregnancy).

modulated amplifier. An amplifier stage during which modulation of the signal is carried out. Also **modulated stage**.

modulated photoelectric system. A *photoelectric alarm* (q.v.) in which the radiation beam is modulated and the detector responds only to a modulated signal. This gives considerable freedom from the effects of varying ambient conditions.

modulation capability. The maximum percentage modulation which can be used without exceeding a specified distortion level.

molecular sieve. Material used in vacuum systems because of its high capacity for the adsorption of gases. It forms the basis of cryogenic pumps as this capacity is increased by cooling and the adsorbed gases can be driven off by subsequent re-heating.

monazite. Mineral occurring principally in sand or gravel, the most important source of thorium which is present as $Th_3 (PO_4)_4$.

Monimax. TN for high-permeability magnetic alloy.

monitoring key. One used for monitoring a telephony channel.

monochrome signal. That part of a TV signal controlling luminance only; cf. *chrominance signal* (in main text). In compatible systems this signal provides the black-and-white picture.

monoformer. A *photoformer* in which the shadow mask is built into the CRT; used where it is only necessary to generate one specific function.

monokinetic. See homokinetic.

monolithic integrated circuit. In this circuit, both active and passive elements are simultaneously formed in a single small wafer of silicon by the diffusion planar technique. Metallic stripes are evaporated on to the oxidized surface of the silicon to interconnect the elements.

monostatic radar. System in which the transmitter and receiver are at the same position and share the same aerial, in contrast to *bistatic radar* (q.v.).

monostatic reflector. One like a *corner reflector* in which the incident beam is reflected back along its incident direction.

moon bounce. The use of the moon to reflect communication signals by using highly-directional aerial systems.

motor controller. Device for remote operation (starting, speed regulation, etc.) of electric motor.

motor converter. A combination of induction motor and synchronous converter with their rotor windings connected in series.

motor torque generator. The motor or drive unit of a servo control system using synchros.

motor torque regulator. An electronic circuit which maintains a constant output torque from a motor despite varying speed.

multichannel television. A TV network which distributes sound and video programmes by line or microwave links to a number of transmitters operating at different frequencies.

multichip integrated circuit. An electronic circuit comprising two or more semiconductor wafers which contain single elements (or simple circuits). These are interconnected to produce a more complex circuit and are encapsulated within a single pack.

multi-electrode valve. One comprising two or more complete electrode systems having independent electron streams in the same envelope. There is usually a common cathode, e.g., double diode.

multipath distortion. That due to direct and indirect paths (arising from ionosphere reflection, etc.) being of different lengths so that in VHF reception the signal becomes amplitude- and phase-modulated.

multiple address. Having several addresses, for the better handling of computer data.

multiple hop transmission. Radio transmission which uses multiple reflection of the sky wave by the ground and ionosphere.

multiprocessor. A computer with multiple arithmetic and logic units, capable of handling two or more programmes simultaneously.

multiprogramming. Technique of handling several programmes simultaneously by interleaving their execution through time-sharing.

NIR colour TV system. Variant of *Secam* system.

natural wavelength. That corresponding to the natural frequency of a tuned circuit, e.g., that of an open aerial by virtue of its distributed capacitance and inductance.

Nautilus. Atomic-powered U.S. Navy submarine.

needle chatter. That arising from vibrations of the gramophone needle being transferred to the tone arm and radiated as noise. Damping pads are often used to reduce the effect.

needle valve. One for controlling gas entry into a vacuum system; it consists of a small aperture partially blocked by a tapered needle which can be advanced or withdrawn by a micrometer screw drive.

negative conductance (*in semiconductors*). The use of 'hot' electrons in some form of a two-terminal negative conductance, forming the basis of both avalanche transit-time devices and those devices relating to the *Gunn effect*. In the former, e.g., using silicon, the negative conductance arises from a phase shift (greater than 90° and preferably near to 180°) between current and voltage. In the Gunn devices it arises within the Ga As crystal in a strong electric field, the local current density decreasing whenever the local electric field exceeds a certain threshold. Other materials showing the Gunn effect are In P, In As, Cd Te, Zn Se, etc., and the fundamental mode of oscillation of such crystals is the *LSA mode* (q.v.).

Neliac. A dialect of the *Algol* computing language developed by the U.S. *Navy Electronics Laboratory*. (NEL *I*nternational *Algol Compiler*.)

neon voltage regulator. Regulator using the property of a neon lamp so that when the applied voltage exceeds the ionization potential the p.d. between the electrodes will remain constant.

nerve current. That arising in a nerve as a result of the *action potential* (q.v.).

nesistor. Transistor depending on a bipolar field effect.

nesting. The programming technique of including shorter loop routines inside longer ones.

neutralizing indicator. One indicating the degree of neutralization present in an amplifier.

neutron logging. The use of neutrons to detect the presence of water or oil in subsoil. These fluids contain an appreciable proportion of hydrogen and are therefore much more effective in slowing down neutrons than the heavy constituents such as calcium and silicon in the soil.

neutron radiography. Radiography by beam of neutrons from nuclear reactor which then produces an image on a photoelectric image intensifier following neutron absorption. It has advantages over *X-ray radiography* where the mass absorption coefficients for neutrons are very different for different parts of the specimen, although the atomic numbers (and hence X-ray absorption) are very similar. This is particularly true of many biological and plastic specimens containing hydrogenous material.

neutron therapy. Medical treatment by use of neutron irradiation—little used because of possible dangers.

Nicaloi. TN for a high-permeability magnetic alloy containing approximately equal amounts of nickel and iron.

nickel-cadmium battery. A secondary battery of 1·2 volt, having a cathode of cadmium and a positive electrode of nickel hydroxide. The battery requires practically no maintenance as water is not lost by electrolysis or evaporation. It is now used as a power supply for transistor TV receivers.

Noctovision. TN for a TV system in which the light-sensitive elements respond to infrared light, and which can therefore be operated in apparent darkness.

nodal lines. Lines on a vibrating surface, e.g., a loudspeaker diaphragm, which do not vibrate.

non-destructive read-out. That of data in a store in such a way that the content of the store is not changed.

non-erasable storage. Data storage in a form (e.g., punched card or tape) which cannot be lost during the normal course of computation.

non-linear network. One for which the behaviour with reference to time cannot be expressed by linear differential equations.

non-operable instruction. One whose only effect is to advance the instruction index counter. Often written as 'continue'.

non-resonant line. One which is perfectly matched and therefore shows no standing-wave pattern.

Norbit. TN for a logic circuit element.

normalize. To adjust the exponent and fraction of a number expressed in floating-point notation to the standard form for the data or system in use.

notation. An agreed set of characters, figures, symbols and abbreviations used to convey information or data.

notch battery. One designed to remove a narrow band of frequencies, i.e., a band-stop filter.

notch filter. One of band-rejection type giving a sharp notch in the frequency response characteristic of the system. Used to provide attenuation at LF end of a channel in a TV transmitter, to minimize interference with sound carrier of lower channel.

noy. Unit of noisiness by which equal-noisiness contours, e.g., for 10 noys, 20 noys, etc., replace equal-loudness contours. These were obtained from relative annoyance judgements of narrow bands of noise. The total annoyance N_t of a noise is deduced by summing the contributions of its various bands, using the relation
$$N_t = N_m + F(\Sigma N - N_m),$$
where N_m is the large contribution (expressed in noys) of any band and ΣN is the sum of all the contributions. The factor F has the following values: 0·15 for third-octave bands; 0·2 for half-octave bands; and 0·3 for full-octave bands.

nuclear battery. One in which the electric current is produced from the energy of radioactive decay, either directly by collecting beta-particles or indirectly, e.g., by using the heat liberated to operate a thermo-junction. In general, nuclear batteries have very low outputs (often only microwatts) but long and trouble-free operating lives. See ripple generator.

nuclear plate. A photoplate for which the emulsion is particularly sensitive to ionizing particles. To obtain a good resolution of the particle tracks, the emulsion forms a very thick coating (>0·1 mm) and has a very fine grain size of 0·1 to 0·2 μ. These are very much used in satellites for radiation measurements because of their low mass and size.

nuclear quadrupole resonance. A technique of analysis similar to *nuclear magnetic resonance* but not requiring a magnetic field since resonance takes place between the electric quadrupole moment of the nucleus and the applied RF field.

nucleon evaporation. Loss of nucleons in the liquid-drop model of the nucleus, similar to evaporation of molecules from liquid surfaces.

nucleonium. Analogous to the *hydrogenoid* (q.v.), but has a proton with a positive charge and an anti-proton with a negative charge. The difference lies in the greater mass of the anti-proton compared with the electron so the two particles are closer together in the nucleonium, which has only a transitory existence.

nude. Said of vacuum gauge pumps, etc., which are supplied without an envelope and are designed for mounting directly into a vacuum chamber.

numerical control system. A system in which digital computers are used to control automatic machines, e.g., machine tools.

object computer. One on which an object deck (or object programme) is processed, if this is different from the one used to compile the object deck from the source deck. Cf. *source computer.*

object deck. A stack of punched cards forming a computer programme in machine language. Usually prepared from an equivalent source deck by the compiler for the machine.

object language. That of the object deck. The basic *machine language* of the computer.

oblique incidence transmission. Short-wave transmission using reflection from ionosphere.

oil switch. One in which sparking at the contacts is eliminated by immersing these in oil.

one-address instruction. An instruction code where normal instructions include one address only.

one-plus-one address instruction. See *two-address instruction.*

open aerial. An aerial which is open-ended and able to support a standing wave of current.

operation code. The code used in a computer instruction to define the operation to be performed in carrying out the instruction.

operation number. One used to indicate the relative position of an operation or subroutine in a computer programme.

operations research. Application of mathematical analysis to solution of problems arising from the operation of plants, offices or factories.

optar. A portable blind guidance device using visible light in combination with radar techniques.

optical ammeter. An instrument used for measuring the rms values of currents with unusual waveforms. The current is passed through a lamp filament and the light output is compared photometrically with that obtained using a d.c. or sine wave supply.

optical character recognition. A method of generating pulse-coded electrical signals directly from the optical scanning of graphic characters by photosensitive devices.

optimization. Development of an *optimum*

programming by a gradual process of improvement.

orthogonal scanning. That subjecting an electron beam simultaneously to axial and transverse magnetic fields. On leaving the combined field, the beam is influenced only by the axial field and emerges parallel to, but displaced from, the axis. Used in low-velocity camera tubes.

ortho-positronium. Positronium in which the positron and electron spin axes are aligned parallel. Cf. *para-positronium.*

Orthonik. TN for magnetic material with rectangular hysteresis loop.

output block. A block of storage locations in a digital computer associated with the information which has to be presented at the output.

output capacitance. The anode-cathode impedance of a thermionic valve or the capacitive component of the output impedance of a transducer.

output variation. That of the peak amplitude of a TV video signal due to noise, etc., and usually taken over a period of one frame.

Ove. Swedish speech-synthesizing computer.

overexcited. Operation of a synchronous motor under conditions such that the back e.m.f. is larger than the terminal voltage.

overlay. The repeated use of the same block of computer storage for successive processing stages of the same problem.

overload. One exceeding the level at which operation can continue satisfactorily for an indefinite period. Overloading may lead to distortion or to overheating and risk of damage—depending on the type of circuit or device. In many cases temporary overloads are permissible. See *overload capacity* (in main text).

Ovshinsky effect. One name for electric switching action in semiconductor loaded glass beads which change sharply from high to low impedance at a given value of the applied electric field. Devices using this effect operate independently of polarity and are more resistant to radiation damage than conventional semiconductors.

oxide isolation. The isolation of the circuit elements in a microelectronic circuit by forming a layer of silicon oxide around each element.

PL1. An attempt to develop a universally acceptable computer language incorporating desirable features from Algol, Cobol and Fortran. (Programme Language One.)

pacemaker. Portable device implanted in the chest wall, used to regulate by means of electrical pulses abnormal heart rhythms in cardiac disease.

packing density. Spacing of data in a store, e.g., bits/cm of magnetic tape.

pad roller. Non-sprocket roller for pressing edges of cinematograph film on to sprockets, so that a sufficient number of teeth are engaged by sprocket holes.

palladium leak. A method of controlling the entry of gas into a vacuum system by temperature control of a porous palladium block. This gives superior control to a needle valve and is widely used in conjunction with ion sources.

Panamac. An extensive electronic computer reservation network introduced by Pan American Airways.

parallel transmission. Simultaneous transmission of the bits forming a word in a computer, e.g., from a parallel memory to a parallel arithmetic unit. Independent conductors are required for each bit.

para-positronium. Positronium in which the positron and electron spin axes are aligned anti-parallel. Cf. *ortho-positronium*.

parastat. A device attached to the tone arm of a record player to clean the record grooves as it tracks and simultaneously to deposit a thin antistatic film on the surface.

Pat. Edinburgh speech-synthesizing computer.

patch cord. Flexible conductor used in setting up problems on a *patch board*.

path attenuation. Fall-off in amplitude of a radio wave with distance from the transmitter.

pattern. (1) That of the radiation field from an aerial system as shown by a polar diagram of field strength and bearing. (2) The luminous trace on the screen of a CRO as traced out by the electron beam.

peak cathode current. The highest instantaneous current drawn from the cathode of a thermionic tube.

peak envelope power. The average power output of a transmitter as measured over one RF cycle at the peak of the modulation envelope.

peak forward voltage. The maximum instantaneous voltage in the forward flow direction of anode current as measured between the anode and cathode of a thermionic rectifier or gas-filled tube.

peak load. The maximum instantaneous rate of power consumption in the load circuit. In a power-supply system the peak load corresponds to the maximum power production of the generator(s).

peak riding clipper. A circuit in which voltage clipping occurs at a level controlled by the peak value of an applied pulse train. Used to limit interference with pulse-amplitude modulated systems.

peak sideband power. The average sideband power of a transmitter over one RF cycle at the highest peak of the modulation envelope.

peaking circuit. One which produces a sharply peaked output from any input wave. See also *peaking transformer* (in main text). Not to be confused with *peaking network*.

peaking network. An interstage coupling circuit which gives a peak response at the upper end of the frequency range that is handled. This is achieved with a resonant circuit and minimizes the fall-off in the

frequency response produced by stray capacitances. Not to be confused with *peaking circuit*.

peaking strip. A device which generates an output pulse at the instant a varying magnetic field has some predetermined value. It consists of a coil wound round a permalloy core, the magnetization of which reverses sharply at the required value of field.

Penning gauge. Cold-cathode ionization vacuum gauge.

Penning ion source. One which uses a magnetic field to introduce charged ions into an acceleration tube in a manner similar to the operation of a *Penning gauge*. It has the drawback, relative to RF ion sources, of producing a smaller proportion of singly-charged ions from the gas source.

Pentone. RTM for a type of *pentode valve*.

percentage regulation. For a stabilized power supply, the value of the output under load expressed as a percentage of that under no load.

perforator. A telegraph instrument for perforating a tape.

peridyne receiver. Early radio tuned by varying inductance of coil through movement of metal plate.

peripheral. Said of units forming part of a computing system but not directly under the control of the main computer. The term applies both to input-output devices and to, e.g., magnetic-tape units.

personal equation. A systematic error of measurement, made by an otherwise skilled experimenter, which may be considered invariable for a length of time, so allowing a correction to be applied to his observations.

phase defect. The phase difference between the actual current in a capacitor and that which would flow in an equivalent ideal (loss-free) capacitor.

phenolic resins. Resins forming the basis of moulding and fabrication industries. May be used up to 150°C. Applied in electrical switchgear construction and, with various fillers, for radio cabinets, etc.

Phoenix. A magnetic mirror plasma containment system built at Culham as part of the thermonuclear research programme.

phonoplug. A connector used on screened cables in signal circuits. It has a single central pin which is 'live' and an outer earth shield.

photochromics. Light-sensitive materials which with the advent of the laser make promising storage and read-out media. In holograph form hundreds of books could be stored in a cm cube of a photochromic material.

photoelectric absorption. That part of the absorption of a beam of radiation associated with the emission of photoelectrons. In general it increases with increasing atomic number but decreases with increasing quantum energy when *Compton absorption* becomes more significant.

photoelectric alarm. A burglar or fire alarm operated by interruption of a beam of ultraviolet or visible light falling on a photocell. Also **photoelectric system.**

photoelectric constant. The ratio of Planck's constant to the electronic charge. This quantity is readily measured by experiments on photoelectric emission and forms one of the principal methods by which the value of Planck's constant may be determined.

photoelectroluminescence. The enhancement of luminescence from a fluorescent screen during excitation by ultraviolet light or X-rays when an electric field is applied, i.e., the field enhancement of luminescence.

photoemissive camera tube. Tube operating on the photoemissive principle, the image falling on its photocathode causing this to emit electrons in proportion to the intensity of the light in the picture elements.

photoproduction. The production of particles by means of photons. Most frequently used for neutron production.

photoresist. Mask for etching semiconductor devices or integrated circuits, produced by photographic reduction process.

photoresist process. Process used to remove selectively the oxidized surface of a silicon slice semiconductor. The photoresist material is an organic substance which polymerizes on exposure to ultraviolet light and in that form resists attack by acids and solvents.

photosource. One used in photoproduction, a typical neutron photosource consisting of radioactive sodium and beryllium. Gamma-rays emitted by the sodium can eject neutrons from the beryllium and they would all have the same energy of about 10^6 eV.

picture black. The signal level which corresponds to the blacked part of the televised scene.

picture slip. Vertical displacement of received picture due to bad synchronism between field frequency of the frame time base and that of the signal.

picture-to-synchronization ratio. The ratio of the total amplitude of the TV waveform assigned to picture information to that assigned to the synchronizing pulses and flyback times.

picture white. The signal level corresponding to the whitest part of a televised scene.

pink-noise generator. A random-noise generator which gives a close approximation to the energy distribution of sound found in nature.

pistonphone. Device in which a rigid piston is vibrated, so that, by measurement of its motion, acoustic pressures and velocities can be calculated.

planar integrated circuit. One produced on a thin silicon wafer; the same technique as is used for *planar transistor.*

planar process. A basic part of the technology of silicon transistors, being a combination of oxidation, selective oxide removal, and then heating to introduce dope materials by diffusion.

planar transistor. Form of transistor con-structed by etching a thin slice of semi-conductor and characterized essentially by a parallel-plane electrode configuration protected by an oxidized surface.

plasma focus. A technique for concentrating hot plasma into a very small confinement volume in order to initiate a thermonuclear reaction. The necessary conditions for this can be very closely approached but the confinement times attained so far are less than 1μs.

plated magnetic wire. Non-magnetic wire plated with a ferromagnetic surface.

playback equalizer. A resistance-capacitance network introduced into an interstage coupling so that all frequencies are repro-duced with equal intensity in the recording of music.

Platinax. TN for an easily machined high-coercivity magnetic material.

player unit. That in which the pickup and turntable are designed as one item.

pleochroism. The dependence of the absorp-tion coefficient on the direction of polariz-ation of the incident light. The phenomenon may give rise to a variation of the colour of the transmitted light with both the direction of transmission and of polarization.

Plumbicon. TN for type of photoconductive colour TV camera tube.

Polaris. A surface-to-surface guided missile, developed by the U.S. Navy, with a range of *ca.* 1500 miles.

Polish notation. A method of representation of algebraic statements which has the advantages of conciseness and unique logical interpretation.

population inversion. The reversal of the normal ratio of populations of two different energy states, i.e., that normally fewer and fewer atoms occupy states of successively higher energies. It forms the basis of laser action.

Potter-Bucky grid. Type of lead grid designed to avoid exposure of film to scattered X-radiation in diagnostic radiography. Mechanical oscillation of the grid eliminates reproduction of the grid pattern on the negative.

power-factor meter. One which reads the power factor of a circuit directly.

power frequency. That of the a.c. supply mains, viz., 50 Hz in the U.K., 60 Hz in the U.S.

power relay. One which operates at a specified power level.

pre-echo. The weak prior impression of an oncoming loud sound which results from a slight deformation of an adjacent groove on a record.

pressure unit. A metal (or plastic) dome forming the diaphragm of a small moving-coil loudspeaker unit, situated in the throat of a horn, for use at intense acoustic pressures.

priority indicators. Code signals which form a queue of data awaiting processing so that this is handled in order of importance.

problem-orientated language. One directed towards the development of a suitable algorithm which could be used to solve a problem, but not to the specific programme required for processing the solution on a particular computer.

programme meter. An instrument used to control the dynamic range of broadcast transmissions.

programme-orientated language. One used in preparing a source programme (*source deck*) for a computer and suitable for direct compilation into machine language. This may be a transcription of a more fundamental algorithm suitable for solving the problem expressed in a *problem-orientated language*. The most widely used programme-orientated languages, e.g., Algol, Cobol, Fortran, PL1, can all be used with a wide range of commercial computing systems.

propagated error. One occurring in one operation and affecting data required for many subsequent operations, so that the error is spread through most of the material processed.

protection ratio. The lowest ratio of the signal strength of a required signal to that of the interfering signals, to obtain satisfactory and consistent radio reception.

protective agent. Any preparation which can be administered to an organism either as food or by injection, and which reduces its sensitivity to the effects of ionizing radiations.

protective gap. A spark gap between conductors on an open transmission line, etc., to protect the system against any excessive high-voltage surges.

proving time. The time used in running a test programme to check if a particular fault in the computer has been corrected.

pseudocode. Computer instructions written in symbolic language which must be translated into an acceptable programme language or direct into machine language before they can be executed.

psycho-acoustics. Phenomena concerned with the psychology of hearing.

pulse bandwidth. The frequency band occupied by Fourier components of the pulse which have appreciable amplitude and which make an appreciable contribution to the actual pulse shape.

pulse detector. One designed for use with modulated pulse signals.

pulse discriminator. Any circuit capable of discriminating between pulses varying in some specific respect, e.g., duration, amplitude or interval.

pulse droop. The exponential decay of amplitude which is often experienced with nominally rectangular pulses of appreciable duration.

pulse flatness deviation. For a nominally rectangular pulse which exhibits pulse droop, the difference between maximum and minimum amplitudes of the top of the pulse, divided by the maximum amplitude.

pulse interrogation. The triggering of a transponder by a pulse or pulse mode.

pulse operation. That of a circuit or device in which energy is supplied in pulses.

pulse stretcher. An electronic unit used to increase the time duration of a pulse.

Q-band. Frequency band used in radar, 36–46 GHz.

Q-switching. The optical pumping of a laser having a low Q-factor and then suddenly increasing the Q of the system. In this way a very short output pulse of high intensity is obtained.

quadrifilar. A transformer in which four windings are used simultaneously to obtain maximum coupling and minimum leakage inductance.

quarter-wave plate. A plate of doubly refracting crystal, cut parallel to the optic axis, of such a thickness that a quarter wavelength path difference between the ordinary and extraordinary light rays is introduced. Used in the analysis of polarized light and to obtain circularly plane-polarized light.

quasi-instruction. A symbolic representation of a computer instruction in a compiler.

quench frequency. The lower frequency signal used to quench intermittently a high-frequency oscillator, e.g., in a super-regenerative receiver. See *squegging* (in main text).

quietening. A receiving circuit device for the suppression of background noise while tuning between radio stations.

quieting. The level reduction (dB) by which background noise is reduced when the tuner of the radio set is fed with an RF signal of a specified voltage.

Quiktran. A programme-orientated computing language based on *Fortran* (q.v.), but designed specifically for users employing a central computing service from a remote terminal, usually via the regular telephone service, although private lines and microwave radio links can also be used.

radar plot. A plan prepared from data displayed on a *radarscope*.

radar storm detection. Use of radar techniques to determine whether precipitation is likely to be associated with clouds.

radio fuse or **radio proximity fuse.** One operated either by a radio signal or by a radar signal interacting with the target.

radio-pill. A miniature broadcasting station of volume about 0·1 cu. in which may be easily swallowed without discomfort. Its contained transducers collect the required information which is transmitted as a modulated radio signal.

rake angle. The angle made by pickup stylus shank with the record disk, as viewed from the side.

raster burn. Deterioration of the scanned area of the screen of a TV picture or camera tube as a result of use.

read-around ratio. The number of times access

may be had to a specific location in a charge storage tube without spill affecting adjacent locations.

Reboul effect. The emission of electrons, soft X-rays and ultraviolet light from the surface of certain semiconductors when very intense electric fields are applied. See also **electro-radiescence.**

receiver gating. Control of sensitivity of a radio receiver by means of gating signals.

receiver quieting sensitivity. *Sensitivity* of FM radio receiver for specified signal noise level.

recirculating loop memory. Small element of magnetic memory in which stored information recirculates continuously to provide rapid access.

Reco. TN for an isotropic magnetic alloy particularly suitable for multipolar applications.

recording channel. One of a number of independent (*a*) tracks on a recording medium or (*b*) recorders in a recording system.

recording spot. The moving element from which a facsimile image is reconstructed.

recording stylus. The cutting tool used in electromechanical recording.

rectilinear scan. A *raster* in which a rectangular area is scanned by a series of parallel lines.

red-tape operations. Those in a computer which do not contribute directly to programme processing.

reference address. One that is used as a locating point for a group of related addresses.

reflectometer. Instrument measuring ratio of energy of reflected wave to that of incident wave in any physical system.

register length. The number of bits stored in a computer register.

relative code. Computer code in which addresses are indicated symbolically in machine language.

relative damage factor. A term used by the ICRP to represent the variation of hazard from bone-seeking internally-absorbed radioisotopes with non-uniformity of distribution. The factor expresses this hazard relative to that for radium.

relocatable programme. A routine prepared in such a way that it can be located and executed in several different parts of the computer memory. The instructions allow for the modification of addresses relating to different fixed-storage locations. This technique assists in developing flexible real-time use of the storage capacity.

rescue dump. The recording of the whole of the contents of a computer memory on to magnetic tape.

resistance lamp. One used to limit the current in a circuit.

resistive component. That part of the impedance of an electrical system which leads to the absorption and dissipation of energy in the form of heat.

resistive coupling. The use of a common resistance to link two otherwise separate circuits.

resistive wall amplifier. A type of travelling-wave tube in which bunching results from interaction between the electron beam and charges induced on an adjacent wall of high-resistivity material.

resonant-gate transistor. A type of *field-effect transistor* with a mechanically-tuned input, obtained by a vibrating cantilever assembly.

right justified. See left justified.

ring-main. A domestic a.c. wiring system in which a number of outlet sockets are connected in parallel to a ring circuit which starts and finishes at a mains supply point.

ring shift. See end-around shift.

ripple factor. The amount of a.c. voltage in the output of a rectifier after d.c. rectification. It is measured, in per cent., by the ratio of the rms value of the a.c. component to the algebraic average of the total voltage across the load. See also *ripple* (1) (in main text).

ripple generator. A thermoelectric generator powered by heat from a radioactive source and suitable for long periods of maintenance-free operation on remote sites, e.g., for lighting marine navigational buoys. (*Radioisotope powered pulsed light equipment.*)

room constant (R). A quantity used to specify the acoustic quality of a room and for small absorption. Given by $R \simeq S\bar{a}$, where S is the total surface area of the room having a mean absorption coefficient $\bar{a}$.

rubidium clock. A compact atomic clock based on an ^{87}Rb gas-discharge lamp and stable after calibration to about 1 part in 10^{10} per year.

S-matrix. One concerned with a scattering process which links the initial and final state vectors ψ_i and ψ_f so that $\psi_f = S\psi_i$. The S-matrix is unitary since total probability must be conserved.

sage. Air defence system whereby information is received from radar and other sources and is processed at a central station to give an evaluation of a situation. (*Semiautomatic ground environment.*)

sanaphant. A circuit which has characteristics between those of a *sanatron* and of a *phantastron.*

Sargent cycle. A type of thermodynamic cycle for an ideal gas.

satellite computer. One used to relieve a central processing device of relatively simple but time-consuming operations, such as compiling, editing and controlling input and output devices.

saturated molecule. One in which each bond of every atom in the backbone of the molecule holds another atom.

scattering loss. The loss of energy from a beam of radiation as a result of various scattering processes arising either at a surface of discontinuity or in the medium traversed.

scintillation cameras. Fixed devices that receive and record the radiation coming from the field of view. They basically consist of a detector, data-transfer system and data-recording system and may be of three types: (1) the gamma-ray, (2) the positron, and (3) the autofluoroscope. The fluorescent detector of the latter may consist of a mosaic of some 300 sodium iodide crystals. Instead of slowly scanning an area of nuclide distribution these cameras visualize the entire concentration at one time.

scrape flutter. Flutter in tape recording created by longitudinal oscillation of the tape arising from scraping against the guides, etc. Although usually of 3 to 4 kHz it may be reduced by several octaves if pre-recorded tapes are dubbed at 4 (or 8) to 1 speed ratio.

sector display. A form of radar display used with continuously rotating antennae, *not* with sector scanning. (So termed because the display uses a long persistence CRT excited only when the antenna is directed into a sector from which a reflected signal is received.)

sector scan. Radar scan in which the antenna rotates only through a limited sector. (Not to be confused with *sector display.*)

seek. Computer process for locating specific data in a random access store. Each memory location inspected is a *seek* and the number of seeks governs the total *search time.*

segment. In computer programming where a long routine has to be subdivided between immediate access and secondary memories, the part retained in the former.

select. The choice by a computer (using logical comparison) of one of several possible *subroutines* or programme *legs.*

select lines. The wires threading and linking the individual cores of a magnetic-core memory.

selective calling. The ability of a transmitting station, by suitable coding, to specify which of several other stations is to receive a message.

selective dump. Recording on tape or printing out the full contents of a specific part of a computer store.

selector. An automatic switching operation in a computer process which enables a logical choice to be based on results of the processing already carried out.

selenophone. Original system of photographically recording sound on paper, the track being reproduced by scanning with a focused slit, the modulated reflected light being received into a photocell.

self-oscillation mixer. A frequency-conversion unit of a superhet receiver, in which a single transistor (or valve) is used to combine the modulated input signal with a local signal of different frequency, to generate a modulated signal of an intermediate frequency.

semantic error. One which results in ambiguous or erroneous meaning of a computer programme. Most programmes have to be *debugged* to eliminate these errors before use.

sepdumag, sepduopt. International codes for motion pictures with two magnetic or optical sound-tracks, each on separate film.

sepmag, sepopt. International codes for motion pictures having magnetic or optical sound-tracks on separate film.

septum. Dividing partition in a waveguide.

serrodyne. Frequency translator using linear sawtooth modulation of transit times.

service routines. Computer routines which enable on-line input and output operations to be performed concurrently with other programmes.

serving. Protective wrapping applied to core of cable.

sextuple play. Magnetic recording tape of one-sixth the standard thickness.

shadowing. A technique for increasing image contrast in electron microscopy, especially valuable with samples of biological material which are relatively transparent to electrons. It consists of evaporating a metal film obliquely on to the sample under vacuum in such a way that where relief occurs one side only of the undulations will be coated (giving the appearance of a shadow in the image).

shared store. One to which two independent computers both have access.

sidereal time. See time (1).

Sidewinder. U.S. air-to-air missile, depending for guidance on infrared detection.

sight check. That of the accuracy of punching or sequence by inspection of a *source deck.*

signal frequency shift. The bandwidth between white and black signal levels in FM facsimile transmission systems.

silicon transistor. One formed from a silicon crystal, sometimes specified in preference to germanium because of its higher temperature stability.

silicones. A family of synthetic materials consisting of a backbone of alternating atoms of silicon and oxygen, usually with associated carbon atoms. They generally have a low vapour pressure and withstand extremely high temperatures.

sine potentiometer. Voltage divider in which the output of an applied direct voltage is proportional to the sine of the angular displacement of a shaft.

slow wave. See space-charge waves.

slow-wave structure. One used to reduce the phase velocity of an electromagnetic wave to a value comparable with the electron beam velocity in a microwave tube. This enables amplification of the wave system to take place: *forward-wave amplification* when the group velocity is in the same direction as the phase velocity and *backward-wave amplification* when in the opposite direction. The former case can give broadband response whereas the latter will be frequency-selective and voltage-tuned.

snapshot. Colloquialism for a *selective dump* (q.v.) at various checkpoints during the running of a computer programme.

software. General term for programming or compiling accessories used for computing or data-processing systems. It covers both actual equipment and library routines. Cf. *hardware*.

solar time. See time (1).

solid logic technology. That relating to the microelectric circuits which are the basic components of solid logic systems. These are so-called because they carry and control the electrical impulses that represent the information within a computer. These extremely small devices operate at speeds between 6 and 300×10^{-12} sec.

source computer. That which is used to compile the source programme, if this is different from the one which processes the object programme. Cf. *object computer*.

source deck. Stack of programme cards ready to insert into compiler of some computers operated by punched cards.

space-charge density. Excess positive or negative charge per unit volume in a space charge.

space-charge region. Any region in a discharge tube, thermionic valve or semiconductor where the space-charge density differs appreciably from zero.

space-charge waves. The type of periodic disturbances imposed on an electron beam in a microwave tube by a bunching device. This always leads to the production of two space-charge waves, a *fast wave* and a *slow wave*, with velocities v_1 and v_2 respectively above and below the beam velocity u. These satisfy the relationship $v_1 v_2 = u^2$.

special modifying factor. See modifying factor.

sphygmomanometer. Instrument for the measurement of blood pressure.

spider. Mount with flexible arms, used for centring coil of dynamic loudspeaker.

spin polarization. Orientation of the spin axis of a nuclear particle.

split-word operation. That of a computer when the number of bits per item of data is a submultiple of the number forming a complete word.

spontaneous emission. In contrast to *stimulated emission* (basis of laser action), process involving the loss of energy in an atomic system without any external stimulation. See *fluorescence* (in main text).

spot punch. Hand punch used for correcting isolated errors on punched cards or tape.

Springfield plant. UKAEA plant for preparing reactor fuel elements.

sprocket pulse. One generated by the sprocket hole in paper tape (or by the equivalent indexing spot on magnetic tape) and used to synchronize the read-in unit with the individual characters recorded on the tape.

stacked ceramic tube. A development of the *lighthouse tube* with electrodes separated by ceramic disks. These tubes operate at such elevated temperatures that additional cathode heating is superfluous.

starting current. The electron-beam current at which oscillations are self-starting in a specific circuit. Such a circuit may be used deliberately as a regenerative amplifier with beam currents kept below this value.

statement number. See instruction number.

static convergence. The adjustment in television of the three-coloured dots so that they converge one on top of the other to give white dots, using the three permanent magnets of the radial assembly and the one magnet of the lateral assembly.

static electricity. Stationary charges on any insulated or insulating body.

Staticon. RTM for a TV camera operating on the photoconducting principle.

Steiver pulses. Those of incorrect amplitude for the energy deposited in a radiation detector because the detector has been triggered during its recovery time.

stepping motor. One in which rotation occurs in a series of discrete steps controlled electromagnetically by individual (digital) input signals.

stepping relay. One with a wiper arm which rotates round a series of stud contacts.

storage allocation. The assignment of data to specific blocks of computer storage.

storage protection. Stored programmes, library subroutines (e.g., evaluation of a square-root) and other data which may have to be retained in a memory indefinitely are all allocated to a protected storage block. With multiprocessing, more complex storage protection is required to ensure that a programme being processed does not attempt to utilize storage associated with a different programme.

stored-programme computer. A digital computer for which the programme instructions are stored in the memory. (This is a somewhat obsolete term as all modern digital electronic computers are of this type. When computers were first being developed some digital machines were programmed on external patch boards similar to those used for analogue computing.)

straight-line coding. Any computer programme which can be completed by carrying out sequentially each programme instruction, i.e., one without any branch points or loop instructions.

stunt box. The unit which controls the response of stations to *selective calling* signals.

sublimation pump. Type of vacuum pump in which gas is effectively adsorbed on to a pad—usually of titanium. It cannot be used to attain a high vacuum unaided but greatly increases the pumping speed obtained when used in conjunction with, e.g., an ion getter pump. Also **titanium pump.**

subscriber trunk dialling. A direct-dialling telephone system in which most of the subscribers in West European towns can dial one another directly.

subscript. Integer numerals or symbols attached to a quantity to indicate its location in an array such as a matrix.

subscripted variable. A variable that must be represented by an array, each element of which carries one or more subscripts for identification.

subtrahend. In computing, a quantity to be subtracted from another quantity.

sum. In computing, the quantity obtained by adding two other quantities (the *addend* and the *augend*).

surface resistivity. That between opposite sides of a unit square inscribed on the surface. Its reciprocal is the *surface conductivity*.

sustained oscillations. Externally-maintained oscillations of a system at or very near its natural resonant frequency. Cf. *forced oscillations, free oscillations* (in main text).

symbolic language. That of a computer programme prepared in any coding other than the specific machine language, and so requiring to be assembled or compiled before it can be carried out.

symmetrical. Said of circuits, networks, or transducers for which the impedance level (image impedance or iterative impedance) is the same in both directions.

symmetry-breaking effects. Effects which violate the invariance of a given symmetry scheme, as exemplified by charge independence of strong interactions, which is broken by the electromagnetic and weak interactions.

synchronizer. (1) A unit used to maintain synchronism when transmitting information between two devices. It may merely control the speed of one (e.g., by *clutch*); or, if the speeds are very different, may include buffer storage. (2) See **synchroscope.**

synchronometer. A device which counts the number of cycles in a given time. If the time interval is unity, the device becomes a *digital frequency meter.*

synchronous communications satellite. One with an orbital period equal to the time of rotation of the earth on its axis and launched so that it remains directly above the same geographical point on the earth's surface. This facilitates the control of the aerial assembly of ground stations operating in conjunction with the satellite.

synchronous converter. Rotary converter with synchronous windings which operates from an a.c. supply at a controlled speed.

synchroscope. Instrument for measuring the phase relationship between two alternating voltages—used particularly with alternators in parallel. Also **synchronizer.**

system check. One on computer performance using external programme checks, not check circuits built into the *hardware.*

table. A collection of data (such as square-root values) laid out in rows and columns for reference, or stored in a computer memory as an array. In computing, elements of such a table would be obtained by direct calculation where possible (*table look at*) to save storage. If this was not possible the whole array would be stored and the element required determined by a comparison search (*table look up*).

Tabsol. Proposed universal computer language. (*Tabular systems oriented language.*)

tantalum-nitride resistor. Resistor consisting of tantalum nitride deposited on a suitable substrate, e.g., industrial sapphire.

tantalum-slug capacitor. One of electrolytic type with a sintered tantalum slug as anode. It permits operation up to 200°C.

tape feed. That part of a data-processing system which controls the input of magnetic or paper tape to reading or writing heads.

tape playing-time formula. Given by

$$T = \frac{12 \times L \times N}{60 \times S},$$

where T = playing time in minutes
L = tape length in feet
N = number of tracks
S = speed in inches/sec.

tape punch. Unit for punching paper tape in a computing or control system.

Tarantula. An experiment on shock-wave heating of a plasma carried out at Culham Laboratories.

target capacitance. That between a camera tube target and backplate.

Teleprompter. RTM for a visual prompter for displaying a script inconspicuously to a speaker in front of a television camera.

television receiver. That part of a TV system in which the picture and associated sounds are reproduced from the input signal.

television system. A complete assembly of units for the production, transmission (by radio or closed circuit) and reproduction of a TV image and associated sound channel.

television transmitter. One which radiates video, aural and synchronizing components of a television signal as a modulated radio-frequency wave.

temperature coefficient of voltage drop. The ratio of the change in voltage drop across a glow-discharge tube to the change in operating temperature.

temperature coefficient of voltage regulator. The rate of change of striking voltage of a glow-discharge tube with ambient temperature.

thermoelectroluminescence. See **electrothermoluminescence.**

thermoluminescent dosemeter. One which registers integrated radiation dose, the read-out being obtained by heating the element and observing the thermoluminescent output with a photomultiplier. It has the advantage of showing very little fading if read-out is delayed for a considerable period, and forms an alternative to the conventional film badge for personnel monitoring.

thin film. Said of a layer which is (or approximates to) monomolecular thickness and is laid down by vacuum deposition. Many types of electronic components and complete microcircuits can be produced in this way.

thin-film capacitor. One constructed by evaporation of two conducting layers and an intermediary dielectric film (e.g., silicon monoxide) on an insulating substrate.

thin-film resistor. Modern high-stability resistor formed by a conducting layer a few hundred ångström thick on an insulating substrate.

three-address instruction. Computer instruction which consists of an operation and a total of three addresses. If these include the address of the next instruction, then it is also known as a $2+1$ **address instruction.**

three-plus-one address instruction. See **four-address instruction.**

through path. The forward path from loop input to loop output in a feedback circuit.

thyratron extinction. Termination of the flow of current through a thyratron by reduction or interruption of the anode supply.

tickler coil. Inductance coil included in the anode current of a valve, magnetically coupled to the grid circuit for reaction.

time. (1) The fundamental unit of time in universal use is the *day* as defined by successive transits of the sun (*solar time*). Unfortunately this is a variable and so-called *universal time* (an internationally accepted value of GMT sometimes known in telecommunications as *zulu time* (in main text)) is based on *mean solar time*. Traditionally mean solar time has been standardized astronomically by observing the transits of distant stars which gives *sidereal time* and making the sidereal year equal to the mean solar year. Variations of a few parts in 10^8 made a time scale based on much longer periods of astronomical observation desirable. This was defined as *ephemeris time* and corresponds approximately to the mean value of universal time over the past 200 years. Modern atomic clocks are now believed to give a more uniform time scale than any possible astronomical observations and, following detailed comparisons made in 1955 and 1958 at the N.P.L., ephemeris time is now defined in terms of the *caesium clock* frequency. (2) In computing, the unit interval for a *digit*, *word* or *instruction*.

time-sharing. A computing technique by which more than one terminal device can use the input, processing and output facilities of a central computer simultaneously. See **automatic interruption.**

titanium pump. See **sublimation pump.**

Tokomak. Russian thermonuclear research device at Kurchatov Institute, Moscow. Tokomaks are toroidal discharge devices employing longitudinal magnetic fields for stabilization. They can realize an ion temperature of up to 2×10^6 °K for about 10 ms.

tone burst. Usually a controlled burst of sine waves for test purposes.

tone-burst generator. One for producing sound pulses of short duration, thus enabling direct audio sound to be separated from the reflected sound in acoustic measurements.

trace interval. The scan time of a time base. *N.B.*—The *period* of the time base is equal to the trace interval plus the *return interval*.

trailer. The final length of a reel of ciné film, or paper or magnetic tape. This may include labels summarizing the contents and control instructions, such as those for rewinding the reel.

transistor current gain. The slope of the output current against input current characteristic for constant output voltage. In a common-base circuit it is inherently a little less than unity but in a common-emitter circuit it may be relatively large. *N.B.*—The current gain is the hybrid parameter h_{21}.

transition card. One at the end of a computer programme deck which signals the end of reading-in and initiates the processing.

trapping. Feature of some computers by which they make an unscheduled jump to some specified location if an abnormal arithmetic situation arises.

triode-heptode. Thermionic valve used as a frequency changer in a similar manner to a *triode-hexode*.

triplener. A filter device to enable three TV bands to use a common VHF download from the aerial system.

tropicalization. The protection of equipment or components from the corrosive effects of hot humid environments.

truncation. (1) Ending of a computational procedure in accordance with some programme rule as soon as a specified accuracy has been reached. (2) Rejection of final digits in a number, thus lessening precision (but not necessarily accuracy).

tuning range. Frequency range over which a mechanically-tuned resonant circuit can be adjusted, e.g., in a reflex klystron or TR tube.

turbomolecular pump. Vacuum pump in which molecules are impelled out of the vacuum chamber by a set of rotating blades similar to a turbine.

twin check. Continuous computer check achieved by duplication of *hardware* and comparison of results.

two-address instruction. Computer instruction which consists of an operation and a total of two addresses. If these include the address of the next instruction then it is also known as a $1+1$ **address instruction.**

two-plus-one address instruction. See **three-address instruction.**

unbranched chain molecule. A large molecule whose component molecules are lined up in a single row.

underwater transducer. One (e.g., a microphone or loudspeaker) designed for operation under water.

unit-distance code. Arrangement of a pattern to avoid ambiguity errors in a computer.

Uniterm system. A data-recording system used by libraries, based on classifying keywords in a coordinate indexing system.

universal time. See time (1).

update. (1) To modify a computer instruction so that the address specified is increased every time the instruction is carried out. (2) To correct entries in a file of data in accordance with a specified procedure.

urea resins. Resins which are light and opaque in colour, used for domestic electrical fittings.

validity check. One based on limits to the data for a specific problem, e.g., a computed time of day would be rejected if outside the range 0-24 hours, but not a computed time interval.

variable. A symbol used in computing to represent a quantity whose value changes during the execution of the programme.

variable address. One which is modified during the course of processing, e.g., by means of an *index register.*

variable resistance pickup. A transducer whose operation depends upon variation of resistance, e.g., a thermistor or strain-gauge element.

variable word length. Said of a computer in which the number of bits per word can be varied in accordance with the data being processed.

varigroove. An arrangement of variable groove spacing on microgroove records, which allows more recording time per radial inch of record to be accommodated.

Veitch diagram. A graphical technique used for the solution of problems arising in logical circuit design.

velocity potential. The function ϕ used in hydrodynamics and acoustics, analogous with gravitational and electrostatic potentials. The fluid velocity in any direction, s, is given by $V_s = -(\delta\phi/\delta s)$.

vertical blanking. The elimination of the vertical trace on a CRT during frame flyback.

vertical hold control. The control in a TV receiver which varies the free-running period of the oscillator providing the vertical deflection.

vertical tracking angle. The vertical stylus motion path in relation to true vertical in a stereo pickup cartridge. Normally the angle is *ca.* 15° forward.

video frequency. Said of circuits or devices which operate over the frequency range required for a video signal—approx. 0–5 MHz.

video integration. In facsimile reception, integration and averaging of successive video signals to improve the overall signal/noise ratio.

video transmitting power. Power radiated in the video band by a TV transmitter.

Vinkor. TN for a class of ferrite cores.

vinyl polymers. A series of copolymers formed from vinyl chloride, polyvinyl alcohol and polyvinyl acetates. The range of materials is extended by the use of plasticizers, etc. Because of their excellent electrical properties and great flexibility, they have a very wide application in the electrical-cable and wire-covering fields. The best known is *polyvinyl chloride (PVC)* which is flexible.

virgin coil. Reel of paper tape before punching.

virtual earth. Live input terminal of a high-grain directly-coupled amplifier which remains approximately at earth potential although not connected to earth.

visibility factor. The ratio of the minimum signal input power to a radar, TV or facsimile receiver for which an ideal instrument can detect the output signal, to the corresponding value when the output signal is detected by an observer watching the CRT.

vocabulary. Set of words for data processing.

wait time. The time interval during which a processing unit is waiting for information to be retrieved from a serial access file or to be located by a search.

waveguide lens. An array of short lengths of waveguide which convert an incident plane wavefront into an approximately spherical one by refraction.

waveguide tee. A T-shaped junction for connecting a branch section of a waveguide in parallel or series with the main waveguide transmission line.

white compression. Reduction of gain of transmitted TV signal at levels corresponding to the highlights of the picture. See also **black compression.**

wired-programme computer. One in which the instructions to be carried out are controlled by flexible cable connections made out on a patch board. Cf. *stored-programme computer.*

X-ray telescope. An instrument used initially to investigate the X-ray emission from the sun. This shows greater variability than might be expected from optical measurements in the visible region. To focus the X-rays by mirrors, grazing incidence techniques had to be used. A gas-filled X-ray proportional counter was placed behind an aperture at the mirror focus and the counts were telemetered to earth, the telescope being rigidly fixed to a rocket which itself was scanned.